# ENGINEERING ECONOMICS

# McGRAW-HILL SERIES IN INDUSTRIAL ENGINEERING AND MANAGEMENT SCIENCE

CONSULTING EDITOR

JAMES L. RIGGS, Department of Industrial Engineering, Oregon State University

GILLETT: Introduction to Operations Research: A Computer-oriented Algorithmic Approach

HICKS: Introduction to Industrial Engineering and Management Science

RIGGS: Engineering Economics

RIGGS AND INOUE: Introduction to Operations Research and Management Science: A General Systems Approach

# ENGINEERING ECONOMICS

JAMES L. RIGGS
Professor and Department Head
Industrial and General Engineering
Oregon State University

McGRAW-HILL BOOK COMPANY New York
St. Louis San Francisco Auckland Bogotá Düsseldorf Johannesburg London
Madrid Mexico Montreal New Delhi Panama Paris São Paulo Singapore
Sydney Tokyo Toronto

This book was set in Optima and Century Schoolbook by Black Dot, Inc.
The editors were B. J. Clark and Annette Hall;
the designer was Joan E. O'Connor;
the production supervisor was Robert C. Pedersen.
New drawings were done by J & R Services, Inc.
R. R. Donnelley & Sons Company was printer and binder.

ENGINEERING ECONOMICS 

1 2 3 4 5 6 7 8 9 0 D O D O 7 8 3 2 1 0 9 8 7

Library of Congress Cataloging in Publication Data

Riggs, James L

Engineering economics.

(McGraw-Hill series in industrial engineering and management science)

Bibliography: p.

Includes index.

1. Engineering economy. 2. Managerial economics. I. Title.

TA177.4.R53 658.1'596 76-44028

ISBN 0-07-052860-8

# CONTENTS

## SECTION THREE MANAGEMENT OF MONEY

## SECTION FOUR RECOGNITION OF RISK AND UNCERTAINTY

# PREFACE

The curricula of most professional schools include a course in applied economics under titles such as *engineering economy, financial management,* and *managerial economics.* Practicing professionals usually rate their "econ" course as one of the most useful subjects taken in college. This book is written to make that econ experience as rewarding as possible and to provide a comprehensive reference for future applications.

Engineering economics is a fascinating subject. Its core is decision making based on comparisons of the worths of alternative courses of action with respect to their costs. Decisions vary from personal investments to corporate capital budgeting, and occur at all organizational levels in both the public and private sectors of the economy. Tools for decision making stretch from standardized worksheets for discounted cash-flow evaluations to refinements for sensitivity and risk analysis. The practices followed are grounded in classical economic theory, yet rely on techniques adapted from financial accounting, decision theory, operations research, and other disciplines. Most of the applications are intuitively logical and computationally simple, but the underlying principles are conceptually demanding.

## CONTENTS

The following brief tour through *Engineering Economics* indicates how this book treats the diverse characteristics described above and how the treatment differs from those in other texts:

- An introductory chapter traces the history of engineering economic thought, relating it to practical problems and associated decision models.

- Section One is called *mesoeconomics* because it combines microeconomic and macroeconomic considerations in evaluating operational problems. Chapter 2 is a digest of the pertinent economic principles that serve as foundations for the allocation of scarce resources. In Chapter 3 attention is directed toward revenue-cost relationships via breakeven analysis of a firm's operations. In the next chapter, calculations to determine minimum-cost levels are explored for the replacement of assets, resource allocations to projects, control of quality, and inventory policies. Ways to economize through value engineering and productivity-improvement programs are investigated in Chapter 5, and methods of evaluating performance, including practical comparison methods for intangible qualities, are discussed.

- The conventional mathematics of money is presented in Section Two. Four chapters are devoted to the mechanics of time-value calculations and comparisons of alternatives based on their equivalent annual worth, present worth, and rate of return. Applications of all three approaches, such as replacement studies, cost of capital, and incremental analysis, are embedded in the chapters rather than treated as separate subjects. Interest factors follow ASEE-suggested functional notations, and interest calculations are displayed on cash-flow diagrams. A wide variety of examples and exercises is provided.

- Section Three examines the practical difficulties of conducting economic comparisons. Data-collection methods, including cause-and-effect and forecasting models, are followed by sensitivity analysis in Chapter 10. The role of accounting and the effects of depreciation and taxes on economic evaluations are extensively treated in Chapter 11. A long chapter on industrial practices features funding sources, inflation effects, customized economic analysis procedures, and capital budgeting. Chapter 13 explores in depth the use of benefit-cost analysis for public projects.

- The last section offers ways to analyze problems when outcomes are in doubt. Risk analysis in Chapter 14 introduces probability concepts and continues with their application to expected-value comparisons, discounted decision trees, minimum-cost models, and investments to avoid risk. Situations in which probabilities cannot be estimated are discussed in Chapter 15, where perspectives for decision making are drawn from game theory and principles of choice.

## ORIENTATION

An author writing a text is like a chef creating a new concoction. Both combine new ingredients with the old, tinker with conventional recipes, season their creations according to their tastes, and hope others will appreciate their contribution. The foregoing survey outlined the ingredients of this text and the recipe for engineering economic studies, both of which are seemingly conventional. My hoped-for contribution arises from the extra emphasis given to certain developments; the flavor of the text should be enhanced by:

- Linking computational methods to the assumptions upon which they are based, particularly the supporting economic principles
- Including financial realities from the business world—both opportunities and restrictions—that influence economic decisions
- Bringing out current concerns confronting today's engineering economists as they face tomorrow's challenges, such as measures for protecting the environment, a possible steady-state economy stagnated by resource limitations, consumer protection, and productivity improvement
- Examining how and why certain evaluation techniques are used in actual applications, even though they are conceptually weak
- Providing special attention to the analysis of public projects and programs, while recognizing the often conflicting issues and responsibilities involved
- Presenting a number of extensions of conventional economic comparisons which use probability-based models, some thought-provoking concepts for decision making under uncertainty, and historical developments that provide a perspective for current practices

## FEATURES

A text on engineering economy is expected to have plenty of detailed examples, comprehensive interest tables, and an ample assortment of practice exercises. This book has them. For instance, there are over 600 questions and problems, with answers provided for about one-third of them. In addition, two unusual features are offered:

1 Short *economy exercises* appear throughout the chapters. These are short problems that pertain to or elaborate on the topic being studied. Worked-out solutions are given at the end of the chapter. An honest effort to solve each economy exercise before flipping pages to find the answer should reveal how well the concepts or techniques are understood. (Another check is offered by chapter summaries; finding anything unfamiliar in the summary is a signal to review.)

2 At the end of each chapter's problem section are *extensions* to subjects presented in the chapter. They are supplementary topics that can be treated as study cases because questions are provided for each one; or, they may be considered as readings because they are outside the regular flow of subject matter. Some of them deal with personal finances and socioeconomic situations (doomsday scenarios, pollution control, distribution of productivity gains, who pays how much tax for what, international investment criteria), whereas others introduce analysis methods that complement those in the chapters (linear programming, critical path scheduling, standardized scoring, learning curves, simulation, queuing theory, utility theory).

The extensions yield added flexibility for classroom studies. There is ample material in the text for a two-term course sequence. When so used, the chapters can be assigned in the given order to secure a logical and integrated development of the subjects. When only one term is available for engineering economics, different chapter arrangements can be utilized to customize the nature of the course. Three possible combinations are tabulated below: (1) a *conventional engineering economy* course which stresses the time value of money and discounted cash-flow comparisons; (2) a survey of *economic decision models* that introduces a wide variety of quantitative analysis techniques with associated qualitative considerations; and (3) a technically oriented *economic analysis of operations* that features practical methodology to evaluate operations and improve operating effectiveness.

SUGGESTED CHAPTERS AND TOPICS FOR ONE-TERM COURSES

| *Engineering Economy* | *Economic Decision Models* | *Economic Analysis of Operations* |
|---|---|---|
| 1 Introduction | 1 Introduction | 1 Introduction |
| 6 Interest formulas | 2 Basic economics | 2 Basic economics |
| 7 Annual worth | 3 Breakeven analysis | 3 Breakeven analysis |
| 8 Present worth | 4 Minimum cost | 4 Minimum cost |
| 9 Rate of return | 5 (last half) Intangibles | 5 Economizing |
| 10 Data and sensitivity | 6 Interest formulas | 6 Interest formulas |
| 11 Taxes and depreciation | 12 Industrial practices | 7 Annual worth |
| 12 Industrial practices | 11 (first half) Benefit-cost | 8 Present worth |
| 5 (last half) Intangibles | 10 (last half) Forecasting | 9 Rate of return |
| 13 Public projects | 14 Risk | 12 Industrial practices |
| 14 (first half) Risk | 15 Uncertainty | 14 (first half) Risk |
| Extensions in Chapters 1, 6 to 9, 11, and 12 | Extensions in all included chapters | Extensions in Chapters 1, 4 to 10, and 12 |

A complete instructor's manual provides solutions to all the problems and additional suggestions for conducting courses.

## TRIBUTE

Money is the modern equivalent of the long-sought "philosopher's stone." For centuries, alchemists vainly sought the "stone" that could transform one type of metal into another. Now we have that capacity, indirectly. With money as the medium of exchange, we can convert one type of resource into other types easily and rapidly, but not always wisely. The worthiness of this transformation is a subtle aspect of the mission of engineering economists. Telltale analyses of alternatives reveal the innermost workings of a project and burden the analyst with ethical responsibilities atop fiscal obligations. Prerequisite to bearing the burden is a thorough knowledge of accepted economic principles and practices. This knowledge facilitates putting a legitimate monetary value on the transformation of each resource and, combined with technical expertise about the subject, allows an accurate and conscientious appraisal of worthiness. Resource commitments monetized as cash flows set a quantitative framework for ensuing qualitative value considerations. This book is dedicated to the economic analysts who contribute to resource-allocation decisions. May you do so wisely.

It can be said that those who manage people manage people who manage works, but those who manage money manage all. I hope you accept the challenge and enjoy a satisfying and profitable experience from *Engineering Economics*.

*James L. Riggs*

# ENGINEERING ECONOMICS

# CHAPTER 1

# PROLOGUE: THE SCOPE OF ENGINEERING ECONOMICS

Which one of several competing designs should be selected?

Should the machine in use be replaced with a new one?

With limited capital available, which investment alternative should be funded?

Would it be preferable to pursue a safer, conservative course of action or follow a riskier one that offers higher potential returns?

Among several proposals for funding that yield substantially equivalent worthwhile results but have different cash flow patterns, which is preferable?

Are the benefits expected from a public-service project large enough to make its implementation costs acceptable?

Such questions are typical of those posed to engineers as they practice their profession, and ways to answer the queries are the contents of this book.

## ENGINEERING DECISION MAKERS

Two characteristics of the above questions should be apparent. The first is that each deals with a choice among alternatives, and the second is that all of them involve economic considerations. Less obvious are the requirements of adequate data and an awareness of technological constraints, to define the problem and to identify legitimate solutions. These considerations are embodied in the decision-making role of engineering economists to

**1** Identify alternative uses for limited resources and obtain appropriate data
**2** Analyze the data to determine the preferred alternative

The breadth of problems, depth of analysis, and scope of application that a practicing engineer encounters vary widely. Newly graduated engineers are regularly assigned to cost-reduction projects and are expected to be cost conscious in all their operations. As they gain more experience, they may become specialists in certain application areas or undertake more general responsibilities as managers. Beginners are usually restricted to short-range decisions for low-budget operations, while engineering managers are confronted with policy decisions that involve large sums and are influenced by many factors with long-range consequences. Both situations are served by the principles and practices of engineering economics.

A decision is simply a selection from two or more courses of action, whether it takes place in construction or production operations, service or manufacturing industries, private or public agencies. Some choices are trivial or largely automatic, but other decisions can be challenging, exciting experiences. Most major decisions, even personal ones, have overtones of economy. This consistent usage makes the subject of engineering economics especially challenging and rewarding.

## ENGINEERING AND ECONOMICS

Prior to about 1940, engineers were mainly concerned with the design, construction, and operation of machines, structures, and processes. Less attention was given to the resources, human and physical, that produced the final products. Many factors have since contributed to expanded engineering responsibilities and concerns.

Besides the traditional work with scientists to develop new discoveries about nature into useful products, engineers now are expected not only to generate novel technological solutions but also to make skillful financial analyses of the effects of implementation. In today's close and tangled relationships among industry, the public, and government, cost and value analyses are supposed to be more detailed and inclusive (e.g., worker safety, environmental effects, consumer protection) than ever before. Without these analyses, an entire project can too easily become more of a burden than a boon.

Most definitions of engineering recognize the mission of engineers to transform the resources of nature for the benefit of the human race. The types of resources susceptible to engineering enrichment include everything from ores and crops to information and energy. A growing awareness of finite limits for earth's resources has added a pressing dimension to engineering evaluations. The focus on scarce resources welds engineering to economics.

Dr. Paul A. Samuelson, Nobel laureate in economics, says that

> Economists today agree on a general definition something like the following: Economics is the study of how men and society end up *choosing*, with or without the use of money,

> to employ *scarce* productive resources that would have alternative uses, to produce various commodities and distribute them for consumption, now or in the future, among various people and groups in society. It analyzes the costs and benefits of improving patterns of resource allocation.*

The relationship of engineering to economics can be likened to that of engineering to physics. The sciences are devoted to the discovery and explanation of nature's laws. Engineers work with the scientists and extend the revelations to practical applications. The "laws" of economics are not as precise as those of physics, but their obvious application to production and the utilization of scarce resources assures increasing attention from engineers.

---

**Economy Exercise 1-1** The past few years have witnessed a remarkable number of events and activities that will undoubtedly have far-reaching effects on engineering practices. Some of the most prominent areas of activity include:

1 The passage of national, state, and local legislation that regulates industrial operations and developments
2 The formation of public pressure groups whose efforts are directed toward improving the quality of life
3 The increasing world market values for raw materials
4 The closer scrutiny of government programs

Select a specific example that illustrates activity in each of the above areas, and relate it to the work of engineers with particular reference to engineering economic considerations.

---

## ECONOMICS: A CAPSULE VIEW

Economics, like engineering, has informal roots deep in history. The construction of the pyramids is considered an engineering marvel; it was also a significant economic accomplishment to have funneled all the necessary resources into monuments rather than consuming them in commerce. The formal roots of economics stretch back two centuries to the publication (in 1776) of Adam Smith's *The Wealth of Nations.*

Early writings deplored government intervention in commerce and promoted a "laissez faire" policy. Thomas Malthus, in *An Essay on the Principles of Population* (1798), conjectured about the causes of economic crises, saying population tends to increase geometrically and the means of subsistence only arithmetically; his forecasts of misery for most of the population predisposed the "dismal science" nickname for economics. Later, John Stuart Mill, in *Treatise on Political Economy* (1800), argued against Malthus' pessimism by suggesting that the laws of distribution are not as immutable as the laws of production. Modern "doomsday" scenarios indicate the issue is still in doubt. (See Extensions 1.1 and 1.2.)

Karl Marx, in *Das Kapital* (1867), argued that capitalism would be superseded by

*P. A. Samuelson, *Economics*, 9th ed., McGraw-Hill, New York, 1973, p. 3.

socialism, which would then develop into communism. According to his views, workers produce more value than they receive in wages. The surplus takes the form of profit and allows capital accumulation, but the capitalist system will eventually fail, owing to cyclic depressions and other inherent weaknesses. About one-third of the world's population agrees with Marx.

"New Economics" evolved from the work of John Maynard Keynes in the 1930s. In his *General Theory of Employment, Interest, and Money,** Keynes clashed with classical economic theory by proclaiming, for example, that interest rates and price-wage adjustments are not adequate mechanisms for controlling unemployment in capitalistic economies. Refinements and extensions of the original work are collectively called "Keynesian economics," which is just one of many current schools of economic thought.

Keynes' and Marx' theories deal with the entire economic system in terms of national income, flow of money, consumption, investment, wages, and general prices. This level of analysis, concerned with the economy as a whole, is called *macroeconomics.* It produces economywide statistical measures such as the national cost of living index and total employment figures.

*Microeconomics* is the study of economic behavior in very small segments of the economy, such as a particular firm or household. It is generally assumed that the objective of a firm is to maximize profit, and the objective of a household is to maximize satisfaction. Measurement statistics for a small economic unit might be the number of workers employed by a firm and income or expenditures of a given firm or family.

Engineering economics, with its focus on economic decision making in an individual organizational unit, is closely aligned with microeconomics.

## ENGINEERING ECONOMY: A SHORT HISTORY

*The Economic Theory of the Location of Railways,* written by Arthur M. Wellington in 1887, pioneered engineering interest in economic evaluations. Wellington, a civil engineer, reasoned that the capitalized cost method of analysis should be utilized in selecting the preferred lengths of rail lines or curvatures of the lines. He delightfully captured the thrust of engineering economy as

> It would be well if engineering were less generally thought of, and even defined, as the art of constructing. In a certain important sense it is rather the art of not constructing; or, to define it rudely but not inaptly, it is the art of doing that well with one dollar which any bungler can do with two after a fashion.†

In the 1920s, J. C. L. Fish and O. B. Goldman looked at investments in engineered structures from the perspective of actuarial mathematics. Fish‡ formulated an investment model related to the bond market. Goldman, in his book *Financial Engineering,* proposed a compound-interest procedure for determining comparative values, and said

*J. M. Keynes, *General Theory of Employment, Interest, and Money,* Harcourt, Brace & World, New York, 1936.

†A. M. Wellington, *The Economic Theory of the Location of Railways,* Wiley, New York, 1887.

‡J. C. L. Fish, *Engineering Economics,* 2d ed., McGraw-Hill, New York, 1923.

> It seems peculiar and is indeed very unfortunate that so many authors in their engineering books give no, or very little, consideration to costs, in spite of the fact that the primary duty of the engineer is to consider costs in order to obtain real economy—to get the most possible number of dollars and cents: to get the best financial efficiency.*

The confines of classical engineering economy were staked out in 1930 by Eugene L. Grant in his text, ***Principles of Engineering Economy***.† Professor Grant discussed the importance of judgment factors and short-term investment evaluation as well as conventional comparisons of long-run investments in capital goods based on compound-interest calculations. His many contributions resulted in the recognition that "Eugene L. Grant can truthfully be called the father of engineering economy."‡

Modern approaches to discounted cash flow and capital rationing were influenced by the work of Joel Dean.§ He incorporated the theories of Keynes and other economists to develop ways to analyze the effects of supply and demand for investment funds in allocating resources.

Current developments are pushing the frontiers of engineering economics to encompass new methods of risk, sensitivity, and intangible analysis. Traditional methods are being refined to reflect today's concerns for resource conservation and effective utilization of public funds.

## PROBLEM SOLVING AND DECISION MAKING

An engineering economist draws upon the accumulated knowledge of engineering and economics to fashion and employ tools to identify a preferred course of action. The tools developed so far are not perfect. There is still considerable debate about their theoretical bases and how they should be used. This concern is wholesome because it promises improved procedures, but the variety of analysis techniques can frustrate practitioners, especially inexperienced ones: There are many aspects to consider, and many ways to consider them.

The fundamental approach to economic problem solving is an elaboration of the time-honored "scientific method." The method is anchored in two worlds: the real, everyday working world and the abstract, scientifically oriented world. As depicted in Figure 1.1, ***problems*** in engineering and managerial economy originate in the real world of economic planning, management, and control. The problem is confined and clarified by ***data*** from the real world. This information is combined with scientific principles supplied by the analyst to formulate a ***hypothesis*** in symbolic terms. The symbolic language aids the digestion of data. By manipulating and ***experimenting*** with the abstractions of the real world, the analyst can simulate multiple configurations of reality which otherwise would be too costly or too inconvenient to investigate. From this activity a ***prediction*** emerges, usually.

The predicted behavior is converted back to reality for testing in the form of hardware,

*O. B. Goldman, ***Financial Engineering***, Wiley, New York, 1920.

†E. L. Grant, ***Principles of Engineering Economy***, Ronald, New York, 1930.

‡A. Lesser, Jr., "Engineering Economy in the United States in Retrospect—An Analysis," ***The Engineering Economist***, vol. 14, no. 2, 1969.

§J. Dean, ***Capital Budgeting***, Columbia, New York, 1951.

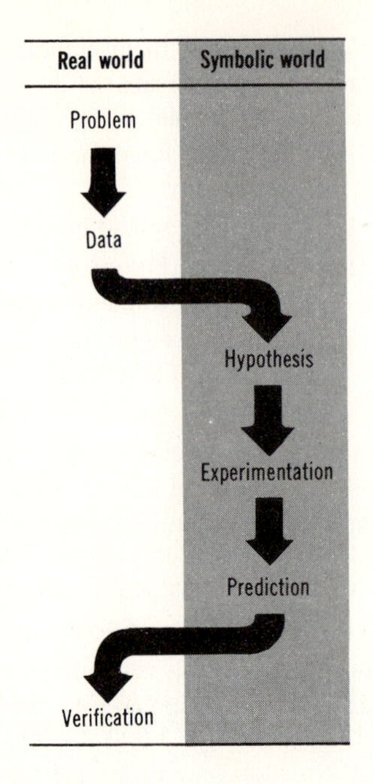


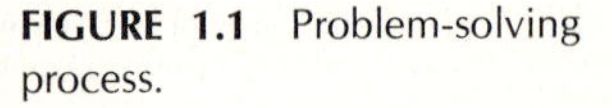
**FIGURE 1.1** Problem-solving process.

designs, or commands. If it is valid, the problem is solved. If not, the cycle is repeated with the added information that the previous approach was unsuccessful. Fortunately, a host of successful approaches have been discovered and validated for economic analyses; the challenge now is to use them wisely.

## Intuition and Analysis

Because engineers generally attack practical problems with solution deadlines instead of engaging esoteric issues for long-term enlightenment, their mission might appear relatively simple. Engineering economic evaluations could even seem mundane, since they usually rely on data from the marketplace and technology from the shelf: Just grab prices from a catalog, plug them into a handy formula, and grind out an answer. Occasionally, such a routine works. Spectacular workbench discoveries and overnight fortunes attest to the fact that plungers sometimes win. There are also innumerable instances where rule-of-thumb, skin-deep evaluations are absolutely unsatisfactory.

As represented in Figure 1.2, a decision made now is based on data from past performances and establishes a course of action that will result in some future outcome. When the decision is shallow and the outcomes are unnotable, a reflex response based upon intuition is feasible. Intuition, immediate knowledge without conscious reasoning, is oriented in the present but informally embraces memories of the past and estimates of what may happen in the future. It is a valid tool for a decision maker's repertoire provided the logic is not flawed by fads and unrecognized bias.

Instinctive judgments are often formalized by *standard operating procedures* (SOPs). In economic analyses, SOPs often take the form of worksheets for the justification of investments. Such short-form justifications are typically limited to smaller investments, say

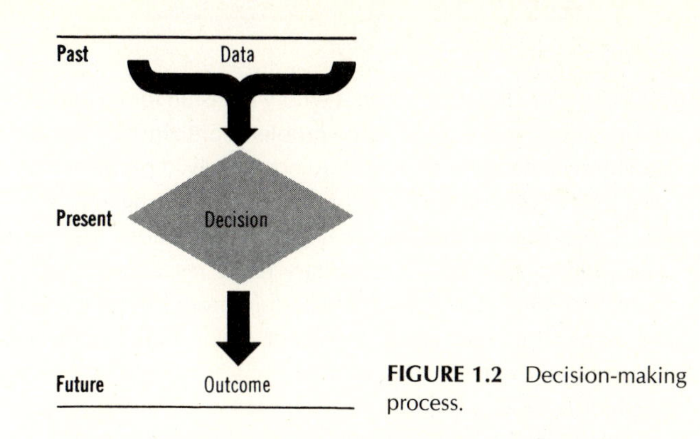


**FIGURE 1.2** Decision-making process.

$2000, which can be recaptured from savings generated by the investment within 6 months or 1 year. These forms or similar SOPs represent collective intuition derived from experience. They have a secure place in economic evaluations, but their use should be tempered by economic principles and a continuing audit to verify that previous judgments are appropriate for current decisions.

Most significant problems require both analysis and personal judgment. Initially, the analyst settles on which evaluation technique to utilize and how to apply it. As the solution procedures progress, factors that are difficult to quantify often arise. These are called ***intangibles;*** they represent aspects of a problem that cannot be translated readily into monetary values. Intuitive ratings are frequently assigned to intangibles to allow them to be included in the decision process. Judgment also enters the process in determining whether a solution is well enough founded to be accepted. Thus intuition and judgment complement analysis methods by contributing to better decisions.

---

**Economy Exercise 1-2** Most decision makers informally set boundaries for routine responses to noncritical problems of a personal and professional nature. Three possible parameters to identify routine responses are shown in Figure 1.3. The level that separates an automatic decision from a problem that requires more investigation varies among decision makers. What are the advantages and dangers of relying on intuition or SOPs for solutions to economic problems?

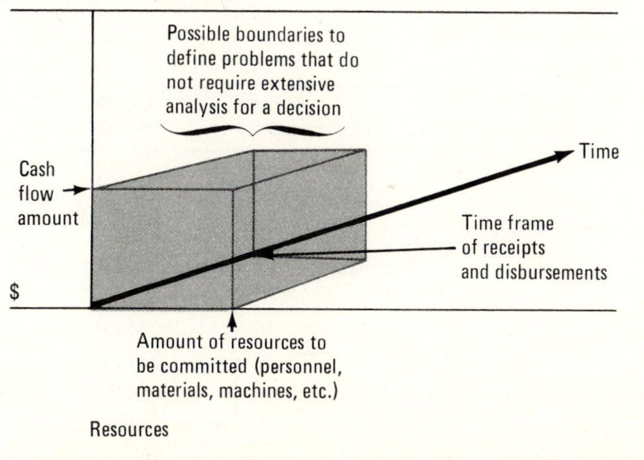


**FIGURE 1.3** Criteria for routine responses to an economic problem.

## Tactics and Strategy

About the only thing more frustrating than a wrong decision for an important problem is the right decision for the wrong problem. Some problems are virtually handed to an analyst on a platter, complete with data trimmings. More commonly, a problem is ill defined, and the analyst is forced to seek the intent of a solution before applying analytical tools. Recognizing the difference between tactical and strategic considerations may clarify the purpose.

*Strategy* and *tactics* historically are military terms respectively associated with broad plans from the high command and specific schedules from lower echelons. Strategy sets ultimate objectives, and the associated tactics define the multiple maneuvers required to achieve the objectives. Strategic and tactical considerations have essentially the same meaning for economic studies.

There are usually several strategies available to an organization. A strategic decision ideally selects the overall plan that makes the best use of the organization's resources in accordance with its long-range objectives. A strategic industrial decision could be a choice from among several different designs to develop or products to promote. In government, strategic evaluations could take the form of benefit-cost analyses to select the preferred method of flood control or development of recreational sites. The measure of merit for strategic alternatives is *effectiveness*—the degree to which a plan meets economic targets.

A strategic plan can normally be implemented in a number of ways. For example, each industrial design or product has tactical alternatives such as which kind of machine to employ or materials to use; tactics for flood control might involve choices among dams, levees, dredging, etc. The relative values of tactical choices are rated according to their *efficiency*—the degree to which an operation accomplishes a mission within economic expectations.

The relationship between strategies and tactics offers some constructive insights. The effectiveness of each strategy is initially estimated from the effect it will have on system objectives. It thus serves as a guide to the area in which tactics will produce the highest efficiency. The actual efficiency of each tactic is determined from a study of the activities required to conduct the tactical operation.

Two strategies, each with three apparent means of accomplishment, are depicted in Figure 1.4. The average efficiency for the tactics associated with strategy 1 (tactics 1-1, 1-2, and 1-3) has a higher value than that for strategy 2. However, it could happen that a strategy

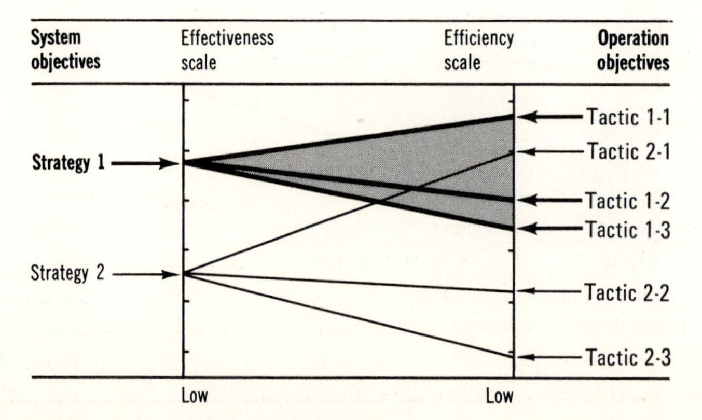


**FIGURE 1.4** Relationship of tactics and strategies.

with a lower effectiveness possesses the tactic with the highest efficiency. Tactic 2-1 is close to the most efficient tactic of strategy 1. If it were the highest on the efficiency scale, it would be the leading candidate for selection, regardless of its strategic origin.

## Sensitivity and Suboptimization

The decision situation related by Figure 1.4 has high sensitivity; that is, it is vulnerable to small changes in the controlling conditions. With tactics 1-1 and 2-1 so close on the efficiency scale, a slight change in operating conditions or external influencing factors could switch the positions of the top tactics, or even the strategies. An insensitive situation occurs when all the tactics for a given strategy have a higher efficiency than the best tactic of any other strategy. The consequence of high sensitivity is to force a complete investigation to assure the validity of the data being evaluated.

A *sensitivity analysis* can be conducted on any problem to explore the effects of deviations from the original problem conditions. Since most engineering economic problems extend over a period of years, future cash flows are necessarily estimated. These estimations may be quite reliable, but it is often enlightening to observe how the attractiveness of alternatives varies as the initial estimates are altered.

Whenever multiple objectives are present in a decision situation, it is probable that there is no single course of action that will optimize all the objectives simultaneously. This conflict situation could lead to *tactical suboptimization*—a solution that optimizes tactical efficiency with little or no regard for strategic effectiveness. "You don't understand the big picture" is a common complaint from harried managers to seemingly sound proposals to alleviate an operating problem, proposals that would, if adopted, clash with policies established for the overall benefit of the organization.

In general, suboptimization occurs when there is a larger problem than the analyst has visualized. It is always tempting to employ intact a classical textbook solution to a real-world problem, whether or not it truly fits the actual conditions. The availability of "canned" computerized solutions to complex problems increases the temptation. Another cause of suboptimal solutions is the legitimate analysis technique of partitioning a large problem into parts during a preliminary investigation to avoid being bogged down in a deluge of details. Trouble enters when tentative solutions to the problem's parts are not integrated. Advances in computer science and operations research may eventually allow analysis of an entire complex system in a single evaluation, but until then it helps to be aware of the areas in which suboptimization is most likely to occur. Three regularly encountered perspectives that lead to suboptimization are described below.

### 1 CROSSEYED VIEW

Both organizations and individuals can be confused by opposing objectives. An example of the danger inherent in focusing on just one parameter while blurring others is what would happen to a company that redeployed its resources to save its ailing flagship product at the expense of the rest of the product line. The rescue could boost sales for the previously eminent product while total sales declined owing to the drain on resources suffered by the rest of the company's products; thus, the battle could be won but the war lost.

Individuals seeking "the good life" also get caught by conflicting goals. If "good" is in-

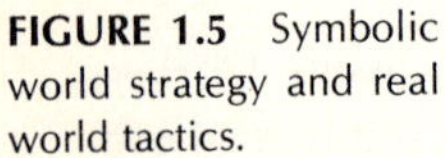
**FIGURE 1.5** Symbolic world strategy and real world tactics.

terpreted as "long and full," then unlimited pleasure seeking for a full life would undoubtedly jeopardize the health needed for a long life. Moderation, however, should produce a temperate plan to satisfy both goals, resulting in a life less full but longer. Of course, there are also irreconcilable objectives such as those pictured in Figure 1.5.

## 2 SHORTSIGHTEDNESS

Tactics based on a planning horizon of 1 or 2 years may not have the same efficiency as those based on a longer span of years. Suppose a manufacturer anticipates using a fixed number of containers each year. The containers can be purchased or the manufacturer can make them by acquiring new production equipment. Costs for the choices are displayed by the breakeven chart in Figure 1.6. A planning horizon under 2 years would indicate that purchasing is the preferable alternative; beyond 2 years it is more attractive to make the

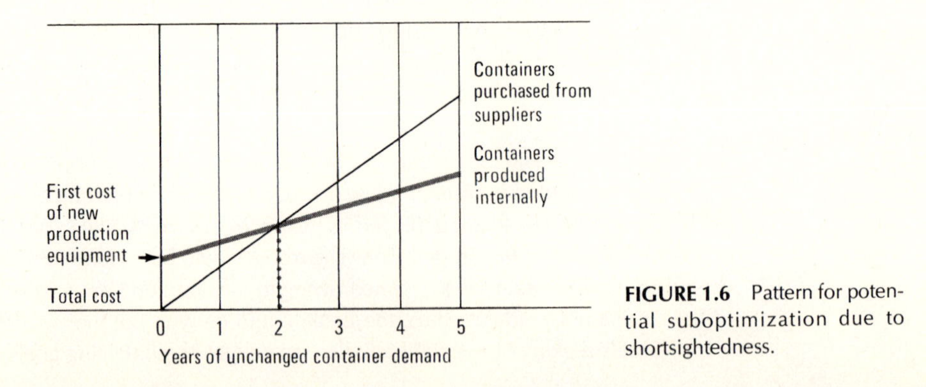


**FIGURE 1.6** Pattern for potential suboptimization due to shortsightedness.

containers. Individuals face the same danger of suboptimization in lease-or-buy decisions for housing and transportation.

### 3 TUNNEL VISION

Organizations are very susceptible to situations in which departments understand the common goal but individually go about working toward the goal in ways that hurt each other. A typical example is the goal to reduce material and inventory costs, as viewed by

- *Purchasing* "Buy in large quantities to get quantity discounts."
- *Comptroller* "Buy in smaller quantities to avoid paying interest on the capital required for purchases."
- *Production* "Larger inventories allow longer production runs which reduce manufacturing costs."
- *Warehousing* "Larger inventories cost more to store and increase the cost of material handling."

If each of the involved departments acts independently, inventory levels will behave like a yo-yo. Obviously, a workable plan will be a compromise, probably satisfying no one completely, but still producing lower total material costs for the organization as a whole.

---

**Economy Exercise 1-3** A subassembly line has been giving the production manager nightmares for months. All kinds of minor modifications have been tried, and all failed to improve output. The current per-unit cost is $4.20, which seems reasonable, but output has failed to reach the required 10,000 units per year. A check with the purchasing department reveals that supplementary units are now being purchased for $4.75 each, but one vendor agrees to provide them at $4.50 each if the entire annual demand is ordered.

The subassembly-line supervisor suggests acquiring three new machines to mechanize successive stages of the production process. Engineers calculate that the machines' purchase price of $100,000 would make their discounted annual cost over a 10-year machine life, coupled with yearly operating costs, amount to $27,000. If needed, the machines have the capacity to double the present output. At the present output level, the remaining subassembly costs using the new machines will annually total $18,000.

While investigating the problem, the engineers uncovered another alternative: All three successive operations could be combined and handled by a single machine. This one combination machine would have the same capacity, speed, life, and remaining production costs as the three-machine alternative, but ownership and operating costs would be reduced by $3000 per year.

Which alternative should be accepted, and why?

---

## THE ENGINEERING ECONOMIC DECISION MAZE

Most important decisions in engineering economics entail consideration of future events. A focus on the future has always had a special and irresistible appeal, but it also encumbers the mission of engineering economists. Not only must they search the past to understand the present and survey the present for hints about the future, but they must consolidate the accumulated results into a pattern that is susceptible to analysis and then select a decision rule to yield a verdict. An indication of the complexities involved is apparent from the maze shown in Figure 1.7.

It would take a much larger maze to portray all the pitfalls and challenges of economic

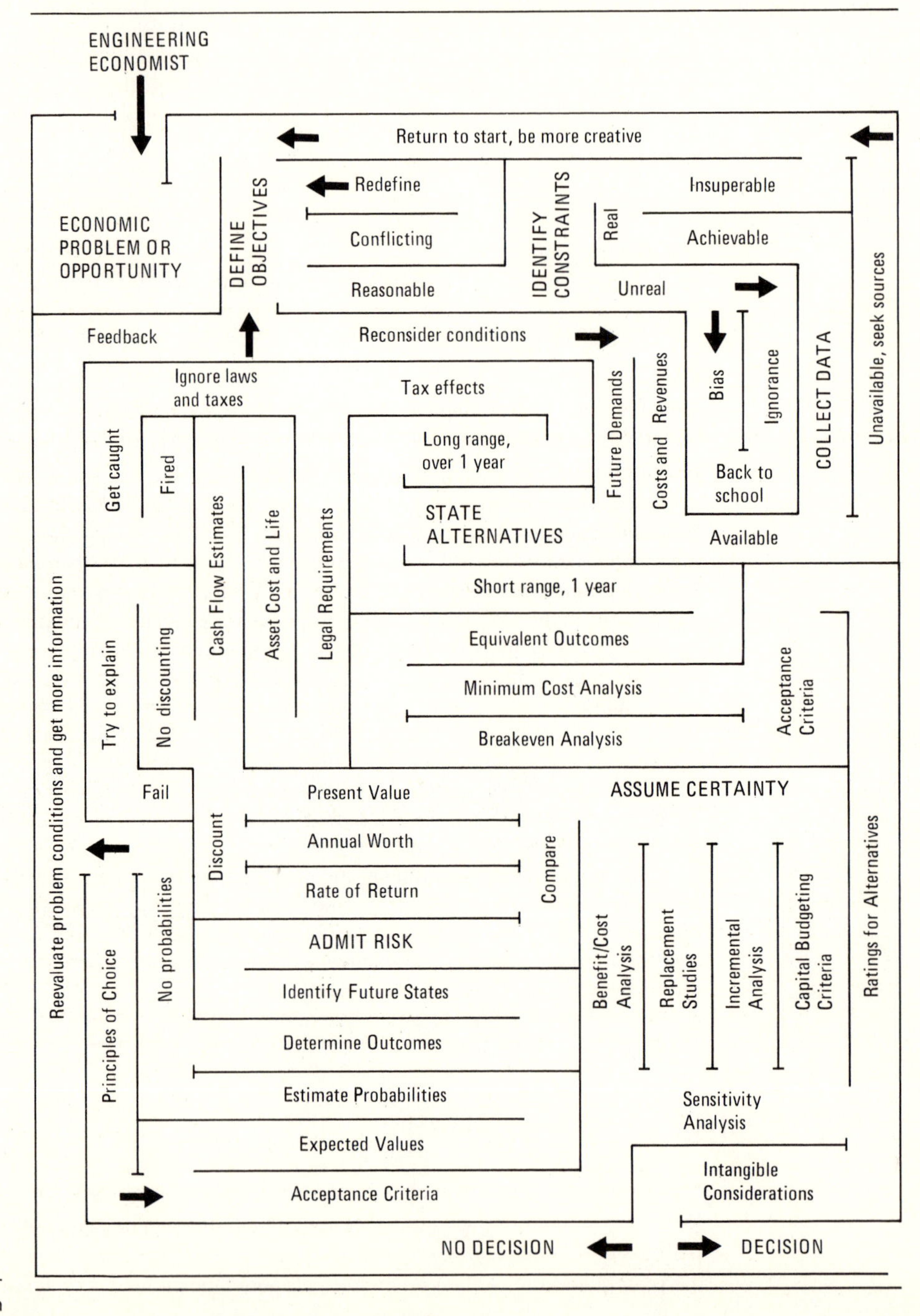


**FIGURE 1.7** Engineering economic decision maze.

analyses, but enough are included to expose the anatomy of engineering economics and to map the contents of this book.

## Feasibility Investigation

Investigation channels.

The maze is entered by invitation from an executive to conduct a study or by the call of regular duties to evaluate options. In both cases the objectives and constraints have to be investigated. An assignment to get the "finest equipment for the least amount of money" is bewildering. It may be possible to decide what is the "finest" equipment, but it is unlikely to be the least expensive. There is no indication of what would be an acceptable compromise. In the same vein, a charge to get the "best equipment for our purposes" is too broad to be useful. Double standards and indefinite objectives lead to second-guessing. A precise statement of realistic goals simplifies the entire decision-making process.

Data act as the fuel for decision making and must be of good quality if the process is to function smoothly. Sources of data are plentiful. Nearly everyone is willing to give an opinion about someone else's problem. Reliable sources are less plentiful. Subjective opinions have a part in information collection, especially in forecasting, but other sources should be thoroughly exploited for "hard" data such as current costs and revenues (see Chapters 10 to 12).

In a preliminary search for a solution, it may be useful to list all possible courses of action. The goal of such a free-wheeling approach is to identify alternatives that might otherwise be overlooked. While it is important to seek novel means to accomplish something, it is just as important to recognize practical limitations. The "laws" of nature and economics (see Chapter 2) cannot be repealed for a pet project. Equivalently, unreal restrictions due to bias or ignorance can constrict options needlessly.

## Adaptivity Analysis

Analysis paths.

Alternatives are compared via economic models that incorporate relationships between cause and effect, supply and demand, objectives and constraints. Representations of the real world may appear as:

1 Scale models that physically look like the real thing
2 Graphical or schematic models that diagram one feature, such as a pie chart to show proportional parts of a total activity or a flowchart to show the sequential steps of an operation
3 Mathematical models which use symbols to represent actual conditions in the form of equations and formulas

The mathematical models used predominantly in engineering economic studies can be categorized according to three analysis areas which correspond to the three main paths through the maze:

1 Analysis of short-range projects involving equivalent outcomes, where the time value of money is not relevant (Section One)
2 Analysis of long-range proposals where inputs to the models are assumed to be known with certainty (Sections Two and Three)

**3** Analysis of long-range proposals where inputs to the models are recognized to be risky or uncertain (Section Four)

Within these categories are several well-accepted methods for comparing alternatives. The preferred method depends on the nature of the problem and the data available. Most long-range proposals involve legal aspects, tax effects, expected receipts and disbursements, and life-cycle data for physical assets (Section Three), but different methods of analysis treat the same inputs in different ways to customize the outputs. For instance, benefit-cost analyses (Chapter 13) are usually associated with government activities, whereas rate-of-return models (Chapter 9) are linked to industrial applications.

## Optimality Determination

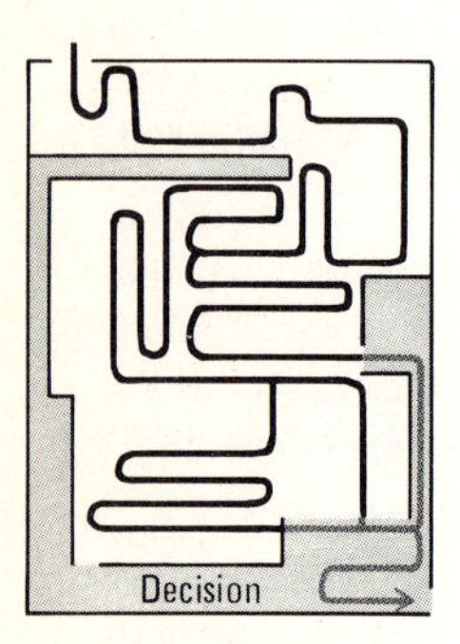


Optimization routes.

An analysis using mathematical models generates a numerical rating for each alternative. These ratings may be dollar values, percentages, or simply numbers that measure attractiveness (see Chapter 5). An analyst's job may be complete after the numerical rating is computed, but someone still has to decide whether that rating is sufficient for the proposal it represents to be accepted.

Most organizations have definite policies to discriminate between acceptable and unacceptable proposals (see Chapter 12). These acceptance criteria will likely vary over the years in response to the organization's financial health and may even vary among divisions of the organization. Intangible considerations (Chapter 5) can sway a decision in subtle ways, although they seldom receive formal recognition in the acceptance criteria. Borderline cases should be subjected to sensitivity analyses (Chapter 10) when there is doubt about any of the influencing factors. If doubts persist, it may be prudent to recycle through the feasibility-adaptivity-optimality process.

Eventually each decision, cultivated by investigation and polished by analysis, is exposed to the unsympathetic workings of the real world. It is worth reviewing both the painful and the pleasurable outcomes because these experiences should improve future decisions. And the pace of technology advances assures there will be no shortage of future economic decisions to engage engineers (see Chapter 15).

## SUMMARY

Proficiency in economic analysis is an essential skill for engineers wherever they practice. Current concerns for economy of operations and conservation of resources emphasize the importance of sound economic decisions, while the complexities of modern technology render such decision making ever more difficult.

***Strategies*** to achieve overall goals are rated according to their effectiveness, and the ***tactics*** available to implement the goals are rated by their efficiency. A decision is considered ***sensitive*** when tactics associated with different strategies have nearly the same efficiencies. ***Suboptimization*** can occur when a solution optimizes tactical efficiency with inadequate regard for its strategic effectiveness. Causes of suboptimization include confusion caused by opposing objectives, planning horizons that are too short, and operating decisions that conflict with organizational policies.

An engineering economist draws upon the accumulated knowledge of engineering and economics to identify alternative uses for limited resources and to select the preferred alternative. Evaluations rely mainly on mathematical models, but judgment and experience are pivotal inputs. Many accepted models are available to conduct analyses of short-range projects where the time value of money is not relevant, and long-range proposals where discounting is required for input data assumed to be certain or admittedly risky. Astute selection and correct use of investigation methods, analysis techniques, and optimality criteria guide an analyst through the engineering economic decision maze.

## Discussion of Economy Exercises

**EE 1-1** Numerous examples can be cited to show how changing societal interests direct the practices of engineers.

1 *Laws* The Occupational Safety and Health Act of 1971 (OSHA) established stricter standards for worker safety. Many states and communities have passed zoning laws that set tighter design requirements for new plants. Most legislative actions require added investments, and engineers are called upon to devise ways to meet new standards with minimal expenditures.
2 *Pressure groups* Environmental groups influenced the design of the Alaska pipeline. Consumer-protection groups successfully campaigned for automobile modifications and control of waste discharges. New technologies have to be developed to protect the quality of air, water, and land; the relationship between control costs and quality levels is largely an engineering function.
3 *Scarce resources* Higher prices for energy, especially gas and oil, sparked worldwide alarm over the future availability of raw materials. Higher prices for certain materials suggest design changes for greater economy. Some resource conservation measures have yielded surprising savings.
4 *Government programs* The abandonment of plans for supersonic transports and declining budgets for space exploration exemplify critical assessments of government activities. Engineers are intimately involved in assessing the effectiveness of government programs. Benefit-cost analyses are routinely conducted by the Corps of Engineers and other government agencies to determine the acceptability of projects.

**EE 1-2** Since there are limits to a decision maker's time and energy whereas the reservoir of problems often seems infinite, guidelines are necessary to confine involvement. SOPs do save time. An intuitive response is quick. Both draw upon experience to yield a reasonable solution. However, handy answers may mask better solutions that could have been exposed by analysis. What was good for yesterday's operations may not be adequate for tomorrow's. Difficulties can occur when the scale of current operations changes without a corresponding revision in the cutoff levels for routine decisions.

**EE 1-3** Apparently, it has been decided that something must be done to improve subassembly production, so the "do-nothing" alternative is eliminated. The single machine is obviously more attractive than the three-machine alternative because the combined operation costs $3000 less. The unit cost for the one-machine alternative is

$$\frac{\$27{,}000 + \$18{,}000 - \$3000}{10{,}000 \text{ subassemblies}} = \$4.20/\text{subassembly}$$

This unit cost is the same as the current cost but promises more reliability. It is also $0.30

per unit less expensive ($4.50 − $4.20) than purchasing all the subassemblies from a supplier. However, more information is needed about the long-range (10-year) expected demand for the subassemblies. Without this information, the decision is subject to the make-or-buy pattern of suboptimization shown in Figure 1.6.

## PROBLEMS

**1.1** There are many general definitions of engineering, and specific ones for different branches of engineering. Look up one of them, and comment on the explicit and/or implied attention paid to economic considerations.

**1.2** It has been said that economists are very busy people because they have to spend full time telling what's going to happen and full time explaining why it didn't. Engineering economists also work with forecasts of the future but are not usually subjected to such joking comments. Why?

**1.3** Efficiency is defined as output divided by input (times 100 percent). Engineering efficiency is commendable when it approaches 100 percent, but financial efficiency must exceed 100 percent before it is considered adequate. Explain.

**1.4** "In the 1930's many economists maintained that the United States economy had reached the apogee of its growth; in the period just after World War II, many predicted an immediate depression; and in late 1969, some economists predicted without qualification that there would be no recession in 1970; but all these assertions proved to be erroneous. This is not to deny, of course, that the predictions made by economists often are accurate; indeed, at a given moment reputable economists make so many conflicting forecasts that one is almost certain to be correct. (To give one illustration, in December of 1969 Milton Friedman said that a recession on the order of that in 1960 seems to be in the cards for 1970; Raymond J. Saulnier said there was a 50-50 chance of a recession; and Pierre A. Rinfret said that 'There ain't gonna be no recession in 1970, period.')"*

**1.4a** Why do economic projections for a future event tend to vary more widely than engineering estimates of performance for a new design?

**1.4b** Why might there be less confidence in the economic performance than the operating performance of a new product?

**1.5** Entropy is a familiar concept in engineering science. Without stretching the concept unduly, it could be applied to the resources of the earth. Ores of value are scattered around the globe, with occasional pockets of higher concentration. Most other raw materials exhibit similar distributions. One of the missions of engineers is to decrease entropy. Relate engineering, production, and consumption to entropy.

**1.6** The following examples illustrate everyday cases in which individuals are required to decide the most advantageous use of limited resources. Identify the strategic or tactical nature of each decision situation, and discuss the factors that should be considered (including sensitivity, if appropriate).

**1.6a** A fisherman stands in his crowded cabin and studies a map. Three fishing grounds are circled on the map. He mentally compares recent reports and gossip concerning the grounds. He also considers the market demands, the weather forecasts, the condition of his ship, and the supplies on board. He has to decide where to fish.

*R. Handy and E. C. Harwood, *A Current Appraisal of the Behavioral Sciences*, rev. ed., Behavioral Research Council, Great Barrington, Mass., 1973.

**1.6b** An engineer contemplates the bulging walls of a large concrete culvert under an interstate highway. The collapsing culvert is the result of thousands of tons of rock recently stockpiled on the roadbed above in preparation for new construction. In a few days the spring thaw will soak the ground and send torrents of water through the culvert. The engineer speculates on possible designs in terms of the conditions and the restrictions imposed by available equipment, materials, and time.

**1.6c** The owner of a wholesale distribution center seeks to improve his delivery service in order to meet competition. To do so he can buy or rent more trucks, subcontract his deliveries, open additional outlets, and/or improve his handling facilities. His capital is limited, and the outlook for increased volume is uncertain. First he must decide if any action is needed. If it is, he must select the most suitable alternative.

**1.6d** The manager of a large manufacturing company surveys a collection of proposals laid out neatly on the long table in the board room. Each proposal represents many hours of staff work. Each is a detailed plan for the development of a new product. She must select the proposal which best serves the interests of the company within the constraints of restricted physical and human resources, competition, legal requirements, available capital, and corporate objectives. Her decision will affect the activities of hundreds of people.

## EXTENSIONS

***1.1 Perpetual-Motion Mystique*** The idea of perpetual motion—something for nothing—seems to be a persistent, insidious dream of humans. It has been around a long time. Over 2000 years ago the Chinese searched for an unpowered "everlasting going." Archimedes tried to find it through hydraulics, and da Vinci experimented with gravity-powered mechanisms. In 1670, John Wilkens, the Bishop of Chester, designed a ramp leading to a pedestal where a magnet was mounted. The magnet was supposed to attract an iron ball up the ramp until it fell off onto a chute that returned it to the bottom of the ramp, again and again and. . . . (What was wrong with the bishop's reasoning?)

A perpetual-motion machine was exhibited in New York in 1813. People paid to see little carriers ceaselessly moving up and down inclined planes to drive a wheel which offered free energy. Robert Fulton, of steamboat fame, exposed the hoax by showing that the contraption was connected by a hidden strand of catgut to a hand-powered crank in an adjacent room.

The infamous John E. W. Keely perpetual-motion machine, unveiled in 1875, enriched its inventor for years without disclosing any practicality. "Whatever other laws he may have violated in his long career," wrote Stanley W. Angrist in *Scientific American,* "Keely had left the first and second laws of thermodynamics inviolate."

The mystique of perpetual-motion machines has a counterpart in economic ventures—rewards without inputs.

### QUESTIONS

**1.1a** Cite an example of a get-rich-quick scheme or a similar hoax that promises a lot for a little.

**1.1b** How could a trip through the engineering economic decision maze expose a financial hoax or at least make a potential investor wary of a scheme such as cited in Question 1?

***1.2 Doomsday Scenarios*** Complicated and controversial mathematical models have been developed to forecast world conditions from now into the next century. The pioneering work was done by Jay W. Forrester* and was followed by *The Limits of Growth: A Report for the Club of Rome's Project on the Predicament of Mankind.*† The major conclusion of both books is that industrial growth will come to an end within a century unless present trends change.

The prediction is based on a computer model that simulates the major ecological forces at work in the world today. All the forces are interlocked. Human population growth increases the need for food. Since the globe's best farmland is already under cultivation, more food can be grown only through the use of more machinery, pesticides, and fertilizers. To produce these products, industry must drain natural resources presently in short supply, and the production causes more pollution. Higher pollution levels and raw-material depletion complete the cycle to interfere with population growth and food harvests.

A computer model for the whole world obviously relies on many simplifying assumptions. No political or geographic subdivisions are included. Data used in the model range from expert opinions to known facts about the world's natural resources, population growth, different forms of pollution, sources of power, investment capital, etc. After the variables were related one to another, numerous computer runs were made to see the effects of altering the numerical values of the variables. A plot of one of the computer runs is shown in Figure 1.8. The given curves are drawn to a variety of scales; significance can be given only to their shape. "Quality of life" is a composite rating for factors such as material standard of living, crowding, and food supply. The somber message of the chart is that scarcity of natural resources will reverse the world's growth trends.

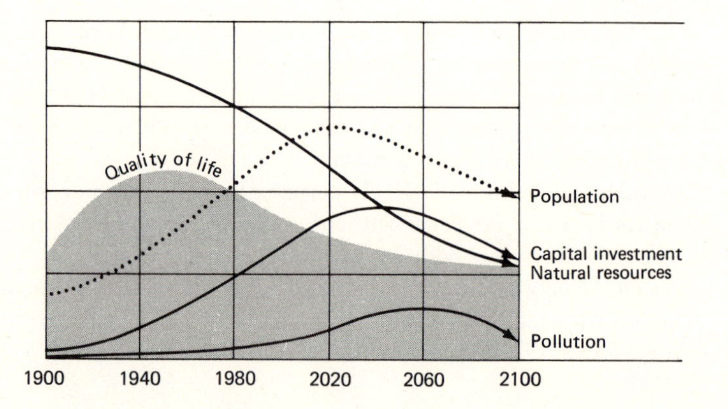


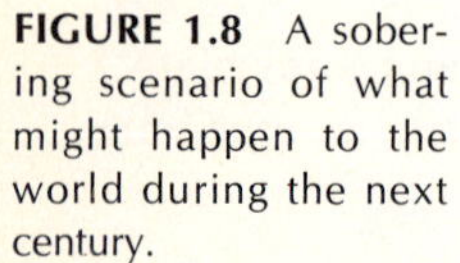
**FIGURE 1.8** A sobering scenario of what might happen to the world during the next century.

A markedly contrasting view based on a different computer model is proposed by

*See *World Dynamics*, Wright-Allen Press, 1971.

†The Club of Rome is an organization of distinguished bankers and scientists from 25 countries. By Donella Meadows, et al., Universe Books, 1972.

Mihajlo Mesarovic and Eduard Pestel.* They extend hope that disaster is not inevitable, but they do feel that "the way to make doomsday prophecy self-fulfilling is to ignore the signs of danger that lie ahead and to rely solely on faith."

## QUESTIONS

**1.2a** Identify future actions designed to counteract the doomsday prediction, and note the role of engineers and engineering economists in implementing them (for example, a requirement that products have a guaranteed minimum service life).

**1.2b** Discuss "quality of life" as an intangible consideration to be included in the evaluation of alternatives that affect the public (for example, a proposal to develop "moving sidewalks" for inner-city transportation).

*In *Mankind at the Turning Point: The Second Report to the Club of Rome,* Dutton, New York, 1974.

# SECTION ONE

# MESOECONOMICS

The term *mesoeconomics* has been applied to the area of economics that lies between and embraces both macro and micro considerations. It draws upon the theories proposed to explain the relationships among *aggregate* units of the economy and upon principles that suggest the behavior of *specific* economic units.

Engineering economists operate in the realm of mesoeconomics. An engineering economist is concerned with aggregate units such as the collection of all industrial firms and agencies engaged in construction, public works, or certain types of production. The ways in which national and international economic forces influence these aggregates are necessary inputs to decisions made at the operating level of a specific organization. Equivalently, the engineering economist must be aware of the interacting forces at the micro level of an organization's activities to be able to evaluate the local effect on decisions.

Generalizations are an inherent danger at the mesoeconomic interface. One firm on one occasion could find that a decision to boost its production of a particular product paid handsomely. But if all firms producing that product also raised their production levels, the result could be a loss for all the producers because excessive supply can depress price. The fallacy to guard against is the generalization that what is valid at one level at one point in time is automatically valid in a somewhat similar situation the next time. Principles do not change, but participants do. Identical situations rarely recur.

Most significant investment decisions have both macroeconomic and microeconomic overtones. The initial decision to buy a new machine is influenced by anticipated market conditions, effects of inflation, national concerns for energy conservation or protection of the environment, and other broad considerations. Once the decision is made to acquire the machine, the selection is based upon narrower considerations including ease of maintenance, reliability, operating costs, and coordination with existing facilities. Thus, aggregate economic considerations are combined with detailed engineering considerations to forge a complete economic analysis.

In this section we shall first observe some of the classic economic principles governing supply, demand, and production (Chapter 2). These principles guide the search for the combination of selling price, fixed and variable costs, and output quantities that promise the most profitable operations (Chapter 3). Then minimum-cost models are explored to determine optimal policies for replacing assets, determining project durations, controlling quality, and minimizing inventory costs (Chapter 4). Finally, attention is given to tactical-level considerations for improving product value, increasing productivity, and comparing alternative designs or courses of action (Chapter 5).

The intent of the mesoeconomics section is to establish a perspective that encompasses both strategic and tactical influences on economic decisions. Analysis methods are limited to those that do not involve long-term effects which rely on the time value of cash flows for evaluation. This limitation is not too restrictive because there are plenty of practical problems for which a direct evaluation of operating conditions will yield a solution. Such evaluations are termed *primary decisions*. The main concern in primary comparisons is that the outcomes of alternatives are essentially equivalent with respect to their effect on the total system. Consequently, knowledge of larger issues and broader principles supports the direct solution methods for workaday problems.

# CHAPTER 2

# SUPPLY, DEMAND, AND PRODUCTION

Most engineers are agents of production. The title or job description for an engineering activity may mask the relationship, but the function performed is part of the transformation process that converts scarce materials into desired products. As diagramed in Figure 2.1, engineering is an input to the transformation process which influences the value of resources and products.

Dashed circles in Figure 2.1 imply the role of engineering in the macro relationship of resources to products. Engineering expertise is itself a resource which is converted into services when utilized to increase the value of other resources. The services are tagged with many names and take many forms:

- Research and development to discover and refine new products and processes
- Design to specify and integrate components for efficient operations
- Coordination and supervision of activities for construction
- Administration of public and private projects and services
- Testing and inspection to collect data and evaluate production processes

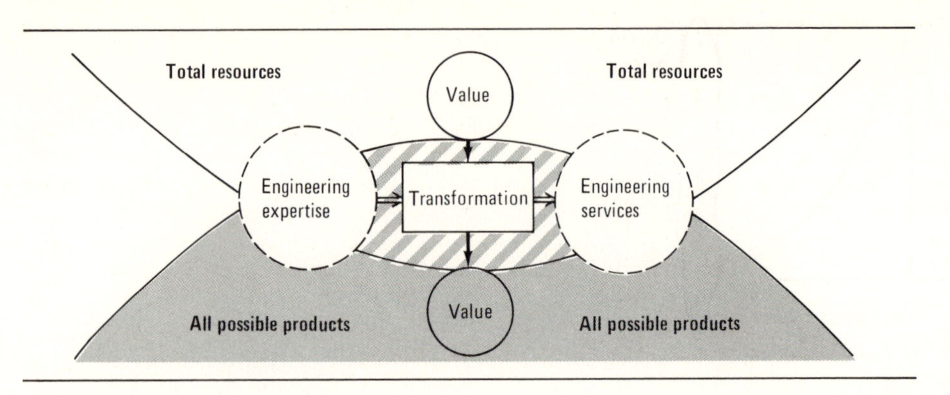


**FIGURE 2.1** Engineering as a part of the resource allocation system.

Engineering services are vital to the technical efficiency of transformation processes, but the effectiveness of transformation depends also on economic and social factors. A highly efficient process to convert foodstuffs to instant garbage would enjoy no more success than a scheme to produce radioactive jewelry. Well-advised mechanics for transformation maximization are futile without regard for the value of the inputs and outputs. Resources have a dynamic economic market value. Product value is determined by the marketplace or by political processes which respond to social values for public benefits such as conservation and recreation. Engineering combines expertise in the physical properties of resources with an awareness of the economic value of resources to optimize the transformation function.

## SUPPLY AND DEMAND

That "supply always adapts to demand" is a commonly accepted truism. What is less apparent in this glib assertion is that a thorough understanding of supply-demand relationships is a major step in understanding the operation of the whole economic system. Features most pertinent to production are highlighted in this section. Additional references to supply-demand concepts are encountered throughout the book because they underlie so many engineering economic decisions.

### Demand Schedule

The relationship between price and demand shown in Figure 2.2 should surprise no one. It displays the readily observed condition that higher prices discourage sales and, equivalently, more units of a product are demanded when the market price is lower. The downward slope of demand curve $D$ reflects buyers' reactions to decreasing prices; a lower price attracts new buyers and coaxes additional purchases from previous customers.

A change in demand for a given product is illustrated in Figure 2.3. A decrease in demand shifts the demand curve $D$ to the left to set curve $D1$. This shift could be the result of any one or a combination of the following reasons:

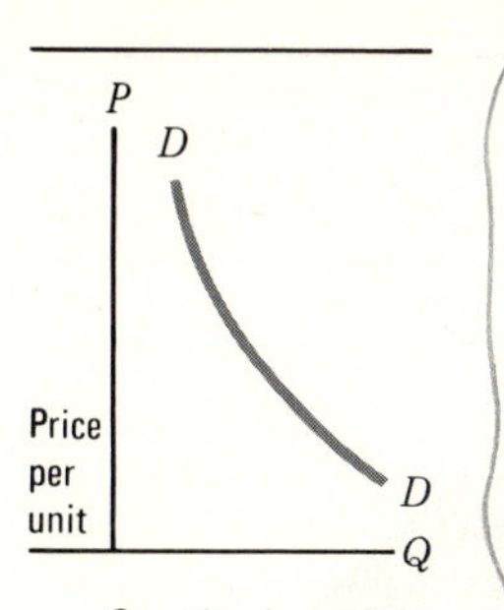

**FIGURE 2.2** Demand schedule.

1. Consumer tastes have switched to favor a competing product; advertising could be an influence.
2. Consumers have switched purchases to a related, less expensive substitute; soybean substitutes can replace meat purchases when the price of meat is deemed to be too high.
3. The number of possible consumers has decreased; a declining birthrate affects the market for prepared baby foods.
4. Expectation of lower prices in the future decreases current demand; conversely, fears of inflation or shortages can cause hoarding which sharply raises current demand.
5. A decline in buyers' incomes normally lowers demand, but the demand for "inferior goods" can increase as incomes go down; equivalently, higher incomes can decrease demand for hamburgers as appetites graduate to steaks.

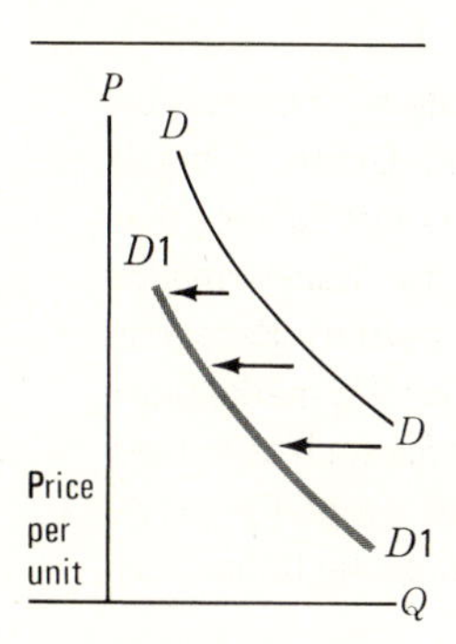

**FIGURE 2.3** Change in demand.

A reversal of the above conditions could cause the demand schedule to shift to the right of curve $D$ in Figure 2.3, signifying that consumers are *willing* and *able* to pay more for a product.

"Change in quantity demanded" is the expression used to describe the movement of demand along a fixed demand curve. It is caused by a change in the price of the product, *not* the reasons listed for a "change in demand" which indicates a displacement of the whole demand curve.

---

**Economy Exercise 2-1.** A highway engineer is preparing a justification for the construction of additional rest stops for motorists along the state's major highways. Each rest stop requires from 3 to 5 acres (12,140 to 20,235 square meters) of land and includes toilet facilities, picnic grounds, tourist information centers, spacious parking, and shaded lawns, but no overnight camping arrangements. Recent data show a significant drop in utilization of existing rest stops over the last 2 years. If additional rest stops would not be used fully by the public, highway funds could be allocated more effectively to other road projects. What are possible reasons for the decreased utilization, and what factors should be evaluated in the justification report?

---

## Supply Schedule

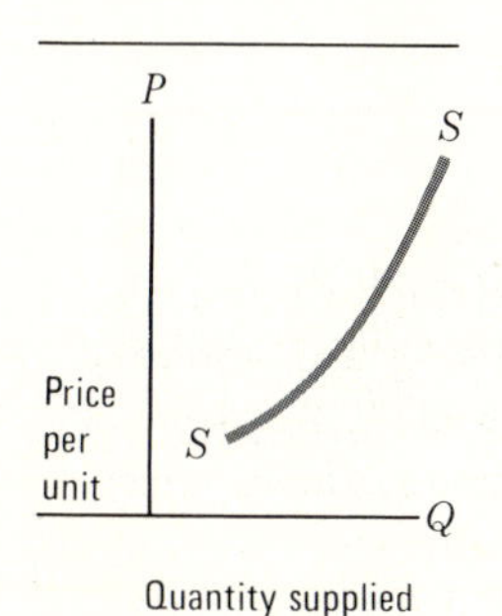

**FIGURE 2.4** Supply schedule

A supply curve as shown in Figure 2.4 displays the relationship between the market price and the amount of a particular product that producers are willing to supply. It confirms the intuitive belief that higher prices are an inducement for greater production when "other things are equal." These "other things" are nonprice determinants which are often important in engineering economic studies.

Advances in technology can shift the supply schedule for a product to the right by lowering production costs and thereby making more units available at lower prices. A shift to the left results from increased prices for raw materials which eventually raise prices for finished products unless the higher material costs can be absorbed elsewhere. Expectations of higher future prices may induce manufacturers to expand production immediately, causing the current supply to increase. When the manufacturers have guessed wrong about higher future prices, the temporary bargains may become lasting, provided the supply schedule still allows a profit. Firms leaving the market obviously decrease supplies. Government actions also affect supplies through tax policies and subsidies. All these nonprice determinants of supply alter supply schedules, ***not*** the "change in quantity supplied" which refers to a movement from one point to another on a stable supply curve.

## Equilibrium

Demand and supply schedules for the same product are brought together in Figure 2.5. The intersection of the two curves reveals the equilibrium price and quantity: the price at which an amount of goods is *willingly* supplied and demanded. Unless other things change, suppliers of product $Z$ can expect a total market of 6000 units, and consumers can expect to pay \$3 per unit.

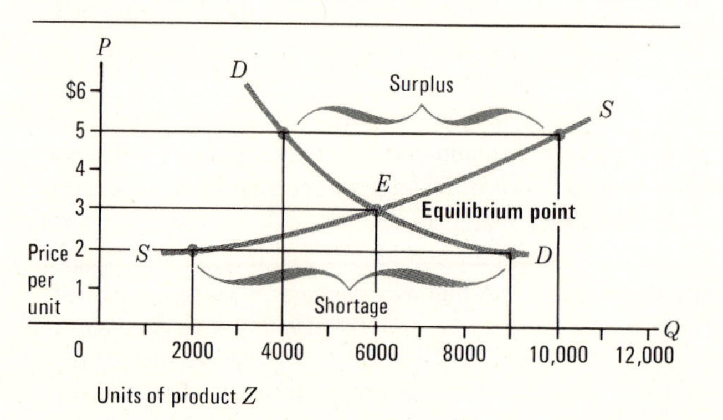


**FIGURE 2.5** Equilibrium point for supply and demand

At the equilibrium point $E$ there is neither a shortage nor a surplus of product $Z$. A price of \$2 per unit creates a shortage of 7000 units because suppliers would be unwilling to meet the 9000 units demanded at the depressed price of \$2. So consumers would drive the price up by their willingness to pay more than \$2 for a unit of product $Z$. A reverse procedure drives prices down. If the suppliers of product $Z$ set a price of \$5 per unit, they would sell only 4000 units and have a surplus of 6000 units. Therefore, the suppliers would rationally decrease the price to sell their surplus. From either position, above or below point $E$, the mechanisms of a competitive market adjust prices until supply equals demand.

**Economy Exercise 2-2** Spot the fallacies in the following observation: "I run a small engineering consulting firm. We have plenty of competition, so our fees are comparable to others. Since we all keep pretty busy, we must be at the equilibrium level. In fact, the equilibrium point must obviously be at whatever price we charge because the amount of services we sell always equals the amount our customers buy. Ergo, equilibrium!"

# ELASTICITY OF SUPPLY AND DEMAND

The behavior of demand with changes in price is a critical consideration to anyone engaged in marketing decisions. Cost-reduction studies by engineers reveal ways to lower production costs. Perhaps these cost savings should be used to cut prices in order to sell more units and thereby increase total profit. And maybe not; it is possible that the lower price would actually decrease the total dollar revenue for the product. Similar conditions arise when a government commission allows a regulated utility to raise its prices to increase returns; the price rise could trigger consumer actions that would frustratingly reduce total returns. Such ventures involve the concept of *elasticity* of demand.

Demand is classified as *elastic* or *inelastic* according to the relative responsiveness of quantity demanded to price changes. Three things can happen to total expenditure (price multiplied by quantity bought: $P \times Q$) for a product along a demand curve. In the first case, demand is *inelastic* if the total expenditure falls when the price falls, and rises when the price rises. Next, if a price rise causes a decrease in total expenditure, or a price decline leads to a higher total expenditure, demand is *elastic*. The third case, in which a change in $P$ and the corresponding change in $Q$ cause no change in $P \times Q$, is *unitary elasticity*. The three cases for elasticity of demand are summarized in Table 2.1.

**TABLE 2.1** Relationship between price $P$ and total expenditure $P \times Q$ for a demand schedule.

| | *Elastic Demand* | *Unitary Elasticity* | *Inelastic Demand* |
|---|---|---|---|
| *P Goes Up* | $P \times Q$ falls | Constant $P \times Q$ | $P \times Q$ rises |
| *P Goes Down* | $P \times Q$ rises | Constant $P \times Q$ | $P \times Q$ falls |

Most demand curves exhibit all three cases of elasticity. Consumers are typically unresponsive to price changes over one range (inelastic demand), while over another range of price changes on the same demand curve there is a significant jump in quantity purchased (elastic demand). Therefore, a complete demand schedule can seldom be classified as elastic or inelastic; only particular ranges can be so categorized.

Concepts of price elasticity also apply to supply. When producers are responsive to price changes, supply is elastic; when they are relatively unresponsive, supply is inelastic. In the limiting case of inelastic supply, the supply schedule is plotted as a vertical line. This is the case when a fishing boat returns to port with a cargo of perishable fish which is sold at whatever price is available. The other extreme is graphed as a horizontal supply line which represents an infinitely elastic supply such that any cut in $P$ causes $Q$ to go to zero. The degree of supply elasticity is generally of less interest than the degree of demand elasticity, but the impact of time on supply elasticity is important to production planning (see page 30).

## Calculation of Elasticity

The measure of elasticity of supply or demand is the ratio of percentage change in $Q$ to the corresponding percentage change in $P$. Specifically, the coefficient of demand elasticity $E_d$ is calculated as

$$E_d = -\frac{\%\Delta Q}{\%\Delta P} = -\frac{\Delta Q/Q}{\Delta P/P} = -\frac{\Delta Q}{\Delta P}\frac{P}{Q}$$

and the supply elasticity $E_s$ is determined by

$$E_s = \frac{\%\Delta Q}{\%\Delta P} = \frac{\Delta Q/Q}{\Delta P/P} = \frac{\Delta Q}{\Delta P}\frac{P}{Q}$$

When the coefficient of elasticity is greater than unity, supply or demand is classified as elastic. Equivalently, $E_d < 1$ or $E_s < 1$ signifies inelasticity, and a coefficient that equals 1 reveals unitary elasticity.

The coefficient of elasticity is a simple calculation, but it involves an annoying feature: $E$ may vary according to the reference point selected for the computation. For example, in calculating the coefficient of elasticity for the segment from $P = \$2$ to $P = \$1$ in the demand schedule given in Table 2.2, different values are obtained when either endpoint is the reference. Thus, using \$2 as the reference point,

$$E_d = -\frac{\Delta Q/Q}{\Delta P/P} = -\frac{-200{,}000/100{,}000}{\$1/\$2} = \frac{2}{0.5} = 4$$

and, using \$1 as the reference point,

$$E_d = -\frac{\Delta Q/Q}{\Delta P/P} = -\frac{-200{,}000/300{,}000}{\$1/\$1}$$

$$= \frac{0.66}{1.00} = 0.66$$

which suggests the segment of the demand curve from $P = \$2$ and $Q = 100{,}000$ to $P = \$1$ and $Q = 300{,}000$ is at once elastic and inelastic. This cannot be.

| *Price, per Unit, P* | $\Delta P$ | *Quantity Demanded, Q* | $\Delta Q$ | *Total Expenditure, P × Q* | *Elasticity* |
|---|---|---|---|---|---|
| \$2.00 | | 100,000 | | \$200,000 | |
| | \$1.00 | | −200,000 | | Elastic |
| 1.00 | | 300,000 | | 300,000 | |
| | 0.50 | | −300,000 | | Unitary |
| 0.50 | | 600,000 | | 300,000 | |
| | 0.30 | | −400,000 | | Inelastic |
| 0.20 | | 1,000,000 | | 200,000 | |

**TABLE 2.2** Demand schedule and factors for the calculation of elasticity.

From Table 2.1 it is clear that the curve segment demonstrates elastic demand because total expenditure increased while the price dropped. This ambiguity is neatly handled by averaging the endpoints of a segment to obtain the formula values for $P$ and $Q$. Again employing the data from Table 2.2 for the first segment of the demand curve,

$$E_d = -\frac{\Delta Q/Q_{av}}{\Delta P/P_{av}} = -\frac{\Delta Q/[(Q_1 + Q_2)/2]}{\Delta P/[(P_1 + P_2)/2]}$$

$$= -\frac{-200{,}000/200{,}000}{\$1.00/\$1.50} = \frac{1.00}{0.67} = 1.5$$

which confirms the demand is elastic for that portion of the curve. The segment at the other end of the demand curve (from $P = \$0.50$ and $Q = 600{,}000$ to $P = \$0.20$ and $Q = 1{,}000{,}000$) has midpoints of $P_{av} = (\$0.50 + \$0.20)/2 = \$0.35$ and $Q_{av} = (600{,}000 + 1{,}000{,}000)/2 = 800{,}000$ to make

$$E_d = -\frac{-400{,}000/800{,}000}{\$0.30/\$0.35} = \frac{0.500}{0.857} = 0.583$$

which indicates inelasticity of demand.

Arc Elasticity

due Pt. Elast

**Economy Exercise 2-3** Only in rare instances can you identify the degree of elasticity just by the slope of the curve. The demand schedule in Figure 2.6 is a straight line that exhibits the same absolute slope for the three line segments *ab, bc,* and *cd.* Calculate the elasticity for each segment. Can you state a rule about the elasticity of demand for a straight-line schedule?

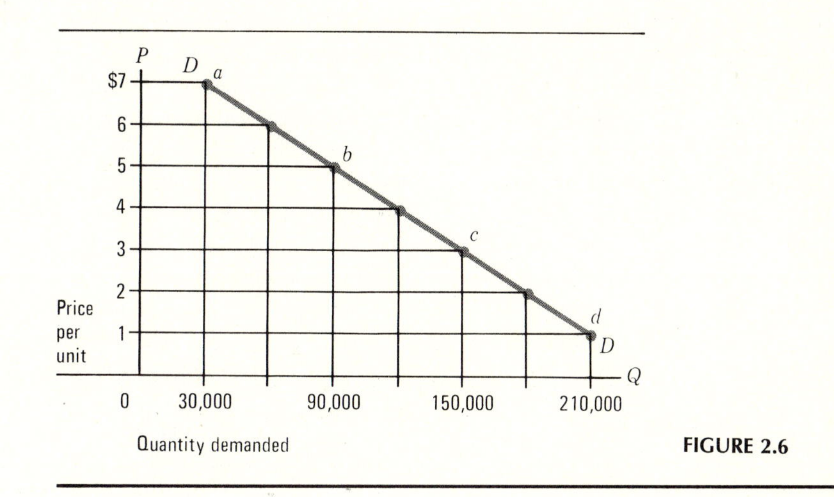


**FIGURE 2.6**

## Supply and Demand Applications

The concepts of supply and demand cannot explain all the mysteries of price movements. But familiarity with the concepts yields an awareness of the complexity of the forces that affect prices and an appreciation for evaluation methods based on supply-demand relationships. These relationships pop up repeatedly in subsequent chapters. Here, a few strategic applications are described.

A decision to automate a process, or to go all the way with robotics, depends in part on the elasticity of demand for the product concerned. Assuming the main reason for automation is to reduce production costs (not, for instance, to avoid labor troubles or improve safety), it is logical to consider passing part of the savings along to consumers in order to increase demand which would allow maximum utilization of the expensive equipment. If demand is elastic, this policy may well succeed. An inelastic demand makes the policy less attractive because any increase in volume resulting from a price cut will be relatively small.

A producer's response to a favorable change in the demand schedule for a company product hinges on the company's ability to shift resources from the production of other products to the newly favored product. Flexibility can be engineered to a large extent. More versatile machines and facilities reduce the response time for production changes and provide a short-term competitive advantage. Over the long run, competing firms will make resource adjustments, and other companies can enter (or leave) the market. This delayed response is typically accompanied by higher prices to compensate for costs incurred to enlarge capacity. Thus, supply elasticity tends to be greater over the longer ranges than in the immediate market period.

Excess demand or supply can be created by actions other than those of the market. For political or social reasons, governments may decide that certain prices are too high or too low. The results are government edicts establishing floor or ceiling prices. Without judging the desirability of these limits, supply and demand relationships reveal why the limits create shortages or surpluses.

Political pressures have occasioned the imposition of wage and price freezes. As emergency actions, freezes may work well for a short period, but continued restrictions lead to distortions. As an illustration, the supply and demand curves in Figure 2.7 indicate that the equilibrium point for a free market is at price $P_e$ for amount $Q_e$. Now assume a ceiling is imposed to limit the price to 80 percent of $P_e$, or $0.8P_e$. If the supply and demand schedules remain unchanged, at $0.8P_e$ the quantity demanded would increase to $Q_d$ while the amount willingly supplied would slip to $Q_s$. This creates a persistent shortage of $Q_d - Q_s$ units. Anxiety over who should go without would likely lead to a rationing plan, or maybe a subsidy for the suppliers. Neither is completely satisfactory because rationing breeds black markets (where the rationed items are sold at prices above legal limits) and subsidies tend to create their own supply-demand distortions.

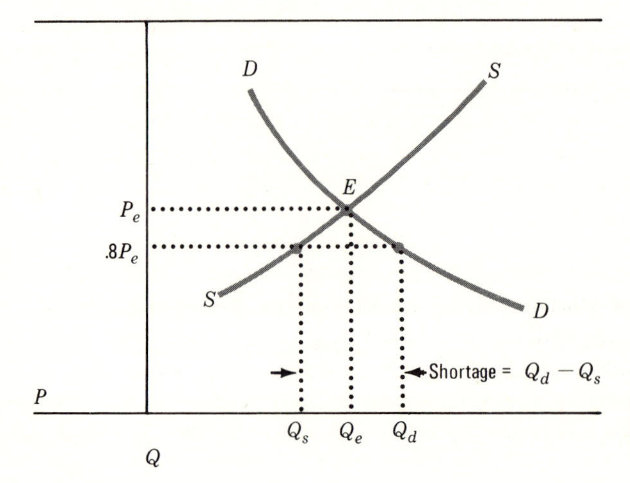


**FIGURE 2.7** Shortage of $Q_d - Q_s$ units resulting from a price ceiling at 80 percent of $P_e$.

## THEORY OF PRODUCTION

Most engineers are involved with some form of *production—the development of any good or service that people will buy.* This definition of production covers both the goods and services people literally purchase (cars, electricity, health care, etc.) and those they indirectly buy through taxes (highways, dams, protection from fire and theft, etc.). The following discussion of the theory of production is directed towards goods produced by privately owned firms; services, mainly those provided by government, are discussed in Chapter 13. However, the principles explored in this chapter apply to both goods and services.

## Production Function

The production of a very simple product, such as a raw apple, requires a wide variety of inputs. A farmer needs land of suitable quality, materials to nurture and protect the trees, capital to buy equipment, and labor to work the orchard and harvest the crop. When the apples become inputs to a production process that yields canned applesauce, more inputs are accumulated. The relationship of inputs to outputs is formally described by a *production function:* a statement (graph, table, or equation) expressing the maximum output that can be produced from any set of inputs at a given state of technical knowledge.

There are innumerable different production functions in United States industry, one or more for every producing unit. For example, an artist might have the following production function that relates daily output in sketches to hours spent at the easel:

| *Output (sketches/day)* | *Input (hours/day)* |
|---|---|
| 3 | 12 |
| 2 | 7 |
| 1 | 3 |

A similar schedule might be developed for the number of hours of study time required to get a certain grade in a course. The schedule would differ among students and courses, just as the artist's schedule varies according to the type of painting being done or as apple growers' schedules differ owing to differences in the quality of the land, age of the trees, etc. And established schedules can change as a result of a technological improvement such as a new spray that better protects an orchardist's apples.

## Fixed and Variable Inputs

Inputs to a production process are classified as fixed and variable. A *fixed input* is one that cannot be immediately changed when greater output is desired. Factories, major pieces of equipment, and key personnel are examples of fixed inputs that cannot be quickly augmented or reduced without incurring excessive costs. Conversely, *variable inputs,* such as common materials and most labor services, may be varied rapidly in response to demand changes.

These two input categories are realistic for *short-run* production periods; over the *long-run* planning horizon, all inputs are variable. For example, a producer could expand output in the short run by enlarging the labor force (variable input) to utilize the existing plant (fixed input) more hours per day. Since the extra labor entails higher overtime or shift-differential costs, it might be more economical in the long run for the producer to enlarge the plant (a fixed input that has become variable) and return to the previous working hours.

A production function relating the inputs of labor $L$ and capital $C$ to the output quantity $Q$ can be expressed as $Q = f(L, C)$. All terms in the expression are rates such as products per day, crew-hours per week, or machine-hours per month. Only variable inputs are included; any input factors not included are regarded as fixed over the range of output being considered. The production function implies how the firm is subject to conditions of supply and demand; it is a supplier of product $Q$ and a customer for its production inputs. It seeks to minimize input costs to achieve a favorable position on its customers' demand curve. It is engaged in a two-front battle in which a misjudgment on either front can waste success on the other.

**Economy Exercise 2-4** A three-dimensional representation of the production function $Q = f(L, C)$ is shown in Figure 2.8. The shaded area is called a *production surface* and reveals how the output $Q$ changes with various combinations of labor $L$ and capital $C$.

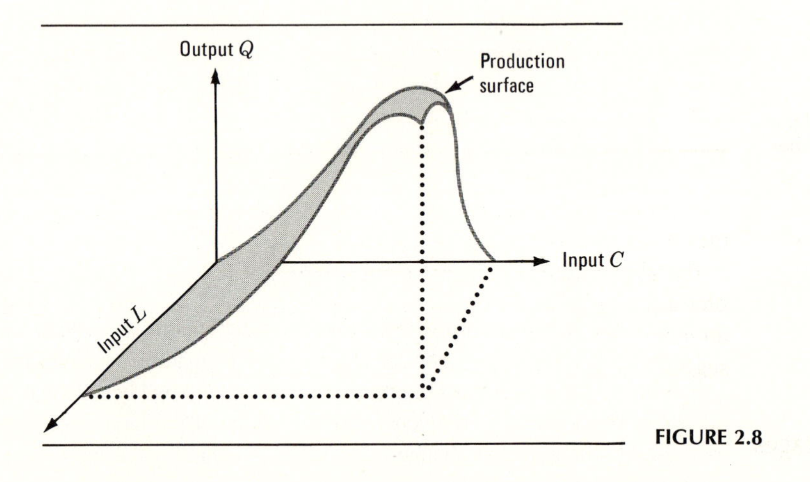


FIGURE 2.8

**a** What assumption is made when $Q = 0$ along both the labor and capital axes?
**b** What trade-off possibilities are indicated by the production surface?
**c** What characteristics of a production function are suggested by the surface around the peak?

## TOTAL, AVERAGE, AND MARGINAL PRODUCT

One of the reasons for investigating the production function of a firm is to determine the operating level that promises the most profit. A single-dimensional production function is shown in Table 2.3: Output depends only on the amount of labor employed. This condition assumes the other factors of production are fixed over the short-run period under consideration.

Columns 1 and 2 of Table 2.3 define a production function. When there is no labor there is no product, because the factory or farm (fixed input) is idle. The total product is the maximum output obtainable from a given amount of labor utilized in the given production facility. Average product (column 2 divided by column 1) reaches a maximum between 300 and 400 worker-years employment and then declines. The marginal-product column shows the change in output associated with each additional level of labor. So long as the marginal product is positive, total product is increasing and reaches a maximum when the marginal product is zero. Note also that when the average product is rising, the marginal

| Input Labor; Worker-Years, in 00's (1) | Output Total Product, Units (2) | Average Product (3) | Marginal Product (4) |
|---|---|---|---|
| 0 | 0 | | |
| 1 | 100 | 100 | 100 |
| 2 | 260 | 130 | 160 |
| 3 | 450 | 150 | 190 |
| 4 | 600 | 150 | 150 |
| 5 | 700 | 140 | 100 |
| 6 | 780 | 130 | 80 |
| 7 | 840 | 120 | 60 |
| 8 | 840 | 105 | 0 |
| 9 | 810 | 90 | −30 |

**TABLE 2.3**
Total, average, and marginal products of labor.

product is greater than the average product; and that when average product reaches its maximum, average product equals marginal product.

The above relationships are not peculiar to Table 2.3; they are representative of any production function in which the average product peaks. Total product at first increases at an increasing rate, then increases at a decreasing rate, and finally decreases. This pattern suggests the most effective ratios for production inputs.

## Production Stages

A production function akin to the data in Table 2.3 is graphed as a continuous total-product (TP) curve in Figure 2.9. Slopes of rays from the origin to the TP curve indicate the average product: Slope equals output divided by labor. Ray 03 is tangent to the TP curve at the level

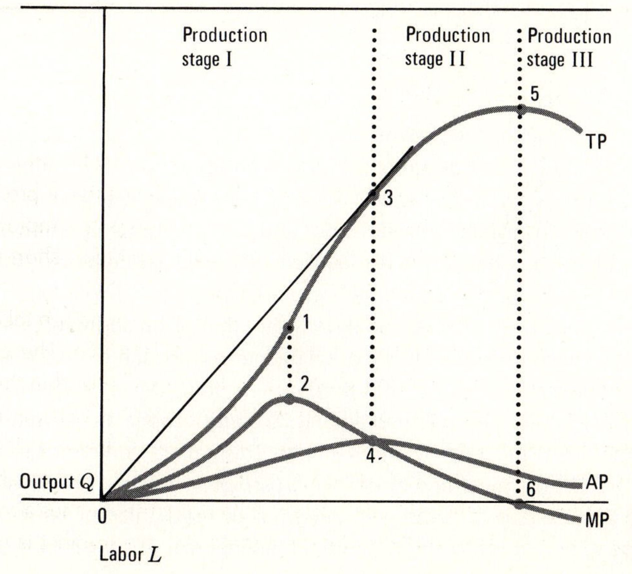


**FIGURE 2.9** Relationship of total product (TP), average product (AP), and marginal product (MP) to production stages for one variable input (labor) of production.

of labor where average product (AP) is at a maximum, point 3. The rate of increase of TP is greatest at point 1, which coincides with the level of labor at which the marginal product (MP) is maximum, point 2. TP increases at a decreasing rate from point 1 until it reaches a maximum at point 5; it declines thereafter, as reflected by the negative MP beyond point 6.

Three stages of production are indicated at the top of Figure 2.9. The first stage corresponds to the range in which average product is increasing as a result of larger applications of the variable input, labor. A rational producer would not operate in this range because the fixed input (plant, land, etc.) is not being utilized to its full capacity. That is, the average output expected from applying more worker-hours is increasing throughout the range of stage I, which indicates the same output could be obtained from a smaller amount of the fixed input.

Stage III is also an uneconomical range for production. Additional units of the variable input actually decrease total product. A larger output than the shown maximum (point 5) can be achieved only by using additional units of the fixed input.

By elimination, stage II is the feasible range for production. If market conditions dictate an output level unavailable in stage II, the fixed input should be raised or lowered rather than operating with the variable input outside the limits of stage II.

## Law of Diminishing Returns

The relationship of inputs to total output displayed in Figure 2.9 is explained by the empirical *"law" of diminishing returns: As amounts of a variable resource are added to fixed resources, beyond some point the marginal product diminishes.* (The law of diminishing returns is also called the *principle of diminishing marginal productivity* and the *law of variable proportions.*) This simple statement captures physical relations that appear repeatedly in the real economic world. It cannot be proven mathematically, but it has not been refuted by experience. Its importance to engineering economists rests on the guarantee it provides that the production function has an optimum input-output ratio.

## Production Isoquants

A production function with two variables as was shown in Figure 2.8 is conveniently analyzed by the use of *isoquants: curves showing all possible combinations of inputs capable of producing a given level of output.* As graphed in Figure 2.10, two isoquants reveal the different combinations of labor and capital required to produce 1000 and 2000 units of output. The isoquants never intersect, and there could be any number of additional isoquants between or beyond the two given levels.

The concave shape of the isoquants indicates that equal additional increments of capital can replace proportionately less labor while maintaining the same output. Increasing capital expenditures from 10 to 20 (from point 1 to point 2 in isoquant I) allows labor to be reduced from 500 to 310 units, but adding 10 units of capital to 40 units (point 3 to point 4) reduces labor requirements by only 40 units. This pattern conforms to the law of diminishing returns. The marginal rate of capital *substitution* for labor diminishes as ever greater amounts are substituted. Investment decisions to mechanize or automate manual operations depend in part on the firm's current marginal rate of substitution of capital for labor.

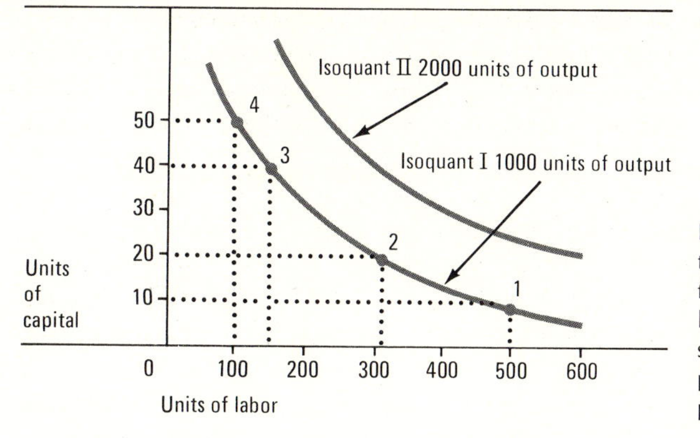


**FIGURE 2.10** Isoquants for two levels of output that result from various combinations of labor and capital. The relationships hold only during a given period of time and for a given production unit.

**Economy Exercise 2-5** The marginal rate of substitution (MRS) of capital for labor is the ratio of the marginal product of labor to the marginal product of capital:

$$\text{MRS}_{C \text{ for } L} = \frac{\text{MP}_L}{\text{MP}_C} \qquad \textit{where } \text{MP}_L = \Delta\text{product}/\Delta\text{labor}$$
$$\text{MP}_C = \Delta\text{product}/\Delta\text{capital}$$

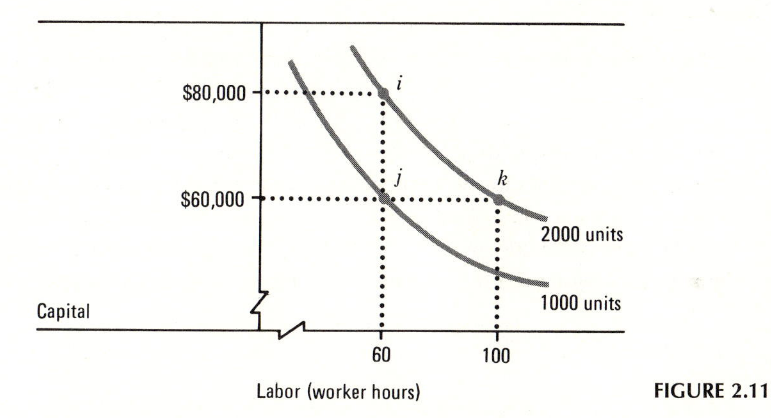


**FIGURE 2.11**

Two isoquants, for a particular product, resulting from various combinations of capital and labor are shown in Figure 2.11. Calculate $\text{MRS}_{C \text{ for } L}$.

# ALLOCATION OF RESOURCES

Supply-demand considerations and a firm's production function are brought together to determine the optimal mix of resources for production operations. The prices of both inputs and outputs are essentially governed by supply and demand in the market. Assuming again that the input variables for production are labor $L$ and capital $C$, and that unit prices for $L$ and $C$ are, respectively, $m$ and $k$, the total cost (TC) of inputs for production is TC $= mL + kC$.

*Isocost* lines as shown in Figure 2.12 define all combinations of $L$ and $C$ that can be obtained from a given total expenditure TC. With a labor wage $m$ of $10,000 per worker-year and a unit of capital $k$ at $5000, a total cost of $100,000 could comprise combinations such as 10 units of $L$ and no $C$; 5 units of $L$ and 10 units of $C$; 20 units of $C$ with no $L$; etc.

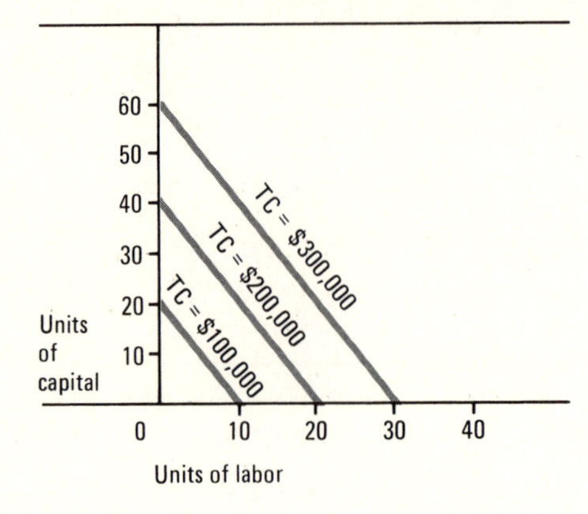


**FIGURE 2.12** Isocost lines for total expenditures (TC) when one unit of labor *(L)* is priced at $10,000 *(m)* and one unit of capital costs $5000 *(k)*.

## Optimality Criterion for Minimizing Production Costs

An isoquant curve for the firm's desired level of output is superimposed on its isocost lines in Figure 2.13 to reveal the least-cost combination of labor and capital. It is clear that it would be uneconomical for the firm to use any combination of capital and labor other than the ratio designated by point $X$, 20 units of capital and 10 units of labor. At this point the isocost line is tangent to the isoquant curve. This indicates the slope of the curve equals the slope of the isocost line at $X$. Therefore, it is logical that the least-cost combination occurs where the marginal rate of substitution is equal to the ratio between the price of labor and the cost of capital. That is, *total cost for a given production output is minimized when*

$$\text{MRS}_{C \text{ for } L} = \frac{\text{MP}_L}{\text{MP}_C} = \frac{m}{k}$$

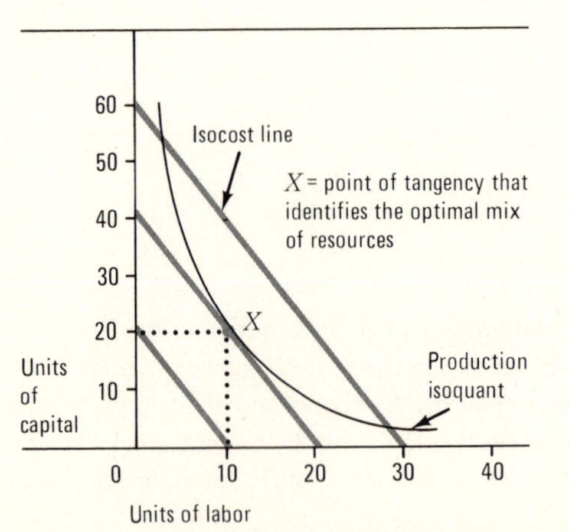


**FIGURE 2.13** Optimal combination of capital and labor to minimize production costs for a given level of output.

## Expansion Path

The level of output for a firm of modest size does not normally affect the prices it must pay for labor and capital inputs. However, changes in output levels do alter the position of isoquants as observed in Figure 2.10. Each isoquant is subject to the optimality condition illustrated by Figure 2.13. These relationships are collectively shown in Figure 2.14. Three isoquants (I, II, and III) are tangent to three isocost lines ($CL$, $C'L'$, and $C''L''$), respectively, and the points of tangency are connected to show the expansion path $OE$.

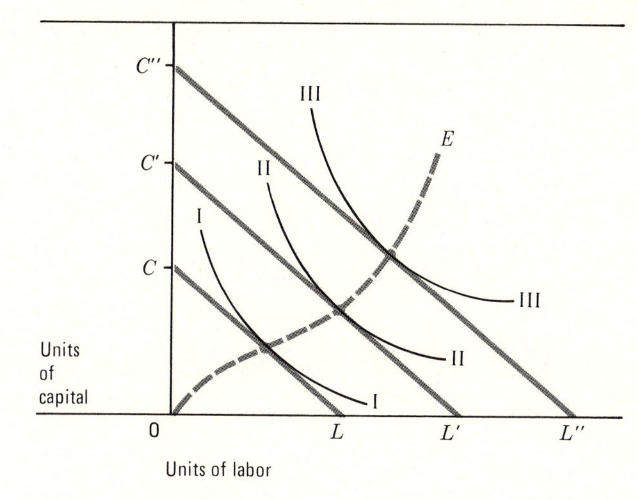


**FIGURE 2.14** The expansion path $O$-$E$ is the curve along which output can be increased with minimum cost when input prices remain constant.

An *expansion path* defines the optimal input combinations for least-cost increases in output when input prices are constant. It is a locus of points along which the marginal rate of substitution is constant and equal to the specific $m/k$ ratio.

The expansion-path concept can guide an expanding economy. The isocost lines can be thought of as budget limits, and the isoquant curves can be operating levels such as different degrees of effectiveness for, say, sanitation, safety, or service provided. Then each point on the path also represents the cost of a function and its effectiveness when operations are optimum. Difficulties do arise in quantifying the values of inputs and output, but the cost-effectiveness concept underlies improvement efforts in both public and private economy.

**Economy Exercise 2-6** Different combinations of workers and equipment are capable of providing the same maintenance service for county roads. Each unit of equipment costs

| *Resource Combination* | *Units of Equipment* | *Number of Workers* |
|---|---|---|
| 1 | 30 | 700 |
| 2 | 31 | 640 |
| 3 | 32 | 600 |
| 4 | 33 | 585 |
| 5 | 34 | 575 |

$200,000 and includes graders, dump trucks, loaders, etc. The wages of maintenance workers average $10,000 each annually. Which of the combinations shown should be utilized to minimize total cost?

## PRINCIPLES AND PRACTICES

The economic principles discussed in this chapter set the framework for engineering economic analyses. Because there are so many factors that can influence the behavior of economic relationships, the general "laws" may seem to have little practical applicability. For example, psychological factors affecting consumers and businesspeople have been blamed when the classic forces of supply and demand have apparently failed to raise or lower prices and wages as expected. The enormous market power of big unions and corporations, actions of international cartels, and even the weather are sometimes held responsible for deviations from economic forecasts. But the economic laws have not been repealed. Eventually, through research and experience, the role of all possible influencing factors may be accounted for, but until then we can still use economic principles to gain a better understanding of relationships between resources and production.

The principles in this chapter reappear many times in following chapters. In most cases the principles are applied under restricted conditions for which optimal solution methods are developed for specific problem situations. In solving confined problems, it is easy to overlook the larger implications. Sometimes the real mesoeconomic issues are neglected in the rush to solve a pressing financial problem: Then, you can't see the economics because of the dollar signs.

## SUMMARY

*Supply* and *demand schedules* show the relationship between price per unit and quantity. The intersection of supply and demand curves for the same product identifies the *equilibrium point* at which goods are willingly exchanged. Price manipulations that distort the equilibrium cause shortages or excesses unless the basic schedules are altered.

Demand is *elastic* if total expenditure falls (rises) as price rises (falls) and is *inelastic* for the inverse relationship. *Unitary elasticity* occurs when price changes do not affect total expenditure. A *coefficient of elasticity* $E = \pm\ \%\Delta P/\%\Delta Q$ greater than unity indicates elasticity; less than unity shows inelasticity; unity reveals unitary elasticity.

A *production function* is an expression of the maximum output that can be produced from a set of inputs at a given state of technology. Inputs are considered *fixed* or *variable* over a *short run,* but over a *long run* all inputs are variable. The *law of diminishing returns* recognizes that as amounts of a variable resource are added to fixed resources, beyond some point the marginal product diminishes.

*Average product* is the average output per unit of input. *Marginal product* is the change in output per additional unit of input. When the average product is rising, the marginal product is greater than the average; when the average reaches its maximum, average equals marginal. The feasible *production stage* stretches over the range of total product from the point where average and marginal products are equal to the point where the marginal product is zero.

*Production isoquants* indicate all possible combinations of inputs capable of producing a given level of output; one input can be substituted for another to maintain the level. When the *marginal rate of substitution* of one input for another equals the ratio of the unit prices of the two inputs, total cost for the given level of production is minimized. An *expansion path* defines the optimal mix of resources for different output quantities when input prices are constant. Graphically, the expansion path is the locus of points where isoquant curves are tangent to *isocost lines*.

## Discussion of Economy Exercises

**EE 2-1** A lower utilization of rest stops probably indicates a change in demand rather than a change in quantity demanded, because the motoring public perceives rest stops as free services; thus the price has not changed. But the number of users and their incomes may have decreased, owing to social and economic conditions. It is also possible that the tastes of travelers have changed to favor commercial rest stops where food is available. The highway engineer should evaluate whether the lower utilization figures represent a temporary change in driving habits or a long-term trend. If it is the latter, the value placed on rest stops has decreased, and the new lower value should be applied to the expected benefits (versus costs) in determining the justification for additional rest stops.

**EE 2-2** The equilibrium point *is* at the price where supply equals demand, but it also represents the point where customers are *willing to keep buying* and suppliers are *willing to go on selling*. If the consultants are not eager to sell more engineering services than the customers will buy at the current price, then *perhaps* the set price is the equilibrium point. However, competition may be imperfect, owing to informal fee setting or an insufficient number of consulting firms.

**EE 2-3** Most demand curves are elastic at the high-$P$ end, inelastic at low $P$, and have unitary elasticity at a point between, where total expenditure is at a maximum. This pattern is present in the given straight-line demand function, as is apparent from the following coefficients of elasticity:

$$E_{d_{(ab)}} = -\frac{\Delta Q}{Q_{av}} \div \frac{\Delta P}{P_{av}} = -\frac{30 - 90}{(30 + 90)/2} \div \frac{7 - 5}{(7 + 5)/2} = \frac{60/60}{2/6} = 3$$

$$E_{d_{(bc)}} = -\frac{90 - 150}{(90 + 150)/2} \div \frac{5 - 3}{(5 + 3)/2} = \frac{60/120}{2/4} = 1$$

$$E_{d_{(cd)}} = -\frac{150 - 210}{(150 + 210)/2} \div \frac{3 - 1}{(3 + 1)/2} = \frac{60/180}{2/2} = 0.33$$

At the midpoint of a straight line, demand is unitary elastic and total expenditure is maximum. Demand is elastic ($E > 1$) above the midpoint, and inelastic ($E < 1$) below.

**EE 2-4**

**a** It is usually assumed that no output is possible unless both inputs are present.
**b** The same level of output is attainable with different combinations of labor and capital.

For instance, an investment in new machines can be utilized to replace human effort.

**c** The peak indicates the highest output obtainable from any combination of capital and labor. Added inputs approaching this point cause output increases at a decreasing rate, and beyond the peak additions actually lower the output. For example, hiring extra workers for an assembly line could reduce the line's output through crowding and confusion.

**EE 2-5** The marginal product of capital is

$$MP_C = \frac{2000 - 1000 \text{ units}}{\$80{,}000 - \$60{,}000} = 0.05 \text{ units/dollar}$$

and the marginal product of labor is

$$MP_L = \frac{2000 - 1000 \text{ units}}{100 - 60 \text{ worker-hours}} = 25 \text{ units/worker-hour}$$

Therefore, the marginal rate of substitution of capital for labor is

$$MRS_{C \text{ for } L} = \frac{MP_L}{MP_C} = \frac{25}{0.05} = \$500\text{/worker-hour}$$

or, using the lettered points in Figure 2.11,

$$MRS_{C \text{ for } L} = \frac{i - j}{k - j} = \frac{\$80{,}000 - \$60{,}000}{100 - 60 \text{ worker-hours}} = \$500\text{/worker-hour}$$

**EE 2-6** Combination 3 is easily calculated by trial and error as the least-cost combination at $12,400,000. Another approach is to note that the $m/k$ ratio is $10,000/$200,000 = 1/20. Therefore, an equipment unit should be substituted for labor so long as it replaces 20 workers. This 1/20 ratio is exceeded for the first three combinations, but not the last two.

## PROBLEMS

**2.1** What are the two demand curves from which the degree of elasticity can be recognized at a glance? State an example of a physical situation that could be represented by each.

**2.2** From the supply and demand relationship shown in Figure 2.5, it is quite easy to visualize why prices rise and quantity falls when supply increases while demand is unchanged. It is not immediately apparent what happens when both supply and demand change simultaneously. Initial supply and demand conditions are shown by curves $S$ and $D$ in Figure 2.15. Curves $S'$ and $D'$ indicate conditions when both supply and demand have increased for the same product; the equilibrium point has gone from $E$ to $E'$ with the increases. State a general rule governing the effect on price and quantity of a simultaneous increase in supply and demand.

**2.3** Draw a diagram similar to the one in Figure 2.15 to illustrate a condition in which both supply and demand have decreased but the price has remained constant while quantity has dropped. Could the quantity possibly increase when both supply and demand decrease?

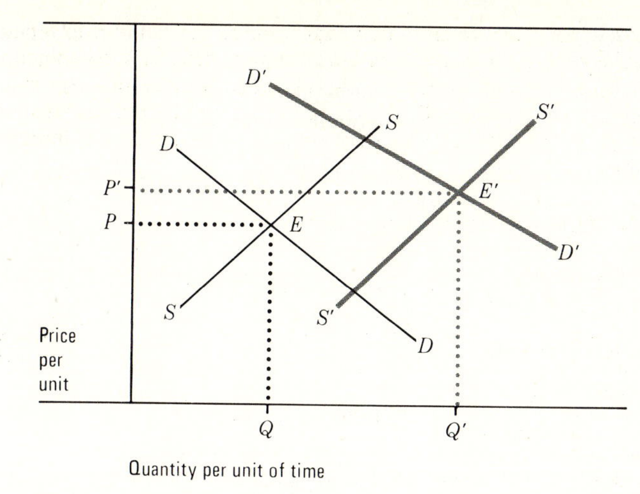


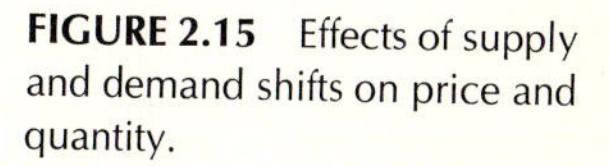
**FIGURE 2.15** Effects of supply and demand shifts on price and quantity.

**2.4** Suggestions to reduce the use of energy include placing a tax on fuel to discourage consumption. Assume the demand curve $D$ in Figure 2.16 represents motorists, and the supply curve applies to petroleum producers. The lower supply curve $S$ depicts conditions before imposition of the new sales tax, and $S'$ is the supply curve after a tax of 30 cents per gallon (7.9 cents per liter) has been added. The new equilibrium point $E'$ indicates that the quantity demanded drops as the price increases. Based on the graph, what portion of the 30 cents per gallon tax is being paid by the producers and the consumers?

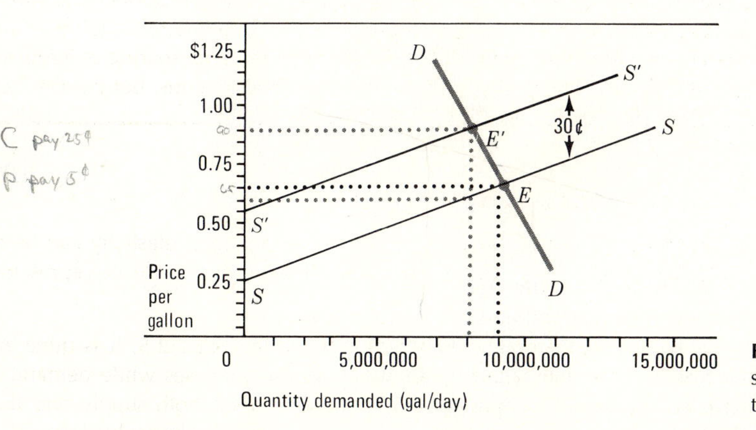


**FIGURE 2.16** Supply curve shift resulting from a tax addition.

**2.5** Draw demand curves representing various degrees of elasticity for the tax situation in Figure 2.16, and observe how the proportion of the tax payment changes between consumers and producers. What general statement can you make about the effects of a higher sales tax on prices and quantities for different degrees of elasticity of demand? How is the tax burden shifted when the supply is relatively more elastic?

**2.6** Productivity (output per worker-hour) gains in the United States have averaged slightly over 2 percent per year during the last eight decades. A major part of this increase is attributable to greater use of capital goods for production. The law of diminishing returns suggests that productivity returns per unit of capital invested in production facilities should be decreasing because each unit of capital has less labor associated with it, assuming that the rate of capital investment is increasing faster than the labor pool. Yet studies show that returns per unit of capital remain about the same. Why?

**2.7** Complete the following table for a one-dimensional production function:

| *Input* | *Total Product* | *Average Product* | *Marginal Product* |
|---|---|---|---|
| 0 | | | |
| 10 | | 300 | |
| 20 | | | 440 |
| 30 | 12,600 | | |
| 40 | | 400 | |
| 50 | | | 250 |

**2.8** A production function relating output to different combinations of capital and labor is given in Table 2.4.

**TABLE 2.4** The production function indicates that 350 units of output can be obtained from an input mix of 3 units of labor and 4 of capital, or 4 units of labor and 3 of capital.

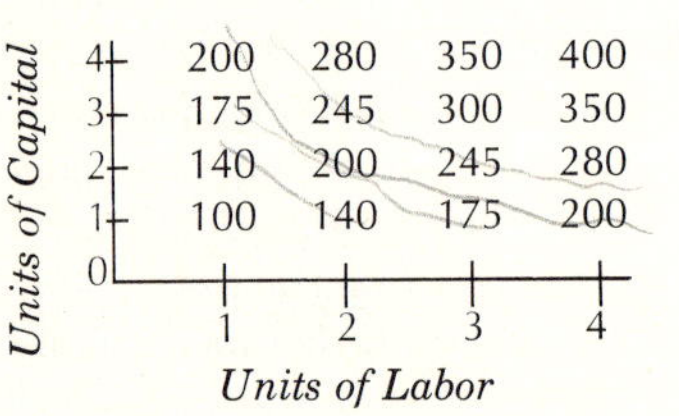

| *Units of Capital* | 1 | 2 | 3 | 4 |
|---|---|---|---|---|
| 4 | 200 | 280 | 350 | 400 |
| 3 | 175 | 245 | 300 | 350 |
| 2 | 140 | 200 | 245 | 280 |
| 1 | 100 | 140 | 175 | 200 |
| 0 | | | | |

*Units of Labor*

**2.8a** Show how the law of diminishing returns is demonstrated by the output pattern when one input is held constant while the other varies.

**2.8b** An explanation of the law of diminishing returns is that each additional increment of the variable input that is added has proportionately less of the fixed variable to work with. For this explanation to be valid, there should be no diminishing returns when both labor and capital are increased in proportion. How do the data in Table 2.4 demonstrate this condition?

**2.8c** Construct an isoquant for 200 units of output from the production function in Table 2.4. Graphically determine the optimal mix of resources for the 200-unit output quantity when the unit prices of land and capital are equal.

**2.8d** Calculate the marginal rate of substitution of capital for labor to determine the $m/k$ ratio that designates an optimal mix of 3 units of labor and 2 units of capital for an output quantity of 245 units.

**2.9** Many production isoquants have the shape shown in Figure 2.10, but some "bend backward," as shown in Figure 2.17. The dashed ridge lines connect the points where the isoquants have vertical or horizontal tangents; outside these points the isoquants have positively shaped segments. The area between the ridge lines indicates the economic range of production equivalent to stage II production in Figure 2.9. Discuss why this region is the only feasible range of production by showing that labor has a negative marginal product between points $L1$ and $L2$, and that capital has a negative marginal product between points $C1$ and $C2$.

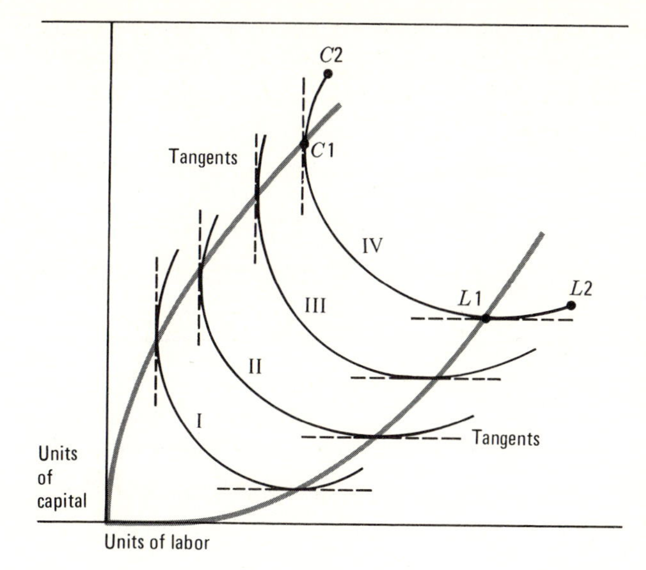


**FIGURE 2.17** Isoquant map for the feasible production range. Each of the isoquants bends backward from the points indicated by horizontal and vertical tangents. The ridge lines connecting these points mark the boundaries of stage II production.

## EXTENSION

***2-1 Linear Programming*** The optimal mix of scarce resources can be determined by a versatile mathematical technique called *linear programming* (LP). In its graphical form it resembles the approach illustrated in Figure 2.13 for identifying the optimal combination of inputs to achieve a given level of output. More complex problems are solved by the *simplex method* originated by G. B. Dantzig in 1947, an iterative routine usually performed by a computer. Since whole books are devoted to explanations of LP intricacies, we shall observe only its relationship to the resource-allocation procedures discussed in this chapter. This brief introduction may whet your intellectual appetite for a whole course in LP because it is a popular and widely used device for economic analyses.

A cost-minimization problem can be graphed when only two inputs are involved; graphing is difficult with three inputs and becomes impossible with more. The production requirements are stated as minimum levels of ingredients required to produce the product, and the cost of production is defined by the per-unit prices of the inputs. Note in the following example how the production requirements form a set of limiting conditions that appear similar to an isoquant when graphed, and the prices of inputs are represented by isocost lines.

A concrete products company has two sources of sand and gravel aggregate. The material from one source is much coarser than the other. The cost from either source depends on pit cost, hauling, and refining. The amount of each grade of rock and sand required for the production of concrete pipe, ready-mix, and other products is fairly well known for 1 month in advance, but it changes from month to month as a function of demand. The objective of determining the proportion of material to take from each source is to *minimize* monthly procurement, handling, and storage costs. Necessary cost and quantity relationships are tabulated below.

| | POUNDS/TON | | |
|---|---|---|---|
| *Aggregate Size* | *Source 1* | *Source 2* | *Tons Required* |
| ½ in to 2 in | 600 | 200 | 10,000 |
| #4 to ½ in | 900 | 600 | 15,000 |
| #50 to #4 | 400 | 900 | 20,000 |
| Less than #50 | 100 | 300 | 5,000 |
| Cost/ton | $1.51 | $1.68 | |

The first step in solving the problem is to convert all measures to the same dimensions. Since most units are in tons, the pounds of each aggregate size per ton can be converted to percentages of tons to establish the following restraint inequations:

$0.30S1 + 0.10S2 \geqslant 10{,}000$ for ½ in to 2 in

$0.45S1 + 0.30S2 \geqslant 15{,}000$ for #4 to ½ in

$0.20S1 + 0.45S2 \geqslant 20{,}000$ for #50 to #4

$0.05S1 + 0.15S2 \geqslant 5{,}000$ for less than #50

These inequations establish the limiting conditions for production, and each appears on a graph as a line connecting all mixes of the inputs that just satisfy the stated requirement; any mix on or above the line is a feasible combination. All the ingredient requirements must be met for an acceptable solution.

The optimal mix minimizes

$\text{Cost} = \$1.51S1 + \$1.68S2$

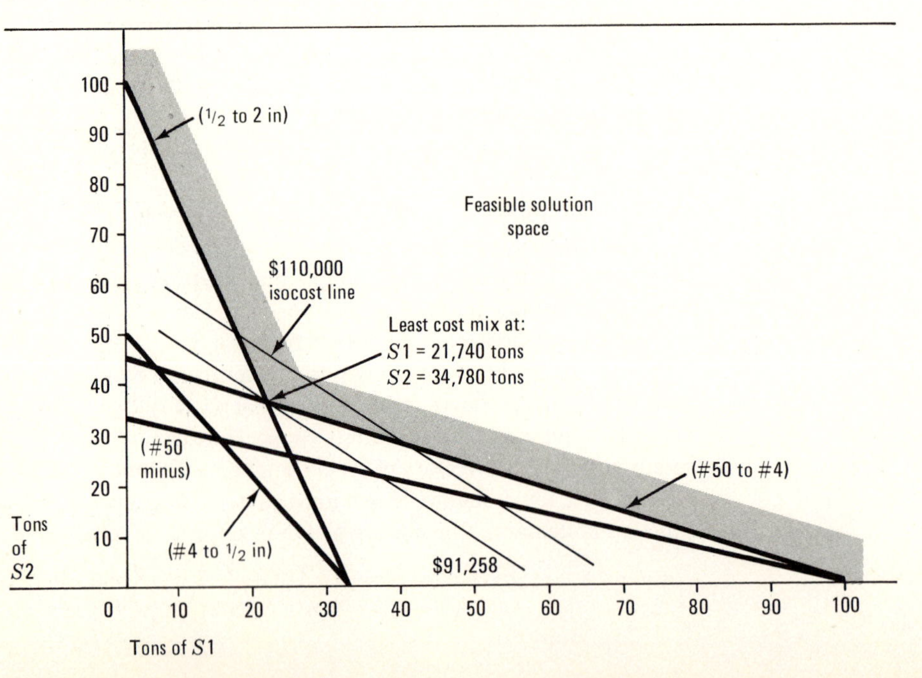


**FIGURE 2.18** Graphical solution of the lowest cost mix of two inputs. The four heavy restraint lines represent minimum requirements for production and the two light isocost lines indicate costs of different mixes.

This cost function is represented by isocost lines for different levels of inputs. The objective is to identify the least cost that fulfills *all* the production requirements. This mix is identified by the lowest isocost line that just touches the feasible-solution space which occurs at the uppermost set of lines representing ingredient limitations. In the example, the quantity of rock and sand must be greater than or equal to the given monthly demand, so the solution space must be on or above all restraint lines.

As shown in Figure 2.18, the lowest point in the shaded feasible solution space occurs where $S1 = 21{,}740$ tons and $S2 = 34{,}780$ tons. The isocost line passing through this point is $91,258. The lowest-cost mix thus exceeds the monthly 50,000 tons demand, but no other combination will satisfy all the minimum rock and sand requirements more economically.

**QUESTION**

**2.1a** Using the graphical method described above, solve the following problem:

Two types of livestock feeds are mixed together to obtain specified minimum requirements of certain ingredients. The percentage of each ingredient per pound of each type of feed, the cost of both feeds, and minimum ingredient requirements are listed in the table:

| | % INGREDIENT/LB | | |
|---|---|---|---|
| *Ingredient* | *F1* | *F2* | *Minimum Daily Requirement, lb* |
| *I1* | 20 | 15 | 2 |
| *I2* | 0 | 25 | 1 |
| *I3* | 25 | 0 | 1.5 |
| *I4* | 20 | 30 | 3 |
| *Cost/lb* | $0.15 | $0.40 | |

What is the optimal combination of the two feeds, $F1$ and $F2$?

*(Mix 9 parts F1 with 4 parts F2.)*

# CHAPTER 3

# BREAKEVEN ANALYSIS

In this chapter the principles from Chapter 2 are applied to operating decisions that face a firm. Should a component of a product be purchased from outside suppliers or should it be produced internally? Would cost-reduction efforts be more likely to succeed by focusing on variable or fixed costs of production? What happens to gross profit when one of a family of products is discontinued?

Most economic comparisons are a form of breakeven analysis. Alternative courses of action are compared to identify which one benefits the organization most. Unless an alternative is clearly dominant for all possible operating situations, there will be certain conditions under which one alternative is preferred and above which a different one is desirable. The level at which preference switches is a *breakeven* or *indifference* point. In future chapters we shall investigate the effects of time and risk on such comparisons. This chapter is limited to short-run evaluations under conditions of *current economy* (the time value of money is not considered) and assumed certainty (data for decisions are believed to be acceptably accurate), and features the economic balance between cost and output.

## COST AND COMPETITIVENESS

Most organizations strive for profits. They do so through a close scrutiny of their internal operating costs and strict attention to their competitive position. Even intentionally non-profit organizations must follow the same policy if they are to achieve excellence. The cost-revenue-profit relationships are exposed by breaking down a unit of output into its component dollar values.

The rectangular block in Figure 3.1(a) represents a unit of output. This output can be a product, such as an automobile, or it can be a service, such as collecting garbage from a subscriber. The unit is divided into three segments which classify the producer's interests. The overall height or price for which it can be sold is a function of the consumer's regard for the item.

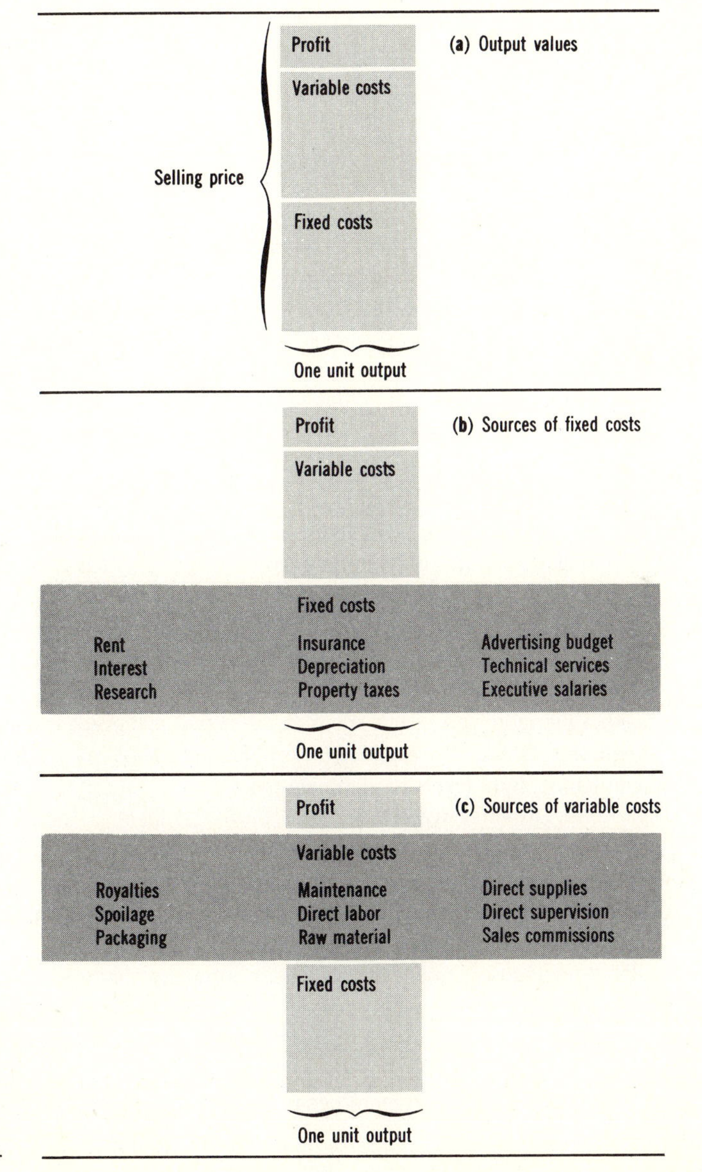


**FIGURE 3.1** Unit costs.

## Fixed Costs

Those costs which remain relatively constant regardless of the level of activity are known as *fixed costs* or *indirect costs*. This description implies that the fixed level is maintained whether output is nil or at 100 percent capacity. In some cases this assumption is not valid; fixed costs may tend to increase as output increases, and they can vary with time. However, the change is usually not significant for short-run studies.

Some of an organization's expenditures which can be considered as fixed are shown in Figure 3.1(b). These costs may be thought of as "preparation" expenses. They arise from measures taken to provide the means to produce a product or service. Before painters can paint a house, they have to have paint brushes. Whether they paint one house or a dozen with the brushes, the expense has already been incurred and shows as a fixed cost. The painter's insurance and advertisements for work would also be indirect costs.

## Variable Costs

Those costs which are generally proportional to output are called *variable costs* or *direct costs*. Such costs are relatively easy to determine, because they are directly associated with a specific product or service. When there is no output, variable costs are zero. The input material and the time required to make a unit give rise to variable costs. For example, the specific type and quantity of paint painters use in painting a house is a variable cost. The more houses they paint, the more paint they use; the quantity used is a function of their output. In a similar manner, the time they spend painting is a direct cost.

## Profit

The dimension of quantity must be included to examine the competitive aspects of profit. A single unit of output is relatively immune to competition. In isolated instances, a fair-sized output distributed in a local area to satisfy a peculiar need is also shielded from competition. However, as output quantity expands, competition is an increasingly apparent factor. Profit is the cause and effect of competitiveness.

A profit (or loss) figure attracts a great amount of attention. It is a handy yardstick of success. Like a thermometer, it only measures the level achieved; it does not control the source it measures. Unlike a thermometer, however, continued low readings may convince the financial temperature takers to eliminate the source.

There are basically three ways to increase profit: (1) increase the selling price, (2) increase the value to increase sales, and (3) decrease the selling price to increase sales. The profit-expansion descriptions are oriented to consumers' interests. The issues become more complicated when we look at them from the producer's viewpoint. Figure 3.2 shows some of the consequences of selling-price manipulations.

The original price-cost-quantity conditions are shown in Figure 3.2(a). *Total revenue* is the product of $N$ units sold at selling price $P$. *Total cost* is the sum of variable and fixed costs incurred in producing $N$ units. *Profit* is the difference between revenue and total cost (when revenue exceeds costs).

Figure 3.2(b) through (d) shows increased profit. The shaded profit areas of (b), (c), and (d)

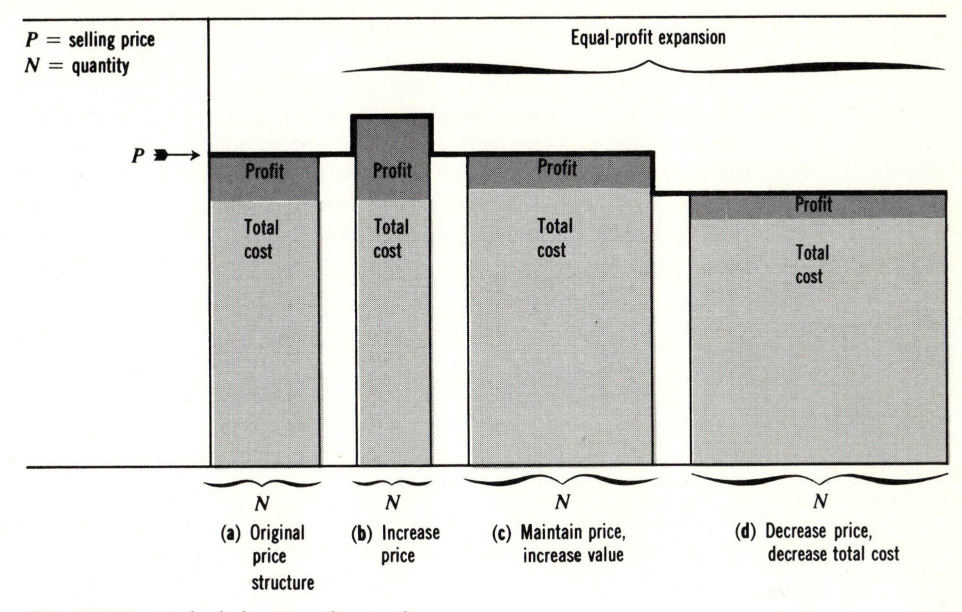


**FIGURE 3.2** Methods for expanding profit.

are equal and are larger than the profit in (a). The dangers and limitations of profit-expansion methods are as follows:

*Increased selling price* Competing products or services set an upper limit to price increases.* Ultimately, this limit is controlled by the consumers. Their willingness to pay is a function of the value they expect to receive and their loyalty to a product. Prices higher than competing products of equivalent value will reduce the number of units sold. The shrinking share of the market eventually causes a decline in total profit.

*Unchanged selling price* One way to increase profit without changing the selling price is to sell more units by increasing the value. The greater value perceived by the consumer can result from better quality, more quantity, or more effective advertising. All these measures increase the total cost for the producer. Higher total cost leads to a lower margin of profit per unit sold. If the market is unstable, a very low profit margin can seriously limit recuperative powers during market fluctuations.

A straightforward means to increase profit while holding prices constant is to reduce total costs. Such a task is the continuous aim of engineers and managers. The obstacle is that it becomes increasingly difficult to make more and more savings in an established operation. At first it is easy. When a product or service is new, it meets a high current demand which compensates for operational inefficiencies. As competition forces the price down, the "fat" is removed from operations. Further effort to reduce costs meets diminishing returns. It is like trying to make a horse run faster. A small whip may help at first, but using ever larger whips fails to force proportional returns in greater speed.

*Reduced selling price* New areas of cost reduction are exposed by changing the level of operations or capacity. A greater output often allows new methods to be incorporated.

*Conditions of *free competition* are assumed: Similar products or services are available from a number of vendors.

Some of the savings resulting from the new methods are passed on to consumers in the form of a lower selling price. In theory, the decreased price should lead to the sale of more units, which in turn satisfies the conditions for incorporating the new methods.

Limitations are inherent throughout the cost reduction-lower price-increased sales-cycle. Cost reductions are limited by minimum levels of quality, maximum levels of expenditure for new equipment, and basic labor or material costs that resist lowering. Reduced prices may be an insufficient incentive to attract enough new sales. However, with reasonable care the cycle rewards the producer and leads to a better standard of living for the consumer.

## BREAKEVEN COMPARISONS

After a decision has been made to pursue a venture that promises to be beneficial, a subsequent decision is usually required to select the best tactic to accomplish the venture's purpose. One method of accomplishment seldom stands out as the best possible when all angles are considered. Assuming the choice among methods has been narrowed to those that can perform the desired function adequately, it is still likely that different alternatives are better over certain ranges of activity. A breakeven comparison detects the range over which each alternative is preferred. Then the decision maker has only to decide the most likely range of future operations to select the proper method to complete the venture.

A "make-or-buy" decision is frequently encountered. It occurs when an item can be made in-house at a lower variable cost than the purchase price for that item from a vendor. The point at which the alternatives are equal depends on the first cost required to begin the in-house production. This situation was illustrated in Figure 1.6 and was used to point out the dangers of suboptimization.

A similar problem arises in a decision to lease or buy an asset. The costs of ownership (purchase price, installation cost, etc.) must be added to operating costs in comparing an owned item to one that is leased. Any fixed-cost investment to lower variable costs with respect to a strictly direct-cost alternative depends on the time period required to pay off the

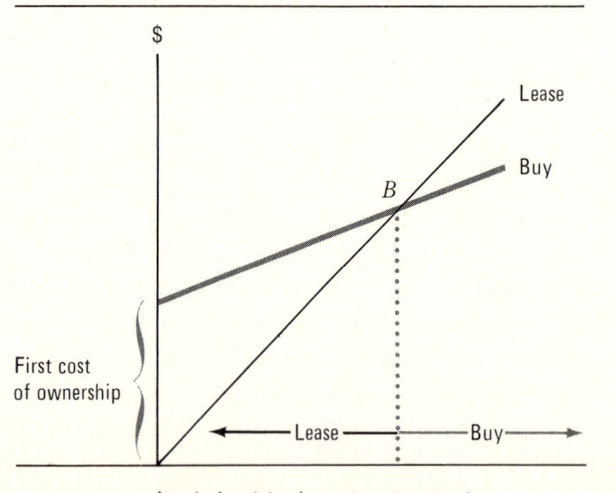


**FIGURE 3.3** Lease-or-buy comparison.

investment from variable-cost savings. One advantage of diagraming a breakeven comparison, as shown in Figure 3.3, is the attention focused on the breakeven point $B$, which designates the level of activity and implies the time period separating the preference for alternatives.

---

**Economy Exercise 3-1** An engineering consulting firm won a contract to design and supervise construction of a sewage-treatment plant at a remote location. The installation phase will last at most 2 years, and two engineers from the firm will supervise on-site operations. They will need both living accommodations and an office. Three alternatives are available, with the costs shown below.

1 Rent a building with furnished living accommodations and an office: $1500 per month including upkeep and utilities.
2 Buy two furnished trailers to live in and rent an office: The purchase price of a house trailer is $12,000 per trailer (the seller will buy back a used trailer for 40 percent of its purchase price any time within 2 years); trailer upkeep, site rental, and utilities are $100 per trailer per month; and office rental is $400 per month.
3 Buy three trailers: Two house trailers as in alternative 2 and a smaller one to serve as an office, purchased for $8000 from the same seller.

If all the alternatives provide adequate facilities, which one do you recommend?

---

## BREAKEVEN CHARTS

The best known breakeven model relates fixed and variable costs to revenue for the purpose of profit planning. The name *breakeven chart* is derived from the concept it depicts, the volume or level at which revenue and total cost of operations exactly break even. At this point, one additional unit made and sold would produce a profit. Until the breakeven point is attained, the producer operates at a loss for the period.

### Standard Format

Properties of a typical breakeven chart are displayed in Figure 3.4. The vertical scale shows the revenue and costs in monetary units. The horizontal scale indicates the volume of activity during the period pictured. The units of volume can be in sales dollars, number of units produced and sold, or the output quantity expressed as a percentage of total capacity.

The horizontal line in the chart shows the fixed costs $F$, which are constant throughout the range of volume. The sloping line originating at the intersection of the fixed-cost line and the vertical axis represents variable costs $V$ plus the fixed costs. For linear relationships, variable costs are directly proportional to volume; each additional unit produced adds an identical increment of cost. The sum of variable and fixed costs is the total cost $C$. The sloping line from the origin of the graph is the revenue line. Revenue $R$ is also assumed to be directly proportional to the number of units produced and sold at price $P$.

The breakeven point $B$ occurs at the intersection of the total-cost and revenue lines. It thus specifies the dollar volume of sales and the unit volume of output at which an operation neither makes nor loses money. The vertical distance between the revenue line and the total-cost line indicates a profit $Z$ to the right of $B$ and a loss to the left.

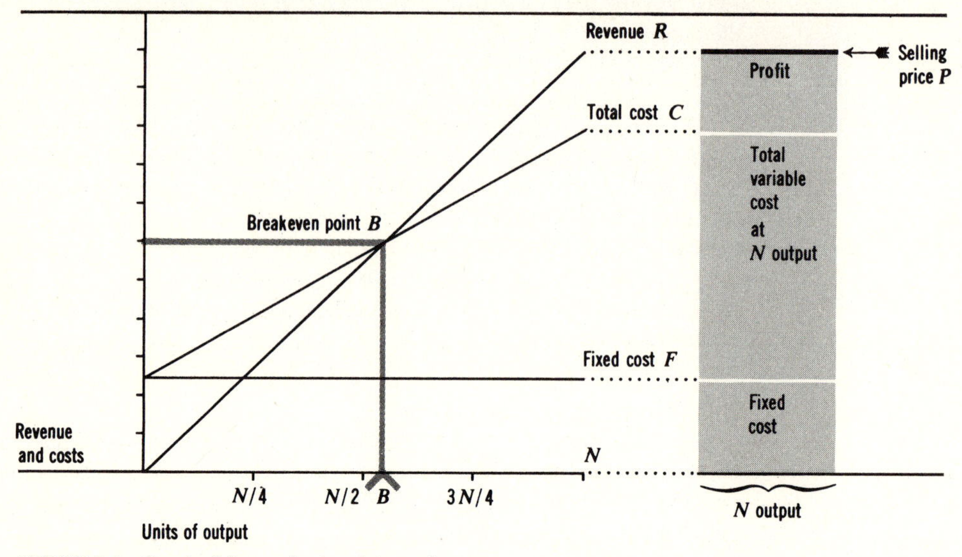


**FIGURE 3.4** Standard format for breakeven charts.

A block cost diagram used in the discussion of profit expansion is shown alongside the breakeven chart in Figure 3.4. For an output volume of $n$, the costs, revenue, and profit are the same in both formats. The breakeven chart further indicates the profit or loss expectation at levels of output other than the specific quantity $N$. This feature helps explain such statements as, "A very low profit margin can seriously limit recuperative powers during market fluctuations." "A very low profit margin" means that the output is barely on the profit side of the breakeven point. An unstable market could easily cause sales to fall below point $B$ and show a loss for the period.

## Algebraic Relationships

The graphic format is convenient for clarifying or presenting economic relationships. It is possible to obtain quantities for particular conditions by scaling values from the chart. However, the same conditions can be easily quantified by formulas. Calculations generally provide greater accuracy. Using the symbols already defined, we have

$$\text{Revenue/period} = R = nP$$
$$\text{Total cost/period} = C = nV + F$$
$$\text{Gross profit/period} = Z = R - C = n(P - V) - F$$

where $n$ can also be a fraction of total capacity when $P$ and $V$ represent total dollar volume at 100 percent capacity.

At the breakeven point, profit equals zero. To determine the output to just break even, we have, at $B$,

$$Z = 0 = R - C = n(P - V) - F$$

and, letting $n = B$,

$$B = \frac{F}{P - V}$$

The term $P - V$ is called *contribution.* It indicates the portion of the selling price that contributes to paying off the fixed cost. At $n = B$ the sum of contributions from $B$ units equals the total fixed cost. The contribution of each unit sold beyond $N = B$ is an increment of profit.

To observe breakeven-chart relationships, assume an airline is evaluating its feeder routes. These routes connect smaller cities to major terminals. They are seldom very profitable themselves, but they feed passengers into the major flights which yield better returns. One feeder route has a maximum capacity of 1000 passengers per month. The contribution from the fare of each passenger is 75 percent of the \$40 ticket price. Fixed costs per month are \$21,000.

To find the average percentage of seats that must be sold on each flight to break even, the cost and revenue data could be converted to the graphical breakeven format shown in Figure 3.5. The same information displayed in the breakeven chart is supplied by the following calculations:

Total revenue per month is

$$nP = 1000 \times \$40 = \$40{,}000$$

$$\text{Total contribution} = 0.75 \times \$40{,}000 = \$30{,}000$$

$$B(\%\text{ of capacity}) = \frac{F \times 100\%}{\text{contribution}}$$

$$= \frac{\$21{,}000}{\$30{,}000}\,100\% = 70\%$$

or,

$$B(\text{passengers}) = \frac{F}{P - V} = \frac{\$21{,}000}{0.75 \times \$40}$$

$$= 700 \text{ passengers/month}$$

Assuming a tax rate of 40 percent, the net profit at full capacity is

$$\begin{aligned} \text{Net profit} &= Z(1 - t) \qquad \textit{where } t = \text{tax rate} \\ &= (R - C)(1 - t) \\ &= [n(P - V) - F](1 - t) \\ &= [1000(0.75 \times \$40) - \$21{,}000](0.60) \\ &= \$9000 \times 0.60 = \$5400 \end{aligned}$$

A gross-profit line is shown in the lower right corner of the chart, starting at $B$. Net profit is a fraction of $Z$ which depends on the tax rate for the total earnings of the organization.

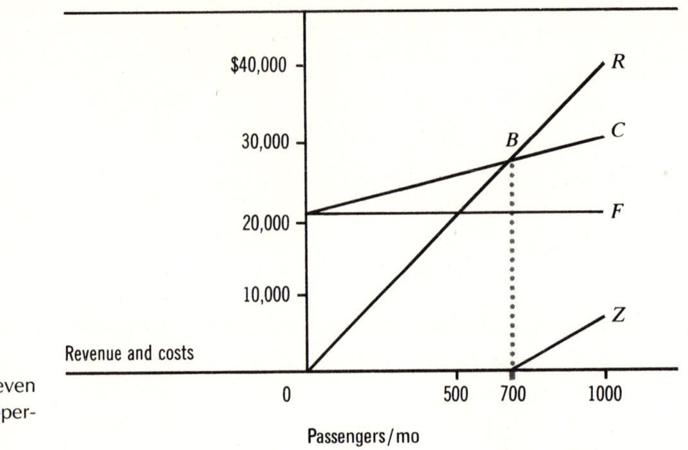


**FIGURE 3.5** Breakeven chart for an airline operation.

## LINEAR BREAKEVEN ANALYSIS

A breakeven chart displays cost-volume-profit relationships that hold only over a short run. Over the long run, the relationships are altered by internal factors (new products, production facilities, etc.) and external impacts (competition, state of the general economy, etc.). Many of the internal activities that affect long-run changes are initiated from analyses of current cost-volume-profit conditions. Thus, a breakeven analysis is like a medical checkup; the physical examination reveals the current state of health and provides clues about what should be done to become or stay healthy.

### Breakeven-Point Alternatives

Any change in costs or selling price affects the breakeven point. We observed the gross effects of profit expansion as a function of selling price. Now we can consider the interaction of revenue, variable costs, and fixed costs in terms of output.

A lower breakeven point is a highly desirable objective. It means the organization can meet fixed costs at a lower level of output or utilization. A sales level well above the breakeven output is a sign of healthiness. Three methods of lowering the breakeven point are shown in Figure 3.6. The original operating conditions are shown as light lines and are based on the following data:

$$V = \$7/\text{unit}$$
$$P = \$12/\text{unit}$$
$$R(\text{at } n = 100 \text{ units}) = \$1200$$
$$C(\text{at } n = 100 \text{ units}) = \$1100$$
$$F = \$400$$
$$B = 80 \text{ units}$$

The bold lines depict the measures necessary to reduce the breakeven point by half, from 80 to 40 units.

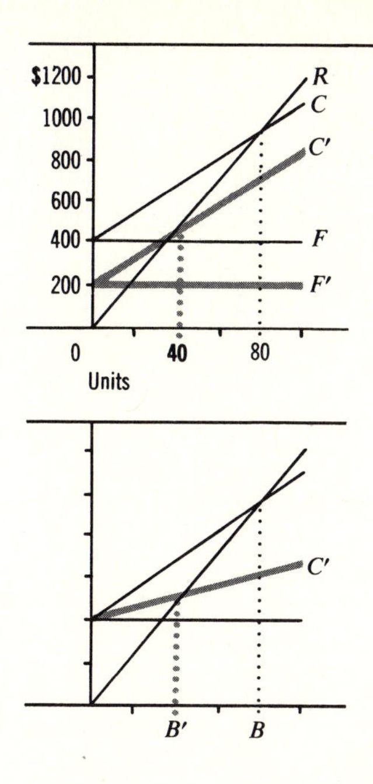


**Fixed-cost reduction:** By reducing the fixed costs by half, $B$ is halved. Therefore the new fixed cost $F'$ is

$$F' = \frac{\$400}{2} = \$200$$

and the new breakeven output $B'$ becomes

$$B' = \frac{F'}{P - V}$$

$$= \frac{\$200}{(\$12 - \$7) \text{ per unit}} = 40 \text{ units}$$

**Variable-cost reduction:** Knowing that $B'$ should equal 40 units, we can solve for the associated $V'$ by

$$B' = \frac{F}{P - V'}$$

or

$$V' = P - \frac{F}{B'} = \$12 - \frac{\$400}{40}$$

$$= \$12 - \$10 = \$2/\text{unit}$$

**Selling-price increase:** Raising the selling price $P'$ increases the slope of the revenue line and augments the contribution. The contribution required for $B'$ to equal 40 units is

$$\text{Contribution} = \frac{F}{B'} = \frac{\$400}{40}$$

$$= \$10/\text{unit}$$

which leads to

$$P' = V + \text{contribution}$$

$$= \$7 + \$10 = \$17$$

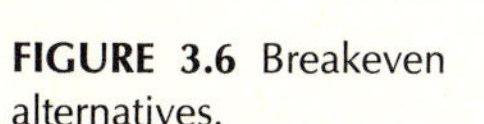

**FIGURE 3.6** Breakeven alternatives.

## Margin of Profit and Dumping

The perspective from which an organization views its breakeven point depends on its particular circumstances at the time of review. A firm struggling to get its products accepted in the market would have concerns different from those of a producer that has already captured a large share of the market. The struggling firm is likely beset with problems of cash availability to meet its fixed costs, while the dominant producer probably seeks investments to maintain or expand its sales.

Consider the alternatives available to a firm producing package waste-disposal units that sell for $35,000 each. Variable costs are $20,000 per unit, and fixed costs are $600,000. The plant can produce a maximum of 80 units per year. It is currently operating at 60 percent capacity. The firm is contemplating the effects of reducing the selling price by $2000 per unit, adding a feature to each unit which will increase the variable costs by $1000, and allocating an extra $120,000 per year for advertising. These actions are designed to sell enough additional units to raise plant utilization to 90 percent.

Under current conditions,

$$B = \frac{F}{P - V} = \frac{\$600{,}000}{\$35{,}000 - \$20{,}000} = 40 \text{ units}$$

Since the company now sells 0.60 × 80 = 48 units per year, the gross annual profit is

$$Z = \text{units sold beyond } B \times \text{contribution/unit}$$
$$= (48 - 40)\$15{,}000 = \$120{,}000$$

The ratio of gross annual profit to fixed costs is

$$\frac{Z}{F} = \frac{\$120{,}000}{\$600{,}000} = 0.20$$

End of Chap 11

and may be thought of as a margin of profit or safety. The same ratio can be obtained by

$$\text{Margin of profit} = \frac{n - B}{B} = \frac{48 - 40}{40} = 0.20$$

where $n$ is the number sold during the period.

Both the current conditions (dotted lines) and the anticipated conditions (solid lines) are displayed in Figure 3.7. As graphed, $B$ increases, as a result of the added expenditures, to

$$B' = \frac{\$600{,}000 + \$120{,}000}{\$33{,}000 - \$21{,}000} = \frac{\$720{,}000}{\$12{,}000} = 60 \text{ units}$$

and the gross profit expected at 90 percent capacity also increases to

$$Z = [(0.90 \times 80) - 60](\$33{,}000 - \$21{,}000)$$
$$= 12 \text{ units} \times \$12{,}000/\text{unit} = \$144{,}000$$

but the margin of profit remains unchanged at

$$\frac{Z}{F} = \frac{\$144{,}000}{\$720{,}000} = 0.20$$

The firm could also follow a course of action in which only one or two of the alternatives are pursued. If the advertising budget is eliminated but price and modifications are retained, the same profit (\$144,000) would be obtained at an output of

$$n = \frac{Z + F}{P - V} = \frac{\$144{,}000 + \$600{,}000}{\$33{,}000 - \$21{,}000}$$

$$= \frac{\$744{,}000}{\$12{,}000/\text{unit}} = 62 \text{ units}$$

and

$$B = \frac{\$600{,}000}{\$12{,}000/\text{unit}} = 50 \text{ units}$$

which makes the margin of safety (62 − 50)/50, or 0.24.

There are, of course, many factors to consider in such a decision. If the market is stable,

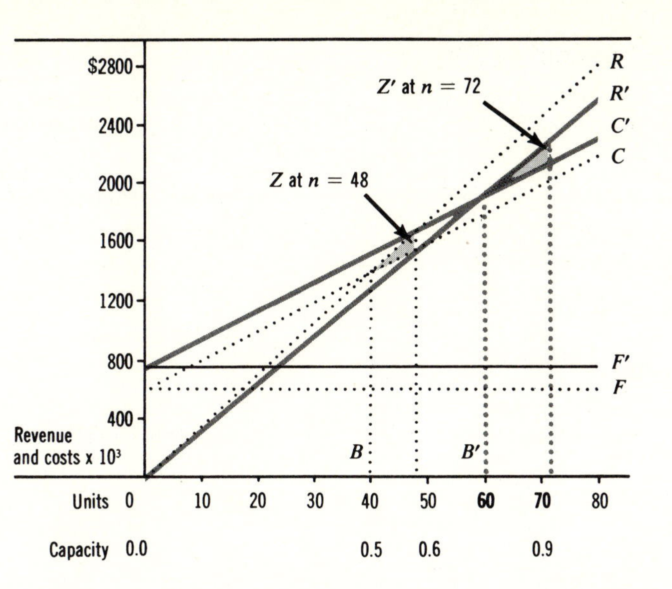


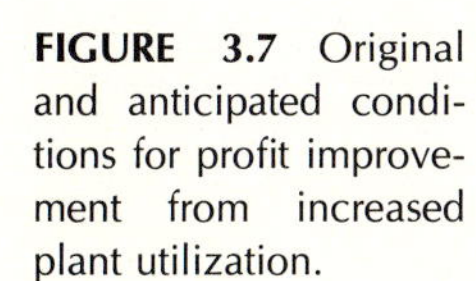

**FIGURE 3.7** Original and anticipated conditions for profit improvement from increased plant utilization.

the margin of profit is less important. Some alternatives are easier to implement than others. Some outcomes are more certain than others.

Still another alternative would be to sell a portion of the output at a reduced price. This practice is called *dumping*. It can be accomplished by selling to foreign markets at a lower price or by selling the same product at different prices under different names. There are many dangers in this practice, but if it works, profit will increase because of the increased plant utilization.

Figure 3.8 illustrates dumping applied to the original data on waste-disposal-unit sales: 48 units are sold at the regular price ($P$ = \$35,000) to account for 60 percent utilization. If 100 percent utilization could be achieved by *dumping* the remaining capacity at a sales price $P'$ of \$25,000 per unit, the gross profit would increase from \$120,000 to \$280,000.

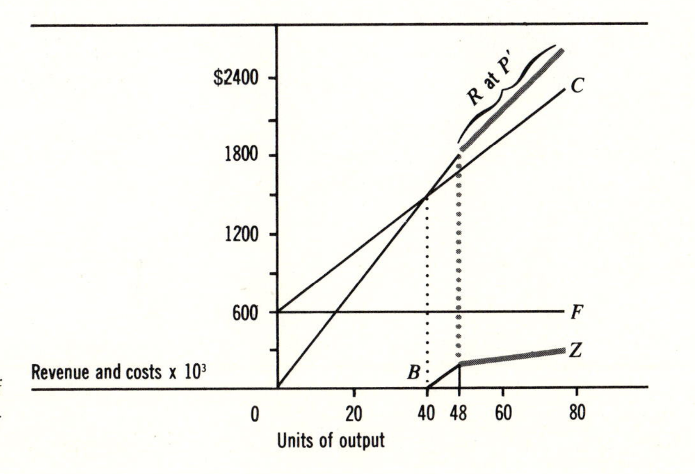


**FIGURE 3.8** Effects of *dumping* on price, profit, and utilization.

**Economy Exercise 3-2** *Operating leverage* reflects the extent to which fixed costs are used in operations, and is a diagnostic measure similar to the margin of profit. Calculation of the degree of operating leverage resembles the elasticity calculation described in Chapter 2:

Breakever Pt, and up

$$\text{Degree of operating leverage} = \text{OL}^\circ = \frac{\%\text{ change in operating income}}{\%\text{ change in quantity sold}}$$

$$= \frac{\Delta Z/Z}{\Delta Q/Q}$$

**a** Develop a formula for OL° composed of breakeven-chart symbols ($n$, $P$, $V$, and $F$).
**b** Apply the formula to the original conditions of the firm producing package waste-disposal units (shown in Figure 3.7) for an output level of 48 units.
**c** Interpret the meaning of operating leverage in terms of elasticity concepts.

## Multiproduct Alternatives

More than one product can be shown on a breakeven chart. Including a whole product line allows the decision maker to evaluate the combined effect of the product mix on plant utilization, revenue, and costs. A slightly different format for the breakeven chart accentuates the effect of multiple products. This type of graph is displayed in Figure 3.9 and is called a *multiproduct profit* or *contribution chart.*

The chart is constructed by plotting fixed costs as a loss on the vertical axis. The horizon-

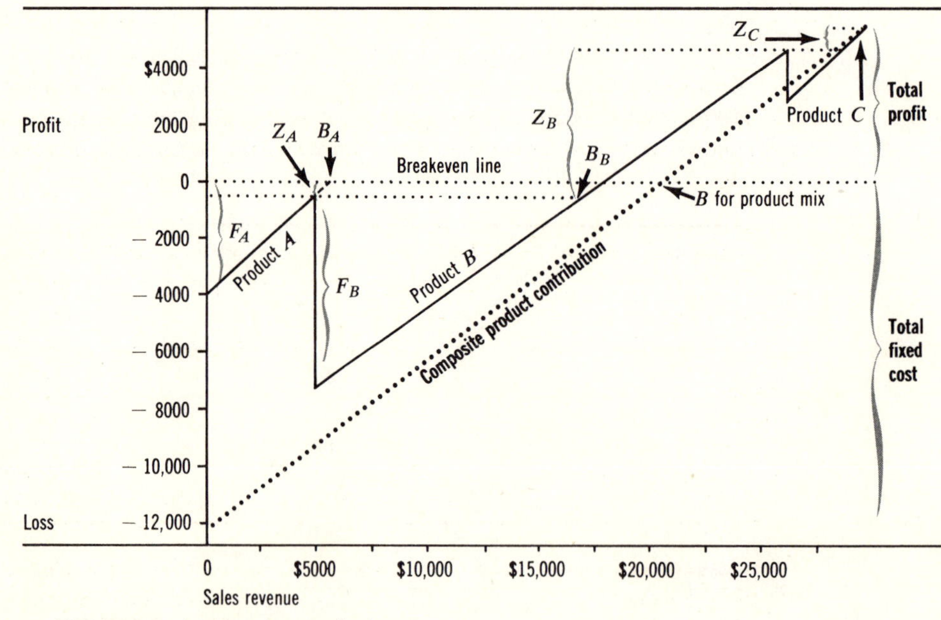


FIGURE 3.9 Multiproduct profit chart.

tal axis denotes sales revenue. At zero sales the only costs associated with a product are the negative preparation costs or fixed costs. As production and sales develop, each unit sold makes a contribution toward paying off the fixed costs. When enough units have been sold to pay these costs, the breakeven point is reached and the contribution from further sales is profit.

Three products, $A$, $B$, and $C$, are represented in Figure 3.9. The plotted values are based on the following assumptions made for the period shown in the chart:

| | PRODUCT | | |
|---|---|---|---|
| | *A* | *B* | *C* |
| Selling price per unit | $8 | $5 | $4 |
| Contribution per unit | $6 | $3 | $3 |
| Fixed cost | $4000 | $7000 | $1000 |
| Number of units sold | 600 | 4000 | 400 |

Product $A$ is entered in the chart by marking its fixed costs ($4000) on the negative side of the breakeven line at zero sales revenue. The total contribution of $A$ is 600 units times $6 per unit equals $3600. Subtracting the fixed costs from the total contribution leaves a loss of $400 from total sales of $4800.

The remaining products are dealt with in a similar manner. Each fixed cost is entered as a vertical line attached to the highest point on the contribution line for the preceding product. The results are cumulative. The distance the last contribution segment extends above the breakeven line is the total plant profit. The dotted line extending diagonally across the chart is the composite contribution line. The point at which this line crosses the plant breakeven line establishes the sales volume at which fixed costs are exactly covered.

The value of multiproduct breakeven charts lies in their use for product comparisons. The portion of fixed costs borne by each product is easily observed. A product is preferred when its contribution line is steeper than the composite contribution line. Such considerations are important in decisions to add new products or drop old ones.

**Economy Exercise 3-3** A fertilizer plant is operating at capacity with production of four mixes which have a total sales volume of $2 million. The sales and production-cost figures for mixes $W$, $X$, $Y$, and $Z$ are as shown:

| | *Mix W* | *Mix X* | *Mix Y* | *Mix Z* | *Totals* |
|---|---|---|---|---|---|
| Percentage of total sales | 10 | 20 | 30 | 40 | 100 |
| Contribution (% of $P$) | 45 | 40 | 45 | 35 | |
| Fixed cost charged | $70,000 | $180,000 | $210,000 | $220,000 | $680,000 |
| Profit | $20,000 | −$20,000 | $60,000 | $60,000 | $120,000 |

Recognizing the loss incurred with product $X$, the company is considering dropping the product or replacing it with another mix. If the product is dropped without a replacement, the new sales and cost figures are estimated to develop as follows:

| | *Mix W* | *Mix Y* | *Mix Z* | *Totals* |
|---|---|---|---|---|
| Percentage of total sales | 15 | 35 | 50 | 100 (for $R$ of $1,800,000) |
| Contribution (% of $P$) | 45 | 45 | 35 | |
| Fixed cost charged | $100,000 | $250,000 | $290,000 | $640,000 |

Construct a multiproduct profit chart to compare the two alternatives.

## NONLINEAR BREAKEVEN ANALYSIS

Cost and revenue functions do not always follow convenient linear patterns. More often than not, realistic cost relationships develop a nonlinear pattern as typified by Table 3.1. The first four columns in the table relate output $n$ to total fixed cost $F$, total variable cost $TV$, and total cost $C$. The right side of the table shows average and marginal costs derived from the figures tabulated on the left.

**TABLE 3.1** Total and average cost data for a firm's operations over a given time period.

| *Total Product,* $n$ | *Total Fixed Cost,* $F$ | *Total Variable Cost,* $TV$ | *Total Cost,* $C$ | *Average Fixed Cost,* $F/n$ | *Average Variable Cost,* $TV/n$ | *Average Total Cost,* $C/n$ | | *Marginal Cost,* $\Delta C/\Delta n$ |
|---|---|---|---|---|---|---|---|---|
| 0 | $3000 | $ 0 | $ 3000 | — | — | — | | |
| 1 | 3000 | 700 | 3700 | $3000 | $700 | $3700 | --------- | $ 700 |
| 2 | 3000 | 1300 | 4300 | 1500 | 650 | 2150 | | 600 |
| 3 | 3000 | 1800 | 4800 | 1000 | 600 | 1600 | | 500 |
| 4 | 3000 | 2400 | 5400 | 750 | 600 | 1350 | | 600 |
| 5 | 3000 | 3100 | 6100 | 600 | 620 | 1220 | | 700 |
| 6 | 3000 | 3900 | 6900 | 500 | 650 | 1150 | | 800 |
| 7 | 3000 | 4900 | 7900 | 429 | 700 | 1129 | | 1000 |
| 8 | 3000 | 6200 | 9200 | 375 | 775 | 1150 | | 1300 |
| 9 | 3000 | 7800 | 10,800 | 333 | 867 | 1200 | | 1600 |

Average and marginal costs behave like the average and marginal product figures for production functions discussed in Chapter 2. This similarity is expected, of course, because costs are just equivalent measures for the use of production inputs such as capital and labor. Therefore, cost relationships also reveal the most economically efficient levels of production. Characteristic patterns of average and marginal costs based on Table 3.1 are pictured in Figure 3.10 and discussed below.

*Average fixed cost* Since fixed costs are independent of output, their per-unit amount declines as output increases. This feature is recognized when business people speak of "higher sales spreading the overhead."

*Average variable cost* The typical saucer-shaped average-cost curve declines at first, reaches a minimum, and then increases thereafter. It reflects the law of diminishing returns. Initially, combining variable resources $TV$ with fixed resources $F$ produces increasing returns, but a point is reached where more and more variable resources must be applied to obtain each additional unit of output. Stated another way, a fixed plant is underemployed when its output is below the minimum average-cost point. As output expands, more complete utilization of the plant's capital equipment will make production more efficient. But continually increasing variable costs will eventually create a condition in which overcrowding and overutilization of equipment impair efficiency.

*Average total cost* Because average total cost is simply the sum of average fixed and average variable costs, it shows the combined effects of spreading out fixed charges and diminishing returns from variable resources.

*Marginal cost* The key to the cost pattern in Figure 3.10 is contained in marginal-cost concepts. Marginal cost is calculated from either $C$ or $TV$ as the extra increment of cost required to produce an additional unit of output. If the last increment of cost is smaller than the average of all previous costs, it pulls the average down. Thus, *average total cost declines until it just equals marginal cost. Equivalently, the rising marginal-cost curve also pierces the average-variable cost curve at its minimum point.*

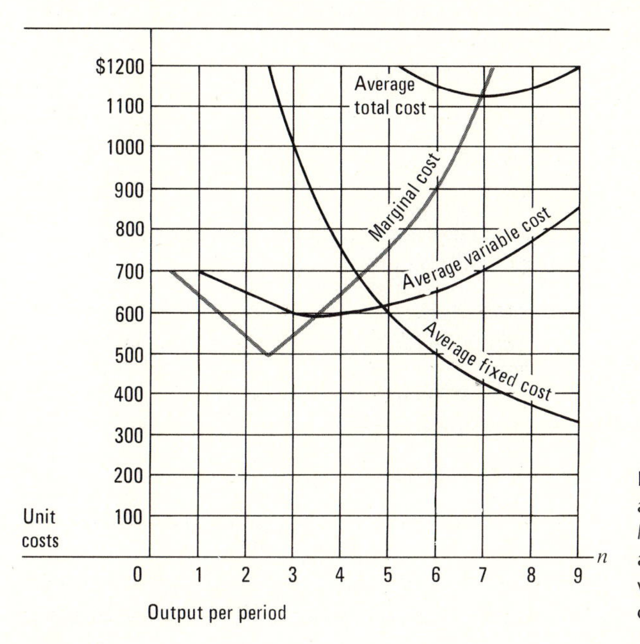


**FIGURE 3.10** Relationship of average and marginal costs. Minimum points on the average and total cost curves occur where they cross the marginal cost curve.

## Marginal Revenue and Profit

Both nonlinear revenue and cost schedules may be expressed as formulas. When such equations are available, their analysis is not much more difficult than that of linear models. Since an assumption of linearity makes all monetary increments constant over an extended range of output, nonlinear models call more attention to marginal relationships.

*Marginal revenue* is the additional money received from selling one more unit at a specified level of output. For linear revenue functions, the marginal revenue is a constant value $P$. That is, for each additional unit sold, the total revenue is increased by $P$ dollars. Consequently, a greater output automatically increases the total profit.

When the linear relationship is replaced by an expression such as

$$\text{Selling price} = P = 21{,}000n^{-1/2} \text{ dollars/unit}$$

the price of each unit is not so obvious. Such expressions are examined with differential calculus. For the price function above, the rate of change of revenue with output is

$$\text{Marginal revenue} = \frac{dR}{dn} = \frac{d(nP)}{dn} = \frac{d(21{,}000n^{1/2})}{dn}$$

$$= 10{,}500n^{-1/2}$$

Figure 3.11(a) shows a decelerating revenue rate and linear costs. Decreasing marginal revenue could result from a policy of lowering prices in order to achieve a higher plant utilization. The nonlinear revenue curve fixes two breakeven points. Between these two points the firm operates at a profit. Outside the breakeven points a loss is incurred, as shown in the profit chart, Figure 3.11(b).

The graphs are based on the following data:

$n$ = 1 unit produced and sold/period
$V$ = \$1000/unit
$F$ = \$100,000/period
$P$ = \$21,000$n^{-1/2}$/unit

Using the formula for total revenue and the selling-price function, $R = nP = 21{,}000n^{1/2}$, leads to a gross-profit equation of

$$Z = R - C = R - (nV + F)$$
$$= 21{,}000n^{1/2} - 1000n - 100{,}000$$

Knowing that $Z = 0$ at a breakeven point allows the value or values for $B$ to be determined by rearranging the terms for $Z$:

$$Z = 0 = -10^3n - 10^5 + 21(10^3n^{1/2})$$

or

$$10^3n + 10^5 = 21 \times 10^3n^{1/2}$$

Squaring each side of the equation and dividing by $10^6$ results in

$$n^2 + 200n + 10^4 = 441n$$

and by collecting terms,

$$n^2 - 241n + 10^4 = 0$$

This expression can be solved with the quadratic formula.*

$$n = \frac{241 \pm \sqrt{241^2 - 4 \times 10^4}}{2} = \frac{241 \pm 134}{2} = 53 \text{ or } 188$$

*For an equation in the quadratic form $Ax^2 + Bx + C = 0$, $x = \frac{-B \pm \sqrt{B^2 - 4AC}}{2A}$

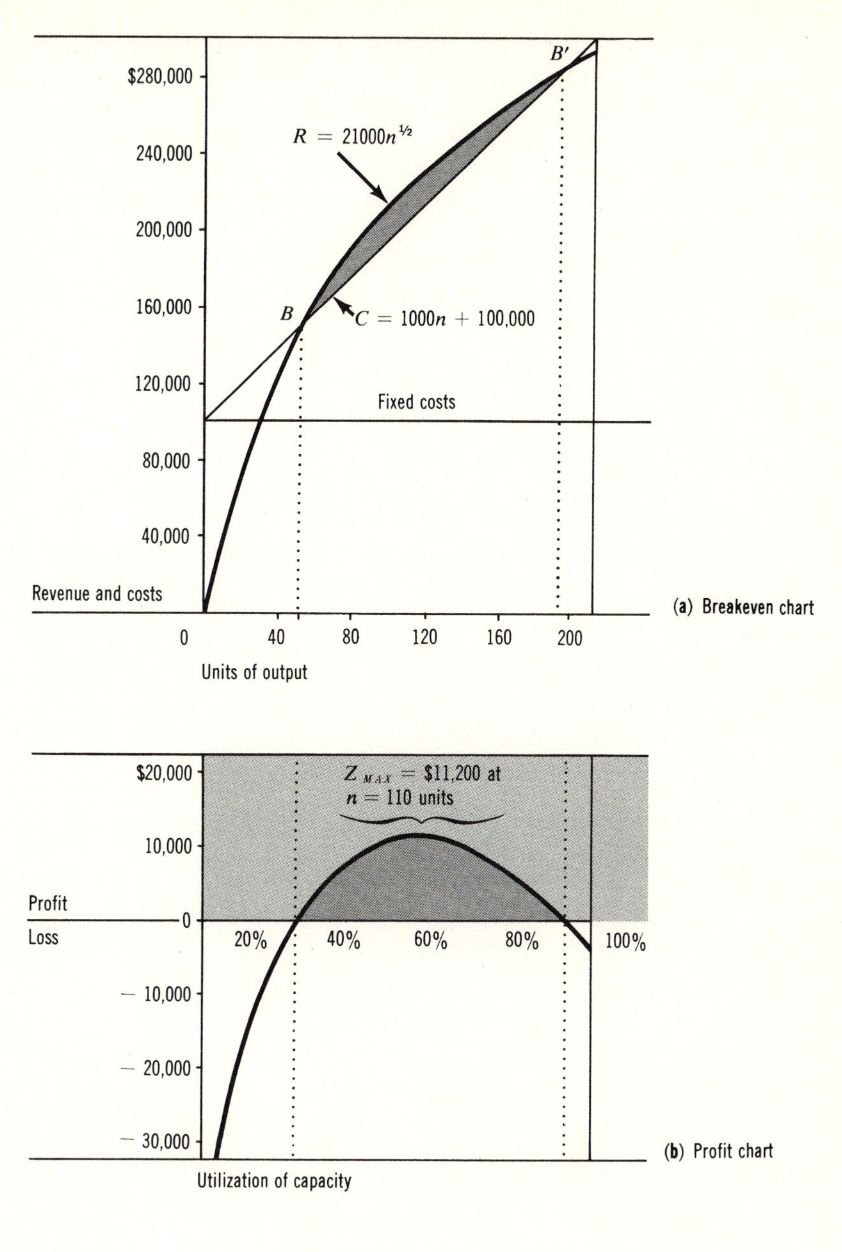


**FIGURE 3.11** Nonlinear breakeven charts for decreasing marginal revenue and linear costs.

or

$$B = 53 \text{ units} \quad \text{and} \quad B' = 188 \text{ units}$$

The point of maximum profit is especially important when two breakeven points are present. As indicated in Figure 3.11(b), the rate of profit is increasing to the left of the point of maximum profit, and decreasing to the right. The rate of change of profit with respect to output is *marginal profit.* At the point of maximum profit, the rate of change (and the marginal profit or slope of the profit line) is zero. Therefore, to find this point, differentiate the profit equation, set the derivative equal to zero, and solve for $n$:

$$\frac{dZ}{dn} = \frac{d(21{,}000n^{1/2} - 1000n - 100{,}000)}{dn} = 0$$

$$= 10{,}500n^{-1/2} - 1000 = 0$$

$$n = \left(\frac{10{,}500}{1000}\right)^2 = 110 \text{ units}$$

## Marginal Cost and Average Unit Cost

As production increases, the total cost per unit may also increase, owing to greater maintenance needs, overtime payments to workers, and general inefficiency caused by congestion during stepped-up operation. Under these conditions there is an increasing ***marginal cost*** (rate of change of total cost with output).

One possible pattern of marginal costs and linear revenue is shown in Figure 3.12. There could be many patterns: two breakeven points (as in Figure 3.11), decreasing marginal costs owing to savings realized from quantity purchases or near-capacity mechanized production, nonlinear functions for both revenue and costs, etc.

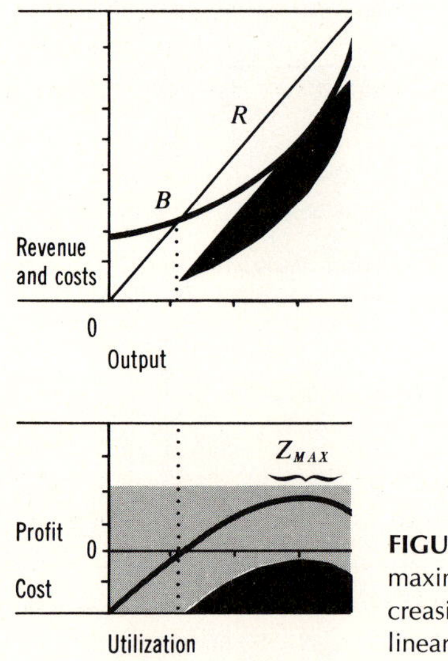


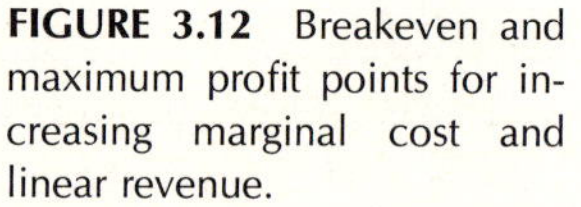


**FIGURE 3.12** Breakeven and maximum profit points for increasing marginal cost and linear revenue.

The feature points of a breakeven analysis are determined in the manner described previously. To find $B$, set the profit equation equal to zero and solve for $n = B$. Differentiate the profit equation, and set the derivative equal to zero to solve for the output that produces maximum gross profit. In doing so, it is interesting to note that

$$\frac{dZ}{dn} = \frac{d(nP - nV - F)}{dn} = 0$$

$$= \frac{d(nP)}{dn} - \frac{d(nV + F)}{dn} = 0$$

or

$$\frac{d(nP)}{dn} = \frac{d(nV + F)}{dn}$$

*marginal revenue* = *marginal cost*

***at the operating level that produces maximum profit.*** This means that when the change in revenue from one additional sale equals the change in cost of producing one more unit, the point of maximum profit is attained. Graphically, this point is at the output at which the vertical distance between the total-cost curve and the revenue curve is greatest. If marginal revenue and marginal cost were plotted, the output for maximum profit would be at the intersection of the two curves.

For linear functions, the *average unit cost* is

$$\frac{nV + F}{n} = V + \frac{F}{n} \qquad \textit{where } n = \text{a specific output}$$

In this case the average unit cost continually decreases with increasing output. For nonlinear costs this condition is not necessarily true. If the average cost goes through a minimum point and then increases, the slope of the curve will be zero at the output for lowest average unit cost:

$$\frac{d(V + Fn^{-1})}{dn} = 0$$

$$\frac{dV}{dn} - \frac{F}{n^2} = 0$$

$$\frac{dV}{dn} = \frac{F}{n^2}$$

From this equation and the maximum-profit equation it is clear that for nonlinear relationships the point of minimum average unit cost does not necessarily coincide with the maximum-profit point.

---

**Economy Exercise 3-4** A monthly record of operating expenses and revenue for a new manufacturing plant is posted in the manager's office. The purpose is to detect any changes in cost-revenue relationships and to establish the plant's operating pattern.

The graphic results of several months' operations are depicted in Figure 3.13. The pattern appears to indicate that as the output becomes greater, marginal costs increase and marginal revenue decreases. The belief is confirmed by formulas developed for curves that fit the data.

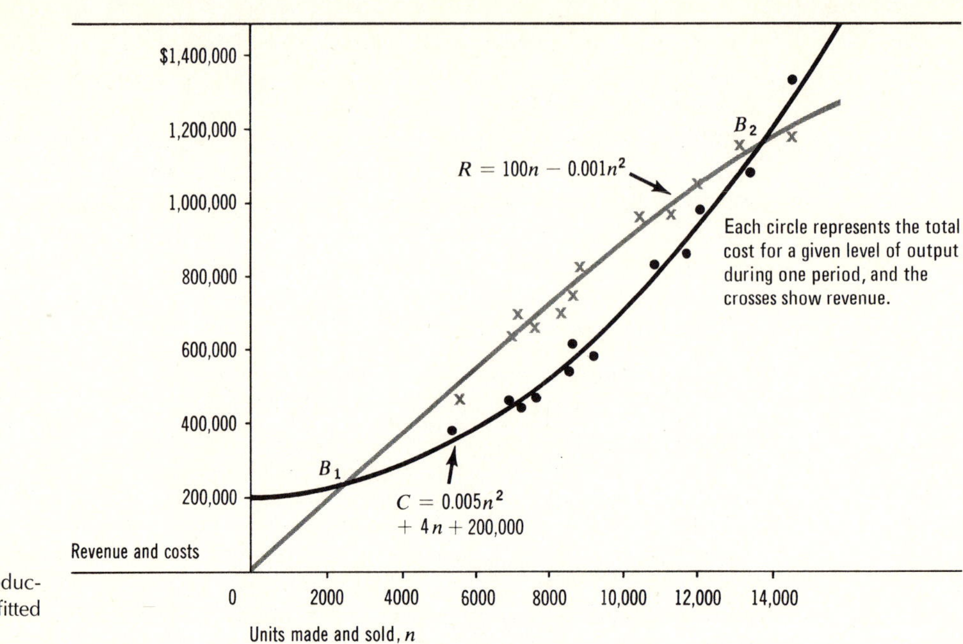


**FIGURE 3.13** Production data and fitted curves.

The selling price of finished units varies according to $P = (100 - 0.001n)$ dollars per unit. The price behavior is attributed to lower per-unit quotes given for large orders. Fixed costs are considered reasonable at $200,000 per month. The variable costs, $V = (0.005n + 4)$ dollars per unit, also appear competitive. The plant is designed to produce 12,000 units per month.

Based on the given data, calculate the level of output which produces the greatest profit, the least average unit cost, and the breakeven points. On a graph, show the curves for average unit cost, marginal cost, marginal revenue, and marginal profit.

## APPLYING BREAKEVEN ANALYSES

A breakeven analysis often tends to oversimplify the decision environment. This is an attribute for presentation purposes and for gross evaluations. It can also be a shortcoming for problems in which detailed measures are needed. A decision to lower the breakeven point for an operation can result from a study of total revenue and costs, but the study alone seldom reveals the in-plant operations that engineers and managers must conduct to implement the decision. The inability to identify tactical procedures is not really a defect of a breakeven analysis; it merely indicates that decision makers should be aware of the limitations of the approach in order to apply it appropriately.

The validity of a breakeven chart is directly proportional to the accuracy of the data incorporated in the chart. When several products are lumped together and represented by one line on a chart, there is a distinct possibility that poor performance by one product may go undetected. A firm should have a good cost-accounting system, but data from past performances are not always indicative of future performances. However, examining graphs of previous breakeven conditions calls attention to developing trends in revenues and costs.

Breakeven relationships imply where engineering efforts can be of most use to an organization. Field or factory-floor engineers can observe the present state of financial affairs and use those observations to guide their cost-control activities. As an engineer's managerial responsibilities increase, the interplay of price, cost, and quantity become of greater concern. Then the combined effect of the system's operations and the underlying economic principles merge to steer strategic decisions.

## SUMMARY

Breakeven analysis organizes current data into a format that assists the presentation and solution of a problem. A ***breakeven chart*** is a snapshot of the relationship of revenue and cost to output quantity during a short period of time.

***Breakeven comparisons*** detect the range over which each alternative solution to a problem is preferred. Typical problems include make-or-buy and lease-or-buy decisions.

Costs which remain relatively constant regardless of the level of a firm's activity are ***fixed costs*** $F$. Per-unit ***variable costs*** $V$ are proportional to ***output*** $n$. ***Profit*** $Z$ is the difference between ***total revenue*** ($R = nP$, where $P$ = price) and ***total cost*** ($C = nV + F$). The ***breakeven point*** $B$ occurs when $Z = 0$. Profit may be increased by raising or lowering $P$, cutting $C$, or combinations thereof.

In linear breakeven charts, $B = F/(P - V)$, where $P - V$ is the ***contribution*** per unit. The $Z/F$ ratio is the ***margin of profit*** or ***safety***, and $(\Delta Z/Z)/(\Delta Q/Q)$ is the degree of ***operating leverage***. ***Dumping*** is the practice of selling a product at a lower price in a separate market to increase plant utilization.

In an nonlinear breakeven analysis, the average total cost is the sum of ever-declining average fixed cost and the typically saucer-shaped average-variable-cost function. The average-total-cost curve is intersected at its lowest point by the rising marginal-cost curve. Similarly, the point of maximum profit occurs where marginal revenue equals marginal cost.

## Discussion of Economy Exercises

**EE 3-1** Total costs $C$ for the three alternatives are calculated as

$C(1) = \$1500$ *where* $N$ is the number of months needed
$C(2) = (2 \times \$12{,}000)0.6 + [(2 \times \$100) + \$400]N = \$14{,}400 + \$600N$
$C(3) = [(2 \times \$12{,}000) + \$8000]0.6 + (3 \times \$100)N = \$19{,}200 + \$300N$

As is apparent in Figure 3.14, $C(1) = C(2)$ at $N = 16$ months; $C(1) = C(3)$ at $N = 16$ months; and $C(2) = C(3)$ at $N = 16$ months. The decision thus narrows to a choice between alternatives 1 and 3, which, in turn, depends on the engineers' estimate of how long the project will take. If it takes longer than 16 months, 3 is preferred. Otherwise, 1 is less expensive. The convenience offered by the rented building (alternative 1 has upkeep and utilities paid, no trade-in hassles, etc.) would likely sway the decision.

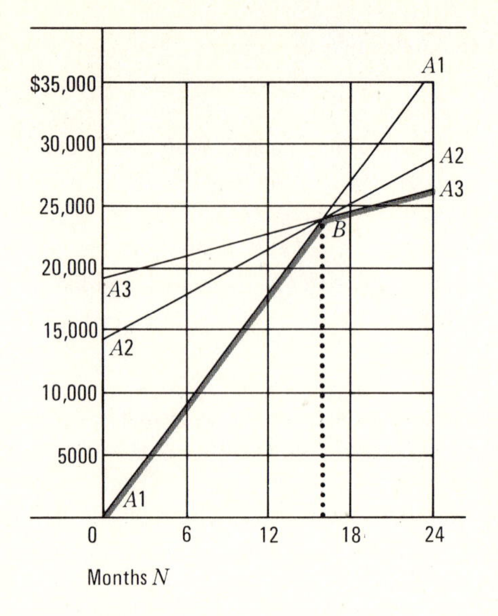


**FIGURE 3.14** Breakeven comparison for three alternatives. $A2$ is dominated by the other two, making the choice between $A1$ and $A3$.

**EE 3-2**

**a** The degree of operating leverage is

$$\text{OL}^\circ = \frac{\Delta Z/Z}{\Delta Q/Q} = \frac{\Delta n(P-V)/n(P-V)-F}{\Delta n/n} = \frac{n(P-V)}{n(P-V)-F}$$

**b** $$\text{OL}^\circ_{48 \text{ units}} = \frac{48(\$35{,}000 - \$20{,}000)}{48(\$35{,}000 - \$20{,}000) - \$600{,}000} = \frac{\$720{,}000}{\$720{,}000 - \$600{,}000} = 6$$

**c** Both operating leverage and elasticity vary according to the range of output being considered. Close to the breakeven point, OL° is greatest because a very small change in volume can produce a very large percentage increase in profits; at $B$, OL° is infinite. Therefore, the degree of operating leverage measures how sensitive operating levels are to changes in sales volume, just as elasticity measures the sensitivity of total product to changes in price.

**EE 3-3** The anticipated and original conditions are displayed in Figure 3.15 in a slightly different form of multiproduct breakeven chart. In this version the first entry is made at zero revenue and the point of maximum fixed cost. Then the contributions of each product are plotted progressively to the right. All other interpretations are the same for the two forms of multiproduct charts.

As is apparent from the chart, the new product line without mix $X$ would be less profitable than the former line:

| | *Mix W* | *Mix Y* | *Mix Z* | *Total* |
|---|---|---|---|---|
| Profit | $21,500 | $33,500 | $25,000 | $80,000 |

The main reason gross profit falls from $120,000 to $80,000 is that most of the fixed costs carried by mix $X$ did not disappear with its elimination from the product line; depreciation,

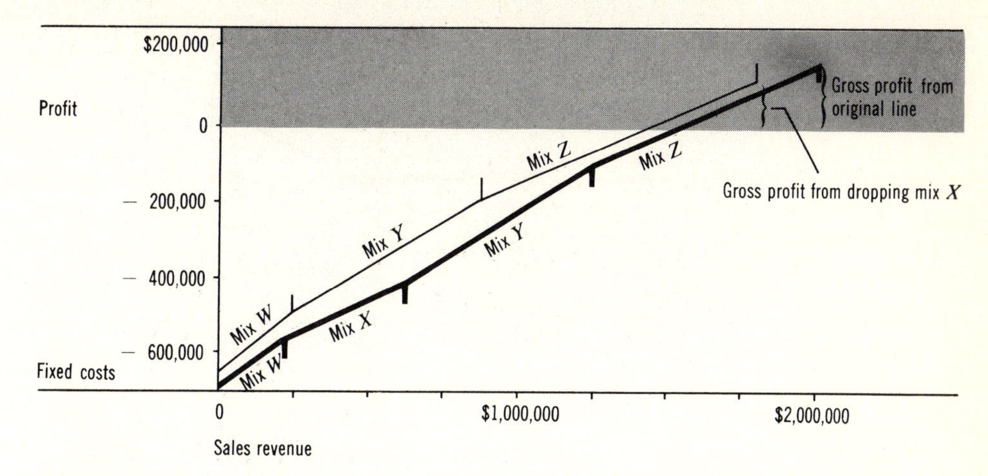


**FIGURE 3.15** Multiproduct contribution chart showing reduced profit from the revised mix.

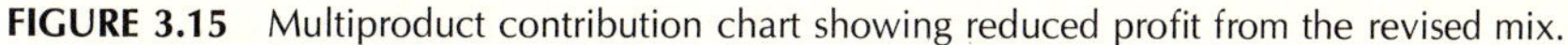

taxes, engineering services, and other indirect costs are still required to run the batching plant, and their magnitude is not reduced appreciably by dropping mix $X$. The situation is further aggravated by having the largest sales increase occur for mix $Z$, which has a lower contribution rate than the product eliminated.

**EE 3-4** Maximum profit occurs where marginal revenue equals marginal cost, as shown in Figure 3.16 and the computations below.

$$\text{Marginal revenue} = \frac{d(nP)}{dn} = \frac{d(100n - 0.001n^2)}{dn} = 100 - 0.002n$$

$$\text{Marginal cost} = \frac{d(nV + F)}{dn} = \frac{d(0.005n^2 + 4n + 200{,}000)}{dn} = 0.01n + 4$$

At maximum profit, $100 - 0.002n = 0.01n + 4$, so

$$n = \frac{96}{0.012} = 8000 \text{ units}$$

For an output of 8000 units,

$$\begin{aligned} Z = R - C &= 100n - 0.001n^2 - 0.005n^2 - 4n - 200{,}000 \\ &= -0.006n^2 + 96n - 200{,}000 \end{aligned}$$

$$Z_{8000} = -0.006(8000)^2 + 96(8000) - 200{,}000 = \$184{,}000$$

The output for maximum profit is also located by the point at which the marginal profit is zero:

$$\text{Marginal profit} = \frac{dZ}{dn} = \frac{d(-0.006n^2 + 96n - 200{,}000)}{dn} = 0$$

$$0 = -0.012n + 96$$

$$n = \frac{96}{0.012} = 8000 \text{ units}$$

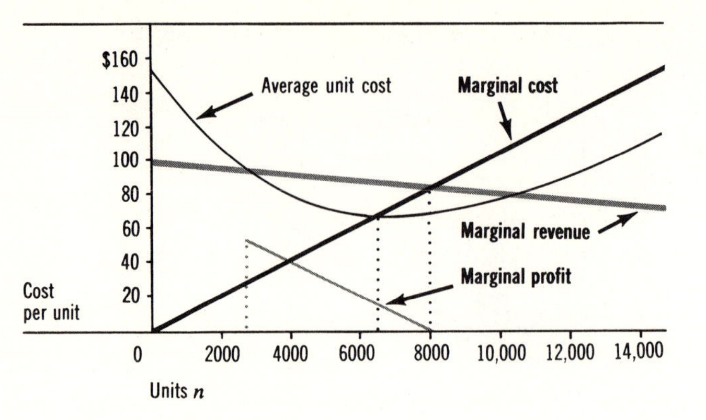


FIGURE 3.16 Average and marginal economic relationships for the data given in Economy Exercise 3-4.

The minimum average cost occurs where the output satisfies the relation

$$\frac{dV}{dn} = \frac{F}{n^2}$$

$$\frac{d(0.005n + 4)}{dn} = \frac{200{,}000}{n^2}$$

$$0.005n^2 = 200{,}000$$

$$n = 6325$$

At $n = 6325$, the average cost is \$67.20 per unit, as determined from the average-cost formula $(0.005n + 4 + 200{,}000/n)$, and as shown in Figure 3.16.

The breakeven output is calculated from the gross profit, where $Z = 0$. Thus,

$$Z = 0 = -0.006n^2 + 96n - 200{,}000$$

$$n = \frac{96 \pm \sqrt{(96)^2 - 4 \times 0.006 \times 200{,}000}}{2 \times 0.006}$$

$$B = n = \frac{96 \pm 66.4}{0.012} = 2467 \text{ and } 13{,}533 \text{ units}$$

## PROBLEMS

**3.1** "Do you mean you would charge me \$32.50 to ride home with you? You're out of your mind! The bus costs only \$23. By air it's just \$31.50. Look at those figures again," implored Gaston.

"Look, Gassie," said Alphonse, "I *know* what it costs me to drive my car. I keep records, complete ones. It's all right here. You look again."

Alphonse handed the paper to Gaston. On it were written the costs of operating a car for 3 years, the length of time Alphonse expected to keep his car before trading it in:

| | |
|---|---|
| Gasoline (40,000 miles at 2.4¢/mile) | $ 960 |
| Oil and grease | 150 |
| Tires | 140 |
| Repairs and maintenance | 400 |
| Insurance | 600 |
| License fees and property taxes | 90 |
| Depreciation in value ($3600–$1700) | 1900 |
| Average interest on investment at 6% interest ($1900/2 + $1700)(0.06)(3) | 477 |
| Total cost for 3 years | $4717 |

Cost per mile = $4717/40,000 miles = 11.8¢/mile

"Ok, Alphie, let's assume your figures are correct. But you haven't used them right. You say the 550-mile trip costs $65. It doesn't. Your car isn't going to lose more value because you drive it an extra 550 miles. Your license fee doesn't go up. Your insurance premium is the same. You still have to pay interest. The only thing this trip will cost you is gas and oil. That's only about 2.8 cents per mile . . . say, $15 for the trip, my share being about $7.50."

"So you say," retorted Alphonse. "I say I use up my tires. I'm going to have to make repairs sooner. Go hitchhike if you want to be cheap."

"I just might do that. I also thought of catching the plane so I could get home in time for a party Saturday night. However, I will forgo the party and give up a relaxing bus ride to keep you company, if you will be reasonable. Now let's look at that list again, Alphie, and remember that car of yours is going to get older whether I ride in it or not. My contribution is clear profit to you. You've got to go home anyway."

"So have you," replied Alphonse, settling more firmly into his bargaining position.

**3.1a** What is Gaston's fair share of the cost if he rides home with Alphonse? Why?

**3.1b** How do you reconcile the counterarguments that Gaston's share of the trip is pure profit to Alphonse and, conversely, whatever Gaston can save compared to the bus price is pure profit to him?

**3.2** Two alternative methods* of crude-oil processing in a producing oil field are available: a manual tank battery (MTB) and an automated tank battery (ATB). Tank batteries are composed of heaters, treaters, storage tanks, and other equipment, which remove salt water and sediment from crude oil prior to its entrance into pipelines for transport to an oil refinery.

Fixed costs for a tank-battery installation include depreciation, taxes, etc. Variable costs that increase in direct proportion to the volume of oil being processed include chemical additives and heating cost. In addition, there are constant costs independent of volume. All these costs are tabulated below.

| | *ATB* | *MTB* |
|---|---|---|
| Annual cost of labor and maintenance | $5485 | $7921 |
| Fixed annual cost of depreciation and taxes | 4064 | 2375 |
| Variable cost per barrel of oil processed | 0.0114 | 0.0081 |

*Adopted from E. J. Ferguson and J. E. Shamblin, "Break-Even Analysis," *Journal of Industrial Engineering*, April 1967.

At what level of oil production is the automatic tank battery preferred?

*(B = 226,000 barrels/year)*

**3.3** An import company buys foreign-made sewing machines for $40 per unit. Fixed costs of the operation are $70,000 per year. The sewing machines are sold on commission by door-to-door sales representatives who receive 40 percent of the selling price for each machine sold. At what price should the machines be sold to allow the importers to break even on a shipment of 5000 sewing machines?

*(P = $90)*

**3.4** A privately owned summer camp for youngsters has the following operating data for a 12-week season:

| | |
|---|---|
| Charge per camper | $60/week |
| Variable cost per camper | $40/week |
| Fixed costs | $24,000/season |
| Capacity | 150 campers |

**3.4a** What is the total number of campers that will allow the camp to just break even?

*(B = 100 campers)*

**3.4b** What is the profit for the 12-week season if the camp operates at 80 percent capacity?

*(Z = $4800)*

**3.4c** What profit would result if the camp stayed open 2 weeks longer and averaged 75 campers paying a reduced rate of $50 per week during the extended season? What might go wrong with the plan?

($Z_{extra}$ = $1500)

**3.5** A manufacturer of hand-operated power tools can produce a convenient attachment for use with his line of tools at an estimated annual fixed cost of $3000 and variable costs of $1.25 per attachment. He can also buy the complete attachment custom-made for his line of tools at a price of $2 per attachment for the first 5000, and $1.05 for all units purchased beyond the 5000-unit breakoff point. At what increments of unit sales should he purchase or manufacture the attachment? What factors other than cost might influence the decision?

**3.6** A lake resort includes 5 three-bedroom cabins which rent for $45 a night, 25 two-bedroom cabins which rent for $35 a night, and 15 single and 15 double rooms in the lodge which rent for $15 and $25 a night, respectively. The cost of cleaning a rental and preparing for new occupants is $10 for a cabin and $6 for a room. Annual fixed costs for operating the resort are $188,000 for the 200-day season. The average stay at the resort is 2 days, and utilization closely follows the proportion of available rooms and cabins.

**3.6a** What percentage of resort capacity must be rented each night to break even?

**3.6b** How many rooms and cabins must be rented each day to make an annual profit of $32,000?

**3.7** Sales of a desk lamp that wholesales at $4 per lamp have been disappointing. The contribution of each lamp sold is $1.50. It is planned to increase the advertising budget by $0.22 per lamp and reduce the price to spur sales. Twice as much will be allocated for advertising as for price reduction. Current gross profit is $50,000 on sales of 100,000 lamps per year. How many lamps must be sold under the proposed conditions to double the profit?

*(n = 171,000 units)*

**3.8** A product currently sells for \$12. The fixed costs are \$4 per unit, and 10,000 units are sold annually for a gross profit of \$30,000. A new design will increase variable costs by 20 percent and fixed costs by 10 percent, but sales should increase to 12,000 units per year. What should the selling price be to keep the same profit (\$30,000)?

*($P' = \$12.17$)*

**3.9** A boat marina now sells motorboats of only one make. The boats are divided into classes by size: *A*, *B*, and *C*. With an average value placed on the accessories sold with each boat, the accounting figures for annual turnover are

| *Type* | *Average P* | *Average V* | *Number Sold* |
|---|---|---|---|
| *A* | \$ 400 | \$ 300 | 300 |
| *B* | 900 | 500 | 175 |
| *C* | 2200 | 1200 | 100 |

Adding a new, fancier line of boats, enlarging the display area, and increasing the sales staff could change the cost and sales figures to

| *Type* | *Average P* | *Average V* | *Number Sold* |
|---|---|---|---|
| *A* | \$ 450 | \$ 350 | 425 |
| *B* | 1000 | 600 | 200 |
| *C* | 2400 | 1200 | 75 |
| *D* | 3600 | 2000 | 50 |

The additional expense of carrying the extra line of boats would double the present \$100,000 fixed-cost charge for the original line.

**3.9a** What is the composite contribution for each line?

**3.9b** Should the new line be added?

**3.10** A consulting engineer was asked by some of her clients to produce a book on "tilt-up" construction. Her costs for preparation—artwork, typesetting, plates, etc.—came to \$6000. For each 1000 books printed, the variable costs—paper, printing, binding, etc.—will be \$2000. She believes the number of books she will sell depends on the price she lists for the book. Her estimates show that the following number of books would sell at the prices shown:

| *Number of Books Sold* | *Price per Book* |
|---|---|
| 2000 | \$10.00 |
| 4000 | 7.00 |
| 10,000 | 3.50 |

**3.10a** What price will give her the greatest profit?

*($P = \$7$)*

**3.10b** An advertising agency claims that a \$10,000 promotion plan for the book would double the sales at any of the prices listed above. If the engineer accepts their forecast, what price should she use?

*($Z = \$24,000$ at $P = \$7$)*

**3.11** A plant produces products 1, 2, and 3 at annual rates of 10,000, 7000, and 5000 units, respectively. Product 1 accounts for 30 percent of the plant's revenue, with a 40 percent contribution on its selling price. Product 2 has fixed costs of $35,000 and a contribution rate of $0.5P$, where $P$ is the purchase price. The fixed costs for product 3 are $20,000, while it accounts for 30 percent of the total sales revenue, with the same contribution rate as the composite contribution rate. Total fixed costs for the plant are $80,000, and total sales amount to $250,000.

**3.11a** What is the plant breakeven revenue?

**3.11b** What is the profit for each product?

**3.12** Another alternative for the fertilizer batching plant described in Economy Exercise 3-3 is to replace mix $X$ with a new mix $XX$. By replacing mix $X$ with mix $XX$, the following operating figures are expected:

| | *Mix W* | *Mix XX* | *Mix Y* | *Mix Z* | *Totals* |
|---|---|---|---|---|---|
| Percentage of total sales | 10 | 15 | 30 | 45 | 100 (on $2 million) |
| Contribution (% of $P$) | 45 | 50 | 45 | 35 | |
| Fixed cost charged | $70,000 | $140,000 | $210,000 | $275,000 | $695,000 |

**3.12a** Calculate the expected profit from the total product line which includes mixes $W$, $XX$, $Y$, and $Z$.

*(Profit = $130,000)*

**3.12b** A *composite contribution rate,* as shown by the sloping dotted line in Figure 3.9, is the sum of each product's contribution, weighted according to its percentage of total sales. That is,

$$\text{Weighted contribution} = \text{product contribution} \times \frac{\text{product sales}}{\text{total sales}}$$

Then, the composite contribution rate is the sum of the weighted contributions, and the total contribution is the composite contribution rate times total sales, which makes the gross profit equal the total contribution less the fixed cost.

Calculate the composite contribution rate, and use it to reaffirm the gross product computed in Problem 3.12*a*.

**3.13** What are some of the dangers faced by a firm that engages in dumping to increase its plant utilization?

**3.14** A small company manufactures rubber matting for the interiors of custom carts. During the past year a revenue of $202,000 from sales was earned with the current costs given below.

| *Current Operating Costs* | |
|---|---|
| Direct material | $51,000 |
| Direct labor | 42,000 |
| Maintenance | 11,000 |
| Property taxes and depreciation | 17,000 |
| Managerial and sales expenses | 35,000 |

The forecast for the next year is a drastic drop in custom-cart sales, which is expected to limit mat sales to $90,000. There is insufficient time to develop new markets before next year. With a skeleton force for the reduced production, anticipated operating costs are shown below:

| *Expected Operating Costs* | |
|---|---|
| Direct material | $28,000 |
| Direct labor | 23,000 |
| Maintenance | 7000 |
| Property taxes and depreciation | 17,000 |
| Managerial and sales expense | 35,000 |

The company can operate at an apparent loss, or mats can be purchased from a large supplier and resold to the custom-cart builders at a price that will just meet purchase and handling costs. Either alternative will retain the market for the company until the following year, when sales are expected to be at least equal to last year. Which of the two alternatives is the better course of action?

**3.15** Assume the variable cost in Table 3.1 is composed completely of labor and represents one week's operation of the firm. Suppose the firm is currently producing 4 units per week with 12 employees (the average wage is $200 per week). The manager recognizes that the firm should operate at a level of 7 or 8 units of output per week to minimize total cost. At a weekly production level of 8 units, variable cost allows an employment level of $6200/$200 per employee = 31 employees at a continued average wage of $200 per week. It thus takes 31 − 12 = 19 employees to double the output produced by the original work force producing 4 units per week. Or, put differently, the original 12 employees accounted for 4/12 = 0.33 units of output apiece, whereas the next 19 employees would account for (8 − 4)/(31 − 12) = 0.21 output units each. Do the original 12 deserve higher pay than the next 19 employees? Discuss wage policy as a function of marginal cost.

**3.16** A company manufactures industrial clips for assembly work. The marginal revenue has been determined to be

$$\text{Marginal revenue} = 100 - 0.02n$$

where $n$ is the number of clips produced. Variable costs plus fixed costs are calculated by the formula

$$\text{Total cost} = 2n^2 \times 10^{-4} + 10{,}000$$

Compute the production in clips per year for the following:

**3.16a** Minimum unit cost of sales

*($n$ = 7071 clips)*

**3.16b** Production for maximum profit

*($n$ = 4902 clips)*

**3.16c** Breakeven volume

*($n$ = 9700 and 100 clips)*

**3.17** Assume that a company can sell all the units it produces, but costs are subject to diminishing returns. Its revenue and cost functions (in thousands of dollars) are

$$R = \frac{3n}{4} \quad \text{and} \quad C = \frac{n^3 - 8n^2 + 25n + 30}{25}$$

**3.17a** Construct a graph of the cost and revenue curves with the breakeven points indicated.

**3.17b** At what output $n$ will profit $Z$ be maximum?

## EXTENSION

***3.1 The Short Run with Pure Competition*** Under conditions of pure competition, no single supplier or purchaser controls enough volume to affect the market price. Since any producing firm can sell all its output at the market price, the elastic demand function is pictured as a horizontal line relating price to output. Marginal revenue coincides with the demand line because a constant selling price means that each additional unit of output yields an additional unit of revenue equal to that market price.

Figure 3.17 shows a healthy producer whose average-total-cost line extends below the marginal-revenue (demand-price) line. To maximize profit, this producer should adjust output to the level at which the marginal cost of producing one more unit equals the marginal revenue. At this point, the profit is represented by the area of the shaded rectangle: The base of the rectangle is the quantity $Q$, and the height is the vertical distance between the horizontal unit-price line and the value of the average cost per unit at $Q$.

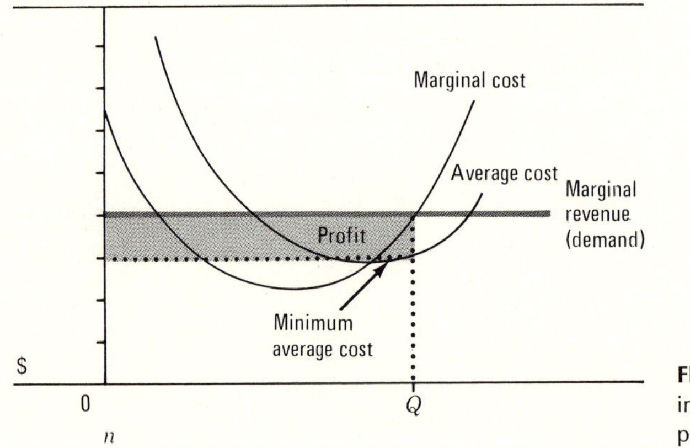


**FIGURE 3.17** Short-run, maximum profit condition for a purely competitive firm.

Note that the profit per unit would be greater in Figure 3.17 if the producer operated at the output at which average cost was minimized, but total profit (the area of the rectangle) is greatest as shown.

Every prudent producer will vary output as the market price varies. If the price drops below the *shutdown point,* the minimum average variable cost at which a producer must operate, production should cease. However, a firm should continue production when the price is between its shutdown point and the point of minimum average cost, even though it means operating at a loss, because this loss is smaller than the fixed-cost loss that would be incurred by shutting down. In this range, loss is minimized by an output level that makes the marginal cost equal the selling price.

A composite marginal-cost curve for all the producers (whether making a profit or not) that supply a certain product at a given time would yield the supply schedule for that product at that time. This is the source and rationale behind the supply schedule shown in Figure 2.4.

Levels of output to optimize profit under short-run, purely competitive conditions are illustrated by the questions below, which utilize the data in Table 3.1.

**QUESTIONS**

**3.1a** For $P = \$1500$, what output produces the maximum profit? Why is this quantity not equal to the output at which average total cost is minimum?

**3.1b** For P = \$900, what output minimizes the loss? How much more would be lost per period if production were halted?

**3.1c** How low can the price drop before the firm reaches its shutdown point? What is the loss per period at this level?

# CHAPTER 4

# MINIMUM-COST OPERATIONS

Engineers are continually confronted with design problems of a type suggested by Figure 4.1: If you vary a certain parameter, the change causes one characteristic to improve and simultaneously causes another characteristic to get worse. It is the engineering equivalent of the adage that there are two sides to every question. This chapter deals with the question of what design level produces the minimum total cost when some of the operating costs vary directly and others inversely to changes in the design variable.

Five factors are considered in a minimum-cost analysis. Their relationship is evident in Figure 4.1 and is expressed by the formula

$$C = Ax + \frac{B}{x} + K$$

A multitude of real-world conditions are represented by this minimum-cost model. The design variable can have dimensions such as money, length, weight, number of parts, and time, or combined units such as speed. Design problems naturally get more complex when several design variables are involved. Only the easiest case of one variable affected by two opposing cost functions will be considered here. However, the relatively simple case covers many problems commonly met in engineering practice, and it sets the framework for

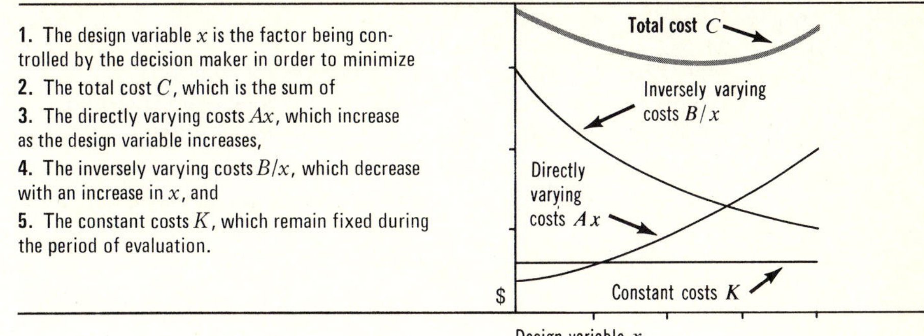


**FIGURE 4.1** Minimum-cost factors.

handling more complicated situations. Operational questions answered by the minimum-cost calculations for a single design variable include:

At what age should an asset be replaced, when operating costs increase with age while capital costs decline?

What project duration minimizes total cost, when activity costs for shorter durations are higher and overhead or penalty costs are lower?

How many items should be ordered at one time, when ordering costs decrease with larger orders while the cost of holding items in storage increases with order size?

## MINIMUM-COST SOLUTION METHODS

There are two ways to calculate the minimum-cost point. Both can be supplemented by graphs, which are less precise for a solution but which excel for explanation and discussion purposes.

### Tabular Approach

A tabular method is appropriate for a design variable that changes in incremental steps. Cumulative direct, inverse, and total costs are tabulated for consecutive changes in the design variable. The tabulation ends when a minimum point is evident in the total-cost column. When the cost patterns consistently rise or fall as a function of the design variable like the curves in Figure 4.1, the minimum-cost level is identified as the point preceding the first upturn in total cost. If the directly or inversely varying costs behave erratically along their general trend line, it is necessary to check several points beyond the first minimum.

**Economy Exercise 4-1** A temporary supply of running water is needed at a construction site situated high above a river. The water is free, but it must be pumped to the site. Larger hose sizes decrease the pumping cost but are initially more expensive. An adequate supply of water can be provided by a 3-, 4-, or 5-inch-diameter (7.62-, 10.16-, or 12.70-centimeter-diameter) hose. Net costs for these sizes (first cost minus resale value) are $460, $590, and $770. Pumping costs during the 6-month project are estimated at $440 for a 3-inch-diameter hose, 20 percent less for a 4-inch size, and half as much for the 5-inch size. What size hose should be purchased?

## Formula Approach

Equations can be used to find the minimum-cost point when the direct cost and inverse cost are proportional to the design variable. The total-cost expression, $C = Ax + B/x + K$, is subjected to the same mathematical manipulations used in nonlinear breakeven analysis. Since the minimum point on a continuous curve occurs where the slope of the curve is zero, the total-cost formula is differentiated to find the slope, and the derivative is set equal to zero:

$$\frac{dC}{dx} = \frac{d(Ax + Bx^{-1} + K)}{dx} = 0 \qquad \textit{where } A, B, \text{ and } K \text{ are constants}$$

$$= A - \frac{B}{x^2} = 0$$

which results in

$$x = \sqrt{\frac{B}{A}}$$

at the point where total cost is lowest. At this point the directly varying costs are equal to the inversely varying costs, as is apparent when the minimum-cost value of $X$ is substituted into the total-cost equation:

$$C_{\min} = AX' + \frac{B}{X'} + K \qquad \textit{where } X' = \text{minimum-cost point} = \sqrt{\frac{B}{A}}$$

$$= A\sqrt{\frac{B}{A}} + \frac{B}{\sqrt{B/A}} + K$$

$$= \sqrt{AB} + \sqrt{AB} + K$$

# CYCLIC REPLACEMENT

Many mechanical items used in service agencies and manufacturing are replaced by essentially the same machine when the original wears out. Informal rules may be used to establish replacement times. One government department replaces a car whenever it exceeds any of the following conditions: total mileage of 65,000 miles (104,605 kilometers), 5 years of service, or the cumulative maintenance cost equals the purchase price. Such rules recognize that automobiles, trucks, typewriters, and similar machines become less efficient and accumulate higher and higher repair bills as they age. Conversely, the longer they are kept in operation, the lower will be their average annual capital cost because the purchase price is spread over more years. The sum of these two types of cost is the total cost of providing the machines' services. As shown in Figure 4.2, total lifetime cost continues to increase with age, but average annual cost passes through a minimum.

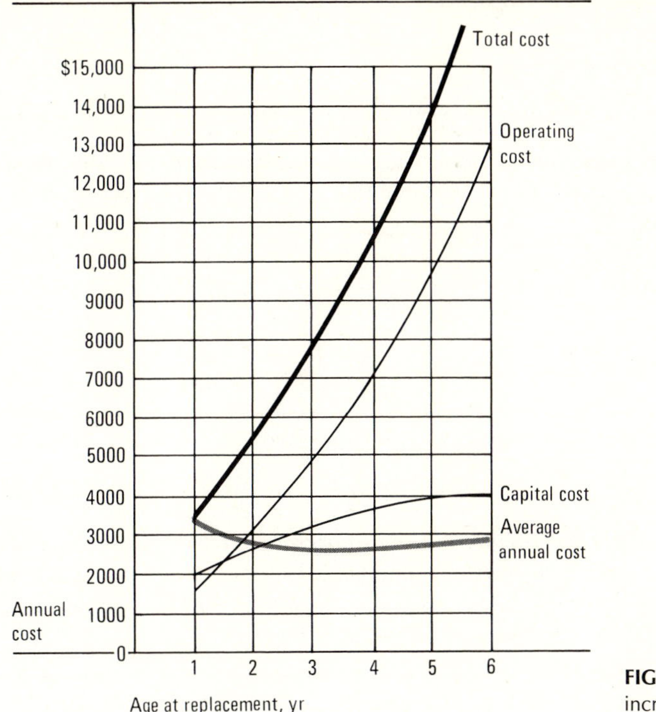


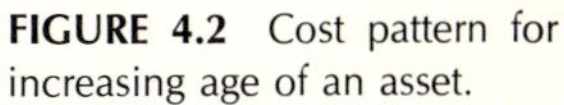
**FIGURE 4.2** Cost pattern for increasing age of an asset.

The objective of a replacement study is to determine the pattern of replacement that will minimize the average annual cost. Data for future costs must be estimated, but cost figures for most assets that are used enough to warrant a cyclic replacement study are available from internal or suppliers' records. Simplifying assumptions made to illustrate the calculation procedure include a replacement price that remains constant and no interest charges for the use of capital. However, relaxing these assumptions does not affect the optimum cycle time very much unless there are dramatic changes in price or exceptionally high interest rates.

Calculations to determine the minimum-cost age follow an iterative routine in which the capital cost (decrease in salvage value to a certain age) is added to the accumulated operating cost (maintenance plus use expenses) for that same period. Each sum is divided by the relevant age to give an average cost for that replacement interval. This routine is illustrated in Table 4.1 for a city delivery service that owns a fleet of panel trucks for store-to-home deliveries. The purchase price per truck is $5000, and the anticipated schedule of future operating costs and salvage values is

| | YEAR | | | | | |
|---|---|---|---|---|---|---|
| | *1* | *2* | *3* | *4* | *5* | *6* |
| Operating cost | $1500 | $1600 | $1900 | $2300 | $2800 | $3400 |
| Resale value | 3000 | 2300 | 1700 | 1300 | 1100 | 900 |

These costs are inputs to the tabular solution method which reveals that trading the trucks in every 4 years for new ones is the minimum-cost replacement cycle. The cost data from Table 4.1 were graphed in Figure 4.2.

**TABLE 4.1** Average-annual-cost calculations for replacing an asset when the purchase price is constant and no interest is charged on capital.

| *Age of Truck at Replacement, Years (1)* | *Cumulative Operating Costs (2)* | *Cumulative Capital Cost* (3)* | *Total Cost [(2) + (3)] (4)* | *Average Annual Cost [(4) ÷ (1)] (5)* |
|---|---|---|---|---|
| 1 | $ 1500 | $2000 | $ 3500 | $3500 |
| 2 | 3100 | 2700 | 5800 | 2900 |
| 3 | 5000 | 3300 | 8300 | 2767 |
| 4 | 7300 | 3700 | 11,000 | 2750 |
| 5 | 10,100 | 3900 | 14,000 | 2800 |
| 6 | 13,500 | 4100 | 17,600 | 2933 |

*Cumulative capital cost equals $5000 less resale value.

Since the 3- and 4-year replacement cycles cost almost the same, the delivery firm might believe the prestige of having new trucks more often is worth $2767 − $2750 = $17 per year. Then the analysis would at least imply the cost of prestige. Additional replacement models involving group versus individual replacements and the most economical use of standby machines are considered in Chapter 14.

**Economy Exercise 4-2** The price $P$ of a piece of equipment is $1000, and the equipment has no salvage value when replaced. Operating cost $O$ is $200 the first year and increases by $O'$ = $100 annually. The average annual cost $AC$ for any replacement cycle $n$ is then

$$AC = \frac{P}{n} + O + (n - 1)\frac{O'}{2}$$

Using this equation, calculate the minimum-cost replacement cycle.

## TIME-COST TRADE-OFFS

In *The Way to Wealth* (July 7, 1757), Benjamin Franklin wrote, "Dost thou love life, then do not squander time, for that's the stuff life is made of." He would likely have relished the thought of buying time. In an informal way, everyone buys time occasionally: hiring someone to do a task that frees you to do something else, or purchasing a machine that cuts the time required to do something. On a larger scale, capital intensive industries make maximum use of high-investment assets by working them 24 hours per day, or a rush order is completed early by having crews work overtime. In all these cases, an extra premium is paid to conserve time. Deciding whether to pay the premium is a time-cost trade-off problem.

Most opportunities to buy time are settled by simply comparing the incremental cost of gaining a time unit against the incremental profit earned from acquiring that time unit. Equivalently, a decision to sell personal time rests on a value judgment for alternative uses of time. When there are time-cost trade-off options, the one providing the most net gain is selected. But identifying that most profitable option gets complicated when activities have to be completed in a certain sequence and the resources involved have different prices for time.

## Network Diagram

A method widely used in construction and maintenance projects is to portray the sequence of required activities in an *arrow network*. This technique is known as *critical path scheduling* (CPS). [CPS is a representative name encompassing the critical path method (CPM), program evaluation and review technique (PERT), and other network-analysis techniques.] It is employed extensively by engineers and managers to plan and control major projects. After its introduction in the late 1950s, it caught on almost immediately as a convenient and effective way to integrate activities and schedule resources. Attention here is focused on its use in determining the most economical schedule for projects which have costs that vary both directly and inversely with time. (See the Extension 4.1 for an explanation of CPS calculations.)

A small maintenance-construction project is described by Figure 4.3. Five activities, represented by labeled arrows, are needed to finish the project. The normal duration for each activity is shown below its arrow. The arrows also designate the order in which the activities must be completed. For instance, the placement of arrow 1,2 (node numbers at the beginning and end of the arrow) shows that old machines must be removed before maintenance (activity 2,4) can begin or site preparation (activity 2,3) can commence. Further, activities 2,3 and 1,3 must be completed before the new machines can be installed (activity 3,4).

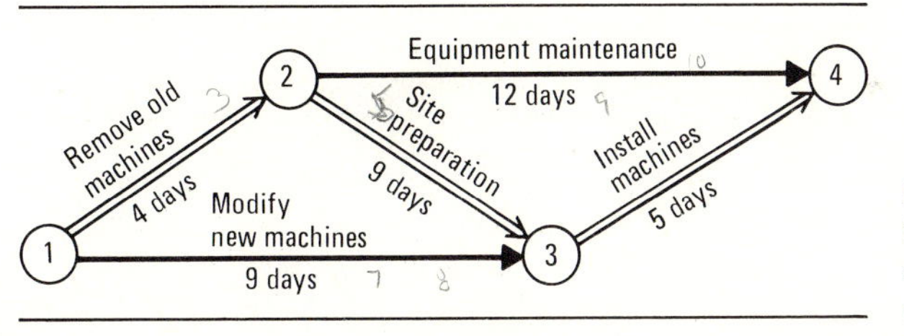


**FIGURE 4.3** Arrow network for the installation of new machines and concurrent maintenance of associated equipment.

The double-lined arrows indicate the *critical path*: the longest chain of activities through the network. The sum of the durations of activities on the critical path gives the total project duration, 4 + 9 + 5 = 18 days.

## Project Costs

The performance cost of an activity is the sum of direct-labor, material, and equipment charges required to complete it. When a separate agency such as a consulting firm or subcontractor undertakes the entire activity, the cost is the amount paid to that agency. Thus, activity time-cost relationships take a variety of forms. Several of the more common configurations are shown in Figure 4.4.

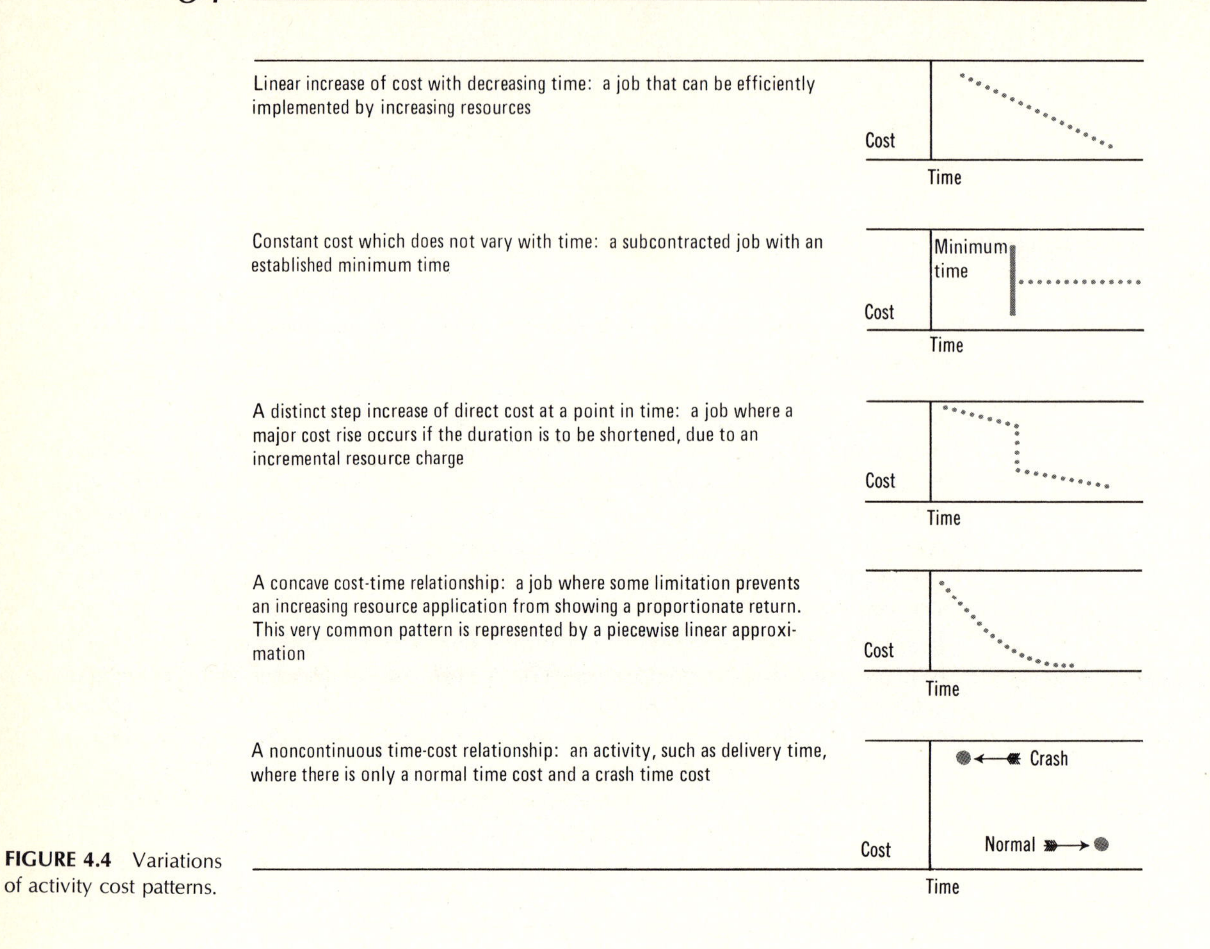


**FIGURE 4.4** Variations of activity cost patterns.

Costs which increase with longer project durations are composed of overhead expenses associated with the project and outage or penalty costs caused by disruption to ongoing operations or by not having a facility available when it is needed.

For the project depicted in Figure 4.3, inversely varying activity costs are assumed to be linearly related to time as indicated in Table 4.2. The cutting cost in the last column of the table is calculated for each activity with the formula

$$\frac{\text{Crash cost} - \text{normal cost}}{\text{Normal duration} - \text{crash duration}} = \text{cutting cost/day}$$

From Table 4.2, the cutting cost for activity 2,4 is

$$\frac{\$1530 - \$1200}{12 - 9} = \frac{\$330}{3 \text{ days}} = \$110/\text{day}$$

Directly varying costs are $220 per day and represent the loss in output from not having the machines available for production during the project duration.

TABLE 4.2 Per-day costs of reducing normal activity durations to their minimum (crash) times.

| | NORMAL | | CRASH | | |
|---|---|---|---|---|---|
| *Activity* | *Day* | *Cost* | *Day* | *Cost* | *Cutting Cost, $/day* |
| 1,2 | 4 | $1200 | 3 | $1400 | 200 |
| 1,3 | 9 | 1800 | 7 | 2100 | 150 |
| 2,3 | 9 | 1440 | 5 | 2080 | 160 |
| 2,4 | 12 | 1200 | 9 | 1530 | 110 |
| 3,4 | 5 | 1650 | 5 | 1650 | |
| *Totals* | | $7290 | | $8760 | |

## Trade-off Analysis

At first glance it might appear that the duration of all activities in the project should be cut to the minimum because all the per-day cutting costs are less than the $220 daily cost of lost production. This impression is incorrect because sequencing requirements dictate that certain activities be done concurrently. The result is that the cost to cut a day from the project when concurrent activities are being performed is the sum of cutting costs for all activities going on at that same time.

The limiting time restrictions are shown on a bar graph in Figure 4.5. The solid portion of each bar is the activity's minimum duration. Costs to cut an activity from its normal to its crash duration are shown in segments to the right of the solid portion. The empty segments on the extreme right of some bars indicate "float" or leeway in scheduling. For instance, activity 1,3 can begin when the project starts and will not interfere with the start of any other activity so long as it is completed in 13 days (activity 3,4 must start at the end of the 13th day to avoid delaying the project's 18-day completion time). Since activity 1,3 normally takes only 9 days to complete, it has 13 − 9 = 4 days of float. The three critical activities with no float are shown at the top of the bar chart and are connected to highlight the critical path.

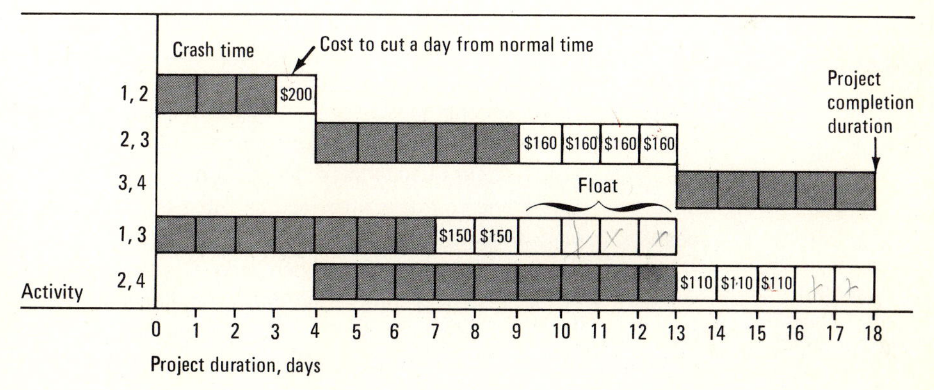


FIGURE 4.5 Bar chart showing the cost and schedule of possible time restrictions for the project diagrammed in Figure 4.4.

From the bar chart, it is possible to develop various project durations and total costs. Since a critical activity must be shortened on each cut, activity 2,3 is the one to cut first.

This cut increases the project cost by $160 but shortens the total duration to 17 days and reduces the float available for activities 1,3 and 2,4 by 1 day. Cutting continues until the cost of the next cut exceeds the savings from directly varying costs of $220 per day:

*16-day schedule* Activity 2,3 is cut again at a cost of $160 to reduce its duration to 7 days. This cut allows activity 3,4 to start at the end of the eleventh day and eliminates the float available for activity 2,4, making it critical.

*15-day schedule* The least expensive cut now available is in activity 1,2 at a cost of $200. This cut allows activities 2,3 and 2,4 to start after the third day, and activity 3,4 to begin after 10 days. The only remaining noncritical activity is 1,3 with 1 day of float.

*14-day schedule* The next cut requires that activities 2,3 and 2,4 be shortened at a cost of $160 + $110 = $270. Since this cut costs more than the savings it produces, 15 days is the lowest-cost project duration. Times and costs for the 15-day duration are shown in Table 4.3.

| *Activity* | *Start* | *End* | *Cost* |
|---|---|---|---|
| 1,2 | Day 0 | Day 3 | $1400 |
| 1,3 | Day 0 or 1 | Day 9 or 10 | 1800 |
| 2,3 | Day 4 | Day 10 | 1760 |
| 2,4 | Day 4 | Day 15 | 1200 |
| 3,4 | Day 11 | Day 15 | 1650 |
| | | Total inversely varying cost = | $7810 |
| Total directly varying cost = 15 days × $220/day = | | | 3300 |
| | | Total project cost = | $11,110 |

**TABLE 4.3** Minimum-cost schedule for the project conditions graphed in Figure 4.5.

In summary, 3 days were cut from the original 18-day project duration given in Figure 4.3, at a cost increase of $160 + $160 + $200 = $520. The shortened schedule reduces directly varying costs by $220/day × 3 days = $660, effecting a saving of $660 − $520 = $140.

## Application of Network Cost Analysis

The preceding example demonstrated that time-cost trade-offs are clumsy to conduct by hand, even for a tiny project. Computer programs are available to relieve manual calculations. Inputs to a program include the sequential ordering of activities obtainable from an arrow network and time-cost data as typified by Table 4.2. Most programs will handle only linear cost relationships as described by the top cost-time pattern in Figure 4.4.

Minimum-cost evaluations have not been used as much as was anticipated when the procedure was introduced in the early 1960s. A major reason is the lack of historical accounting data from which to obtain time-cost relationships for individual activities. Accounting records are traditionally maintained for line items, not collections of cost categories associated with different activities needed to perform a maintenance or construction project.

"Crashing" is about the only salvation for a project that has fallen seriously behind schedule and faces stiff penalty costs for late completion. Then the project engineers are exhorted to find ways to capture time. Each remaining activity is examined to find out if it can be shortened, how, and at what cost. In this situation, manual time-cost trade-off calculations are feasible because only a relatively small portion of the project's activities need be investigated, and their cutting costs can be estimated accurately with respect to current conditions. The advantage of the systematic CPS cost-analysis method is that it limits time-reduction efforts to the activities where they are most profitable and thereby avoids the panic-button approach of trying to buy time wholesale.

**Economy Exercise 4-3** Continue to reduce the duration of the project charted in Figure 4.5. Calculate the cost of the minimum duration, and show the resulting schedule on a bar chart.

## VIGILANCE VERSUS ERRORS

Most individuals and organizations take pride in the quality of their output. Higher quality tends to be automatically equated with superior performance, which is assumed to produce more profit. This relationship usually holds true, but there are occasions when the cost of higher quality is not repaid from the receipts earned by the achievement.

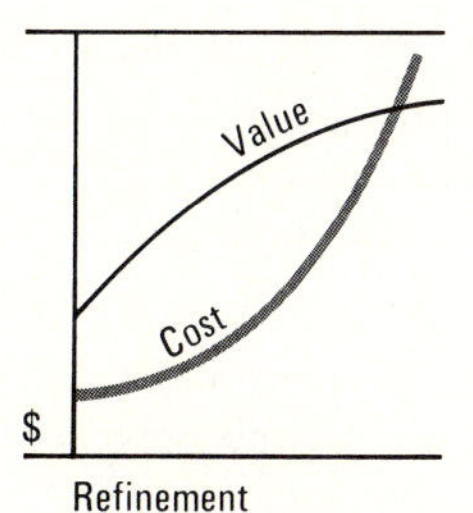


**FIGURE 4.6** Product value and cost as a function of refinement.

Both service and manufacturing quality start with a process capable of producing to the design specifications and continue with a quality-assurance program that ascertains if standards are being met. The initial decision concerning specifications is based on what customers supposedly want. It is moderated by the realization of what the supplier can provide. These two considerations can take the shapes shown in Figure 4.6, in which the customer's perceived value of a service or product levels off at higher degrees of refinement while the cost of providing refinements continues to increase.

A supplier's calibrated-to-profit perspective focuses on the difference between perceived value and the cost-refinement capabilities of the process. Suppliers with high refinement capabilities might choose to advertise their highly refined offerings with the intent of educating customers to appreciate them. Wine, automobile, and clothing advertisements abound with examples of persuasions to buy high quality at a high price.

Another view of value-cost relations centers on cost. For a given degree of refinement there are many contributing factors affecting cost. One of the most critical is the expense of quality assurance. A minimum-cost quality-assurance program balances the cost of maintaining a desired level of quality (vigilance) against the cost of failure (errors). In most production situations the cost of vigilance varies directly with the excellence of output, and the cost of errors varies inversely as shown in Figure 4.7.

### Cost of Inferior Quality

Failure to meet quality standards creates internal costs for the production process and external costs associated with the output. Internal costs include labor, material, and overhead charges on scrapped products; the expense of activities to rework products to make them acceptable; disruptions in the production process caused by quality deficiencies; and

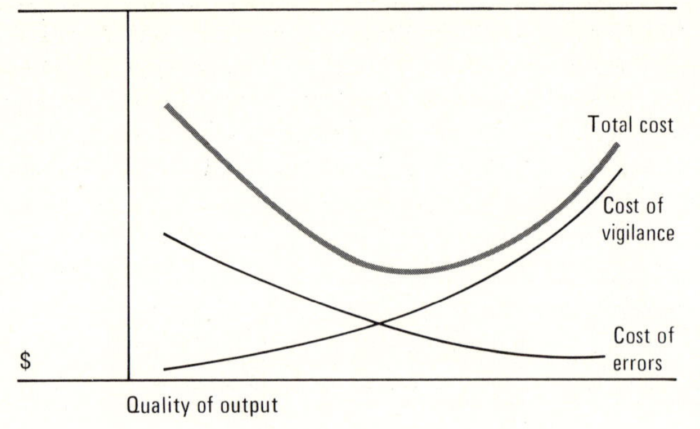


**FIGURE 4.7** Relationship of vigilance to errors in quality assurance.

the engineering time to correct deficiencies. External costs are due mainly to loss of consumer confidence and the complaint or field services designed to placate customers.

## Cost of Prevention

Errors can be prevented throughout the production process. The first opportunity is in the design of the product. Designers must beware of tolerances and standards that are unnecessarily restrictive; tight tolerances introduce more chances for error, and unrealistic standards can raise enforcement costs prohibitively. Clever engineering designs may even reduce the cost of production errors by making their correction and prevention easier.

Deviations from design specifications are detected by inspection. Materials purchased for the production process are inspected before they are accepted; in-process inspections confirm quality during the production process; and output is inspected before it is released to customers. Higher inspection costs normally provide a greater degree of protection from the effects of defective output, but the marginal cost of detection rises sharply as the percentage of undetected errors approaches zero. A point is usually reached where further improvement is not worthwhile.* For instance, if 98 percent viability of grass seeds is considered acceptable, decreasing the proportion of nonviable seeds below 2 percent by inspection would be uneconomical unless the higher quality is rewarded with a higher price.

Quality-control and design engineers are thus exposed to economic questions of vigilance versus errors. Would it be better to set a standard that can be held within stated tolerances 90 percent of the time at a lower cost, or accept a higher cost to have 95 percent meet the standard? Should a higher-cost process be used to consistently produce to a certain degree of excellence, or should a lower-cost process be coupled with intensive inspection to produce the same degree of quality? Answers are straightforward when complete information is available, but costs often involve human value systems and performances that are difficult to verify and subject to abrupt changes. This subject is discussed further in Chapters 5 and 14.

*Exceptions include projects in which extreme reliability is required, e.g., space vehicles and nuclear power plants.

**Economy Exercise 4-4** A manufacturing plant makes 600 Wambos each day to get 400 good ones. Labor costs are $3 per good unit of output, and daily overhead is $1200. A Wambo goes through two stages of processing which require an equal amount of labor. Inspection currently takes place after the second stage. The 200 defective units are melted and returned to the first stage as a raw-material input at approximately the same cost as other raw materials.

A new inspection plan is proposed to add an intermediate inspection point at the end of the first stage, to detect defectives before they are processed through the second stage. The plan will add $150 per day to overhead costs and is expected to reject one-sixth of the flow (another one-sixth will be rejected at the final inspection). If plant capacity is limited to the present level of material input, should the plan be implemented?

## INVENTORY ANALYSIS

Every organization needs supplies in order to operate. But stored supplies (accumulated inventory) are an idle resource. That the resource is idle does not mean it is serving no purpose. It is available if needed. It serves as an insurance policy to protect the organization from breakdowns, delays, unexpected surges in demand, and other disturbances that could upset ongoing activities. Insurance is not free. The idle resource can be damaged or become obsolete before it serves any purpose. The objective is again to secure an economic balance between the cost of loss and the cost of preventing it.

Engineering activities are affected by inventory policies in a number of ways. Designers should be aware of what is already stocked and what the delivery time is for items that have to be ordered. Engineers in production are concerned with supply of materials to the production process and the storage of finished products. Sales engineers need to know which of their products are immediately available to buyers and which have delivery delays. Engineers themselves are occasionally an "idle resource" when organizations hire more engineers than are immediately needed for operations in order to be prepared for the award of a large contract.

Inventory analysis is a classical minimum-cost evaluation. One of the first formulas was developed by an engineer, F. W. Harris, in 1915. Since that time, especially in the last two decades, an immense volume of literature has been written about inventory problems and solution methods. The content ranges from crude one-line rules to elaborate computer models. So much emphasis is owed to the universality of inventory usage; supplies are consumed in all kinds of organizations from family units to government agencies, and from one-person companies to multinational conglomerates.

Basic inventory cost relationships are explored in this section. The discussion is limited to short-run, current conditions, and all data are assumed to be known accurately. These restrictions are eased in Chapter 14, where analyses recognize the uncertainties of future events. Under conditions of either assumed certainty or risk, two important questions are:

How often should supplies be replenished?

What amount should be replenished each time?

The two questions are related, and models for their solution seek to minimize the sum of directly and inversely varying costs.

## INVENTORY FACTORS

An organization orders, receives, stores, and uses a great variety of supplies, but a relatively few classes of supplies usually account for most of the inventory costs.* Therefore, analytical evaluations are applied only to those key classes in which close control will promote the most savings. After an inventory item has been selected for analysis, the following information is needed:

$D$ = *annual demand or usage for the item.*

$M$ = *annual manufacturing rate for the organization producing the item.*

$O$ = *order cost or set-up cost to procure the item.* Order costs to obtain supplies from a vendor include the fixed cost of maintaining an order department and the variable costs of preparing and executing purchase requisitions. Set-up cost accounts for the expenses incurred in physically preparing for a production run to produce the item and includes the overhead costs for shop orders, scheduling, and expediting. Order and set-up costs remain relatively constant regardless of the order size.

$H$ = *holding cost for keeping each item in storage 1 year.* Costs originating from many sources are consolidated under the heading of holding cost. In general, holding cost exhibits a fixed charge for a very low inventory level and then varies directly with additional quantities stored. Costs collected under $H$ include operating and ownership expenses for a storage facility; handling charges composed of wages and equipment costs; protection and insurance for stored items; taxes on the items stored and the storage facilities; and damage, pilferage, or obsolescence of inventory during the storage period.

$P$ = *price of the item.* The value of an item is its unit purchase price if it is obtained from an outside supplier, or its unit production cost if it is produced internally. The amount invested in an item being manufactured is a function of its degree of completion as it passes through the production process. The price per piece for outside purchases can vary as a function of quantity discounts for larger orders.

$i$ = *interest rate for money invested in inventory.* The amount of money invested in an item is an amount of capital not available for other purposes. If the money were invested in a bank, a return on the investment would be expected. Therefore, a charge to inventory expense is made to account for this unreceived return from money represented by stored items. The interest rate charged, $i$, is the percentage return expected from other investments and is applied as an annual capital cost based on the price of the item, $iP$.

### Design Variable

The number of items procured for each inventory replenishment is the design variable denoted as $Q$. If all the items needed for an entire year are procured at one time, then

*This condition demonstrates the Pareto principle: A few activities in a group of activities always account for the major share of the resources used or gained.

$Q = D$. When inventory is replenished more than once a year,

$$Q = \frac{D}{N}$$

where $N$ is the number of procurement periods per year, or $N = D/Q$.

## Directly Varying Costs

The cost $H$ of holding items in storage and the interest on money invested in stored items are directly proportional to the inventory level. Immediately after a procurement is made, there are $Q$ units in storage. If a constant usage rate is assumed and all the items in storage are depleted before another procurement is received, the average inventory stored is $Q/2$. Then the total annual holding cost is

$$\text{Holding cost} = H\frac{Q}{2}$$

and the annual interest charge is the value of the average inventory, $PQ/2$, times the interest rate $i$, or

$$\text{Interest charge} = \frac{iPQ}{2}$$

which makes the total carrying cost

$$\text{Carrying cost} = (H + iP)\frac{Q}{2}$$

## Inversely Varying Costs

The cost $O$ associated with placing an order or preparing to produce an order occurs once each procurement period. Total annual preparation cost is then the order cost for each procurement period times the number of periods per year $N$, or

$$\text{Procurement cost} = \frac{OD}{Q}$$

## Constant Costs

If $P$ is also assumed constant throughout a year, the total annual purchase cost is

$$\text{Purchase cost} = DP$$

## Total Cost

The sum of the above costs is the total inventory cost $C$, or

$$C = \frac{OD}{Q} + \frac{(H + iP)Q}{2} + DP$$

Total cost = procurement cost + carrying cost + purchase cost

The total-cost formula can now be adapted to represent different real-world inventory situations. The terms in the formula are changed to fit specific inventory patterns, but the enduring objective is to minimize $C$ by calculating the optimal quantity $Q$.

## ECONOMIC ORDER QUANTITY

An organization ordering items from an internal or external source can usually minimize its costs by adhering to the economic order quantity $Q$. The calculations for $Q$ are based on the assumptions and inventory pattern shown in Figure 4.8.

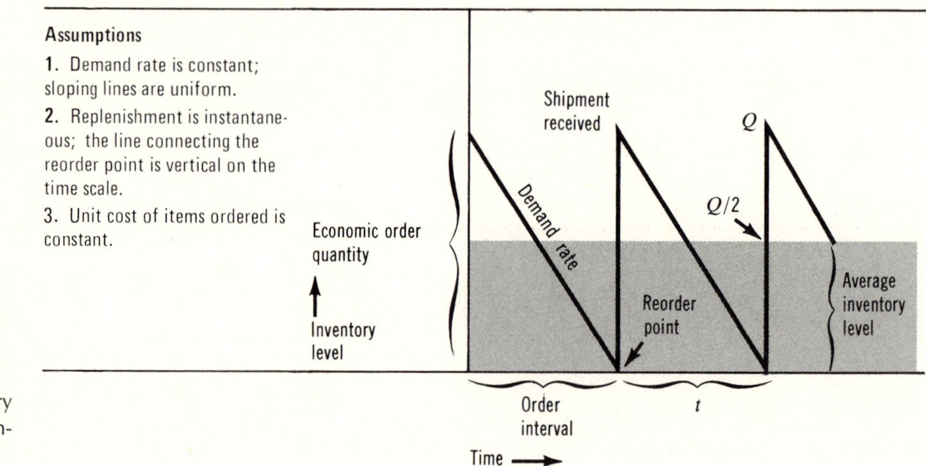


**FIGURE 4.8** Inventory usage pattern and controlling assumptions.

Under the stated assumptions, the previously encountered minimum-cost formula appropriately designates the economic order quantity (EOQ) as

$$\text{Design variable} = Q = \sqrt{\frac{\text{inversely varying costs}}{\text{directly varying costs}}}$$

$$= \sqrt{\frac{2OD}{H + iP}}$$

When the interest charge is included as part of the total holding cost $H$, as is often the case, the order-size formula reduces to

$$Q = \sqrt{\frac{2OD}{H}}$$

and related inventory values based upon $Q$ are

$$\text{Number of orders per year} = N = \frac{D}{Q} = \sqrt{\frac{HD}{2O}}$$

$$\text{Order interval} = t = \frac{\text{working time per year}}{N}$$

$$\text{Total annual cost} = C = \frac{OD}{\sqrt{2OD/H}} + \frac{H}{2}\sqrt{\frac{2OD}{H}} + DP$$

$$= \sqrt{2ODH} + DP$$

**Economy Exercise 4-5** A retail paint distributor with several outlets in a large city annually sells 8000 gallons (30,280 liters) of Do-or-Dye stain. The costs of determining the order amount of each shade of stain, preparing the order forms, and transportation charges amount to \$150. The firm has its own warehouse, which is used to store only its own merchandise. Warehousing and handling costs are prorated at \$0.21 per gallon (\$0.055 per liter) per year. The average price paid to the stain manufacturer is \$3.15 per gallon (\$0.832 per liter). Insurance and interest charges are 12 percent. What is the economic order quantity for Do-or-Dye stain, and how often should orders be placed? (*Hint*: Warehousing costs are based on the maximum level of inventory held because warehousing and handling expenses are fixed costs and remain the same whether the warehouse is full or empty.)

## QUANTITY DISCOUNTS

Price discounts are offered by some suppliers to encourage larger orders. "Cheaper by the dozen" is a household slogan which reflects availability of discounted prices for large volume purchases. Lower freight rates and handling charges for bigger shipments create the same effect as price discounts. These benefits have to be measured against the incremental increase in carrying costs required to accommodate purchases larger than the economic order quantity.

Total-cost curves for a commodity subject to a price discount are shown in Figure 4.9. The top curve $C$ is the annual inventory cost at price $P$ for different order quantities. The lower curve $C'$ shows annual inventory cost based on a lower price $P'$. However, price $P'$

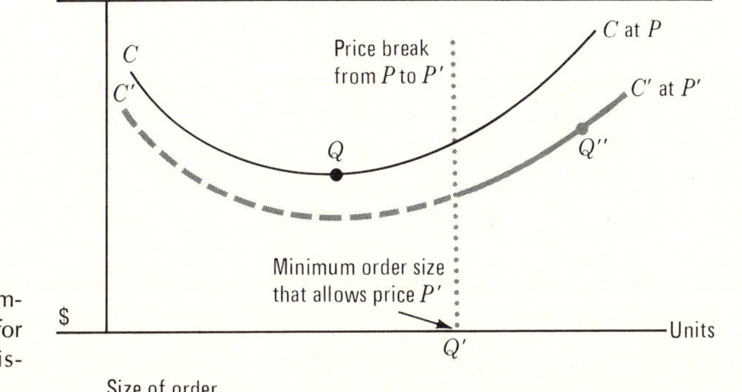


**FIGURE 4.9** Minimum-cost order quantities for purchase price discounts.

is applicable only for order size $Q'$ or larger, as indicated by the solid portion of curve $C'$ to the right of $Q'$. Because $C'$ is lower at $Q'$ than $C$ is at its minimum point $Q$, the minimum-cost ordering policy is to order $Q'$ units each time replenishments are needed. Note that if the cutoff quantity for price $P'$ were at $Q''$ on curve $C'$, it would be less costly to order quantity $Q$ at price $P$.

The procedure outlined in Figure 4.10 discloses the minimum-cost order quantity for most common quantity-discount patterns. It is applicable to a commodity with several price breaks that would be graphed as a family of curves for different prices similar to the two-curve family in Figure 4.9.

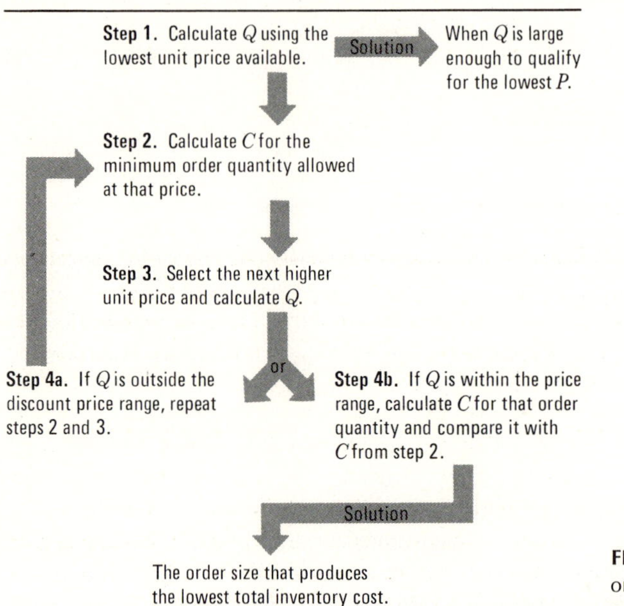


**FIGURE 4.10** Flow chart for order-size calculations when quantity discounts are allowed.

To amplify the quantity-discount procedure, consider the following costs for a product with a large annual demand:

$D$ = 400,000 units/year
$O$ = \$25/order
$H$ = \$0.06 + 0.2$P$/unit, based on average inventory
$P1$ = \$0.50/unit on orders up to 9999
$P2$ = \$0.47/unit on orders from 10,000 to 24,999
$P3$ = \$0.45/unit on orders greater than 24,999

For the lowest unit price ($P3$ = \$0.45), the order quantity is (step 1)

$$Q_{P3} = \sqrt{\frac{2 \times 25 \times 400{,}000}{0.06 + 0.2(0.45)}} = 11{,}547 \text{ units}$$

which is less than the 25,000 order size needed to qualify for the lowest unit price. The annual total cost at the \$0.45 price break is (step 2)

$$C_{P3} = \frac{(25)(400{,}000)}{25{,}000} + [0.06 + 0.2(0.45)]\frac{25{,}000}{2} + 400{,}000(0.45)$$

$$= 400 + 1875 + 180{,}000 = \$182{,}275$$

The next higher price is $P2$ = \$0.47, and $Q$ at this price is (step 3)

$$Q_{P2} = \sqrt{\frac{2 \times 25 \times 400{,}000}{0.06 + 0.2(0.47)}} = 11{,}396 \text{ units}$$

This time the order size is within the price discount range. The total cost is then calculated as (step 4b)

$$C_{P2} = \frac{25 \times 400{,}000}{11{,}396} + [0.06 + 0.2(0.47)]\frac{11{,}396}{2} + 400{,}000(0.47)$$

$$= 877.50 + 877.50 + 188{,}000 = \$189{,}755$$

The solution is to select the lowest-total-cost order size: Order 25,000 units at \$0.45 per unit. In this example the solution was obtained by passing once through the steps outlined in Figure 4.10. If $P2$ had been less than 10,000 units, it would have been necessary to pass again through steps 2 and 3 using the next higher price, $P1$. That the EOQ at $P1$ equal to \$0.50 does not yield a lower total inventory cost is demonstrated by

$$C_{P1} = \sqrt{2ODH} + DP = \sqrt{2 \times 25 \times 400{,}000\,[0.06 + 0.2(0.50)]} + 400{,}000(0.50)$$

$$= 1789 + 200{,}000 = 201{,}789$$

Thus the largest price break provides the lowest cost, but it should be remembered that bigger orders always increase the dangers associated with carrying higher inventory levels. The degree of risk is gauged by the stability of past demand, the resale value of stock, and market trends.

## ECONOMIC PRODUCTION QUANTITY

An economic production quantity (EPQ) is associated with a manufacturing environment, whereas the economic order quantity (EOQ) is more common in retail situations. The difference between the two quantities is due to delivery time. EOQ calculations are based on an assumption of instantaneous delivery time. EPQ calculations realistically show that inventory is gradually built over a period of time, as illustrated in Figure 4.11.

If the manufacturing rate $M$ is just equal to the usage rate $D$, the items will be used as fast as they are produced. If $M$ is greater than $D$, inventory is accumulated at the daily rate of $(M - D)$ divided by the number of operating days per year. The inventory level reaches a maximum after $Q/M'$ days, where $M'$ is the daily production rate. This level is equal to $(M - D)Q/M$, and the average inventory is half this amount, or $(Q/2)(1 - D/M)$. Thus, the total cost for a production situation is

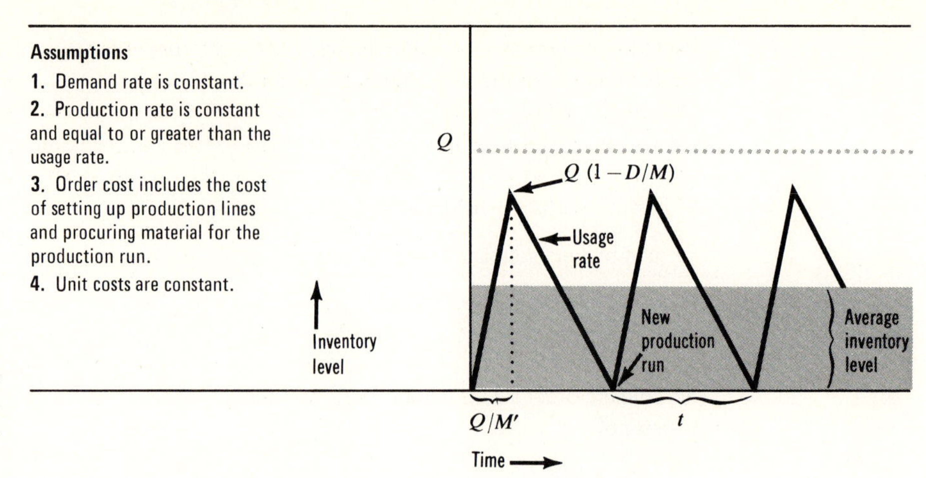


**FIGURE 4.11** Inventory accumulation-usage pattern and controlling assumptions.

$$C = \frac{OD}{Q} + \frac{HQ}{2}\left(1 - \frac{D}{M}\right) + DP$$

and the economic production quantity is

$$Q = \sqrt{\frac{2OD}{H(1 - D/M)}}$$

**Economy Exercise 4-6** One department of a plant produces switch plates which are used in the assembly department of the same plant. Switch plates can be produced at the rate of 4000 units per day. The assembly department needs only 182,000 per year. The cost of authorizing switch-plate production and setting up the machines is $300 for each run. The plant operates 260 days per year. The annual cost of holding one unit in storage is $0.10. How many switch-plate production runs should be scheduled each year?

## INVENTORY PRACTICES

The methods and formulas presented in this chapter to determine minimum-cost inventory policies are indicative of what can be done. They should not be used indiscriminately because each formula was based on specified assumptions. Only the tip of the inventory iceberg was exposed. A multitude of adaptations have been developed to fit different situations. For instance, more realistic inventory models include the cost of running out of stock and the effects of variable delivery times for ordered materials. (Risk analysis applied to inventory policy is discussed in Chapter 14.)

Inventory control involves much more than just the ordering policy. ***Safety stock***, a reserve supply beyond the amount consumed by average usage, is usually maintained to cushion the consequences of running out of needed items. This reserve obviously increases the carrying cost, but the buffer it provides may be worth the extra holding expense when demands or delivery times are uncertain. Even when economic order quantities and acceptable safety-stock levels are known, the administration of material controls is a demanding and critical task. Records must be maintained to know when stock levels have dropped to a reorder point, the condition of reserve supplies needs to be checked, and materials have to be physically handled to make room for new deliveries while sustaining access to existing stores.

Many engineering problems of material handling and storage for specific conditions have enjoyed ingenious solutions. Some warehousing systems are automated to mechanically fill orders on command and to place orders for replenishments automatically. Tables of economic order quantities are stored in computers, on call from the shop floor for order placements or to set the size of production runs. Subsequent chapters enlarge upon factors affecting investment decisions to improve inventory practices.

## SUMMARY

Replacement periods, project durations, quality-assurance plans, and inventory order quantities typify problems that can be solved with a *minimum-cost model* which includes costs that vary directly and inversely to a design variable. The model is based on short-run operations for which data are assumed to be accurately known.

The replacement age which minimizes the cost of ownership of an asset, subject to maintenance costs that increase with age and capital costs that decrease with longer ownership, is determined through a tabular approach. The optimum age provides the lowest average annual cost.

Construction works and other projects which integrate activities that utilize different resources are often analyzed by *critical path networks*. Earlier than normal completion of an activity increases its cost, but a shorter project duration decreases directly varying costs. *Time-cost trade-offs* are conducted by comparing the cost of incremental activity-duration cuts to project savings generated by the cuts. Manual methods guided by graphic displays are feasible for limited-size applications.

A preferred level for the quality of goods and services balances the cost of attaining that quality against the value placed on the quality by consumers. Process capabilities, product design, and inspection practices are considered in minimizing the total cost of quality assurance.

Basic inventory costs include order or set-up costs $O$ that decrease with larger order quantities $Q$, and holding costs $H$ that increase with increases in $Q$. When demand and delivery times are constant, the economic order quantity (EOQ) is calculated by $Q = \sqrt{2OD/H}$, where $D$ is annual demand. The use of quantity discounts is evaluated by comparing the lowest total inventory cost $C$ possible at each price level $P$, where

$$C = \frac{OD}{Q} + \frac{HQ}{2} + DP$$

When items are consumed as they are produced, the economic production quantity (EPQ) formula is

$$Q = \sqrt{\frac{2OD}{H(1 - D/M)}} \qquad \textit{where } M = \text{annual manufacturing rate}$$

## Discussion of Economy Exercises

**EE 4-1** A typical tabular approach lists the design variable (in this case the hose size) and respective costs as shown below.

| *Design Variable (hose diameter)* | *Directly Varying Cost (hose purchase)* | *Inversely Varying Cost (pumping cost)* | *Total Cost* |
|---|---|---|---|
| 3 in (7.62 cm) | \$460 | \$440 | \$900←minimum cost |
| 4 in (10.16 cm) | 590 | 352 | 942 |
| 5 in (12.70 cm) | 770 | 220 | 990 |

The smallest hose size exhibits the lowest total cost.

**EE 4-2** A tabular approach gives about the same solution as differentiating the average-annual-cost formula with respect to $n$, as

$$\frac{dAC}{dn} = -\frac{P}{n^2} + \frac{O'}{2} = 0$$

solving for $n$,

$$n = \sqrt{\frac{2P}{O'}}$$

and substituting in the cost data:

$$n = \sqrt{\frac{2(1000)}{100}} = \sqrt{20} = 4.47 \text{ years}$$

The resulting average annual cost is

$$AC = \frac{\$1000}{4.47} + \$200 + (4.47 - 1)\frac{\$100}{2} = \$223.71 + \$200 + \$168.50$$

$$= \$592.21/\text{year}$$

**EE 4-3** Two additional cuts can be made to reduce the project from the described 15-day duration to a minimum 13-day duration. The first cut takes a day off activities 2,3 and 2,4 at a cost of \$160 + \$110 = \$270. The next cut involves activities 2,3, 2,4, and 1,3 at a cost of \$160 + \$110 + \$150 = \$420. No further cuts are possible according to the conditions stated in Table 4.2. The minimum-duration schedule and associated costs are shown in Figure 4.12, where all activities are now critical. The project cost for a minimum duration is greater than the minimum-cost schedule (\$7810 + 15 × \$220 = \$11,110) by \$250.

**EE 4-4** Under present conditions, material for 600 Wambos is processed daily, 400 units from new raw material and 200 units from reject material. The new inspection plan will not

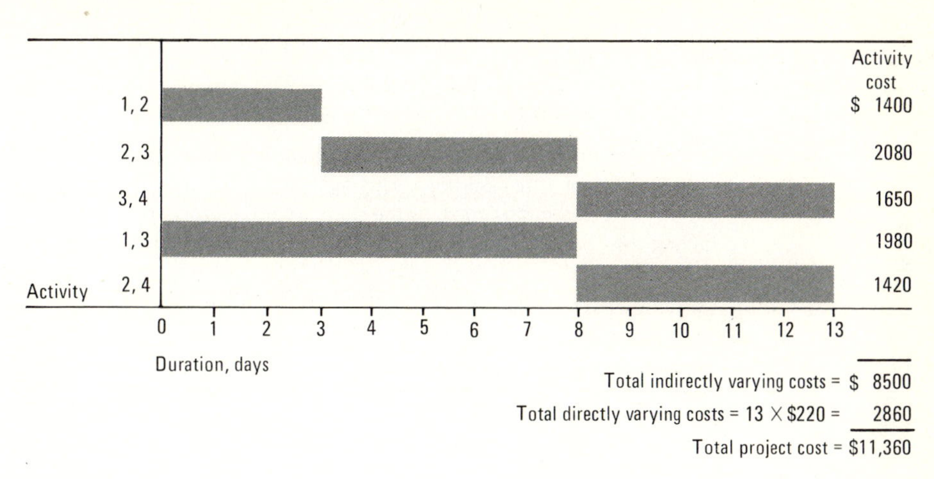

FIGURE 4.12

improve the capability of the process, but it lowers labor costs because less labor is spent on rejected units. The present flow pattern for raw-material sources, process stages, and output is shown below.

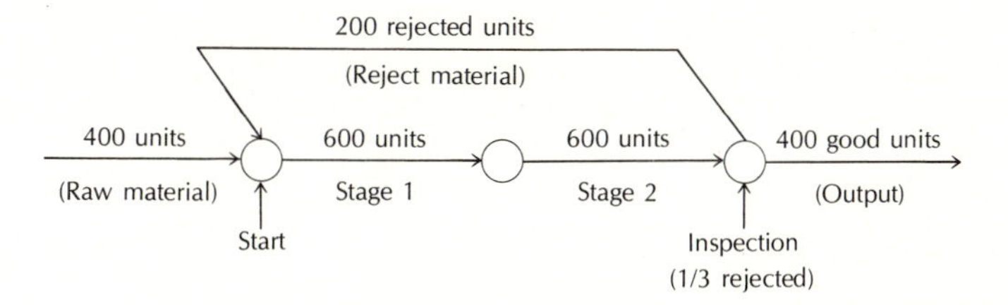


Since labor costs are evenly divided between the two stages, 400 × \$3/2 = \$600 is spent at each stage, or \$1 per unit processed. If the output is maintained at 400 units per day, the input $X$ when one sixth of the flow is rejected at each of two stages is

$$X \frac{5}{6} \frac{5}{6} = 400 \qquad \text{and} \qquad X = 576 \text{ units}$$

which means only 5/6 (576) = 480 go through the second stage. At \$1 per unit, the labor savings are 400 × \$3 − (576 + 480)\$1 = \$144 per day. Under this arrangement the plan would be turned down.

But if the flow was kept at the maximum level of 600, 600 × 1/6 = 100 units would be rejected at the first inspection and (600 − 100)1/6 = 83 at the second. Therefore the raw material input would equal the output at 600 − 100 − 83 = 417 units per day. In this pattern labor costs at \$1 per unit processed decline by \$100, but the 17 units of increased output valued at \$3 per unit would contribute 17 × \$3 = \$51 to make total savings of \$100 + \$51 = \$151. The cost of the extra inspection is thus recovered, barely.

**EE 4-5** Relating the costs in the exercise to formula symbols, we have

$D$ = 8000 gal/year (30,280 l/year)
$O$ = \$150/order

$H$ = \$0.21/gal/year (\$0.055/l/year) on maximum inventory
$P$ = \$3.15/gal (\$0.832/l) wholesale price
$i$ = 12% of $P$ on the average inventory level

The statement of the problem indicates that the storage space allowed for paint and stain is not used for any other purpose. This means space is provided for the maximum number of gallons received in an order. As the stain is sold, less space is occupied, but the total cost of having the space available is constant. Therefore, the directly varying costs are

$$\text{Carrying costs} = HQ + \frac{iPQ}{2} = \frac{(2H + iP)Q}{2}$$

Then the economic-lot-size formula is

$$Q = \sqrt{\frac{2OD}{2H + iP}}$$

and substituting figures into the EOQ expression, we get

$$Q = \sqrt{\frac{2 \times \$150 \times 8000}{2(\$0.21) + 0.12(\$3.15)}} = \sqrt{\frac{\$2,400,000}{\$0.798}}$$

$$= 1734 \text{ gal/order (6556 l/order)}$$

The number of orders per year is $N = 8000/1734 = 4.6$ orders per year, which makes the order interval in weeks $t$ = (weeks/year)/$N$ = 52/4.6, or about 11 weeks between orders.

**EE 4-6** The EPQ for the given data is calculated as follows:

$D$ = 182,000 units/year
$O$ = \$300/production run
$H$ = \$0.10/unit/year
$M$ = 4000 × 260 = 1,040,000 units/year

$$Q = \sqrt{\frac{2 \times 300 \times 182,000}{0.10 \times [1 - (182,000/1,040,000)]}} = 36,400 \text{ switch plates/run}$$

$$N = \frac{182,000}{36,400} = 5 \text{ runs/year}$$

## PROBLEMS

**4.1** A concrete contractor uses different sizes of crews for commercial and residential work. A typical residential job requires 3 hours on site and ½ hour travel each way, with a minimum crew of two, a finisher and a laborer. A finisher is paid \$7 per hour, and laborers get \$4 per hour.

Adding another pair (finisher and laborer) to the minimum crew cuts the working time required for a typical job by 60 percent. Adding just one laborer to the minimum crew reduces the time by 30 percent, and adding two laborers cuts it by 50 percent. Because of the sequence of work and space limitations, the maximum crew size is four.

Transportation costs are unchanged by the size of the crew. An overhead charge of \$20 per hour for equipment is also independent of crew size and is levied for both working and travel time. What makeup of crew will provide the most economical performance?

**4.2** For a rough analysis of different bridge designs, it is assumed that costs are proportional to the span length $X$ between piers. As the span length increases, a greater amount of steel is required to support the superstructure. However, fewer piers are required when span lengths are longer. The cost of each span for a certain bridge design is

$$C_s = 50X^2 + 5000X - 100{,}000$$

and the cost for a pier or an abutment is given by

$$C_p = 200{,}000 + 1000X$$

**4.2a** Use the tabular approach to find the minimum cost for a bridge crossing a 300-foot (91.44-meter) bay.
**4.2b** Use the formula approach to find the number of spans that minimize the cost for bridging a 300-foot (91.44-meter) bay.

**4.3** Lord Kelvin developed a classic minimum-cost analysis for the most economical wire size to conduct electricity. He proposed that the investment cost in wire diameter (directly varying cost) should be equal to the energy loss due to wire resistance (inversely varying cost). The energy loss in watts is equal to $I^2R$, where $I$ is the current to be conducted, and $R$ is the electric resistance which is inversely proportional to the cross-sectional area of the wire. Using the following symbols for the cost factors, determine the formula for the wire size which produces the minimum total cost:

$A$ = cross-sectional area of the wire
$I$ = current conducted
$R$ = resistance of the wire
$C_c$ = rate charged for the use of invested capital, %
$C_t$ = rate charged to cover taxes and insurance, %
$C_A$ = cost of wire
$C_I$ = cost of energy

$[(A = \sqrt{I^2C_I/C_A(C_c + C_t)}]$

**4.4** A mechanical testing machine with an initial cost of $8000 closely follows the cost pattern shown below.

| *Year* | *Operating Cost* | *Salvage Value* |
|---|---|---|
| 1 | $3500 | $6000 |
| 2 | 3800 | 5000 |
| 3 | 4200 | 4300 |
| 4 | 4600 | 3900 |
| 5 | 5100 | 3500 |
| 6 | 5800 | 3200 |
| 7 | 6700 | 3000 |

Using 0 percent interest and assuming that replacement equipment will follow the same cost pattern, determine the economic life.

*(4 years; average annual cost = $5050)*

**4.5** A new type of testing machine to perform the same function as that depicted in Problem 4.4 is now available at a purchase price of $14,000. The new machines cost more, but they have twice the capacity of the older models. A company plans to replace four of the older type with the new design. The four candidates for replacement are now 2 years old. They will be replaced when the expected cost of keeping them in service one more year is greater than the average annual cost for new machines of comparable capacity. The expected cost pattern for the new type of testing machine is:

| *Year* | *Operating Cost* | *Salvage Value* |
|---|---|---|
| 1 | $ 7200 | $10,800 |
| 2 | 7600 | 8600 |
| 3 | 8400 | 7500 |
| 4 | 9500 | 5500 |
| 5 | 11,000 | 4000 |

When should the replacement occur?

*(at 3 years)*

**4.6** A dump truck with a first cost of $8000 has the depreciation and service pattern shown below

| | END OF YEAR | | | | | |
|---|---|---|---|---|---|---|
| | *1* | *2* | *3* | *4* | *5* | *6* |
| Depreciation during year | $2800 | $2000 | $1400 | $ 500 | $ 400 | $ 400 |
| Annual service cost | 1800 | 2100 | 2500 | 2900 | 3400 | 4000 |

**4.6a** Assuming no interest charges are necessary for the evaluation, how many years should the truck be kept in service before replacement?

**4.6b** Assume the truck presently owned is 2 years old. It is known that a truck will be needed for only 6 more years. When should another truck having the same cost pattern be purchased in order to minimize ownership and operating costs during the 6 years?

**4.7** A production machine has an initial cost of $3750 and has no salvage value at any time. The costs of inferior performance and maintenance are $300 the first year and increase by $300 each successive year the machine is used. Develop a total-cost formula, different than the one given in Economy Exercise 4-4, for the special case in which operating expenses increase each year by the amount of the first year's cost. Then use the formula to solve for the minimum-cost ownership period.

*($n = 5$ years)*

**4.8** Plans are being made to remodel a shipping and receiving area. The critical path network for the project is shown on p. 103. The dashed "dummy" arrow indicates that both activities $C$ and $B$ must be completed before activity $G$ or $F$ can begin. (See Extension 4.1 for an explanation of network construction and use.) The critical path is denoted by double arrows. The earliest start (ES) and latest start (LS) times for each activity based on normal activity durations are shown in Table 4.4. Also shown in the table are the normal and crash costs.

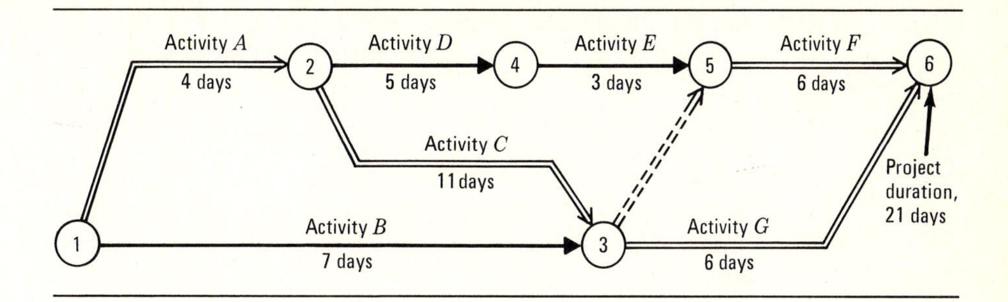


The remodeling project has a \$1100 fixed burden cost for overhead and supervision charges, plus a \$100 per day directly varying cost that results because extra help and additional storage space are needed while the old area is torn up.

TABLE 4.4

| | NORMAL TIMES | | | | CRASH TIMES | |
|---|---|---|---|---|---|---|
| *Activity* | *Days* | *ES* | *LS* | *Cost* | *Days* | *Cost* |
| $A$(1,2) | 4 | 0 | 0 | \$200 | 3 | \$250 |
| $B$(1,3) | 7 | 0 | 8 | 300 | 5 | 450 |
| $C$(2,3) | 11 | 4 | 4 | 800 | 8 | 950 |
| $D$(2,4) | 5 | 4 | 7 | 100 | 2 | 180 |
| $E$(4,5) | 3 | 9 | 12 | 300 | 3 | 300 |
| $F$(5,6) | 6 | 15 | 15 | 180 | 4 | 240 |
| $G$(3,6) | 6 | 15 | 15 | 560 | 4 | 800 |

**4.8a** Develop a bar chart similar to Figure 4.5 to show the sequence, float, and cutting costs of project activities.

**4.8b** Determine the duration and total cost for the project that minimizes the sum of directly and inversely varying costs.

**4.9** A closer look at the activity duration costs for the project described by the network for Problem 4.8 leads to a conclusion that all the activity costs do not linearly increase from normal to crash amounts. A revised cost schedule is shown in Table 4.5, where an intermediate duration and associated cost are given for each activity.

TABLE 4.5

| | NORMAL | | INTERMEDIATE | | CRASH | |
|---|---|---|---|---|---|---|
| *Activity* | *Days* | *Cost* | *Days* | *Cost* | *Days* | *Cost* |
| $A$ | 4 | \$ 200 | 4 | \$ 200 | 3 | \$ 250 |
| $B$ | 7 | 300 | 6 | 350 | 5 | 450 |
| $C$ | 11 | 800 | 10 | 870 | 8 | 950 |
| $D$ | 5 | 100 | 3 | 140 | 2 | 180 |
| $E$ | 3 | 300 | 3 | 300 | 3 | 300 |
| $F$ | 6 | 180 | 5 | 210 | 4 | 240 |
| $G$ | 6 | 560 | 5 | 600 | 4 | 800 |
| *Totals* | | \$2440 | | \$2630 | | \$3170 |

Assuming the given costs vary linearly between each pair of the three designated values, and that the directly varying costs given in Problem 4.8 are still accurate, determine the minimum-cost project duration and its total cost.

*(Minimum-total-cost project duration = 16 days with a cost of $5410.)*

**4.10** During the fruit harvest a group of orchard growers estimate they can reduce the number of culls sent to the packing plant by hiring field inspectors. Their crop averages 16 percent culls without inspection on 80 tons picked per day. They expect that for every two inspectors hired the percentage of culls would be cut in half (that is, two inspectors, 8 percent culls; four inspectors, 4 percent culls; etc.). An inspector would be paid $25 per day. The growers will save $0.40 per ton for each percent the culls are reduced. How many inspectors should be hired?

*(6 inspectors)*

**4.11** Where products are inspected during a production process depends on cost and the nature of the process. Inspections are logically scheduled in front of operations that are costly, irreversible, or masking. A study of an assembly process reveals the following rejection percentages and cumulative manufacturing costs for five consecutive operations:

| | OPERATION | | | | |
|---|---|---|---|---|---|
| | 1 | 2 | 3 | 4 | 5 |
| Percentage rejected | 5.1 | 0.6 | 4.8 | 3.2 | 1.2 |
| Cumulative production-process costs | $2.75 | $3.98 | $4.27 | $4.85 | $5.12 |

The standard unit of inspection is 100 products, and inspectors are paid $4 per hour. Inspection times in minutes per product are, in order of operation, 1.2, 0.6, 0.5, 2.0, and 1.8.

**4.11a** If two inspection stations are to be established, where in the sequence of operations should they be located? Why?

**4.11b** If three inspection stations are allowed, where should they be placed? Why?

**4.12** Depending on their responsibilities, different engineers in a production organization would likely have different opinions about the most advantageous quantity of a certain item to produce in each production run. State whether the following individuals would probably prefer longer or shorter production runs and what reasons they would give to support their contentions:

**4.12a** Sales engineer concerned with customer relations

**4.12b** Material-handling engineer concerned with warehousing

**4.12c** Design engineer concerned with purchasing

**4.12d** Engineering economist concerned with capital budgeting

**4.12e** Production engineer concerned with the production line

**4.13** A company buys $100,000 worth of a certain material each year. Order costs are 1 percent of the amount of each order, and carrying costs are 10 percent of the average inventory. If annual order and carrying costs are equal, how many weeks of supply should be ordered at one time?

*(Every 10 weeks for a 50-week year)*

**4.14** A type of raw material is used at the rate of 80,000 pounds (36,288 kilograms) per year. The current price is $1.00 per pound ($2.205 per kilogram). Storage costs are estimated at $0.09 per pound ($0.198 per kilogram) per year on the maximum inventory value. The cost of placing an order is $10, and the lead time is negligible. What is the EOQ?

*(Q = 2982 pounds/order)*

**4.15** "I propose that we apply EOQ calculations to all our inventory procurements," said the machine shop's all-purpose engineer. "At the present time we have one all-purpose purchasing agent who operates in a very informal manner, so informal that we don't even find out how much it costs us to place an order. We know he gets paid $16,000 a year, occupies an office that we figure costs $2000 a year, has a telephone bill averaging $500 per year, and incidental expenses of about $125 per month. Since our small company has no major material requirements, he just circulates around the building to see what supplies are needed. Then he phones the orders to local merchants and writes the checks himself to make all the payments. Apparently he places about 2500 orders each year, ranging in size from a few dollars to $300 or $400. That comes out to about $8 ordering cost per order. We can probably cut the number of orders down to 1500, which means a savings of 2500 − 1500 orders) $8/order = $8000/year."

Discuss three attributes of the present system that should receive consideration before changes are made to formalize the inventory policy.

**4.16** A wholesaler forecasts annual sales of 200,000 units for one product. Order costs are $75 per order. Holding costs of $0.04 per unit per year and interest charges of 12 percent are based on the average inventory level. The cost to the wholesaler is $0.80 per unit acquired from the factory.

**4.16a** What is the EOQ?

**4.16b** What is the annual inventory cost?

**4.16c** What is the time between orders based on 250 working days per year?

**4.17** A heavy equipment manufacturing company produces components which are later fabricated into finished products. One product is produced at the rate of 6 units per day. A component used in this product can be manufactured at the rate of 30 units per day and has annual carrying costs per unit of $1.50 based on the average inventory level. Administrative and set-up costs total $240 per order. If the company operates every day of the year, what is the value of the EPQ?

*(EPQ = 936 units)*

**4.18** Determine the economic production quantity for the following conditions:

| | |
|---|---|
| Set-up costs per production run | $80 |
| Variable production costs per unit produced | $0.30 |
| Percentage charge for interest and insurance | 25 |
| Selling price per finished unit | $0.60 |
| Manufacturing rate per year, units | 110,000 |
| Usage rate per year, units | 20,000 |

*(EPQ = 7221 units)*

**4.19** A manufacturer of children's toys uses approximately 200,000 nuts and bolts of one size each year. The nuts and bolts are purchased in standard quantities of 5000 at a cost of $50 per package. The storage costs of $0.001 per year to hold one bolt and matching nut are based on the maximum expected inventory level. Interest and insurance charges are 13 percent of the average value of inventory on hand. Typical order costs for preparing a purchase order, mailing, receiving, inspecting, and transporting are $25 for each purchase. The company operates 250 days per year.

**4.19a** What is the minimum-cost order quantity?

**4.19b** What is the time interval between orders?

**4.20** Determine the economic order size for the following conditions, where $n$ refers to the number of units, and holding costs are based on the average inventory size.

$$O = \$10 + \$0.01n$$

$$H = \$2n + \frac{0.01P}{n^2}$$

$$P = \$40/\text{unit}$$

$$D = 800 \text{ units/year}$$

[$n$ = *15.9 (16) units/order*]

**4.21** The Hour Glass Girdle Works has been offered a 2 percent discount if it purchases its annual material requirements in equal quarterly amounts. The costs are given below. Note that the carrying cost is based on maximum amount in stock at any time.

Ordering cost = $30/order
Carrying cost = 10% of maximum inventory level
Annual demand = $270,000

Should the girdle works use the discount?

**4.22** Four thousand tons (3,628,800 kilograms) of raw material are used each year. Order costs are $20 per order; carrying costs are $8 per ton ($0.0088 per kilogram) based on the maximum storage requirement. The supplier has offered to reduce the $40-per-ton ($0.044-per-kilogram) price by 5 percent if the minimum order is 500 tons (453,600 kilograms) and by another 5 percent if the minimum order is 1000 tons (907,200 kilograms). The capacity of the present storage facility is 250 tons (226,800 kilograms). Any increase in capacity will increase the carrying costs in direct ratio. Show the calculations and assumptions needed to prove whether or not the quantity discount should be utilized.

**4.23** Determine the economic lot size under the following conditions:

Annual usage rate is 3000 units.

Cost of acquisition is $8 per order.

Interest, insurance, and damage charges are 12 percent of the value of the average inventory.

Annual warehousing cost is $0.08 per unit, based on the maximum inventory level of 500; if the inventory level exceeds the 500 limit, an additional warehouse must be leased to make the cost $0.12 per unit for the maximum number of units stored.

The purchase price of $3 per unit is subject to a

4 percent discount on orders of 200 or over
7 percent discount on orders of 400 or over
9 percent discount on orders of 500 or over
10 percent discount on orders of 600 or over

## EXTENSION

***4.1 Critical Path Scheduling*** Most engineers will work with network management diagrams in their careers, especially civil and industrial engineers. The diagrams take a variety of forms, as they are used to analyze and control projects ranging from fund-raising drives and theatrical productions to construction projects and space programs. The simplicity and versatility of CPS techniques assures their continuing stature as one of the most used management tools.

The development of a critical path network starts with a listing of all activities required to do a project and the order in which the activities must be accomplished. This is the most demanding and critical part of the whole application. Without accurate activity definition and restriction (order of performance) lists, anything that follows is suspect. Durations for the activities are estimated from previous experience in conducting the same or similar activities. When equivalent activities have never been conducted before, as in research and development work, a range of completion times may be used. This is the PERT (program evaluation and review technique) method in which the *expected activity time* $t_e$ is the weighted average of optimistic $a$, most likely $m$, and pessimistic $b$ time estimates, calculated as $t_e = (a + 4m + b)/6$. The expected times thus calculated are treated as single time durations for critical path computations.

The information required to construct a network is shown in Table 4.6. Alphabetic symbols are used here to represent activity descriptions to save space, but a verb and noun description written on the network improves communications. The $<$ sign in the restriction column is read "precedes" and indicates that the activities to the right of $<$ *immediately* follow the activity on the left.

**TABLE 4.6** Activity list showing descriptions, durations, and sequencing restrictions for all activities required to complete a project.

| *Activity Description* | *Symbol* | *Duration, Weeks* | *Restrictions* |
|---|---|---|---|
| Design prototype | $A$ | 6 | $A < B,C$ |
| Obtain materials | $B$ | 2 | $B < D$ |
| Order parts | $C$ | 2 | $C < E,F$ |
| Manufacture parts | $D$ | 3 | $D < E$ |
| Assemble component no. 1 | $E$ | 2 | $E < G$ |
| Assemble component no. 2 | $F$ | 4 | $F < G$ |
| Assemble prototype | $G$ | 1 | |

As introduced in Figure 4.3, activities are represented by directed arrows in a network diagram. The arrow network which conforms to the activity list in Table 4.6 is shown in Figure 4.13. The diagram faithfully portrays the logic and information from the activity list. Circled nodes at both ends of the arrows designate *events* that define the beginnings and/or ends of activities. Thus, the event at the end of arrow $A$ when $A < B,C$ marks the end of $A$ and the beginning of $B$ and $C$. All activities merging at a node must be completed before activities bursting from that node can begin.

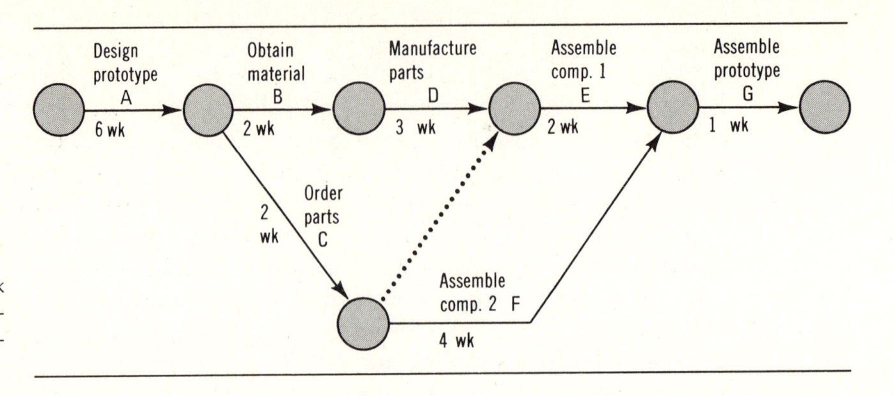


**FIGURE 4.13** Network diagram for the development of a prototype.

A dashed "dummy" arrow is used to show a one-way restriction. It is treated like an activity arrow with zero duration. As shown in Figure 4.14, the dummy from node 4 to node 5 indicates that activity $C$ must be completed before activity $E$ can begin.

Time calculations are carried out manually on a network or by computer from network data. Manual calculations, appropriate for smaller projects of, say, 100 activities or fewer, begin with the first node, the project start time. Crosses are added atop the node circles to assist the arithmetic. With a project start at time zero, the *earliest finish* (EF) time for an initiating activity is its duration added to zero. An EF value is inscribed at the head end of an arrow. The EF then becomes the *earliest start* (ES) for any immediately following activities. When two or more activities merge at a node, the *largest* EF is entered on the left bar of the cross at the merge node. This limiting EF becomes the ES for postrequisite activities.

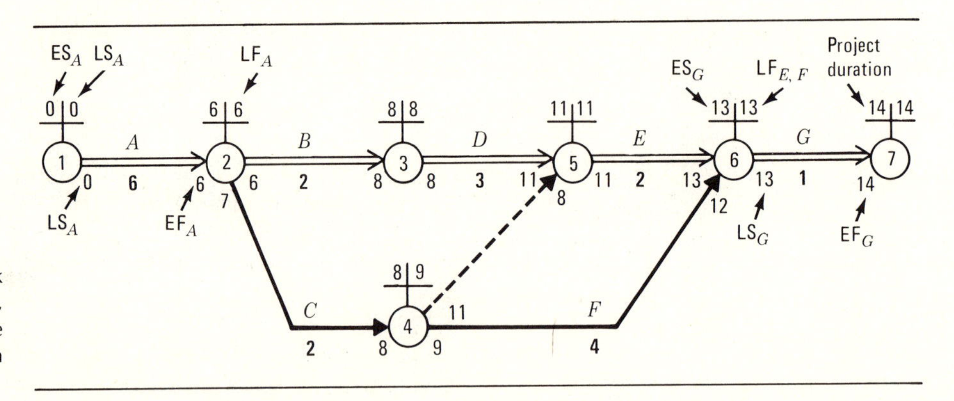


**FIGURE 4.14** Network calculations for ES, EF, LS, and LF times for the project described in Table 4.6.

As diagrammed in Figure 4.14, the project start is set by a zero on the left bar of the first cross. Then $0 + 6 = 6$ is the EF for the first activity (entered below the arrowhead of activity $A$), and this value in turn is entered on the left bar of the cross at node 2 to give the ES for activities $B$ and $C$. Note that dummy 4,5 is treated as an activity. The calculation of EF and ES values continues as a forward pass through the network until the cross on the last node is filled. This last entry is the earliest completion of the project, and the earliest start for a following project that relies on the same resources.

A backward pass, right to left, through the network allows the calculation of *latest start* (LS) and *latest finish* (LF) times. These latest-time calculations are the reverse of ES and EF calculations: An activity's duration is subtracted from the entry on the right cross of its arrowhead node and is then entered below the activity's arrow at its tail end. When two or more activities burst from a node, the smallest LS in the burst is entered as the limiting value on the cross at that node. Note that the limiting LS is the latest finish for prerequisite activities.

For example, in Figure 4.14 at node 7, $LS_G = 14 - 1 = 13$ is entered below the tail of arrow *G* and on the right bar of the cross at node 6, because 13 is the latest time activities *E* and *F* can be completed in order not to delay the project completion time of 14. At node 2, $LS_B = 8 - 2 = 6$, and $LS_C = 9 - 2 = 7$ are shown as tail-end numbers below outgoing arrows *B* and *C*, and the smaller LS of the two, 6, is entered on the cross as the latest finish of activity *A*.

After the start and finish times are inscribed on the network, they can be collected in a boundary timetable as shown in Table 4.7, or be used to develop a time schedule chart such as the one given in Figure 4.5. The difference between the latest and earliest times is the *total float* (TF), the allowable leeway in scheduling; TF = LS − ES = LF − EF. The critical activities that comprise the critical path are those with no float, TF = 0.

**TABLE 4.7** Boundary timetable developed from the network in Figure 4.14.

| *Activity* | *Duration* | *ES* | *LS* | *EF* | *LF* | *TF* |
|---|---|---|---|---|---|---|
| *A* | 6 | 0 | 0 | 6 | 6 | 0 |
| *B* | 2 | 6 | 6 | 8 | 8 | 0 |
| *C* | 2 | 6 | 7 | 8 | 9 | 0 |
| *D* | 3 | 8 | 8 | 11 | 11 | 0 |
| *E* | 2 | 11 | 11 | 13 | 13 | 0 |
| *F* | 4 | 8 | 9 | 12 | 13 | 1 |
| *G* | 1 | 13 | 13 | 14 | 14 | 0 |

Both network construction and calculations get easier with practice because they are largely mechanical. The measure of merit in network analysis is the validity of the input data and what is done with the boundary time output to produce a better project plan.

## QUESTIONS

**4.1a** Construct an arrow network segment for each set of restrictions given below:

| *a* | *b* | *c* | *d* |
|---|---|---|---|
| *A* < *C* | *A* < *D* | *A* < *C* | *A* < *C*,*D* |
| *B* < *C* | *B* < *D*,*E* | *B* < *D*,*E* | *B* < *C*,*D* |
| *C* < *E* | *C* < *E* | *C* < *D* | *C* < *E* |
| *D* < *E* | *D* < *F* | *D* < *F* | *D* < *E* |
| | *E* < *F* | *E* < *F* | |

**4.1b** The activities and restrictions for a fabricating project are given below. Construct an arrow network, and develop a boundary timetable for the project.

| *Activity* | *Description* | *Duration* | *Restrictions* |
|---|---|---|---|
| *A* | Make parts list | 2 | $A < B,C$ |
| *B* | Prepare routings | 3 | $B < D$ |
| *C* | Order and procure materials | 4 | $C < E,F$ |
| *D* | Make schedule | 3 | $D < E,F$ |
| *E* | Process parts for subassembly 1 | 3 | $E < G$ |
| *F* | Process parts for subassembly 2 | 2 | $F < H$ |
| *G* | Assemble subassembly 1 | 2 | $G < I$ |
| *H* | Assemble subassembly 2 | 4 | $H < I$ |
| *I* | Final assembly | 1 | |

**4.1c** Given the following activity list, time estimates, and restrictions, draw an arrow network, and complete a boundary timetable.

| *Activity* | *Duration* | *Postrequisites* | *Activity* | *Duration* | *Postrequisites* |
|---|---|---|---|---|---|
| *A* | 2 | *B* | *H* | 6 | *M* |
| *B* | 4 | *G,F,L* | *I* | 11 | *K* |
| *C* | 3 | *E,I,J* | *J* | 7 | *K* |
| *D* | 4 | *I,J* | *K* | 12 | *M* |
| *E* | 7 | *G,F,L* | *L* | 9 | *K* |
| *F* | 16 | *M* | *M* | 5 | |
| *G* | 6 | *H* | | | |

# CHAPTER 5

# ECONOMIZING AND COMPARING

Engineering economy begins with a state of mind, a reference frame that seeks more efficient ways to utilize resources: materials, machines, processes, and people. Good intentions to conserve resources have a much better chance of succeeding when the mission is guided by proven procedures. Fresh sets of procedures are regularly unveiled. Often they are just a rehash of old methods packaged under catchy new slogans, yet the promotions are valuable. The "new" programs may simply recast old concepts in rejuvenated language, but they are reminders of the worthiness of economizing and add vigor to its pursuit.

A few of the programs and methods to encourage economy are described in this chapter. All draw from the same tool bin of cost-conscious concepts. Minimum-cost and breakeven models presented in the previous two chapters appear in customized versions. Many suggestions could be classified as common sense, although they rely on uncommon ability to employ them effectively.

Value-improvement programs focus on short-run actions to better operating performance. They emphasize motivation to seek out improvements and propose ways to hasten accomplishments. They lead eventually to the question of which improvements deserve implementation. Will the new way detract from traditional values as it reduces cost? Is the new way good enough, or should further improvement be sought? How should difficult to quantify but influential factors be included in the analysis?

Such questions always seem to cloud economic evaluations. Future chapters face the questions from a variety of viewpoints; in this chapter productivity measures are considered, and a comparison method is introduced for the evaluation of alternatives with mixed dimensions.

## VALUE ENGINEERING

The design of a new product, process, or service offers an exceptional opportunity to economize. It is a chance to start anew, build better from experience, rectify old mistakes with refinements, utilize the latest technological developments, and break away from stagnating traditions or customs. The search for and selection of new means to reduce cost and improve value during the design phase is the forte of ***value engineering*** (VE). The concepts involved grew from government-sponsored cost-prevention campaigns in the 1940s to become a recognized discipline with a national organization called American Society for Value Engineering, cleverly initialed SAVE.

Value engineering is an organized effort to identify and eliminate unnecessary costs without sacrificing quality or reliability. Some larger companies have slots in their tables of organization for value-engineering staffs. In other companies the VE function is less formally performed by engineers in research and development, design groups, and plant-engineering departments. The innovative approaches forwarded by value engineering are equally applicable to service organizations and individual economizing.

### Value

The methodology of value engineering is intended to (1) identify the function of a product or service, (2) establish a value for that function, and (3) develop a means to provide value. Value is established by comparison, and by no other means. It is not assumed to be an automatic ingredient of any design. Since many of today's products and systems are designed by teams, design groups cannot always devote detailed attention to all subassemblies. This leads to suboptimization and subassemblies designated by standard or rote practices. The end result may be a decrease in economic value and an increase in unneces-

| | |
|---|---|
| *Use value* | The properties or qualities which accomplish a use, work, or service |
| *Esteem value* | The properties of an object which make its ownership desirable |
| *Cost value* | The sum of labor, material, overhead, and other costs required to produce economic value |
| *Exchange value* | The qualities of an object that make it possible to procure other items in its place |

**FIGURE 5.1** Types of economic value.

| | |
|---|---|
| *Incomplete information* | Failure to gather all the facts |
| *Lack of an idea* | Failure to explore all possible ways of performing a service or making a product |
| *Honest wrong beliefs* | Decisions made on what is believed to be true, and not on the facts |
| *Habits and attitudes* | Habits which take us where we were yesterday and attitudes which tend to keep us there |

**FIGURE 5.2** Reasons for unnecessary costs.

sary costs. Figure 5.1 lists four types of economic value, and Figure 5.2 offers reasons for unnecessary costs.*

Value engineering emphasizes *use* value. The value of anything can be no more than the value of the function it performs. Unnecessary costs arise when there is a less expensive method that performs the same function without downgrading quality.

## Function

A value-engineering study starts with the identification of an item's primary function by means of a verb and a noun. According to this technique, the primary function of a table would be to "support weight," and that of a shipping container would be to "provide protection." Any secondary functions are also identified; for example, a shipping container might also "attract sales." All subassemblies, such as a zip opener for a container, should contribute to the value of either the primary or secondary function.

Next, a value is assigned to the function by comparing it with something else that will do the same task. Areas of costs are evaluated to determine where current quality standards or expenses are excessive. A search is made for alternative methods that preserve the desired quality. For instance, a manufacturer had a contract for engines ranging from 10 to 20 horsepower. Under the terms of the contract the engines were to be packed in individual crates. The requirement appeared to be unnecessarily costly. After a study of the problem, a proposal was submitted to the buyer to use returnable steel racks in lieu of the crates and to cover the six engines on each rack with a plastic dust cover. The proposal was accepted, and significant savings resulted.

If the noun in the function description has a measurable parameter, such as weight or length, a mathematical evaluation is feasible. The key to comparison is an assignment of dollar values to the parameters of equivalent quality. When the comparison looks favorable, there is still the problem of implementation. As a rule, any change will meet resistance. The effort required to produce a usable idea often pales into insignificance when compared with the effort required to sell it. In a sound study, objections are anticipated and means to overcome them are included.

Another example of value engineering concerns a lock on a small compartment door that appeared unnecessarily expensive. The door was on the outside of a newly designed truck

***Value Engineering Methods Manual*,* The Boeing Company, Seattle, Wash., 1962.

cab and led to a tool compartment. The original design called for a tool-actuated lock which cost $8.27. Its functional description was "secures door." The search for alternatives revealed a spring lock that served the purpose equally well and cost $3.11. A further search led to a clip fastener that cost only $0.42. Finally, the whole assembly was questioned and it was determined there was no need even to have a door, because the tool-storage compartment was readily accessible from inside the cab. The study led to a total saving of $26.42 per unit.

The cost of a VE study must be deducted from the savings derived from the study. In order to receive maximum return for the effort expended, studies should be concentrated first on items for which the largest expenditures are planned. There may be many candidates. A selection is made by examining the factors involved in a study by means of the formula

$$\frac{\text{Potential savings}}{\text{Study and implementation costs}} \times \text{probability of implementation} = \text{rating factor}$$

Items with the highest ratings are given first attention, and studies with lower cost-reduction potential are made when time permits.

---

**Economy Exercise 5-1** Unusual solutions to usual problems are characteristic of value engineering. Techniques such as brainstorming and synectics are employed to enhance engineering creativity. A brainstorming session could very well generate a hundred different ways to perform a certain function. Some of the ideas are probably so wild and impractical that they can be easily discarded. Other ideas are filtered through a screening process to test their feasibility. One method of screening is to compare each new alternative to the existing or conventional method by means of a *T-chart*. The T-chart consists of a list of criteria that an acceptable solution should satisfy and a place to mark whether the new design is better or worse than another design in reference to each criterion. A sample T-chart is shown

| *Design X: Present Design* | *Better* | *Worse* |
|---|---|---|
| Initial cost | ✓ | |
| Operating cost | | ✓ |
| Reliability | | ✓ |
| Appearance | ✓ | |
| Comfort | | ✓ |

Check marks, as used in the sample T-chart, are usually adequate to eliminate less desirable ideas in a preliminary screening. Design *X* in the T-chart above would probably be rejected because its lower first cost and more attractive appearance are outweighed by lower reliability, less comfort, and higher operating costs than the existing design. A finer T-chart screen requires that percentages replace check marks in the comparison.

Develop a T-chart to evaluate the following suggestion for a company-sponsored plan to conserve energy and assist employees commuting to and from work:

The company will purchase 12-passenger deluxe vans and recruit employees to drive them. The fees paid by the first 10 passengers will be allocated to gasoline, upkeep, and paying for the van over a 5-year period. The driver will get to keep the fees paid daily by the eleventh and twelfth passengers, and have free use of the van at night and on weekends. The company will assist in signing people up for this *Vanpool* service and will provide reserved parking adjacent to the plant. There is no public transportation available at the plant location.

## VALUE-IMPROVEMENT PROGRAMS

Any operation can be improved. What was thought of as being the ultimate in efficiency a few years ago is now commonplace or outdated. Further improvements are not always dramatic leaps ahead, but multiple small refinements can gradually fashion a breakthrough. Greater value might result from a minor design change that makes a product more useful to the consumer or easier to build by the producer. A small change in the layout of the workplace can make operations more convenient or safer for the operator. A larger investment in robots can reduce the exposure of humans to dangerous or distasteful work. The criterion of acceptability of all these engineered improvements is that the value gained from each change must be greater than the cost incurred.

### Materials: Value Analysis and Design to Cost

The control of material costs is becoming more significant as higher rates of inflation boost prices from one period to the next and resources become scarce owing to either temporary shortages or dwindling world supplies. The potentials for material savings are manifold; the challenge is to discover those that retain the required output quality while lowering cost.

Minimum-cost inventory policies were discussed in Chapter 4. These policies are important but comprise only one aspect of material cost control. *Value analysis* is a name given to studies that determine how appropriate purchased materials are for the functions they are to perform. The analysis is very similar to value engineering, as it seeks answers to questions such as: Could a less expensive substitute be used that would still perform the required task?

Weight, strength, machineability, and appearance are typical selection alternatives for materials. For instance, a building contractor comparing galvanized-steel gutters with aluminum gutters faces a material-selection decision. Both materials are adequately strong and rust or corrosion resistant. The desired box-type gutter is available in both materials. Appearance is less important because the gutters will be painted. The main differences are in weight and price. Weight affects installation and shipping costs. On-site costs for a 10-foot (3.05-meter) trough section would amount to:

| | Basic Cost | Weight, lb | Difference in Cost due to Weight | Cost Comparison |
|---|---|---|---|---|
| Aluminum | $4.40 | 4 (1.81 kg) | | $4.40 |
| Steel | 3.80 | 7.5 (3.40 kg) | $0.38 | 4.18 |

This cost comparison could change abruptly if the gutters had to be shipped a greater distance. The 3.5-pound (1.59-kilogram) difference per section would cause a major cost contrast. A building site in Alaska could change the outcome ratings to:

| | Basic Cost | | Shipping Cost | | Installation Cost | | Total-Cost Comparison |
|---|---|---|---|---|---|---|---|
| Aluminum | $4.40 | + | $0.64 | + | $0.82 | = | $5.86 |
| Steel | 3.80 | + | 1.19 | + | 1.14 | = | 6.13 |

*Design to cost* is a program that seeks to avoid cost overruns for major projects by starting with the amount of funds available and working backward to design the best possible product within the funding limits. Material selection is an essential part of most designs. Government bodies make frequent use of the design-to-cost concept when they know the funds available from an appropriation or bond levy; they then juggle material costs and production bid options to design the weapon, facility, or system that yields the greatest value for the budgeted amount. Equivalent output-to-input reasoning is suitable when market conditions set the top limit on the price that vendors can charge for a product or service; they strive to provide the most value within the imposed limits to protect or build their reputations.

**Economy Exercise 5-2** A homeowner has budgeted $1000 for a new driveway and is adamant about not budging from her budget. She wants your construction company to do the job because she respects your reputation. Use the design-to-cost concept to prepare specifications for the driveway.

## Machines: Unit Cost and Man-Machine Analyses

To get the maximum value from a machine, it has to be used for the proper purpose and be kept in operation. Idle machines, like idle materials, represent capital that is not earning any returns. And machines, like people, have seemingly individual characteristics. Some machines do more precise work than others. Some take a long period to start or set up. Others have frequent breakdowns and are unreliable. Operating costs vary widely. Because of these diverse characteristics, the assignment of work to different machines can be frustrating.

Even the relatively simple question of whether to perform a set of calculations on a hand calculator or on a computer involves several considerations:

| | |
|---|---|
| What accuracy is needed? | Both are probably accurate enough for most problems. |
| Is machine time available? | Hand calculations can be performed almost any time, anywhere. |
| How complex is the problem? | The memory capacity of a computer makes it more suitable for multistep solutions. |
| Is a program available? | It may take as long to write a program for the computer as it would to do the entire problem step by step on a hand calculator. |
| Will the problem occur again? | If the answer is yes, it makes writing a program more attractive. |
| When is the answer needed? | A computer is faster for complex problems if a machine and a program are ready when needed. |

Most of the answers to such questions involve time: machine running time, set-up time, checking time, and operator's time.

Ways to minimize idle time of resources have long been of interest, probably because an idle resource is such an obvious exemplification of waste. Solution methods range from queuing theory (Chapter 14) to block diagrams. ***Man-machine charts*** fit the latter category. These charts are used to display the relationship between operator and machine time when one operator or a crew operates a machine or a battery of machines. The operations required to keep the machines producing form a repetitive cycle. Each cycle is represented in a chart by a schedule of operations for the operators and the machines.

Assume that a large number of semiautomatic machines producing identical products are serviced by operators who follow the repetitive operating sequence given below.

| *Operation* | *Standard Time, min* |
|---|---|
| Load machine | 1.2 |
| Remove finished unit | 0.4 |
| Inspect finished unit | 2.3 |
| Package finished unit | 3.1 |
| Walk to next machine | 0.3 |

One unit of output is produced during each machine cycle of 17.4 minutes. As shown in Figure 5.3, the proportion of idle time between operator and machine varies with the number of machines serviced by an operator.

The most economical man-machine ratio is determined from the relative costs of idle time. If the machine burden rate (maintenance, depreciation, operating cost, etc.) is $10 per hour and the labor rate for an operator is $4.20 per hour, the minimum cost per unit is calculated as

$$\begin{aligned}\text{Minimum machine cycle time} &= \text{loading-unloading time} + \text{running time}\\ &= 1.6 + 17.4 = 19 \text{ min}\end{aligned}$$

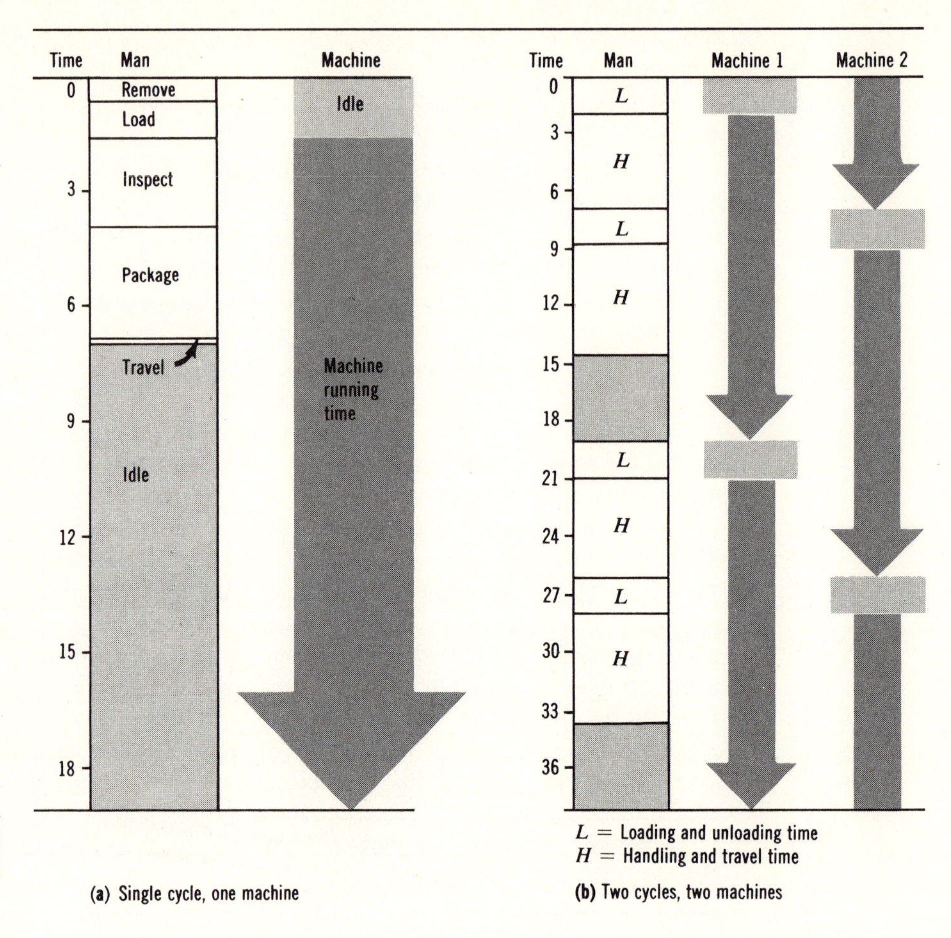


**FIGURE 5.3** Man machine charts for one and two machines serviced by one operator.

$$\begin{aligned}\text{Minimum operator cycle time} &= \text{loading-unloading time} + \text{handling and travel time}\\ &= 1.6 + 5.7 = 7.3 \text{ min}\end{aligned}$$

which can be used to roughly fix the solution area:

$$\text{Approximate number of machines} = \frac{19}{7.3} = 2.6$$

If one operator services two machines, the cycle time is set by the machine's running time, 19 minutes. The unit cost is

$$\textit{Labor:} \qquad (19 \text{ min})\frac{\$4.20}{60 \text{ min}} = \$1.33$$

$$\textit{Burden:} \quad (19)(2)\frac{\$10.00}{60 \text{ min}} = \$6.33$$

$$\textit{Total cost per cycle} = \$7.66$$

$$\textit{Unit cost} = \frac{\$7.66}{2} = \$3.83$$

If one operator services three machines, the cycle time is set by the required operator's time, $3 \times 7.3 = 21.9$ minutes. The unit cost is

$$\textit{Labor:} \quad (3 \times 7.3)\frac{\$4.20}{60} = \$\ 1.53$$

$$\textit{Burden:} \quad (21.9)(3)\frac{\$10.00}{60} = \$10.95$$

$$\textit{Total cost per cycle} = \$12.48$$

$$\textit{Unit cost} = \frac{\$12.48}{3} = \$4.16$$

As is apparent in the preceding example, ***unit cost*** is the periodic total cost divided by the units of output. Unit costs are a convenient measuring scale by which to compare different machines, methods, and services. Such evaluations are similar to average-cost analyses (discussed in Chapter 3) applied to individual assets or operations.

**Economy Exercise 5-3** Two types of machines with the capacities and costs shown in Table 5.1 are capable of producing a certain kind of unit. The annual demand for these units is 45,000. If all production is done during the normal working year of 250 eight-hour days and the machines are not needed for any other production, what type of machine or combination of types should be used?

**TABLE 5.1**

| | *Type S* | *Type L* |
|---|---|---|
| Depreciation, taxes, insurance/year | $7500 | $16,800 |
| Labor rate of operator/hour | $10.50 | $11.20 |
| Power consumption/hour | $0.30 | $0.50 |
| Production rate/hour, finished units | 10 | 15 |

## Processes: Sequencing and Life Cycle Analyses

A process is an organized method of doing something. Processes of most interest to engineers are multistep and involve several resources. Time is the critical resource that integrates the activities of people utilizing materials and machines.

Impressive strides have been made in developing mathematical models of processes. They are usually complex because the processes modeled are complex. A very simple

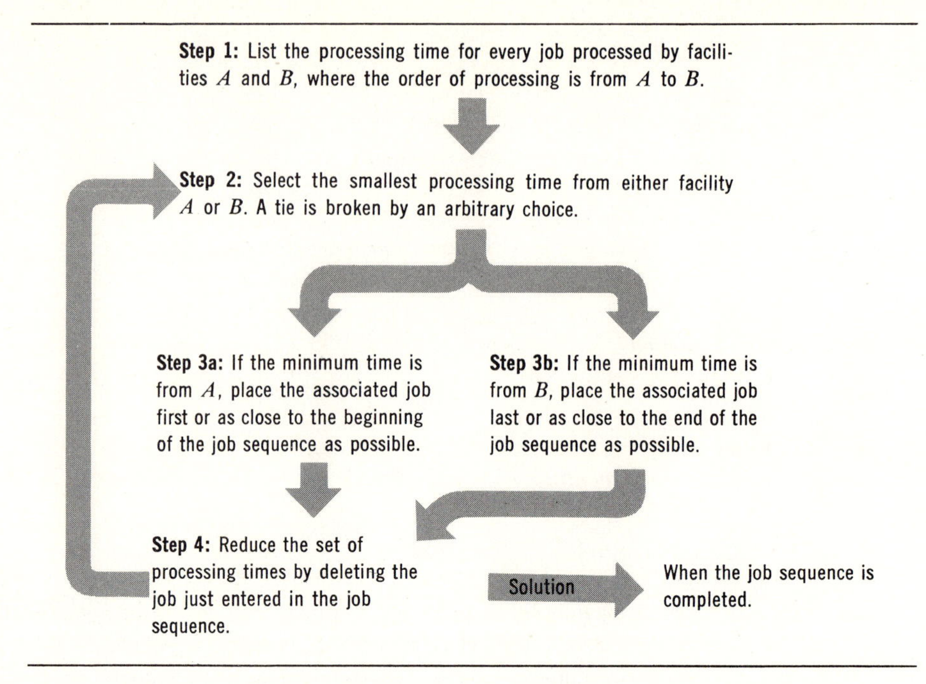


**FIGURE 5.4** Flow chart for processing $n$ jobs through two facilities.

model to minimize process times, known as ***Johnson's rule,*** is shown in Figure 5.4. It is an algorithm to optimize the flow of any number of jobs through two processes. Time units take the place of monetary values as inputs in the model, yet the solution is the minimum-cost sequence because it minimizes idle time, a costly and irretrievable resource.

Johnson's rule is applicable to a situation in which $n$ jobs are to be processed through two work stations, always in the same order from station $A$ to station $B$. The algorithm specifies the sequence of jobs which minimizes the elapsed time from the beginning of the first job to the end of the last job. Such sequencing problems quickly increase in size as more jobs and work stations are included; five products going through five work stations can be sequenced in $(5!)^5$ = 25,000,000,000 ways.

**Economy Exercise 5-4** Four reports are to be prepared for reproduction in the typing and drawing department. They are then to be sent to the printing shop, where several hundred copies will be printed, collated, and bound. Preparation and printing times vary among reports, as shown by the hours required for each in the table. Apply the steps given in Figure 5.4 to determine the minimum time to process the four reports.

| | REPORT | | | |
|---|---|---|---|---|
| | $W$ | $X$ | $Y$ | $Z$ |
| Preparation process $(A)$ | 6 | 5 | 4 | 10 |
| Printing process $(B)$ | 2 | 6 | 8 | 5 |

Time is also a critical parameter in determining whether a process is initially acceptable. ***Life cycle analysis*** is a term used to describe an economic evaluation of expected costs over the entire life of a proposed project. Its purpose is to anticipate the effects of various activities with respect to the total cost of producing the product or service. For instance, additional expenditures for the preliminary design might lower operating costs and thereby reduce total costs.

Expenditures during the life of most projects roughly follow the pattern shown in Figure 5.5. The first stage, design, accounts for research, engineering design, administration, and financing costs. The development stage takes the basic plan and converts it to hardware or services through charges for fabrication, installation, delivery, training, trial runs, and material purchases. After the process is established, operating costs required to keep it going include personnel, consumable supplies, overhead, maintenance, and services.

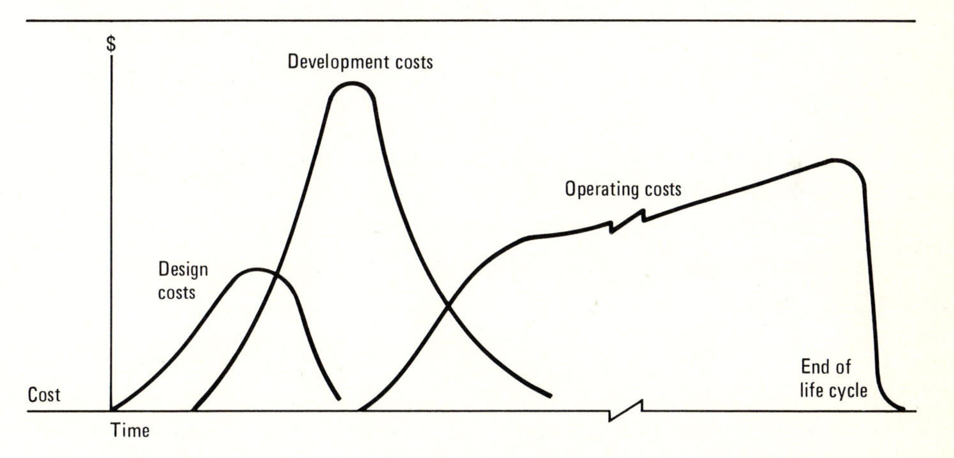


**FIGURE 5.5** Typical stages of life cycle costs for a product or service.

Life cycle analysis is informally done for many purchases. In buying an automobile, most buyers consider gas mileage, maintenance costs, and resale value as well as the sticker price. Some government agencies require life cycle costing for purchasing. The intent is to evaluate trade-offs between cost categories. For example, a government contract might inform potential suppliers that a product will be evaluated according to the life cycle model that minimizes the total cost for

$$\frac{\text{Custom design}}{\text{Project life}} + \frac{\text{unit price} + \text{maintenance}}{\text{service life}} + \text{fuel consumption}$$

or, the life cycle cost for an education program as

$$\text{Course development} + \frac{\text{instruction and supplies}}{\text{number of students}} + \text{follow-up}$$

Separate measures of quality are necessary to determine the final bid preference.

## People: Training and Motivation

The most conspicuous, best studied, yet least comprehensible resource in any system is the human resource. Materials and machines obey the rules of physics and are predictable.

People are not. Since the behavior of large groups of people is more predictable than individual behavior, collective responses to particular conditions allow the formulation of macroeconomic "laws." Deviations become more pronounced as the group gets smaller. The response by any given individual to an economic incentive may fall anywhere in the spectrum of possibilities. Because engineering economists deal with both individuals and groups under shifting economic circumstances, they should never be bored by lack of variety.

Just as individual opinions vary, capabilities also vary. Unit costing can measure what an operator has done before, but it may not be accurate for tomorrow's performance. Uniformity and consistency are encouraged by training and motivation programs.

Improvement programs are packaged under many names and are backed by a varied host of supporters. Most programs do spur improved performance, at least for a short time, in the way a pep talk arouses go-get 'em emotions before a contest. Some programs rely on researched principles of information feedback on past performance to improve future performance, while others focus on changing the environment to affect worker behavior. Regardless of the methods used, the value of the program ultimately rests on the difference between the output before the program was instituted and the output afterward.

Figure 5.6 indicates the general pattern of costs for conducting a performance-improvement program and savings owed to it. Savings generated by training or exposure usually lag the presentation expenses but drop off more slowly.

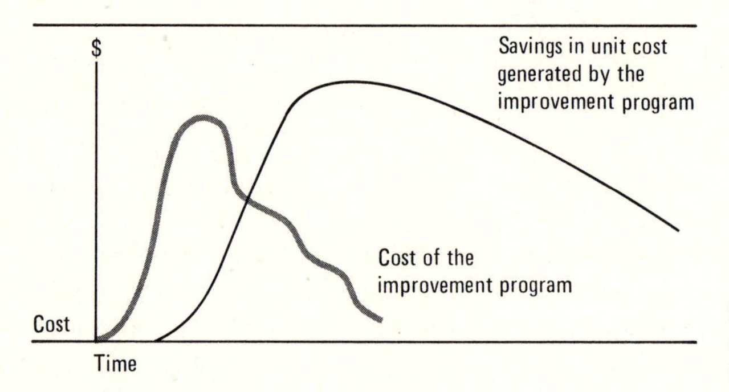


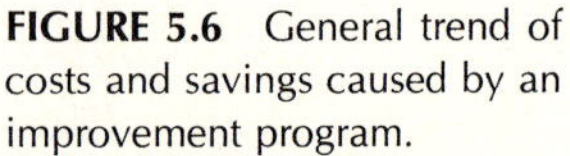

**FIGURE 5.6** General trend of costs and savings caused by an improvement program.

Variations in the cost-savings pattern for value-improvement programs make economic analyses difficult because measurements taken at different periods exhibit different saving-to-cost ratios as do ratios of total savings to total costs over different time intervals. A yardstick of effectiveness for programs to improve personnel practices is how long it takes for the savings resulting from a program to equal its cost. This payback rating can be used to compare trials of different programs to determine which ones deserve continuation. In selecting which programs to try initially, an adaptation of the formula used for selecting value-engineering studies is useful:

$$\frac{\text{Potential savings per period}}{\text{Cost of improvement program per period}} \times \begin{matrix}\text{probability of}\\ \text{success within}\\ \text{the period}\end{matrix} = \text{rating factor}$$

**Economy Exercise 5-5** New improvement programs for quality and safety are being considered.

A "Down with Defects" promotion would have a big kickoff rally with banners and speeches where workers would pledge how much they expect to reduce defective output during the 6-month program. Individual pledgers who successfully achieve their defect-reduction goals would receive publicity and rewards. Potential savings from defect reduction are $1 million per year, but it is estimated that there is only a 50 percent chance of reaping half the potential. The promotion would cost $100,000.

An "Avoid Accidents" program would involve hiring another safety engineer and increasing the safety budget. New classes would be developed to explain unsafe acts associated with each type of job, and all the workplaces would be visited to detect unsafe conditions. A better safety record would reduce insurance costs and avoid work disruptions to save $265,000 annually, but slower production due to workers' anxiety about being careful might reduce output by $25,000 per year. The accident-reduction program would cost $60,000 per year, and the probability of success is 80 percent.

Assuming the estimates are not overly optimistic, what would you recommend?

## COMPARABILITY

Decision makers can outsmart themselves. In a rush to get the best deal, they can be distracted from the main issue by the attraction of a "bargain." A practice of making comparisons strictly on the basis of immediate cost will inevitably lead to some long-run disappointments. Consider the following three examples of routine decisions:

- A family plans to visit relatives on the other side of the country. They plan to drive and are concerned about overnight lodging costs. Motel or hotel accommodations are estimated to cost $40 per night for 5 nights each way. A sleeping trailer can be rented for $80 per week. They will be gone 4 weeks. Other costs associated with pulling the trailer should amount to about $70. The nearly equal cost of the two alternatives ($400 versus $390) makes a decision difficult.
- A homemaker reads that steak sells for $0.40 per pound ($0.882 per kilogram) less at a market across town from her usual neighborhood shopping area. She spends an additional $2 traveling to and from the distant market to save $2.80 on 7 pounds (3.2 kilograms) of steak.
- A shopper discovers a coat in a discount store that is almost identical to a far more expensive coat in a stylish clothing shop. Both coats fit, both are attractive, and both appear to have the same wearing qualities; the difference is that one costs only half as much as the other. The expensive coat is purchased.

In the above decision situations, cash outlays are influential and conspicuous factors, but they are not the only factors that should be considered. Comparisons limited to input factors are not complete unless the outputs resulting from all the alternatives are essentially equivalent. Consider the examples again:

- The two types of lodging for the traveling family should be compared for convenience, comfort, and safety, as well as price.
- The homemaker is purchasing quality as well as quantity. Even if the less expensive meat is the same quality as regularly priced meat, the comparison is still inadequate because the value of the homemaker's time is not included.
- Price appears to be the only difference between the two coats of apparently equal quality, but a subtle, qualitative factor probably prejudiced the decision: pride of "Name brand" ownership. Though vanity or snob appeal may be deplored, it remains a real factor affecting long-term satisfaction.

Such examples of "false economy" have counterparts in organizational decisions. Offices have been located in low-rent districts and then relocated owing to the inconveniences caused. Protective clothing and devices have been purchased for workers who resisted wearing them because they were unattractive.

Wage payments demonstrate two sides of comparability. The rewards for extreme proficiency are most apparent in the "star" ratings of the entertainment industry. Famous people receive great remuneration from motion pictures, television, recordings, book writing, and athletics. Stars deserve extravagant rewards if their presence is repaid by proportional increases in revenue. A major-league baseball player could expect to earn at least \$200,000 if his batting average were .400, and his presence might increase the club's ticket sales by an amount equal to his big salary. Yet the star cannot play the game alone. Should the rest of the players on the team be compensated in proportion to their proficiency measured against the star's? If so, a .100 hitter would earn (.100/.400)(\$200,000) = \$50,000. That a .100 hitter cannot even stay in the league is evidence that comparisons based on proficiency alone are inadequate. Yet it warrants consideration along with other factors.

Consider a similar example from manufacturing. A worker receives an hourly wage of \$6, and the indirect costs of employment are 50 percent of wages. The worker operates a machine with a burden rate of \$21 per hour and averages 10 pieces of output per hour. The resulting unit cost is

$$\text{Unit cost} = \frac{\$6 + (0.5 \times \$6) + \$21}{10} = \$3.00/\text{piece}$$

Another worker uses the same type machine but produces only 7 pieces per hour. An equivalent pay scale calculation shows that this worker's wage $W$ should be such that

$$\frac{W + 0.5W + 21}{7} = \$3/\text{piece}$$

or

$$W = \frac{\$21 - \$21}{1.5} = \$0/\text{hour}$$

which means that, on a relative basis, the second worker deserves no pay.

While this example exaggerates the equivalent-pay concept, it does highlight the problem of proper rewards for greater productivity. Standardized hourly-wage contracts, step-graded annual salaries, guaranteed annual wages, minimum wages, and other payment plans that are not tied to output are deemed socially beneficial by custom, unions, and Congress. But they severely restrict incentives for productivity improvement. However, when wage adjustments for proficiency are possible, it is always difficult to measure the *value added* by a specific operator because all workers depend on an extensive support system which has to function correctly to sustain any output. Productivity measures are discussed in the next section.

---

**Economy Exercise 5-6** Operator 1 has an hourly wage of \$6 and produces 50 parts each hour on a machine with a \$14-per-hour burden rate. The raw material in each part costs \$0.50, and if spoiled its scrap value is \$0.09. This operator has a spoilage rate of 10 percent.

Operator 2 does the identical job for the same pay and has the same spoilage rate, but works slower and always detects a ruined part by the time the operation that produces it is 50 percent complete.

How many parts per hour should operator 2 produce to have the same unit cost as operator 1?

---

# PRODUCTIVITY

Output divided by input is the trademark of engineering studies. Physical efficiency is a dimensionless number that approaches but never quite reaches 100 percent. A comparable measure of performance used by engineering economists is ***financial efficiency***. This ratio of input to output is also usually dimensionless, but it is expected to exceed 100 percent. It can take the form of

$$\text{Financial efficiency} = \left(\frac{\text{worth}}{\text{cost}} \quad \text{or} \quad \frac{\text{receipts}}{\text{disbursements}}\right)100\%$$

which shows why the ratio has to be greater than 100 percent for an undertaking to be profitable.

## Measurement

Productivity measurements are similar to efficiency ratings, except the ratios are not necessarily dimensionless. Productivity is a criterion of performance for an individual operation in a micro analysis and a measure of performance for an entire nation in a macro analysis, or it can be applied to collected entities between the extremes. Ratios used for relatively small systems take the general form of total returns divided by the conversion costs responsible for the returns:

$$\text{Productivity} = \frac{\text{total sales}}{\text{sales expenses}} \quad \text{or} \quad \frac{\text{patient bed-days}}{\text{staffing hours}}$$

$$= \frac{\text{papers processed}}{\text{worker-hours worked}} \quad \text{or} \quad \frac{\text{materials handled}}{\text{worker-hours worked}}$$

The productivity of larger systems is measured by statistics that best represent the segment of the economy being analyzed. A ratio that recognizes the contribution of capital investments to output is

$$\text{Productivity} = \frac{\text{total annual sales}}{\text{number of employees} + \dfrac{\text{capital investment}}{\text{average annual earnings/employee}}}$$

Service industries can be measured by a ratio that discloses the net annual output per employee:

$$\text{Productivity} = \frac{\text{value added/year}}{\text{number of employees}}$$

where "value added" is the increased worth of a commodity due to the service performed.

An example of productivity analysis for a firm is shown in Figure 5.7. The purpose of the analysis is to observe the effects of changes, from one year to the next, of cost categories in relation to revenue. Receipts and disbursements for two consecutive years' operations are shown. All figures are converted to ***base-year*** values to cancel the effect of inflation on the comparison. With 1974 as the base year which has a designated value of 100, a 23 percent increase in the price of certain materials between 1974 and 1976 leads to a price index of 123 for those materials. That is, the 1976 price for the materials would be divided by 123/100 = 1.23 to reveal how much the same materials would have cost in 1974. Price-index figures for different cost categories are published to facilitate constant-dollar comparisons.

PRODUCTIVITY ANALYSIS OF ANNUAL OPERATIONS

| | 1975 | | | 1976 | | | |
|---|---|---|---|---|---|---|---|
| | *Current Dollars* | *Price Index (1974 = 100)* | *Constant (1974) Dollars* | *Constant Dollars* | *Price Index (1974 = 100)* | *Constant (1974) Dollars* | *Productivity Rise (1975 to 1976)* |
| Net sales (1) | $2,459,000 | 111 | $2,215,000 | $3,043,000 | 123 | $2,474,000 | |
| Labor (2) | 681,000 | 109 | 625,000 | 852,000 | 122 | 698,000 | |
| Materials (3) | 1,065,000 | 112 | 951,000 | 1,304,000 | 128 | 1,019,000 | |
| Services (4) | 344,000 | 108 | 319,000 | 388,000 | 120 | 323,000 | |
| Depreciation (5) | | | 104,000 | | | 131,000 | |
| Total inputs (6) [(2) + (3) + (4) + (5)] | | | 1,999,000 | | | 2,171,000 | |
| Labor-factor effect (7) [(1) − (3) − (4)] | | | 945,000 | | | 1,132,000 | |
| Factor input (8) [(2) + (5)] | | | $729,000 | | | 829,000 | |
| Labor productivity (9) [(7)/(2)] | | | 1.51 | | | 1.62 | 7.3% |
| Factor productivity (10) [(7)/(8)] | | | 1.30 | | | 1.37 | 5.4% |
| Total productivity (11) [(1)/(6)] | | | 1.11 | | | 1.14 | 2.7% |

**FIGURE 5.7** Breakdown of input and output data from annual financial statements used for productivity analysis.

The firm described by the financial data in Figure 5.7 experienced a better year in 1976 than in 1975. Sales rose at a rate faster than inputs to provide a total productivity improvement of 2.7 percent, calculated as

$$\frac{(1976 - 1975) \text{ constant dollars}}{1975 \text{ constant dollars}} 100\% = \frac{1.14 - 1.11}{1.11} 100\% = 2.7\%$$

The high rate of labor productivity improvement is notable and suggests attention now be given to the steeply climbing material costs which held the total productivity rise below that of the labor productivity gain. A better perspective is obtained by viewing each year's performance in relation to previous years. Shifts either up or down from the trend should be checked to find the source, either good or bad.

## Improvement

From 1890 to 1945, productivity in the United States rose at a fairly uniform annual rate of 2 percent. The rate increased to 3 percent for the next 25 years but declined to about the former level in the 1970s. The first full-year productivity decline, recorded in 1974, was followed by a strong rise in 1975. The output per worker-hour ratio used in the national index is a composite of representative productivity figures from the manufacturing, service, and government sectors.

Efforts to improve productivity are usually directed toward personnel practices and technological advances. (See Distribution of Productivity Gains in Extension 5.1.) New technology has contributed a major share of the productivity gains in the last three decades; the continuing problem is to find investment capital to take advantage of existing technology and to support budgets for further research and development. There is less agreement on how to improve personnel productivity. But there is no shortage of suggestions. Among the practices advocated to improve worker productivity are to:

- Have shorter 3- or 4-day workweeks by lengthening shifts to 10 to 13 hours
- Tie wages more closely to output and use merit awards
- Develop and utilize more standard times in service industries
- Redesign the content of jobs to make them more interesting and challenging
- Improve communications to encourage everyone to work toward the same desired objectives

The list could go on and on because there are so many factors that affect on-job performance. Engineers cannot ignore these "managerial problems" because they design the work environment, set the way jobs should be done, arrange schedules, and supervise operations. Even more convincing is the recognition that most engineering positions have management responsibilities, and engineers are a prime source of managerial talent.

---

**Economy Exercise 5-7** "By 1980, over half of payrolls are likely to be white collar workers. The general level of efficiency in white collar areas averages about 50%, with the clerical areas notably poorer than the average. A further statistic suggests that in many clerical departments, 75% of the work is done by 25% of the people—close to Pareto's 80/20 law."*

The quoted conditions portray an alarming situation, one with a huge potential for productivity improvement. Can engineering activities help tap this source of productivity gains?

## RATING INTANGIBLE FACTORS

Numbers have been called the "language of engineering." Every engineering student is well aware of the emphasis given to mathematics and the sometimes bewildering variety of number manipulations used in solving engineering problems. What tends to be overlooked is the difficulty of acquiring those numbers. "Hard" data, ratio scale numbers for receipts and disbursements, can usually be obtained, although diligent digging may be necessary. "Soft" data, composed of subjective opinions, are normally easy to obtain, but their reliability is questionable. Both hard and soft data are involved in most comparisons.

Many factors that affect a decision have no natural measures. How can you measure the relative *attractiveness* of two designs—by the number of beauty points they possess? If so, how do you define a "beaut"? The evaluation of intangibles is important because the final decision could hinge on the value placed on a factor such as attractiveness.

No ideal method has yet been devised to quantify intangibles. The choice is to use imperfect methods or ignore the intangible aspects in quantitatively evaluating economic decisions. Where intangibles have little influence, neglect is reasonable. When there are important subjective factors involved, imperfect methods are better than nothing because they at least expose the opinions to formal scrutiny.

### Ordinal Scaling

A simple order scale ranks every item in a list in order of preference. It works fine when there are just two alternatives to be subjectively rated for a single criterion. Then it is merely a choice of judging which alternative is more satisfying. However, an order scale does not measure the degree of preference; it is only a listing in preferential order. If three alternatives, $X$, $Y$, and $Z$, are respectively ranked in order 1, 2, and 3, the ranker expresses preference for $X$ over $Y$ and $Y$ over $Z$. But there is no clue as to how much $X$ is preferred to $Y$ and $Z$, or why it is preferred.

**Economy Exercise 5-8** Try ranking your three favorite sports. Then stop and ask yourself why you ranked them as you did. For the sake of argument, say the ranking came out skiing, tennis, and jogging. You reason that skiing is more thrilling than tennis, tennis is more fun than jogging, but jogging is handier and cheaper than skiing. How do you get out of this circular reasoning, called *intransivity*, where $X$ is preferred to $Y$ which is preferred to $Z$, and $Z$ is preferred to $X$?

*From G. McBeath, ***Productivity through People,*** Wiley, New York, 1974.

After narrowing the purpose of ranking to a single merit such as more useful than, more convenient than, or easier to understand than, all methods of ordinal ranking are about the same. Whether starting from a "most preferred" level and working down the list or starting from the bottom up, ranking is essentially by pairs.

Assume there are five alternatives to be ranked for the same criterion: $V, W, X, Y,$ and $Z$. As shown in Figure 5.8, all possible pairings are listed on the "steps." There will always be one step fewer than the number of items to be compared. The sequence in which the items are listed is purely arbitrary. In the first step, $V$ is paired with $W, X, Y,$ and $Z$. In the second step, $V$ is dropped, and the next item, $W$, is compared in turn to the remaining items. The procedure is continued until a single pair remains.

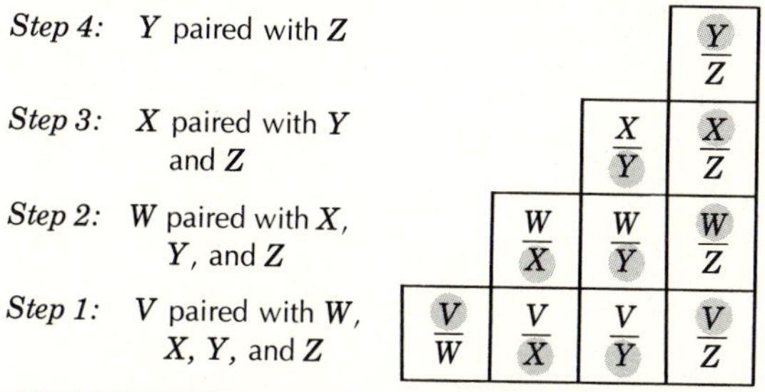

|  |  | | | |
|---|---|---|---|---|
| *Step 4:* $Y$ paired with $Z$ | | | | Y/Z |
| *Step 3:* $X$ paired with $Y$ and $Z$ | | | X/Y | X/Z |
| *Step 2:* $W$ paired with $X$, $Y$, and $Z$ | | W/X | W/Y | W/Z |
| *Step 1:* $V$ paired with $W$, $X$, $Y$, and $Z$ | V/W | V/X | V/Y | V/Z |

| *Alternative* | *Number of Circles* | *Rank* |
|---|---|---|
| V | 2 | 3 |
| W | 1 | 4 |
| X | 3 | 2 |
| Y | 4 | 1 |
| Z | 0 | 5 |

**FIGURE 5.8** Step-by-step comparison of alternatives to develop an ordinal scale ranking. Circles indicate the preference among pairs at each step. The tabulation shows that the ordinal scale for the five alternatives is $Y$ over $X$ over $V$ over $W$ over $Z$.

Ranking is accomplished by circling the preferred item in each comparison. In the first step, if $V$ is preferred to $W$ and $Z$, it is circled for these pairs. Meanwhile, $X$ and $Y$ are circled to show their rank above $V$. The formal ordering results from a tabulation of the number of times each item is preferred in the complete pairwise comparison. In the example, item $Y$ was circled four times to receive the top rank, and the never preferred item $Z$ is relegated to the last position.

The tempting violation of ordinal scaling is to read into an ordering a certain mathematical spacing between entries. The simple order of $X > Z > W > Y$ (where $>$ is read "is preferred to") could represent a number set of $100 > 99 > 98 > 2$ or $41 > 10 > 3 > 2$. The lack of specific intervals rules out any arithmetic operations.

## Interval Scaling

An interval scale is the next improvement from an order scale. This type of scale provides a relative measure of preference in the same way a thermometer measures relative warmth. An interval scale is a big improvement over ranking, but it still cannot be used like a ratio scale of distance or weight. This limitation stems from the lack of a natural zero. A zero in a ratio scale has a universal meaning; a zero distance or a zero weight means the same thing to everyone. A zero temperature can convey different meanings according to the type of interval scaling employed: Fahrenheit or Celsius. But once this zero value is understood, both temperature scales use standardized units of measurements which allow certain arithmetic operations, such as averaging, to be performed with the scaled values. Three methods for developing an interval scale are described on the next four pages.

## 1 CHURCHMAN-ACKOFF METHOD*

Churchman and Ackoff offer a procedure for quantifying intangibles in which the developed values are assumed to be additive. The rater is asked first to rank the items and then to assign numbers between 1.0 and 0.0 to alternative outcomes according to the approximate intensity of preference. Thus, a rating for outcomes from alternatives $W$, $X$, $Y$, and $Z$ might appear as

| $X$ | $Z$ | $W$ | $Y$ |
|---|---|---|---|
| 1.0 | 0.8 | 0.4 | 0.3 |

Now the sum of the values for $Z$, $W$, and $Y$ ($0.8 + 0.4 + 0.3 = 1.5$) is compared with the rating for $X$ (1.0). In order to show a distinct preference for $X$, its rating must exceed the sum of all lower-ranked ratings ($X > Z + W + Y$).† If the ratings do not conform to the rule, they are changed as little as possible in making them conform. The new value assignment might be

| $X$ | $Z$ | $W$ | $Y$ |
|---|---|---|---|
| 1.0 | 0.6 | 0.2 | 0.1 |

where $1.0 > 0.6 + 0.2 + 0.1$. Next the value for $Z$ is compared to the sum of $W$ and $Y$. The values above confirm a preference for $Z$, since $0.6 > 0.2 + 0.1$. The sequence ends with a preference shown for $W$ over $Y$, with $0.2 > 0.1$.

There are many sets of numbers that conform to the procedure and show the same order of preference but different intervals:

| $X$ | $Z$ | $W$ | $Y$ |
|---|---|---|---|
| 1.00 | 0.97 | 0.02 | 0.01 |
| 1.00 | 0:34 | 0.32 | 0.01 |
| 1.00 | 0.04 | 0.02 | 0.01 |

The procedure by itself does not assure that a legitimate interval scale has been developed. It systematizes the judgment process, but accuracy is still a function of the rater's conscientiousness.

## 2 STANDARD GAMBLE METHOD

Another procedure designed to yield an interval scale is called the *standard gamble method*. The top and bottom levels of the scale are mentally fixed by visualizing the perfect outcome of the criterion for a 1.0 rating, and the worst possible outcome for a 0.0 rating. Then the alternative being rated is compared to the extreme examples. The comparison is

*Discussed in R. L. Ackoff and M. W. Sasieni, *Fundamentals of Operations Research*, Wiley, New York, 1968.
†> is used here in its usual mathematical sense, to indicate a quantity relationship.

made like a lottery: The rater selects acceptable odds for a gamble between having a perfect outcome (1.0) against the worst outcome (0.0) *or* having the certain outcome of the alternative. The mental gymnastics required to conduct this mental lottery are difficult to master, but the scale boundaries for the best and worst outcomes make the ratings comparable for all alternatives.

To further describe the standard gamble method, assume graduate schools are being compared. One of the criteria is ***prestige,*** an attribute with no natural measurements. The first step is to select the most prestigious school imaginable, and give it a rating of 1.0. Next select a school with the least possible prestige for the 0.0 rated outcome. The best and worst limits do not have to be practical outcomes; that is, it might be impossible to attend the most prestigious institution but it still sets the upper limit. Then a theoretical lottery matches the preference for the top school (1.0) over the lowest (0.0) against surely attending the school being rated.

The lottery takes the form of a specific query aimed at each school being rated: "What probabilities of going to the 1.0 rated school instead of the 0.0 school would I accept to make the gamble equivalent to surely going to school $X$ ($X$ is the school being rated)." An answer of 0.4 indicates indifference between attending school $X$ and having 4 chances in 10 of attending the top school (which means there are 6 chances in 10 of attending the worst school). A rating of 0.5 shows no preference between school $X$ and a 50 percent chance of going to either the top or bottom school. The selected probabilities become the ratings for each alternative. As shown in Figure 5.9, multiversity, with a rating of 0.9, is preferred over the other two alternatives by the intervals given by the lotteries.

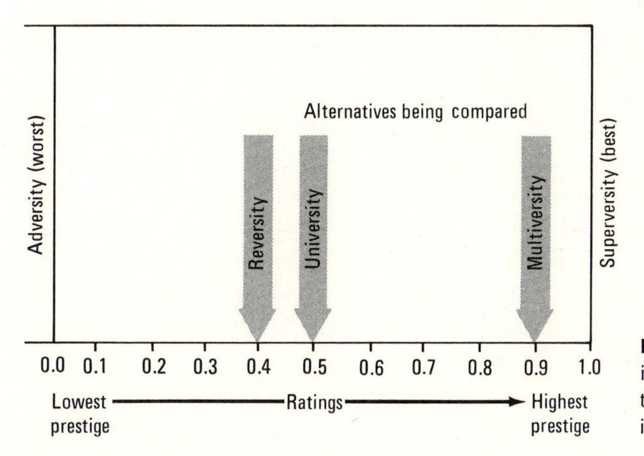


**FIGURE 5.9** Interval scale ratings developed to compare three graduate schools according to their prestige.

## 3 RATING FORMS

A standardized rating form which has written descriptions of each level of desirability is the most commonly used method for rating intangibles. The scales typically run from 0 to 10 with explanations of the attributes expected at each interval. Well-composed rating forms define, in easily understood language, the outcome that qualifies an alternative for each numbered rating.

Personnel ratings are often made on standardized appraisal forms. A satirical form is innocently offered in Figure 5.10. More serious versions are widely used. A rater is asked to assign a number or choose a proficiency level that describes the person being rated. The selection of a fitting description fixes a number to a particular criterion of performance. Resulting numbers are collectively taken as a representative measure of stature for the person being ranked.

EMPLOYEE APPRAISAL GUIDE

| *Ranking* | *Phenomenal* 4 | *Marvelous* 3 | *Good* 2 | *Not so Good* 1 | *Pathetic* 0 |
|---|---|---|---|---|---|
| Competitiveness | Slays giants | Holds his own against giants | Holds his own against equals | Runs from midgets | Gets caught by midgets |
| Personal appearance | Could be a professional model | Could model but wouldn't be paid much | Could model as the "before" | Goes unnoticed in a crowd | Panics a crowd if noticed |
| Leadership | Walks on water consistently | Walks on water in emergencies | Wades through water | Gets caught in hot water | Passes water in emergencies |
| Intelligence | Knows everything | Knows a lot | Knows enough | Knows nothing | Forgets what he never knew |
| Communication | Talks to big shots | Talks to little shots | Talks to himself | Argues with himself | Loses those arguments |

**FIGURE 5.10** A spoof of an employee-ranking system that makes just enough sense to be frightening.

Rating forms with similar characteristics have been developed to evaluate recurring decision situations. For example, government agencies engaged in research solicit bids from internal and outside investigators for conducting studies. A request for proposals (RFP) contains a statement of the technical requirements of the work and requests bidders to provide cost estimates, time schedules, and proof of competence. The replies are then evaluated by a board according to how well they meet the criteria of acceptance. A typical guideline for assigning numerical ratings for each criterion is given below.

| *Score* | | | | *Description* |
|---|---|---|---|---|
| 10 | 9 | | Very good | Has a high probability of exceeding all the requirements expressed in the RFP for the criterion |
| 8 | 7 | 6 | Normal | Will most likely meet the minimum requirements and scope of work established in the RFP |
| 5 | 4 | 3 | Below normal | May fail to meet the stated minimum requirements but is of such a nature that it has correction potential |
| 2 | 1 | | Unacceptable | Cannot be expected to meet the stated minimum requirements and is of such a nature that drastic revision is necessary for correction |

While using an appraisal form, it is important to keep referring to a mental standard that conforms to each level. In the RFP evaluation, the standards are defined in writing. In personnel rating forms the standards result from experiences with the performance of people who were previously rated in each category. Each rater likely has a different interpretation of what constitutes perfection, based on personal views and past exposures. It is not vital that all raters have the same absolute limits for their interval scale; it is vital that they are consistent in applying their own scale among alternatives. Some raters believe their consistency is improved by initially giving each criterion a rating of 10 and then subtracting points as the alternatives are compared.

A review of past ratings can improve future ratings. A numerical rating system is only as good as the rationale exercised in its use. A rater should be prepared to convince a questioner that the judgment was correct. Since intangible judgments are necessarily fragile, they deserve to be handled with care.

## CONDITIONAL DECISION-TREE COMPARISONS

Certain characteristics of new designs cannot be expressed in ratio scale numbers. Such characteristics are seldom independent; that is, the "charm" of a design for, say, an apartment, could depend on its furnishings which depend on its size. When a rating is based on a chain of choices with interacting effects, the rating is called a ***conditional decision.***

Several alternatives are usually available at each stage of development or in each area of design. Preliminary evaluations of various combinations of methods or components are particularly suitable to evaluation by interval scales. This approach acts as a filter. The most promising designs are identified for a more thorough examination.

A ***decision-tree*** format is helpful in making conditional decisions. In this section only immediate decisions are considered. Chapter 14 explains the use of decision trees that display the effect of time on outcomes. Figure 5.11 illustrates possible combinations available for the specification of a design component. There are three characteristics to be specified: material, shape, and capacity. The choices within the limits of these characteristics have been narrowed to three types of material, two configurations of shape, and three

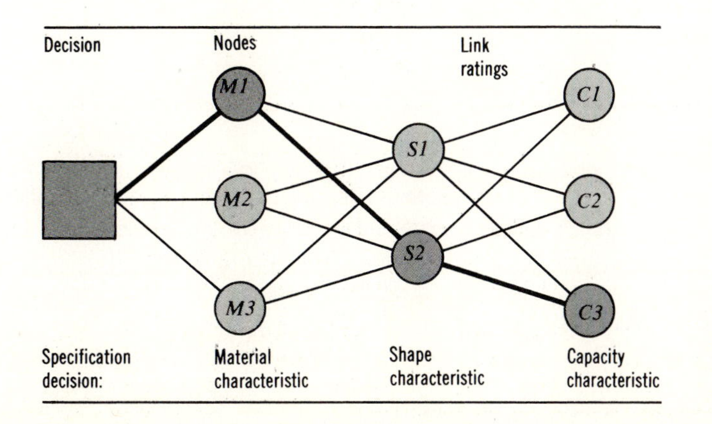


**FIGURE 5.11** Design characteristics displayed on a conditional decision tree.

capacity sizes. The square symbol at the left indicates that a decision is to be made among the paths leading to the right. Each node (circle) represents a design quality. The nodes are grouped in columns according to common characteristics. A complete path from the square to a node in the last column represents one alternative.

A conditional rating is given to each link between nodes. That is, each characteristic is rated with respect to its value in relation to the other characteristics in its path. One specification for the component could be material 1 ($M1$), shape 2 ($S2$), and capacity 3 ($C3$). The ratings for the material characteristic would depend only on a preference between types $M1$, $M2$, and $M3$, but the link ratings between material and shape would be based on the value of a given shape produced from a given material. Thus the link rating between $M1$ and $S2$ indicates the relative value of shape 2 made from material 1. In turn, the link value between $C3$ and $S2$ is the rating for capacity 3, given that $M1$ and $S2$ are part of the path. There are 18 ($3 \times 2 \times 3$) unique combinations or paths from which to select a specification.

The bold-lined path in Figure 5.11 represents one design alternative. Rearranging the compact but intertwined branches delimits the alternatives. Figure 5.12 shows an *extensive* form of the decision tree for the same component with the addition of link ratings. The three characteristics are still segregated by columns, but nodes are provided for each link. This arrangement produces a cluster of nodes for each quality within the characteristic. Link ratings, developed by interval scaling procedures, add up to 1.0 for each cluster. Whenever a quality has a zero rating (completely unacceptable), that path is obviously terminated ($M1$-$S2$, for example).

The evaluation procedure is simply a forward pass through each branch to determine the path with the largest product. The comparison thus uses a *multiplicative* model. According to the expressed ratings, the path $M2$-$S1$-$C1$ would be the preferred specification, with a product rating of 0.168.

The task of determining the link values is both the main advantage and the main limitation of this approach to primary comparisons. Each conditional rating must reflect the value of that quality in relation to all preceding qualities. Even for a relatively small problem, the rating procedure becomes onerous and complicated. Still, this effort makes the procedure worthwhile, in that it encourages identification and consideration of interdependencies. This close scrutiny may reveal new alternatives, or it may eliminate some branches from further consideration. The final evaluation tends to indicate which alternatives warrant further investigation.

## MIXED-RATING COMPARISONS

A legitimate question at this stage of our rating discussion is: "What do we do if some of the ratings are on an interval scale and others are on a ratio scale?" A corollary to this question is: "How are alternatives compared when there are several bases for equivalent outcomes?" The answers are found through dimensionless numbers.

Again, several models are available for dimensionless comparisons. The best-known version is an *additive* model. It is a procedure in which all ratio scale ratings are converted to dimensionless numbers; these and interval scale numbers are multiplied by an impor-

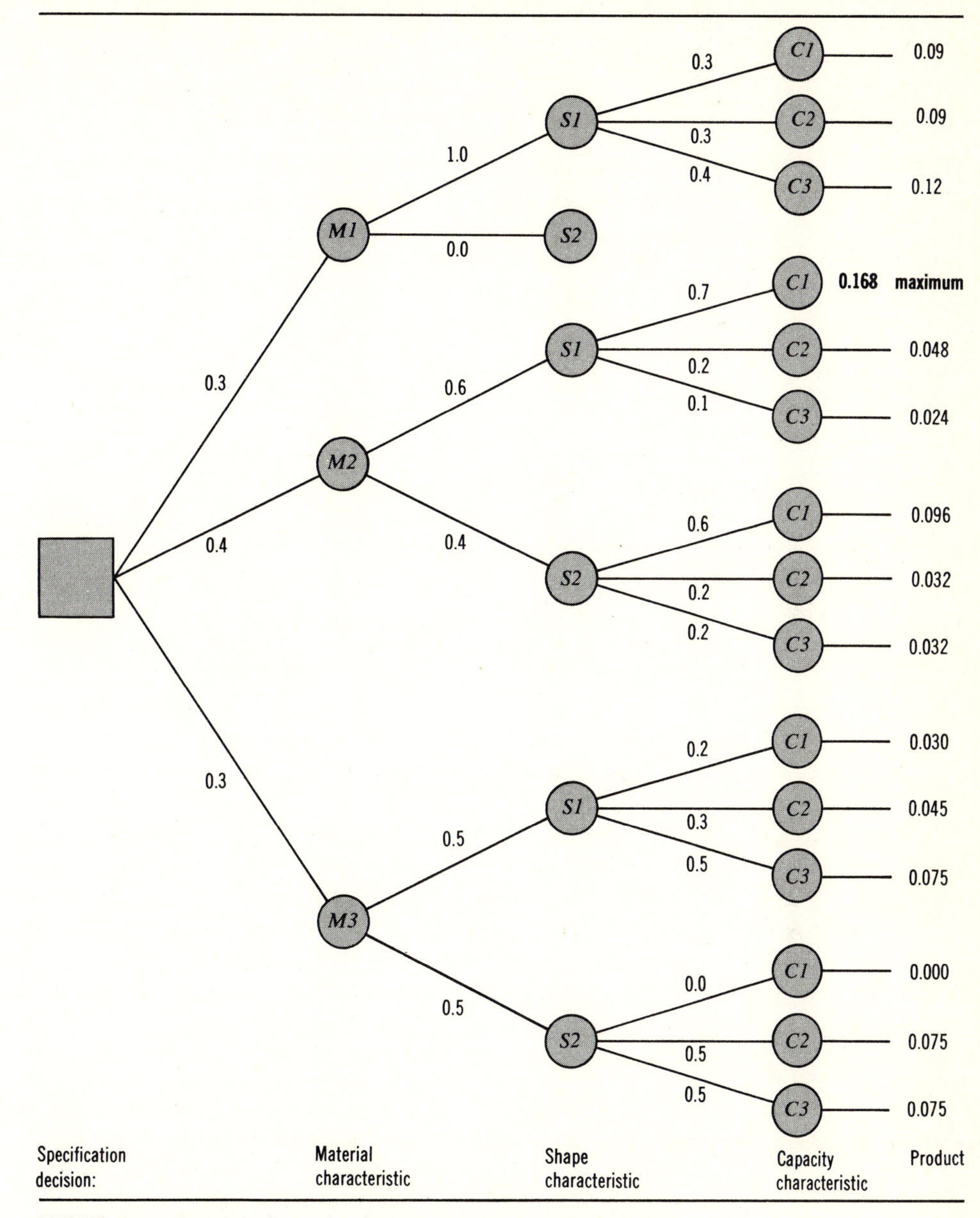


**FIGURE 5.12** Extensive form of a decision tree for a design evaluation.

tance factor associated with each criterion; and the products are summed to give a single number that scores the alternative.

Consider the comparison of three prototypes for a new bumper-jack design. Five independent criteria have been selected for evaluation: safety, cost, appearance, weight, and reliability. The relative importance of each criterion is determined as an interval scale rating. With 10 representing the utmost importance, the criteria have the ratings shown in Table 5.2. While rating the importance, a cutoff score may be recognized that makes an alternative unacceptable. For example, $20 is considered the top limit for a jack's cost, and any alternative that exceeds this cutoff level is eliminated regardless of how well it scores on the other criteria.

**TABLE 5.2** Criteria and comparison values for evaluating two bumper-jack designs.

| | *Safety* | *Appearance* | *Cost* | *Weight* | *Reliability* |
|---|---|---|---|---|---|
| Jack 1 | 8 | 4 | $17.56 | 9.7 | 0.96 |
| Jack 2 | 7 | 9 | $ 9.95 | 6.2 | 0.81 |
| Jack 3 | 7 | 7 | $14.47 | 6.0 | 0.90 |
| Importance | 9 | 4 | 8 | 3 | 6 |
| *Cutoff* | | | < $20 | | |

Cost, weight, and reliability are ratio scale measurements. Lower cost and weight are preferred, while a higher reliability is desired. All criteria measurements are converted to dimensionless numbers by taking a ratio of each criterion value to the best value available among the alternatives for that criterion. The best value is assumed to have a 10 rating. Ratios used to convert the reliability figures to a 10-point scale are

| *Jack 1* | *Jack 2* | *Jack 3* |
|---|---|---|
| $\frac{0.96}{0.96} 10 = 10$ | $\frac{0.81}{0.96} 10 = 8.4$ | $\frac{0.90}{0.96} 10 = 9.4$ |

since 0.96 is the highest reliability rating among the alternatives.

The converted cost scores are calculated as

| *Jack 1* | *Jack 2* | *Jack 3* |
|---|---|---|
| $\frac{\$9.95}{\$17.56} 10 = 5.7$ | $\frac{\$9.95}{\$9.95} 10 = 10$ | $\frac{\$9.95}{\$14.47} 10 = 6.9$ |

since $9.95 is the lowest cost for any alternative. Note that none of the costs exceeded the $20 cutoff level which would have eliminated an alternative from further consideration. Safety and appearance need no conversion because they are already rated on an interval scale with a top score of 10.

The final step is to multiply the criteria ratings by their respective importance factors and add the resulting products for each alternative. The alternative with the highest total score is the winner. As shown in Table 5.3, jack 2, with a score of 258, is preferred over the other two prototypes.

## SUMMARY

Ways to economize are as diverse as they are numerous. ***Value engineering*** involves a systematic search for ways to increase value by discovering less expensive means to fulfill necessary functions. More efficient use of materials is encouraged by ***value analysis*** of

| ALTERNATIVES | | JACK 1 | | | JACK 2 | | | JACK 3 | | |
|---|---|---|---|---|---|---|---|---|---|---|
| *Criteria* | *I* | *Observation* | *Rate* | *R × I* | *Observation* | *Rate* | *R × I* | *Observation* | *Rate* | *R × I* |
| Safety | 9 | | 8 | 72 | | 7 | 63 | | 7 | 63 |
| Appearance | 4 | | 4 | 16 | | 9 | 36 | | 7 | 28 |
| Cost (< $20) | 8 | $17.56 | 5.7 | 46 | $9.95 | 10 | 80 | $14.47 | 6.9 | 55 |
| Weight, lb | 3 | 9.7 | 6.2 | 19 | 6.2 | 9.7 | 29 | 6.0 | 10 | 30 |
| Reliability | 6 | 0.96 | 10 | 60 | 0.81 | 8.4 | 50 | 0.90 | 9.4 | 56 |
| *Totals* | | *Jack 1* | | 213 | *Jack 2* | | 258 | *Jack 3* | | 232 |

**TABLE 5.3** Typical format by which alternatives are scored according to the sum of the products from dimensionless criterion ratings $R$ multiplied by their relative importance $I$.

purchases and *design-to-cost* programs that constrain designs to conform to budgetary limits. Calculation of unit cost reveals the most advantageous combination of operators and machines. Sequencing rules identify work flows that conserve time, and *life-cycle analysis* seeks to minimize total cost over the life of a project or product. Programs to improve personnel practices are evaluated by the savings they produce compared to presentation expenses.

Comparisons based on inputs are definitive only if the outputs from all the alternatives are essentially equivalent.

Productivity is usually measured by the ratio of output value to the labor required to produce the output. Year-to-year comparisons utilize base-year computations to cancel the effects of inflation. Efforts to improve productivity are primarily directed toward personnel practices and the utilization of technological advances.

Giving the proper weight to intangible factors is one of the toughest problems in making economic comparisons. Rating factors on an ordinal scale gives only the order of preference. An interval scale shows the relative positions of the factors rated. Three procedures designed to yield interval scaling are the Churchman-Ackoff method, standard gamble, and rating forms.

A *conditional decision tree* is a multiplicative model that utilizes ratings based on a chain of relationships. Dimensionless numbers are developed in the *mixed-rating comparison* model by using fractions to convert ratio scale numbers to a 10-point scale. Importance factors are assigned to each criterion by which the alternatives are being evaluated. The sum of the 10-point scaled criteria ratings multiplied by their respective importance ratings gives a score for measuring alternatives.

## Discussion of Economy Exercises

**EE 5-1** The Vanpool plan has been exceptionally well received in actual practice. Initial reports indicate the vans are popular because they are comfortable (air conditioned, carpeted, stereo equipped, etc.) and the passengers enjoy the companionship (some drivers even installed a bar in the back to serve coffee in the morning and cocktails on the way home). The T-chart comparison of commuting by individual cars against Vanpool commuting might appear as follows:

| *Vanpool Plan versus Private Cars* | *Better* | *Worse* |
|---|---|---|
| Conservation of energy (fuel and materials) | ✓ | |
| Cost of commuting (to commuters) | ✓ | |
| Flexibility of transportation (emergencies) | | ✓ |
| Condition of commuters arriving for work | ✓ | |
| Convenience (to commuters) | | ✓ |
| Parking facilities (adequacy and upkeep) | ✓ | |

**EE 5-2** Off the top of the $1000 comes the usual profit margin for the company, say 10 percent. With the remaining $900 you have to pay for materials and construction. Design variables include asphalt, concrete, and gravel for materials, the needed width of the driveway, and timing that would allow lower equipment and labor costs by delaying the project until additional roadwork was needed in the same area. The options would be explained to the homeowner to let her select the trade-off she prefers between, say, a narrower driveway made of concrete and done immediately versus a wider driveway of asphalt completed sometime within a month.

**EE 5-3** Unit costs for the three ways to meet the annual demand for 45,000 units with the two types of machines are calculated as

$$\text{Unit cost (three } S\text{-type machines)} = \frac{(\$10.80)(3)}{10 \times 3} + \frac{(\$7{,}500)(3)}{45{,}000} = \$1.58/\text{unit}$$

$$\text{Unit cost (two } L\text{-type machines)} = \frac{(\$11.70)(2)}{15 \times 2} + \frac{(\$16{,}800)(2)}{45{,}000} = \$1.53/\text{unit}$$

The unit cost using one $S$-type machine plus one $L$-type machine is determined by first checking which machine has the lowest unit cost when used to capacity:

$$\text{Unit cost}_{S\text{-type}} = \frac{\$10.80}{10} + \frac{\$7{,}500}{10 \times 8 \times 250} = \$1.455/\text{unit}$$

$$\text{Unit cost}_{L\text{-type}} = \frac{\$11.70}{15} + \frac{\$16{,}800}{15 \times 8 \times 250} = \$1.34/\text{unit}$$

Thus, the $L$-type machine should produce 30,000 of the 45,000 units required, and the $S$-type machine the remainder, to give

$$\text{Average unit cost (one machine of each type)} = \frac{1}{2}\left(\frac{\$10.80}{10} + \frac{\$7{,}500}{15{,}000} + \$1.34\right)$$

$$= \$1.46/\text{unit}$$

which is the lowest-cost alternative.

**EE 5-4**

Step 1 is included in the problem statement.

Step 2 leads to the smallest processing time of 2 hours for report $W$ in the printing process.

Step 3 assigns report $W$ to the last position in the sequence of jobs: | | | | W |

Step 4 yields the reduced set of processing times by eliminating report $W$:

| *Report* | *X* | *Y* | *Z* |
|---|---|---|---|
| Preparation | 5 | 4 | 10 |
| Printing | 6 | 8 | 5 |

Step 2 is repeated to select the smallest processing time of 4 hours for report $Y$.

Step 3 assigns report $Y$ to the first position in the sequence, which now has designated beginning and end jobs: | Y | | | W |

Step 4 reduces the set to two processing times, and step 2 is repeated again to reveal a tie between 5 hours for preparing report $X$ and 5 hours for printing report $Z$. The tie is broken by an arbitrary choice of job $X$ in the first available sequence slot to give the final job sequence: | Y | X | Z | W |

In this example the same job sequence results from breaking the tie between $X$ and $Z$ in favor of $Z$ rather than $X$. In any case, the way a tie is broken will not alter the total elapsed time, although it can produce alternative optimal sequences. A detailed schedule for the reports is given in Table 5.3. The minimum elapsed time for completing the reports is 27 hours. If the facilities have no work other than the reports, the idle time would be 27 − 25 = 2 hours for the preparation, and (4 − 0) + (19 − 18) + (25 − 24) = 6 hours for the printing facility.

**TABLE 5.3** Total elapsed time for an optimal sequence of reports $W$, $X$, $Y$, and $Z$ through two processes.

| | PREPARATION | | | PRINTING | | |
|---|---|---|---|---|---|---|
| *Report* | *Total Time* | *Time In* | *Time Out* | *Total Time* | *Time In* | *Time Out* |
| *Y* | 4 | 0 | 4 | 8 | 4 | 12 |
| *X* | 5 | 4 | 9 | 6 | 12 | 18 |
| *Z* | 10 | 9 | 19 | 5 | 19 | 24 |
| *W* | 6 | 19 | 25 | 2 | 25 | 27 |

**EE 5-5** Using the program-selection formula to develop rating factors for the two proposals, we get

$$\textit{Down with Defects:} \quad \frac{\$1{,}000{,}000 \times 0.5}{\$100{,}000} 0.5 = 2.5$$

$$\textit{Avoid Accidents:} \quad \frac{\$265{,}000 - \$25{,}000}{\$60{,}000} 0.8 = 3.2$$

which indicates a preference for the safety program even though the potential savings are over twice as great for the defect-reduction program. The lower presentation cost and higher probability of success suggest the safety program be done first. It is usually not as productive to have more than one major improvement program underway at one time.

**EE 5-6** The equation for unit cost is

$$\text{Unit cost} = \frac{W}{N} + \frac{M}{N} + SR\frac{RM - SV + (W/N + M/N)K}{100\% - SR}$$

*where* $W$ = wage = \$6/hour
$M$ = machine burden = \$14/hour
$SV$ = scrap value = \$0.09/part
$K$ = percent of operation completed when scrap is detected = 100% for operator 1 and 50% for operator 2
$SR$ = scrap rate = 10%
$RM$ = raw materials = \$0.50/part
$N$ = units/hour = 50 for operator 1

The unit cost for operator 1 is

$$\frac{6}{50} + \frac{14}{50} + 0.10\frac{0.50 - 0.09 + (6/50 + 14/50)(1.0)}{1.0 - 0.10} = 0.40 + \frac{0.10}{0.90}(0.41 + 0.40)$$

$$\text{Unit cost} = \$0.40 + \$0.09 = \$0.49$$

Then, for operator 2 to have the same unit cost,

$$\$0.49 = \frac{6}{N} + \frac{14}{N} + 0.10\frac{0.50 - 0.09 + (6/N + 14/N)(0.5)}{1.0 - 0.10}$$

$$= \frac{\$20}{N} + \frac{0.10}{0.90}\left[0.41 + \left(\frac{\$20}{N}\right)(0.5)\right]$$

or

$$N \doteq \frac{21.111}{0.445} \doteq 47.5 \text{ parts}$$

which is about 97 percent of the output of operator 1.

**EE 5-7** It should first be realized that engineers are usually classified as white-collar workers. Whether they are subject to the liberal interpretation of Pareto's law that says 80 percent of the activities associated with any system are owed to 20 percent of the members of that system is debatable, but possible.

New data-handling devices and better office equipment are examples of engineering advances that assist clerical work. "Management engineers" are involved in setting work standards and improving work flow.

**EE 5-8** The exit from circular reasoning is found in more precise objectives. The intangible values of sports could be ranked according to fun per minute of activity, thrills per outing, fellowship enjoyed, satisfaction obtained, etc. Each of these qualities could be the basis for a different ranking episode. They cannot be mixed effectively within one rank. Only one dimension is implicit in a ranking episode. Each quality deserves its own trial.

## PROBLEMS

**5.1** A portable concrete batching plant is to be set up near a bridge construction site. Two sand sources which provide sand of the same fineness modulus are available. One source is 2 miles (3.22 kilometers) south of the bridge site, and the other is 4 miles (6.44 kilometers) north of the site. Coarse aggregate can be obtained from a pit 3 miles (4.83

Mix x miles north of bridge

kilometers) north of the bridge site. The design mix calls for twice as many pounds of coarse aggregate as sand for 1 cubic yard (0.765 cubic meter) of mixed concrete. Dump trucks haul rock and sand from their source to the batching plant, and mixer trucks haul the mixed concrete from the plant to the bridge site. The cost of hauling mixed concrete is twice the cost of hauling a similar weight of sand or coarse aggregate in a dump truck. There is room to set up the batching plant anywhere between the two sand sources. Where is the most economical location for the plant, and which sand source should be used?
*(Use north sand source with the batching plant at the bridge site.)*

**5.2** Value engineers are expected to think positively in order to encourage the generation of new ideas for improvements. Negative attitudes prevail among people who are satisfied with present conditions and get disturbed when someone suggests change. Some customary comments that discourage changes are listed below. What response to each one could a positive-thinking value engineer make so that the countercomment would make the negative thinker think again? For instance, the negative comment that "We're making a profit now, why change?" could elicit a VE response of "That's no reason not to make more."

**5.2a** Why change it? It has been done this way for years.
**5.2b** The customer will never agree to it!
**5.2c** It costs too much.
**5.2d** It's against company policy!
**5.2e** Plastics are too weak.

**5.3** A tie clasp issued to all new recruits in a branch of the military is chrome-plated, has a pivoted clasp, and contains a replica of the insignia of the military branch in the center. The description of the working function of the clasp could be "hold parts," where the parts are the tie and shirt. A secondary (or perhaps the basic) function is "create impression." Show how the creation of substitute means and the method of evaluation depend on the functional description of the item.

**5.3a** If the present tie clasp costs $2.93, determine a less expensive alternative based on the "hold parts" function.

**5.3b** How does the alternative from Problem 5.3*a* satisfy the function "create impression"?

**5.4** Four items have been singled out by the communications department as possibilities for value-engineering studies. The data accumulated for these items are shown in the table below.

| *Item* | *Estimated Yearly Savings* | *Estimated Study and Implementation Cost* | *Probability of Implementation* |
|---|---|---|---|
| 1 | $14,000 | $4000 | 0.9 |
| 2 | 6,500 | 1000 | 0.6 |
| 3 | 41,000 | 9000 | 0.7 |
| 4 | 3,300 | 600 | 0.8 |

**5.4a** According to the VE rating expression, which alternative should be selected?
*(Item 4)*

**5.4b** What are some other considerations that are not included in the data that could influence the decision?

**5.5** An appliance store has received a franchise to distribute a new type of recorder. Plans are being made to advertise the recorder by a direct-mail campaign of at least 14,000 flyers. A mimeograph machine is available at the store. If it is used, the flyers can be produced for \$0.02 apiece. However, an artist will have to be employed to make the line drawings of the recorder and to do the lettering. The art charge is \$195. Franked envelopes can be purchased for \$100 per thousand.

Another alternative is to have the flyers printed. Plates are supplied by the manufacturer, but printing costs will still amount to \$615. A slight change in the format will allow the flyers to be sent without envelopes. The alterations will cost \$60, and mailing expense will be \$0.06 per flyer.

A suitable and equivalent advertisement will result from either method. The addressing costs will be the same for both alternatives. Which one has the lowest unit cost?

*(Use printed flyers at \$0.108/pc.)*

**5.6** An operator earns \$6.25 per hour running a plastic molding machine that has an hourly burden rate of \$6.50. The production rate is 16 units per hour, of which 10 percent are rejected. Rejection takes place during a midprocess inspection when the operation is 60 percent complete. Raw-material costs are \$0.90 per piece, and there is no scrap value.

A plan is being considered under which the operator will receive a \$1 per hour pay increase to operate two machines. Average production should be 14 units per hour from each machine, but the spoilage rate will double and the operator will not have time for the midprocess inspection.

**5.6a** What is the unit cost for each situation?

**5.6b** What is the cost of producing a rejected unit when two machines are being operated?

**5.7** Develop a graph of unit costs versus units produced according to the data and conditions given in Economy Exercise 5-3. How does the graph refute the traditional belief that "the more you produce, the less expensive it is per unit"?

**5.8** The 4-day, 40-hour workweek has been proposed as an alternative to improve worker motivation and morale. It has been tried in several industries and many government agencies with mixed success. Develop a T-chart to compare the traditional 5-day week with a 4-day pattern, both for a 40-hour workweek in a manufacturing industry. Select criteria that are important to employees and employers, and evaluate the alternative by checking whether the 4-day pattern is better or worse from the standpoint of operating effectiveness for the firm, realizing that satisfied employees are usually more productive.

**5.9** An engineering architect is preparing plans for a prefabricated vacation home. The decision tree in Figure 5.13 shows the ratings placed on the characteristics of two tentative roof and ceiling designs. Ratings are based on expected consumer preferences as to simplicity of do-it-yourself construction, ease of maintenance, and final appearance. A rating of 1.0 is the highest possible. Which combination of design, structural materials, and finish materials is preferred?

*(D1-SM3-FM1, with a rating of 0.168)*

**5.10** A testing laboratory has seven samples to test for the same properties. Each test is in two parts which must follow the same order, because the samples are destroyed during the tests. A sample is always subjected to test 1 first. The laboratory has only one machine of

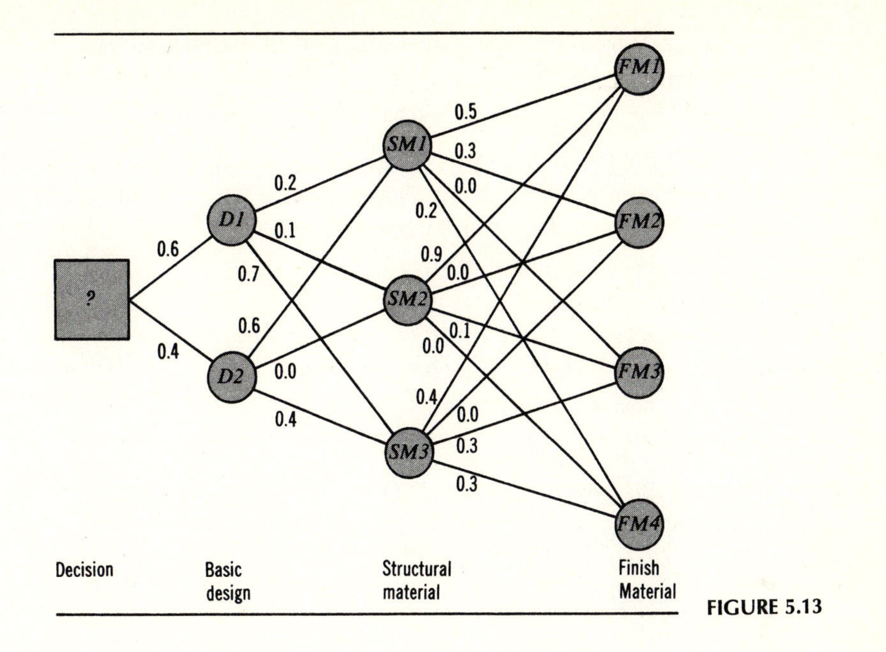


FIGURE 5.13

the type required for each part of the test. Based on the testing times below for each test, determine the order in which the samples should be tested to minimize the total time needed to complete all seven tests.

| *Sample* | *Test 1* | *Test 2* |
|---|---|---|
| 1 | 2 | 3 |
| 2 | 7 | 9 |
| 3 | 10 | 8 |
| 4 | 5 | 2 |
| 5 | 6 | 11 |
| 6 | 4 | 3 |
| 7 | 8 | 11 |

*(1-5-2-7-3-6-4; 54 time units)*

**5.11** A company received four rush orders to be processed through two departments. All other work for the departments is stopped while the rush orders are being completed. Each order passes from Department $A$ to Department $B$ with the following expected processing times in hours:

| | *1* | *2* | *3* | *4* |
|---|---|---|---|---|
| Department $A$ | 3 | 7 | 6 | 6 |
| Department $B$ | 2 | 5 | 10 | 4 |

**5.11a** What is the minimum elapsed time to complete all four orders?

**5.11b** What would the elapsed time be if Department *A* actually took 10 hours to complete order 2, and Department *B* finished order 2 in 2 hours?

**5.12** An operator earning $3 per hour runs a machine with a burden rate of $4.80 per hour. What is the unit cost when one operator runs two and three machines? The activity times, in minutes, for producing one piece are:

| | |
|---|---|
| Insert piece in machine | 0.60 |
| Remove finished piece | 0.30 |
| Inspect piece | 0.50 |
| File burr and set aside | 0.20 |
| Walk to next machine | 0.05 |
| Machine running time | 3.95 |

*(2 machines = $0.51/piece; 3 machines = $0.47/piece)*

**5.13** Two operators, each paid $5.50 per hour, operate two semiautomatic machines. Both operators must work together at all times. The element times, in minutes, for the work cycle are as follows:

| | |
|---|---|
| Load | 1.2 |
| Run | 3.4 |
| Unload | 0.7 |
| Inspect prior to loading | 0.4 (the part is on the worktable during inspection) |
| Inspect after completion | 0.8 (done while the machine is running) |

The machines are located next to each other, and transportation times are included in the above elements. If material costs $15.80 per part, direct machine cost is $24 per hour, and overhead is calculated to be 25 percent of labor, materials, and machine costs, what is the cost of each completed part?

**5.14** An electronic specialty company employs 60 people to fabricate miniature resistor assemblies. The workers average 2000 hours of work per year, with an average production of 16.2 assemblies per hour, of which 6 percent are defective. The top third of this group produce 19.1 assemblies per hour with only 3.4 percent defective. The cost to the company to rework each defective unit is $0.30.

All the work is paid at a straight piecework rate of $0.25 per assembly. Each employee uses a scope and special equipment which has an annual fixed cost of $612. Indirect costs and supervision amount to $740 per year per worker.

For the same total production, how much could be spent per year to select, train, and motivate the workers so that all of them would produce at the level of the top 20? (*Hint:* The original top 20 are part of the new, smaller assembly crew.)

*($26,023)*

**5.15** One year ago, a city-center bakery was close to going out of business, owing to competition from bakery sections built into the new shopping centers in the suburbs. Fewer people visited the downtown site of the bakery, and those who did were purchasing less to take home. The turnaround in business originated at a meeting of all employees called by the manager.

The purpose of the meeting was to urge the employees to further reduce costs or face the possibility of having the shop close. As the meeting progressed and the employees became aware of how serious the situation was, they started offering many ideas that the manager had never considered before. Suggestions ranged from an easier way to make butterflake rolls to opening a coffee-and-rolls sidewalk bar or a discount booth inside the store for day-old products.

The manager shrewdly recognized the productivity improvement potential of the suggestions and gave them full backing. Highlights of the resulting year's activities compared with the previous year and price-index data are listed below.

Total sales increased by 60 percent (including a 5 percent general increase due to inflation in the overall economy) to $480,000.

Cost of materials used increased by 50 percent (including a 10 percent industrywide rise in material expenses) to $150,000.

Labor costs increased by 10 percent (the same rate as the industrywide wage averages) to $110,000.

Annual depreciation charges jumped by $15,000 (owing to investments made for facility innovations) to $65,000.

Other expenses increased by 40 percent (including inflation) to $140,000.

The increased output with essentially the same number of worker-hours was due to improved methods instituted by the bakers and partially to new equipment. Sales increased more than material costs because waste was reduced.

Calculate the percentage increase in different productivity measures to determine the effect of the revised operations, and interpret the significance of the different ratios.

**5.16** The most widely used mousetrap is the spring-operated type that uses impact to kill the mouse. This kind has been used for many years and is quite efficient. Both the spring and triggering systems are almost the ultimate in simplicity and economy. However, there are certain disadvantages to this type of trap. It is dangerous because it is not selective. It can kill kittens and puppies, and it can hurt babies or adults who happen to touch the sensitive bait trigger. It can make quite a mess if the mouse bleeds or is cut in half. Setting the trap and removing the dead victim is not to the liking of many sensitive people.

In an attempt to have the world beat a path to your door by inventing a better mousetrap, assume the three designs shown on p. 146 are your best creations and you want to select one for a market trial.

**5.16a** List the criteria for evaluating the designs.

**5.16b** Weight the importance of the criteria, using either the Churchman-Ackoff or standard gamble method.

**5.16c** Determine a rating for each outcome of the criteria.

**5.16d** Decide which design is the most promising. Comment on the results of your calculations (sensitivity, confidence, etc.).

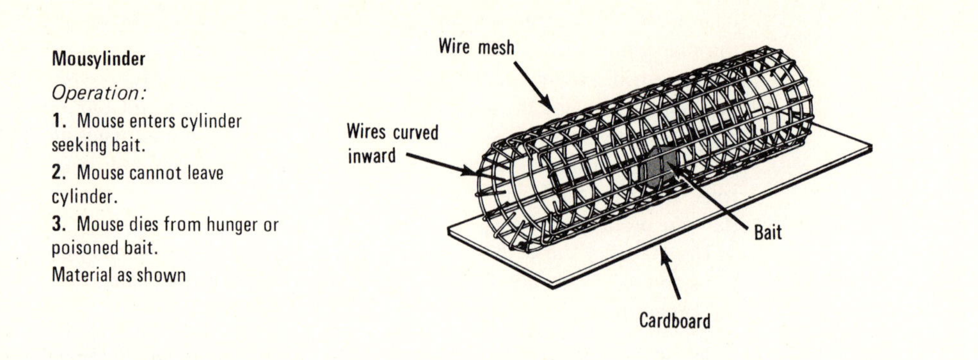


**Mousylinder**

*Operation:*

1. Mouse enters cylinder seeking bait.
2. Mouse cannot leave cylinder.
3. Mouse dies from hunger or poisoned bait.

Material as shown

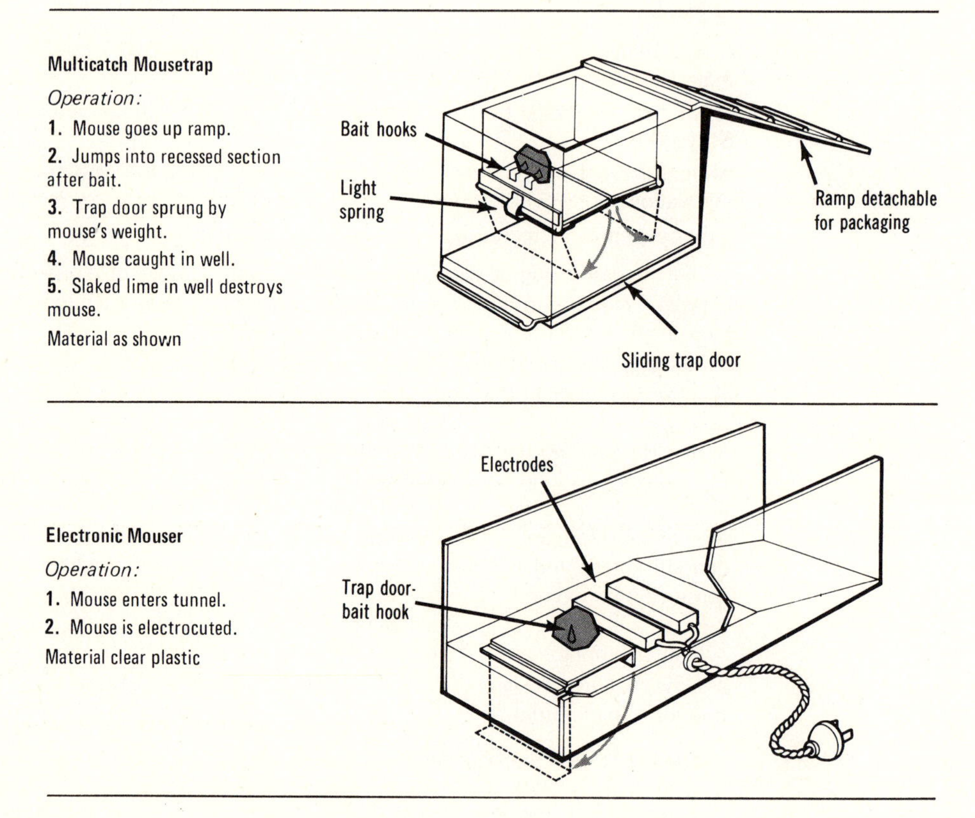


**Multicatch Mousetrap**

*Operation:*

1. Mouse goes up ramp.
2. Jumps into recessed section after bait.
3. Trap door sprung by mouse's weight.
4. Mouse caught in well.
5. Slaked lime in well destroys mouse.

Material as shown

**Electronic Mouser**

*Operation:*

1. Mouse enters tunnel.
2. Mouse is electrocuted.

Material clear plastic

**5.17** Three designs have been proposed for a new type of can opener. Careful studies have been conducted to evaluate the degree to which each design meets the desired criteria. The characteristics with familiar measurements were determined by design engineers. Opinions of many people were collected to obtain interval-scaled ratings for intangible characteristics. The results are shown on p. 147.

| *Criteria* | *Design A* | *Design B* | *Design C* |
|---|---|---|---|
| Cost (minimize) | $3.42 | $5.84 | $9.88 |
| Cleanability (minimize time) | 3.3 min | 1.8 min | 3.0 min |
| Reliability (maximize) | 0.78 | 0.91 | 0.99 |
| Size (minimize) | 102 in³ | 102 in³ | 320 in³ |
| Appearance (maximize) | 6 | 7 | 9 |
| Safety (maximize) | 7 | 9 | 9 |

**5.17a** Assuming all criteria are rated equally important, use the dimensionless comparison model to determine the preferred design.

**5.17b** Select importance ratings for the criteria using a scale of 1 to 10, and apply the additive, mixed-rating model to determine the preferred design. Compare the results from Problem 5.17*a* with calculations based on your importance ratings.

**5.18** Prospective sites for a new chemical plant have been narrowed to three locations. The criteria for each alternative and the importance of the criteria are shown below. Higher interval-scaled ratings show a preference. Which site is apparently more attractive? Comment on the sensitivity of the choice.

| | ALTERNATIVE | | | |
|---|---|---|---|---|
| *Criteria* | *Site 1* | *Site 2* | *Site 3* | *Importance* |
| Labor supply | 2 | 8 | 9 | 9 |
| Raw materials | 3 | 10 | 5 | 8 |
| Transportation | 8 | 7 | 9 | 7 |
| Cost of land | $200,000 | $700,000 | $400,000 | 5 |
| Building costs | $2,000,000 | $3,800,000 | $1,800,000 | 6 |
| Annual taxes and utility costs | $60,000 | $120,000 | $80,000 | 7 |
| Climate | 7 | 6 | 4 | 5 |

**5.19** Contestants in the "Miss All" competition are judged on beauty, personality, and talent. Beauty is considered twice as important as personality and four times as important as talent. Girls are rated on a simple order scale (1, 2, 3, . . ., $n$) with 1 being the highest rating. One girl has been judged first in beauty but tenth in talent. She knows the girl rated second in beauty is sixth in talent.

**5.19a** What is the minimum rating in personality the girl first in beauty needs to assure being selected as "Miss All"?

**5.19b** How could the beauty judging system be improved?

## EXTENSIONS

***5.1 Distribution of Productivity Gains*** People do not get too excited about a sterile statement such as "productivity increased by 2.7 percent this year." They say "That's nice," and drop the subject when they are not personally affected by the results. But the discussion can get hot when it concerns the emotional issue of who should get the benefit of the 2.7 percent increase, and it stays hot when the subject is what to do to maintain the gains in the future.

A productivity gain means resources are being used more effectively. One of the resources is labor. It is logical that workers should receive a share of the wealth gained by a productivity rise if part of the gain comes from their contribution. Another resource is capital. Investments in productivity-improving machines and facilities came from stockholders or taxpayers who deserve a share of the gains they finance. Consumers also have a claim on the gains because they buy the output, and greater output is meaningless if no one acquires it.

**QUESTIONS**

**5.1a** Assume your company enjoyed a 3 percent productivity gain last year but decreased the dividends to stockholders. What reasons might be behind the dividend decline during a period of productivity prosperity?

**5.1b** A *synergistic* relationship exists when the total effect is greater than the sum of the parts causing it. For example, an additional unit of water added to a field might raise output by 2 bushels, another unit of fertilizer increases output by 3 bushels, another unit of labor adds 4 bushels, to provide a total increase of 9 bushels when these inputs are applied independently. But when they are all applied together, the output might be increased by 16 bushels, owing to interactions among the inputs that mutually support each other. Describe an industrial example in which synergism could contribute to productivity improvement.

***5.2 Standardized Scoring*** Alternatives are often evaluated by a team of appraisers. Each person evaluates all alternatives using the same scaling technique, and the resulting scores for each alternative are summed to provide a ranking among alternatives. Unless all appraisers have about the same range of scores, bias is introduced. A single score for an alternative that is far outside the range used by the rest of the appraisers can bias the selection toward the alternative favored by one enthusiastic backer. A simple procedure for standardizing the scores reduces this type of biasing.

The procedure described below is used by several organizations to convert each appraiser's ratings to an average value of 500 with a standard deviation of 100. After the scores are standardized, they are averaged for a final evaluation.

To standardize the scores given by each evaluator, calculate:

$$\text{Mean} = \bar{x} = \frac{\Sigma x_i}{N}$$

*where* $x_i$ = rating for alternative $i$
$N$ = number of alternatives

$$\text{Standard deviation} = s = \sqrt{\frac{\Sigma(x_i - \bar{x})^2}{N - 1}}$$

$$\text{Standardized score} = 500 + 100\frac{x_i - \bar{x}}{s}$$

To illustrate the procedure, assume four people have rated six alternatives, $A$ through $F$. The raw scores for the alternatives (sum of the criteria ratings multiplied by importance ratings) given by each appraiser are shown below.

| | APPRAISER | | | |
|---|---|---|---|---|
| *Alternative* | *1* | *2* | *3* | *4* |
| *A* | 490 | 680 | 610 | 730 |
| *B* | 515 | 625 | 530 | 610 |
| *C* | 480 | 640 | 560 | 760 |
| *D* | 505 | 670 | 500 | 685 |
| *E* | 460 | 700 | 590 | 750 |
| *F* | 800 | 600 | 480 | 660 |

The mean score for appraiser 1 is (490 + 515 + 480 + 505 + 460 + 800)/6 = 542. The sum of the squares of the differences between each score and the mean is $(490 - 542)^2 + (515 - 542)^2 + (480 - 542)^2 + (505 - 542)^2 + (460 - 542)^2 + (800 - 542)^2 = 81{,}934$. Then the standard deviation is

$$s = \sqrt{\frac{81{,}934}{6 - 1}} = 128$$

which makes the standardized score given alternative *A* by appraiser 1 equal to

$$500 + 100\,\frac{490 - 542}{128} = 459$$

Standardized scores for the rest of appraiser 1's ratings are *B* = 479, *C* = 452, *D* = 471, *E* = 436, and *F* = 702.

**QUESTIONS**

**5.2a** Calculate the standardized scores for the other three appraisers depicted in the table, and sum the scores for the six alternatives to determine their ranking.

**5.2b** Alternative *F* would have had the top ranking based on raw scores, but it ranks fifth after the scores have been standardized. How do you account for this switch?

**5.2c** Collect ratings from three or more people for the three mousetrap designs given in Problem 5.16 (all based on the same criteria for evaluation), and standardize the scores to determine a group preference.

# SECTION TWO

# MATHEMATICS OF MONEY

*Interest* is the cost of using capital. Its history extends as far back as the recorded transactions of mankind. In earliest times, before money was coined, capital was represented by wealth in the form of personal possessions, and interest was paid in kind. For example, a loan of seed to a neighbor before planting was returned after harvest with an additional increment. We can surmise that the concept of interest in its modern sense arose from such loans for productive purposes.

Capital and credit have been important to human progress since about 5000 B.C. At that time Neolithic man was engaged in agriculture and animal culture to provide his own food. Capital was counted by seeds, tools, and herds of animals. Cattle were probably the first true productive assets and are the origin of many financial terms. *Pecuniary* stems from *pecus* meaning "flock" in Latin, and the Egyptian term *ms*, meaning "interest," is derived from the verb *msj*, which means "to give birth." Early Greeks measured wealth in terms of cattle; in the Odyssey, Ulysses was promised a contribution "of bronze and gold to the value of twenty oxen."

By the time the Greek and Roman empires were in their ascendancies, interest rates were somewhat standardized and occasionally legislated. The amount charged for loans to the most reliable borrowers was around 10 percent, with the range from 4 percent in first-century Rome to about 50 percent for grain loans in Egypt during the same period.*

Along with the development of money and credit came abuses. Aristotle pointed out that money was "barren," and it was unfair to charge interest for loans. Early Israelites did not permit lending at interest. Romans permitted credit but limited the rate of interest to about 5 to 12 percent. Greeks encouraged credit without limit but forbade personal bondage for debt. Biblical utterances against usury were aimed at loans for consumption rather than production, yet interest was forbidden by canon law through the Middle Ages.

*S. Homer, *History of Interest Rates,* Rutgers, New Brunswick, N.J., 1963.

The concept of interest has not changed much through the centuries, but the modern credit structure differs markedly from that of antiquity. Lending or investing was relatively inconvenient in ancient days because transactions were made directly between individuals. There were no banking organizations to act as intermediaries, and no credit instruments in the money market. Governments were not often able to float loans since they could not pledge the private resources of their people. And they had not discovered the practice of deficit financing.

Today there are many credit instruments, and most people use them. Business and government are the biggest borrowers. Businesses seek the use of capital goods to increase productivity. Governments borrow against future tax revenues to finance highways, welfare programs, and public services. Households also borrow to make purchases in excess of their current cash resources. Such borrowers, and the corresponding lenders, must acknowledge the time value of their commitments.

- The purchase of a home is the largest investment most people make. The two tables below vividly portray the impact of interest rates and loan periods: A shorter repayment period at a given interest rate or a lower interest rate for a given loan period begets a conspicuous saving.

| *Repayment Period, Years* | *Monthly Payment* | *Total Interest* |
|---|---|---|
| 15 | $295.50 | $23,190 |
| 20 | 260.50 | $32,520 |
| 25 | 241.75 | 42,525 |
| 30 | 230.75 | 53,069 |

A $30,000 loan at 8½ percent interest for four repayment periods.

| *Interest Rate, %* | *Monthly Payment* | *Total Interest* |
|---|---|---|
| 7½ | $210.00 | $45,600 |
| 8½ | 230.75 | 53,069 |
| 9½ | 252.50 | 60,899 |
| 10½ | 274.50 | 68,820 |

A $30,000 loan for 30 years at four interest rates.

- During the War of 1812, New York City loaned $1 million to the nation's capital. During the period of New York's financial crises in 1975, it was suggested that Washington might be billed for the original loan and the accumulated interest. At 6 percent interest compounded annually, the $1 million loan would have increased to an $11.2 billion debt.

Laws have been passed to encourage "truth in lending" by requiring standardized statements of interest charges. The need for such laws arises from the many ways interest can be calculated and the unique vocabulary associated with the subject. The profusion breeds confusion.

In this section we shall first explore interest vocabulary and calculations (Chapter 6) and then consider three methods for comparing the economic time value of alternatives: equivalent annual amount (Chapter 7), present worth (Chapter 8), and rate of return (Chapter 9). In the previous section we observed that price (or cost) is a natural rationing device. Interest rates, being the price of loanable funds, ration capital to the most rewarding alternatives. Interest calculations are definitive because they are objective, and the mathematics of money is absorbing because it is so practical.

# CHAPTER 6

# TIME-VALUE MECHANICS

As different economic situations are investigated, one indispensable but uncontrollable factor is continually involved: *time*. In the previous chapters we did not directly consider the prevailing effect of time. The problems in the first section were characterized by immediate outcomes or alternatives affected similarly by time. In this chapter interest calculations are utilized to evaluate the time value of economic factors.

In our present-day culture nearly everyone is directly exposed to interest transactions occasionally, and is indirectly affected regularly. Credit cards are a mainstay of commerce; they have an interest load for delayed payments. Key parts of a contract for purchasing an automobile or home are the interest stipulations. The rate of interest paid on municipal bonds directly affects tax rates for property in the affected area. Businesses borrow to expand or just maintain operations, and the cost of their borrowing must be repaid from more profitable operations allowed by loans. All this borrowing, taken together, adds up to an enormous debt, and it all has interest charges. The actual rates charged vary from the lowest granted to least-risk institutions to prohibitive loan-shark rates for desperate individuals. (See Chapters 12 and 13 for a discussion of interest rates.)

Two examples demonstrate the significance of interest charges:

1 "One/ten-a-week" loans are offered by shadowy characters to tide borrowers over until the next paycheck. The lenders charge $1 interest for each $10 borrowed for a week.

Thus, $140 would have to be repaid for the use of $100 for 4 weeks. That amounts to a 520 percent annual rate!

**2** The famous purchase of Manhattan Island from the Indians for $24 is often referred to as an exceptional bargain. This incident reputedly occurred in 1626, when Peter Minuit of the Dutch West India Company bought the rights to the island from local residents. Was it a bargain? For the sake of argument, suppose the Indians could have invested the money at a reasonable interest rate of 6 percent compounded annually. Over the years since then, the original $24 investment would have grown by the following proportions:

| *Year* | *Value of the Original $24 Investment* |
|---|---|
| 1626 | $24.00 |
| 1676 | 442.08 |
| 1726 | 8,143.25 |
| 1776 | 149,999.92 |
| 1826 | 2,763,021.69 |
| 1876 | 50,895,285.76 |
| 1926 | 937,499,015.11 |
| 1976 | 17,268,876,484.38 |

The two examples might reinforce the adage that "a fool and his money are soon parted," or an observation that "all you have to do to be rich is live long enough." Both are partly right, but the underlying message is a call for basic economic astuteness.

## REASONS FOR INTEREST

The significance of interest is obvious in the previous examples. The reasons for this effect become more apparent when we examine the uses of capital. In our economic environment, capital is the basic resource. It can be converted into production goods, consumer goods, or services. It has the power to earn and to satisfy wants.

From a lender's viewpoint, capital is a fluid resource. Capital can be spent on goods expected to produce a profit or on personal satisfaction. It can be hoarded or given away. It can also be loaned. If it is loaned, the lender will normally expect some type of compensation. The common compensation is interest. Interest compensates for the administrative expense of making the loan, for the risk that the loan will not be repaid, and for the loss of earnings which would have been obtained if the money had been invested for productive purposes.

From a borrower's viewpoint, a loan is both an obligation and an opportunity. A borrower must expect to repay the loan. Failure to repay leads to a damaged reputation, loss of possessions, and other consequences. The loan offers an opportunity to do something immediately that would otherwise have to be delayed. In some cases an objective would no longer exist after a delay. In order to take advantage of an existing course of action or to fulfill a current need, the borrower agrees to pay a certain amount in addition to the sum immediately received. This premium is the interest paid to avoid waiting for the money.

Implied in both the lender's and borrower's viewpoints is the earning power of money. For money to earn something, the owner or user must wait (*waiting-earning* is obviously opposed to the *spending-owing* use of money to gratify immediate desires). Interest payments have been likened to the reward for waiting, but it is more appropriate for engineering economists to view interest as the productive gain from efficient use of the money resource. The prevailing interest rate is essentially a measure of the productivity to expect from the resource. An owner of money can lend it at the prevailing rate and wait to be repaid the original amount plus an extra increment. Equivalently, the borrower could reloan the money at a higher rate to acquire a gain larger than the amount to be repaid, or the money could be converted to productive goods that would be expected to earn more than the amount needed to repay the loan. In both cases the prevailing interest rate sets the minimum level of expected productivity, and both cases involve time between receipt and return of the loan to secure the earnings: the *time value of money*.

---

**Economy Exercise 6-1** The money earned from making a loan is evident in the contract (sometimes in fine print). A loan of $10,000 for 1 year at an interest rate of 10 percent earns the lender $10,000 × 0.10 = $1000. A borrower usually does not know in advance exactly how much will be gained from a loan to buy productive goods. It is often impossible to segregate precisely the receipts due a certain production operation when that operation is a small part of a much larger production system. In such cases an evaluation study may be made on the amount production costs of the system are decreased by improvements to the given operation, assuming the operation must be performed to maintain the total process. Then the earnings are in the form of "cost savings" in the system which are compared to the investment cost of acquiring and using assets to improve the operation.

Assume a machine is purchased for $10,000 with the loan mentioned above. The machine will be completely worn out by the end of the year, and its operating costs will be $100 per month more than the costs of the present operation. How large a cost reduction must be provided by the machine for its purchase to earn a 15 percent return for the borrower?

---

## SIMPLE INTEREST

When a *simple interest rate* is quoted, the interest earned is directly proportional to the capital involved in the loan. Expressed as a formula, the interest earned $I$ is calculated with

$$I = PiN$$

*where* $P$ = present amount or principal

$i$ = interest rate/period

$N$ = number of interest periods

Since the principal or amount borrowed $P$ is a fixed value, the annual interest charged is constant. Therefore, the total amount a borrower is obligated to pay a lender is

$$F = P + I = P + PiN = P(1 + iN)$$

where $F$ is a future sum of money. When $N$ is not a full year, there are two ways to calculate the simple interest earned during the period of the loan. Using *ordinary simple interest,* the year is divided into twelve 30-day periods, or a year is considered to have 360 days. In *exact simple interest* a year has exactly the calendar number of days, and $N$ is the fraction of the number of days the loan is in effect that year.

An example of simple interest as the rental cost of money is a loan of \$1000 for 2 years at 10 percent, where interest is charged only on the principal. Then the interest owed at the end of the 2 years is $\$1000 \times 0.10 \times 2 = \$200$, and the total amount due after 2 years is $\$1000 + \$200 = \$1200$. Note that the principal has earned $\$1000 \times 10\% = \$100$ at the end of 1 year, but no interest is charged on this \$100 increment.

**Economy Exercise 6-2** A loan of \$200 is made for a period of 13 months, from January 1 to January 31 the following year, at a simple interest rate of 8 percent. What future amount is due at the end of the loan period?

# COMPOUND INTEREST

Again we use the example of \$1000 loaned for 2 years, this time at an interest rate of 10 percent compounded annually; the pattern of interest compounding is shown in Table 6.1.

**TABLE 6.1** Future value of a \$1000 loan when interest is due on both the principal and unpaid interest.

| *Year* | *Amount Owed at Beginning of Year* | *Interest on Amount Owed* | *Amount Owed at End of Year* |
|---|---|---|---|
| 1 | \$1000 | $\$1000 \times 0.10 = \$100$ | $\$1000 + \$100 = \$1100$ |
| 2 | 1100 | $\$1100 \times 0.10 = \$110$ | $\$1100 + \$110 = \$1210$ |

The amount to be repaid for the given loan is thus $\$1210 - \$1200 = \$10$ greater for compound than for simple interest. The \$10 difference accrues from the interest charge on the \$100 earned during the first year that was not accounted for in the simple-interest calculation. The formula approach for the calculations in Table 6.1, using previously defined symbols, is

$$\text{Compound amount due in 2 years} = \text{amount borrowed} + \text{year-1 interest} + \left(\begin{matrix}\text{amount bor-}\\ \text{rowed plus}\\ \text{interest due}\end{matrix}\right)\left(\begin{matrix}\text{interest}\\ \text{rate}\end{matrix}\right)$$

$$\begin{aligned} F_2 &= P + Pi + (P + Pi)i \\ &= P(1 + i + i + i^2) \\ &= P(1 + i)^2 \\ &= \$1000(1 + 0.10)^2 \\ &= \$1000(1.21) = \$1210 \end{aligned}$$

The key equation in the above development is $F_2 = P(1 + i)^2$. Generalized for any number of interest periods $N$, this expression becomes $F = P(1 + i)^N$, and $(1 + i)^N$ is known as the *compound-amount factor*. It is one of several interest factors derived in this chapter for which numerical values are tabulated in Appendix B.

**Economy Exercise 6-3** Let the interest on a \$1000 loan for 2 years at an interest rate of 10 percent compounded annually be paid when it comes due. That is, at the end of the first year an interest payment of $\$1000 \times 0.10 = \$100$ is paid, and at the end of the second year the principal (\$1000) plus the interest earned during the second year (\$100) is paid. The total interest charge is thus $\$100 + \$100 = \$200$ at 10 percent compounded annually, which is the same as the interest charge at 10 percent simple interest ($I = PiN = \$1000 \times 0.10 \times 2 = \$200$). How come?

## Nominal Interest Rates

Interest rates are normally quoted on an annual basis. However, agreements may specify that interest will be compounded several times per year: monthly, quarterly, semiannually, etc. For example, a year divided into four quarters with interest at 2 percent per quarter is typically quoted as "8 percent compounded quarterly." Stated in this fashion, it is called a *nominal interest rate.* The future value at the end of 1 year for \$200 earning interest at 8 percent compounded quarterly is developed as

$$
\begin{aligned}
F_{3\text{ mo}} &= P + Pi = \$200 + (\$200)(0.02) \\
&= \$200 + \$4 = \$204 \\
F_{6\text{ mo}} &= \$204 + (\$204)(0.02) \\
&= \$204 + \$4.08 = \$208.08 \\
F_{9\text{ mo}} &= \$208.08 + (\$208.08)(0.02) \\
&= \$208.08 + \$4.16 = \$212.24 \\
F_{12\text{ mo}} &= \$212.24 + (\$212.24)(0.02) \\
&= \$212.24 + \$4.24 = \$216.48
\end{aligned}
$$

The result of the nominal interest rate is to produce a higher value than might be expected from the 8 percent figure stated in its expression. At 8 percent compounded annually, the \$200 mentioned above would earn, in 1 year, $F_{12\text{ mo}} = \$200 + \$200(0.08) = \$216$, which is 48 cents less than the amount accrued from the nominal rate of 8 percent compounded quarterly. An interest of 1½ percent per month is also a nominal interest rate that could appear to the uninitiated as being quite reasonable. Using the compound-amount factor to calculate how much would have to be repaid on a 1-year loan of \$1000 at a nominal interest rate of 18 percent compounded monthly (1½ percent per period with 12 interest periods per year) gives

$$
\begin{aligned}
F_{12} &= \$1000(1 + 1\tfrac{1}{2}\%)^{12} = \$1000(1.015)^{12} \\
&= \$1000(1.1956) = \$1196
\end{aligned}
$$

This can be compared to the future value of the same loan at 18 percent compounded semiannually (9 percent per period with 2 interest periods per year):

$$
\begin{aligned}
F_{12} &= \$1000(1 + 9\%)^{2} = \$1000(1.09)^{2} \\
&= \$1000(1.1881) = \$1188
\end{aligned}
$$

Thus, more frequent compounding within a nominally stated annual rate does indeed increase the future worth.

## Effective Interest Rates

Confusion about the actual interest earned is eliminated by stating the charge as an ***effective interest rate***. Efforts to protect borrowers from exotic statements of interest charges was the thrust behind the national "truth in lending" law passed in 1973. The effective interest rate is simply the ratio of the interest charge for one year to the principal (amount loaned or borrowed). For the \$1000 one-year loan at a nominal interest rate of 18 percent compounded monthly,

$$\text{Effective interest rate} = \frac{F - P}{P} = \frac{\$1196 - \$1000}{\$1000}$$

$$= \frac{\$196}{\$1000} 100\% = 19.6\%$$

For the same loan at 18 percent compounded semiannually,

$$\text{Effective interest rate} = \frac{\$1188 - \$1000}{\$1000}$$

$$= \frac{\$188}{\$1000} 100\% = 18.8\%$$

The effective interest rate can be obtained without reference to the principal. Based on the same reasoning utilized previously, and with

$i$ = effective interest rate
$r$ = nominal interest rate
$m$ = number of compounding periods per year

the effective interest rate for a nominal interest rate of 18 percent compounded semiannually is

$$i = \left(1 + \frac{r}{m}\right)^m - 1 = \left(1 + \frac{0.18}{2}\right)^2 - 1$$

$$= (1 + 0.09)^2 - 1 = 1.188 - 1$$

$$= 0.188 \quad \text{or} \quad 18.8\%$$

which means that a nominal interest rate of 18 percent compounded semiannually is equivalent to a compound interest rate of 18.8 percent on an annual basis.

The ultimate limit for the number of compounding periods in a year is called ***continuous compounding***. Under this accrual pattern, $m$ approaches infinity as interest compounds continuously, moment by moment. The effective interest rate for continuous compounding is developed as follows:

The interest periods are made infinitesimally small:

$$i = \lim_{m \to \infty} \left(1 + \frac{r}{m}\right)^m - 1$$

The right side of the equality is rearranged to include $r$ in the exponent:

$$\left(1+\frac{r}{m}\right)^m - 1 = \left[\left(1+\frac{r}{m}\right)^{m/r}\right]^r - 1$$

The bracketed term is recognized as the value of the mathematical symbol $e$ [$e = 2.718$ is the value of $(1 + 1/n)^n$ as $n$ approaches infinity]:

$$\lim_{m\to\infty}\left(1+\frac{r}{m}\right)^{m/r} = e$$

By substitution,

$$i = \lim_{m\to\infty}\left[\left(1+\frac{r}{m}\right)^{m/r}\right]^r - 1 = e^r - 1$$

As an example of continuous compounding, when the interest rate is $r = 18.232$ percent,

$$i = e^r - 1 = e^{0.18232} - 1 = 0.20 \quad \text{or} \quad 20\%$$

and, correspondingly, when the effective interest rate is $i = 22.1$ percent (see Table 6.2),

$$0.221 = e^r - 1$$
$$1.221 = e^r$$
$$r = 20\%$$

**Economy Exercise 6-4** A loan can be arranged at a nominal rate of 12 percent compounded monthly, or 13 percent compounded semiannually. Which arrangement provides the lower debt at the end of the loan period?

## Continuous Compounding

Occasionally, economic studies are conducted with continuous compounding rather than conventional discrete interest rates. The most obvious computational effect of using continuous interest is that a given nominal rate, say 20 percent, produces a significantly larger future amount than the same figure as an effective interest rate; as demonstrated above, a continuously compounded $r$ of 20 percent compared to $i = 20$ percent yields an annual return greater by

$$\frac{0.221 - 0.20}{0.20}100\% = 10.5\%$$

Values of the continuous compound-amount factor for various periods and nominal interest rates are given in Table 6.2.

The rationale for using continuous interest in economic analyses is that the cash flow in certain situations is best approximated by a continuous pattern; that is, cash transactions tend to be spread out over a year in more or less of an even distribution, rather than being concentrated at particular dates. Some mathematical models are also facilitated by the assumption of continuous compounding rather than periodic compounding.

CONTINUOUS COMPOUND-AMOUNT FACTOR AT INTEREST RATE $r$, $e^{rN}$

| $N$ | $r=2\%$ | $r=4\%$ | $r=6\%$ | $r=8\%$ | $r=10\%$ | $r=15\%$ | $r=20\%$ | $r=30\%$ |
|---|---|---|---|---|---|---|---|---|
| 1 | 1.020 | 1.041 | 1.062 | 1.083 | 1.105 | 1.162 | 1.221 | 1.350 |
| 2 | 1.041 | 1.083 | 1.128 | 1.174 | 1.221 | 1.350 | 1.492 | 1.822 |
| 3 | 1.062 | 1.128 | 1.197 | 1.271 | 1.350 | 1.568 | 1.822 | 2.460 |
| 4 | 1.083 | 1.174 | 1.271 | 1.377 | 1.492 | 1.822 | 2.226 | 3.320 |
| 5 | 1.105 | 1.221 | 1.350 | 1.492 | 1.649 | 2.117 | 2.718 | 4.482 |
| 6 | 1.128 | 1.271 | 1.433 | 1.616 | 1.822 | 2.460 | 3.320 | 6.050 |
| 7 | 1.150 | 1.323 | 1.522 | 1.751 | 2.014 | 2.858 | 4.055 | 8.166 |
| 8 | 1.174 | 1.377 | 1.616 | 1.896 | 2.226 | 3.320 | 4.953 | 11.023 |
| 9 | 1.197 | 1.433 | 1.716 | 2.054 | 2.460 | 3.857 | 6.051 | 14.880 |
| 10 | 1.221 | 1.492 | 1.822 | 2.226 | 2.718 | 4.482 | 7.389 | 20.086 |
| 15 | 1.350 | 1.822 | 2.460 | 3.320 | 4.482 | 9.488 | 20.086 | 90.017 |
| 20 | 1.492 | 2.226 | 3.320 | 4.953 | 7.389 | 20.086 | 54.598 | 403.429 |
| 30 | 1.822 | 3.320 | 6.050 | 11.023 | 20.086 | 90.017 | 403.429 | 8103.084 |
| 40 | 2.226 | 4.953 | 11.023 | 24.533 | 54.598 | 403.429 | 2980.958 | |
| 50 | 2.718 | 7.738 | 20.086 | 54.598 | 148.413 | 1808.042 | | |

**TABLE 6.2** Values for the factor needed to find $F$ given $P$ at a nominal interest rate $r$ compounded continuously for $N$ periods: $F = Pe^{rN}$.

In actual practice, however, interest rates are seldom quoted on a continuous basis, and the vast majority of organizations use discrete compounding periods in their economic studies. The reason for this is probably custom or the familiarity that makes it easier to understand periodic interest charges. Accounting practices that categorize receipts and disbursements as end-of-year values, and financial experiences with annual tax, insurance, or mortgage payments, contribute to thinking in terms of discrete periods. Yet continuous and discrete compounding are both only approximations of true cash flow, because cash neither flows like a free stream of water nor gushes like a geyser at given intervals. Receipts and disbursements are irregular in amount and in timing.

In the following discussion of the time value of money and in subsequent chapters on economic comparison methods, end-of-year compounding is utilized. All interest statements are for effective rates unless specified otherwise. Tables of interest factors are provided in Appendix B only for discrete compounding. However, the development of comparable interest factors for continuous compounding is explained in case they might be needed for special applications.

**Economy Exercise 6-5** A sum of money doubles in size after 9 years when invested at 8 percent interest compounded annually:

$$F = P(1 + 0.08)^9$$
$$2P \doteq P(1.999)$$

At what rate of continuous compounding will an amount double in just half the time taken at the 8 percent effective rate?

## TIME-VALUE EQUIVALENCE

Two things are equivalent when they produce the same effect. The effective interest rate computed for a nominally stated interest rate is an equivalent expression of the interest charge. Both interest charges produce the same effect on an investment. In considering time-value conversions, the equivalent numerical values of money are determined, ***not*** values with equivalent ***purchasing power***. The amount of goods that can be purchased with a given sum of money varies up and down (more often down) as a function of special localized circumstances and nationwide or worldwide economic conditions. Ways to include inflation effects are discussed in Chapter 12. In this chapter about time-value mechanics, attention is directed toward calculations based on the ***earning power*** of money, which relates time and earnings to locate time-equivalent money amounts.

If $1000 were sealed and buried today, it would have a cash value of $1000 when it was dug up 2 years from now. Regardless of changes in buying power, the value remains constant because the earning power of the money was forfeited. It was observed earlier that $1000 deposited at 10 percent interest compounded annually has a value of $\$1000(1 + 0.10)^2 = \$1210$ after 2 years. Therefore, $1000 today is equivalent to $1210 in 2 years from now if it earns at a prevailing rate of 10 percent compounded yearly. Similarly, to have $1000 in 2 years from now, one need only deposit

$$\$1000 \frac{1}{(1 + 0.10)^2} = \$826.44$$

today. In theory then, if 10 percent is an acceptable rate of return, an investor would be indifferent between having $826.44 in hand or having a trusted promise to receive $1000 in 2 years.

The $1000 could also be used to pay two equal annual $500 installments. The buried $1000 could be retrieved after 1 year, an installment paid, and the remaining $500 interred again until the second payment became due. If, instead, the $1000 is deposited at 10 percent, there would be $1100 available at the end of the first year. After the first $500 installment was paid, the remaining $600 would draw interest until the next payment. Paying the second $500 installment would leave

$$\$600(1.10) - \$500 = \$660 - \$500 = \$160$$

in the account. Because of the earning power of money, the initial deposit could have been reduced to $868 to pay out $500 at the end of each of the two years:

*First year:* $\$868(1.10) - \$500 \doteq \$955 - \$500 = \$455$

*Second year:* $\$455(1.10) = \$500 \doteq$ second installment

Thus, $868 is equivalent to $500 received 1 year from now plus another $500 received 2 years from now:

*First year:* $\dfrac{\$500}{1.10} \doteq \$455$

$$\textit{Second year:} \qquad \$455 + \frac{\$500}{(1.10)^2} = \$455 + \frac{\$500}{1.21}$$

$$\doteq \$455 + \$413 \doteq \$868$$

The concept of equivalence is the cornerstone for time-value-of-money comparisons. To have a precise meaning, income and expenditures must be identified with time as well as with amount. A decision between alternatives having receipts and disbursements spread over a period of time is made by comparing the equivalent outcomes of the alternatives at a given date. Figure 6.1 shows the translation of $1000 at time zero (now) into equivalent alternative expressions of cash flow.

$1000 today is equivalent to $1791 received 10 years from now.

$1000 today is equivalent to $237.40 received at the end of each year for the next 5 years.

$1000 today is equivalent to $317.70 received at the end of years 6, 7, 8, 9, and 10.

$237.40 received at the end of each year for the next 5 years is equivalent to a lump sum of $1791 received 10 years from now.

$317.70 received at the end of years 6, 7, 8, 9, and 10 is equivalent to $1791 received 10 years from now.

$237.40 received at the end of each year for the next 5 years is equivalent to $317.70 received at the end of years 6, 7, 8, 9, and 10.

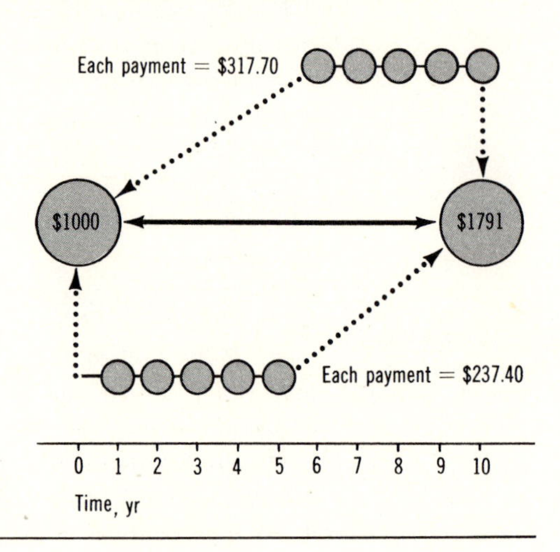


**FIGURE 6.1** Equivalent outcomes with an interest rate of 6 percent compounded annually.

## COMPOUND-INTEREST FACTORS

Cash flow is translated to a given point in time by determining either its present worth or its future worth. A present-worth calculation converts a single future sum or a series of future values to an equivalent amount at an earlier date. This date is not necessarily the present time. Future-worth calculations convert values occurring at any time to an equivalent amount at a later date.

Equivalent values could be determined by calculating the compound amount of each sum for each period. This tedious routine is avoided by using compound-interest tables for different present- and future-worth factors. There are two basic types of factors. The one we have already considered converts a single amount to a present or future value. The other type is for a series of uniform values called an *annuity*. The tables in Appendix B are based on an annuity characterized by (1) *equal payments A,* (2) *equal periods between payments N,* and (3) *the first payment occurring at the end of the first period.* Annuity factors are used to convert a series of payments to a single future or present sum and to translate single sums into a series of payments occurring in the past or future.

## Conversion Symbols

There are seven basic interest factors for discrete compounding. Names and notations for these factors are those suggested by the Engineering Economy Division of the American Society for Engineering Education.* Each factor is described by a name (for instance, one is the "compound-amount factor" used previously to find the future worth of a single payment) and two notational forms: (1) a mnemonic symbol (to assist memory by association) as in ($CA$-$i$%-$N$) for the compound-amount factor, and (2) a functional symbol (to suggest the use of the interest factor) as in ($F/P$, $i$%, $N$), again for the compound-amount factor which is used to find $F$ given $P$. Since the functional notation is most descriptive of the operation to be performed, it will be utilized for all the interest factors. Time-value conversions and associated factors are summarized in Table 6.3.

**TABLE 6.3** Interest factors for discrete cash flow with end-of-period compoundings.

| *Factor* | *To Find* | *Given* | *Symbol* |
|---|---|---|---|
| Compound amount | Future worth, $F$ | Present amount, $P$ | ($F/P$, $i$%, $N$) |
| Present worth | Present worth, $P$ | Future amount, $F$ | ($P/F$, $i$%, $N$) |
| Sinking fund | Annuity amounts, $A$ | Future amount, $F$ | ($A/F$, $i$%, $N$) |
| Series compound amount | Future worth, $F$ | Annuity amounts, $A$ | ($F/A$, $i$%, $N$) |
| Capital recovery | Annuity amounts, $A$ | Present amount, $P$ | ($A/P$, $i$%, $N$) |
| Series present worth | Present worth, $P$ | Annuity amounts, $A$ | ($P/A$, $i$%, $N$) |
| Arithmetic gradient conversion | Annuity amounts, $A$ | Uniform increase in amount, $G$ | ($A/G$, $i$%, $N$) |

The symbols for the first six time-value conversions are abbreviations for the equivalent values sought (future worth $F$, present worth $P$, or uniform series amounts $A$) and the data given ($F$, $P$, or $A$ with its associated interest rate $i$ and number of compounding periods $N$). The arithmetic gradient conversion factor is used to convert a constantly increasing series into a uniform series of amounts $A$ which can then be an input to other interest factors. In the equation

$$F = \$1000(F/P, 10, 2)$$

$1000 is the known present amount, the interest rate is 0.10 per period ($i$ is 10 percent), and $F$ is the equivalent future worth after two periods ($N = 2$). The whole symbol stands for the numerical expression $(1 + 0.10)^2$, and the numerical value is found in Appendix B. To find the value for ($F/P$, 10, 2), look for 2 in the $N$ column of the 10 percent table, and then read across to the compound-amount-factor column to find 1.2100.

The conversion descriptions and symbols connote that certain factors are reciprocals of one another:

$$(F/P, i, N) = \frac{1}{(P/F, i, N)}$$

*See *The Engineering Economist*, vol. 14, no. 2, Winter 1969.

$$(A/F, i, N) = \frac{1}{(F/A, i, N)}$$

$$(A/P, i, N) = \frac{1}{(P/A, i, N)}$$

Other relationships are not so apparent from the abbreviations but are useful in understanding conversion calculations. The following equalities are verified during the development of the conversion symbols:

$$(F/P, i, N) \times (P/A, i, N) = (F/A, i, N)$$
$$(F/A, i, N) \times (A/P, i, N) = (F/P, i, N)$$
$$(A/F, i, N) + i = (A/P, i, N)$$

## Development of Interest Formulas

A better understanding of the conversion process is achieved by studying the development of interest formulas. The symbols employed are the same as those described previously. The numerical values of the formulas for different interest rates and compounding periods are tabulated in Appendix B.

### 1 COMPOUND-AMOUNT FACTOR (SINGLE PAYMENT)

*Use:* To find $F$, given $P$
*Symbols:* $(F/P, i\%, N)$
$(CA\text{-}i\%\text{-}N)$
*Formula:* $P(1 + i)^N$
$P(F/P, i, N)$

The effect of compound interest on an investment was demonstrated in previous examples. The future worth of a present amount when interest is accumulated at a specific rate $i$ for a given number of periods $N$, where $F1$ is the future worth at the end of the first period and $F_N$ is the future worth at the end of $N$ years, is

$$F1 = P + Pi = P(1 + i)$$
$$F2 = P[(1 + i) + (1 + i)i]$$
$$= P(1 + i)(1 + i) = P(1 + i)^2$$
$$F3 = P[(1 + i)^2 + (1 + i)^2 i]$$
$$= P(1 + i)^2(1 + i) = P(1 + i)^3$$
$$F_N = P(1 + i)^N$$

The ratio of future worth to present amount is then expressed as

$$\frac{F}{P} = (F/P, i, N) = (1 + i)^N$$

### 2 PRESENT-WORTH FACTOR (SINGLE PAYMENT)

*Use:* To find $P$, given $F$
*Symbols:* $(P/F, i\%, N)$
$(PW\text{-}i\%\text{-}N)$
*Formula:* $P = F \frac{1}{(1 + i)^N}$
$= F(P/F, i, N)$

$P$ is the present worth of a sum $N$ periods in the future. Rearranging the single-amount future-value formula $F = P(1 + i)^N$ to express $P$ in terms of $F$ gives

$$P = F \frac{1}{(1 + i)^N}$$

Then the ratio of present worth to future value is:

$$\frac{P}{F} = (P/F, i, N) = \frac{1}{(1 + i)^N}$$

## 3 SINKING-FUND FACTOR

*Use:* To find $A$, given $F$
*Symbols:* $(A/F, i\%, N)$
$(SF\text{-}i\%\text{-}N)$
*Formula:* $A = F \frac{i}{(1 + i)^N - 1} = F(A/F, i, N)$

A fund established to accumulate a given future amount through the collection of a uniform series of payments is called a ***sinking fund.*** Each payment has a constant value $A$ and is made at the end of an interest period.

The growth pattern of a sinking fund is illustrated in Table 6.4. Each end-of-year payment $A$ is equal to \$1000, and payments continue for 5 years. Interest is 8 percent compounded annually. It is assumed that each payment begins to draw interest as soon as it is deposited in the sinking-fund account. Thus, the first payment draws interest for 4 years, and the last payment receives no interest.

**TABLE 6.4** Compound amount of a uniform series of payments.

| *Time of Payment (end of year)* | *Amount of Payment, A* | *Future Worth at the End of Each Year* |
|---|---|---|
| 1 | \$1000 | $\$1000(1.08)^4$ = \$1360 |
| 2 | 1000 | $1000(1.08)^3$ = 1260 |
| 3 | 1000 | $1000(1.08)^2$ = 1166 |
| 4 | 1000 | $1000(1.08)^1$ = 1080 |
| 5 | 1000 | $1000(1.08)^0$ = 1000 |
| | | Annuity value $F$ at the end of year 5 = \$5866 |

A more general expression for the future worth of an annuity develops from the use of symbols to represent the values in Table 6.4. The first payment, earning interest for $N - 1$ periods, where $N$ is 5 years in the example, increases to a future worth of

$$F = A(1 + i)^{N-1}$$

Each of the payments is treated in the same manner and collected to obtain the total amount $F$:

$$F = A(1 + i)^{N-1} + A(1 + i)^{N-2} + A(1 + i)^{N-3} + A(1 + i)^{N-4} + A(1 + i)^{N-N}$$

Factoring out $A$ and letting the exponent $N - N = 0$, we have

$$F = A[(1 + i)^{N-1} + (1 + i)^{N-2} + (1 + i)^{N-3} + (1 + i)^{N-4} + 1]$$

Multiplying this equation by $1 + i$ results in

$$F(1 + i) = A[(1 + i)^N + (1 + i)^{N-1} + (1 + i)^{N-2} + (1 + i)^{N-3} + (1 + i)]$$

Subtracting the original equation from the last equation gives

$$F(1 + i) - F = -A + A(1 + i)^N$$
$$Fi = A[(1 + i)^N - 1]$$

Solving for $A$,

$$A = F \frac{i}{(1+i)^N - 1}$$

we see the sinking-fund factor expressed as

$$(A/F, i, N) = \frac{i}{(1+i)^N - 1}$$

## 4 SERIES COMPOUND-AMOUNT FACTOR (UNIFORM SERIES)

*Use:* To find $F$, given $A$
*Symbols:* $(F/A, i\%, N)$
$(SCA\text{-}i\%\text{-}N)$
*Formula:* $F = A \frac{(1+i)^N - 1}{i}$
$= A(F/A, i, N)$

From the development of the sinking-fund-factor formula,

$$Fi = A(1+i)^N - 1$$

which is expressed in terms of $F$ as

$$F = A \frac{(1+i)^N - 1}{i}$$

Then the time value for the future worth of an annuity is

$$(F/A, i, N) = \frac{(1+i)^N - 1}{i}$$

**Economy Exercise 6-6** A manufacturing firm in a foreign country has agreed to pay $25,000 in royalties at the end of each year for the next 5 years for the use of a patented product design. If the payments are left with the foreign company, interest on the retained funds will be paid at an annual rate of 15 percent.

What total amount will be available in 5 years under these conditions?

How large would the uniform annual payments have to be if the patent owners insisted that a minimum of $175,000 must be accumulated in the account by the end of 5 years if they are to leave it with the company?

## 5 CAPITAL-RECOVERY FACTOR

*Use:* To find $A$, given $P$
*Symbol:* $(A/P, i\%, N)$
$(CR\text{-}i\%\text{-}N)$
*Formula:* $A = P \frac{i(1+i)^N}{(1+i)^N - 1}$
$= P(A/P, i, N)$

The capital-recovery factor is used to determine the amount of each future annuity payment required to accumulate a given present value when the interest rate and number of payments are known. For instance, the amount of each annual payment made for 5 years in order to repay a debt of $3993 bearing 8 percent annual interest can be determined through the use of the capital-recovery factor. Table 6.5 shows that it would take five $1000 payments to repay the $3993 debt.

Using symbols to represent the conversions shown in Table 6.5, the present worth of an annuity is

$$P = A[(1+i)^{-1} + (1+i)^{-2} + (1+i)^{-3} + (1+i)^{-4} + (1+i)^{-N}]$$

| Time of Payment (end of year) | Amount of Payment, A | Present Worth of Payments at End of Year |
|---|---|---|
| 1 | \$1000 | $\$1000(1.08)^{-1} = \$\ 926$ |
| 2 | 1000 | $1000(1.08)^{-2} =\ 857$ |
| 3 | 1000 | $1000(1.08)^{-3} =\ 794$ |
| 4 | 1000 | $1000(1.08)^{-4} =\ 735$ |
| 5 | 1000 | $1000(1.08)^{-5} =\ 681$ |
| | | Present worth $P$ of the 5-year annuity = \$3993 |

**TABLE 6.5** Present worth of a uniform series of payments.

and multiplying both sides of the equation by $(1 + i)^{-1}$ results in

$$P(1 + i)^{-1} = A\,[(1 + i)^{-2} + (1 + i)^{-3} + (1 + i)^{-4} + (1 + i)^{-N} + (1 + i)^{-N-1}]$$

Subtracting the first equation from the second equation gives

$$P\,[(1 + i)^{-1} - 1] = A\,[(1 + i)^{-N-1} - (1 + i)^{-1}]$$

Converting $(1 + i)^{-1} - 1$ to $-i/(1 + i)$, multiplying both sides by $-(1 + i)$, and rearranging yield

$$P\,\frac{i(1 + i)}{1 + i} = A\,[(1 + i)(1 + i)^{-1} - (1 + i)(1 + i)^{-N-1}]$$

$$Pi = A\,[1 - (1 + i)^{-N}]$$

$$= A\,\frac{(1 + i)^N - 1}{(1 + i)^N}$$

or $$A = P\,\frac{i(1 + i)^N}{(1 + i)^N - 1}$$

from which comes the expression for the capital-recovery factor,

$$(A/P, i, N) = \frac{i(1 + i)^N}{(1 + i)^N - 1}$$

The relationship among time-value annuity factors is apparent from the way the capital-recovery factor can be converted to the sinking-fund factor by substituting $P = F(1 + i)^{-N}$ in the capital-recovery formula, as

$$A = P(A/P, i, N) = \frac{F}{(1 + i)^N}\,(A/P, i, N) = \frac{F}{(1 + i)^N}\,\frac{i(1 + i)^N}{(1 + i)^N - 1}$$

$$A = F\,\frac{i}{(1 + i)^N - 1} = F(A/F, i, N)$$

or

$$(A/P, i, N) = (A/F, i, N) + i$$

as indicated by

$$\frac{i(1+i)^N}{(i+i)^N-1} = \frac{i}{(1+i)^N+1} + i = \frac{i + i(1+i)^N - 1}{(1+i)^N - 1} = \frac{i(1+i)^N}{(1+i)^N-1}$$

## 6 SERIES PRESENT-WORTH FACTOR (UNIFORM SERIES)

*Use:* To find $P$, given $A$
*Symbols:* $(P/A, i\%, N)$
$(SPW\text{-}i\%\text{-}N)$
*Formula:* $P = A\dfrac{(1+i)^N-1}{i(1+i)^N} = A(P/A, i, N)$

The present value of a series of uniform end-of-period payments can be calculated in the cumbersome fashion shown in Table 6.5. The present worth is more readily determined by use of the series present-worth factor.

Expressing the known relationship

$$A = P\frac{i(1+i)^N}{(1+i)^N-1}$$

in terms of $P$ yields

$$P = A\frac{(1+i)^N-1}{i(1+i)^N} = A(P/A, i, N)$$

which is the time value expression of the present worth of an annuity.

---

**Economy Exercise 6-7** The management of a wearing-apparel firm is considering a proposal from a consulting group to introduce a new method of training inexperienced sewing-machine operators. The consultants claim that their program will produce savings of \$7000 per year over the planned 5-year life of the project. Immediate costs to implement the program are \$12,000. Annual training expenses will be \$4000. The company uses 6 percent annual interest for cost comparisons. Do the anticipated savings warrant the expense of hiring the training consultants?

---

## 7 ARITHMETIC-GRADIENT CONVERSION FACTOR (TO UNIFORM SERIES)

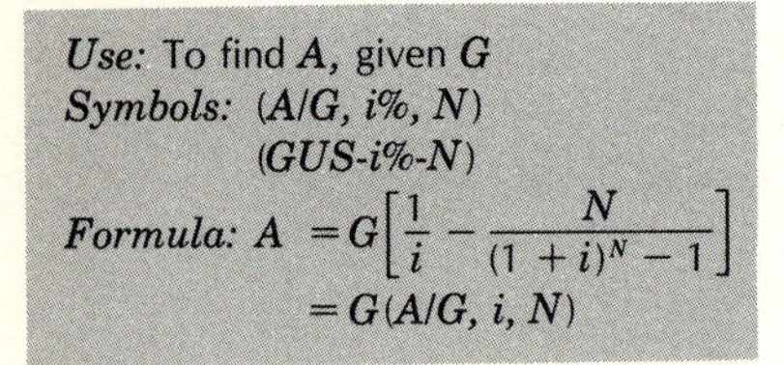


Enough situations occur in which series of payments increase at equal increments to warrant a special conversion factor. A series of payments that increases at a rate of \$200 per year is illustrated in Figure 6.2. The \$200 periodic change is the gradient $G$, and the payment at the end of the first period is the base annuity value $A'$. The pattern of an arithmetic gradient is then

$$A', A' + G, A' + 2G, \ldots, A' + (N-1)G$$

where $N$ is the duration of the series ($N = 5$ in Figure 6.2).

A uniformly increasing series can be evaluated by calculating $F$ or $P$ for each individual payment and summing the collection. Calculation time is reduced by converting the series into an equivalent annuity of equal payments $A$. The formula for this translation is developed by separating the series shown in Figure 6.2 into two parts: a base annuity designated $A'$ and an arithmetic-gradient series increasing by $G$ each period. The future worth of

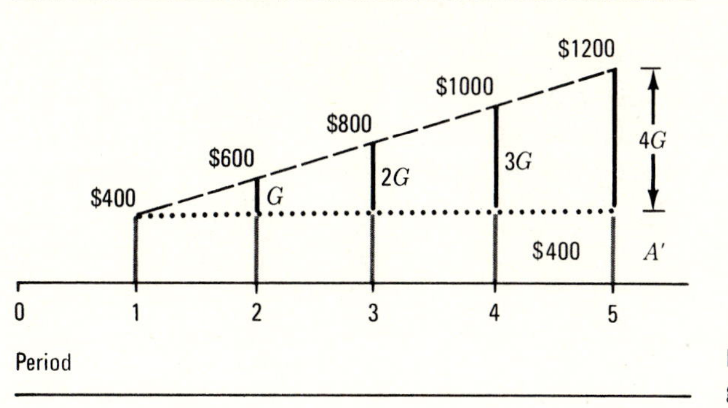


**FIGURE 6.2** A uniform gradient series for five periods.

the $G$ values in Figure 6.2 is calculated as

$$F = \$200(F/P, i, 3) + \$400(F/P, i, 2) + \$600(F/P, i, 1) + \$800$$

or

$$F = G(1 + i)^3 + 2G(1 + i)^2 + 3G(1 + i)^1 + 4G$$

Multiplying the latter expression by $(1 + i)^1$ gives

$$F(1 + i) = G(1 + i)^4 + 2G(1 + i)^3 + 3G(1 + i)^2 + 4G(1 + i)^1$$

Subtracting the last equation from the one above it yields

$$F - F(1 + i) = -G(1 + i)^4 - G(1 + i)^3 - G(1 + i)^2 - G(1 + i)^1 + 4G$$

Letting $4G = NG - 1$ and multiplying both sides of the equation by $-1$ give

$$F(1 + i) - F = G[(1 + i)^4 + (1 + i)^3 + (1 + i)^2 + (1 + i)] - NG$$

We have, in brackets, the series compound-amount factor $(F/A, i, 5)$, so

$$Fi = G(F/A, i, N) - NG$$

To convert $F$ to an annuity, both sides of the equation are multiplied by the sinking-fund factor $(A/F, i, N)$, which is the reciprocal of $(F/A, i, N)$, to get

$$Fi(A/F, i, N) = G - NG(A/F, i, N)$$

and since $A = F(A/F, i, N)$,

$$A = \frac{G}{i} - \frac{NG}{i}(A/F, i, N)$$

or

$$A = G\left[\frac{1}{i} - \frac{N}{i}(A/F, i, N)\right]$$

in which the bracketed expression is called the *arithmetic gradient conversion factor* with the symbol $(A/G, i, N)$.

For the cash flow diagramed in Figure 6.2, the equivalent uniform annuity calculated at an effective interest rate per period of 10 percent is

$$\begin{aligned} A &= A' + G(A/G, i, N) \\ &= \$400 + \$200(A/G, 10, 5) \\ &= \$400 + \$200(1.8100) = \$762 \end{aligned}$$

which means that five end-of-period payments of \$762 are equivalent to five payments starting at \$400 and increasing by \$200 each period.

The gradient factor may also be applied to a pattern of payments that decrease by a constant increment each period. The formula would then be

$$A = A' - G(A/G, i, N)$$

As an example, assume an endowment was originally set up to provide a \$10,000 first payment with payments decreasing by \$1000 each year during the 10-year endowment life. What constant annual payment for 10 years would be equivalent to the original endowment plan if $i = 8$ percent? We have

$$\begin{aligned} A &= \$10{,}000 - \$1000\,(A/G, 8, 10) \\ &= \$10{,}000 - \$1000(3.8712) = \$6128.80 \end{aligned}$$

---

**Economy Exercise 6-8** A contract has been signed to lease a building at \$20,000 per year with annual increases of \$1500 for 8 years. Payments are to be made at the end of each year, starting 1 year from now. The prevailing interest rate is 7 percent. What lump sum paid today would be equivalent to the 8-year lease-payment plan?

---

## Continuous-Interest Factors

The compound-amount factor for a single payment with continuous interest, $F = Pe^{rN}$, was described previously; values for the factor are listed in Table 6.2. From the development of other interest factors, it is apparent that the present-worth factor for a single amount is the reciprocal of the compound-amount factor. Therefore, the ***continuous-compounding present-worth factor*** is $e^{-rN}$ and is used as

$$P = F\,\frac{1}{e^{rN}}$$

Continuous-compounding factors with ***discrete payments*** can be similarly developed from the end-of-period annuity formulas. First recall from the discussion of effective interest that $e^{rN}$ corresponds to $(1 + i)^N$, and that the effective interest rate is $i = e^r - 1$. Substitution of these expressions into the sinking-fund formula yields the fact that the discrete cash flow, end-of-period compounding, sinking-fund formula is equal to the discrete cash flow, continuous-compounding, ***sinking-fund*** formula when $i = e^r - 1$:

$$A = F\,\frac{i}{(1 + i)^N - 1} = F\left[\frac{e^r - 1}{e^{rN} - 1}\right]$$

and the reciprocal of the bracketed expression is the *series compound-amount factor* for continuous compounding.

By similar reasoning, the *capital-recovery* formula with continuous compounding and discrete payments is

$$A = P\left[\frac{e^{rN}(e^r - 1)}{e^{rN} - 1}\right]$$

and the reciprocal of the bracketed expression is the continuous-compounding *series present-worth factor.*

In every formula the $A$, $F$, and $P$ values resulting from computations using either end-of-period or continuous compounding are identical for discrete payments when the continuous interest rate is equivalent to the effective interest rate. This relationship is apparent when numbers are substituted into the sinking-fund formulas displayed above. Letting $i$ = 22.1 percent, which corresponds to a nominal continuous interest rate of 20 percent ($0.221 = e^{0.2} - 1$), and applying the equivalent interest rates in the two sinking-fund formulas with $N = 2$, give

| *End-of-period compounding* | | *Continuous compounding* |
|---|---|---|
| $A = F\dfrac{0.221}{(1.221)^2 - 1}$ | = | $F\dfrac{e^{0.2} - 1}{e^{(0.2)(2)} - 1}$ |
| $= F\dfrac{0.221}{1.491 - 1}$ | = | $F\dfrac{1.221 - 1}{1.492 - 1}$ |
| 0.45 | = | 0.45 |

This shows that the factors are equal when $i = e^r - 1$ and payments are discrete.

Another version of continuous compounding occurs when the *total payment for a period is received in continuous, small, equal payments during that period.* Letting $\overline{A}$ designate this total amount of each payment in the series, which continues for $N$ periods at interest rate $r$ per period, we find the future worth directly by integration as

$$F = \overline{A}\int_0^N e^{rt}\,dt = \overline{A}\left[\frac{e^{rN} - 1}{r}\right]$$

The expression in brackets is called the *continuous-compounding series compound-amount factor* for continuous, uniform payments.

Functional notations and formulas for the six basic continuous-compounding factors for continuous-flow payments are given in Table 6.6. The first two in the table are for a single continuous payment for one period, and the last four are annuity payments. Bars over the symbols for future, present, and annuity payments (respectively, $\overline{F}$, $\overline{P}$, and $\overline{A}$) represent the total amount accumulated from continuous small payments throughout each compounding duration.

The assumption of a continuous flow of disbursements and incomes throughout a year is rare as compared to the end-of-year payment pattern. However, the continuous-flow assumption is more revealing and applicable than continuous compounding of discrete

| Functional Notation | Application Formula |
|---|---|
| $(F/\overline{P}, r\%, N)$ | $F = \overline{P}\,\dfrac{e^{rN}(e^r - 1)}{re^r}$ |
| $(P/\overline{F}, r\%, N)$ | $P = \overline{F}\,\dfrac{e^r - 1}{re^{rN}}$ |
| $(\overline{A}/R, r\%, N)$ | $\overline{A} = F\,\dfrac{r}{e^{rN} - 1}$ |
| $(F/\overline{A}, r\%, N)$ | $F = \overline{A}\,\dfrac{e^{rN} - 1}{r}$ |
| $(\overline{A}/P, r\%, N)$ | $\overline{A} = P\,\dfrac{re^{rN}}{e^{rN} - 1}$ |
| $(P/\overline{A}, r\%, N)$ | $P = \overline{A}\,\dfrac{e^{rN} - 1}{re^{rN}}$ |

**TABLE 6.6** Symbols and formulas for continuous compounding of a continuous flow. $\overline{F}$, $\overline{P}$, and $\overline{A}$ designate total amounts accumulated in small equal payments during one compounding period $N$ for $\overline{A}$, and over $N$ periods for $\overline{F}$ and $\overline{P}$. The continuous interest rate is $r$ per period.

payments because the main reason for using continuous interest in an economic evaluation is to determine the effects of continuous cash flows, not just the continuous interest on discrete payments.

**Economy Exercise 6-9** The inventor of an automatic, coin-operated newstand, called "Automag," believes the economic evaluation of the invention should be based on continuous compounding of continuous cash flow because income from the Automag will be essentially continuous, and disbursements for services and purchases of materials (newspapers, paperbacks, news magazines, etc.) will occur regularly and frequently. The expected life of an Automag is 5 years. Annual income should average \$132,000 at a good location, and total expenses for servicing and materials are expected to average \$105,000 per year. What initial price could be paid for an Automag (delivered and ready to operate) to allow a buyer to earn 20 percent on the investment if the income and disbursement estimates are accurate?

## CALCULATION OF TIME-VALUE EQUIVALENTS

The purpose of time-value calculations is to translate receipts and disbursements of various amounts occurring at various times to a cash-flow pattern that assists an economic evaluation. The translation is essentially mechanical in the same fashion a vector is routinely decomposed into component forces; rules of geometry direct vector operations, and time-value relationships direct cash-flow translations. Although errors in translation can arise from carelessness, mistakes due to incorrect problem formulations are the ones to guard against.

*Cash-flow diagrams* are pictorial aides for problem formulation and cash-flow calculations. They serve the same purpose as the "free-body" diagrams used to portray the effect of forces acting on a body. The intent is to isolate the factors pertinent to a problem and to display them clearly to observe what data are available and what calculations are needed. They also help the analyst to keep track of data during manipulations.

The format of a cash-flow diagram is left to the discretion of the user. A rough sketch showing dates and amounts is adequate for simple problems. A more systematized format is useful for more complex problems, to avoid misinterpretations. Style is unimportant because a cash-flow diagram does not yield a graphical solution; it just displays the data in a form that contributes to correct mathematical operations. The suggested version followed in this text has a time line with a scale large enough for clarity and lines or arrows representing given cash flows or equivalent translations. Positive cash flows are registered above the time line, and negative flows below.

The notable statements in a time-value problem are the interest elements: $P$, $F$, $A$, $N$, and $i$. Generally, three of the elements are known for each cash flow, and the problem entails solving for a fourth element. Several money-time translations may be required in one solution. A variety of typical cash-flow patterns are treated in the following pages. Problems are presented to put cash-flow questions in a realistic setting, show different perspectives for the same type of problem, demonstrate the use of cash-flow diagrams, and provide familiarity with the use of interest factors and tables.

## Single-Payment Cash Flow

Translation of a future amount to its present worth, or the reverse from present to future, has been demonstrated previously for both discrete and continuous compounding. Sometimes both present and future amounts are known, and the problem is to find the value of $i$ or $N$ that makes them equivalent.

1 *Unknown interest rate* At what annual interest rate will \$1000 invested today be worth \$2000 in 9 years?
*Solution*

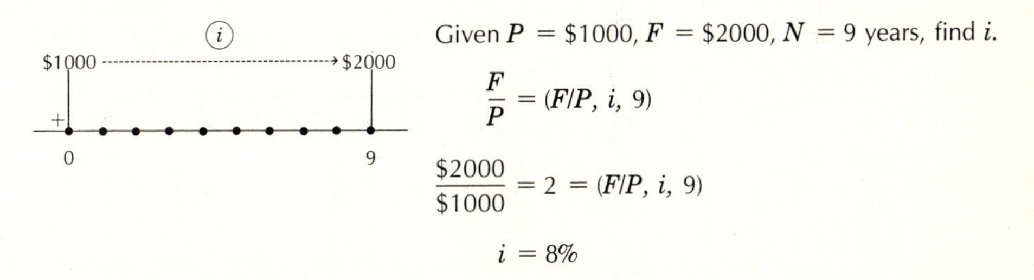


Given $P = \$1000$, $F = \$2000$, $N = 9$ years, find $i$.

$$\frac{F}{P} = (F/P, i, 9)$$

$$\frac{\$2000}{\$1000} = 2 = (F/P, i, 9)$$

$$i = 8\%$$

The interest rate $i$ is determined by locating the interest rate at which the single-payment compound-amount factor is equal to 2.0 at $N = 9$ (a reciprocal relationship using a present-worth factor could serve just as well). The numerical value of $i$ is found by leafing through the pages of interest rates and noting the appropriate factor values for the given number of periods.

It is usually necessary to interpolate from table values when $N$ or $i$ is unknown. The error introduced by linear interpolation is relatively insignificant for most practical applications. If the investment period for Example 1 had been 10 years instead of 9, the interest-rate calculation would have been

$$(F/P, i, 10) = \frac{F}{P} = \frac{\$2000}{\$1000} = 2.0$$

At $i = 7$ percent, $(F/P, 7, 10) = 1.9671$; and $i = 8$ percent, $(F/P, 8, 10) = 2.1589$. Then, by interpolation,

$$i = 0.07 + 0.01\frac{2.0000 - 1.9671}{2.1589 - 1.9671}$$

$$= 0.07 + 0.01\frac{0.0329}{0.1918}$$

$$= 0.07 + 0.0017 = 0.0717$$

or,

$$i \doteq 7.2\%$$

**2** *Unknown number of interest periods* A loan of \$1000 is made today under an agreement that \$1400 will be received in payment some time in the future. When should the \$1400 be received if the loan is to earn interest at a rate of 8 percent compounded quarterly?
*Solution*

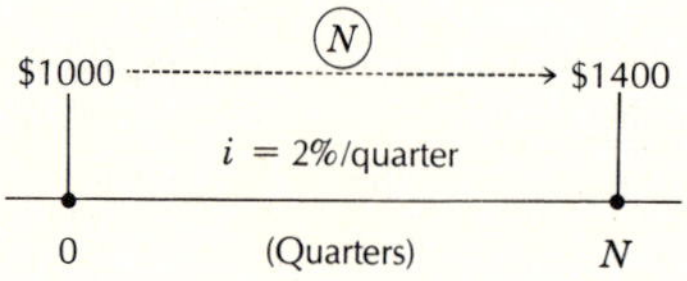

\$1000 ---- (N) ----→ \$1400

$i$ = 2%/quarter

0 (Quarters) $N$

Given $P = \$1000$, $F = \$1400$, $i = r/m = 8\%/4 = 2\%$, find $N$ (in quarters).

$$\frac{F}{P} = \frac{\$1400}{\$1000} = 1.4 = (F/P, 2, N)$$

$N = 17$ quarters or 4 years, 3 months

**3** *Unknown N at continuous interest* When should the loan in Example 2 be repaid if it is to earn 8 percent compounded continuously?
*Solution*

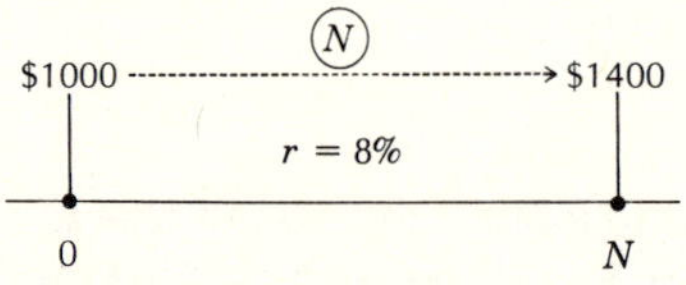

\$1000 ---- (N) ----→ \$1400

$r$ = 8%

0 $N$

Given $P = \$1000$, $F = \$1400$, $r = 8\%$, find $N$.

$$\frac{F}{P} = 1.400 = (F/P, 8, N)$$

$$1.377 = (F/P, 8, 4)$$
$$1.492 = (F/P, 8, 5)$$

$$N = 4 + 1\,\frac{1.400 - 1.377}{1.492 - 1.377} = 4 + 0.2 = 4.2 \text{ years}$$

Examples 2 and 3 both involve nominal interest rates. A statement of 8 percent compounded quarterly means there are four compounding periods per year and the interest rate for each period is $r/m = 0.08/4 = 0.02$. The problem could also have been solved by converting the nominal rate to its equivalent effective interest rate,

$$i = \left(1 + \frac{0.08}{4}\right)^4 - 1 = 1.082 - 1 = 0.082$$

and solving the following equation by interpolation between the 8 percent and 9 percent interest tables and $N$ values between 4 and 5:

$$1.4000 = (F/P, 8.2, N)$$
$$1.3604 + 0.2(1.4115 - 1.3604) = 1.3706 = (F/P, 8.2, 4)$$
$$1.4693 + 0.2(1.5386 - 1.4693) = 1.4831 = (F/P, 8.2, 5)$$

$$N = 4 + 1\,\frac{1.4000 - 1.3706}{1.4831 - 1.3706} = 4.265 \text{ years}$$

The result differs from 4 years, 3 months only by "round-off" errors in the interest factors.

## Multiple-Payment Cash Flows

Practical problems customarily involve both single payments and annuities. For instance, determining the equivalent present worth (cost) of owning a car for 3 years involves a series of payments to purchase the car and provide gas, repair and maintenance costs at irregular intervals, and a single payment (receipt) when the car is sold. All these receipts and disbursements would be translated to "now" and summed to find the present worth—a single sum equivalent to ownership costs for 3 years.

**4** *More compounding periods than payments* "Now" is June 30, 1977. Three payments of $500 each are to be received every 2 years starting 2 years from now and deposited in a bank where they will earn interest at 7 percent per year. How large will the bank account be on June 30, 1985?

*Solution*

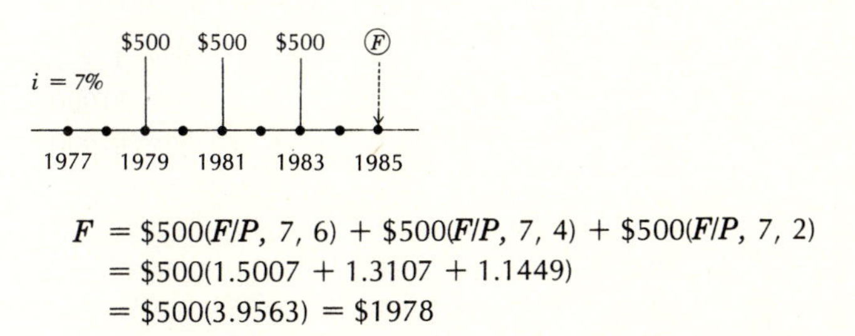


$$F = \$500(F/P, 7, 6) + \$500(F/P, 7, 4) + \$500(F/P, 7, 2)$$
$$= \$500(1.5007 + 1.3107 + 1.1449)$$
$$= \$500(3.9563) = \$1978$$

The equal payments in this example do not constitute an ordinary annuity because there are fewer payments than compounding periods. Therefore, each payment must be translated individually to the 1985 date and added to the worth of the other payments at that date to obtain the equivalent future worth of all three payments.

**5** *Annuity with an unknown i* A new machine that can be purchased for $8065 will annually reduce production costs by $2020. The machine will operate for 5 years, at which time it will have no resale value. What rate of return will be earned on the investment? (Alternative statement of the problem: At what interest rate will a cash flow of \$2020 per year for 5 years equal a present value of \$8065?)

***Solution***

$$\frac{P}{A} = \frac{\$8065}{\$2020} = 3.993 = (P/A,\ i,\ 5)$$

$$i = 8\%$$

**6** ***Continuous-flow, continuous-compounding annuity*** A savings plan offered by a company allows employees to set aside part of their daily wages and have them earn 6 percent compounded continuously. What annual amount withdrawn from pay will accumulate \$10,000 in 5 years? (Alternative statement of the problem: What continuous-flow annual annuity will yield a future value of \$10,000 in 5 years when compounded continuously at 6 percent?

***Solution***

$$\bar{A} = F(\bar{A}/F,\ 6,\ 5) = F\frac{r}{e^{rN} - 1}$$

$$= F\frac{0.06}{e^{0.3} - 1} = \$10{,}000\frac{0.06}{1.35 - 1}$$

$$= \$10{,}000(0.1714) = \$1714$$

A series of payments made at the beginning instead of the end of each period is sometimes referred to as an *annuity due.* Rather than create a special factor for this annuity pattern, the series is divided into two parts. If the first payment is translated separately, the remaining payments fit the pattern for an ordinary annuity beginning at the time of the first payment. The present worth of the series is the sum of the first payment plus the product of one payment times the series present-worth factor, where $N$ is the number of payments minus 1.

**7** ***Annuity due*** What is the present worth of a series of 15 year-end payments of \$1000 each when the first payment is due today and the interest rate is 5 percent?

***Solution***

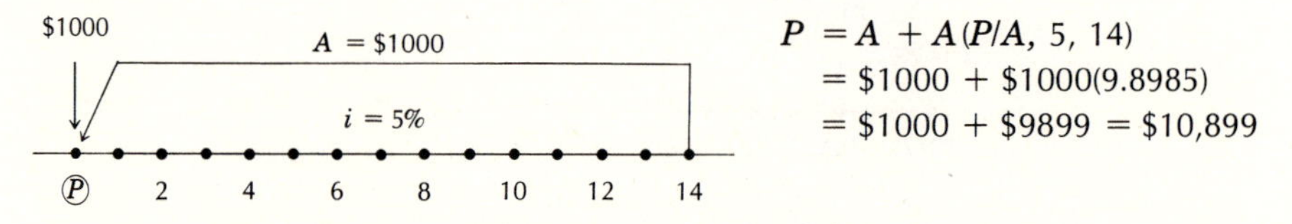


$$P = A + A(P/A,\ 5,\ 14)$$
$$= \$1000 + \$1000(9.8985)$$
$$= \$1000 + \$9899 = \$10{,}899$$

Another pattern for a series of payments, in which the first payment does not begin until some date later than the end of the first period, is called a *deferred annuity.* Like an annuity due, a deferred annuity is evaluated by dividing the time period into two parts. One portion is $N$, the number of payment periods, plus 1. This portion forms an ordinary annuity of $N$ periods. The second portion is the number of periods left after subtracting $N + 1$ periods. A solution results from determining the present worth of the ordinary annuity and then discounting this value through the deferred period.

**8** ***Deferred annuity*** With interest at 6 percent, what is the worth on June 30, 1979, of a series of end-year payments of $317.70 made on June 30 from 1985 through 1989?
***Solution***

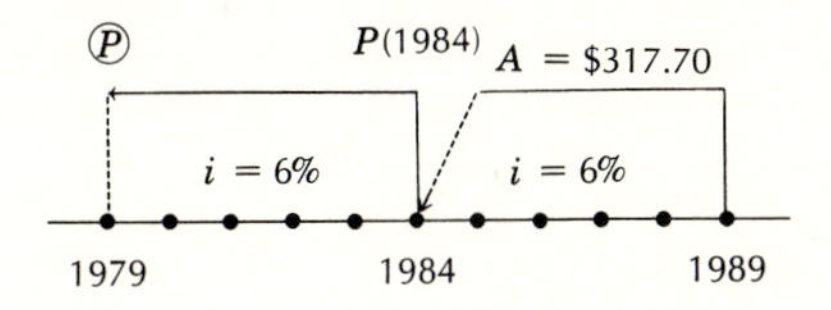


Starting from the known values of $A$, $i$, and $N$,

$$P(1984) = A(A/P, 6, 5)$$

$P(1984)$ becomes the $F$ value in calculating $P(1979)$:

$$P(1979) = P(1984)(P/F, 6, 5)$$

Collecting terms,

$$P(1979) = A(P/A, 6, 5)(P/F, 6, 5)$$
$$= \$317.70(4.2123)(0.74726) = \$1000$$

The results from this example may be recognized as one of the equivalent outcomes presented without proof in Figure 6.2. Another of the outcomes from the same figure is shown in Figure 6.3.

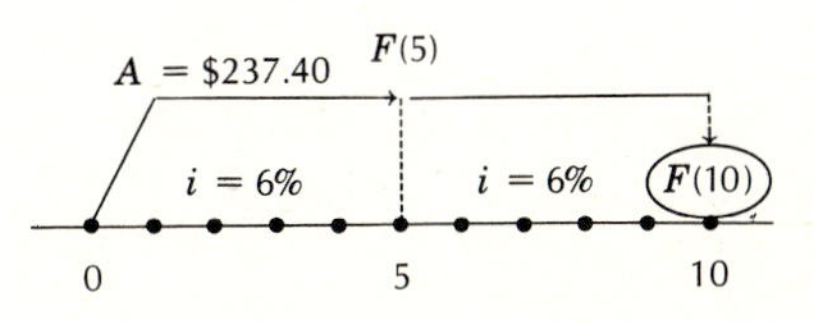


**FIGURE 6.3** Cash flow diagram displaying the data required to calculate the future worth of a 5-year annuity. The data shown are based on cash flows given in Figure 6.2.

The value of the annuity after five payments of $237.40 is labeled $F(5)$. By letting the end of the fifth year be "now," $F(5)$ is treated as a present value, and the compound-amount factor is used to find the future value at the end of year 10, or $F(10)$:

$$F(5) = A(F/A, 6, 5) \quad \text{and} \quad F(10) = F(5)(F/P, 6, 5)$$
$$F(10) = A(F/A, 6, 5)(F/P, 6, 5)$$
$$= \$237.40(5.6370)(1.3382) = \$1791$$

More extensive economic situations often include both income and outlay. Such situations are evaluated by calculating the net outcome at a certain point in time. A cash-flow diagram incorporates receipts and disbursements by displaying income above the time line and outlays below the line. Other payment categories can be handled similarly.

**9** ***Income and outlay*** A boy is now 11 years old. On his fifth birthday he received a gift of $4000 from his grandparents which was invested in 10-year bonds bearing interest at 4 percent compounded semiannually. His parents plan to have $3000 available for the boy's nineteenth, twentieth, twenty-first, and twenty-second birthdays to help finance a college education. To assist the financing, the grandparents' gift will be reinvested when the bonds mature. How much should the parents allocate to invest each year on the boy's twelfth through eighteenth birthdays to complete the education plan? All future investments will earn 6 percent annually.

*Solution*

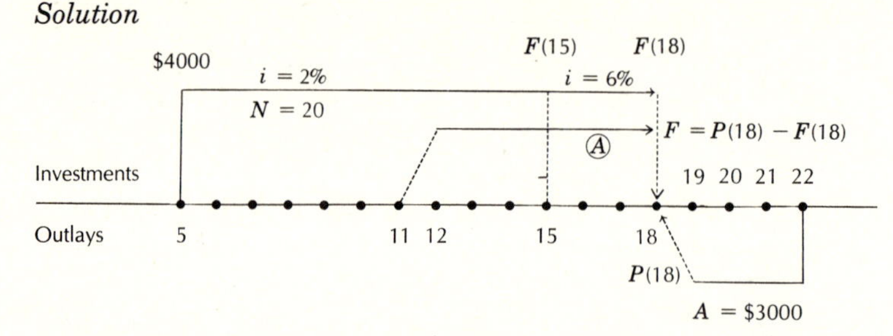


The cash-flow diagram for the problem is based on an evaluation date of the boy's eighteenth birthday. It would be equally valid to use any of his birthdays for the comparison date. The same value for the parents' payments, $A$, would result for any chosen date, but there is usually one time which reduces the number of required calculations.

As shown in the diagram, there are three payment plans involved: a series of outlays, a lump-sum investment, and a series of annual investments. After computing the difference between the amount available from the lump-sum investment and the amount required for the education expenses at a common date, the annual payment necessary to accumulate this difference can be determined.

The present worth of the education annuity on the boy's eighteenth birthday is

$$P(18) = A(P/A, 6, 4) = \$3000(3.4650) = \$10{,}395$$

The future worth of the grandparents' gift is a function of two investment rates: 2 percent per period for 20 periods (a nominal interest rate of 4 percent compounded semiannually for 10 years), followed by 6 percent per period for three periods:

$$\begin{aligned} F(18) &= P(F/P, 2, 20)(F/P, 6, 3) \\ &= \$4000(1.4859)(1.1910) = \$7079 \end{aligned}$$

The difference $F$ between the amount required and the amount available at year 18 is

$$F = P(18) - F(18) = \$10{,}395 - \$7079 = \$3316$$

which is the required future value for the series of $18 - 11 = 7$ payments of amount $A$ beginning on the boy's twelfth birthday:

$$A = F(A/F, 6, 7) = \$3316(0.11914) = \$395$$

---

**Economy Exercise 6-10** A very successful engineer plans to endow a chair of Engineering Economics at his alma mater. The endowment will last for 7 years at $20,000 per year, commencing 3 years from now. In addition, he wants to award annually a graduate Engineering Economist scholarship for the last 4 years of the endowment period. The 1-year scholarships will start at $5000 and increase by $500 each year. Funds for the endowment are to be provided by three equal payments starting 1 year from now, and the transfer to the university of a $90,000 bond the engineer owns that will mature in 6 years. How large will each of his payments be for the next 3 years to fund the Engineering Economics chair and scholarships, if all funds earn 7 percent interest?

---

## SUMMARY

Interest represents the earning power of money. It is the premium paid to compensate a lender for the administrative cost of making a loan, the risk of nonrepayment, and the loss of use of the loaned money. A borrower pays interest charges for the opportunity to do something now that would otherwise have to be delayed or never done. ***Simple interest*** $I$ is a charge directly proportional to the capital (principal $P$) loaned at rate $i$ for $N$ periods, so that $I = PiN$. ***Compound interest*** includes charges for the accumulated interest as well as the amount of unpaid principal.

A ***nominal interest rate*** $r$ of 8 percent compounded quarterly, for example, indicates an interest charge of 2 percent per period compounded four times a year. If $m$ is the number of compounding periods per year, the equivalent ***effective interest rate*** $i$ of a nominal rate is $i = (1 + r/m)^m - 1$. ***Continuous interest*** is the nominal interest rate as $m$ approaches infinity, and its equivalent effective interest rate is $i = e^r - 1$.

Time-value mechanics involves the use of compound-interest factors to translate payments of various amounts occurring at various times to a single equivalent payment. Interest factors are symbolized by notations based on $i, N, P$ = present worth, $F$ = future worth, and $A$ = annuity payment. An ordinary annuity is a series of equal payments, at equal intervals, with the first payment at the end of the first period. When payments in an annuity increase by a constant increment $G$ each period, an equivalent ordinary annuity is determined through use of the arithmetic-gradient factor.

Interest factors for discrete compounding are listed in Table 6.3, and the values of the factors for different interest rates and numbers of compounding periods are tabulated in Appendix B. Continuous-flow, continuous-compounding factors are listed in Table 6.6. The $e^{rN}$ values from Table 6.2 can be substituted into the factor formulas of Table 6.6 to determine factor values.

Time-value calculations begin with recognition of the type of interest compounding required and continue with identification of the timing and amounts of payments. Cash-flow diagrams display the available data on a time scale and suggest what calculations are appropriate. With practice, different cash-flow patterns are recognized and associated with certain interest factors. Then the challenge is to determine the correct cash flows for economic comparisons.

## Discussion of Economy Exercises

**EE 6-1** The costs involved include repaying the loan plus interest charges for the loan, extra operating costs, and investment earnings resulting from the purchase of the machine:

| | |
|---|---|
| Loan repayment (purchase price of the machine) | \$10,000 |
| Interest paid on the loan for 1 year = \$10,000 × 0.10 | 1000 |
| Additional operating cost incurred = \$100/month × 12 | 1200 |
| 15% earnings on the \$10,000 borrowed = \$10,000 × 0.15 | 1500 |
| Necessary cost reduction to support the investment | \$13,700 |

If the cost reduction turns out to be only \$13,700 − \$1500 = \$12,200, it would just cover expenses and nothing would be gained from the machine's purchase. However, if the company could afford \$10,000 of its own money for the machine rather than borrowing it, a cost reduction of \$12,200 would yield a 10 percent return on the investment, and a cost reduction of \$13,700 would produce a return of

$$\frac{\$13{,}700 - \$10{,}000 - \$1200}{\$10{,}000} = \frac{\$2500}{\$10{,}000}\,100\% = 25\%$$

**EE 6-2** Using ordinary simple interest, the total amount to be repaid after 13 months is

$$\begin{aligned} F &= P + PiN \\ &= \$200 + (\$200)(0.08)(1 + 1/12) \\ &= \$200 + (\$200)(0.0867) \\ &= \$200 + \$17.34 = \$217.34 \end{aligned}$$

If exact simple interest is used, the future value (assuming the year in question is not a leap year) is

$$\begin{aligned} F &= P + PiN \\ &= \$200 + (\$200)(0.80)\left(1 + \frac{31}{365}\right) \\ &= \$200 + (\$200)(0.0868) \\ &= \$200 + \$17.36 = \$217.36 \end{aligned}$$

**EE 6-3** A borrower obviously pays the same amount of interest in both cases, but the timing is different. When interest is compounded, \$100 is due at the end of *each* year. With simple interest charged, the first \$100 interest charge is delayed until the end of the second year. In effect, the lender is deprived of the use of \$100 for one year under the simple-interest arrangement.

**EE 6-4** The more attractive arrangement is the one with the lowest effective interest rate. At 12 percent compounded monthly, $r = 0.12$ and $m = 12$, and

$$\text{Effective interest rate} = \left(1 + \frac{0.12}{12}\right)^{12} - 1$$

$$= (1.01)^{12} - 1 = 1.127 - 1 = 0.127$$

At 13 percent compounded semiannually, $r = 0.13$ and $m = 2$, and

$$\text{Effective interest rate} = \left(1 + \frac{0.13}{2}\right)^{2} - 1$$

$$= (1.065)^{2} - 1 = 1.134 - 1 = 0.134$$

The loan at 12 percent compounded monthly is thus seen to have the lower $F$ value.

**EE 6-5** Given $F = 2P$ at $N = 9/2 = 4.5$ years. Substituting these values into the future-worth equation using continuous compounding gives

$$F = 2P = P(e^{rN}) \quad \text{or} \quad 2 = e^{4.5r}$$

The value of $r$ can be spotted in Table 6.2 as being slightly greater than 15 percent by going down the $N$ column to 4.5 and passing along that row until a value close to 2 is detected in the 15 percent column. Or, the right-hand equation can be solved directly for $r$ to get $r =$ 15.4 percent.

**EE 6-6** The annual payments form an annuity. Knowing that $A$ = \$25,000 per period, $i$ = 15 percent per period, and there are five periods, the future worth $F$ is calculated as

$$F = A(F/A, 15, 5) = \$25{,}000(6.7423) = \$168{,}558$$

If the patent owners insisted on an accumulated value of \$175,000, the five end-of-year royalty payments would have to be

$$A = F(A/F, 15, 5) = \$175{,}000(0.14832) = \$25{,}956$$

**EE 6-7** Assuming that the costs and savings occur at the end of the year,

$$A = \text{annual savings} - \text{annual costs} = \$7000 - \$4000 = \$3000$$

For the proposal to be acceptable, the net return must be greater than the \$12,000 initial cost. By translating the 5-year annuity to the present time, the initial cost and the present worth of savings $P$ are compared directly:

$$P = A(P/A, 6, 5) = \$3000(4.2123) = \$12{,}637$$

The indicated total net savings exceeds the initial cost by

$$\$12{,}637 - \$12{,}000 = \$637$$

which gives very little leeway for any error in cost or savings estimates.

The same conclusion results from a different approach with the same data. From the capital-recovery formula, the annual return on the gross savings $A$ required for 5 years to meet a current obligation of \$12,000 is

$$A = P(A/P, 6, 5) = \$12{,}000(0.23740) = \$2849$$

Comparing the required return to the expected annual gross savings shows annual net savings of

$$\$3000 - \$2849 = \$151$$

**EE 6-8** The first step in solving for the present worth of the lease-payment plan is to convert the increasing annual payments to a uniform series:

$$A = A' + G(A/G, 7, 8) = \$20{,}000 + \$1500(3.1463)$$
$$= \$20{,}000 + \$4719.45 = \$24{,}719.45$$

Then the annuity is translated to its present worth as

$$P = A(P/A, 7, 8) = \$24{,}719.45(5.9712) = \$147{,}604.78$$

which is the amount in today's dollars equivalent to the lease contract that provides yearly payments of \$20,000 with annual increases of \$1500 for 8 years.

**EE 6-9** The net annual receipts from the Automag are expected to average $132,000 − $105,000 = $27,000. The acceptable price of an Automag is the present-worth equivalent of the receipts for 5 years at 20 percent continuous compounding of continuous cash flow:

$$P = \overline{A}(P/\overline{A}, 20, 5) = \$27{,}000\frac{e^{(0.2)(5)} - 1}{0.2e^{(0.2)(5)}} = \$27{,}000\frac{e^1 - 1}{0.2e^1}$$

$$= \$27{,}000\frac{2.718 - 1}{(0.2)(2.718)} = \$27{,}000(3.160)$$

$$= \$85{,}320$$

Values for the $e^{rN}$ terms in continuous-interest factors can be taken from Table 6.2 or computed directly with certain hand-held calculators.

**EE 6-10** Assuming the bond does not pay dividends before its maturity and all university disbursements occur at the first of each funding year, the cash-flow diagram takes the pattern shown in Figure 6.4, where the engineer's investments are considered positive cash flows and the university's disbursements for the chair and scholarships are negative cash flows.

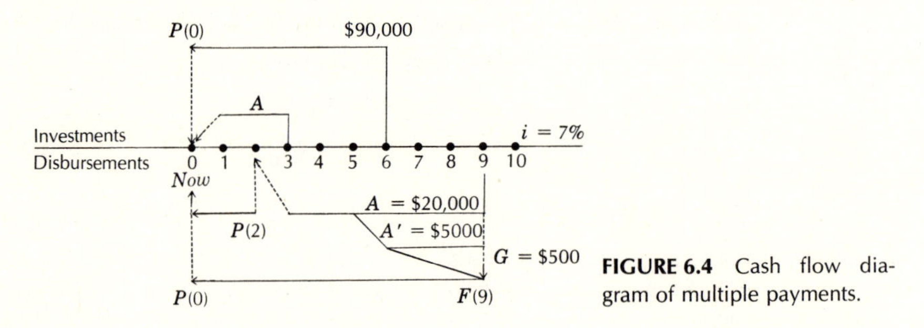


**FIGURE 6.4** Cash flow diagram of multiple payments.

The future worth of the scholarships is calculated by determining the equivalent uniform annuity of the gradient and then its $F$ value:

$$F(9) = [A' + G(A/G, 7, 4)](F/A, 7, 4)$$
$$= [\$5000 + \$500(1.4153)](4.4398) = \$25{,}340.82$$

The equivalent cost at the end of the second year for the seven annual disbursements to the holder of the Engineering Economics chair is

$$P(2) = A(P/A, 7, 7) = \$20{,}000(5.3892) = \$107{,}784$$

The current present worth of the two annuities less the present worth of the bond is

$$P(0) = F(9)(P/F, 7, 9) + P(2)(P/F, 7, 2) - \$90{,}000(P/F, 7, 6)$$
$$= \$25{,}340.82(0.54394) + \$107{,}784(0.87344) - \$90{,}000(0.66635)$$
$$= \$13{,}774 + \$94{,}143 - \$59{,}972 = \$47{,}945$$

Then the annuity of three annual payments to get the endowment started is calculated to be

$$A = P(0)(A/P, 7, 3) = \$47{,}948(0.38105) = \$18{,}271$$

## PROBLEMS

Problems 1 through 15 are adapted from ***Mathematics of Finance*** by L. L. Smail (McGraw-Hill, New York). This college text was published in 1925, and it shows that basic interest problems have not changed much over the years.

**6.1** What sum must be loaned at 6 percent simple interest to earn $47 in 2 years?
*($391.67)*

**6.2** How long will it take $800 to yield $72 in simple interest at 4 percent?
*(2¼ years)*

**6.3** At what rate will $65.07 yield $8.75 in simple interest in 3 years, 6 months?
*(3.8%)*

**6.4** How long will it take any sum to double itself at a 5 percent simple interest rate?

**6.5** Find the ordinary and exact simple interest on $3300 at 6 percent for 56 days.
*($30.80; $30.38)*

**6.6** If the interest on a certain sum for 3 months is $63.87 at 5 percent simple interest, what would it be at 6 percent?

**6.7** Find the compound amount of $100 for 3 years at 5 percent compounded annually.
*($115.76)*

**6.8** What is the compound amount of $750 for 5 years at 6 percent compounded quarterly?

**6.9** Accumulate a principal of $600 for 5 years, 9 months at 6 percent compounded monthly. How much interest is earned?
*($246.42)*

**6.10** Find the difference between the amount of $100 at simple interest and at compound interest for 5 years at 5 percent.
*($2.62)*

**6.11** Find the compound amount of $5000 at 6 percent for 4, 6, 8, and 10 years, and compare the results. Does doubling the time double the amount?

**6.12** A company was incorporated for $50,000; at the end of 10 years the capital had increased to $90,000. What was the average rate of increase?
*(6.052%)*

**6.13** By the conditions of a will, a sum of $25,000 is left to a girl to be held in trust by her guardian until it amounts to $45,000. When will the girl receive the money if the fund is invested at 8 percent compounded quarterly?

**6.14** Find the amount of an annuity due of $240 per year payable annually for 5 years, money being worth 6 percent.

**6.15** At the beginning of 10 successive years, a premium of $120 is paid on a certain insurance policy. What is the worth of the sum of these payments at the end of the tenth year if computed at 4 percent compound interest?

*($1440.64)*

**6.16** Find the present value of a deferred annuity of $500 per year for 12 years, deferred 6 years, if the current rate of interest is 5 percent annually.

*($3306.91)*

**6.17** What is the effective interest rate for:

**6.17a** 12 percent compounded semiannually?
**6.17b** 12 percent compounded quarterly?
**6.17c** 12 percent compounded monthly?

($i = 12.7\%$)

**6.18** What nominal interest, compounded quarterly, is required to provide a 6 percent effective interest rate? A 12 percent effective interest rate?

**6.19** A personal loan is made at an interest rate of 3/4 percent per month on the unpaid balance.

**6.19a** What is the effective interest rate?
**6.19b** How much interest is earned if the duration of the loan is 18 months?

*(Interest earned = 14.4%)*

**6.20** How long will it take for $1 to double in value (disregarding any change in the buying power of the dollar) if:

**6.20a** The interest rate is 10 percent compounded annually?
**6.20b** The interest rate is 10 percent compounded semiannually?
**6.20c** The interest rate is 10 percent ordinary simple interest?

*(7.1 years)*

**6.21** How is it possible to determine the numerical value of a capital-recovery factor ($A/P$) if the only table available is:

**6.21a** ($P/F$)?
**6.21b** ($A/F$)?
**6.21c** ($F/A$)?

**6.22** Determine the value of ($F/A$, 4, 8) using only the table for ($P/A$).

**6.23** Develop a formula for the present worth of an annuity due.

**6.24** What is the future worth of each of the following investments?

**6.24a** $6300 in 6 years at 15 percent compounded annually
**6.24b** $2000 in 4 years at 4¼ percent compounded annually
**6.24c** $200 in 27 years at 6 percent compounded annually
**6.24d** $4300 in 6 years at 7 percent compounded semiannually
**6.24e** $500 in 17 years at 9 percent compounded quarterly

*($2305)*

**6.25** What annual interest rate increases an investment of $1400 to $2000 in 9 years?

*(4.037%)*

**6.26** How many years will it take for the balance left in a savings account to increase from $1000 to $1500 if interest is received at a nominal rate of 6 percent compounded semiannually throughout the period?

*(6.86 years)*

**6.27** Solve the following problems with an annual interest rate of 5 percent:

**6.27a** What amount must be invested today to secure a perpetual income of $6000 per year?

**6.27b** A present expenditure of $50,000 is justified by what annual saving for the next 12 years?

**6.27c** What is the annual payment that will provide a sum of $20,000 in 20 years?

**6.27d** How much money can be loaned today on an agreement that $700 will be paid 6 years from now?

**6.27e** What payment can be made today to prevent a series of year-end expenses of $2800 lasting 15 years?

**6.27f** If the down payment on a piece of land is $7000 and the annual payments for 8 years are $2000 per year, what is the value of the land now?

**6.27g** What single payment 10 years from now is equivalent to a payment of $5500 in 3 years?

**6.27h** If $12,000 must be accumulated by equal annual payments in 15 years, but the first end-of-year payment cannot be made until 4 years from now, how much will each payment be?

**6.28** What semiannual cost beginning today is equivalent to spending lump sums of $2000 in 2 years, $4000 in 4 years, and $8000 in 8 years if the nominal interest rate is 8 percent compounded semiannually?

*($703.73)*

**6.29** Derive the equation for calculating the future worth of a series of discrete payments when interest is compounded continuously. Start from the expression

$$F = A + Ae^{r} + Ae^{2r} + \cdots + Ae^{(n-2)r} + Ae^{(n-1)r}$$

and follow the procedure used in developing the sinking-fund factor for end-of-period compounding.

**6.30** Solve the following problems using a continuous interest rate of $r = 8$ percent:

**6.30a** What total amount must be invested each year to accumulate $10,000 in 10 years if cash flow is continuous?

**6.30b** What total amount invested continuously during the first year will be worth $10,000 in 10 years?

**6.30c** What is the value of $A$ in Problem 6.30b if the cash flow is in discrete payments?

**6.30d** How long will it take a $10,000 single payment to double in amount?

**6.30e** Continue the comparison in Problem 6.10 by finding the compound amount of $100 for 5 years. Also compare the answer to Problem 6.16 under the assumptions that:

**6.30f** Cash flow is continuous.

**6.30g** Cash flow is discrete.

**6.31** Check the annual payments calculated for the engineer's endowment of an Engineering Economics chair in Economy Exercise 6-10 to see whether they are adequate to cover the annual disbursement to the recipient of the chair, $20,000 payable at the first of each year starting on the day the final payment is received from the engineer.

**6.32** If the population of a certain suburb is 29,000 in 1979 and the average annual rate of increase is estimated at 7 percent, what will its population be in 1989, if the growth rate remains constant?

**6.33** A person owes debts of $3380 due in 4 years, with annual interest included at 9 percent, and $1200 plus interest due in 2 years on which the interest charge is 10 percent compounded annually. She now wishes to discharge these debts with two equal payments, payable 1 and 2 years from now. What is the value of the installments if money is now worth 6 percent?

**6.34** An orchard will come into full bearing in 6 years and is expected to yield a net income of $30,000 per annum. If it can be expected to maintain a constant average for 20 years, what is an equitable cash value for the orchard today if money is worth 8 percent?

**6.35** A beachcomber bought a dune buggy, paying $1000 in cash and agreeing to pay $500 every 6 months for 3 years. As she was driving her buggy down the beach one day, she spotted a bottle that held five moldy but cashable $500 bills. She had not yet made her first semiannual installment for the dunebuggy. When should she make a payment of $2500 to discharge her obligation for the purchase in a lump-sum settlement, if the interest rate is 12 percent compounded semiannually?

**6.36** Net receipts from a continuously producing oil well add up to $120,000 over 1 year. What is the present worth of the well if it maintains steady output until it runs dry in 8 years, if $r = 10$ percent?

**6.37** If investments of $2000 now, $2500 in 2 years, and $1000 in 4 years are all made at 4 percent effective interest, what will be the total in 10 years?

*($7647)*

**6.38** An inventor has been offered $12,000 per year for the next 5 years and $6000 annually for the following 7 years for the exclusive rights to his invention. At what price could he afford to sell his rights to earn 10 percent, disregarding taxes?

*($63,625)*

**6.39** A family is planning to buy a vacation cabin for $19,000. They intend to keep the cabin for 6 years and expect the annual upkeep and taxes to amount to $900 per year. Without any major repairs, the cabin should have a resale price of $16,000. What is the equivalent annual cost of owning the cabin if the family's acceptable interest rate is 6 percent?

**6.40** The family depicted in Problem 6.39 could also do some renovations on the cabin during their vacation periods which would increase the resale value to $26,000. How much could they afford to invest in materials each year if they hope to receive $500 per year for their labor?

**6.41** Maintenance records of a certain type of machine indicate that the first-year maintenance cost of $80 increases by $30 per year over the 10-year replacement period of the machine. Answer the following, if the maintenance cost is considered to occur at the end of the year and the firm's interest rate is 12 percent:

**6.41a** What equal annual payments could the firm make to a service organization to carry out the maintenance for 20 machines?

**6.41b** How much additional could be paid for a new type of machine with the same service life that required no maintenance during its life?

**6.42** Three years ago a student borrowed \$4000 to help pay for his education, agreeing to repay the loan in 100 payments at an interest rate of 12 percent compounded monthly. He has just received a bonus to play professional tennis and desires to pay the principal in a lump sum. How much does he owe?

*(\$2989)*

**6.43** What annual expenditure for 10 years is equivalent to spending \$1000 at the end of the first year, \$2000 at the end of the fourth year, and \$3000 at the end of the eighth year, if interest is at 8 percent per year?

*(\$598.62)*

**6.44** The village hotshot borrowed \$4000 to buy a chrome-plated peanut stand. He agreed to pay the loan back in 45 equal payments at 18 percent interest compounded monthly. A year after he took out the loan, he asked you to calculate how much of the \$4000 debt he had paid off. Determine not only how much he has paid on the principal, but also how much interest he has paid so far.

**6.45** A postal clerk wants to accumulate \$20,000 in order to take a year's vacation 10 years from now. She now has \$6000 in savings certificates which earn interest at 6 percent compounded semiannually. She plans to invest an equal amount each year in an account that earns 6 percent annual interest. How large should this amount be to give her \$20,000 in 10 years when it is combined with the future value of the savings certificates?

**6.46** The postal clerk in Problem 6.45 figures that in addition to the annuity calculated to give her \$20,000 in 10 years, she will also be able to save part of her annual raises in pay. If the portion saved increases by \$300 each year starting at the end of the second year from now (no additional increment is possible at the end of this year), how much extra money will she have when she is ready to take her year's vacation?

**6.47** A shady individual engaged in making small loans offers to lend \$200 on a contract under which the borrower must pay \$6.80 at the end of each week for 35 weeks in order to pay off the debt. What are:

**6.47a** The nominal interest rate per annum?
**6.47b** The effective interest rate?

*(67.8%)*

**6.48** A lender offers loans under the following conditions: The total amount of interest owed is the difference between the future worth and the present worth of the loan, $I = P(F/P, i, N) - P$. This amount $I$ is then subtracted from the amount given to the borrower. That is, amount $P$ is borrowed, but only $P - I$ is received from the lender. If the interest rate used in the calculation is 9 percent and the duration of the loan is 3 years, what effective interest is charged?

## EXTENSION

***6.1 More Payments than Compounding Periods*** **Many banks and financial institutions calculate the interest earned on savings accounts by applying the stated interest rate to the *minimum balance* maintained in the account during the interest**

period. This means that funds deposited during the interest period and funds withdrawn during the period receive no interest credit. Sometimes a grace period is allowed for the first 10 days of an interest period; deposits made during the grace period are considered to have been in the account for the full interest period.

**QUESTION**

**6.1a** How much interest is credited to the account for the cash flow displayed in Figure 6.5, if the bank pays 8 percent compounded quarterly and a grace period is allowed during the first 10 days of each quarter?

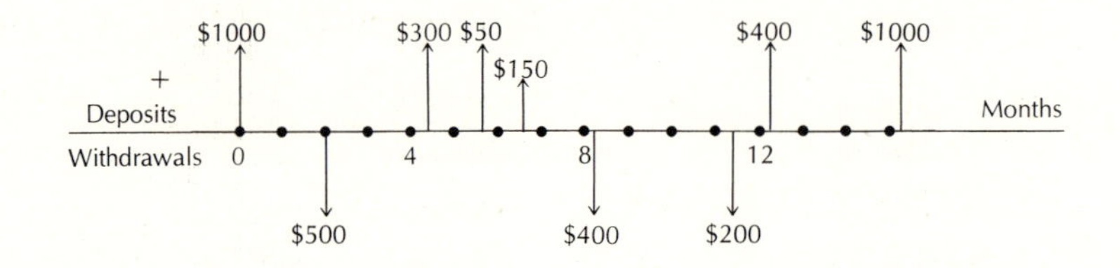


FIGURE 6.5

A procedure that is somewhat the reverse of the interest-payment procedure is used to determine how much is owed on loans in which the balance changes between loan-charge calculations. The formulas used to determine the amount owed on a credit account make a major difference in the financing charge. There are two basic methods of computing the interest owed on "revolving" accounts such as bank credit cards or charge accounts issued by stores. Actual practices and interest rates charged vary among lending institutions and the states within which they operate.

The first is called the *adjusted-balance* method. An example of how it works is as follows: Suppose you charge a purchase for $200 on March 1, and the billing date for your account is March 2. You are given 25 days from the billing date to pay your full bill without incurring any interest charges. If the bill is not paid in full, you are charged interest on the unpaid portion figured from the billing date. Assume you pay $100 on March 24, leaving a balance due of $100. On your April bill you will be charged interest, on the adjusted balance, which amounts to ($200 − $100)(0.015) = $1.50, if the monthly interest rate is the typical 1½%.

The second method is the *average-daily-balance* system that works as follows: The purchase of $200 on March 1 and the $100 payment on March 24 with the billing date of March 2 are interpreted to mean that you owed $200 from March 1 to March 24 and $100 from the latter date until the next billing on April 2. That is 22 days at $200 and 9 days at $100, which result in

$$\text{Average daily balance} = \frac{22(\$200) + 9(\$100)}{31} = \$170.97$$

Again assuming an interest rate of 1½% per month,

$$\text{Interest owed} = \$170.97 \times 0.015 = \$2.56$$

which is $2.56 − $1.50 = $1.06 greater than the interest charge using the adjusted-balance method, a significant difference.

Thus, with the adjusted-balance method, the interest charges are determined solely by the amount you pay. With the average-daily-balance system, the charges are also determined by when in the billing period you pay; the earlier you pay, the lower your interest. If you pay your entire bill within the first 25 days of the billing period, there is no interest charge in either system.

**QUESTION**

**6.1b** Let interest be 18 percent compounded monthly, and the billing date be the second day of each month. Compare the total interest charges for April, May, and June, using both the adjusted-balance and average-daily-balance methods, for the following purchases on credit (−) and repayments (+):

| | | | |
|---|---|---|---|
| March 1 | −\$200 | April 29 | −300 |
| March 24 | +100 | May 1 | +200 |
| March 29 | −300 | May 7 | −150 |
| April 1 | +200 | May 15 | +100 |
| April 14 | −150 | May 21 | −50 |
| April 21 | +200 | June 1 | +350 |

# CHAPTER 7

# EQUIVALENT ANNUAL-WORTH COMPARISONS

The time-value mechanics introduced in the previous chapter revealed that cash flows can be translated to equivalent amounts at any point in time in accordance with the selected interest rate. Three ways to compare cash flows are suggested by the mechanics of translation:

1. *Annual worth* Convert the cash flow to an equivalent annual annuity over duration $N$ at interest $i$.
2. *Present worth* Convert the cash flow to a single equivalent amount (usually the present value) based on the duration of the study period $N$ and interest rate $i$.
3. *Rate of return* Equate the equivalent value of positive cash flow to the negative cash flow over the study period $N$ and determine the value of $i$.

All three methods are suitable for comparing the effects of cash-flow transactions. The choice of which one to use depends on the type of problem being analyzed, the personal preference of the analyst or the organizational preference, and which one yields results that are most understandable to the decision makers involved.

The mission of discounting cash flows to time equivalents is purely to establish the relative attractiveness of alternative courses of action. All alternatives must be compared on the basis of equivalent outcomes, as discussed in Chapter 5. In this chapter, the formulation and interpretation of annual-worth calculations are explored. In doing so, three assumptions are made:

1. Evaluations are made on *before-tax* cash flows. The effects of taxes are discussed in Chapter 11, where methods for including tax considerations in all three comparison methods are given.
2. Evaluations do not include intangible considerations. Equivalent values can be inputs to the total decision process, where they are evaluated in conjunction with difficult-to-quantify aspects by procedures such as the mixed-rating comparison method described in Chapter 5.
3. Evaluations do not include consideration of the availability of funds to implement alternatives. It is explicitly assumed that funds will be found to finance a course of action if its benefits are large enough. Although financing is not a direct input for computations, the output computed can be appraised with respect to available funding. For instance, an old, inefficient machine could be kept in operation because there appears to be insufficient capital available in the organization to afford a replacement, but an engineering economic analysis might point out that the savings from replacing the bungling machine would be so great that the organization cannot afford not to find funds for a replacement. Such capital-budgeting considerations are examined in Chapter 12.

The limitations listed above apply to the discussions in this and the next two chapters on economic comparison methods.

## UTILIZATION OF THE EQUIVALENT ANNUAL-WORTH COMPARISON METHOD

With an annual-worth method all the receipts and disbursements occurring over a period of time are converted to an equivalent uniform yearly amount. It is a very popular method because of the widespread inclination to view a year's gains and losses as a yardstick of progress. Cost-accounting procedures, depreciation expenses, tax calculations, and other summary reports are annual in nature. These yearly cost tabulations generally make the annual-worth method easier to apply and comprehend than the other comparison methods.

Equivalent annual-worth comparisons produce results compatible with present-worth and rate-of-return comparisons. For a set of common assumptions, a preference for an alternative exhibited by one method will be mirrored by the other two. Annual-worth calculations are frequently a part of the computations required to develop present-worth and rate-of-return values, and parallel computations by different methods are useful for complementary comparisons that improve the clarity of an analysis.

### Approximate Annual-Cost Calculation

A readily understood and very simply applied procedure for calculating the uniform yearly cost of owning an asset is called *straight-line depreciation plus average interest.* The main deficiency of this method is that it produces inaccurate results. However, it nicely relates annual cost to accounting practices that show how an asset loses value over its lifetime; it is thus introduced to illustrate concepts, ***not*** as the recommended method for calculating annual worth.

A machine, home, automobile, or other asset has a purchase price $P$, expected life $N$, and salvage value $S$. The purchase price is the present worth of the asset at its time of purchase, and the resale value is its future value at the end of the ownership period. The resale value may be negative (if it costs more to get rid of the asset than is returned by its scrap value) or positive ($S$ may even be greater than $P$ under unusual conditions of scarcity or rareness). During the period of ownership the asset's value gradually declines from $P$ to $S$. This decline is known as ***depreciation.***

One way to picture the devaluation of an asset is to assume it loses an equal amount of value each year, as shown in Figure 7.1. This pattern is named ***straight-line depreciation*** and is widely used for accounting purposes to charge ownership costs against operating revenue and to compute taxes. (The types and uses of depreciation accounts are discussed in Chapter 11.) In effect, an annual depreciation charge is made to recover the loss in value suffered by an asset from the income that asset helped produce during the year. The amount of the charge, using straight-line depreciation, is $(P - S)/N$. For the asset depicted in Figure 7.1, which has an initial cost $P$ of \$60,000 and an expected resale value $S$ of \$20,000 after 4 years of ownership, the annual loss in value using straight-line depreciation is

$$\text{Annual depreciation charge} = \frac{\$60{,}000 - \$20{,}000}{4 \text{ years}} = \$10{,}000/\text{year}$$

Note that the devaluation is assumed to occur at the end of each year of ownership. This accounting practice agrees with the payment pattern of an ordinary annuity.

In addition to the devaluation cost, an owner is deprived of the money that the capital invested in the asset could have earned if it had been invested elsewhere, unless a ***capital cost*** is also charged. This charge for the use of capital committed to an asset's purchase is

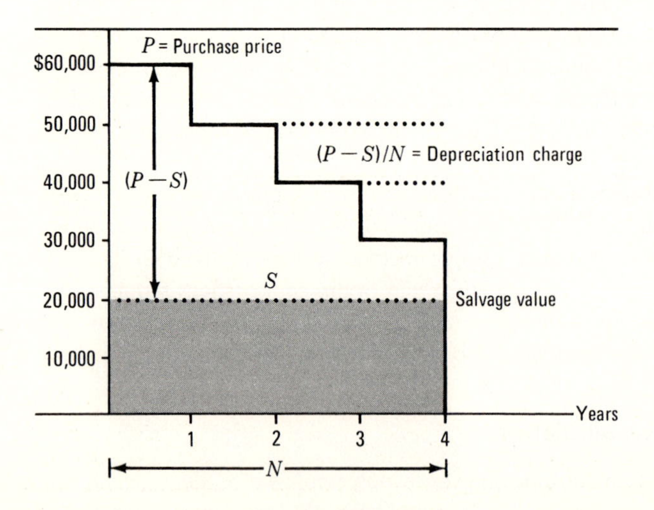


**FIGURE 7.1** Annual loss of value of an asset as a linear function of age as represented by straight-line depreciation when $P > S$.

identical to the charge for the use of loaned capital; in both cases the money is expected to earn a return. Thus the income produced by an asset beyond its operating expenses should be sufficient to repay the asset's loss in value ***plus*** a return on the value it still possesses. Its year-to-year value normally decreases, as indicated in Figure 7.1. Consequently, the capital cost decreases with age also. Capital costs for the first and last years of an asset's life can be averaged to obtain a uniform annual charge.

If $i$ is the interest rate for capital, then the first-year capital cost is $Pi$ or, equivalently, $(P - S)i + Si$, because interest is expected on the full purchase price. Capital cost during the last year of ownership is

$$\frac{P - S}{N} i + Si \qquad \text{where } N = \text{years of ownership}$$

and the average capital cost is

$$\frac{1}{2}\left[(P - S)i + Si + \frac{P - S}{N} i + Si\right] = (P - S)\frac{i}{2}\frac{N + 1}{N} + Si$$

Then the cost of ownership, not including operating expenses, is the sum of the depreciation charge to recover the capital invested plus the interest charge to pay for the use of the capital:

Annual cost for straight-line depreciation plus average interest =

$$\frac{P - S}{N} + (P - S)\frac{i}{2}\frac{N + 1}{N} + Si$$

Table 7.1 shows how straight-line depreciation plus average interest is applied (the interest rate is 10 percent) to recover the capital invested in the asset represented in Figure 7.1.

| *End of Year* | *Capital Not Recovered by End of Year* | *Interest Due on Unrecovered Capital* | *Straight-Line Depreciation Charge, (P − S)/N* | *Interest on Salvage Value ($20,000)(0.1)* |
|---|---|---|---|---|
| 0 | $40,000 | | | |
| 1 | 30,000 | $ 4000 | $10,000 | $2000 |
| 2 | 20,000 | 3000 | 10,000 | 2000 |
| 3 | 10,000 | 2000 | 10,000 | 2000 |
| 4 | 0 | 1000 | 10,000 | 2000 |
| Totals | | $10,000 | $40,000 | $8000 |

Comparable uniform payments (from table) $= \frac{\$40{,}000 + \$10{,}000}{4} + \frac{\$8000}{4}$
$= \$12{,}500 + \$2000 = \$14{,}500$

Straight-line depreciation plus average interest $= \frac{\$60{,}000 - \$20{,}000}{4} +$
$(\$60{,}000 - \$20{,}000)\frac{0.10}{2}\frac{4 + 1}{4} + \$20{,}000(0.1)$
$= \$10{,}000 + \$40{,}000(0.05)(5/4) + \$2000$
$= \$10{,}000 + \$2500 + \$2000 = \$14{,}500$

Total interest charged during asset's life $= \$10{,}000 + \$8000 = \$18{,}000$

**TABLE 7.1** Comparison of tabular and formula calculations of straight-line depreciation plus average annual interest when $P = \$60{,}000$, $S = \$20{,}000$, $N = 4$, and $i = 10$ percent.

## Exact Annual-Cost Calculation

Examination of the formula for straight-line depreciation plus average interest reveals why the annual cost so computed is inaccurate; the time value of compound interest is not recognized. The true annual cost results from applying the ***capital-recovery factor*** $(A/P, i, N)$. The total capital to recover is still $P - S$, but the capital-recovery factor accounts for both the repayment of capital and an interest charge on the unrecovered portion of the investment in one annual payment. Although the payments are uniform, the proportion of capital recovered and interest charged changes each year, as is apparent in Table 7.2. The annual interest charged on the salvage value is again equal to $Si$, because $S$ is assumed to be a constant amount of unrecovered capital throughout the study period. Therefore, using the capital-recovery factor,

*Equivalent annual cost* $= (P - S)(A/P, i, N) + Si$

This formula is the cornerstone of annual-worth computations.

| *End of Year* | *Capital Not Recovered by End of Year* | *Interest Due on Unrecovered Capital* | *Amount of Capital Recovered* | *Annual Capital-Recovery Charge* | *Interest on Salvage Value ($20,000)(0.1)* |
|---|---|---|---|---|---|
| 0 | $40,000 | | | | |
| 1 | 31,381 | $ 4000 | $ 8619 | $12,619 | $2000 |
| 2 | 21,900 | 3138 | 9481 | 12,619 | 2000 |
| 3 | 11,471* | 2190 | 10,429 | 12,619 | 2000 |
| 4 | 0 | 1147 | 11,472* | 12,619 | 2000 |
| Totals | | $10,476 | $40,001* | $50,476 | $8000 |

Annual cost = annual capital-recovery charge + interest on salvage value
$= (P - S)(A/P, i, N) + Si$
$= (\$60{,}000 - \$20{,}000)(0.31547) + (\$20{,}000)(0.10)$
$= \$12{,}619 + \$2000 = \$14{,}619$

Total interest charged $= \$10{,}476 + \$8000 = \$18{,}476$

*Differences are due to rounding errors.

**TABLE 7.2** Pattern of capital recovery and interest charges when the capital-recovery factor is applied to the data used in Table 7.1.

**Economy Exercise 7-1** Make a table to compare the percentage difference between annual payments for capital recovery plus interest by the exact and approximate methods for interest rates of 4, 8, 12, and 20 percent for durations of 5, 10, 15, and 20 years. Use the capital-recovery factor from the tables in Appendix B and the formula for straight-line depreciation plus average interest, $[1/N + (i/2)(N + 1)/N]$.

What general observation can you make about the trend of errors resulting from the approximate method as a function of increases in $i$ and $N$?

## SITUATIONS FOR EQUIVALENT ANNUAL-WORTH COMPARISONS

The term ***annual worth*** suggests a positive value, but the calculations can just as well produce a negative value. A negative annual worth indicates that the equivalent value of negative cash flow for disbursements is greater than the corresponding positive flow of receipts. Negative worths usually mean that an alternative is unacceptable. Exceptions occur when projects must be undertaken to satisfy certain requirements such as safety citations or building codes. Then the objective is to identify the alternative with the least equivalent cost (negative cash flow).

It is often very difficult, and not worth the required study time, to discover the income derived from one component in a complex system. For instance, the income produced by a copying machine is troublesome to derive exactly since its output is utilized by many people, often from different departments, working on many projects. In this type of situation, alternatives to satisfy the copying needs are evaluated on the basis of their relative costs, because each alternative capable of meeting the requirements of the system will produce the same income to the system. When it is apparent that *only* costs are involved in an evaluation, it is convenient to ignore the negative sign convention and let comparison figures represent the absolute value of costs.

Several situations for applying equivalent annual-worth calculations are described in the following examples.

**Example 7.1** **Equivalent Net Worth of Cash Flows**

A consulting firm proposes to provide "self-inspection" training for clerks who work with insurance claims. The program lasts 1 year, costs $2000 per month, and professes to improve quality while reducing clerical time. A potential user of the program estimates that savings in the first month should amount to $800 and increase by $400 per month for the rest of the year. However, operation confusion and work interference are expected to boost clerical costs by $1200 the first month, but this amount should decline in equal increments to zero by the end of the training year. If the required return on money is 12 percent compounded monthly and there is a stipulation that the program must pay for itself within 1 year, should the consultants be hired?

**Solution 7.1**

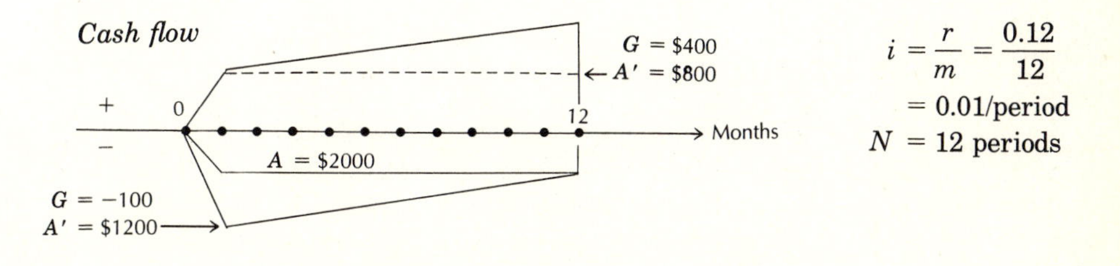


$$i = \frac{r}{m} = \frac{0.12}{12} = 0.01/\text{period}$$
$$N = 12 \text{ periods}$$

$$\begin{aligned}\text{Equivalent monthly worth of savings} &= \$800 + \$400(A/G, 1, 12)\\ &= \$800 + \$400(5.3682)\\ &= \$2947\end{aligned}$$

$$\begin{aligned}\text{Equivalent monthly worth of costs} &= -\$2000 - [\$1200 - \$100(A/G, 1, 12)]\\ &= -\$3200 + \$100(5.3682)\\ &= -\$2663\end{aligned}$$

Equivalent net monthly cash flow = \$2947 − \$2663 = \$284

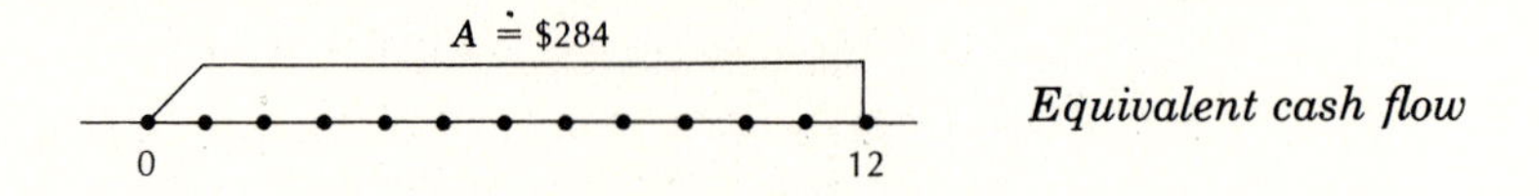


The program looks very promising because the equivalent monthly worth is positive during the first year, and savings generated by the training should continue into the future.

Gradient factors are utilized in Example 7.1 to convert uniformly varying cash flows to their equivalent constant worths. Recovery of capital is not an issue, since no property ownership is involved. The comparison is made directly on the basis of expected income versus outgo.

**Example 7.2** **Net Annual Worth of a Single Project**

The purchase of a truck with an operator's platform on a telescoping hydraulic boom will reduce labor costs for sign installations by \$10,000 per year. The price of the boom truck is \$57,000, and its operating costs will exceed those of the present equipment by \$100 per month. The resale value is expected to be \$6000 in 12 years. Should the boom truck be purchased when the prevailing interest rate is 12 percent?

**Solution 7.2**

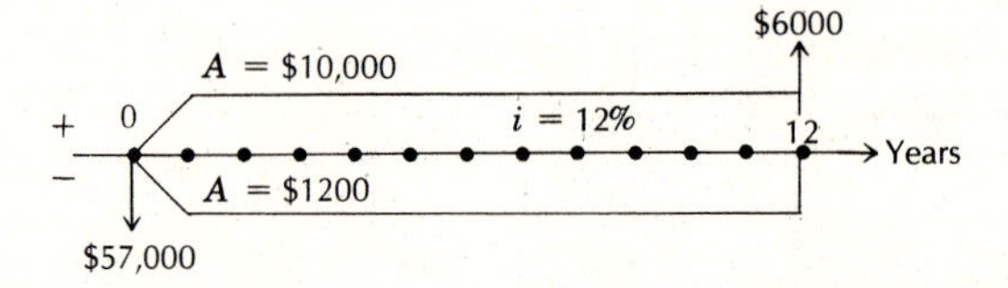


*Net annual worth*

$$\begin{aligned} AW &= -\$57{,}000(A/P, 12, 12) + \$6000(A/F, 12, 12) - \$100(12) + \$10{,}000 \\ &= -\$57{,}000(0.16144) + \$6000(0.04144) - \$1200 + \$10{,}000 \\ &= -\$9202 + \$249 + \$8800 = -\$153 \end{aligned}$$

Equivalent annual-worth calculations indicate that the purchase and use of the boom truck will cause a loss equivalent to \$153 per year for 12 years, compared with other investments that could earn a 12 percent return.

The solution to Example 7.2 was developed from the accompanying cash-flow diagram. The capital-recovery factor leads to the same solution when capital-recovery costs are registered negatively and the net annual savings are positive.

$$\begin{aligned} AW &= \text{annual savings} - \text{capital-recovery costs} \\ &= \$10{,}000 - \$1200 - [(P - S)(A/P, 12, 12) + Si] \\ &= \$8800 - [(\$57{,}000 - \$6000)(0.16144) + \$6000(0.12)] \\ &= \$8800 - (\$8233 + \$720) = \$8800 - \$8953 = -\$153 \end{aligned}$$

Whether to use $(P - S)(A/P, i, N) + Si$ or $P(A/P, i, N) - S(A/F, i, N)$ to calculate the cost of recovering capital is simply a personal choice, although the former method is most popular.

**Example 7.3** **Comparison of Net Annual Worths**

**A supplier of laboratory equipment estimates that profit from sales should increase by \$23,000 per year if a mobile demonstration unit is built. A large unit with sleeping accommodations for the driver will cost \$71,000, while a smaller unit without sleeping quarters will be \$55,000. Salvage values for the large and small units after 5 years of use will be, respectively, \$8000 and \$3500. Lodging costs saved by the larger unit should amount to \$3000 annually, but its yearly transportation costs will exceed those of the smaller unit by \$1300. With money at 15 percent, should a mobile demonstration unit be built, and if so, which size is preferable?**

**Solution 7.3** *AW of large mobile demonstration unit*

| | |
|---|---|
| Annual increase in profit | \$23,000 |
| Savings in lodging costs over smaller unit per year | 3,000 |
| Extra transportation costs over smaller unit, per year | −1,300 |
| Capital-recovery cost: | |
| (\$71,000 − \$8000)($A/P$, 15, 5) + \$8000$i$ | |
| (\$63,000)(0.29832) + \$8000(0.15) | −19,994 |
| Net AW = | \$4,706 |

*AW of small mobile demonstration unit*

| | |
|---|---|
| Annual increase in profit | \$23,000 |
| Capital-recovery cost: | |
| (\$55,000 − \$3500)($A/P$, 15, 5) + \$3500$i$ | |
| (\$51,000)(0.29832) + \$3500(0.15) | −15,888 |
| Net AW = | \$7,112 |

**The net annual worths indicate that both alternatives will produce positive cash flows while concurrently repaying investment costs. The small mobile demonstration unit is preferred because it promises a larger annual profit.**

Since the profit increase expected from building either of the mobile labs is the same, the comparison could have been conducted by considering just the costs first:

| | Small Unit | Large Unit |
|---|---|---|
| Capital-recovery cost: | \$15,888 | \$19,994 |
| Net extra cost of smaller unit: \$3000 − \$1300 | 1,700 | |
| Total annual cost | \$17,588 | \$19,994 |

and then evaluating the best resulting alternative in terms of the expected income.

**Example 7.4** **Alternatives with Equal Annual Costs**

A machine needed for 3 years can be purchased for \$77,662 and sold at the end of the period for about \$25,000. A comparable machine can be leased for \$30,000 per year. If a firm expects a return of 20 percent on investments, should it lease or buy the machine?

**Solution 7.4**

$$\begin{aligned}\text{Equivalent annual cost to buy} &= (\$77{,}662 - \$25{,}000)(A/P, 20, 3) + \$25{,}000(0.20)\\ &= \$52{,}662(0.47473) + \$5000\\ &= \$30{,}000\\ \text{Annual cost to lease} &= \$30{,}000\end{aligned}$$

If the salvage value is considered reasonably accurate, the machine should be purchased because the funds invested in it will earn 20 percent (assuming 20 percent is an attractive rate of return for the firm).

The two alternatives in Example 7.4 are compared on the basis of their costs because the income resulting from their contribution is not available and, it is believed, both are capable of producing that contribution. The question is not whether to get a machine. It is known that the machine is necessary, so it is a choice between buying and leasing it. In selecting between the alternatives with equal equivalent annual worth, it should be recognized that they are equal only after ownership of the machine has earned 20 percent on the capital invested in it. Therefore, if the owners are satisfied with a 20 percent return on their money and no other considerations are involved, it is prudent to purchase.

**Example 7.5** **Comparison of Assets with Unequal Lives**

Two models of machines can be purchased to perform the same function. Type I has a low initial cost of \$3300, high operating costs of \$900 per year, and a short life of 4 years. The more expensive type II costs \$9100, has annual operating expenses of \$400, and can be kept in service economically for 8 years. The scrap value from either machine at the end of its life will just cover its removal cost. Which is preferred when the minimum attractive rate of return is 8 percent?

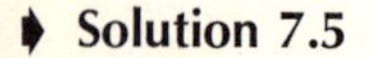
**Solution 7.5**

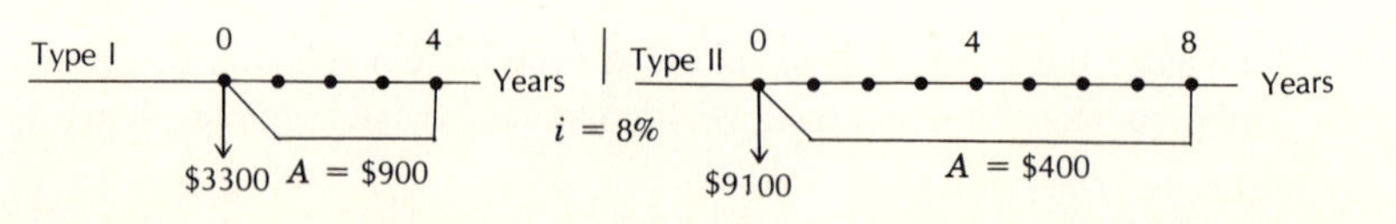

| *Type I Machine* | | *Type II Machine* | |
|---|---|---|---|
| Operating expenses | \$ 900 | Operating expenses | \$ 400 |
| Capital recovery: | | Capital recovery: | |
| $P(A/P, 8, 4)$ = \$3300(0.30192) | 996 | $P(A/P, 8, 8)$ = \$9100(0.17402) | 1584 |
| | AW = \$1896 | | AW = \$1984 |

**The type I machine has a lower annual cost for service during the next 4 years and is therefore preferred.**

The machines described in Example 7.5 exhibit the common feature that more expensive models, designed to serve the same function as less expensive versions, are expected to operate more economically and/or last longer. (If a costlier machine also produces better-quality products, the benefits from improved quality must be included in the analysis to make the outcomes comparable.) The difficulty in comparing alternatives with unequal lives is to account for the service provided during the period in which one outlasts the other.

The implied assumption in Solution 7.5 is that two machines of type I will be purchased consecutively to provide the same length of service as one type II machine. The equivalent annual cost for 8 years of service from two type I machines is, of course, the same as calculated in the solution above:

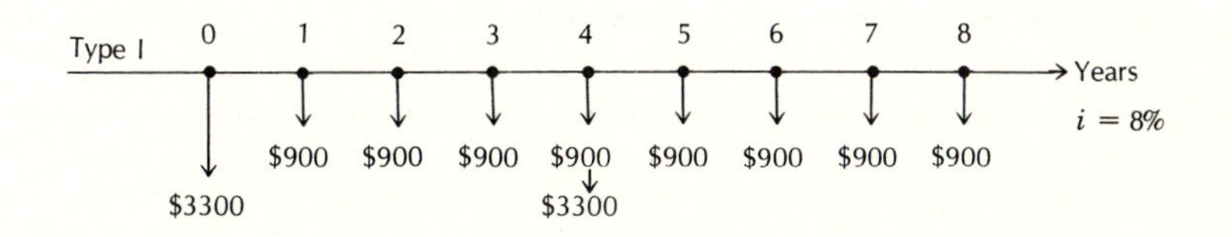


$$\begin{aligned} AW &= \$3300(A/P, 8, 8) + \$3300(P/F, 8, 4)(A/P, 8, 8) + \$900 \\ &= \$3300(0.17402) + \$3300(0.73503)(0.17402) + \$900 \\ &= \$574 + \$422 + \$900 = \$1896 \text{ (as calculated previously)} \end{aligned}$$

The "repeated projects" assumption for evaluating assets with different lives is reasonable and widely used; other interpretations are explored in the next section. Unless reliable forecasts can be made about future operating conditions and the probability of technical advances, the assumption that today's conditions will exist in the future is plausible. When future conditions can be estimated with confidence, these valuations are the data for equivalent annual-worth calculations. For instance, a confident prediction that current developmental work on the type I machine will produce refinements within 4 years to reduce operating costs by one-third while increasing the purchase price by one-half leads to a revised economic analysis:

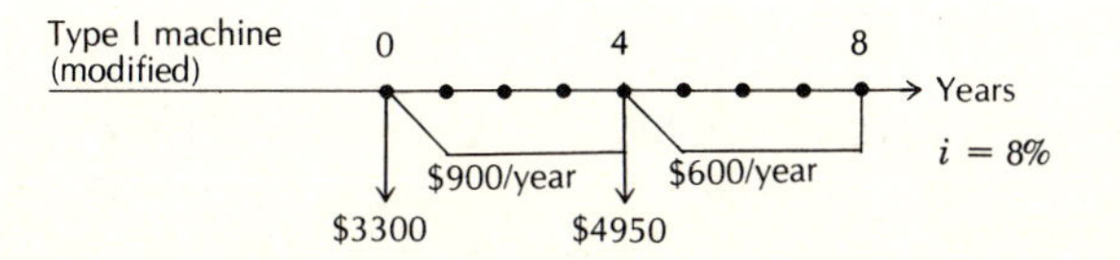

The annual worth is calculated by (1) translating the cash flow in the last 4 years to a present value at year 4 and discounting this value to year 0; (2) computing the present value of the first 4 years' cash flow; (3) adding the two present worths; and (4) converting the total to an 8-year annuity:

$$\begin{aligned} \text{AW (type I modified)} &= \{[\$600(P/A, 8, 4) + \$4950](P/F, 8, 4) + \$900(P/A, 8, 4) + \$3300\}(A/P, 8, 8) \\ &= \{[\$600(3.3121) + \$4950](0.73503) + \$900(3.3121) + \$3300\}(0.17402) \\ &= [\$6937(0.73503) + \$2981 + \$3300](0.17402) \\ &= (\$5099 + \$6281)(0.17402) = \$1980 \end{aligned}$$

The modified type I machine now has nearly the same equivalent annual cost as the type II model. If a decision maker has confidence in the forecasted data, the choice between types could go either way, but it will more likely swing to type I because the lower purchase price and shorter life mean less capital is committed for a shorter period. This conservative philosophy provides some protection from unexpected developments. It was assumed in the comparison that the need for the machine would exist for 8 years. Many things can happen in 8 years to thwart the most carefully conceived plans—new designs, changing markets, successful competition, etc. The opportunity to reevaluate tactics in 4 years, as possible with the type I machine, is a subtle but valued attribute.

**Economy Exercise 7-2** The five investment proposals shown below have been investigated carefully and are deemed to be equally safe. An investor will be satisfied if a minimum before-tax return of 9 percent is realized.

| *Proposal* | *Investment P* | *Life N, years* | *Salvage Value S* | *Net Annual Cash Flow* |
|---|---|---|---|---|
| Alpha | $30,000 | 5 | $ 0 | $ 7,500 |
| Beta | 60,000 | 5 | 10,000 | 13,755 |
| Gamma | 20,000 | 5 | 0 | 5,000 |
| Delta | 40,000 | 5 | 10,000 | 10,000 |
| Epsilon | 30,000 | 5 | 5,000 | 7,500 |

**a** Which proposal is preferred if only one can be selected? Why?
**b** How much should be invested if unlimited capital is available? Why?
**c** Which proposals should be utilized if an investor wishes to allocate exactly $60,000 of investment capital, and any funds not put into one or more of the proposals will be used to purchase bonds that pay annual dividends of 7½ percent on the amount invested?

## ASSETS' LIVES IN ECONOMIC COMPARISONS

Translating cash flows to equivalent annuities is a mechanical process that becomes almost automatic with practice. Understanding the meaning of an economic comparison and being able to explain its significance to others are the critical skills. The issues raised in this section apply to other comparison methods besides annual worth, and related considerations are sprinkled along the rest of the text.

## Definitions of Asset Life

In time-value mechanics, $N$ is simply the number of compounding periods appropriate for the analysis of cash flows. $N$ takes on a special meaning when it represents the life of an asset that loses value as a function of use or time. The more frequently applied terms to describe the life of an asset are listed and defined as follows:

*Ownership life* or *service life* is the period of time an asset is kept in service by an owner or owners. Implied is a period of useful service from the time of purchase until disposal. Actually, under the vague expectation that it might somehow again prove useful, equipment is often retained beyond the point where it is capable of satisfying its intended function. A machine can have a *physical life* longer than its service life; the machine is still physically sound, but there is no useful function for it to perform.

*Accounting life* is a life expectancy based primarily on bookkeeping and tax considerations. It may or may not correspond to the period of usefulness and economic desirability. As will be seen in Chapter 11, the accounting records for the life and depreciation pattern of an asset affect taxes which in turn affect the net income derived from employing the asset.

*Economic life* is the time period that minimizes the asset's total equivalent annual cost or maximizes its equivalent net annual income. This period terminates when the asset is displaced by a more profitable replacement or the asset's service is no longer required. Economic life is also referred to as the *optimal replacement interval* and is the condition appropriate for most engineering economic studies.

Land is not subject to a specified life or to capital recovery because it historically appreciates in value rather than depreciating with age. The cost of land ownership is the interest not received on funds invested in the property.

**Economy Exercise 7-3** Experience with a conventional mechanical log debarker suggests that its service life is 4 years. A newly designed hydraulic debarker costs one-third more than a mechanical debarker but makes much less noise. Both debarkers have about the same operating costs and no salvage value. What will the optimal replacement interval of the new hydraulic debarker have to be to make its cost comparable to the mechanical debarker at $i = 12$ percent?

## Evaluations Involving Unequal Lives

The *repeated-projects method* of analysis forwarded in Example 7.5 assumes that assets are to be replaced by similar assets. The implication is that the function performed by the asset remains the same indefinitely, and no better way to perform the function will be developed. Many kinds of equipment historically fit the assumption, and similar replacements are usually valid for short-lived assets. When there is good reason to suspect that the ratio of prices or operating costs among alternatives will change in the future, comparisons should include the prospective differences.

The possibility of modifying costs during subsequent life cycles was illustrated in the discussion of Example 7.5, where costs revised from the first 4-year life were used in the second 4-year life cycle of an alternative to compare its annual worth with an 8-year-life alternative. When such revisions are made in repeated life cycles, the study period should extend to a point at which all alternatives have a common endpoint. For instance, three alternatives with lives of 2, 3, and 4 years would be evaluated over a 12-year period, the smallest interval divisible by complete life cycles for all the alternatives being compared.

The *study-period method* for evaluating alternatives with unequal lives begins with the selection of a time period for the comparison. The choice is based on analysis convenience, custom, and data availability. Then all cash flows of an asset beyond the study period are translated to the endpoint of the period and converted to their equivalent annual worth. For example, using the data for the type I and type II machines from Example 7.5 and a study period of 4 years, the annual cost of the type I machine is

$$\begin{aligned}\text{AW(type I)} &= P(A/P, 8, 4) + \$900 \\ &= \$3300(0.30192) + \$900 = \$1896\end{aligned}$$

and the annual cost of the type II machine, which has a life of 8 years, prorates costs during the last 4 years of life over the 4-year study period as

$$\begin{aligned}\text{AW(type II)} &= P(A/P, 8, 8)(P/A, 8, 4)(A/P, 8, 4) + \$400 \\ &= \$9100(0.17402)(3.3121)(0.30192) + \$400 \\ &= \$1984\end{aligned}$$

which are the same annual worths calculated in Solution 7.5.

That both the study-period and repeated-projects methods give the same AW is reassuring. The methods just approach the question differently. The study-period procedure implies that asset costs beyond the study period are ignored. These costs are actually replaced by a "salvage" value at the end of the study period equivalent to their worth at that time. For the sample data, the implied salvage value of the type II machine is the present value at the end of year 4 of the unrecovered balance of the purchase price:

$$\begin{aligned}&\text{Implied salvage value of the} \\ &\text{type II machine at end of year 4} = P(A/P, 8, 8)(P/A, 8, 4) \\ &\qquad = \$9100(0.17402)(3.3121) = \$5245\end{aligned}$$

Using this salvage value in the capital-recovery formula once again results in the same AW for the type II machine:

$$\begin{aligned}\text{AW(type II)} &= (P - S)(A/P, 8, 4) + S(0.08) + \$400 \\ &= (\$9100 - \$5245)(0.30192) + \$5245(0.08) + \$400 \\ &= \$3855(0.30192) + \$420 + \$400 \\ &= \$1984\end{aligned}$$

A third, slightly different, approach is to assume ***an asset will be succeeded by the most economical replacement then available.*** Recognizing the problem of prediction accuracy, one rationalizes that the replacement will be at least as good as the best available today. Calculations abiding by this line of reasoning do not normally change the preference indicated by the repeated-projects method, but differences in annual worth between alternatives may be less (see Problem 7.13). In the special case where the alternatives being compared are known not to be the best possible but are nonetheless the best available (perhaps owing to urgent need, there is insufficient time to secure the most economical asset and the intent is to determine which substitute is best), the known cost for the best possible replacement is used in the analysis.

### Perpetual Life

Occasionally, an asset is treated as if it will last forever. The assumption of infinite life in terms of capital recovery is slightly more reasonable than the physical interpretation. Nothing made by humans, even Egyptian pyramids or the great wall of China, lasts forever, but the difference between infinity and 100 years in the numerical value of the capital-recovery factor is quite small:

$$(A/P, 10, \infty) = \frac{0.10(1.1)^{\infty}}{(1.10)^{\infty} - 1} = 0.10 \frac{\infty}{\infty} = 0.10$$

$$(A/P, 10, 100) = 0.10001$$

Therefore, in an economic comparison involving an asset with an infinite life, ***such as land***, the interest rate replaces the capital-recovery factor. The man-made assets most closely approaching perpetual life are dams, tunnels, canals, aqueducts, and monuments. The nature of very long-lived assets relegates them mostly to public projects, and there the trend has been to set a study period of 50 years or so in recognition of changing public needs and technological advances that generate new ways to fulfill the needs.

---

**Economy Exercise 7-4** A short concrete canal can be constructed as part of a flood-control project; the placement of a large galvanized culvert will serve the same function. The cost of the canal, which will last indefinitely, is $75,000, and maintenance costs will average $400 per year. A culvert, which will have to be replaced every 30 years, will cost $40,000 and have annual maintenance costs of $700. Salvage values are negligible for both alternatives, and the government interest rate is 6 percent. Which alternative has the lowest equivalent annual cost?

---

## USE OF A SINKING FUND

The sinking-fund factor was used as an alternative means of calculating capital-recovery costs in the discussion of Example 7.2: $(P - S)(A/P, i, N) + Si = P\,(A/P, i, N) - S\,(A/F, i, N)$. As is apparent in the equation, the sinking-fund factor is applied to compute the annuity required to accumulate a certain future amount. Organizations are sometimes obligated by legislated or contractual agreements to establish a fund, separate from their internal operations, to accumulate a specified amount by a specified time. This accumulation is called a *sinking fund.*

A provision for a sinking fund requires that an organization set aside a portion of the income derived from sales or taxes each year in order to retire a bond issue (or, in some cases, an issue of preferred stock). Failure to meet the sinking-fund payments forces the bond issue to be thrown into default, causing serious credit and credibility problems. The payments are a direct cash drain on the organization. That is the purpose of the sinking-fund—to protect investors by enforcing an orderly retirement of debt from current income.

A debt can be retired by regular payments from the sinking fund or by letting money accumulate in the sinking fund until the debt is due and then paying it in full (plus interest). Periodic payments are associated with *callable bonds.* A call provision in a bond gives the issuing organization the right to pay off a bond before its maturity date. Funds reserved from income are allocated each year to retire a portion of the bond issue. Sinking-fund payments thus utilized earn at the rate at which interest payments are avoided on the bonds retired.

When a sinking fund is established to accumulate sufficient money to meet the bond cost at maturity, annual payments are normally required to be invested in a savings institution. The interest rate for these "savings" typically is at a rate lower than the organization earns on its own capital, but the sinking fund is less an earning device than a way to assure that funds will not be diverted to other ventures.

For example, a firm borrows $1 million at 9 percent simple interest because it believes that the amount borrowed can be utilized within the firm to earn double the borrowing rate, 18 percent. If the $1 million is acquired by issuing 20-year bonds with the stipulation that a sinking fund be set up, the firm has to set aside a payment each year and put it in an external account. This account probably earns less than the 18 percent internally earned on investment funds, but it represents a very secure investment. Assuming the account pays 6 percent annual compound interest, the annual payments into the sinking fund would be

$$\begin{aligned} A &= \$1{,}000{,}000(A/F, 6, 20) \\ &= \$1{,}000{,}000(0.02719) = \$27{,}190 \end{aligned}$$

to make total annual debt repayment on the principal plus interest of $27,190 + $1,000,000(0.09) = $117,190.

## REPLACEMENT STUDIES

*Replacement* refers to a broad concept embracing the selection of similar but new assets to replace existing assets, and evaluation of entirely different ways to perform an asset's function. For instance, old trucks could be replaced with new models that operate similarly but have advanced features that improve their performance. The trucks could also be replaced with a conveyor system, an overhead crane, a subcontract for hauling, or even manual labor, if any of these methods serves the needed function at a lower total cost.

Replacement decisions are critically important to a firm. A hasty decision to "get rid of that junk" because a machine is temporarily malfunctioning, or a decision to faddishly buy the latest model because "we take pride in being very modern," can be a serious drain on operating capital. A firm hard pressed for operating funds may go to the other extreme by adopting a policy that postpones replacements until there is no other way to continue production. A policy of postponement places a firm in the dangerous position of becoming noncompetitive. Reliance on inefficient equipment and processes that lead to higher long-run operating costs or low quality, while competitors enjoy declining costs and better quality gained from modern machinery, is a delaying action that eventually must be paid for, perhaps in bankruptcy. Engineers bear responsibility to recognize when an asset is no longer employed efficiently, what replacements should be considered, and when replacement is economically feasible.

A replacement decision is a choice between the present asset, sometimes called the ***defender***, and currently available replacement alternatives, sometimes called ***challengers***. The defender may or may not be at the end of its economic life. An asset is ***retired*** when its owner disposes of it, but it may still serve other owners as second-hand equipment before it is scrapped. Unsatisfactory performance and inability to meet current capacity needs are the main causes of retirement. The challenger may or may not perform the function of the defender in the same way.

Replacement studies are usually made as annual-worth calculations to take advantage of data traditionally collected as annual charges: depreciation, maintenance costs, operating expenses, salaries, taxes, etc. In this chapter the annual-worth comparisons are made on a before-tax basis, and the book values of assets as derived from depreciation accounting are not considered. These considerations and the use of present-worth and rate-of-return methods are developed in subsequent chapters. Here, challengers are compared to the defender according to their annual worth (or cost).

## Replacement due to Deterioration

Deterioration is manifested through excessive operating costs, increased maintenance cost, higher reject rates, or a combination of added equipment costs. As costs climb, it soon becomes apparent that a replacement study is warranted. Successive studies may be required to determine when the costs for operating another period without change become greater than the average cost expected from the replacement.

In many replacement studies it is appropriate first to check the annual cost of retaining the existing asset one more year. This 1-year cost of ownership is compared to the equivalent annual cost for the challenger, taken over its economic life. The assumption is that the challenger, when acquired, will be kept its full economic life, but it will not be acquired until its equivalent annual cost is lower than next year's cost for the defender. Before final acceptance, the challenger must also pass the test that its equivalent annual cost is smaller than the equivalent annual cost of the defender, based on the defender's remaining years of service.

**Example 7.6** **Annual Costs for Next Year and Remaining Years of Ownership of an Existing Asset Compared to the Equivalent Annual Cost of a Proposed Replacement**

An existing machine is worth \$2500 today and will lose \$1000 in value by next year plus \$500 per year thereafter. Its \$8000 operating cost for this year is predicted to increase by \$1000 annually, owing to deterioration. It will be retired in 4 years, when its salvage value will be zero.

A new, improved machine that satisfactorily performs the same function as the existing machine can be purchased for \$16,000 and is expected to have relatively constant annual operating costs of \$6000 to the end of its 7-year economic life, at which time the salvage value will be \$1500. No major improvements are expected in designs for machines of this type within 7 years.

If the minimum attractive rate of return is 12 percent, should the existing machine be replaced? If so, when?

**Solution 7.6**

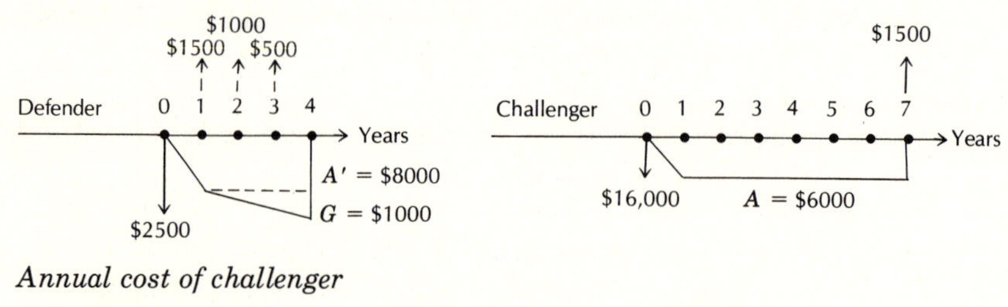


*Annual cost of challenger*

| | |
|---|---|
| Operating cost | \$6000 |
| Capital recovery: $(P - S)(A/P, 12, 7) + Si$ | |
| (\$16,000 − \$1500)(0.21912) + \$1500(0.12) | 3357 |
| | \$9357 |

*One-year cost of defender*

| | |
|---|---|
| Operating cost | \$8000 |
| Capital recovery: $(P - S)(A/P, 12, 1) + Si$ | |
| (\$2500 − \$1500)(1.12) + \$1500(0.12) | 1300 |
| | \$9300 |

Since the defender has a lower annual cost for next year, it should be retained. To anticipate the future replacement budget, conditions for service during year 2 should also be checked. At the beginning of year 2 the existing asset has a value of \$1500, which declines to \$1000 at the end of the year. Operating costs for the year are \$8000 + \$1000 = \$9000.

*Annual cost of defender during year 2*

| | |
|---|---|
| Operating cost | \$9000 |
| Capital recovery: (\$1500 − \$1000)(1.12) + \$1000(0.12) | 680 |
| | \$9680 |

If the cost estimations are deemed reasonable, purchase of the challenger should be anticipated 1 year from now.

It would be a mistake to base the decision on comparison of the equivalent annual cost of the existing machine for its four remaining years before retirement, as follows:

*Annual cost of defender based on a 4-year life*

| | |
|---|---|
| Operating cost: $A' + G(A/G, 12, 4)$ | |
| $8000 + $1000(1.3588) | $ 9,359 |
| Capital recovery: $(P - S)(A/P, 12, 4) + Si$ = $2500(0.32924) | 823 |
| | $10,182 |

The challenger would then have been purchased while the defender's cost of one more year of ownership was lower than the replacement's equivalent annual cost. A comparable calculation for the conditions expected 1 year from now confirm the decision to make the replacement then:

*Annual cost of defender based on the last 3 years before retirement*

| | |
|---|---|
| Operating cost: $A' + G(A/G, 12, 3)$ | |
| $9000 + $1000(0.9246) | $ 9,925 |
| Capital recovery: $P(A/P, 12, 3)$ = $1500(0.41635) | 625 |
| | $10,550 |

A *cyclic replacement* pattern was discussed in Chapter 4 for a minimum-cost evaluation using a zero interest rate. The purpose of the analysis was to determine when an existing asset should be replaced with a new asset with the same capital and operating costs. The same evaluation procedure is suitable when interest is charged on the cash flow. In effect, the analysis determines the economic life of the asset.

The data utilized in Table 4.1 to calculate the average annual cost for replacing a truck when the pattern of operating and capital costs remain unchanged for successive cycles are repeated below.

| | YEAR | | | | | | |
|---|---|---|---|---|---|---|---|
| | 0 | 1 | 2 | 3 | 4 | 5 | 6 |
| Operating cost | | $1500 | $1600 | $1900 | $2300 | $2800 | $3400 |
| Resale value | $5000 | 3000 | 2300 | 1700 | 1300 | 1100 | 900 |

Assuming each year is the economic life $N$ of the asset, the equivalent annual cost for owning the asset $N$ years is calculated as

$$\begin{aligned} \text{AW} &= \text{capital recovery} + \text{equivalent annual operating cost} \\ &= (P - S)(A/P, i, N) + Si + \text{FW(operating costs for } N \text{ years)}(A/F, i, N) \end{aligned}$$

The annual cost for 1 year of ownership when $i$ = 10 percent is

$$AW_{N=1} = (\$5000 - \$3000)(A/P, 10, 1) + \$3000(0.10) + \$1500$$
$$= \$2000(1.10) + \$300 + \$1500 = \$4000$$

For successive years,

$$AW_{N=2} = (\$5000 - \$2300)(A/P, 10, 2) + \$2300(0.10) + [\$1500(F/P, 10, 1) + \$1600](A/F, 10, 2)$$
$$= \$2700(0.57619) + \$230 + [\$1500(1.10) + \$1600](0.47619)$$
$$= \$1786 + \$1548 = \$3334$$

$$AW_{N=3} = (\$5000 - \$1700)(A/P, 10, 3) + \$1700(0.10) + [\$1500(F/P, 10, 2) + \$1600(F/P, 10, 1) + \$1900](A/F, 10, 3)$$
$$= \$3300(0.40212) + \$170 + [\$1500(1.21) + \$1600(1.10) + \$1900](0.30212)$$
$$= \$1497 + \$1654 = \$3151$$

$$AW_{N=4} = (\$5000 - \$1300)(A/P, 10, 4) + \$1300(0.10) + [\$1500(F/P, 10, 3) + \$1600(F/P, 10, 2) + \$1900(F/P, 10, 1) + \$2300](A/F, 10, 4)$$
$$= \$3700(0.31547) + \$130 + [\$1500(1.331) + \$1600(1.21) + \$1900(1.10) + \$2300](0.21547)$$
$$= \$1297 + \$1793 = \$3090$$

$$AW_{N=5} = (\$5000 - \$1100)(A/P, 10, 5) + \$1100(0.10) + [\$1500(F/P, 10, 4) + \$1600(F/P, 10, 3) + \$1900(F/P, 10, 2) + \$2300(F/P, 10, 1) + \$2800](A/F, 10, 5)$$
$$= \$3900(0.26380) + \$110 + [\$1500(1.4641) + \$1600(1.331) + \$1900(1.21) + \$2300(1.10) + \$2800](0.16380)$$
$$= \$1139 + \$1958 = \$3097$$

which indicates that the economic life of the truck and the minimum-cost replacement period is 4 years. This is the same replacement interval indicated in Table 4.1, where the analysis was conducted at $i = 0$. Note that equivalent annual capital costs decline as the ownership period increases, while the equivalent annual operating costs increase owing to deteriorating performance. The pattern of changes in capital and operating costs should be checked periodically, when like-for-like replacements are in effect, to confirm that the economic life remains the same.

## Replacement due to Obsolescence

Each new development or refinement of an older asset makes the previous way of accomplishing an objective less appealing. Improved methods and machines affect replacement studies in two opposing ways. Improvements usually result in higher first costs for a challenger and reduce the salvage value of a defender; together these two capital costs tend to favor retention of the existing asset. However, dramatic operating-cost reductions or impressive quality gains provided by technologically advanced challengers are the typical causes of obsolescence, and these annual savings tend to exceed the capital-cost advantage of a defender.

Progressive design enrichments can outdate an existing asset well before its economic life has expired. When this happens, the owned asset probably retains considerable resale value, unless the new design is revolutionary. The replacement study should be based on the current *market* value of the defender, *not* on the accounting records or original estimates of resale value because these figures rely on data acquired before the existence of the improved challenger became known. The remaining years of service life for an obsolete asset usually conform to the original estimate of $N$.

**Example 7.7** **Selection of a Salvage Value for an Obsolete Asset**

A low-volume office copying machine was purchased 2 years ago for \$700. At the time of purchase it was believed that the machine would have an economic life of 5 years and a salvage value of \$100. Operating costs over the first 2 years for material, labor, and maintenance have averaged \$4200 annually and are expected to continue at the same level. Some type of copier will be needed for the next several years.

The same company that manufactured the presently used copying machine has a new model which costs \$1000 but will perform the current workload with operating costs of \$3500 per year. They are offering \$500 for the old model as a trade-in on the new machine. The expected salvage value for the new model is \$200 at the end of 10 years.

Another company has a different type of copier which is available only on a lease basis. The company claims that leasing their copier at \$750 per year will reduce the operating expense for the present amount of work to \$2750. Since they do not accept trade-ins, the machine now in use would have to be sold in the open market, where it is expected to bring only \$250.

If the minimum acceptable rate of return is 10 percent before taxes, should the defending copier be replaced by one of the challengers?

**Solution 7.7** The market value of the present copier seems to depend on the alternative to which it is being compared, but in either case the remaining period of expected service is 3 years and the salvage value is \$100. Trade-in values are often placed unrealistically high to enthuse a buyer. What the bloated trade-in really amounts to is a discount from the selling price. For the given data, it appears that the true market value is \$250 rather than \$500. The difference is a discount of \$500 − \$250 = \$250 which makes the effective purchase price = \$1000 − \$250 = \$750. In order to compare both challengers to the defender under equitable conditions, the apparent market value (\$250) is used, and the price for purchasing a copier is discounted accordingly.

*Annual cost of defender*

| | |
|---|---|
| Operating costs | \$4200 |
| Capital recovery: $(P - S)(A/P, 10, 3) + S(0.10)$<br>(\$250 − \$100)(0.40212) + \$100(0.1) | 70 |
| | \$4270 |

*Annual cost of the challenger from the same vendor*

| | |
|---|---|
| Operating costs | \$3500 |
| Capital recovery: $(P - S)(A/P, 10, 10) + S(0.10)$ | |
| [(\$1000 − \$250) − \$200](0.16275) + \$200(0.1) | 110 |
| | \$3610 |

*Annual cost of the leased challenger*

| | |
|---|---|
| Operating costs | \$2750 |
| Annual lease contract | 750 |
| | \$3500 |

The annual-cost calculations indicate both challengers are preferred to the defender. Leasing is more attractive than buying, if it can be assumed that lease charges will not increase. Another consideration is the possibility that future improvements will make the current \$1000 challenger obsolete in the near future. A lease is less expensive, for next year at least, and provides flexibility to take advantage of future cost-reduction developments if they occur.

Using the trade-in value for the existing asset in Example 7.7 does not alter the solution appreciably:

*Annual cost of defender when P = trade-in value = \$500*

| | |
|---|---|
| Operating costs | \$4200 |
| Capital recovery: (\$500 − \$100)(0.40212) + \$100(0.1) | 171 |
| | \$4371 |

*Annual cost of purchased challenger when P = \$1000*

| | |
|---|---|
| Operating costs | \$3500 |
| Capital recovery: (\$1000 − \$200)(0.16275) + \$200(0.1) | 150 |
| | \$3650 |

The disparity between annual worths using the trade-in value (\$4371 − \$3650 = \$721) and the market value for the defender (\$4270 − \$3610 = \$660) is \$721 − \$660 = \$61. It is due to the difference in the capital-recovery periods of the defender and challenger. Equal reductions in the present values of the challenger and defender favor the defender in a comparison because it almost always has a shorter life. Thus the use of effective value is a more conservative approach.

Viewing a replacement decision from a stranger's outlook may clarify the situation and provide impartiality. A stranger viewing the choice of a copier described in Example 7.7 could place the decision in the following context: "I need a copying machine. From one manufacturer I can buy a new machine for \$1000 which will have a \$200 salvage value at the end of 10 years. The annual operating costs for this machine will be \$3500. From the same manufacturer I can also get a used copier for \$500 which will last 3 years. It will have a $100 salvage value and annual operating expenses of $4200. On the other hand, I can lease a machine for $750 a year, with expected operating costs of $2750, or buy a used machine for $250 which will last 3 years and have annual operating costs of $4200. I will select the alternative which provides the lowest equivalent annual cost."

## Replacement due to Inadequacy

When current operating conditions change, an older asset occasionally lacks the capacity to meet new requirements. Sometimes a similar asset can be purchased to supplement the old asset, as in the case of placing a new generator alongside an old one to meet new power demands. New layouts, building additions, and design changes are examples of possible modifications to increase capacity.

Alternatives or supplements to an existing asset are usually compared with a challenging new asset which may perform an equivalent function in an entirely different manner. Replacing a wood stove with an oil furnace, for example, provides an entirely different way to heat a home. Even though a challenger is deemed a desirable replacement, the defender may still have value as usable equipment. In such cases it can be sold or retained for standby purposes. It is also possible that secondary uses can be found for assets replaced from their primary function; the wood stove replaced by an oil furnace might be used as an incinerator.

**Example 7.8** **Upgrading versus Demolition and Replacement**

A small bridge leading to a proposed industrial park has a load limit of 10,000 pounds (4,536 kilograms). A manufacturing firm will lease a building site in the park if the capacity of the bridge is raised to 60,000 pounds (27,216 kilograms). The developers of the land have two alternatives. They can reinforce the old bridge, or they can tear it out and fill in the low area, leaving a culvert to carry away surface water.

The present bridge has no realizable salvage value. Reinforcement would cost $30,000 and should provide adequate access for 10 years without any major additional work. The salvage value from added materials would be $8000 in 10 years.

A culvert-and-fill approach to the park would cost $60,000 and should meet all requirements for the next 50 years. There would be no salvage value. In addition, it will cost $2000 to remove the old bridge. Maintenance costs are expected to be $2200 per year less than upkeep for a bridge.

Annual property taxes and insurance on the improvements will be 1 percent of the first cost. The required return on investments is 8 percent before taxes. If the developer feels that a new approach to the park is required, which alternative should be selected?

**Solution 7.8**

Since a replacement for the old bridge is definitely necessary, there is no distinct defender-challenger relationship. The lower-initial-cost alternative, reinforcement, would best fit the role of defender. It's equivalent annual cost is the sum of capital recovery, extra maintenance costs, and taxes plus insurance.

*Equivalent annual cost of reinforcing the bridge*

| | |
|---|---|
| Additional maintenance costs | \$2200 |
| Capital recovery: $(P - S)(A/P, 8, 10) + S(0.08)$ <br> (\$30,000 − \$8000)(0.14903) + \$8000(0.08) | 3919 |
| Taxes and insurance: \$30,000(0.01) | 300 |
| | \$6419 |

*Equivalent annual cost of the culvert and fill*

| | |
|---|---|
| Capital recovery: $P(A/P, 8, 50)$ <br> (\$60,000 + \$2000)(0.08174) | \$5068 |
| Taxes and insurance: \$60,000(0.01) | 600 |
| | \$5668 |

From strictly a cost viewpoint, the culvert and fill has a clear advantage of \$6419 − \$5668 = \$751 per year. Other management considerations could influence the final choice. For instance, the developers might be willing to forgo the annual \$751 benefit in order to be allowed an opportunity to change plans in 10 years when the reinforced bridge will again need a replacement study. They might be short of capital at the present time or have other possible investments with a greater potential rate of return than that earned on the extra increment of investment (\$62,000 − \$30,000 = \$32,000) required for the culvert and fill. (Note that the extra increment earns 8 percent *plus* \$751 a year.)

---

**Economy Exercise 7-5** The headquarters building owned by a rapidly growing company is not large enough for current needs. A search for enlarged quarters revealed only two alternatives that provide sufficient room, enough parking, and the desired appearance and location. One can be leased for \$144,000 per year, and the other can be purchased for \$800,000, including a \$150,000 cost for land.

The study period for the comparison is 30 years, and the desired rate of return on investments before income taxes is 12 percent. It is believed that land values will not decrease over the ownership period, but the value of a structure will decline to 10 percent of the present worth in 30 years. Property taxes are 4 percent and rising. For comparison purposes, annual tax payments should be uniformly close to 5 percent of the purchase price.

The present headquarters building is already paid for and is now valued at \$300,000. The land it is on is appraised at \$60,000. An engineer suggested that consideration be given to remodeling the present structure. The engineer estimates that an expenditure of \$300,000 will provide the necessary room and improve the appearance to make it comparable to the other alternatives. However, the remodeling will occupy part of the existing parking lot. An adjacent privately owned parking lot can be leased for 30 years under an agreement that the first year's rent of \$9000 will increase by \$500 each year. If upkeep costs are the same for all three alternatives, which one is preferable?

---

## PRECAUTIONS FOR REPLACEMENT STUDIES

The value of a replacement study, like other economic evaluations, is directly proportional to the validity of the data. Some costs have more effect in a comparison than others. Replacement decisions are very sensitive to recurring cash flows, especially operating costs. Unfortunately, operating expenses are more difficult to extrapolate into the future than other periodic cash flows such as taxes and insurance.

A conservative approach when estimates are highly uncertain is to give every advantage to the defender. This is accomplished by assuming that the present differential between the operating costs of the defender and challenger will remain constant during the study period, and that there are no capital-recovery costs for the defender. If a challenger still looks good under these handicaps, it is truly a valid contender.

Salvage values are necessarily subject to question because they occur at the most distant point in a replacement study. The basic principle is to use the best *current* estimate of the future, regardless of previous estimates. Appraisals may change from one study to the next, owing to price fluctuations, availability, and needs. An often-neglected cost associated with salvage is the expense of getting an old asset ready for the new purchaser. These expenses could include dismantling, overhaul, painting, crating, cartage, and repairs to the area vacated by the disposed asset. When such costs exceed the disposal price, the salvage value is a loss and is treated as a minus quantity ($S$ is negative) in capital-recovery calculations.

Another commonly ignored cost is the expense associated with putting a new asset in operating order. Special wiring, piping, guard rails, foundations, and other facilities may be needed before new equipment can operate. Radically different or complex equipment often requires more "debugging" than is provided by the supplier. In a replacement study these once-only, operational-type costs should be treated as capital costs.

The bridge-replacement example demonstrated a case in which there was no choice about making a replacement study. More commonly an asset performs its intended function without obvious financial loss. If an acceptable challenger goes unnoticed, the accumulated yearly losses from failing to recognize the need for a replacement can be substantial. One clue to the replaceability of an asset is its economic life. As an asset nears the end of its original life estimation, it becomes a more likely candidate for a replacement study. Other clues include an awareness of new developments which could lead to asset obsolescence and the deterioration of performance, as indicated by reject rates or frequent repairs. Since machines age less obviously than humans and cannot complain about their frailties, it is an engineering function to diagnose infirmities and prepare remedies.

## SUMMARY

Cash flows are converted to an equivalent annual annuity for *annual-worth comparisons.* An annual charge to account for the repayment of invested capital plus interest earnings on the unrecovered balance of the investment is obtained as

$$\text{Equivalent annual cost} = (P - S)(A/P, i, N) + Si$$

*Where* $P$ = present value or purchase price
$S$ = salvage value or resale price
$N$ = study period or economic life of the asset
$i$ = interest rate expected by the organization

Alternatives are compared on the basis of ***net annual worth*** when both income and disbursements are known, or on their ***equivalent annual costs*** when they serve the same function but the income resulting from that function is not known.

When assets in a comparison have different lives, their annual cost can be calculated by the repeated-projects method, study-period method, or best-available-replacement approach. It is assumed that land does not lose value during ownership, and some man-made assets have perpetual life.

A ***sinking fund*** is sometimes established to protect investors by enforcing an orderly retirement of debt from current income. Periodic payments are made by an organization to an outside institution where the payments earn interest at the institution's rate.

A replacement decision is between the existing asset (***defender***) and currently available replacement alternatives (***challengers***). Existing assets may be replaced because of deteriorating performance, obsolescence, or inadequate capacity. The equivalent annual cost of a challenger should be lower than the defender's next year's cost ***and*** the equivalent annual cost for the remaining years of its life before replacement due to deteriorating performance.

The present value $P$ of a defender should be based on its current market value, and $P$ for a challenger should include all first costs required to make it operational. Salvage values are based on current estimates of an asset's future worth, which includes all costs associated with its disposal. The economic life of a challenger is the time period that minimizes its total equivalent annual cost.

## Discussion of Economy Exercises

**EE 7-1**

**TABLE 7.3** Percentage error for the approximate method compared to the exact method for calculating capital recovery; the error in percent is

$$\frac{1/N + i/2(N+1)/N - (A/P, i, N)}{(A/P, i, N)} 100\%$$

| *Recovery Period, N* | INTEREST RATE, $i$ *4%* | *8%* | *12%* | *20%* |
|---|---|---|---|---|
| 5 | −0.3 | −1 | −2 | −4 |
| 10 | −1 | −3 | −6 | −12 |
| 15 | −2 | −6 | −11 | −19 |
| 20 | −4 | −10 | −16 | −24 |
| | % error | | | |

The trend of errors revealed in Table 7.3 indicates that the approximate method is always too low (when $P > S$), and the percentage difference increases with higher interest rates and longer durations. About the best that can be said about the approximate method is that it can be used when no interest tables are available.

**EE 7-2** The annual worth of each proposal is calculated as follows:

$$\begin{aligned}\text{AW (Alpha)} &= \$7500 - \$30{,}000(A/P, 9, 5) = \$7500 - \$30{,}000(0.25709)\\ &= \$7500 - \$7713 = -\$213\end{aligned}$$

$$\begin{aligned}\text{AW (Beta)} &= \$13{,}755 + (-\$60{,}000 + \$10{,}000)(A/P, 9, 5) + (-\$10{,}000)(0.09)\\ &= \$13{,}755 - \$50{,}000(0.25709) - \$900\\ &= \$13{,}755 - \$13{,}755 = 0\end{aligned}$$

$$\begin{aligned}\text{AW (Gamma)} &= \$5000 - \$20{,}000(A/P, 9, 5) = \$5000 - \$20{,}000(0.25709)\\ &= \$5000 - \$5142 = -\$142\end{aligned}$$

$$\begin{aligned}\text{AW (Delta)} &= \$10{,}000 - (\$40{,}000 - \$10{,}000)(A/P, 9, 5) - \$10{,}000(0.09)\\ &= \$10{,}000 - \$30{,}000(0.25709) - \$900\\ &= \$10{,}000 - \$8613 = \$1387\end{aligned}$$

$$\begin{aligned}\text{AW (Epsilon)} &= \$7500 - (\$30{,}000 - \$5000)(A/P, 9, 5) - \$5000(0.09)\\ &= \$7500 - \$25{,}000(0.25709) - \$450\\ &= \$7500 - \$6875 = \$625\end{aligned}$$

**a** If only one proposal can be accepted, Delta is preferred because it has the largest net annual worth on the capital invested.

**b** With unlimited capital, Delta and Epsilon are obvious selections, owing to their positive net annual worths. Beta should also be accepted, even with its zero annual worth, because the $60,000 invested will still earn 9 percent; any positive AW indicates a return *beyond* the minimum attractive rate, 9 percent. Therefore, $60,000 + $40,000 + $30,000 = $130,000 should be invested from the unlimited treasury.

**c** An investment capital limitation of $60,000 allows three logical choices:

**(1)** Invest the full amount in Beta where it will earn exactly 9 percent.

**(2)** Invest in Delta ($40,000) plus Gamma ($20,000) to earn 9 percent plus $1387 − $142 = $1245.

**(3)** Invest in Delta plus a $20,000 bond. This combination is obviously superior to the similar combination of Epsilon plus a $30,000 bond. The annual worth of Delta plus the bond is calculated from the cash flow diagram below where annual returns from the bond are 0.075 × $20,000 = $1500.

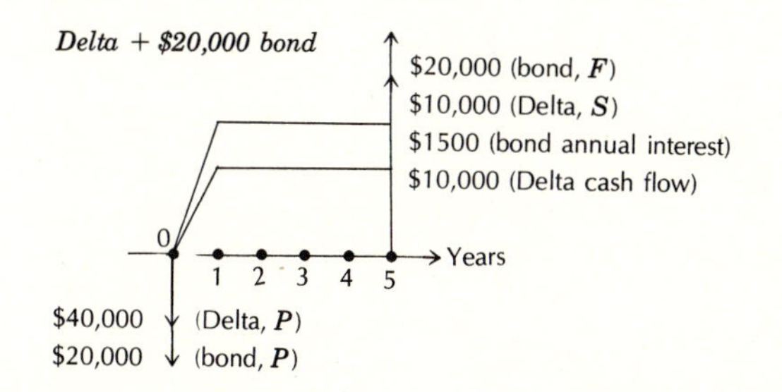


$$\begin{aligned}\text{AW} &= \$10{,}000 + \$1500 - (\$40{,}000 + \$20{,}000)(A/P, 9, 5) + (\$20{,}000 + \$10{,}000)(A/F, 9, 5)\\ &= \$11{,}500 - \$60{,}000(0.25709) + \$30{,}000(0.16709)\\ &= \$11{,}500 - \$15{,}425 + \$5013 = \$1088\end{aligned}$$

Thus the AW of the $60,000 investment in Delta plus the bond is better than the $60,000 Beta investment proposal, but less rewarding than investing in Delta and Gamma.

**EE 7-3**

$$\text{AW (hydraulic debarker)} = \text{AW (mechanical debarker)}$$
$$P(\text{hd})(A/P, 12, N) = P(\text{md})(A/P, 12, 4) \quad \text{and} \quad P(\text{hd}) = 1.3P(\text{md})$$
$$(A/P, 12, N) = \frac{P(\text{md})}{1.33P(\text{md})}(0.32924) = 0.24693$$
$$N = 5 + 1\,\frac{0.27741 - 0.24693}{0.27741 - 0.24323} = 5 + 0.89 = 5.89 \text{ years}$$

The less tangible advantage of noise reduction associated with the hydraulic debarker should also be considered to make the alternatives' outcomes comparable.

**EE 7-4** Annual-cost comparison of a canal with perpetual life and a culvert with an economic life of 30 years:

*Canal*

| | |
|---|---|
| Annual maintenance | \$ 400 |
| Interest on investment: \$75,000(0.06) | 4500 |
| Equivalent annual cost | \$4900 |

*Culvert*

| | |
|---|---|
| Annual maintenance | \$ 700 |
| Capital recovery: \$40,000($A/P$, 6, 30) = \$40,000(0.07265) | 2906 |
| Equivalent annual cost | \$3606 |

The culvert has the advantage of a lower equivalent annual cost.

**EE 7-5** The cash flows are summarized in the following three diagrams:

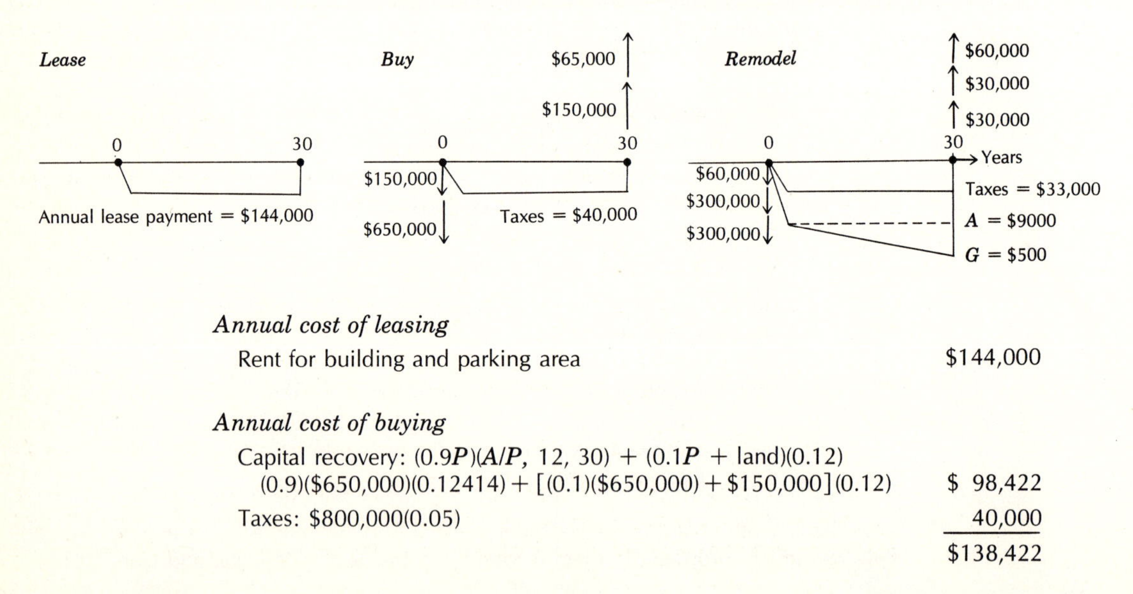


*Annual cost of leasing*

| | |
|---|---|
| Rent for building and parking area | \$144,000 |

*Annual cost of buying*

| | |
|---|---|
| Capital recovery: $(0.9P)(A/P, 12, 30) + (0.1P + \text{land})(0.12)$ | |
| $(0.9)(\$650{,}000)(0.12414) + [(0.1)(\$650{,}000) + \$150{,}000](0.12)$ | \$ 98,422 |
| Taxes: \$800,000(0.05) | 40,000 |
| | \$138,422 |

*Annual cost of remodeling:*

| | |
|---|---|
| Parking-lot rent: $A' + G(A/G, 12, 30) = \$9000 + \$500(7.2974)$ | \$ 12,649 |
| Capital recovery: (0.9)(\$300,000 + \$300,000)($A/P$, 12, 30) + [(0.1)(\$300,000 + \$300,000) + \$60,000](0.12) = \$540,000(0.12414) + \$120,000(0.12) | 81,436 |
| Taxes: \$660,000(0.05) | 33,000 |
| | \$127,085 |

Remodeling the presently occupied building is the preferred course of action. Note that the only capital recovery on land value is for interest on the amount invested in land, and that the already-paid-for value of the present headquarters building is included in the capital-recovery charge.

## PROBLEMS

**7.1** Example 7.1 describes a training program which has uniformly increasing savings and uniformly decreasing costs.

**7.1a** At what point during the year does the training program break even when $i = 0$?
**7.1b** At what point in time do savings just equal costs when the interest rate is 12 percent compounded monthly?
**7.1c** Explain why the breakeven point moved in the direction it did as the cash flow was discounted.

**7.2** A pilot plant has been constructed to convert garbage to oil, natural gas, and charcoal. The plant cost \$4 million to build and will have a 20-year life with no salvage value. The cost of processing 1 ton (907 kilograms) of garbage is \$14. From each ton are derived products which can be sold for \$21 (\$0.023 per kilogram). However, 20 percent of the end products are consumed as energy during the processing. The cost of delivery, sorting, and shredding a ton of garbage is \$2.50 (\$0.0028 per kilogram), after deducting the value of salvageable materials recovered. How many tons (kilograms) of garbage must be processed each year to recover the cost of the plant when the interest rate is 9 percent?

**7.3** A standby generator was purchased 6 years ago for \$4200. Similar equipment had shown an economic life of 15 years with a salvage value of 15 percent of the first cost. The generator is no longer needed and is to be sold for \$1800. The interest rate is 8 percent. What is the difference between the actual and anticipated equivalent annual capital-recovery costs?

*(\$196/year)*

**7.4** Five years ago, a car owner bought an automobile for \$5600. A trade-in of \$600 was allowed on the purchase of the new car. The old one had been driven 70,000 miles (112,651 kilometers). If the owner's other investments earn 6 percent annually, what was the cost per mile (kilometer) for capital recovery plus interest during the period of ownership?

*(\$0.087/mile; \$0.054/kilometer)*

**7.5** A grain elevator was built 15 years ago at a cost of \$43,000. It was supposed to have a salvage value after 30 years of 10 percent of its first cost. Depreciation is by a sinking fund held by a local bank which has paid interest during the period at 4 percent compounded annually. The owners now want to add a second grain elevator which will cost $60,000. How much additional capital will they need if they apply the depreciation reserves from the first elevator toward construction of the second?

*($46,184)*

**7.6** A family that enjoys outdoor activities can purchase a small, run-down cabin in a desirable mountain location for $6000. To make it livable they will have to repair the roof and construct a new outhouse at a cost of $1800 for materials and transportation. If the cabin is to be usable for skiing in the winter, they will have to add a fireplace or stove and insulation that will cost at least $2000. Other annual expenses for electricity, taxes, maintenance, etc., will average $550. They expect to own the cabin 5 years before a job transfer will force them to sell it. Assuming the repairs are done at once, heating arrangements are added by the end of the second year, and the cabin can be sold for $7000, how much could the family afford to spend each year for other outings in place of going to their cabin? Their savings earn 7 percent. What intangible factors might affect their decision to buy the cabin?

**7.7** Today is January 24, 1978. Suppose that 10 years ago you had started putting $10 per month in the bank. You made payments continually for 6 years and then stopped, but the accumulated deposit was left in the bank. On January 24, 1982, you plan to open a mousetrap factory. The money in the savings account will be used to advertise your new mousetrap. If you use $100 per month in advertising, how many months can you continue before the fund is exhausted? The money is invested at 6 percent compounded monthly.

*(14.51 months)*

**7.8** An investment in a cherry orchard is being considered. The asking price for the orchard is $120,000, half of which is accounted for by the land value. The cherry trees and harvesting equipment are past their prime and are expected to have no realizable value after 10 years. Income from the orchard, after deducting operating expenses and taxes, should be about $20,000 the first year, declining by $1000 each year. To obtain this net income, a working fund of $15,000 is required to maintain operations (the working fund is essentially invested capital because it is not available for other investments). Since the orchard is near a growing city, the land is expected to increase in value by 50 percent within 10 years. What is the annual worth of the orchard investment over 10 years when money is expected to earn 9 percent annually?

*($2078)*

**7.9** A factory uses 11 machines of a special design which are custom made. Each one costs $5000. In operation, maintenance costs are $1000 the first year and increase by 20 percent each year (for example, the third-year maintenance cost is $1000 × 1.2 × 1.2). Depreciation is figured on a straight-line basis, and a machine has no salvage value at any time. Interest for the use of capital is charged at 10 percent. At what service life will the average annual cost be a minimum if:

**7.9a** Capital recovery is calculated by the straight-line depreciation plus average interest method?

**7.9b** Capital recovery is calculated by the exact method using the $(A/P, i, N)$ factor?

**7.10** Laser beams are to be used on a major construction project to assure the exact alignment of components. Two types of laser alignment systems, with the costs shown below, are suitable for the project.

| | *IC System* | *UC System* |
|---|---|---|
| First cost | $5000 | $3200 |
| Salvage value | 1000 | 0 |
| Annual operating cost | 600 | 950 |
| Additional taxes and insurance | 180 | 0 |

**7.10a** If both systems have a life of 4 years and the minimum rate of return is 15 percent, which offers the lowest equivalent annual cost?

*(IC = $2331; UC = $2071)*

**7.10b** How much longer would the economic life of the IC system have to be in order to make the equivalent annual costs of the two systems equal?

*(1.4 years)*

**7.11** A necessary function can be performed by any of the assets described by the accompanying data:

| *Alternative* | *P* | *N* | *S* | *Annual Cost* |
|---|---|---|---|---|
| I | $6000 | 6 | $2000 | $ 800 |
| II | 3000 | 3 | 1000 | 1000 |
| III | 2000 | 3 | 0 | 1200 |

When investment funds are expected to earn 12 percent annually, what is the equivalent annual cost of the alternatives according to each of the following comparison procedures?

**7.11a** Repeated-projects method.
**7.11b** Study-period method (3-year period).
**7.11c** Best possible replacement as determined in a 9-year evaluation period.
**7.11d** Compare the answers obtained in using the three comparison methods, and discuss them as you would in explaining the results to someone unfamiliar with discounted cash-flow analysis. Which one seems to you to be most realistic? Why?

**7.12** Two methods of supplying water and sewage treatment for a housing development outside the districts where water and sewage-disposal services are provided by the city are described by the accompanying cash flows for a 40-year study period:

| *Years* | *Method 1* | *Method 2* |
|---|---|---|
| 0 | −$350,000 | −$735,000 |
| 1–10 | −11,000 | −8000 |
| 10 | −25,000 | +100,000 |
| 11–20 | −13,000 | −13,000 |
| 20 | −150,000 | +100,000 |
| 21–30 | −15,000 | −15,000 |
| 30 | −25,000 | −75,000 |
| 31–40 | −18,000 | −15,000 |

Both methods have the same absolute cash-flow amounts over the study period ($1,120,000), and both provide comparable quality of service. At an interest rate of 8 percent, which method has the lower equivalent annual cost?

**7.13** Machine 1 will do a required operation adequately; it can be delivered immediately at a price of $11,250. Its operating costs are $9500 annually, and it will have no realizable salvage value at the end of its 5-year economic life. This machine is being compared with another presently available machine; machine 2 has a first cost of $30,000, annual operating costs of $6500, and a salvage value of $3000 in 10 years.

The much improved machine 3 performs the same function as the other two contenders; it has a first cost of $14,500 with an expected salvage value of $4250 after 5 years. Its major advantage is a lower operating cost of $5000 per year. The only drawback is that the machine is so much in demand that there is a 2-year wait for delivery. Since the firm cannot delay acquiring one of the machines, it must choose between the currently available alternatives 1 and 2. It uses a 15 percent rate of return before taxes for economic comparisons.

**7.13a** Compare the alternatives using the repeated-projects method of analysis.

[*AC(1) = $12,856; AC(2) = $12,331*]

**7.13b** Compare the alternatives using a 10-year study period.

**7.13c** Compare the alternatives using the best-possible-replacement approach.

[*AC(1 + 3) = $10,639; AC(2) = $12,331*]

**7.14** Based on the data in Economy Exercise 7-2 and the investment conditions in part c (where the investment capital is limited to $60,000), explain the following characteristics of the solution given on page 215.

**7.14a** The reasoning that eliminated the Epsilon proposal without recourse to calculations when it is compared to the Delta proposal as part of the $60,000 investment in a proposal plus a bond.

**7.14b** The reason why the annual worth of Delta plus a $20,000 bond paying 7½ percent interest is less than the AW of the Delta proposal alone.

**7.15** A bond issue for $75,000 has been passed by voters to buy six minibuses for a senior citizens' transportation service. It is anticipated that the revenue from the bus service will yield a rate of return of 7 percent on the investment. A provision in the bond issue was that a sinking fund be established through a local bank to accumulate enough money to recover the $75,000 in 6 years. The bond issue is to pay 8 percent simple interest (due in a lump sum at maturity), and the local bank pays annual interest of 6 percent.

**7.15a** What annual payment is required for the sinking fund at the bank?

**7.15b** What annual return is required to recover the capital invested plus interest?

**7.16** A company borrows $100,000 under an agreement to set up a sinking fund to pay back the amount owed in 5 years. The interest earned on the sinking fund is 7 percent. Annual interest payments of $100,000 × 0.09 = $9000 are to be paid by the company to the lenders. The company expects internally invested funds to earn twice the interest rate paid on the external sinking fund. What is the annual cost to the company to repay the loan?

**7.17** An earth compactor costs $18,000 and has an economic life of 9 years. However, the purchaser needs it only for one project that will be completed in 3 years. At the end of the project, it can be sold for half its purchase price. What is the annual cost to the owner if the required rate of return is 20 percent?

**7.18** At the end of half its expected economic life, a 4-year-old machine has a book value of $5800 from its original cost of $9200. Estimated operating costs for the next year will amount to $6000. An equipment dealer will allow $3600 if the machine is traded in now, and $2800 if it is traded in a year later. The dealer proposes the purchase of a new machine to perform the same function; it will cost $14,000 installed. This machine will have an estimated operating cost of $4500 per year, and at the end of 4 years will have a salvage value of $3000. Is it profitable to replace the existing machine now if the minimum return on investments is 15 percent before taxes?

[*Next year's cost for defender = $7340; AC (challenger) = $8803*]

**7.19** Machine *A* was installed 6 years ago at a total cost of $8400. At that time it was estimated to have a life of 12 years and a salvage value of $1200. Annual operating costs, excluding depreciation and interest charges, have held relatively constant at $2100. The successful marketing of a new product has doubled the demand for parts made by machine *A*. The new demand can be met by purchasing an identical machine which now costs $9600 installed. The economic life and operating costs for the two machines will be the same. The salvage value for the second *A*-type machine will be $1600.

Machine *B*, a different type, costs $17,000 installed but has twice the capacity of machine *A*. Its annual operating costs will be about $3100, which should be relatively constant throughout its 10-year economic life. Salvage is expected to be $4000. The present machine can be used as a trade-in on the new machine *B*. It is worth $3000.

Compare the two alternatives on the basis of equivalent annual cost when the interest rate is 10 percent.

**7.20** An estimating service for contractors plans to provide an additional service for its customers. Besides the normal routine for making estimates, an additional follow-up service will be offered to review the records of each project and analyze actual costs versus estimated costs. Based on the expected workload for the next 3 years, the proposal could be conducted by:

1 One secretary and a clerk with total direct and indirect wages of $14,000 per year, using hand-operated equipment with an initial cost of $1400
2 One secretary with total annual wages of $7500, using computerized equipment which is priced at $8000 and has maintenance costs of $150 per month
3 Subcontracting part of the work at a monthly cost of $700 and hiring a part-time secretary for $5000 per year

Any investment must be written off in 3 years with no salvage value, and an 8 percent rate of return is required. Assuming all three alternatives provide work of equal quality, which one should be selected?

**7.21** A new delivery truck has a sticker price of $6895, but the dealer will sell it for $6300 cash. A 3-year-old van can be traded for $2700 on the new truck purchased at the sticker price, or it can be sold to a used-car dealer for $2500. The resale value of the van is expected to decrease annually by 40 percent of the previous year's value each year. Operating costs for the van will be $1020 next year and will increase each following year by $400.

A new truck is expected to have operating costs of $700 per year for the next 2 years; these costs will then increase by $300 each year. After 6 years a truck can likely be sold for 10 percent of its sticker price. It is the policy to retire trucks and vans after 6 years of service and to earn 8 percent annually on invested funds.

Assuming the price of a new truck will not increase in the next few years, when should the van be replaced?

# EXTENSION

***7.1 Amortization plus Interest*** Some government agencies and utility companies use a method of capital recovery referred to as *amortization plus interest.* "Amortization" means money put aside at intervals for gradual payment of a debt. The agencies utilizing this method normally employ the sinking-fund factor to calculate depreciation on their assets; hence, amortization is associated with a sinking fund in this context. The "interest" part of the method occurs as a charge on the first cost of the asset.

The rationale for applying the "sinking fund plus interest on first cost" method runs somewhat as follows: "I pay out a certain sum $P$ for an asset, and this asset decreases in value to $S$ by the end of its economic life $N$. To account for this devaluation, I will make payments into a fund that earns interest at rate $i$ to accumulate $P - S$ dollars in $N$ years. However, during this period I am deprived of the use of amount $P$ because this amount is invested in the asset and unavailable to me. Therefore, I will charge interest at rate $i$ on $P$ for the entire period as compensation for not having the use of that money."

**QUESTIONS**

**7.1a** For an asset described by $P = \$15{,}000$, $S = \$5000$, $i = 5$ percent, and $N = 20$ years, calculate its annual cost of ownership according to amortization plus interest: $\text{AC} = (P - S)(A/F, i, N) + Pi$.

**7.1b** Apply the capital-recovery factor to obtain the annual cost for the asset in Question 7.1a. Use the formulas for the interest factors to prove that the two methods for recovering capital plus a return must produce the same annual cost.

**7.1c** An advantage sometimes cited for the "amortization plus interest" method is that different interest rates can be used in the calculation. For instance, when the sinking fund for an asset earns interest at a rate of 5 percent, the asset's owner might want 10 percent compensation for the investment funds that are tied up in the asset. Consequently, $i = 5$ percent when used in the sinking fund, and $i = 10$ percent when applied to the first cost. With reference to Questions 7.1a and b and the discussion of the use of sinking funds in the chapter, comment on the use of two different interest rates in making an economic comparison.

# CHAPTER 8

# PRESENT-WORTH COMPARISONS

The previous chapter, on equivalent annual-worth comparisons, serves as an entry to present-worth comparisons. The steps in annual-worth calculations are the ingress to the much more widely applied present-worth evaluations. Although time-value mechanics authenticate the equality of

$$\textit{Present worth} = \textit{annual worth}(P/A, i, N)$$

the lump-sum present value is a more frequently used measure of economic merit than the comparable annual amount, probably because it provides a better grasp of the total commitment to, or potential of, a course of action.

Most of the analysis possibilities investigated for AW calculations apply to PW comparisons. Diagraming helps in keeping track of the timing and amount of cash flows. Assumptions about the continued use of assets with different lives must be carefully appraised, and the appropriate values for $P$ and $S$ should be assessed in replacement studies. In effect, all considerations pertinent to AW comparisons apply also to PW comparisons; only the perspective changes. Therefore, analysis techniques in this chapter build upon the concepts developed previously.

A student once answered the question, "What is the present worth?" as, "The value the person receiving the present attaches to it." Though hardly apropos for an engineering economics class, the answer deserves at least some credit because it translates anticipation of future pleasures to the present's present worth. The answer even implies interest, albeit a personal value rather than a rate of return. This chapter answers "What is the present worth?" by presenting the reasons for and methods of calculating the equivalent present value of cash flow and a discussion of the appropriate interest rate for economic evaluations.

## UTILIZATION OF THE PRESENT-WORTH COMPARISON METHOD

Present-worth calculations are associated with ***coterminated projects:*** comparisons in which the study periods or lives of all alternatives end together. Many analyses fall naturally into this pattern. Whenever assets are purchased to provide service for a specified period, all alternatives have the same service life. Large-scale public projects that provide benefits over a long span of years are usually evaluated by the present-worth method or a variant, called ***benefit-cost analysis,*** in which the present worth of benefits is divided by the present worth of costs to obtain a measure of acceptability. (Benefit-cost analysis is discussed in detail in Chapter 13.)

Another use of present-worth calculations is to determine the value today of an anticipated stream of cash flow. The computation places a ***valuation*** on the prospective net receipts. Valuation studies typically reveal the amount that could be paid today to receive a given rate of return for future receipts expected from ownership of property or bonds.

The present worth of a future amount drops off rapidly as the time between "now" and "then" increases, particularly at higher interest rates. The pattern of this decrease in present value is shown in Figure 8.1. Examination of the curves supports an interesting rule of thumb: the ***72-rule*** indicates the number of years $N^*$ over which a future value loses half its present worth at the annually compounded interest rate $i$:

$$\textit{72-rule}: \quad N^* = \frac{72}{i}$$

Another way to view the 72-rule is that $(1 + i)^N$ doubles about every $N^*$ years. Thus, as is apparent in Figure 8.1, the present worth is halved in about 15 years at $i = 5$ percent, 7 years at $i = 10$ percent, and 4 years at $i = 20$ percent.

## SITUATIONS FOR PRESENT-WORTH COMPARISONS

Given that an equivalent annual worth can be translated to a present worth by simply multiplying AW times a present-worth uniform series factor at a common interest rate, it is sometimes more convenient to approach a problem in terms of net AW. For instance, Example 7.1 dealt with a training proposal that promised uniformly increasing savings while causing uniformly decreasing operational costs. The first step in evaluating the cash flow was to calculate the uniform series equivalents to the two gradient series. Then the uniform costs per month ($2663) were subtracted from the uniform monthly savings

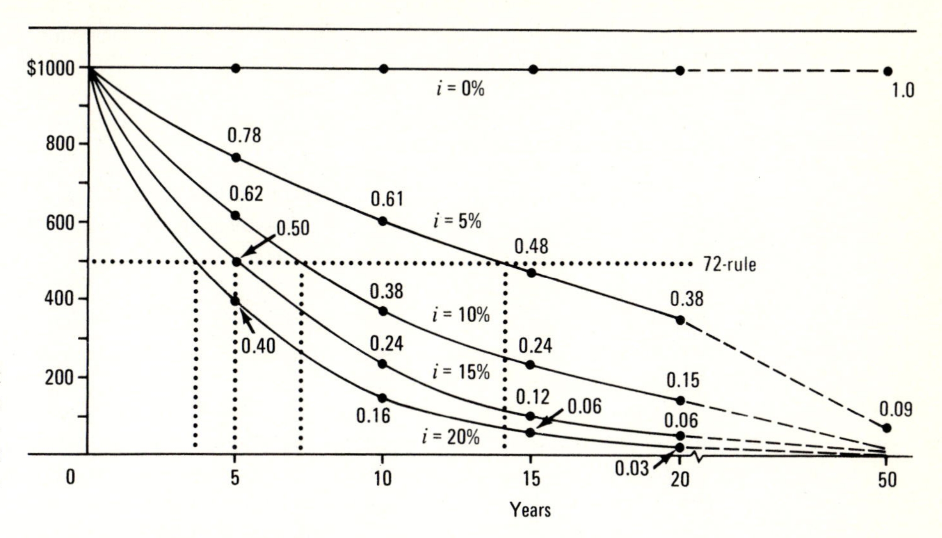


**FIGURE 8.1** Present worth of \$1000 as a function of time and interest rates when interest is compounded annually.

(\$2947) to obtain the equivalent net monthly cash flow (\$284). If the solution would be more meaningful in terms of the equivalent worth today of one year's receipts from the training program, it could be translated to

$$\text{Net present worth} = \$284(P/A, 1, 12) = \$284(11.253) = \$3196$$

where the interest rate ($i = 1$ percent per month) is necessarily the same in the translation. The interpretation of the PW is that the training program will produce net receipts during the year equivalent, at the expected interest rate, to a single payment of \$3196 today.

Example 7.4 illustrates other features shared by AW and PW comparisons. The "worth" of an alternative can be either positive or negative. With the usual convention of negative values for costs and positive for income, a negative net present worth indicates a loss and an unattractive alternative. When the alternatives for a certain situation all produce the same effect on income, each alternative's present worth is a cost which is often handled, for convenience, without minus signs. The solution to Example 7.4 revealed that the two alternatives being compared (lease or buy) had the same annual cost. They equivalently have the same present worth: Given

1. Alternative to buy at $P = \$77{,}662$, $S = \$25{,}000$, $N = 3$, and $i = 20$ percent
2. Alternative to lease at \$30,000 per year
3. Both alternatives provide equal service and have the same operating costs

we have

$$\text{PW(buy)} = \$77{,}662 - \$25{,}000(P/F, 20, 3) = \$77{,}662 - \$25{,}000(0.5787) = \$63{,}195$$

$$\text{PW(lease)} = \$30{,}000(P/A, 20, 3) = \$30{,}000(2.1065) = \$63{,}195$$

The economic interpretation of this equality is that an investment of \$77,662 today and a receipt of \$25,000 in 3 years to save three annual \$30,000 payments earns 20 percent on the invested capital.

## Comparison of Assets Having Unequal Lives

The utilization of present-worth comparisons for *coterminated projects* implies that the lives involved have a common endpoint. The necessity for cotermination is readily apparent in considering a familiar decision such as the choice between paying \$30 for a 3-year subscription to a magazine and paying \$40 for a 5-year subscription to the same publication. A simple comparison of \$30 to \$40 for a subscription is inaccurate, because the extra \$10 buys two more years of issues. *Alternatives must be compared on the basis of equivalent outcomes.*

The three methods described for comparing the annual costs of alternatives with differing lives are applicable to PW comparisons. The following example illustrates the comparability of the three methods.

**Example 8.1** **PW Comparisons of Alternatives with Unequal Economic Lives**

Assets $A1$ and $A2$ have the capability of satisfactorily performing the required function. $A2$ has an initial cost of \$3200 and an expected salvage value of \$400 at the end of its 4-year economic life. Asset $A1$ costs \$900 less initially, with an economic life 1 year shorter than that of $A2$, but it has no salvage value and its annual operating costs exceed those of $A2$ by \$250. When the required rate of return is 15 percent, which alternative is preferred when compared by:

**a** The repeated-projects method
**b** A 2-year service period (assuming the assets are needed for only 2 years)
**c** The best-possible-replacement or study-period method

**Solution 8.1**

**a** The *repeated-projects method* is based on the assumption that assets will be replaced by identical models possessing the same costs. Equivalent service results from comparing costs over a period divisible evenly by the economic lives of the alternatives; in this case the least common multiple is 12 years.

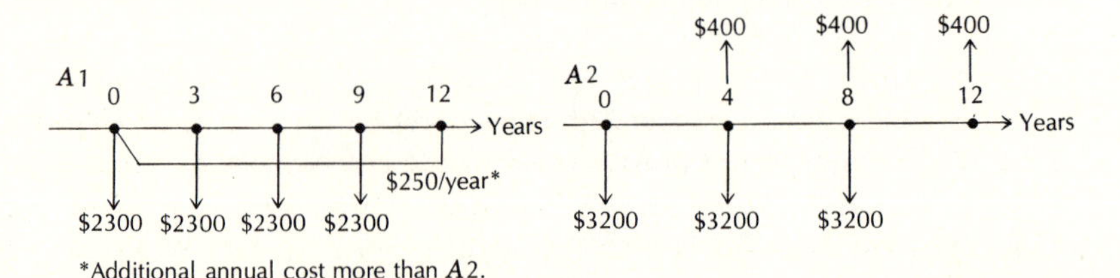


*Additional annual cost more than $A2$.

$$\text{PW}(A1) = \$2300 + \$2300(P/F, 15, 3) + \$2300(P/F, 15, 6) + \$2300(P/F, 15, 9) + \$250(P/A, 15, 12)$$

$$= \$2300 + \$2300(0.65752) + \$2300(0.43233) + \$2300(0.28426) + \$250(5.4206) = \$6816$$

$$\begin{aligned}\text{PW}(A2) &= \$3200 + \$2800(P/F,\ 15,\ 4) + \$2800(P/F,\ 15,\ 8) - \$400(P/F,\ 15,\ 12)\\ &= \$3200 + \$2800(0.57175) + \$2800(0.3269) - \$400(0.18691) = \$5642\end{aligned}$$

The calculations above could be shortened by first computing the annual cost as

$$\begin{aligned}\text{PW}(A1) &= \text{AC}(A1)(P/A,\ 15,\ 12)\\ &= [\$2300(A/P,\ 15,\ 3) + \$250](P/A,\ 15,\ 12)\\ &= [\$2300(0.43798) + \$250](5.4206) = \$6816\end{aligned}$$

$$\begin{aligned}\text{PW}(A2) &= \text{AC}(A2)(P/A,\ 15,\ 12)\\ &= [(\$3200 - \$400)(A/P,\ 15,\ 4) + \$400(0.15)](P/A,\ 15,\ 12)\\ &= [\$2800(0.35027) + \$60](5.4206) = \$5642\end{aligned}$$

The present-worth advantage of $A2$ over $A1$ for 12 years of service is $6816 − $5642 = $1174.

**b** A *service-period comparison* is utilized when a limited period of ownership of assets is set by specific operational requirements. A 2-year study period for $A1$ and $A2$ indicates that the service required from either asset will be needed only 2 years, and it will be disposed of at that time. If possible, estimates of the worth of assets at the end of the study period should be secured. These salvage values may be quite large when the service period is only a small fraction of the economic life. When it is difficult to secure reliable estimates of worth after use during the ownership period, minimum resale levels can be calculated to make the alternatives equivalent. Then only a judgment is needed as to whether the market value will be above or below the minimum level.

For instance, assuming $S = 0$ for alternatives $A1$ and $A2$ after 2 years of service,

$$\begin{aligned}\text{PW}(A1) &= \$2300 + \$250(P/A,\ 15,\ 2) = \$2300 + \$250(1.6257)\\ &= \$2707\end{aligned}$$

$$\text{PW}(A2) = \$3200$$

shows that $A1$ has the lower present worth of costs for the 2-year service period. The salvage value for $A2$ that would make PW($A1$) = PW($A2$) is

$$\$2707 = \$3200 - S(P/F,\ 15,\ 2)$$

$$S = \frac{\$3200 - \$2707}{(P/F,\ 15,\ 2)} = \frac{\$493}{0.75614} = \$652$$

which means that $A2$ is preferred to $A1$ when the resale value of $A2$ at the end of 2 years is more than $652 greater than the resale value of $A1$ at the same time. The advantage of calculating this *aspiration level* is to avoid making an estimate of $S$; only a judgment is required about whether $S$ will exceed a certain amount, the aspiration value, which is $652 in this case.

**c** The *best possible replacement* for $A1$ is $A2$, as is evident from previous calculations, unless it is known that an even better replacement will be available within 3 years (the economic life of $A1$). Assuming $A2$ is the best available, the present worth can be calculated for any study period that allows equivalent service of the two assets. An infinite study period takes care of all possible life multiples as

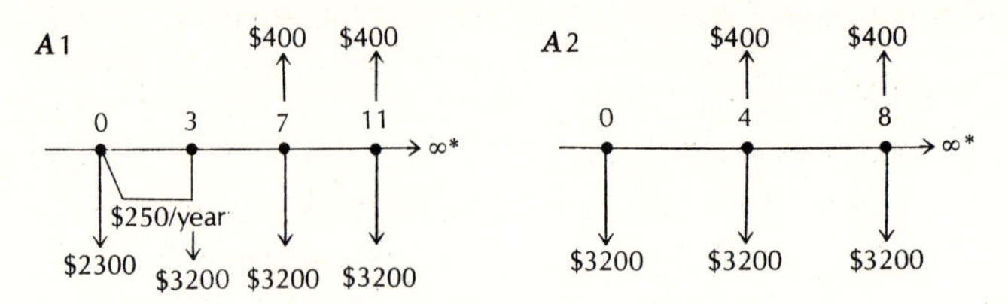


*The uniform series present-worth factor at $N = \infty$ has a value of $1/i$. For $i = 15$ percent, $(P/A, 15, \infty) = 1/0.15 = 6.6667$.

$$\begin{aligned}
\text{PW}(A1) &= \$2300 + \$250(P/A,\ 15,\ 3) + \$3200(P/F,\ 15,\ 3) + \\
&\qquad \$2800(A/P,\ 15,\ 4)(P/A,\ 15,\ \infty)(P/F,\ 15,\ 7) \\
&= \$2300 + \$250(2.2832) + \$3200(0.65752) + \\
&\qquad \$2800(0.35027)(6.6667)(0.37594) \\
&= \$2300 + \$571 + \$2104 + \$2458 = \$7433
\end{aligned}$$

$$\begin{aligned}
\text{PW}(A2) &= \$3200 + \$2800(A/P,\ 15,\ 4)(P/A,\ 15,\ \infty)(P/F,\ 15,\ 4) \\
&= \$3200 + \$2800(0.35027)(6.6667)(0.57175) \\
&= \$3200 + \$3739 = \$6939
\end{aligned}$$

The difference between the present worth of perpetual service for $A1$ and $A2$ indicates that, if future assets are no better than those available now, the advantage of $A2$ over $A1$ can be no greater than $7433 − $6939 = $494. The same difference is obtained by a 3-year *study period* in which the annual costs of the assets for only one replacement period of the shortest-lived asset are considered:

$$\begin{aligned}
\text{PW}(A1_{3\text{ years}}) &= [\$2300(A/P,\ 15,\ 3) + \$250](P/A,\ 15,\ 3) \\
&= [\$2300(0.43798) + \$250](2.2832) &= \$2871
\end{aligned}$$

$$\begin{aligned}
\text{PW}(A2_{3\text{ years}}) &= [(\$3200 - \$400)(A/P, 15, 4) + \$400(0.15)](P/A, 15, 3) \\
&= [\$2800(0.35027) + \$60](2.2832) &= \underline{2376} \\
&\text{Present-worth advantage of } A1 \text{ over } A2 &= \$\ 495
\end{aligned}$$

This difference ($495) is less than the difference ($1174) obtained by the repeated-projects method. The reason, of course, is that the higher annual costs for $A1$ occur only during the first replacement period (3 years) when it is assumed $A2$ replaces $A1$ after that first period. Thus all three methods of evaluating assets with different lives indicate the same preference among alternatives, but the numerical value of the advantage depends on which replacement assumption is utilized:

*Repeated-project assumption* An asset is replaced by a like asset with the same costs.

*Study-period assumption* Costs outside the study period are ignored.

*Best-possible-replacement assumption* The asset with the lowest annual cost replaces any asset with a higher annual cost.

The best-possible-replacement approach to a coterminated study provides the most conservative and realistic present-worth comparison.

## Capitalized-Cost Comparisons

The sum of first cost plus the present worth of disbursements assumed to last forever is called a ***capitalized cost.*** This type of evaluation is essentially limited to long-lived assets. It is not used as extensively as it once was, but it is still favored by some for studies of dams, railway rights of way, tunnels, and similar structures which provide extended service.

The calculation of capitalized cost is conducted like a present-worth comparison, where $N$ equals infinity. This makes the analysis very sensitive to the selected rate of return. Like all present-worth calculations, the final figure is usually an impressive amount. As such, it could appear discouragingly high unless properly interpreted. Expressed as a formula,

$$\text{Capitalized worth} = P + \frac{A}{i}$$

where $A$ is the uniform difference between annual receipts and disbursements. When there is no revenue, the formula becomes

$$\text{Capitalized cost} = P + \frac{\text{disbursements}}{i}$$

---

**Economy Exercise 8-1** A $500,000 gift was bequeathed to a city for the construction and continued upkeep of a music hall. Annual maintenance for a hall is estimated at $15,000. In addition, $25,000 will be needed every 10 years for painting and major repairs. How much will be left for the initial construction costs after funds are allocated for perpetual upkeep? Deposited funds can earn 6 percent annual interest, and the returns are not subject to taxes.

---

## Comparison of Deferred Investments

An occupational hazard for engineers is the habitual appeal to "get it done today, or preferably yesterday." The accustomed response to such an appeal is a workable solution that admittedly may not be the most economical long-run course of action. But keeping an operation going with a less-than-optimum solution is often less costly than the wait caused by a search for something better. For instance, suppose flooding at a construction site was the provocation for an engineer to remedy the situation quickly. An emergency purchase and installation of a pump plus sandbags for revetments cost $4000. The pumping facility was used off and on during the 2-year construction project for a total cost discounted to the time of purchase (assuming no salvage value and $i$ = 12 percent) of

| | |
|---|---|
| First cost of pump and revetment | $4000 |
| Pumping cost and maintenance ($460/year): | |
| $460($P/A$, 12, 2) = $460(1.690) | 770 |
| PW of emergency pumping operation | $4770 |

A post-project review shows that a smaller pump could have been purchased to perform the same work adequately at

| | |
|---|---|
| First cost of smaller pump and revetment | \$3100 |
| Pumping cost and maintenance (\$640/year): | |
| \$640($P/A$, 12, 2) = \$640(1.690) | 1082 |
| PW of a lower-cost solution to flooding | \$4182 |

However, if it had taken only a week longer to select and get delivery of the lower-cost pump, it is likely that the cost of 7 days of flooding would have exceeded the \$4770 − \$4182 = \$588 potential saving.

A more typical analysis of deferred investments is to determine the timing of capital expenditures to meet anticipated activity increases. Piecemeal additions to existing capacity, infrequently of an emergency nature, almost promise to be an inefficient expansion program. It is a fine example of suboptimization due to shortsightedness. A planning horizon should extend far enough into the future to accommodate growth.

Designs to accommodate growth usually involve the question of whether to acquire a full-size facility now and absorb the temporary cost of unused assets, or to acquire a smaller facility with a later addition and accept the extra cost of duplicated effort and dislocation inconvenience. For a given capacity, one large facility inherently has a lower-per-unit cost because it is designed specifically for that level of operation, but it increases the chance of technical obsolescence and idleness, owing to changing future conditions. The economic analysis of a deferred addition is usually conducted by a present-worth comparison of the options.

### Example 8.2 Immediate and Deferred Investments for Identical Capacity

A small novelty manufacturing company needs to acquire storage space in order to reduce production costs by stabilizing employment. Ninety percent of the products produced are sold during the Christmas holiday season. A resource utilization study has shown that producing at a constant rate during the year and storing output will reduce the overall manufacturing costs.

The products produced by the novelty company have been well received, and sales have increased each year. Increased capacity will be needed in the future, and two alternatives have been identified. A large warehouse with sufficient space to meet all needs for 10 years can be leased for that period at \$23,000 per year. Since there is some doubt about how much business will increase in the future, and the company is reluctant to go into debt deep enough to build a warehouse as large as the one available for leasing, the other feasible alternative is to build a small warehouse now for \$110,000 and make an addition to it in 3 years for \$50,000. Annual costs for taxes, insurance, maintenance, and repairs are expected to be \$1000 for the first 3 years and \$2000 for the next 7 years. The added-to warehouse should have a resale value of \$50,000 in 10 years. Based on a study period equal to the lease contract and a 12 percent cost for capital, which alternative is preferable?

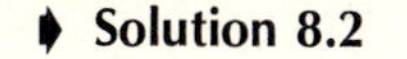

**Solution 8.2**

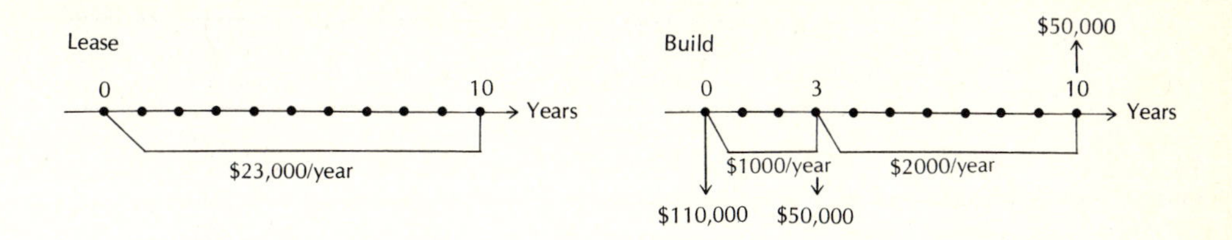


$$\text{PW(lease)} = \$23{,}000(P/A,\ 12,\ 10) = \$23{,}000(5.6502) = \$129{,}955$$

$$\begin{aligned}\text{PW(build and add)} &= \$110{,}000 + \$50{,}000(P/F,\ 12,\ 3) - \$50{,}000(P/F,\ 12,\ 10) + \\ &\qquad \$1000(P/A,\ 12,\ 3) + \$2000(P/A,\ 12,\ 7)(P/F,\ 12,\ 3) \\ &= \$110{,}000 + \$50{,}000(0.71178) - \$50{,}000(0.32197) + \\ &\qquad \$1000(2.4018) + \$2000(4.5637)(0.71178) \\ &= \$138{,}389\end{aligned}$$

The analysis indicates that the present worth of storage costs for the next 10 years will be $138,389 − $129,955 = $8434 less from leasing than building the required warehousing facility.

This same problem is examined from a different perspective in Chapter 14, where it is no longer assumed future demands are known with certainty. If reliable probabilities for future capacity needs can be generated, they should be included in the analysis. The difficulty is in finding a reliable way to peek into the future.

---

**Economy Exercise 8-2** A bank has offered to loan the novelty company described in Example 8.2 the sum of $145,000 at 8 percent interest compounded annually, to build a large warehouse immediately. The warehouse would be the same size as the small one plus the addition and would have annual expenses of $1500 per year. If its resale value at the end of 10 years is $50,000, would it be a better alternative than the other two in Example 8.2?

---

## CLASSIFICATION OF ALTERNATIVES

On the first page of this text the role of engineering economists was defined to include the identification of economic alternatives and analysis to determine their worth. Since then, alternatives have been evaluated for many different decision situations. Alternatives that share certain features can be categorized together to facilitate analysis. The alternatives in Table 8.1 will be utilized to illustrate the classifications; for computational convenience all proposals have the same life and no salvage value.

| Proposal | Investment | Life | Net Annual Cash Flow | Net PW at $i = 5\%$ |
|---|---|---|---|---|
| $N$ | −$1000 | 5 | +$300 | +$299 |
| $E$ | −2000 | 5 | +400 | −268 |
| $W$ | −3000 | 5 | +900 | +896 |
| $S$ | −4000 | 5 | +1000 | +329 |

**TABLE 8.1** Cash flows for proposals having the same life and no salvage value.

## Individual or Multiple Alternatives

The difference between individual and multiple alternatives results from physical properties or characteristics of the alternatives. Consequently, there is no real difference in methods of economic analysis, but the physical relationships could bias a decision maker and influence the way comparisons are interpreted.

Multiple alternatives are associated with levels of development for the same asset or activity. Each level is an alternative, and the purpose of analysis is to select the most promising level. For instance, the proposals in Table 8.1 would be considered multiple alternatives if they represented different sizes of the same design (e.g., diameters of pipes or thicknesses of insulation) or incremental levels of resource application (e.g., number of crews assigned or number of units ordered). Most of the evaluations in Chapter 4 dealt with multiple alternatives. The purpose of the comparison of multiples is to select the level that minimizes cost or maximizes profit.

Let the proposals in Table 8.1 represent investments in insulation to reduce heat loss (equivalently, to conserve energy), and the cash flows the savings from various thicknesses compared with the cost of heating without insulation. The net present worth of proposal $W$,

$$\begin{aligned} \text{PW}(W) &= -\$3000 + \$900(P/A, 5, 5) \\ &= -\$3000 + \$900(4.3294) = -\$3000 + \$3896 = \$896 \end{aligned}$$

is greater than any of the other proposals and indicates that the thickness of insulation it represents is the one to use.

It is often instructive to observe the incremental returns for multiple alternatives. Savings resulting from each additional $1000 of investment are indicated in Table 8.2.

| Levels of Investment Proposals | Increment of Investment, $\Delta I$ | Incremental Savings per Year, $\Delta S$ | $\Delta S/\Delta I$ |
|---|---|---|---|
| No investment to $N$ | $ 0–$1000 | $ 0–$ 300 | $300/$1000 |
| Investment $N$ to $E$ | 1000– 2000 | 300– 400 | 100/ 1000 |
| Investment $E$ to $W$ | 2000– 3000 | 400– 900 | 500/ 1000 |
| Investment $W$ to $S$ | 3000– 4000 | 900– 1000 | 100/ 1000 |

**TABLE 8.2** Incremental annual savings in relation to increments of investment.

It appears that proposal $W$ is again far superior. However, the increment of investment from $N$ ($I = \$1000$) to $W$ ($I = \$3000$), from one acceptable level of investment to the next *acceptable* level, increases the savings from $300 to $900 per year to produce a ratio of

$$\frac{\Delta S}{\Delta I} = \frac{\$900 - \$300}{\$3000 - \$1000} = \frac{\$600}{\$2000} = \$300/\$1000$$

which is identical to the return earned on $N$. This means that money invested in alternative $N$ earns at the same rate as an investment in $W$. The only reason $W$ would be preferred to $N$ is that a larger investment at a given rate of return earns a larger total amount, ignoring other influencing factors such as the future availability of energy. The actual percentage rate of return earned on incremental investments will be discussed in the next chapter.

*Individual investment alternatives* are compared in the same manner as multiple alternatives. For example, now assume alternatives $N$, $E$, $W$, and $S$ are four different ways to conserve energy. Investments in $N$, $W$, and $S$ are better than doing nothing; $E$ has a negative PW, which is worse than the zero present worth of doing nothing. $W$ is the preferred alternative, owing to its highest PW. The meaning of its $896 present worth is that the $900 annual installment repays the $3000 investment *and* provides a series of payments worth $896 today, all at 5 percent interest. The part of the $900 that repays the $3000 investment earning 5 percent annually is

$$\$3000(A/P, 5, 5) = \$3000(0.23098) = \$693$$

which leaves $\$900 - \$693 = \$207$ as the balance received each year for 5 years that has a present value of

$$\$207(P/A, 5, 5) = \$207(4.3294) = \$896$$

Occasionally, individual investment opportunities are linked to other alternatives through legal, administrative, political, or physical requirements. Then the acceptance of one alternative depends on the simultaneous acceptance of one or more related alternatives. Cash flows are also affected by relational factors, as in a flood-control proposal where the benefits of a levee of a given height depend on the acceptance of proposed dams on the headwaters of the river.

Returning again to Table 8.1, assume proposal $W$ could be implemented only if $E$ were also funded. This linkage simply consolidates two investments into one alternative with an initial cost of $\$2000 + \$3000 = \$5000$ which is expected to earn $\$400 + \$900 = \$1300$ annually. The evaluation then proceeds with $E + W$ treated as an individual alternative. The end effect of relational considerations is a listing of grouped, internally dependent alternatives which are collectively unrelated. An impetus for investigating relational connections is to avoid the selection of a course of action that cannot be implemented without additional commitment of resources.

## Dependent or Independent Alternatives

A *dependent alternative* is influenced by the outcome or selection of other alternatives being considered. The most common dependencies fall into the following two categories:

1 *Conditional dependency* The final outcome of an alternative depends on the successive results from a chain of evaluations. Conditional decision-tree comparisons in Chapter 5 were made to evaluate the combined effect of a sequence of dependent design

variables not affected by time. "Discounted decision trees" are developed in Chapter 14 to realistically portray a sequence of time-dependent decisions affecting the cash flow of alternatives. A conditional comparison involves a step-by-step procedure that discloses the present condition of each alternative with respect to previous considerations, and then selects the single course of action that is preferred for the stated conditions.

2 *Mutual exclusiveness* The selection of one alternative eliminates the opportunity to accept any other alternative. Most of the comparisons made in previous chapters were among mutually exclusive alternatives. Operational problems normally fit into this category, because a single course of action is sought to solve a particular, often urgent, problem. When the best solution is determined, the problem is theoretically resolved by implementing the indicated course of action.

An *independent alternative* is not affected by the selection of another alternative. Each proposal is evaluated on its merit and will be accepted if it meets the criteria of acceptability. Comparisons of independent investment proposals are designed to determine which proposals satisfy a minimum level of economic value. All those that surpass the minimum level may be implemented so long as sufficient capital is available. For example, if the proposals in Table 8.1 were independent and the criterion of acceptability were a return of 5 percent on any investment, proposals $N$, $W$, and $S$ would be satisfactory investments because they possess positive net PW at the required interest rate. The choice of which proposals to accept if investment funds are insufficient to undertake all the proposals meeting the minimum requirements is a capital-budgeting decision as discussed in Chapter 12.

**Economy Exercise 8-3** Data for three alternative investment plans are listed in Table 8.3.

**TABLE 8.3**

| *Alternative* | *Investment* | *Salvage Value* | *Life, Years* | *Annual Net Cash Flow* |
|---|---|---|---|---|
| $X$ | $ 6,000 | $ 0 | 3 | $2600 |
| $Y$ | 12,000 | 3000 | 6 | 2500 |
| $Z$ | 18,000 | 0 | 6 | 4000 |

When the minimum attractive rate of return is 10 percent, which alternative(s) should be selected under each of the following decision conditions?

**a** Individual alternatives are mutually exclusive
**b** Multiple alternatives are mutually exclusive.
**c** Individual alternatives are independent.

## VALUATION

Value is a measure of the worth of something in terms of money or goods. A barter system uses a personally directed trade of goods to establish equivalent values. In an auction the measure of worth is established by competitive monetary bids. To some extent, selling price in an openly competitive market sets a value on goods as a result of the amount customers are willing to pay. Value is more difficult to determine before a transaction occurs. Expert appraisers are familiar with prices from previous transactions and interpolate to set values on goods that have not been exposed to the market. Works of art and land properties are subject to sometimes controversial evaluations.

Appraisals regularly encountered in financial practice are known by specific names. A *going value* is how much the assets of an organization are worth as an operating unit. It is opposed to a *liquidating value,* which is the amount that could be realized if the assets were sold separately from the organization that is using them. The going value is normally greater than the liquidating value, in recognition of the "organizational" value of a unit still in operating condition; accountants term this difference *goodwill.* The worth of an asset for accounting purposes is its *book value,* which may be quite different from its *market value,* the price at which it can be sold. Book value reflects historical cost, whereas market value is dependent upon earnings.

When the value of property depends upon its earning capacity, valuation results from discounting future probable earnings to their present worth. When risk is ignored, the "value" of such property is simply its present worth at the interest rate deemed appropriate by the appraiser. Examples of properties whose value tends to be a function of future cash flows are bonds, stocks, and rental assets.

## Bond Valuation

A bond is sold by an organization to raise money. Bonds represent a debt to the bondholders rather than a share of ownership in the organization. Most bonds bear interest semiannually and are redeemable for a specified maturity value at a given date. There are many variations designed to make bonds more attractive to purchasers, such as an option to convert them to common stock under specified conditions, or to make the bond debt more manageable for the issuing organization, as in *callable* bonds that may be paid off prior to maturity according to a printed repurchase schedule. Some public organizations are allowed to issue bonds for which the interest payments are not taxable income to the bondholders.

The value of a bond of a given denomination depends on the size and timing of the periodic dividends and the duration before maturity. The bond valuation is thus the present worth* of the cash-flow stream of dividends plus the discounted value of the redemption payment. The key to the valuation is the rate of return expected by the bond purchaser. Lower rates are reasonable when there is very little risk of default. For instance, a United States Treasury security would have less risk of nonrepayment than one issued by the Fly-by-Nite Corporation; consequently, a lower discount rate would be appropriate. The calculation procedure is illustrated by the following example.

**Example 8.3** **Evaluation of a Bond Purchase**

**A utility company sold an issue of 4 percent bonds 6 years ago. Each bond has a face value (value at maturity) of $1000, is due in 14 years, and pays interest twice a year**

*The procedure for calculating the value is sometimes called *capitalization of income.*

(2 percent per period). Because interest rates on savings have climbed in recent years, the bond can now be sold on the bond market for only $760. If buyers expect their money to earn 8 percent compounded semiannually and they must pay a brokerage charge of $20 to purchase each bond, is the current selling price reasonable?

**Solution 8.3**

Semiannual interest payments amount to 2 percent of the face value of the bond, or $0.02 \times \$1000 = \$20$, and $1000 will be redeemed in 14 years, or 28 half-year periods. Therefore, the present value of the cash flow when the desired nominal interest rate is 4 percent per 6-month period is

$$\begin{aligned} PW &= \$20(P/A, 4, 28) + \$1000(P/F, 4, 28) \\ &= \$20(16.662) + \$1000(0.33348) = \$666.72 \end{aligned}$$

Since the price of the bond is $\$760 + \$20 = \$780$, prospective bond purchasers should look elsewhere to obtain their desired rate of return on investments.

Assuming an original purchaser of the utility bond in Example 8.3 paid the face value ($1000) 6 years ago and accepted $760 for it now, only the original investment was recovered and no interest was earned during the 6 years it was held:

$$\begin{aligned} \$1000 &= \$20(P/A, i, 12) + \$760(P/F, i, 12) \\ &= \$20(P/A, 0, 12) + \$760(P/F, 0, 12) \\ &= \$20(12) + \$760(1) = \$1000 \end{aligned}$$

The current market rate of interest strongly affects bond prices. Higher market rates tend to lower bond prices by decreasing the present worth of the future stream of payments promised by the bond. Figure 8.2 displays the valuation at different market rates of interest for a 20-year bond as described in Example 8.3 and a bond which pays the same dividends but matures in 3 years.

As is apparent in Figure 8.2, the longer the maturity of a security, the greater its price change in response to a change in the market rate of interest. This pattern explains why short-term bonds with the same risk of default as long-term bonds usually have lower

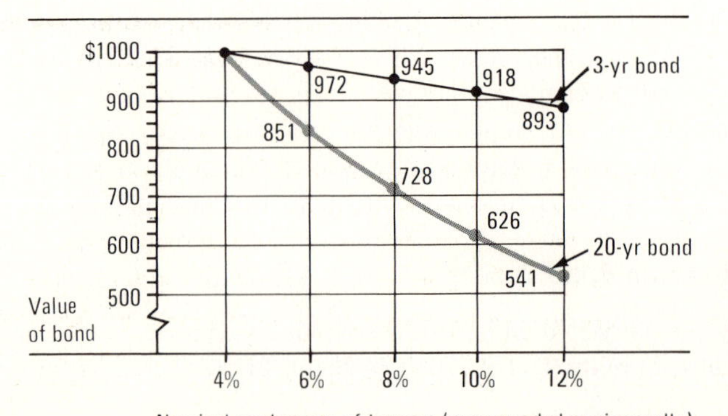


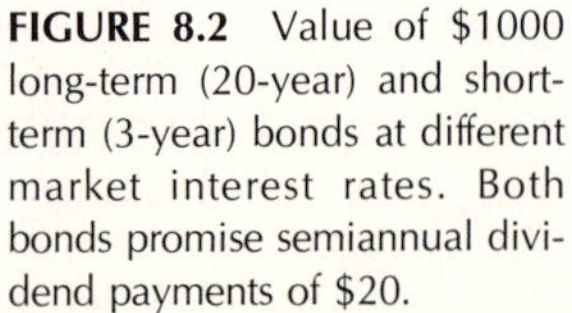
**FIGURE 8.2** Value of $1000 long-term (20-year) and short-term (3-year) bonds at different market interest rates. Both bonds promise semiannual dividend payments of $20.

dividend rates. Investments in short-term securities expose the investor to less chance of severe fluctuations in the market rate of interest than comparable long-term investments.

## Stock Valuation

Stock in a company represents a share of ownership, as opposed to a bond which is essentially a promissory note. There are many types of stocks and bonds, varying as to their degree of security and associated special privileges. In general, all bonds have claims on a company's assets before stock in case of a business failure. Stock is still a popular investment because it has the potential of increasing in value and may pay higher dividends than bonds when the company is very successful. Some stockholders have voting privileges which allow major investors to have some say in company policy.

*Preferred stocks* usually entitle owners to regular, fixed dividend payments similar to bond interest. Characteristically, preferred stock has no voting rights and initially has a higher yield than bonds. The *par value* of a preferred stock is the amount due the stockholder in the event of liquidation, and the annual yield is often expressed as a percentage of the par value. Since preferred stock has no maturity date, it may be treated as a perpetuity whose value is

$$\text{PW} = \frac{\text{annual dividend on the preferred stock}}{\text{annual rate of return expected by the investor}}$$

*Common stocks,* the most common form of equity shares in a company, are more difficult to value than preferred stocks or bonds because dividends and prices of common stocks are not constant; investors hope they will increase over time. It is therefore necessary to forecast future earnings, dividends, and stock prices. If reliable forecasts could be made (and that is a highly questionable assumption), stock valuation would result from discounting the forecast cash flow.

To illustrate, suppose a share of Sumplex, Inc., has a current market price of \$40. It is expected to earn \$5 per share and to pay a \$3 dividend by the end of 1 year. Since the company went public in 1947, the value of its stock has been rising at an average rate of 4 percent. Using a 1-year study period,

$$\text{Present worth} = \frac{\text{dividend}}{(P/F, i, 1)} + \frac{\text{market price at end of year}}{(P/F, i, 1)}$$

$$\text{Present price} = \frac{\text{dividend}}{1+i} + \frac{\text{present price} \times 1.04}{1+i}$$

Rearranging and substituting numerical data give

$$1 + i = \frac{\$3 + \$40(1.04)}{\$40} = \frac{\$44.60}{\$40} = 1.115$$

to reveal the discount rate

$$i = 1.115 - 1 = 0.115 \text{ or } 11.5 \text{ percent}$$

If an investor is satisfied with an 11.5 percent rate of return after considering the risk involved, shares in Sumplex, Inc., could be purchased. The same expected rate of return could be calculated as

$$i = \text{dividend rate} + \text{growth rate} = \frac{\$3}{\$40} + 0.04 = 0.075 + 0.04 = 0.115$$

The partition of expected returns into dividends and growth is important when taxes are considered. Capital gains, the increase in market value between the purchase and sale of a stock, are taxed at a lower rate than ordinary income. Thus, a very prosperous investor would be attracted to stocks that promise greater growth over more generous dividends, given equal risk and rate of return.

**Economy Exercise 8-4** An investor has been investigating the stock performance of two companies: Withit and Righton. The Withit Corporation has consistently paid dividends that increase 10 cents per year while the selling price of the stock has averaged a 2 percent annual rise. Righton is a new glamour company that has paid no dividends because all earnings are retained for expansion, but its market price is expected to increase by $10 per year. Further, in about 5 years it is expected to start paying dividends equal to 2 percent of its price per share (a price-to-earnings ratio of 50:1). Current data about the two companies are summarized below.

| | *Withit corporation* | *Righton Company* |
|---|---|---|
| Dividend | $2.25 (10¢/year increase) | 0 (2% of market price after 5 years) |
| Market price | $28 (2% annual increase) | $65 ($10/year increase) |
| Capitalization rate | 9% | 12% (risk adjusted) |

Since it is generally believed that Righton stock is less stable than Withit's, the extra risk of investing in Righton is recognized by requiring a higher rate of return for the valuation, 12 percent versus 9 percent.

Disregarding tax effects and brokerage commissions to buy or sell, which stock has the greater valuation for an anticipated 10-year ownership?

## COST OF CAPITAL

The cost of capital is a composite interest rate representing the cost of providing money for an organization. Sources of funds include debt (sale of bonds, or promissory notes), equity (sale of stock, or shares of ownership), and retained earnings (profit not distributed to owners). The cost of capital is not necessarily the minimum rate of return an organization expects to receive on its own investments; the expected rate is probably higher than the cost of capital, to compensate for the risk involved and because the organization would retire its own debts if its funds could not be invested more profitably elsewhere. Still, it is useful to know the cost of capital when deciding what interest rate to use in economic evaluations. Other considerations involved in determining the minimum attractive rate of return are discussed in the next chapter.

One way to measure the cost of capital is to calculate the weighted average of the cost of funds acquired from all sources. The theory behind this approach is that the return on any project funded at the weighted average rate will not change the market price of the firm's stock. That is, if the yield from a project equals the composite cost of funds from other sources, the money generated will neither raise nor lower the stock price. Since changes in the price of stock act as a rough index of a firm's economic health, and managers do not care to see the index dip, the cost of capital effectively sets a bottom limit for the rate of return required from future investments.

With $k$ representing the discount rate that aligns future cash flows with the present value of debt and equity funding, the cost of capital from each source can be calculated according to the valuation procedures for stocks and bonds. If 20 percent of a firm's capital comes from \$1000 bonds, now selling for \$926, which mature in 10 years and pay annual dividends of \$50, then

$$\$926 = \$50(P/A, k, 10) + \$1000(P/F, k, 10)$$

from which $k$ is computed by trial and error to be 6 percent.

Next, if 50 percent of the firm's capital is represented by common stock which now sells for \$65 per share, pays no dividends, but is expected to continue to increase in price by \$7 per year for the next 5 years,

$$\$65 = [\$65 + \$7(5)](P/F, k, 5)$$

from which $k$ is found to be 9 percent.

With the remaining 30 percent of total capital derived from retained earnings which, at a minimum, should be valued the same as common stock because stockholders are being denied dividends from retained earnings, the weighted cost of capital can be determined as shown in Table 8.4.

**TABLE 8.4** Calculation of the cost of capital as a weighted average of the funding from debt, equity, and retained earnings.

| *Source of Capital* | *Percentage of Total (1)* | *Cost, % (2)* | *Weighted Cost, % [(1) × (2)]* |
|---|---|---|---|
| Debt (bonds) | 20 | 6 | 1.2 |
| Equity (stock) | 50 | 9 | 4.5 |
| Retained earnings | 30 | 9 | 2.7 |
| *Cost of capital* | | | 8.4 |

The process of determining the cost of capital is greatly oversimplified in Table 8.4. The approach is valid, but the practical difficulties of estimating future cash flows and determining which value to use for current worth are formidable. These difficulties and ways to overcome them are explored in succeeding chapters.

## SUMMARY

A present-worth calculation reveals the sum today that is equivalent to a future cash-flow stream. When alternatives are compared according to their present worths, the study periods must coterminate. Equivalent outcomes for assets with unequal lives are developed by assuming the same assets will be repeatedly employed, specifying a limited study period to conform to the duration of needed service, or insisting each asset is replaced at the end of its economic life with the best replacement then available. Deferred investments, usually associated with capacity decisions, involve different timings for cash flows. The present worth of disbursements over an infinite study period is called the ***capitalized cost.***

The analysis of alternatives may be clarified by categorizing them according to the following characteristics:

- ***Multiple*** Different levels of development for the same asset or activity
- ***Individual*** Distinguishably different options
- ***Dependent*** Influenced by the outcome or selection of other alternatives; often alternatives are mutually exclusive
- ***Independent*** Not affected by the selection of any other alternative

In a ***valuation*** study, the present worth is the discounted value of future earnings; it indicates how much can be invested for a given rate of return to obtain the earnings. Such appraisals are suitable for stocks, bonds, and rental properties.

The ***cost of capital*** is a composite interest rate which represents the cost of all acquired funding for an organization. It is usually lower than the minimum acceptable rate of return expected for investments. One way to measure the cost of capital is to determine the weighted average of funding costs from all sources of capital.

## Discussion of Economy Exercises

**EE 8-1** The total capitalized cost is known to be $500,000. From the capitalized-cost formula,

$$\text{First cost} = \text{capitalized cost} - \frac{\text{annual disbursements}}{i}$$

The annual disbursements are $15,000 for maintenance plus the annual payments necessary to accumulate $25,000 every 10 years. The funds will earn interest at 6 percent, so we have

$$\text{First cost} = \$500{,}000 - \frac{\$15{,}000 + \$25{,}000(A/F, 6, 10)}{0.06}$$

$$= \$500{,}000 - \frac{\$15{,}000 + \$25{,}000(0.07587)}{0.06}$$

$$= \$500{,}000 - \$281{,}613 = \$318{,}387$$

which means that the interest earned on the amount left after allowing $318,387 for construction will cover all the anticipated upkeep indefinitely, provided the interest rate continues at 6 percent or more.

**EE 8-2** The interest rate offered by the bank should not be used in the comparison, because the required rate of return for an organization includes more than just the cost of borrowing. At $i = 12$ percent, the present worth of 10 years of storage from the construction of a large warehouse comparable in size to the other alternatives is

$$\text{PW} = \$145{,}000 - \$50{,}000(P/F, 12, 10) + \$1500(P/A, 12, 10)$$
$$= \$145{,}000 - \$50{,}000(0.32197) + \$1500(5.6502) = \$137{,}375$$

This cost is lower than that of the build-small-and-add alternative, but is more than the leasing cost. Note that the cost of construction in two phases of the same-sized building exhibits the typical relationship that the absolute cost (at $i = 0$ percent) of acquisition by parts is greater than acquisition all at one time. However, a deferral may make the time value of acquisition by parts less expensive.

**EE 8-3** The present worths of the three alternatives displayed in Table 8.3 could be calculated as

$$\text{PW}(X) = -\$6000 + \$2600(P/A, 10, 3) = -\$6000 + \$2600(2.4868)$$
$$= \$465$$

$$\text{PW}(Y) = -\$12{,}000 + \$2500(P/A, 10, 6) + \$3000(P/F, 10, 6)$$
$$= -\$12{,}000 + \$2500(4.3552) + \$3000(0.56448)$$
$$= \$582$$

$$\text{PW}(Z) = -\$18{,}000 + \$4000(P/A, 10, 6) = -\$18{,}000 + \$4000(4.3552)$$
$$= -\$579$$

**a** The deception that alternative $Y$ is more attractive than alternative $X$ occurs because the present worths are not based on equal time durations. When both investments are compared over a 6-year study period, implicitly assuming that a second investment in $X$ can be made at the end of 3 years at the original cost, the present worth of repeated investments in $X$ ($X_1X_2$) is

$$\text{PW}(X_1X_2) = -\$6000 - \$6000(P/F, 10, 3) + \$2600(P/A, 10, 6)$$
$$= -\$6000 - \$6000(0.75132) + \$2600(4.3552) = \$816$$

Therefore, when the individual alternatives are mutually exclusive, $X$ is preferred.

**b** When $X$, $Y$, and $Z$ are three levels of development for a certain asset or function, alternative $X$ is again preferred for the reasons given in part a. A visual incremental analysis confirms the selection when it is observed that no increase in net cash flow accompanies the added increment of investment going from $X$ to $Y$.

**c** Under the assumption that funds are available for all alternatives that meet the minimum attractive rate of return, investments in $X$ and $Y$ are indicated by their positive present worths. Both satisfy the 10 percent return criterion.

**EE 8-4** A 10-year study period is used to calculate the present worth of each stock alternative, assuming dividends are paid at the end of the year. The valuation of Withit stock at $i =$ 9 percent is

$$\text{PW(Withit)} = [\$2.25 + \$0.10(A/G, 9, 10)](P/A, 9, 10 + \$28(F/P, 2, 10)(P/F, 9, 10)$$
$$= [\$2.25 + \$0.10(3.7976)](6.4176) + \$28(1.2189)(0.42241)$$
$$= \$16.87 + \$14.42 = \$31.29$$

which makes the current market price of \$28 appear attractive.

The valuation of Righton Company's stock based on a 12 percent desired rate of return and the assumption that dividends are paid annually after a 5-year wait is diagramed and calculated as

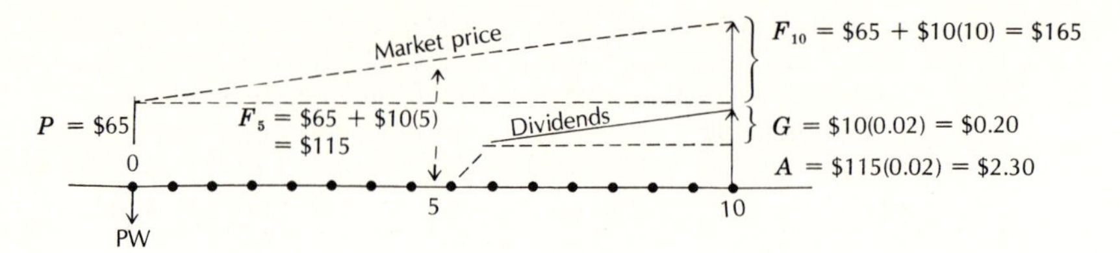


$$\begin{aligned}\text{PW(Righton)} &= [\$2.30 + \$0.20(A/G,\ 12,\ 5)](P/A,\ 12,\ 5)(P/F,\ 12,\ 5) + \$165(P/F,\ 12,\ 10)\\ &= [\$2.30 + \$0.20(1.7745)](3.6047)(0.56743) + \$165(0.32197)\\ &= \$5.43 + \$53.13 = \$58.56\end{aligned}$$

Since the \$60 market price for Righton Stock exceeds the calculated valuation, it appears that Withit Stock is a better investment opportunity, assuming the risk-return ratings and forecast cash flows are reasonably accurate.

## PROBLEMS

**8.1** What is a contract worth today that promises you \$50 per month for 20 months with the first payment to be received immediately, if 12 percent compounded monthly is expected on such contracts?

*(\$911.30)*

**8.2** What would the contract in Problem 8.1 be worth if the amount received increased by \$5 per month?

**8.3** Determine the present worth of the given end-of-year cash-flow stream for a discount rate of 15 percent.

| *Year* | *Flow* | *Year* | *Flow* |
|---|---|---|---|
| 0 | −\$47,000 | 4 | −\$2,000 |
| 1 | \$12,000 | 5 | \$12,000 |
| 2 | \$10,000 | 6 | \$10,000 |
| 3 | \$8,000 | 7 | \$8,000 |
| | | 8 | \$19,000 |

**8.4** A company borrowed \$100,000 to finance a new product. The loan was for 20 years at a nominal interest rate of 6 percent compounded semiannually. It was to be repaid in 40 equal payments. After half the payments were made, the company decided to pay the remaining balance in one final payment at the end of the tenth year. How much was owed?

*(\$64,357.90)*

**8.5** A proposed improvement in an assembly line will have an initial purchase and installation cost of \$67,000. The annual maintenance cost will be \$3500; periodic overhauls once every 3 years, including the last year of use, will cost \$6000 each. The improvement will have a useful life of 12 years, at which time it will have no salvage value. What is the present worth of the lifetime costs of the improvement at $i$ = 8 percent?

**8.6** Machine $A$ has a first cost of $9000, no salvage value at the end of its 6-year useful life, and annual operating costs of $5000. Machine $B$ costs $16,000 new and has a resale value of $4000 at the end of its 9-year economic life. Operating costs for machine $B$ are $4000 per year. Compare the two alternatives on the basis of their present worths, using the repeated-projects assumption at 10 percent annual interest.

**8.7** A commercial rental property is for sale at $100,000. A prospective buyer estimates the property would be held for 12 years, at the end of which it could be sold for $90,000. During the ownership period, annual receipts from rentals would be $15,000, and average disbursements for all purposes in connection with ownership would be $6000. If a rate of return of 9 percent is expected, what is the maximum bid the prospective purchaser should make to buy the property?

*($96,444)*

**8.8** A manufacturer requires an additional 10,000 square feet (929 square meters) of warehouse space. A reinforced-concrete building added to the existing main structure will cost $220,000, whereas the same amount of space can be constructed with a galvanized building for $106,000. The life of the concrete building is estimated at 50 years with a yearly maintenance cost of $1200. The life of the galvanized building is estimated to be 30 years, and the annual maintenance cost is $1800. Average annual property taxes are 1.2 percent for the concrete building or 0.5 percent for the metal building. Another building having approximately the same floor space and located a half mile (0.805 kilometer) from the plant can be leased for $10,800 per year. In addition to the lease expense, management estimates it will cost an additional $480 per month in material-handling expenses to transport materials to and from the warehouse.

**8.8a** Compare the present worths of the two warehouse additions, using 12 percent interest.

**8.8b** Compare the annual cost of the preferred addition to the lease expense, again at $i$ = 12 percent.

**8.9** Two alternative structures are being considered for a specific service. Compare the present worth of the cost of 24 years' service, using an interest rate of 5 percent, when neither structure has a realizable salvage value. The pertinent data are as follows:

| | *Structure I* | *Structure II* |
|---|---|---|
| First cost | $4500 | $10,000 |
| Estimated life | 12 years | 24 years |
| Annual disbursements | $1000 | $720 |

*[PW(I) = $20,805; PW(II) = $19,935]*

**8.10** A firm purchased a pump and motor for $1925 installed. It was later discovered that the pump had been improperly selected for the required head and discharge. As a result, the power bill for operating the pump was $900 for a year.

A new pump, suited to the requirements, is available for $2450 installed, with a guarantee that power costs will not exceed $500 annually. The original pump and motor can be sold for $375.

Assume a 10-year study period with zero salvage value for both pumps at the end of the period. The firm uses a minimum attractive rate of return of 10 percent before taxes. Based on present-worth calculations, should the pump be replaced?

*[Yes, as PW(old) = $5904.60; PW(new) = $5522]*

**8.11** A marina has two alternative plans for constructing a small-boat landing on a lake behind their sales building; one is a wooden dock, and the other is a metal and concrete wharf. Data for the two plans are shown.

| | *Metal* | *Metal and Concrete* |
|---|---|---|
| First cost | \$25,000 | \$50,000 |
| Period before replacement | 20 years | 30 years |
| Salvage value | \$5000 | 0 |
| Annual maintenance | \$5500 | \$2800 |

Using a minimum attractive rate of return of 10 percent, compare the present worths of the two plans. Assume both will provide adequate service and that replacement costs will be the same as the original cost.

**8.12** A refining company entered into a contract for raw materials with an agreement to pay \$600,000 now and \$150,000 per year beginning at the end of the fifth year. The contract was made for 10 years. At the end of the third year, because of unexpected profits, the company requested that it be allowed to make a lump-sum payment in advance for the rest of the contract. Both parties agreed that 7 percent compounded annually was a fair interest rate. What was the amount of the lump sum?

**8.13** A machine can be repaired today for \$2000. If repairs are not made, the operating expenses will increase by \$200 each year for the next 5 years. Assume that the expenses will occur at the end of each year, and the machine will have no value under either alternative at the end of the 5-year period. The minimum acceptable rate of return is 12 percent.

**8.13a** Compare the present worths of the two alternatives.

**8.13b** Compare the annual costs of the two alternatives.

[*AC(both) = \$554.82*]

**8.14** The following alternatives are available to accomplish an objective of 12 years' duration:

| | *Plan A* | *Plan B* | *Plan C* |
|---|---|---|---|
| Life cycle | 6 years | 3 years | 4 years |
| First cost | \$2000 | \$8000 | \$10,000 |
| Annual cost | \$3200 | \$700 | \$500 |

Compare the present worths of the alternatives using an interest rate of 7 percent.

**8.15** A firm finds that it will be necessary to air-condition a rather large area for its computers and data-processing equipment. An engineering study revealed that the more money spent on insulating the walls and ceiling area, the less money is required for the air-conditioning unit. The engineer's estimates are as follows:

| | 1 | 2 | 3 | 4 |
|---|---|---|---|---|
| First cost of insulation | \$35,000 | \$45,000 | \$60,000 | \$80,000 |
| First cost of air-conditioning equipment | 52,000 | 45,000 | 38,000 | 32,000 |
| Annual power cost | 6,500 | 5,100 | 4,100 | 3,500 |

The study also estimated that the insulating material would have a life of 20 years with zero salvage value, and the air-conditioning equipment would have a life of 10 years with no salvage value. Taxes and insurance are expected to be 2 percent of the first cost per year. Which alternative should be selected if the firm requires a 15 percent rate of return before taxes?

**8.16** The lining of a chemical tank must be replaced every 3 years at a cost of \$1800. A new type of lining is available that is more resistant to corrosion. The new lining costs \$3100. If the minimum rate of return required is 12 percent, and taxes and insurance are 4 percent of the first cost annually, how long must the improved lining last to be more economical than the present lining?

**8.17** A question has arisen whether it is more economical to replace a forklift truck or to rebore the engine and completely rebuild the present one. The original cost of the present truck 10 years ago was \$6000. To rebore the engine and completely rebuild the truck will extend its life another 5 years and will cost \$4800. A new forklift truck of the same capacity will have a first cost of \$8200 and will have an estimated life of 10 years. Fuel and lubricants for the rebuilt truck will be about \$4000 per year. Similar costs for the new truck will be about 15 percent less. Repairs for the new truck are expected to be about \$300 per year less than for the rebuilt truck. Neither unit will have any realizable value when retired. State your assumptions, and determine the equivalent present worths of the two alternatives using an interest rate of 8 percent.

**8.18** A single underground transmission circuit is needed immediately, and load studies indicate the need for a second circuit in 6 years. If provision is made for a second conduit when the conduit for the first circuit is installed, there will be no future need for reopening, trenching, backfilling and repaving. The cost of installing a single circuit with minimum preparation for the eventual second circuit is \$850,000. The installation of the second circuit will be considered to cost \$800,000 at the end of year 5, in order to be in operation by the end of year 6. If the second circuit is installed immediately, the total cost will be \$1.4 million.

Levelized annual operating and maintenance costs of the circuits are 8 percent of the first cost and begin 1 year after the first costs are incurred. The average life of a circuit is 32 years. The required rate of return on such investments is 10 percent before taxes.

**8.18a** Compare the deferred investment with the immediate investment using a 32-year study period.

**8.18b** Compare the two conduit plans on a study period taken to eternity. Then compare this solution to the one in Problem 8.18*a*. Which is the most reasonable? Why?

**8.19** After a recent visit by OSHA inspectors, a company must install better ventilation and filtering equipment or replace existing machines with newer versions having built-in controls. Costs of two alternatives that will satisfy safety codes are listed below.

| | *Add Ventilation Equipment to Existing Plant* | *Install New, Improved Machines* |
|---|---|---|
| First cost | \$80,000 | \$420,000 |
| Present value of old machines | \$150,000 | |
| Economic life, years remaining | 4 years | 9 years |
| Salvage value at end of life | | \$70,000 |
| Added maintenance cost/year | \$7,000 | |
| Total annual operating costs | \$36,000 | \$25,000 |

The company desires a rate of return of 9 percent on investments to improve operations and, owing to current cash-availability problems, seeks to minimize major capital allocations. Therefore, a temporary priority system has been established to rate the desirability of alternative investments based on need, immediate cash outlays (net first cost), and total present worth, with relative-importance ratings of 10, 10, and 5, respectively. Both alternatives have the same score for need, 10.

**8.19a** Under the assumption of a before-tax economic evaluation for a 9-year study period and using the additive comparison model described in Chapter 5, which alternative is preferred?

**8.19b** Discuss the dangers of temporal suboptimization involved in the given importance ratings. What other method can you suggest to include consideration of the conservation of capital in economic evaluations?

**8.20** Perpetual care for a small shrine in a cemetery is estimated to be available for \$500 per year. The long-term interest rate is expected to average about 5 percent. If the capitalized cost is estimated at \$15,000, what amount is anticipated for the first cost of the shrine?

*(\$5000)*

**8.21** A proposed mill in an isolated area can be furnished with power and water by a gravity-feed system. A stream high above the mill will be tapped to provide flow for water needs and power requirements by connecting it to the mill with a ditch-and-tunnel system or with a wood-and-concrete flume that winds its way down from the plateau. Either alternative will meet current and future needs, and both will utilize the same power-generating equipment.

The ditch-and-tunnel system will cost \$500,000 with annual maintenance of \$2000. The flume has an initial cost of \$200,000 and yearly maintenance costs of \$12,000. In addition, the wood portion of the flume will have to be replaced every 15 years at a cost of \$100,000.

Compare the alternatives on the basis of capitalized costs with an interest rate of 6 percent.

*[CC(d&t) = \$533,333; CC(flume) = \$471,600]*

**8.22** Cash flows from five investment proposals are itemized below.

| | | ANNUAL NET CASH FLOWS EACH YEAR | | | | |
|---|---|---|---|---|---|---|
| *Proposal* | *Investment* | *1* | *2* | *3* | *4* | *5* |
| *A* | −$15,000 | $ 4,500 | $ 4,500 | $ 4,500 | $ 4,500 | $ 4,500 |
| *B* | −25,000 | 12,000 | 10,000 | 8,000 | 6,000 | 4,000 |
| *C* | −20,000 | 2,000 | 4,000 | 6,000 | 8,000 | 10,000 |
| *D* | −30,000 | 0 | 0 | 15,000 | 15,000 | 15,000 |
| *E* | −10,000 | 4,500 | 4,500 | −2,500 | 4,500 | 7,500 |

All investment opportunities are evaluated using a 15 percent rate of return and are based on a 5-year study period.

**8.22a** If the proposals are mutually exclusive, which one is the best investment based on its present worth?

**8.22b** If investment capital is limited to $30,000, which proposals should be accepted?

**8.23** Everwhite, a small community with big ideas, envisions development of a nearby winter recreational area. The idea seems reasonable, owing to the area's location near a major population center and because it normally has snow of exceptional quality. The state government has agreed to guarantee bonds, which pay 6 percent interest and mature in 20 years, to be sold to fund the project. The main decision remaining is whether to build a modest facility at first and enlarge it later or to acquire all the land and do all the construction at one time.

Costs for a small lodge with an open ice rink and two ski runs with lifts on 400 acres of land will be $2 million. Construction will take 2 years. Once the resort opens, net revenues from this size operation are expected to average $300,000 per year, exclusive of bond interest and debt retirement.

A year-round resort with provisions for golf, tennis, and two more ski runs will boost the first cost to $5 million, require 3 years to build, and will occupy 1500 acres. Net revenue (exclusive of debt payments) for a resort of this size should average $400,000 per year for the first four years of operation until the resort becomes well known, and then should increase by $100,000 per year for the next 13 years, at which time net revenue will level off.

The enlargement of the resort can be deferred 5 years, but the extra land required (1100 acres) should be purchased immediately for $500,000, to avoid the increased valuation expected when the resort opens. Construction can be done while the original facilities continue to operate. Money for the $3-million addition should be available at the end of the fifth year. When the additions are ready for use by the end of the seventh year, net revenue should at once jump annually by $150,000 over the revenue for the smaller version. This gradient increase should continue for 5 years and then drop to an annual increase of $100,000 over the previous year until the end of the twentieth year.

Depending on the development plan followed, bonds will be sold to accumulate $2, $2.5, or $5 million immediately, and $3 million at the end of year 5. Analyze the alternatives using a desired rate of return of 8 percent. Discuss your preferred solution with respect to the other options available.

**8.24** What is the maximum amount you could afford to bid for a $3000 (face value) bond with a simple interest rate of 4 percent payable quarterly if your minimum attractive rate of return were 8 percent compounded quarterly? The bond matures in 5 years.

*($2510)*

**8.25** A sum of $200,000 has been invested in bonds which yield a return of 2 percent of the face value every 6 months. The bonds mature in 20 years. The dividends received from the bonds are invested at a rate of 10 percent compounded semiannually. What amount invested today and earning 7 percent annual interest would produce a 20-year cash flow equivalent to the bond and its reinvested dividends?

**8.26** A $2 million school-bond issue bearing interest at 3 percent payable annually and maturing in 25 years was sold at a price which provided a 4 percent annual rate of return to the investors. The brokerage fee for handling the sale was 0.3 percent of the total bond issue. What amount was realized from the sale to actually use for school construction?

*($1,681,500)*

**8.27** A corporation sold callable 10-year bonds with 5 percent semiannual interest payments (2½ percent per period). Each bond sold at the $500 face value was guaranteed a 10 percent premium if called before maturity. The extra $50 was added to the redemption value of the bonds bought back by the corporation 6 years from the time of the original issue. What was the yield to the bondholders of these bonds?

**8.28** Interest on a 3½ percent, $1000 bond due in 25 years is payable semiannually by clipped coupons, with the first payment possible 6 months from now. What must be the price of the bond to provide a rate of return to the buyer of 5 percent compounded semiannually, assuming the coupons are cashed promptly?

*($787.30)*

**8.29** Bonds of the Overightors Corporation are perpetuities bearing 7 percent annual interest. Their par value is $1000.

**8.29a** If bonds of this type currently are expected to yield 6 percent, what is the market price?

**8.29b** If interest rates rise to the level at which comparable bonds return a yield of 8 percent, what would the market price be for Overightor bonds?

**8.29c** How would the prices change in Problems 8.29*a* and 8.29*b* if the bonds had a definite maturity date of 20 years?

**8.30** The XX Company is currently earning $2 million per year after taxes. Shareholders own 1 million shares. If investors require a rate of return of 15 percent on stocks in the same risk class as XX stock, and the previous dividend was $1, what price will the stock sell for when future dividends are expected to grow at a constant rate of 5 percent per year? What is the price-to-earnings ratio? Since XX is not considered to be in a legitimate growth industry, any increases in price per share will result only from improved dividends.

**8.31** An engineer working on a construction project has reason to believe that the stock of one of the companies supplying materials to the project is undervalued at the current selling price of $60 per share. The last dividend was $2.50. She believes both the value of the stock and the dividends will double in 10 years. Assuming the increase occurs in constant increments each year, and the engineer requires a 10 percent return on such investments, how much could she afford to pay per share for the stock?

**8.32** Four basic forms of debt and equity financing are described in Table 8.5.

TABLE 8.5

| *Name of Security* | *Type* | *Market Value* | *Life* | *Obligation to Pay Return* | *Return* | *Vote* |
|---|---|---|---|---|---|---|
| First-mortgage bond | Mortgage on physical assets | Relatively stable | 30–35 years | First | Lowest (fixed) | None |
| Debenture bond | Unsecured obligation of the company | Relatively stable | 10–50 years | Second | Second lowest (fixed) | None |
| Preferred stock | Part owner of company | Variable | Usually perpetual | Third | Second highest (usually fixed) | Usually yes |
| Common stock | Part owner of company | Variable | Perpetual | Last | Highest (variable) | Yes |

**8.32a** Discuss the relative merits of investing in each security from the viewpoint of a private investor.

**8.32b** Discuss the relative merits of each source of funds from the viewpoint of a small company that recently went public (owners sold stock to the public), is in desperate need of cash, has not been able to sell all the initial stock offering, and has mortgages on 10 percent of its physical assets.

**8.33** A company's current capitalization structure is 35 percent equity and 65 percent debt. The estimated weighted average of all debt financing over an extended future period is 4.2 percent. The corresponding cost of all outstanding equity financing is 10.77 percent.

**8.33a** What is the cost of capital?

*(6.5%)*

**8.33b** Using the firm's cost of capital as its required rate of return, how much could it afford to pay for an investment that promised to return $10,000 for three consecutive years starting 2 years from now?

**8.33c** Should the firm make the investment (in Problem 8.33*b*) if the risk is considered quite small? Why?

## EXTENSION

*8.1* ***Engineering, Economics, and Pollution*** Life style in highly industrialized nations has been parodied as the "effluent society." The more developed countries do not have a monopoly on pollution problems. All large population concentrations face the danger of overloading nature's self-regenerating capacity to absorb and recycle wastes. However, pollution is acute and most obvious in industrialized countries because they process more resources per capita and use more advanced technologies that may damage the ecological system on a grand scale.

Alarming examples of pollution practices and frightening forecasts of eventual outcomes from continued contamination are regularly reported in the press. Progress toward cleansing some parts of our environment is also being reported. The Great Lakes cleanup has brought marine life back to "dead" Lake Erie. By 1975, exhaust from automobiles had 85 percent less carbon monoxide and hydrocarbons than did uncontrolled cars in the 1960s, and the level of smoke, soot, and dust in the air dropped by 14 percent from the levels of 5 years earlier. Advances have also been made in reducing discharges of untreated wastes into waterways, control of pesticides and industrial chemicals, and trash disposal. But the sweeping drive to clean up the environment appears to be slowing, owing in large part to related engineering and economic difficulties.

Technology to drastically curtail most pollutants is available, and it is expensive. Even if engineering advances can substantially cut the cost of pollution control, the remaining costs must be balanced against other national priorities. What is the cost of a cleaner environment in terms of jobs lost or created, energy consumed, personal inconvenience, and changes in the standard of living—up, down, or just different?

*The present worth of a pollution-control investment is the net cash outlays (−) from public and private sources to prevent a particular form of pollution minus the reduction in the costs (+) created by that form of pollution.* Examples of pollution-prevention costs (−) include expenditures of public funds for municipal sewage-treatment plants, expenditures by industrial firms to remove contaminants from waste discharges, and expenditures by individuals for pollution-control devices on automobiles. Reduction in costs (+) to overcome pollution that already exists could include lower expenditures for water-purification systems, medical bills for illnesses due to polluted conditions, and remedies to combat mental and physical discomfort associated with pollution. The costs for pollution prevention are obviously easier to quantify than the benefits resulting from less pollution.

The relationship between pollution costs and the benefits of reduced pollution generally follows the pattern shown in Figure 8.3. The message in this graph centers on the *economically feasible* level $X$ of pollution control for a given state of technology and set of economic conditions. Ideally, the pollution level would be zero, but the sacrifice in terms of economic disruptions to reach that level is probably prohibitive. However, if the current level of pollution is $Y$, it would be economically sound to spend enough on pollution prevention to lower the level to $X$.

**QUESTIONS**

**8.1a** Assume that Figure 8.3 represents a sewage-disposal situation in a certain locality. The pollution-prevention costs are expenditures to construct and operate a waste-treatment plant. Pollution damages include downstream water-purification expenses, human health-care costs, and the loss of recreational activities.

(1) How could these costs be determined?

(2) Once the costs are obtained, how should they be used to convince the polluters to allocate resources for prevention when most of the benefits accrue downstream?

(3) Should the downstream beneficiaries contribute to the pollution-prevention costs?

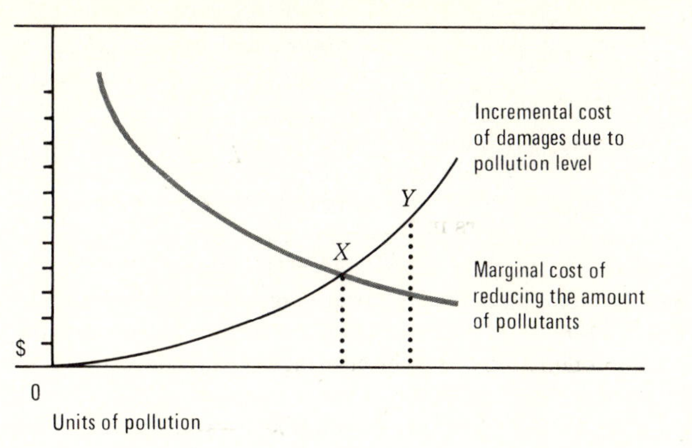


**FIGURE 8.3** Relationship of costs and benefits of pollution control.

**8.1b** *Spillovers* (or *externalities* or *diseconomies*) is an economic term that describes the situation in which some costs or benefits associated with production or consumption spill over to third parties not directly involved with the transaction. Spillover benefits are discussed in Chapter 13. Environmental pollution is a spillover cost. When an industrial plant or municipality dumps its untreated wastes into a lake or river, the people exposed to the contamination bear the spillover costs. The producers of the wastes are avoiding treatment expenses by making those exposed to the wastes suffer the damage or discomfort without compensation. Thereby, the producers are able to market their goods or services at illegitimately low prices.

The market effect of a spillover cost for a given product is to set its supply curve in a supply and demand diagram (see Figure 2.5) too far to the right. That is, the supply curve does not include all the legitimate costs of production, and this understatement allows the product to be sold at a lower price. This lower selling price then encourages greater consumption (the equilibrium point moves further to the right) which tends to allocate more resources to the polluting production which in turn increases the amount of pollution. There is no natural braking mechanism in a freely competitive market to stop this pollution-growth pattern.

An antipollution policy attempts to "internalize" the spillover costs of polluting. Laws have been passed to prohibit or to regulate the amount of pollutants dispersed. This legislation is designed to force polluters to bear the costs of avoiding pollution that would otherwise be afflicted on society; actually, the cost is just shifted to the consumers of products produced by the former polluter, rather than falling on the residents near the pollution source. Special fees or taxes based on the amount of pollution damage can serve the same purpose as laws to regulate standards of emissions. The fees or taxes assessed are costs of production which must be reflected in the product's price and, if the costs are high enough, will encourage capital investments to purchase pollution abatement equipment to reduce or eliminate emission fees.

(1) Draw supply and demand curves similar to Figure 2.5 to show how a supply curve that is shifted to the left by pollution-abatement costs is reflected in product prices and switches the pollution-control costs to the consumer.
(2) Discuss the equity of the shift.
(3) Discuss the economic implications of regulation versus the imposition of fees or taxes to control pollution.

(4) What are the economic advantages and disadvantages of subsidies (federal grants) given by the government to polluting municipalities and industries to correct their pollution problems?

**8.1c** A novel approach to control pollution, suggested by J. H. Dales,* is the sale of "pollution rights." The proposal presumes the creation of an appropriate pollution-control agency which sets the amount of pollutants that can be discharged into a certain portion of land, water, or air. An upper limit of, say, 1000 tons (907,185 kilograms) of wastes per year at a given treatment level, might be the quota for a certain section of a stream. The limit is based on the stream's capacity to absorb or recycle the wastes. The rights to use the stream's regenerative abilities are made available for sale each year to organizations that need to dispose of wastes.

The price of pollution rights provides an explicit monetary value for waste disposal and, commensurately, provides an incentive to potential polluters to develop their own waste-disposal arrangements. The established upper limit for discharges would also assure a specific standard of quality for the environment and would even allow conservation-minded groups to buy up a portion of the pollution rights to reduce the actual pollution level below the governmentally set standards.

When a public body enforces a limit on the amount of pollution that a segment of the environment can safely recycle, the supply of pollution rights becomes a scarce resource that is perfectly inelastic (see page 27).

(1) Draw a graph of an inelastic supply of 1000 units of pollution rights and a linear demand ranging from $400 per right for 500 rights to $0 for 1500 rights. Assume the demand curve represents the current situation. Next, construct a parallel demand curve in which the price per pollution right increases by 25 percent, and assume this is the condition 10 years from now. What could cause this increase?

(2) Discuss the advantage of having the price of pollution rights sensitive to demand.

(3) What problems might be encountered in applying the pollution-rights concept?

*In *Pollution, Property, and Prices,* University of Toronto Press, Toronto, 1968.

# CHAPTER 9

# RATE-OF-RETURN COMPARISONS

*Rate of return* is the most celebrated method of comparing investment alternatives. It is also known as *internal rate of return* (IRR) and *return on investment* (ROI). Under whatever name it is applied, the rate-of-return (RR) concept provides a percentage figure that indicates the relative yield on different uses of capital. Since interest rates are well understood throughout the world of commerce, there should be little danger of misinterpretation of RR figures. Another minor advantage is that it avoids the necessity of knowing a required or minimum rate of interest before calculations can be conducted; the calculations produce a percentage figure that can be compared directly with other investment proposals. These features are achieved at the expense of more tedious calculations, a minor but frustrating drawback.

Annual-worth and present-worth formulations are the foundations for rate-of-return calculations. It was observed in the previous two chapters that AW and PW are readily convertible at a common interest rate, and some problems are more easily formulated from one or the other approaches. Equivalently, the rate of return can be calculated by equating either the annual or present worths of cash flows to zero and solving for the interest rate

(RR) that allows the equality. Although both the AW and PW approaches are legitimate, the rate of return is generally defined in terms of present worth:

> The rate of return over cost is that rate which, employed in computing the present worth of all costs and the present worth of all the returns, will make these two equal. Or, as a mathematician would prefer to put it, the rate which employed in computing the present worth of the whole series of differences between two income streams (some differences being positive and others negative) will make the total zero.*

> The interest rate at which the present worth of the cash flows on a project is zero.†

## UTILIZATION OF THE RATE-OF-RETURN METHOD

Because rate-of-return computations begin with a problem expressed in terms of present worth or annual worth, it is necessary to heed the guidelines for the AW and PW methods. In particular, alternatives must be compared on the basis of equivalent outcomes. As in the previous discussions of discounted cash flow, we initially investigate the rate-of-return method without considering the effects of income taxes. Tax considerations are introduced in Chapter 11.

The rate of return for a single proposal is determined by setting the present worth (or AW) of receipts equal to the present worth (or AW) of disbursements. Then an interest rate is sought that makes the discounted flows conform to the equality:

Find $i$ so that PW(receipts) = PW(disbursements)

The same relationship obviously occurs when the discounted flows are subtracted from each other to equal zero:

Find $i$ so that PW(receipts) − PW(disbursements) = 0

When a single proposal is for a cost-reduction project, the receipts take the form of net savings from the method of operation used before the cost-reduction investment. For either PW formulation, the calculation of $i$ is usually a trial-and-error procedure.

**Example 9.1** **Income-producing Proposal**

**A parcel of land adjacent to a proposed freeway exit is deemed likely to increase in value. It can be purchased now for $80,000 and is expected to be worth $150,000 within 5 years. During that period it can be rented for pasture at $1500 per year. Annual taxes are presently $850 and will likely remain constant. What rate of return will be earned on the investment if the estimates are accurate?**

**Solution 9.1**

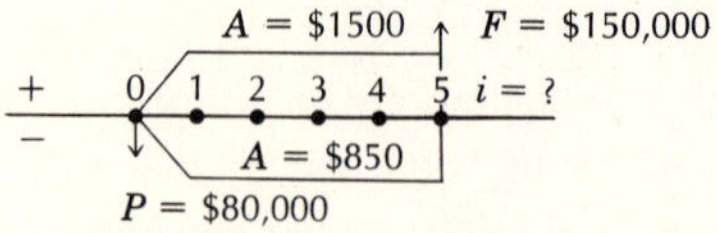

$A$ = $1500  $F$ = $150,000

+ 0 1 2 3 4 5 $i = ?$

−

$A$ = $850

$P$ = $80,000

*I. Fisher, *The Theory of Interest,* Kelley and Millman, New York, 1930.

†*The Engineering Economist,* Spring 1972.

The income (positive cash flows) and disbursements (negative cash flows) can be equated according to their equivalent present worths as

$$\$150{,}000(P/F, i, 5) + \$1500(P/A, i, 5) = \$80{,}000 + \$850(P/A, i, 5)$$

or the positive and negative cash flows can be subtracted as

$$\$150{,}000(P/F, i, 5) - \$80{,}000 + \$1500(P/A, i, 5) - \$850(P/A, i, 5) = 0$$

which reduces to

$$\$150{,}000(P/F, i, 5) - \$80{,}000 + \$650(P/A, i, 5) = 0$$

The value of $i$ that conforms to the above equation is the rate of return on the \$80,000 investment. Its value is determined by trial and error.

Now, a quick preliminary check to see if the relationship has a positive rate of return results from letting $i = 0$. At $i = 0$,

$$\$150{,}000 - \$80{,}000 + \$650(5) = \$70{,}000 + \$3250 = \$73{,}250$$

The positive value indicates the investment will produce a positive rate of return because the total income is much greater than the outgo. The check also gives a very rough idea of how large the rate of return might be. For instance, the *72-rule* suggests that a sum doubles in value every $72/i$ years. Since the \$80,000 almost doubles in value in 5 years, $i$ should be near $72/5 = 14.4$ percent.

Letting $i = 15$ percent as the first trial, we have

$$\$150{,}000(P/F, 15, 5) - \$80{,}000 + \$650(P/A, 15, 5) \stackrel{?}{=} 0$$
$$\$150{,}000(0.49718) - \$80{,}000 + \$650(3.3521) = -\$3244.14$$

The negative value indicates that the interest rate used was too large. Now it is known that $i$ lies between 0 and 15 percent.

Letting $i = 14$ percent gives

$$\$150{,}000(P/F, 14, 5) - \$80{,}000 + \$650(P/A, 14, 5) \stackrel{?}{=} 0$$
$$\$150{,}000(0.51937) - \$80{,}000 + \$650(3.4330) = \$136.95$$

which shows that 14 percent $< i <$ 15 percent. The approximate value of $i$ is determined by linear interpolation from

| $i$ | *PW* |
|---|---|
| 14% | \$136.95 |
| ? | 0 |
| 15% | −\$3244.14 |

$$\text{Range of } i = 15\% - 14\% = 1\%$$

$$\text{Range of PW} = \$136.95 - (-\$3244.14) = \$3381.09$$

The amount by which $i$ is greater than 14 percent is equal to the proportion of the PW range to the point where PW = 0:

$$i = 14\% + 1\%\frac{\$136.95 - 0}{\$3381.09} = 14\% + 1\%(0.041)$$

$$\text{RR} \doteq 14\%$$

A characteristic worth noting in the previous calculations is that whenever the present worth turns out to be positive, the next trial should employ a higher interest rate to approach the desired zero outcome. Conversely, lowering the interest rate in the present-worth formulation increases the resulting outcome.

---

**Economy Exercise 9-1** A $1000 utility bond with 14 years remaining before maturity can now be purchased for $760. It pays interest of $20 each 6-month period. What rate of return is earned by purchasing the bond at the current market price plus a brokerage charge of $20?

---

## Interpretation

A small degree of error is introduced by linear interpolation between interest-table values that are not linearly related. To keep the error as small as possible, interpolations should be conducted between adjacent interest tables. The error is naturally less between lower-interest-rate tables, separated by ½ percent, than at increments of 10 percent for the largest interest rates. For the purposes of this book, interpolated rates of return computed to the nearest tenth of a percent are adequate. The slight error that may be thus introduced will very seldom influence the choice among alternatives; this error is probably much less significant than actual deviations from the cash flows estimated in the comparisons.

**Example 9.2** **Cost-Reduction Proposal**

Subassemblies for a model IV scope are purchased for $71 apiece. The annual demand is 350 units, and it is expected to continue for 3 years, at which time the model V scope now under development should be ready for manufacturing. With equipment purchased and installed for $21,000, the production costs to internally produce the subassemblies should be $18,500 for the first year and $12,250 each of the last 2 years. The equipment will have no salvage value. Should the company make or buy the subassemblies?

**Solution 9.2** The savings expected in a cost-reduction proposal are treated as income. Assuming the transactions occur at the end of each year, the cash-flow diagram appears as shown below.

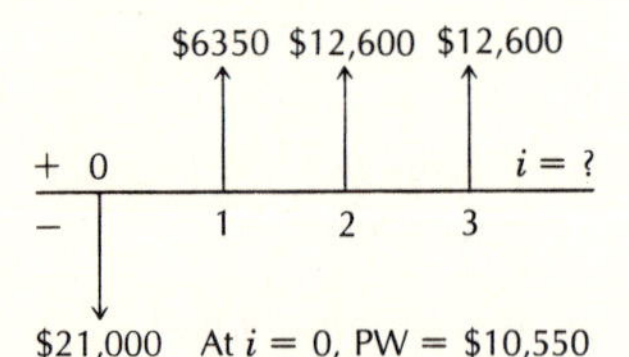


$$\text{Present annual cost} = 350 \times \$71 = \$24,850$$

$$\text{Net savings (year 1)} = \$24,850 - \$18,500 = \$6350$$

$$\text{Net savings (years 2, 3)} = \$24,850 - \$12,250 = \$12,600$$

$$\text{PW} = -\$21{,}000 + \$6350(P/F, i, 1) + \$12{,}600(P/F, i, 2) + \$12{,}600(P/F, i, 3) = 0$$

Trying successively higher rates of return gives at $i = 10$ percent,

$$\text{PW} = -\$21{,}000 + \$6350(0.90909) + \$12{,}600(0.82645) + \$12{,}600(0.75132) = \$4652.62$$

at $i = 15$ percent,

$$\text{PW} = -\$21{,}000 + \$6350(0.86957) + \$12{,}600(0.75614) + \$12{,}600(0.65752) = \$2333.89$$

at $i = 20$ percent,

$$\text{PW} = -\$21{,}000 + \$6350(0.83333) + \$12{,}600(0.69445) + \$12{,}600(0.57870) = \$333.78$$

and at $i = 25$ percent,

$$\text{PW} = -\$21{,}000 + \$6350(0.80000) + \$12{,}600(0.64000) + \$12{,}600(0.51200) = -\$1404.80$$

By interpolation, the rate of return on the \$21,000 investment is

$$\text{RR} = 20\% + 5\%\frac{\$333.78 - 0}{\$333.78 - -\$1404.80} = 20\% + 0.96\% \doteq 21\%$$

The answer to the make-or-buy question depends on how large a return the firm expects on its invested capital. Conditions for accepting the proposal to manufacture the subassemblies internally are displayed by the graph in Figure 9.1, where the

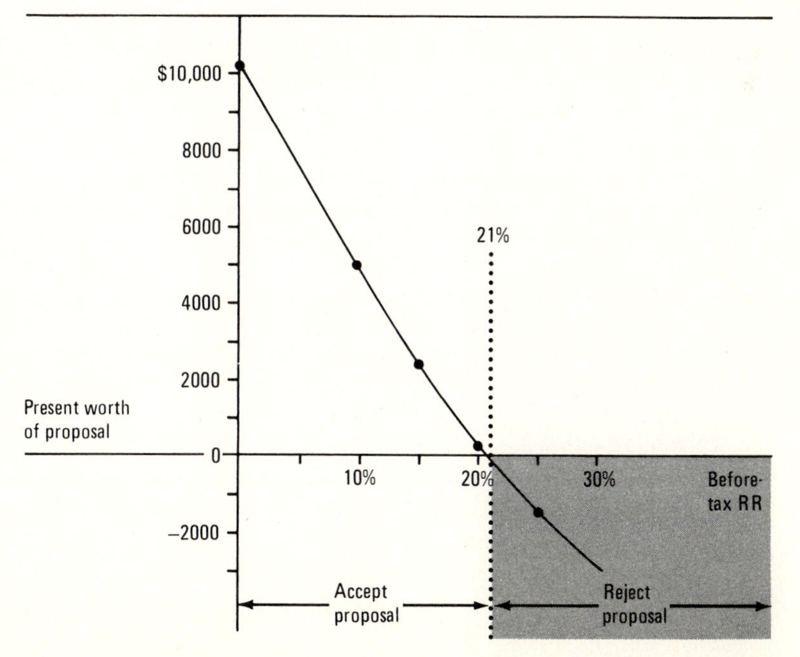


**FIGURE 9.1** Present worth of cash flow for the proposal in Example 9.2 at different interest rates. At any required RR less than 21 percent, the present worth of the proposal is positive and therefore acceptable. For a minimum acceptable RR greater than 21 percent, the proposal exhibits a negative present worth and is thereby rejected.

present worth of the proposal is shown as a function of the required RR. For any required rate of return lower than 21 percent, the firm should view the proposal favorably.

### Clues for RR Calculations

There is no way to avoid the trial-and-error search procedure for determining the RR in problems with complex cash flows. But a little preplanning may narrow the area of search. An even neater tactic, if available, is to use a computer with a canned rate-of-return program.

A quick proximity fix on the RR is possible for problems that have their major cash flows at the beginning and end of the study period, or those consisting largely of constant cash-flow streams. When the salvage value is close to 100 percent of the first cost, the net annuity divided by the first cost gives a close approximation to $i$; that is, $A/P \doteq i$. As demonstrated in Example 9.1, when the total net income at $i = 0$ sums to about twice the initial outlay, $i \doteq 72/N$.

The more variation in cash flows, the more difficult the guessing game gets. Sometimes irregular cash flows can be rounded off to approximate an ordinary annuity, or individual transactions within short time intervals can be lumped together to allow gross preliminary calculations that suggest the vicinity of the RR. For instance, in Example 9.2 the irregular receipts could be approximated by an average $A$ of, say, \$10,000. Then, $A/P = \$10{,}000/\$21{,}000 = 0.4761$. With this figure as an entry to an interest table for a capital-recovery factor at $N = 3$, $(A/P, 20, 3) = 0.47473$ gives a good place to begin RR trial computations.

## CONSISTENCY OF COMPARISON METHODS

In practically all situations, the acceptability of alternative courses of action will be identical whether evaluated according to their annual worths, present worths, or rates of return. (Some exceptions are noted on page 269.) The important aspect is to recognize the meaning of the measures of acceptability and the assumptions upon which they are based. Sample calculations applying all three comparison methods to the data in Table 9.1 reveal the consistency of results. They also illustrate how the rate of return is calculated to compare investments when only the disbursements are known and how to interpret the outcomes.

Suppose a certain function is currently being performed at an annual labor expense of \$20,000. One alternative, plan $A$, is to leave the operation unchanged. In effect, this is the "do-nothing" alternative which is almost always present in a decision situation. A second alternative, plan $B$, is to invest \$30,000 in layout modifications which will allow the function to be performed at a reduced labor cost of \$15,000. The expense of the renovations must be recovered in 10 years according to operating policy. Plan $C$ is a proposal to install a labor-saving device which will cut the labor cost to \$12,000. The device will be worn out in 5 years and has no salvage value. Table 9.1 is a year-by-year tabulation of the cash flow for the three plans. If the company's minimum acceptable rate of return is 8 percent, which plan offers the greatest economic benefit?

| Year | Plan A | Plan B | Plan C |
|---|---|---|---|
| 0 | | $30,000 | $25,000 |
| 1 | $20,000 | 15,000 | 12,000 |
| 2 | 20,000 | 15,000 | 12,000 |
| 3 | 20,000 | 15,000 | 12,000 |
| 4 | 20,000 | 15,000 | 12,000 |
| 5 | 20,000 | 15,000 | 37,000 |
| 6 | 20,000 | 15,000 | 12,000 |
| 7 | 20,000 | 15,000 | 12,000 |
| 8 | 20,000 | 15,000 | 12,000 |
| 9 | 20,000 | 15,000 | 12,000 |
| 10 | 20,000 | 15,000 | 12,000 |

**TABLE 9.1** Estimated cash flows for alternative operating plans.

## AW and PW Comparisons

Using the already familiar procedures for computing the equivalent annual worth (cost) of a cash-flow stream, the annual cost for the current operating method, plan $A$, is read directly from the table:

AW(plan $A$) = labor expense = \$20,000

In plan $B$ the initial investment is spread over the 10-year study period and added to the annual labor expense to get

AW(plan $B$) = \$30,000($A/P$, 8, 10) + \$15,000
= \$30,000(0.14903) + \$15,000 = \$19,471

Since the study period comprises two cycles of the 5-year economic life of the labor-saving device in plan $C$, the annual cost will be the same over each 5-year period and is equal to

AW(plan $C$) = \$25,000($A/P$, 8, 5) + \$12,000
= \$25,000(0.25046) + \$12,000 = \$18,262

Thus, plan $C$, with the lowest annual cost, is preferred. Compared with the currently existing plan $A$, an investment of \$25,000 in a labor-saving device will yield a return of 8 percent per year plus the equivalent receipt of \$20,000 − \$18,262 = \$1738 each year from savings in labor expense for 5 years.

The equivalent present worth (cost) of the three plans is calculated by simply multiplying each AW by the uniform-series present-worth factor, ($P/A$, 8, 10) = 6.710:

PW(plan $A$) = \$20,000(6.710) = \$134,200

PW(plan $B$) = \$19,471(6.710) = \$130,650

PW(plan $C$) = \$18,262(6.710) = \$122,538

Since a lower present cost is preferred, plan $C$ again gets the nod, as advertised. This means that over a 10-year period when money is worth 8 percent, plan $C$ is expected to cost \$134,200 − \$122,538 = \$11,662 less in ***today's dollars*** to accomplish the same operation now being done under plan $A$. This total saving is, of course, the present worth of the annual gain beyond the 8 percent return calculated in the AW comparison:

$\$1738(P/A, 8, 10) = \$1738(6.710) = \$11{,}662$

## RR Comparison

When one and only one alternative is to be selected by the rate-of-return method, ***each increment of capital expended must justify itself.*** That is, every increment of capital expended should meet or exceed the criterion for an acceptable investment, usually expressed as a minimum required rate of return. The initial increment of investment is typically the difference between the alternative of doing nothing (no change) and applying the alternative with the lowest first cost. If this investment meets the minimum acceptability criterion, the additional increment of investment required for the alternative with the next highest first cost is evaluated. Each evaluation increment is the difference between the higher investment alternative and the next lower ***accepted*** alternative.

In Table 9.1, the do-nothing alternative is plan $A$, and the alternative with the lowest initial investment is plan $C$. The "earnings" from a \$25,000 investment in a labor-saving device are the annual reductions in labor expense, \$20,000 − \$12,000 = \$8000. The rate of return can be calculated directly from

Annual cost for plan $C$ = annual cost for plan $A$ at RR

or,

First cost to initiate plan $C$ − present worth of annual savings using plan $C$ = 0 at RR

$$\$25{,}000 - \$8000(P/A, i, 5) \stackrel{?}{=} 0$$

$$(P/A, i, 5) = \frac{\$25{,}000}{\$8000} = 3.125$$

which, by interpolation between the 15 percent and 20 percent interest tables, furnishes

$$\text{RR} = 15\% + 5\% \frac{3.352 - 0}{3.352 - 2.991} = 15\% + 5\% \frac{0.227}{0.361} = 18.1\%$$

proving plan $C$ is an acceptable alternative when the required rate of return is 8 percent.

Although it is already known from the AW and PW comparisons that plan $C$ is preferable to plan $B$, it will be checked again using the annual-worth formulation:

Annual cost for plan $B$ = annual cost for plan $C$ at RR

$$\$30{,}000(A/P, i, 10) + \$12{,}000 \stackrel{?}{=} \$25{,}000(A/P, i, 5) + \$15{,}000$$

or, rearranging,

$$\$30{,}000(A/P, i, 10) - \$25{,}000(A/P, i, 5) - \$3000 \neq 0$$

at any positive value of $i$. That plan $B$ is unattractive by comparison even when no interest is charged is confirmed by letting $i = 0$ in the above equation:

$$\$30{,}000(0.1) - \$25{,}000(0.2) - \$3000 = -\$5000$$

Since it is necessary to lower the value of $i$ to make a negative outcome approach zero, plan $B$ could equal plan $C$ only if a negative rate of return were applied to the extra increment of investment involved. From this relationship it is easily inferred that plan $B$ has a lower RR with respect to plan $A$ than 18.1 percent, and the suspicion is confirmed by

$$\text{Annual cost for plan } B = \text{annual cost for plan } A \quad \text{at RR}$$
$$\$30{,}000(A/P,\ 10.5\%,\ 10) + \$15{,}000 = \$20{,}000$$

The acceptance of plan $C$ means that the investment of $25,000 is expected to be recovered by annual savings that allow an 18.1 percent return on capital. This interpretation involves an important implication: *The RR method assumes that all cash flows can be reinvested at the calculated rate of return.* In PW and AW calculations, an explicit interest rate is used. This rate generally represents a logical expectation for capital earnings, and the funds recovered each year should have a reasonable chance of being reinvested at that given rate. A proposal with a very high rate of return may create unrealistic expectations if the monies recovered for depreciation purposes cannot be reinvested at the calculated RR.

## RR Reinvestment Assumption

To delve deeper into the reinvestment assumption, assume the monies recovered from savings each year by investing in plan $C$ *could not* be reinvested; that is, $i = 0$. Then the annual capital-recovery charge would be $25,000/5 = $5000, and the ***explicit reinvestment rate of return*** would be

$$\frac{\text{Net return}}{\text{Investment}} = \frac{\$8000 - \$5000}{\$25{,}000} = 0.12 \quad \text{or} \quad 12\%$$

If the payments to recover the original investment were made into a sinking fund which could earn 8 percent,

| | |
|---|---|
| Annual savings from plan $C$ | $8000 |
| Annual sinking-fund payment to recover initial investment = $25,000($A/F$, 8%, 5) | 4263 |
| *Net annual return* | $3737 |

$$\text{Explicit reinvestment rate of return} = \frac{\$3737}{\$25{,}000} = 0.149 \quad \text{or} \quad 14.9 \text{ percent}$$

Finally, if the sinking fund for capital recovery paid interest at an annual rate of 18.1 percent, the calculated RR would be

| | |
|---|---|
| Annual savings from plan $C$ | $8000 |
| Annual sinking-fund payment to recover the initial investment = $25,000($A/F$, 18.1%, 5) | 3498 |
| *Net annual return* | $4502 |

$$\text{Explicit reinvestment rate of return} = \frac{\$4502}{\$25{,}000} = 0.18 \quad \text{or} \quad 18\%$$

This varies only by "rounding error" from the previously determined RR. It is thus apparent that the calculated RR will be accurate in practice only if the funds recovered each period can be reinvested at the computed interest rate. This assumption does ***not*** affect the order of acceptability of competing alternatives, and it is therefore important mainly in understanding the meaning of rate-of-return comparisons.

---

**Economy Exercise 9-2** An old hotel was recently damaged by a fire. Since it has a desirable location in the old part of the city that is currently being rejuvenated by an urban-renewal project, it will be rebuilt and renovated as either a showroom and office building or a modern apartment building. Estimated receipts and disbursements for the 30-year life of the refurbished structure are shown.

| | *Offices* | *Apartments* |
|---|---|---|
| First cost of renovation | $340,000 | $490,000 |
| Increase in salvage value from renovation | 120,000 | 190,000 |
| Annual receipts | 212,000 | 251,200 |
| Annual disbursements | 59,100 | 88,000 |
| Present value of fire-damaged building | 485,000 | 485,000 |
| Expected salvage value of the fire-damaged building after 30 years | 266,000 | 266,000 |

If the required rate of return is 12 percent, which renovation plan is preferable?

---

## RR COMPARISON OF SEVERAL MUTUALLY EXCLUSIVE ALTERNATIVES

Economic decisions usually involve several possible courses of action. During preliminary evaluations, less attractive alternatives may be eliminated by the obviously similar but superior outcomes of different courses of action. Others are eliminated by lack of the funds, personnel, or equipment required to conduct them. The remaining eligible alternatives are then evaluated by comparing both tangible and intangible qualities. When they are multilevel alternatives designed for essentially the same function, as in how many floors a new building should have, the intangible differences may be minor. When they are mutually exclusive and different in nature, intangibles likely differ also and may be highly influential in the final choice.

The aim of a cash-flow evaluation is to put all competing alternatives into a comparable investment perspective. It acts as a screen. The alternative with the most promising future from one evaluation is then pitted against winners of other evaluations where intangible and financial considerations become critical. No organization has sufficient capital to fund all conceivable worthwhile proposals within its province. An advantage of RR comparisons is the laddering effect it provides for the final decision. Each alternative has a spot on the ladder according to its RR (see ***capital budgeting***, page 406). Care must be taken in all evaluations to assure correct and consistent selections. Two logical-sounding selection criteria that sometimes lead to inaccurate conclusions are

1 Selecting the alternative that offers the highest rate of return on total investment
2 Selecting the alternative with the largest investment that meets the minimum required rate of return

The first criterion may bypass alternatives that earn lower rates of return which are still higher than other options available. The second criterion could lead to a larger investment than desirable, which prevents a portion of the funds from earning higher returns available through substitute investments. A decision procedure to avoid inaccurate selections is shown by the flowchart in Figure 9.2. The procedure implicitly assumes that any size investment is possible. Whether the selected proposal is actually funded is a financial

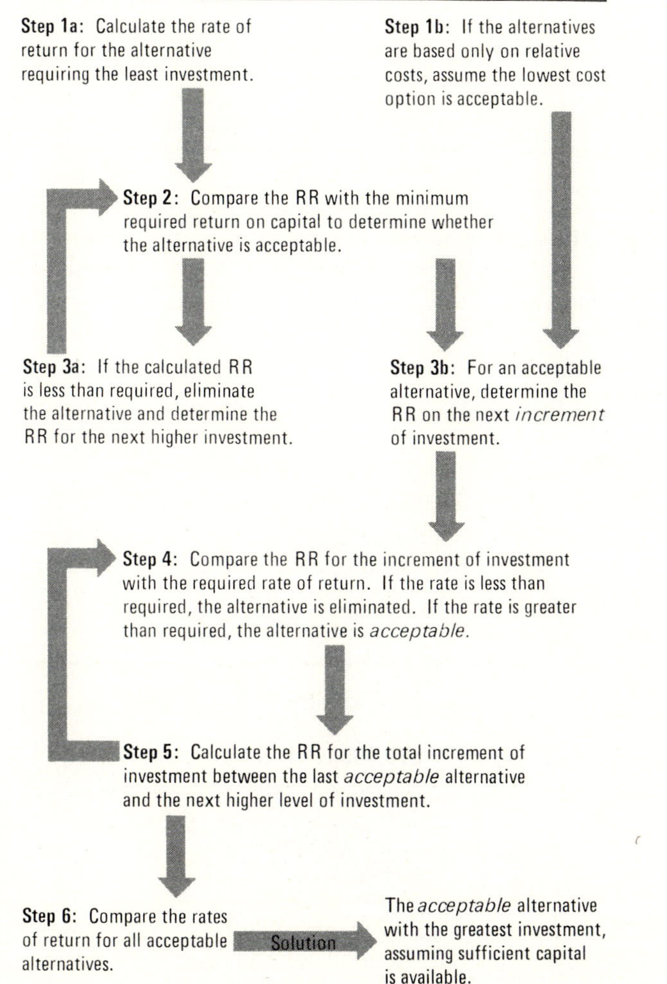


**FIGURE 9.2** Flowchart for a selection among several mutually exclusive investment alternatives when a minimum rate of return is required and capital is assumed to be unlimited.

decision that relates available capital to all types of investment proposals and their effect on the funding organization. The RR comparison rates the alternatives as inputs to the final decision process.

**Example 9.3** **Comparison of Several Mutually Exclusive Alternatives**

Four designs for a product with their associated revenue and cost estimates have been presented to top management for a decision. A 10-year study period was used. A minimum rate of return of 10 percent before taxes, a rate expected from other investments with similar risk, is required. Based on the following projected cash flows, which of the four alternative designs appears the most attractive?

**TABLE 9.2** Ten-year cash flows.

| | DESIGN | | | |
|---|---|---|---|---|
| | *A* | *B* | *C* | *D* |
| Initial investment | \$170,000 | \$260,000 | \$300,000 | \$330,000 |
| Annual receipts | 114,000 | 120,000 | 130,000 | 147,000 |
| Annual disbursements | 70,000 | 71,000 | 64,000 | 79,000 |

**Solution 9.3** The alternatives are arranged in order of increasing investment requirements in Table 9.2. A supplementary table of incremental values is given below. The annual net returns are calculated by subtracting the annual disbursements from the annual receipts for each alternative and then obtaining the differences between alternatives. For example,

$$\text{Annual net return of increment } A \rightarrow B = (\$114{,}000 - \$70{,}000) - (\$120{,}000 - \$71{,}000) = \$5000$$

| | INCREMENT | | |
|---|---|---|---|
| | $A \rightarrow B$ | $B \rightarrow C$ | $C \rightarrow D$ |
| Additional investment | \$90,000 | \$40,000 | \$30,000 |
| Annual net return | 5,000 | 17,000 | 2,000 |

Using the steps in Figure 9.2, the RR of design *A* is calculated from

$$-\$170{,}000 + (\$114{,}000 - \$70{,}000)(P/A, i, 10) \stackrel{?}{=} 0$$

and

$$(P/A, i, 10) = \frac{\$170{,}000}{\$44{,}000} = 3.8636$$

which, by interpolation, gives a rate of return of 22.5 percent. It is thereby an acceptable alternative because RR = 22.5 percent > 10 percent required.

The increment of investment from design $A$ to design $B$ is \$260,000 − \$170,000 = \$90,000, and the net annual returns are \$5000. These annual returns will obviously not pay back the investment in 10 years: \$5000 × 10 < \$90,000. Therefore, design $B$ is eliminated from contention because its incremental RR is lower than the required 10 percent.

Design $C$ is evaluated by comparing it with the *last acceptable alternative,* which in this case is design $A$. Then the incremental comparison $A \rightarrow C$ progresses as

$$-\$300{,}000 - (-\$170{,}000) + (\$130{,}000 - \$64{,}000 - \$44{,}000)(P/A, i, 10) \stackrel{?}{=} 0$$

$$(P/A, i, 10) = \frac{\$130{,}000}{\$22{,}000} = 5.9090$$

to obtain

$$\text{RR} = 10\% + 1\% \frac{6.1445 - 5.9090}{6.1445 - 5.8892} = 10.9\%$$

With its rate of return greater than 10 percent on the indicated extra increment of investment, design $C$ becomes an acceptable alternative.

The next additional increment of investment to afford design $D$ shows

$$(P/A, i, 10) = \frac{\$30{,}000}{\$2000} = 15$$

which is unacceptable by inspection. Therefore, assuming that sufficient capital is available, design $C$ is preferred. Its rate of return on total capital invested is calculated from

$$(P/A, i, 10) = \frac{\$300{,}000}{\$130{,}000 - \$64{,}000} = 4.5454$$

which leads to an RR of 18.9 percent.

The results of the evaluation of alternative designs in Example 9.3 are supplemented and summarized in Table 9.3.

**TABLE 9.3** RR relationships for Example 9.3.

| *Design* | *Total Investment* | *Annual Return* | *RR on Total Investment, %* | *Incremental RR* |
|---|---|---|---|---|
| $A$ | \$170,000 | \$44,000 | 22.5 | |
| | | | | $A \rightarrow B$: (−) |
| $B$ | 260,000 | 49,000 | 13.5 | |
| | | | | $A \rightarrow C$: 10.9% |
| $C$ | 300,000 | 66,000 | 18.9 | |
| | | | | $C \rightarrow D$: (−) |
| $D$ | 330,000 | 68,000 | 15.9 | |

One type of mistake would have been to select design $A$, which has the greatest rate of return on total investment, as the preferred alternative. This choice would prevent the additional investment of \$300,000 − \$170,000 = \$130,000 in design $C$, which returns 10.9 percent. Since 10.9 percent is higher than the 10 percent expected from other investments with similar risks, a loss of about 1 percent on $130,000 would occur. However, it is prudent to keep in mind the high returns possible through an investment in design $A$ in case there is insufficient capital to fund design $C$ or if there is another independent opportunity for investing $130,000 at an RR greater than 10.9 percent.

Another type of error would be to select the largest investment that still meets the 10 percent rate-of-return requirement. The unsatisfactory RR for the extra investment in design $D$ over design $C$ was apparent in the incremental analysis. Therefore, putting $330,000 − $300,000 = $30,000 into design $D$ forces this amount of capital to earn less than the 10 percent it could receive if invested elsewhere.

## COMPARISON OF INDEPENDENT ALTERNATIVES

The demarcation between dependent and independent alternatives gets fuzzy at times. Consider the plans for different product designs in Example 9.3. When only one new product design is sought, the four product proposals are certainly mutually exclusive. Similarly, four plans for remodeling an office are surely mutually exclusive. The best product proposal and the best office plan appear to be completely independent of each other. Yet both contribute to the profit status of the organization.

The product proposal is for a new product that will bring in revenue in excess of costs in order to increase profit, and the office plan is designed to produce improved working conditions that will lower costs to increase profit. At least tangentially, both plans are related through their contribution to the firm's financial standing and by their reliance for initial funding on the firm's capital. In effect, they may be mutually exclusive.

### Interdependency

It is seldom worthwhile to trace tenuous links that may relate various investment proposals, particularly so during initial comparison screenings. Diligent detective work could probably uncover links between any and all alternatives, but their effect on early economic comparisons would be negligible. Most mutually exclusive options are clearly evident by coinciding functions; and conditional dependencies are conspicuous by physical relationships, such as the condition for adding a second floor to a building is the construction of the first floor. Such obvious relationships are natural and necessary recognitions for most economic evaluations. The question of how to finance the alternatives may or may not enter the initial comparisons.

When the amount of capital available to fund proposals is restricted, all the alternatives that have passed minimum-required-return screening are linked by the need for funding from a common capital pool. In practice, different divisions of an organization such as manufacturing, research and development, sales, and regional branches usually have separate budgets, and funds are not readily transferable between divisions. If each of the four product proposals in Example 9.3 were funded by a separate budget, they would be independent. Then their acceptance would depend largely on the figures displayed in Table 9.4. At a minimum required return of 10 percent, all would be acceptable and all

**TABLE 9.4** Designs from Example 9.3 considered as independent proposals with a minimum required rate of return of 10 percent.

| *Design* | *Investment* | *RR, %* | *PW* | *PW (receipts)** / *Investment* |
|---|---|---|---|---|
| *A* | $170,000 | 22.5 | $100,336 | 1.59 |
| *B* | 260,000 | 13.5 | 41,056 | 1.16 |
| *C* | 300,000 | 18.9 | 105,504 | 1.35 |
| *D* | 330,000 | 15.9 | 87,792 | 1.27 |

*The ***PW index*** is based on the required 10 percent rate of return.

would be funded, assuming capital were available and other independent alternatives were not more attractive.

A close look at Table 9.4 reveals a "preference among equals." All the designs meet the minimum-return requirement and are therefore acceptable. But design $A$ definitely appears more "acceptable" than design $D$, in which the investment is almost twice as large as in design $A$ while the present worth of net returns is smaller. In the same vein, is the extra $130,000 investment in design $C$ worthwhile when compared with the slight advantage in PW it has over a $170,000 investment in design $A$, or should a search be made to find a proposal with an RR approaching that of design $A$? The question is akin to a choice between a net $500 gain in 2 years on investments of $1000 and $10,000; both have a positive PW of $500, but the $1000 is a much more comfortable commitment. This type of reasoning places more reliance on the "laddering" effect of ordered RR or PW index (last column in Table 9.4) rankings. The same reasoning suggests the possibility of making several smaller investments, drawn from alternatives that have already passed the minimum-required-return screen, in place of one large "acceptable" investment when capital is limited.

## Conversion of Independent Proposals to Mutually Exclusive Combinations

Independent proposals can be collected in various combinations and evaluated as grouped to determine how well each combination meets the investment objectives. The groupings selected depend on the conditions set for the evaluation. For instance, one division of a firm might be allowed to fund two proposals while other divisions are allowed only one. Then each combination would include two proposals from the favored division with one from each of the other divisions.

Several programming techniques, such as linear programming, are adaptable to systematic selections among numerous combinations. A direct search method is used to illustrate the conversion of independent proposals to mutually exclusive combinations because it is conceptually simple and adequate for most routine situations. Consider the four independent proposals listed in Table 9.5, from which any combination with first costs totaling $5000 or less can be selected. After appropriate groupings that satisfy the specified capital limitation are identified, each combination is evaluated by one of the discounted cash-flow comparison methods. Since only one combination can be selected, the decision is between mutually exclusive alternatives.

A check-off procedure for identifying appropriate combinations of proposals is shown in

**TABLE 9.5** Three-year cash flows for four independent proposals, all of which have passed a screening based on a minimum required rate of return of 10 percent.

| *Proposal* | END-OF-YEAR CASH FLOW *First Cost* | *Year 1* | *Year 2* | *Year 3* |
|---|---|---|---|---|
| I | −$1000 | $ 550 | $ 550 | $ 550 |
| II | −2000 | 875 | 875 | 875 |
| III | −3000 | 1400 | 1400 | 1400 |
| IV | −4000 | 1665 | 1665 | 1665 |

Table 9.6. The leftmost columns indicate by a 1 when a proposal is included in a combination, and by a 0 when it is not. The first combination is the do-nothing alternative; the four zeros in the first row indicate no proposals are included. Altogether there are nine combinations that have total first costs of $5000 or less.

**TABLE 9.6** Binary numbers 0 and 1 represent the proposals included in a combination. An RR is calculated for each combination to rank the alternatives. Note that the extra $1000 investment required for alternative 9 over alternative 7 earns less than 0 percent rate of return.

| PROPOSALS | | | | | END-OF-YEAR CASH FLOW | | | | |
|---|---|---|---|---|---|---|---|---|---|
| *I* | *II* | *III* | *IV* | *Alternative* | *0* | *1* | *2* | *3* | *RR, %* |
| 0 | 0 | 0 | 0 | 1 | 0 | 0 | 0 | 0 | 0 |
| 1 | 0 | 0 | 0 | 2 | −$1000 | $ 550 | $ 550 | $ 550 | 30 |
| 0 | 1 | 0 | 0 | 3 | −2000 | 875 | 875 | 875 | 15 |
| 0 | 0 | 1 | 0 | 4 | −3000 | 1400 | 1400 | 1400 | 18.8 |
| 1 | 1 | 0 | 0 | 5 | −3000 | 1425 | 1425 | 1425 | 20 |
| 0 | 0 | 0 | 1 | 6 | −4000 | 1665 | 1665 | 1665 | 12 |
| 1 | 0 | 1 | 0 | 7 | −4000 | 1950 | 1950 | 1950 | 21.7 |
| 1 | 0 | 0 | 1 | 8 | −5000 | 2215 | 2215 | 2215 | 15.7 |
| 0 | 1 | 1 | 0 | 9 | −5000 | 2275 | 2275 | 2275 | 17.3 |

Tangible ratings for a managerial decision on the allocation of capital are clearly presented in the table. The highest RR is earned by allocating just $1000 to fund proposal I. The next highest RR results from an investment of $4000 in proposals I and III. If the full $5000 is to be allocated, 17.3 percent is the best RR that can be expected. The final allocation is a management prerogative relying on economic and political objectives.

Much more complex comparisons can be conveniently accomplished by the iterative approach represented by Table 9.6. However, without constraints such as a limit on investment capital, the number of combinations grows rapidly as more proposals are included, and computations grow correspondingly cumbersome. Computerized and programmed assistance then becomes a near necessity.

**Economy Exercise 9-3** The four independent proposals in Table 9.5 are still under consideration, and new conditions have been added. Proposal I is so attractive that it must be included in any accepted combination. Another proposal with the following cash flow has been added to the list:

| *Proposal* | END-OF-YEAR CASH FLOW *0* | *1* | *2* | *3* |
|---|---|---|---|---|
| IIA | −$2500 | $1150 | $1150 | $1150 |

In addition to proposal I, either proposal II or II*A* must be in any acceptable combination. If the management decision is to fund the combination possessing the highest RR, regardless of the amount of capital required, which proposals will be funded?

## RR IRREGULARITIES

The consistency of AW and PW comparisons is above reproach, and both *generally* agree with RR evaluations. However, there are two situations in which RR calculations obscure or contradict the preferences shown by the other two comparison methods. Both situations involve distinctive cash-flow patterns that provide a clue to possible confusion. The apparent disagreements can be reconciled by reference to the reinvestment-rate assumptions inherent in each comparison method.

### Ranking Reversal

Let two projects have the cash flows indicated in Table 9.7. Both proposals require the same $1000 initial investment. The contrasting net annual returns are conspicuous; project $X$ starts low and increases, whereas project $Y$ has a high first-year flow followed by constant lower flows.

**TABLE 9.7** Cash flows for two projects with 4-year lives and no salvage value.

| | END-OF-YEAR CASH FLOW | | | | |
|---|---|---|---|---|---|
| *Project* | *0* | *1* | *2* | *3* | *4* |
| $X$ | −$1000 | $ 100 | $350 | $600 | $850 |
| $Y$ | −1000 | 1000 | 200 | 200 | 200 |

The two projects are first compared by their present worths when the minimum required rate of return is 10 percent:

$$\begin{aligned} \text{PW}(X) &= -\$1000 + [\$100 + \$250(A/G,\ 10,\ 4)]\,(P/A,\ 10,\ 4) \\ &= -\$1000 + [\$100 + \$250(1.3810)]\,(3.1698) \\ &= \$411.56 \end{aligned}$$

$$\begin{aligned} \text{PW}(Y) &= -\$1000 + [\$1000 + \$200(P/A,\ 10,\ 3)]\,(P/F,\ 10,\ 1) \\ &= -\$1000 + [\$1000 + \$200(2.4868)]\,(0.90909) \\ &= \$361.27 \end{aligned}$$

This *ranks project X higher than project Y.*

When an RR comparison is made, the rankings switch, as shown by the following calculations: For project $X$,

$$\text{PW} = -\$1000 + [\$100 + \$250(A/G,\ i,\ 4)]\,(P/A,\ i,\ 4) \stackrel{?}{=} 0$$

At $i$ = 20 percent,

$$\text{PW} = -\$1000 + [\$100 + \$250(1.2742)]\,(2.5887) = \$83.51$$

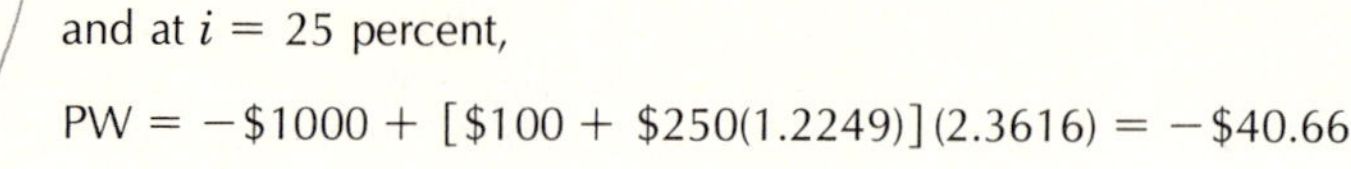

and at $i = 25$ percent,

$$\text{PW} = -\$1000 + [\$100 + \$250(1.2249)](2.3616) = -\$40.66$$

to give

$$\text{RR}(X) = 20\% + 5\% \frac{\$83.51 - 0}{\$83.51 - -\$40.66} = 23.4\%$$

For project $Y$,

$$\text{PW}(Y) = -\$1000 + [\$1000 + \$200(P/A, i, 3)](P/F, i, 1) \stackrel{?}{=} 0$$

is solved by trial and error to obtain RR($Y$) = 34.5 percent, which ***ranks project Y ahead of project X.***

The net present-worth profiles of the two projects are shown in Figure 9.3. Note that at 10 percent the PW curve for project $X$ is above the curve for project $Y$. The positions are reversed for required rates of return above 13 percent.

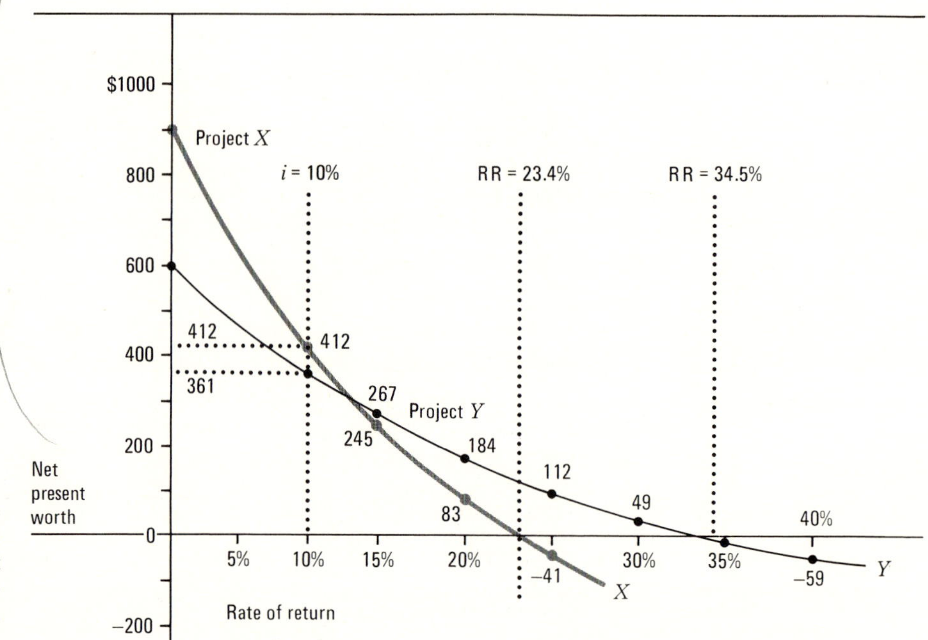


**FIGURE 9.3** Relationship of net present worth and different discount rates, showing how the rankings for two projects can switch.

The reversed rankings are the result of assuming the annual returns can be reinvested at 10 percent in the PW comparison versus 23.4 percent for project $X$ and 34.5 percent for project $Y$ in the RR comparison. The influence of a higher reinvestment rate is exaggerated by the markedly different cash-flow patterns of the two projects. In project $X$ the larger receipts occur near the end of the project's life, as opposed to the recovery of the initial $1000 investment by the end of the first year in project $Y$. This $1000 inflow is assumed to be reinvested at the calculated RR, and it consequently returns handsome dividends over the remaining 3 years of the project.

If the two alternatives in Table 9.7 were independent, both would be acceptable at the 10 percent minimum required rate of return, and both would continue to be acceptable by either PW or RR evaluations up to a discount rate of 23.4 percent. Thus, the discrepancy in ranking is more significant for mutually exclusive alternatives.

Critics of the RR comparison method claim the reinvestment assumption is unrealistic. They suggest that it is unlikely that another project can be found for reinvestment purposes that will provide returns equivalent to the RR of the highest-ranked proposal. The PW method is considered more conservative when the discount rate employed is a function of the cost of capital, and it is therefore believed by many to be theoretically superior.

## Multiple Rates of Return

When the *cumulative* cash flow of a project switches from negative to positive (or the reverse) *more than once,* the project may have more than one rate of return. In such situations, relatively rare in practice, no single percentage is immediately available to rank the alternative; two or more RR figures are equally correct, as demonstrated in Example 9.4.

**Example 9.4** **Two Solutions for an RR Evaluation**

One of the alternatives for improving an operation is to do nothing to it for 2 years and then spend \$10,000 on improvements. If this course of action is followed, the immediate gain is \$3000 followed by two years of breakeven operations. Thereafter, annual income should be \$2000 per year for 4 years. What rate of return can be expected from following this course of delayed action?

**Solution 9.4**

| *End of Year* | *Cumulative Cash Flow* |
|---|---|
| 0 | +\$3000 |
| 1 | +3000 |
| 2 | −7000 |
| 3 | −5000 |
| 4 | −3000 |
| 5 | −1000 |
| 6 | +1000 |

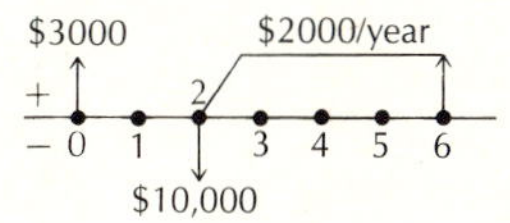


Converting the given data to a cash-flow diagram suggests there may be more than one RR because the cumulative cash flow reverses from positive to negative at year 2 and again reverses signs at year 6. Since a double sign reversal is not always accompanied by dual rates of return, trials can be conducted at arbitrarily selected interest rates to determine the general PW profile. Based on

$$PW = \$3000 - \$10{,}000(P/F, i, 2) + \$2000(P/A, i, 4)(P/F, i, 2) \stackrel{?}{=} 0$$

at $i = 0$,

$$PW = \$3000 - \$10{,}000 + \$2000(4) = \$1000$$

(also indicated by the cumulative cash-flow pattern shown above); at $i = 10$ percent,

$$PW = \$3000 - \$10{,}000(0.82645) + \$2000(3.1698)(0.82645) = -\$25$$

(which indicates by the sign reversal from PW at $i = 0$ that one RR is slightly less than 10 percent); at $i = 51$ percent,

$$PW = \$3000 - \$10{,}000(0.43906) + \$2000(1.5856)(0.43906) = \$2$$

(where the second sign reversal confirms there is a second RR.)

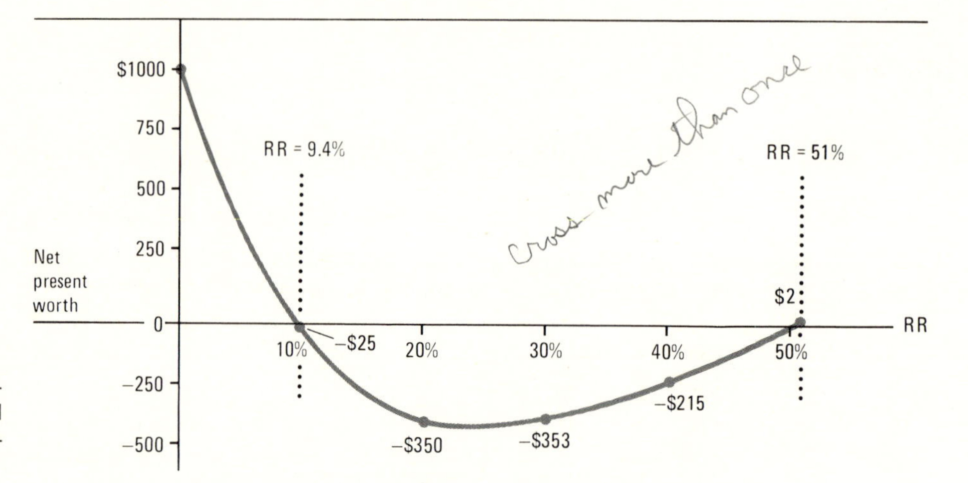


**FIGURE 9.4** Net present worth of a proposal with multiple rates of return.

**The profile of present worths over the discounting range of 0 percent to 51 percent is shown in Figure 9.4. The equivocal answer for the proposal's RR is that it is either 9.4 percent or 51 percent when returns can be assumed to be reinvested at either rate.**

A more specific answer to the RR question in Example 9.4 is developed by applying an *explicit reinvestment rate of return* to a limited portion of the cash flow that will disturb the total cash-flow pattern as little as possible while eliminating one of the sign reversals. The explicit reinvestment rate may be the minimum attractive rate of return employed by the organization or a rate suggested by the PW profile. The significance of the choice is apparent from comparisons in Table 9.8, where the $3000 receipt is assumed to be invested at an explicit interest rate for 2 years. Under this assumption, one sign reversal is avoided and the present worth of the modified cash flow is

$$PW = \$3000(F/P, i\%, 2) - \$10{,}000 + \$2000(P/A, i, 4) \stackrel{?}{=} 0$$

$3000(*F/P*, *i*%, 2)
$2000/year
+
−
0 1 2 3 4
Years
$10,000

*where i*% = explicit reinvestment rate

| *Explicit Reinvestment Rate Applied to the $3000 Payment for 2 Years, %* | *RR on Net Investment when an Explicit Reinvestment Rate Is Utilized, %* |
|---|---|
| 0 | 5.5 |
| 5 | 7.6 |
| 9.4 | 9.4 |
| 15 | 12.4 |
| 20 | 18.9 |
| 30 | 22.7 |
| 40 | 33.4 |
| 51 | 51 |
| 60 | 82 |

**TABLE 9.8** Different RR percentages resulting from explicit reinvestment rates used on a limited portion of the cash flow to convert a dual rate of return to a single RR.

From Table 9.8 it appears that the proposal is attractive only when the expected reinvestment rate is below 9.4 percent, and above 51 percent. Thus the translation of dual RRs to a single RR is contingent upon the reinvestment rate assumed. The closer the reinvestment rate of return is to the calculated RR, and the less manipulation needed to remove a sign reversal, the more closely the single RR approaches one of the dual RRs.

**Economy Exercise 9-4** Expected cash flows for a strip-mining project are estimated as shown in the cash-flow diagram below.

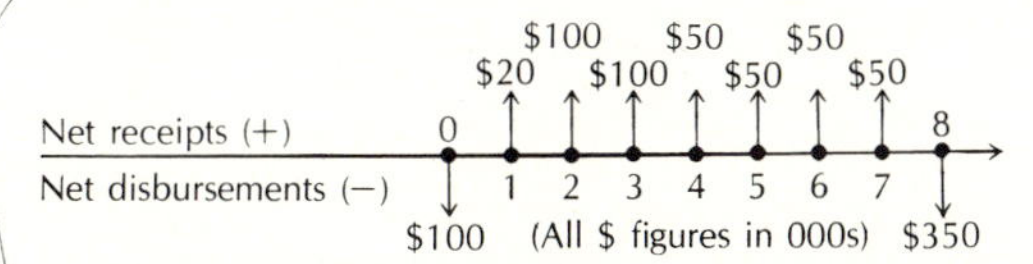


A start-up cost is incurred immediately. Then income exceeds outlays for the next 7 years. During the eighth year the major cost is for landscape improvement. Does the strip-mining project appear to be a profitable investment?

## MINIMUM ATTRACTIVE RATE OF RETURN

The *minimum attractive*, or *required*, or *minimum acceptable* rate of return is a lower limit for investment acceptability set by organizations or individuals. It is a device designed to make the best possible use of a limited resource, money. Rates vary widely according to the type of organization, and even within the organization. Historically, government agencies and regulated public utilities have utilized lower required rates of return than competitive industrial enterprises. Within a given enterprise, the required rate may be different for various divisions or activities. These variations usually reflect the risk involved. For instance, the rate of return required for cost-reduction proposals may be lower than that required for research and development projects in which there is less certainty about prospective cash flows.

There is a wealth of literature on the subject, but a poverty of agreement. It is generally accepted that the lower bound for a minimum required rate of return should be the *cost of capital.* The constitution of this cost is also subject to controversy. As discussed in Chapter

8, the cost of capital for competitive industries must reflect the expense of acquiring funds from various sources; and, as discussed in Chapter 13, the cost of safe government bonds is a basis for the lower bound of interest rates used to evaluate public investments.

How much above the cost of capital to set the minimum required rate of return depends on an organization's circumstances and aspirations. A small company strapped for cash and burdened by a low credit rating must have a very, very attractive proposal before it can consider investing. Larger established companies tend to view the rate as a realistic expectation of how much their capital can earn when reinvested.* This *reinvestment rate* is a typical figure promised (and later substantiated) for a large number of high-quality investment proposals available to the firm; it is assumed that the proceeds earned from current projects can be reinvested at comparable rates in future proposals. The rate so derived is sometimes called the *opportunity cost of capital* because any proposal funded to earn a lower rate precludes the opportunity to earn the minimum attractive rate of return.

The effect of establishing a minimum required rate of return is to ration capital (examples are provided in Chapter 12). It is rationed to divisions of an organization and to the whole organization as a function of time. The purpose is to avoid unproductive investments in marginal activities, perhaps favored for political reasons, and to conserve capital during periods when fewer "attractive" proposals are submitted. Capital rationing is implemented by adhering to the cutoff rate unless powerful intangible factors are boldly involved, or, as an engineering economics function, rationing is accomplished by selecting the most advantageous alternative, when one of the alternatives is to do nothing.

## SUMMARY

*Rate of return* is the last of the discounted cash-flow comparison methods considered. It usually produces the same rankings as the annual-cost and present-worth methods. The exception occurs when alternatives with heavy cash flows in early years are compared with those having the major flows in later years. Multiple rates of return can also confuse comparisons. These may occur when the *cumulative* cash flow reverses signs *more than once.* Both multiple rates of return and contradictory rankings are a result of the reinvestment-rate assumption inherent in RR evaluations; it is assumed that funds recovered during the life of a project can be reinvested at the calculated RR. This assumption causes some criticism of the rate-of-return method and the claim that present-worth comparisons are theoretically superior.

The rate-of-return method is favored by some who claim it provides the most easily understood comparisons and that it produces rankings that are more useful in rationing capital. RR calculations can begin with AW or PW formulations, the latter being more common. There is no way to avoid trial-and-error computations for complex formulations, but the structure offers clues as to where to begin.

Each increment of additional investment in mutually exclusive alternatives must justify itself in relation to a minimum required rate of return. Increments are always derived from the next-lowest-cost *acceptable* alternative. Independent alternatives can be converted to

*Returns on equity for *Fortune's* 500 largest United States corporations were 10.3 percent in 1972, 12.4 percent in 1973, and 13.6 percent in 1974.

mutually exclusive options composed of legitimate combinations of independent proposals. Then the combinations are evaluated for acceptability. Most alternatives share an interdependency resulting from their reliance, for funding, on an organization's available capital.

A *minimum acceptable* or *required* rate of return is the lowest level at which an alternative is still attractive. It varies among and within organizations. Although there are a wide variety of recommendations for determining this lowest level of acceptability, it is generally agreed that it should be no lower, and most likely considerably higher, than the cost of capital. How much higher depends on the circumstances, objectives, and policies of the organization. The purpose of establishing a minimum attractive rate of return is to ration capital to the most deserving proposals.

## Discussion of Economy Exercises

**EE 9-1** This is a variation of the bond problem in Example 8.3. In the original version the question was how much could be paid for the described bond in order to earn a rate of return of 8 percent compounded semiannually. The solution revealed that the PW of the bond at $i = 4$ percent per period was \$666.76. So we know the actual rate of return must be smaller than 8 percent compounded semiannually.

Trying $i = 5$ percent compounded semiannually, or 2½ percent per period for 28 periods, gives

$$\begin{aligned} PW &= \$1000(P/F, 2\tfrac{1}{2}, 28) + \$20(P/A, 2\tfrac{1}{2}, 28) - (\$760 + \$20) \overset{?}{=} 0 \\ &= \$1000(0.50089) + \$20(19.964) - \$780 = \$120.17 \end{aligned}$$

which indicates $i$ should be greater than 2½ percent per period. At $i = 6$ percent compounded semiannually, again for 28 periods,

$$\begin{aligned} PW &= \$1000(P/F, 3, 28) + \$20(P/A, 3, 28) - \$780 \overset{?}{=} 0 \\ &= \$1000(0.43709) + \$20(18.763) - \$780 = \$32.38 \end{aligned}$$

which is closer, but $i$ is still too low. The present worth of the bond at $i = 4$ percent per period is \$666.76; the difference between this PW and the selling price is \$666.76 − \$780 = −\$113.24. By interpolation between the PWs at 3 percent and 4 percent,

$$i = 3\% + 1\%\frac{\$32.38 - 0}{\$32.38 - -\$113.24} = 3\% + 0.2\% = 3.2\%$$

which means the bond purchased for \$780 will earn 6.4 percent compounded semiannually.

**EE 9-2** Investigating first the lowest-cost alternative, we check to see if an office building will be profitable at $i = 0$.

$$PW = \underbrace{-(\$485{,}000 + \$340{,}000)}_{P = -\$825{,}000} + \underbrace{(\$120{,}000 + \$266{,}000)}_{S = \$386{,}000}(1) + \underbrace{(\$212{,}000 - \$59{,}100)}_{A = \$152{,}900}(30) = \$4{,}148{,}000$$

Knowing a positive cash flow exists, a rough estimate of the RR is determined from the more significant flows of $P$ and $A$:

$$(A/P, i, 30) \doteq \frac{A}{P} \doteq \frac{\$152{,}900}{\$825{,}000} \doteq 0.1853$$

which falls between the 15 percent and 20 percent interest tables. Then, by trial and error,

$$\text{PW} = -\$825{,}000 + \$386{,}000(P/F, i, 30) + \$152{,}900(P/A, i, 30) \stackrel{?}{=} 0$$

At RR = 18 percent,

$$\begin{aligned}\text{PW} &= -\$825{,}000 + \$386{,}000(0.00698) + \$152{,}900(5.5168)\\ &= \$21{,}213\end{aligned}$$

At RR = 19 percent,

$$\begin{aligned}\text{PW} &= -\$825{,}000 + \$386{,}000(0.00560) + \$152{,}900(5.2478)\\ &= -\$20{,}450\end{aligned}$$

From these, by interpolation, RR = 18.5 percent.

The conversion of the fire-damaged hotel into a showroom and office building is thus an acceptable alternative; the 18.5 percent RR is greater than the required 12 percent.

The alternative plan to convert to an apartment has incremental additional values of

*First cost:* $490,000 − $340,000 = $150,000
*Salvage value:* $190,000 − $120,000 = $70,000
*Net annual returns:* $251,200 − $88,000 − $152,900 = $10,300

The incremental rate of return is calculated as

$$\text{PW} = -\$150{,}000 + \$70{,}000(P/F, i, 30) + \$10{,}300(P/A, i, 30) \stackrel{?}{=} 0$$

At RR = 6 percent,

$$\begin{aligned}\text{PW} &= -\$150{,}000 + \$70{,}000(0.17412) + \$10{,}300(13.764)\\ &= \$3958\end{aligned}$$

At RR = 7 percent,

$$\begin{aligned}\text{PW} &= -\$150{,}000 + \$70{,}000(0.13137) + \$10{,}300(12.409)\\ &= -\$12{,}991\end{aligned}$$

By interpolation,

$$\text{RR} = 6\% + 1\% \frac{\$3958 - 0}{\$3958 - (-\$12{,}991)} = 6.2\%$$

The RR lower than the required 12 percent rate disqualifies the additional investment needed to proceed from the office plan to the apartment plan. It should be noted that the office plan still has a total RR greater than the minimum required 12 percent:

$$\text{PW} = -\$975{,}000 + \$456{,}000(P/F, i, 30) + \$163{,}200(P/A, i, 30) \stackrel{?}{=} 0$$

At RR = 17 percent,

$$PW = -\$975{,}000 + \$456{,}000(0.0107) + \$163{,}200(5.927) = -\$2834$$

At RR = 16.9 percent,

$$PW = -\$975{,}000 + \$456{,}000(0.0110) + \$163{,}200(5.963) = \$3178$$

Even with a 17 percent rate of return, the apartment plan is not an acceptable alternative because the additional investment required for its implementation does not meet the minimum rate of return standard. The results are summarized in the table.

| | *Office Plan* | → | *Increment* | → | *Apartment Plan* |
|---|---|---|---|---|---|
| First cost | $825,000 | | $150,000 | | $975,000 |
| Salvage value | $386,000 | | $70,000 | | $456,000 |
| Annual returns | $152,900 | | $10,300 | | $163,200 |
| Rate of return | 18.5% | → | 6.2% | → | 17% |

**EE 9-3** All possible combinations that satisfy the investment criteria are delineated in Table 9.9.

**TABLE 9.9**

| | ALTERNATIVES | | | | | | | | | | | |
|---|---|---|---|---|---|---|---|---|---|---|---|---|
| *Proposal* | *1* | *2* | *3* | *4* | *5* | *6* | *7* | *8* | *9* | *10* | *11* | *12* |
| I | 1 | 1 | 1 | 1 | 1 | 1 | 1 | 1 | 1 | 1 | 1 | 1 |
| II | 1 | 1 | 1 | 1 | 0 | 0 | 0 | 0 | 1 | 1 | 1 | 1 |
| II*A* | 0 | 0 | 0 | 0 | 1 | 1 | 1 | 1 | 1 | 1 | 1 | 1 |
| III | 0 | 1 | 0 | 1 | 0 | 1 | 0 | 1 | 0 | 1 | 0 | 1 |
| IV | 0 | 0 | 1 | 1 | 0 | 0 | 1 | 1 | 0 | 0 | 1 | 1 |

From knowing the RRs for individual proposals already listed in Table 9.6, and by calculating RR(II*A*) = 18 percent, all combinations which include proposal II can be eliminated because RR(II) = 15 percent; compared to proposal II*A*, the inclusion of proposal II has to lower any combination's RR below 18 percent. This narrows the choice to alternatives 5, 6, 7, and 8 in Table 9.9.

Again noting from Table 9.6 that the RR of proposal III is above 18 percent while that of proposal IV is below, the best combination must be alternative 6. This combination has a first cost of $1000 + $2500 + $3000 = $6500 and annual returns of $550 + $1150 + $1400 = $3100, which lead to an RR of 20.3 percent. Utilizing dominating relationships can bypass an lot of computations.

**EE 9-4** The − to + to − cash-flow pattern suggests dual rates of return. The suspicion is verified by trial-and-error calculations based on

$$\text{PW} = \$100{,}000 + \$20{,}000(P/F, i, 1) + \$100{,}000(P/F, i, 2) + \$100{,}000(P/F, i, 3) + \$50{,}000(P/A, i, 4)(P/F, i, 3) - \$350{,}000(P/F, i, 8) \stackrel{?}{=} 0$$

to identify the RRs shown below.

| | | PW AT 2.9% | | PW AT 44.5% | |
|---|---|---|---|---|---|
| *End of Year* | *Cash Flow* | *Factor* | *Amount* | *Factor* | *Amount* |
| 0 | −$100,000 | 1.00 | −$100,000 | 1.00 | −$100,000 |
| 1 | 20,000 | 0.97 | 19,000 | 0.69 | 14,000 |
| 2 | 100,000 | 0.94 | 94,000 | 0.47 | 47,000 |
| 3 | 100,000 | 0.92 | 92,000 | 0.33 | 33,000 |
| 4 | 50,000 | 0.89 | 45,000 | 0.22 | 11,000 |
| 5 | 50,000 | 0.86 | 43,000 | 0.15 | 7,500 |
| 6 | 50,000 | 0.83 | 42,000 | 0.10 | 5,000 |
| 7 | 50,000 | 0.81 | 40,000 | 0.07 | 3,500 |
| 8 | −350,000 | 0.78 | −275,000 | 0.06 | −21,000 |
| *Net PW* | | | 0 | | 0 |

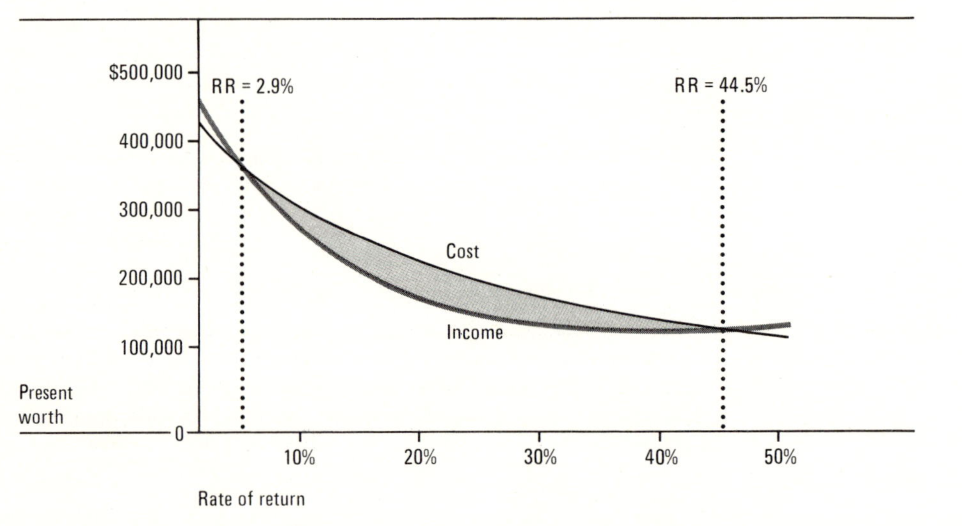


**FIGURE 9.5** Present worths of income and cost streams at different rates of return. The curves intersect where net PW = 0.

A more comprehensive picture of the proposed strip-mining venture is provided by the discounted income and cost curves shown in Figure 9.5. It is apparent that the project can be considered profitable only when the interest rate used in the present-worth calculations is below 3 percent or above 45 percent. This means that the strip-mining project will be attractive only when the organization is satisfied with a modest 3 percent rate of return or is optimistic enough to accept the premise of a reinvestment rate of 45 percent or greater. Between the points where the curves intersect in Figure 9.5, the PW is negative. Thus, a minimum attractive rate of return of 10 percent rejects the project when it is used as an explicit reinvestment rate to remove the sign reversal in the cash flow, as demonstrated below.

Let $i$ = 10 percent to compound the values of the first two transactions (−\$100,000 and \$20,000); the cash-flow diagram is revised as

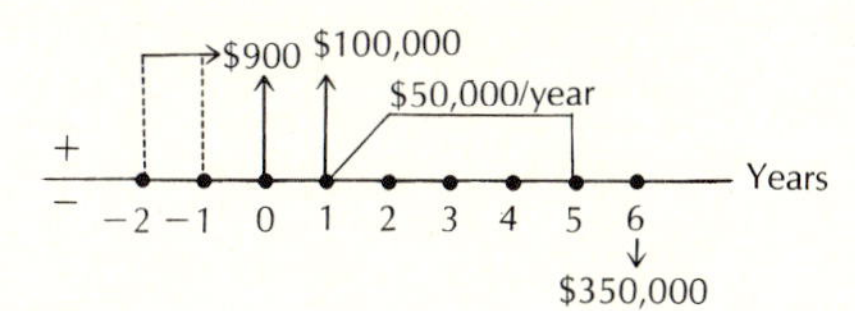


where, at year 0,

$$\$900 = \$100{,}000 + \$20{,}000(F/P,\ 10\%,\ 1) + -\$100{,}000(F/P,\ 10\%,\ 2)$$
$$= \$100{,}000 + \$22{,}000 - \$121{,}100$$

to reveal that the net PW is negative even when $i$ = 0 percent:

$$\text{PW} = \$900 + \$100{,}000(P/F,\ 0,\ 1) + \$50{,}000(P/A,\ 0,\ 4)(P/F,\ 0,\ 1) - \$350{,}000(P/F,\ 0,\ 6)$$
$$= \$900 + \$100{,}000 + \$200{,}000 - \$350{,}000 = -\$49{,}100$$

## PROBLEMS

**9.1** Sometimes objects of art are respectable investments. In 1975 a marble bust of Benjamin Franklin, from which the engraving was made for \$100 bills, was auctioned for \$310,000. It was sculpted in France in 1778 by Jean-Antoine Houden. In 1939 the same bust sold for \$30,000. What rate of return was earned by the collector who owned the statue from 1939 to 1975?

**9.2** Occasionally, rates of return are stated in ways that are deceptive; they sound intuitively logical, but the effective rate may be much higher than the instinctive impression. The age-old warning to buyers could aptly be reworded as "let the borrower beware." Examples of characteristic false impressions involving interest rates are provided in the following problems:

**9.2a** A gullible borrower went to the Fast-Buck Loan Company expecting to borrow money at the advertised rate of "one percent per month." He agreed to repay a loan of \$500 in 12 equal end-of-month payments of \$47.50 each. The amount of the monthly payments was calculated as follows:

| | |
|---|---|
| Principal to be repaid (amount borrowed) | \$500 |
| Interest at 1%/month (12 months × 0.01 × \$500) | 60 |
| Credit investigation (check on credit rating) | 10 |
| *Total cost of loan* | \$570 |

$$\text{Monthly payments} = \frac{\$570}{12\ \text{months}} = \$47.50/\text{month}$$

What effective interest rate was charged?

*(27.9%)*

**9.2b** A prepaid interest plan has the interest deducted from the amount of the loan. For a 1-year loan of \$1000 at a 1½ percent interest charge per month, the interest of 12 months × 0.015 × \$1000 = \$180 is deducted from the amount loaned to give the borrower just \$1000 − \$180 = \$820 to spend. If monthly payments are \$1000/12 = \$83.33, what is the actual nominal interest rate?

**9.2c** "Balloon" payments occur when the last loan-repayment charge is exceptionally large compared with the constant payments up to the final one. This repayment plan may not be as troublesome with respect to unusually high interest charges as with respect to the difficulty of making that last balloon payment; inability to meet the final payment may lead to repossession of the item purchased or the necessity to refinance, often at a higher interest rate. For example, a loan of \$1000 for a used car could lead to a repayment plan of \$60 per month for 11 months with a final, year-end payment of \$460: that is, $\$460 + \$60 \times 11 = \$1000 + 0.01 \times \$1000 \times 12$. What effective interest rate is charged? Discuss why such plans might be offered, and why they may be accepted by the borrower.

**9.2d** Sometimes, it is possible to buy an item for less by paying the purchase price in one lump sum because the seller prizes immediate cash over a time contract for a series of future payments. Assume you can purchase a \$10,000 building lot on a 6-year contract with annual payments at 7 percent compounded annually on the unpaid balance. As you consider the purchase, you learn that someone else purchased an essentially identical lot from the same seller for \$8500 in cash. In effect, what total interest rate are you paying if you elect to make the purchase on a 6-year contract?

**9.3** Assume *Hotstuff* magazine offers two types of subscriptions, payable in advance, as follows:

| | |
|---|---|
| 1-year subscription | \$10 |
| 2-year subscription | \$16 |

**9.3a** In comparing the economy of a 1-year subscription with that of a 2-year subscription, what is the rate of return on the extra investment in the 2-year subscription?

*(66.7%)*

**9.3b** As an added bonus, a nature-lover's calendar is included with each subscription to *Hotstuff*. However, only one calendar is included, even with a 2-year subscription. If you figure your money is worth 10 percent, how much could you afford to pay for next year's calendar so that both subscription plans can be considered equal in value?

**9.4** A \$5000 bond matures in 10 years and pays 3 percent interest twice a year. If the bond sold for \$5050, what is the actual investment rate?

*(5.9% compounded semiannually)*

**9.5** What was the interest rate paid by the school district for the money raised by the bond issue described in Problem 8.26?

**9.6** A construction firm can lease a crane required on a project for 3 years for \$180,000 payable now, with maintenance included. The alternative is to buy a crane for \$240,000 and sell it at the end of 3 years for \$100,000. Annual maintenance costs are expected to be \$5000 the first 2 years, and \$10,000 the third year (payable at the end of each year). At what interest rate would the two alternatives be equivalent?

*(9.4%)*

**9.7** Compare the alternatives of replacing two type-*A* machines with one type-*B* machine, as described in Problem 7.19, by the rate-of-return method. Does the resulting RR have any advantage over the results obtained from an annual-worth comparison when it is necessary to explain the evaluation to someone not familiar with discounted cash-flow comparisons? Explain.

**9.8** A spare-parts service department for a construction-equipment supplier must be established in Alaska to meet contractual commitments. The firm has eliminated the option of building a new field office because it expects to conduct business in Alaska for just the next 10 years, during the construction boom. At best, the field office will meet expenses and make a small profit contribution. Only three structures that are available for purchase meet the space needs and have the desired location. A major portion of the investment will likely be recovered by appreciation in land values. A minimum acceptable rate of return of 15 percent before taxes is expected on such investments. What rate of return will be earned by each increment of investment from the anticipated cash flows shown?

| | *Site 1* | *Site 2* | *Site 3* |
|---|---|---|---|
| Purchase price | $140,000 | $190,000 | $220,000 |
| Resale value | 125,000 | 155,000 | $175,000 |
| Net revenue | 24,000 | 31,000 | 41,000 |

[*RR*(1) = 16.9%; *RR*(1→2) = 11.7%; *RR*(1→3) = 19.8%]

**9.9** Compare the flume and ditch-and-tunnel alternatives to provide power and water for the mill described in Problem 8.21. Base the RR calculation on a 60-year economic life for both alternatives.

(*RR on extra investment increment* = 4.6%)

**9.10** A temporary water line is required to supplement the water supply at a plant until city water becomes available in a new industrial area. Three alternative pipe sizes with associated pumping facilities will satisfy the water requirements:

| | PIPE SIZE | | |
|---|---|---|---|
| | *14 in (35.6 cm)* | *16 in (40.6 cm)* | *18 in (45.7 cm)* |
| First cost | $18,000 | $25,000 | $34,000 |
| Annual pumping cost | 6,400 | 4,400 | 2,800 |

The pipeline and pumping stations will be in the same locations for all three alternatives. The planning period is 5 years, and pipe can be recovered at the end of the period. It is expected to yield 40 percent of its first cost when recovered, and the cost of recovery will be $2000 regardless of pipe size. Compare the rates of return for the alternatives when a 9 percent return before taxes is desired.

**9.11** Solve Problem 8.15 by the rate-of-return method. What is the RR for the preferred level of insulation?

**9.12** Owing to perennial complaints by students and faculty about the lack of parking spaces on campus, a parking garage on university-owned property is being considered. Since there are no university funds available for the project, it will have to pay for itself from parking fees over a 15-year period. A 10 percent minimum rate of return is deemed reasonable in deciding how large the structure should be. Based on the income and cost data shown below, how many levels should be built?

| *Number of Levels* | *Cumulative Construction Costs* | *Annual Operating Cost* | *Income per Year* |
|---|---|---|---|
| 1 | $ 600,000 | $35,000 | $100,000 |
| 2 | 2,200,000 | 60,000 | 350,000 |
| 3 | 3,600,000 | 80,000 | 570,000 |
| 4 | 4,800,000 | 95,000 | 810,000 |

**9.13** Compare the three alternatives for providing the additional estimating services described in Problem 7.20 according to their rates of return.

*(Select alternative 2 at an incremental RR of 25%.)*

**9.14** Compare the rates of return for the following plans, and select the preferable alternative. The minimum acceptable rate of return is 6 percent.

| | *Plan 1* | *Plan 2* | *Plan 3* |
|---|---|---|---|
| First cost | $40,000 | $32,000 | $70,000 |
| Salvage value | $10,000 | $6,000 | $20,000 |
| Economic life | 7 years | 7 years | 7 years |
| Annual receipts | $18,000 | $18,000 | $24,500 |
| Annual disbursements | $11,000 | $14,500 | $14,500 |

**9.15** Additional parking space for a factory can be either rented for $6000 per year on a 10-year lease or purchased for $90,000. The rental fees are payable in advance at the beginning of each year. Taxes and maintenance fees will be paid by the lessee. The land should be worth at least $60,000 after 10 years. What rate of return will be earned from the purchase of the lot?

*(RR = 4.2%)*

**9.16** A business property can be purchased today for $90,000; the expected resale value after 20 years is $60,000. If annual rental income is $11,800 and expenses are $4700, what before-tax rate of return would be earned by purchasing the property?

**9.17** Six mutually exclusive alternatives are being considered. They are listed in order of increasing first costs in the following table, where the rate of return for the overall investment in each and the incremental RR for every increment are given. All the alternatives have the same lives and comparable intangible values.

| | | RR ON INCREMENTS OF INVESTMENT COMPARED WITH ALTERNATIVE: | | | | |
|---|---|---|---|---|---|---|
| *Alternative* | *RR on Overall Investment* | *I* | *II* | *III* | *IV* | *V* |
| I | 1% | | | | | |
| II | 8% | 21% | | | | |
| III | 11% | 15% | 12% | | | |
| IV | 15% | 22% | 19% | 17% | | |
| V | 13% | 19% | 16% | 15% | 9% | |
| VI | 14% | 21% | 18% | 16% | 14% | 21% |

**9.17a** If one of the alternatives *must* be implemented but there are insufficient funds available to afford any of the last three alternatives, which one should be selected? Why?
**9.17b** What would you recommend if none of the alternatives is mandatory (do nothing is an acceptable alternative); there are still insufficient funds for alternatives IV, V, and VI; and the minimum attractive rate of return is 12 percent? Why?
**9.17c** If the selection of one of the alternatives is mandatory, at what range of RR is alternative I the proper choice?
**9 17d** At what minimum required rate of return is alternative IV the correct choice?
**9.17e** If funds are almost unlimited and no minimum rate of return is required (although the cost of capital is 9 percent), which alternative would you select? Why?
**9.17f** If the minimum attractive rate of return is 12 percent, what would be the opportunity cost, expressed as a rate-of-return percentage, when a shortage of funds causes alternative III to be selected?

**9.18** A half dozen cost-reduction proposals have been forwarded by the industrial engineering department. A 20 percent rate of return is expected, and all equipment investments are to be written off in 5 years with no salvage value.

| *Proposal* | *Equipment Cost* | *Net Annual Savings* |
|---|---|---|
| Combine | $25,000 | $ 9,500 |
| Rearrange | 60,000 | 21,500 |
| Modify | 20,000 | 7,000 |
| Eliminate | 20,000 | 9,000 |
| Up-quality | 40,000 | 17,000 |
| Make-safer | 30,000 | 10,000 |

**9.18a** Which proposal should be selected if only one can be accepted?
**9.18b** Which proposal should be selected if all proposals are independent and there is effectively unlimited capital available for cost-reduction projects?
**9.18c** What is the highest RR that can be earned from any combination of independent alternatives that has a total investment cost of at least $50,000 but less than $100,000?
**9.18d** What combination of alternatives would produce the greatest total savings on an investment of not over $100,000?
**9.18e** Would you recommend the combination identified in Problem 9.18*c* or 9.18*d*? Why?

**9.19** A cash-flow pattern shows an income of $250 at the end of year 1 between expenditures of $100 now and $156 at the end of year 2.
**9.19a** Calculate the dual rates of return by trial and error.
**9.19b** Use the quadratic formula to solve for the two rates. As a reminder, the quadratic formula is $(-b \pm \sqrt{b^2 - 4ac})/2a$, and the coefficients in this case are $a = -100$, $b = 250$, and $c = -156$.

**9.20** The owner of a truck-weighing and lumber-scaling station has agreed to lease the facility for 15 years at $10,000 per year under an agreement that the scales and other equipment will be overhauled and repaired by the owner at the end of the eighth year at a cost not to exceed $150,000.
**9.20a** What rate(s) of return will the owner receive for the station lease with the equipment-repair agreement?
**9.20b** After negotiations on the above lease, caused by concern that the equipment

needed overhauling before 8 years, it was agreed that the owner would pay up to $105,000 for repairs at the end of year 4 instead of making the repairs at the end of year 8. What rate(s) of return will the owner receive under the revised agreement?

**9.20c** Check the revised agreement to see if it is acceptable at an RR of 20 percent. Use the 20 percent as an explicit reinvestment rate to determine a single RR for the lease and repair agreement in Problem 9.20*b*. How would you explain the expected RR to the owner?

**9.21** Three independent proposals have passed a preliminary screening to confirm that all are acceptable at a minimum RR of 15 percent. Each one has an economic life of 4 years. The cash flows are given in the table, with the salvage values included in the final year's income.

| | | END-OF-YEAR CASH FLOWS | | | |
|---|---|---|---|---|---|
| *Proposal* | *First Cost* | *1* | *2* | *3* | *4* |
| *A* | $17,000 | $10,000 | $ 8,000 | $ 6,000 | $ 4,000 |
| *B* | 22,200 | 4,000 | 7,000 | 10,000 | 13,000 |
| *C* | 20,700 | 8,000 | 8,000 | 8,000 | 8,000 |

**9.21a** Which combination of proposals should be selected if sufficient capital is available to fund any choice, so long as the rate of return is 20 percent or greater? Why?

**9.21b** Which combination of proposals should be selected when the minimum RR must be 15 percent, but an income of at least $14,000 per year is necessary?

**9.22** An interesting article* about the reasons why Continental Oil Company switched in 1955 to the discounted-cash method for evaluating investments was written by John G. McLean, then vice-president for international and financial operations. One of the applications described was a water-flood project that exhibited dual rates of return.

The problem was to determine the profitability of acquiring a small oil-producing property in which the primary reserves were nearly exhausted. The owner of the property would receive a royalty of 12½ percent on all oil produced from the property, and the company would agree to water-flood the reservoir at an expected cost of $2.5 million. The injection of water into a reservoir is a method of "secondary recovery" that often increases the total amount of oil recovered after the free-flowing oil supply has diminished.

Estimated cash flows for the 10-year project are shown in the second column of Table 9.10. The primary reserve of oil yields returns the first 3 years. Then the water flood is expected to boost the company's income to $700,000 in the fourth year while it invests $2.5 million, for a net outlay of $1,800,000 that year. Thereafter, income decreases annually by $100,000. The present worths of the yearly cash flows are indicated for different discount rates. At $i = 28$ percent and $i = 49$ percent, the present worth of the cash flow is zero. Between these rates the present worth of the venture is negative.

Instead of settling for two rates of return, assume the cash flows from the first 3 years can actually be reinvested at an annual interest rate of only 15 percent. What is the RR on the resulting net investment when this portion of the total is reinvested at the explicit 15 percent

*J. G. McLean, "How to Evaluate New Capital Investments," *Harvard Business Review*, November-December 1958.

TABLE 9.10 Present worth of a secondary oil-recovery project at different discount rates (cash flows are in thousands of dollars).

| | | PRESENT WORTH OF CASH FLOW AT | | | | | | |
|---|---|---|---|---|---|---|---|---|
| *Year* | *Cash Flow* | *10%* | *20%* | *28%* | *30%* | *40%* | *49%* | *50%* |
| 1 | $ 200 | $ 182 | $167 | $156 | $154 | $143 | $134 | $133 |
| 2 | 100 | 83 | 69 | 61 | 59 | 51 | 45 | 44 |
| 3 | 50 | 38 | 29 | 24 | 23 | 18 | 15 | 15 |
| 4 | −1800 | −1229 | −868 | −671 | −630 | −469 | −365 | −356 |
| 5 | 600 | 373 | 241 | 175 | 162 | 112 | 82 | 79 |
| 6 | 500 | 282 | 167 | 114 | 104 | 66 | 46 | 44 |
| 7 | 400 | 205 | 112 | 71 | 64 | 38 | 24 | 23 |
| 8 | 300 | 140 | 70 | 41 | 37 | 20 | 12 | 12 |
| 9 | 200 | 85 | 39 | 21 | 19 | 10 | 5 | 5 |
| 10 | 100 | 39 | 16 | 8 | 7 | 3 | 2 | 2 |
| *Total* | $650 | $198 | $ 42 | 0 | −$ 2 | −$ 8 | 0 | $ 1 |

rate? How would you explain your solution to a group of investors not very familiar with discounted cash-flow analysis?

# EXTENSION

***9.1 Behind Interest Rates*** A minimum attractive rate of return sets a price on money. Prices are rationing devices. The rate (or price) an organization sets for the use of its money reflects its internal economic conditions, such as the amount of capital accumulated from successful operations, and external factors such as the total supply of loanable capital in the whole economy and current or anticipated government fiscal policies.

## SOME INTEREST-RATE THEORY

A supply-and-demand relationship is responsible for a "market" rate of interest under traditional capital theory. This single theoretical rate is the percentage return yielded by any riskless bond or other security. Since capital is assumed to be subject to the law of diminishing returns, its demand curve $D$ takes the general shape shown in Figure 9.6. At a certain point in time, the demand curve is intersected by an inelastic supply line $S$ that represents the amount of capital then available. The intersection point is the temporary interest rate at which projects yielding that rate of productivity (measured as a percentage) are funded. This short-run equilibrium point implies that all projects with greater net productivity have already been funded, and those that will yield a lower rate must await the accumulation of additional capital. The graph also suggests that interest rates will decline as the pool of capital increases; this contention has not been supported by recent history.

The relationship of supply and demand in Figure 9.6 is probably more representative of an individual firm than the economy as a whole. Shifting expectations, technological innovations, inflation, and other factors impinge on traditional capital

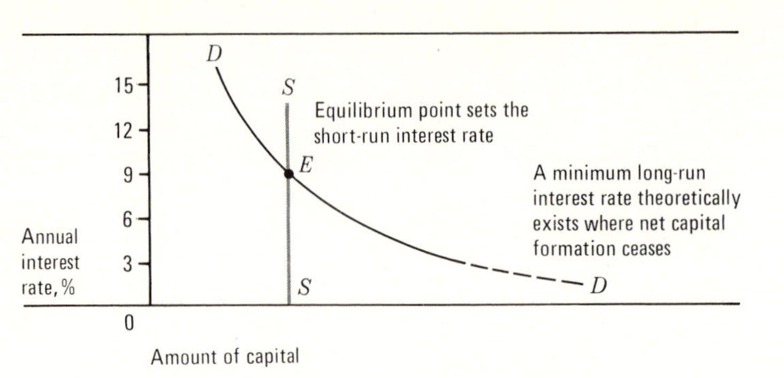


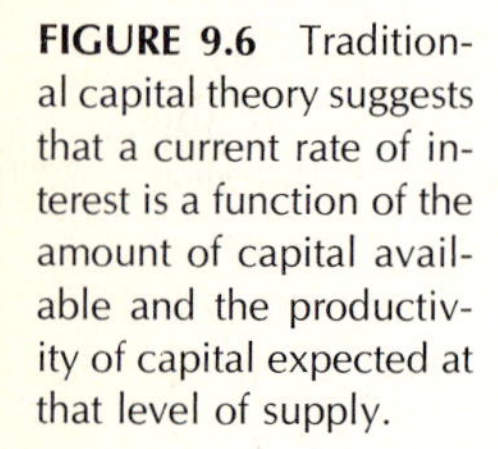
**FIGURE 9.6** Traditional capital theory suggests that a current rate of interest is a function of the amount of capital available and the productivity of capital expected at that level of supply.

theory to thwart the stability of conceived investment relationships. Yet the theory does offer a simplified explanation of the purpose served by interest rates. Money is "hired" by businesses to buy capital goods. The productivity of these goods must at least pay the "wages" of the money used to acquire the goods, and attaining this wage level sets a lower limit on accepting investment proposals.

Although it is convenient to theorize a single interest rate for the whole economy, there exist, of course, many different interest rates. Reasons for the differences among loans include:

1. Higher interest charges are applied when there is greater risk that a loan will not be repaid.
2. Charges for long-term loans are usually greater than for short-term loans because a lender forgoes the opportunity to take advantage of alternative uses of the money for a longer period and incurs greater risk from potential changes in the economic environment.
3. Administrative costs of lending are a higher percentage of the loan value of a smaller than a larger loan.
4. Local money markets vary, owing to regional differences; interest rates are often higher in small towns than in large cities because borrowers find it less convenient to shop for better rates.

**QUESTION**

**9.1a** Determine the interest charges or rate of return earned in your locality for each of the following loans or investments. Discuss reasons for the differences.

- $1000 government bond that matures in 20 years
- $1000 Treasury note due in 90 days
- $4000 borrowed from an auto-maker's credit corporation to buy a new car, with repayment over 3 years
- $4000 borrowed from a local commercial bank to buy a new car, with repayment over 3 years
- $30,000 FHA mortgage loan for a new house, with repayment over 20 years
- $50 pawn-shop loan for 1 month
- $300 borrowed from a small-loan company to be repaid in 6 months

**SOME INTEREST-RATE PRACTICES**

Whatever theory one adopts to explain how interest rates depend on the supply of and demand for capital, no one can deny the dominant influence of government on those rates. In effect, an interest rate is an administered price. Fiscal policies of government affect the supply of capital through the power to issue money, and the demand through massive borrowing. The reason government purposely influences interest rates is to direct the economy toward national priorities. For instance, other things being equal, a lowering of interest rates will stimulate business expenditures and create more employment. However, other things are seldom equal, so fiscal policies do not always achieve the desired objectives in themselves. Interacting forces in the national economy (and indirect international influences) are awesomely complex; money management is only one contributor.

During periods of severe inflation, as happened in 1974, The U.S. Federal Reserve Board ordinarily tries to minimize the increase in money supply. This is done in part by the sale of Treasury bills which indirectly drains money out of the banking system. It also drives up the interest rate for borrowing. In 1974, the interest rate for 90-day Treasury bills ranged from about 7 percent to over 9 percent and then settled to an average around 6 percent during 1975, when government emphasis switched to a policy aimed at encouraging greater employment.

As opposed to the wild gyrations of short-term interest rates, long-term rates are relatively well behaved. It is generally agreed that the *money* interest rate varies with business cycles, reflects profitability of capital, and incorporates an inflation premium. The comparison of a 5-year *moving average* (see Chapter 10) of the Consumer Price Index and yields on new issues of high-grade corporate bonds in Figure 9.7 suggests the *real interest rate* is only about 3 percent, where the real interest rate is defined as the money interest rate minus the percentage price rise.

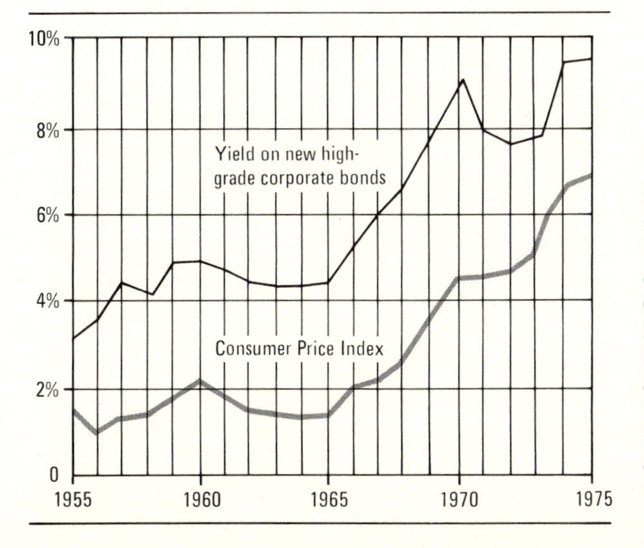


**FIGURE 9.7** The rise of interest rates for new issues of high-grade corporate bonds, compared with the rise in the Consumer Price Index over a 20-year span. The CPI trace is based on a 5-year moving average to smooth out seasonal variations. (Adapted from A. F. Ehrbar, "The Only Certainty about Interest Rates Is Their Uncertainty," *Fortune*, Jan. 1976.)

**QUESTION**

**9.1b** A surprising revelation of the 1973–74 business boom was a shortage of plant capacity in United States basic industries. The possible need for plant expansion has led to much discussion of a future capital shortage. This potential shortage could affect business needs as well as social wants such as low-income housing. A study by the New York Stock Exchange projected capital demands through 1985 of almost $4.7 trillion against capital supplies of $4 trillion.

Assuming the New York Stock Exchange projection is fulfilled, discuss the $0.7 trillion capital shortage in terms of its possible effect on interest rates, sources of capital, and the sorting out of proposed projects (such as plant-capacity expansions) and national priorities (such as energy independence or housing improvements).

# SECTION THREE

# MANAGEMENT OF MONEY

"The name of the game is business, and the score is kept with money" is a catchy old saying that contains a lot of truth. Perhaps business is too serious to be called a game and money is too restrictive a measure, but there are plenty of players and they are surely interested in finances. Even nonprofit organizations and governments are subject to economic yardsticks. Very few individuals or organizations are immune to the economic evaluation criteria.

Professor Han of the University of Novi Sad in Yugoslavia observed an intriguing semantic quality associated with the quest for money. In English we say that we "make money" in the notion of manufacturing it, or "earn money" as just compensation for labor. Oriental languages roughly translate as "seeking" money. In German one "serves" for money, and in Russian one "works" for it. Hungarians "search" for money. The French and Italian versions are inspired by the notion of "winning" money. An appropriate but apparently unused phrase would be to "manage" money. Engineering economists manage money by directing and controlling its movement.

Only the very brash or very naïve would enter a chess competition without knowing the rules and practices of the game. Similar inadequacies may fail to bar the unprepared and the unsuspecting from economic competition. Many are lulled into complacency by an honest but often erroneous belief that they possess an innate money sense that will carry them through. In the last section we dealt with some of the mechanics of economic competition. In this section we shall investigate the data needed to compete, the special rules such as taxes, and the practices used by experienced competitors.

# CHAPTER 10

# DATA AND SENSITIVITY ANALYSIS

Economic analyses are built from data as a house is built of bricks, but an accumulation of data is no more an analysis than a pile of bricks is a house. There are piles of data everywhere. They fill the shelves of libraries and accumulate limitlessly from the unending parade of statistics and information issued by all kinds of agencies, public and private. Yet, when a particular fact is crucial to an analysis, it may be elusive or impossible to procure. Data difficulties range from an overabundance, when the problem is to digest information into a usable form, to a scarcity that may force guesses to substitute for facts.

Information is both the raw material and the finished product of an engineering economic evaluation. The evaluation process is similar to familiar production processes. Bits of data, the raw materials, are received from many sources and inspected for accuracy before being accepted. They are fed into the engineering economics "machinery," where they are translated and transformed into the desired product: an economic comparison. As in any production process, it is necessary to know what raw materials are needed, the best sources, and ways to determine whether they are adequate. In this chapter, tools to tune the process include

*Cause-and-effect diagrams* to determine what data are needed
*Estimating techniques* to collect existing data
*Forecasting methods* to predict future data
*Sensitivity analysis* to observe the effect of data used

University graduates have commented that the biggest difference between applying engineering economics in the real world and studying it in the confined world of a textbook is that real-world data and problems seldom match so well as in text examples. The methods and techniques in this chapter and the other chapters in Section Three are aimed at mitigating the transition from idealized to real-world problem solving.

## CAUSE-AND-EFFECT ANALYSIS*

Often, the toughest part of problem solving is getting started. Inertia is a physical law that seems to have a counterpart in mental motion. Once mental activity gets under way and directed, things happen. It is comparatively easy to keep them happening after that first movement, even when there is an uncomfortable suspicion that not enough is known about the problem. As the solving process progresses, the uneasiness may be accompanied by frustrations in correlating facts and drawing meaningful interpretations from data. Better organization of data alleviates some of the difficulties by highlighting pertinent information. Cause-and-effect (C&E) diagrams assist in defining a problem by categorizing related factors to make their influence more observable.

### Structure of a C&E Diagram

A C&E diagram is a portrait of a problem. It serves much the same purpose as the black-box concept in engineering studies. A black-box diagram, as shown in Figure 10.1, represents an undefined transformation process that converts known inputs into specified outputs. Examination of the inputs and outputs is supposed to expose the workings of the black-box conversion process. In a similar fashion, also displayed in Figure 10.1, a C&E diagram elaborates on a problem (black box) through the identification of its causes (inputs) and effects (outputs).

The cause-and-effect concept is especially fitting for economic studies. In most situations the basic problem is known. The question is not how a mechanism works to convert inputs to outputs, but which inputs (causes) are needed to solve the problem or accomplish the mission to produce the wanted outputs (effects). Specific alternatives are *not* identified in the initial investigation. The characteristics associated with the problem are identified and noted on the diagram. Then the monetary values corresponding to each alternative's reaction to the problem characteristics are examined.

The mechanics of constructing a C&E diagram are elementary, but the mental effort preceding the drawing is demanding. The first step is to develop a pithy statement of the

*Adapted from J. L. Riggs and M. S. Inoue, *Introduction to Operations Research and Management Science*, McGraw-Hill, 1975.

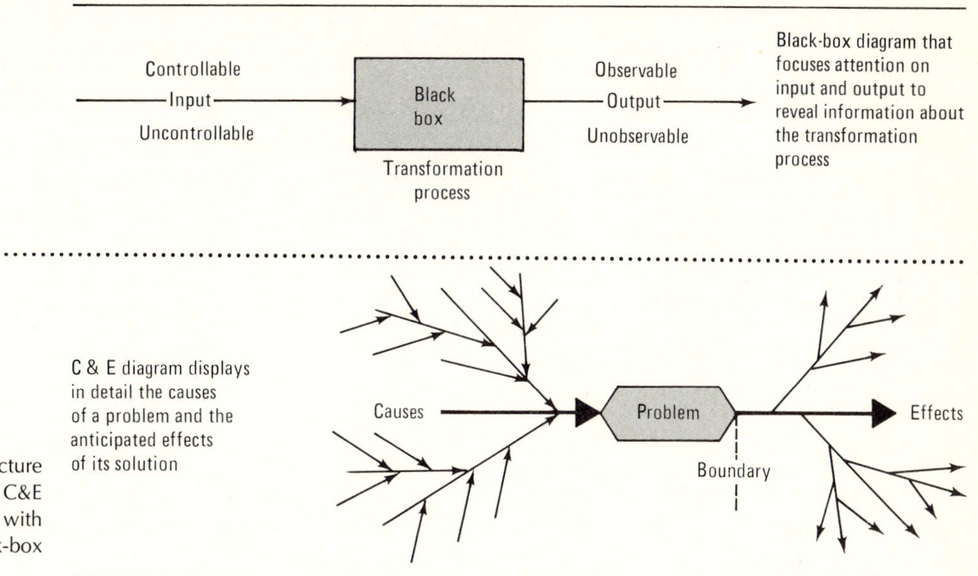


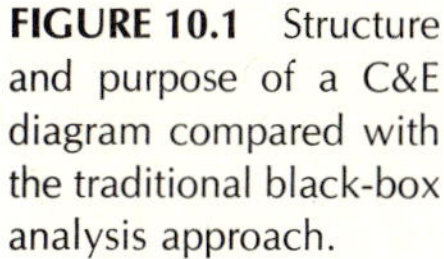
**FIGURE 10.1** Structure and purpose of a C&E diagram compared with the traditional black-box analysis approach.

problem, situation, or objective. This is entered in a six-sided box. Below the box is a boundary line that sets a limit (such as the point in time when the solution is to be implemented) that denotes when the causes take effect. Then, labeled ribs are drawn to or from the cause-and-effect spine (bold arrows in Figure 10.1) to specify the main causes and effects. Short definitions in ovals at the ends of the rib arrows identify these main factors. Finally, smaller arrows identifying subfactors run to or from the main ribs and, in turn, sub-subfactor arrows can lead to or from these to provide greater detail. The construction procedure is illustrated in the following example.

**Example 10.1**

### C&E Diagram Developed to Examine the Practicality of Adopting a 4-Day Workweek

An engineering economist in the Public Works Department of a large city is investigating the relative costs and benefits to expect from replacing the conventional 5-day workweek with a 4-day workweek. Both workweeks comprise 40 hours; only the lengths of the workdays differ. No wage differential is involved.

The main categories of factors involved in "adopt a 40-hour, 4-day workweek" are shown by the labeled ribs in Figure 10.2. These titles serve the same function as headlines or chapter headings, attention-getters for related data. The boundary line below the problem description is labeled "start" because this is a feasibility study, not an action plan.

Specific conditions of interest are next listed on the diagram through arrows attached to the titled ribs. Either an individual or group can discuss the main factors to pinpoint the considerations that are most important to a decision. Elaboration of successively more detailed factors creates the branching effect of arrows shown in

**FIGURE 10.2** Main factors affecting the adoption of a 40-hour, 4-day workweek are shown as ribs joined to the spinal cause-and-effect arrows. Brief descriptions of the prime considerations are enclosed in ovals for easy recognition.

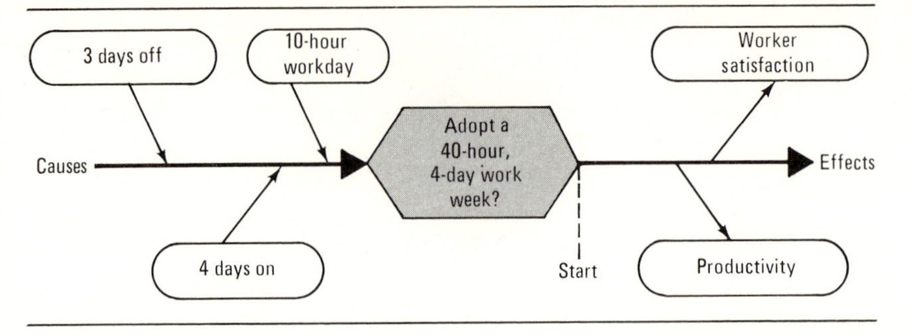


Figure 10.3. This preliminary sketch can be circulated among affected persons to solicit additional considerations.

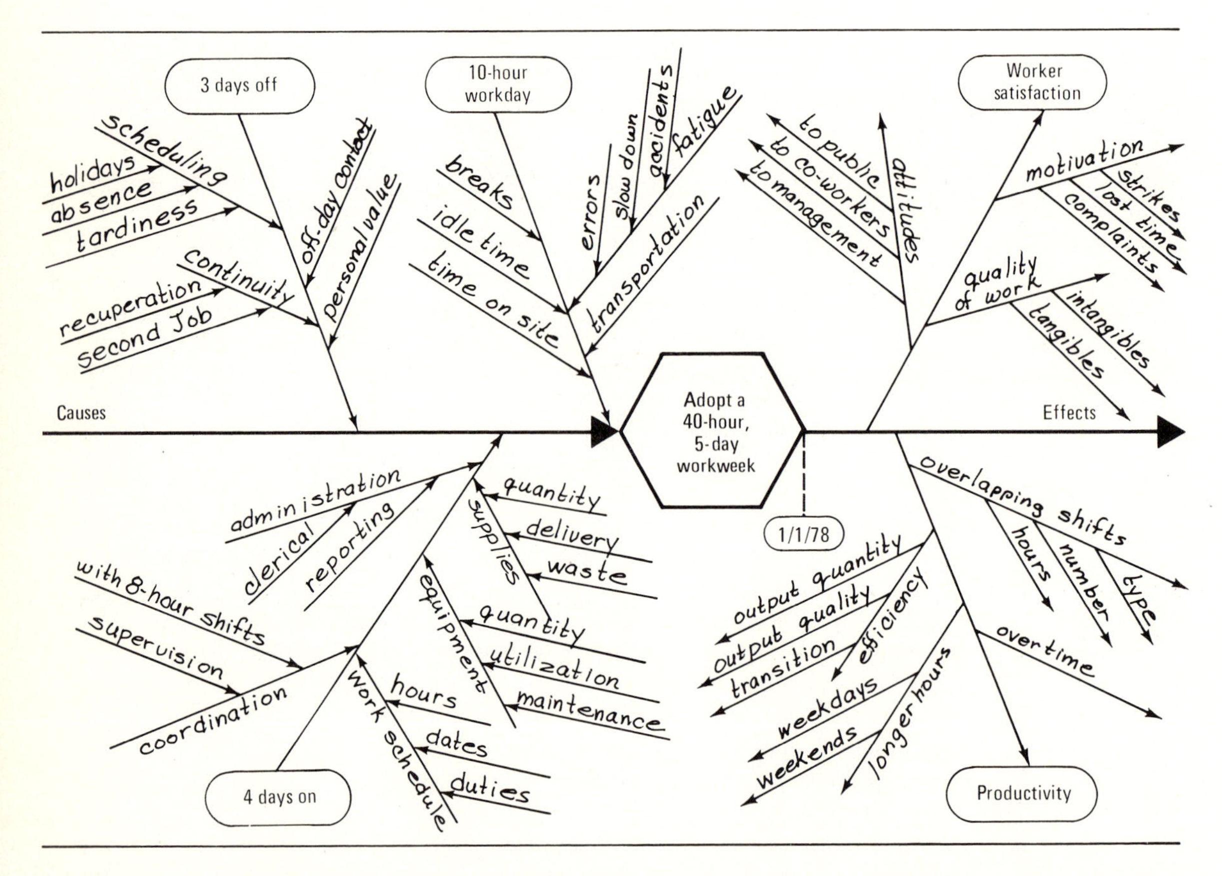


**FIGURE 10.3** A C&E sketch of considerations involved in switching to a 4-day workweek. Additional subfactors can be added during the investigation to determine alternative courses of action. After alternatives have been itemized, measurements are inserted in a refined diagram to represent specific proposals.

*After* the comprehensive structure has been refined, numerical values for appropriate factors are estimated to evaluate specific alternatives. Although the same considerations are involved, the advantages and disadvantages of adopting a 4-day workweek are quite different for distinct departments or segments of the work force. The police and fire departments would benefit from overlapping shifts, made more convenient by the 10-hour days, during their busiest hours. More hours would be available each day during the early part of the week for sanitation workers, when they are most needed. Equipment utilization would be better for some types of work and worse for others, as would several other factors.

Once the receipts and disbursements are collected, the evaluation continues according to the most suitable comparison method. In the analysis of a problem that has numerous intangible factors, such as "motivation" and "attitudes" in the 4-day-workweek question, a mixed-rating comparison model may be most appropriate.

**Economy Exercise 10-1** A glance at the general shape of a C&E diagram may be sufficient to spot the most troublesome aspects or to judge the adequacy of the representation. This

1. *Ailment*: Fatigue. *Symptoms*: Data not sorted, classified, or ordered. Too crowded and regimented.

2. *Ailment*: Unbalanced diet. *Symptoms*: Too much experience or concern for one area and not enough for others.

3. *Ailment*: Feverish. *Symptoms*: Apparent singlemindedness yields exotic detailing, but breadth is obviously lacking.

4. *Ailment*: Stupor. *Symptoms*: Classifications are too mechanical (either/or) and seem to ignore other possibilities.

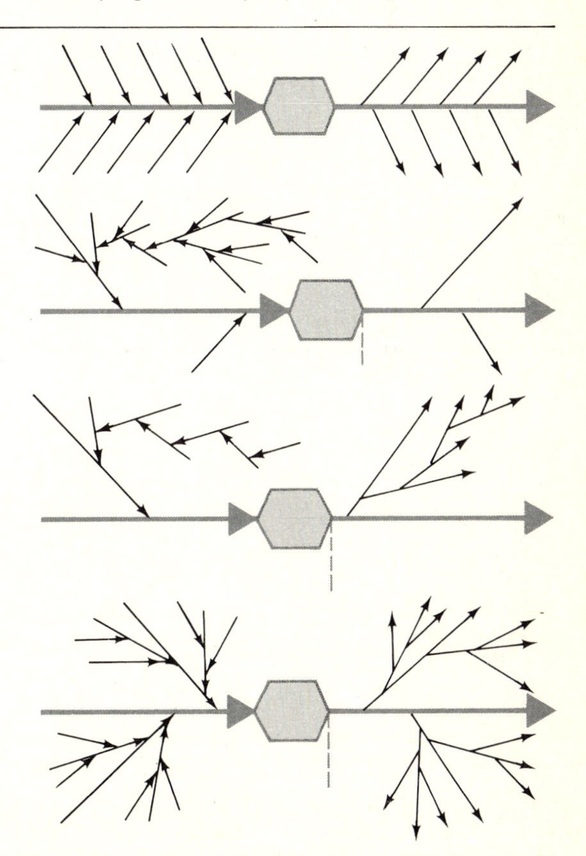

quick-inspection procedure is especially useful to the leader of a group discussion which is being diagramed. The logic and analytic abilities of a group will be made apparent by the basic structure of the diagram, while the wealth of knowledge will show up as the details of the entries. An unbalanced structure reveals specific areas in which the individual or group either lacks experience or is overly concerned. The patterns on page 295 might emerge as a preliminary evaluation of a subject. Each is described by an apparent ailment. Prescribe the cure.

## C&E Applications

A cause-and-effect analysis is conducted to explore a situation. The visual impact tends to strengthen an investigation by suggesting additional considerations associated with the factors already recorded. Familiarity gained by the diagraming exercise reveals what data are needed to compare alternative courses of action.

There are no rigid rules for the construction of a C&E diagram. The finished figure can be a sketch on a napkin or a colorful poster for presentation purposes. Innumerable modifications can be made to suit the situation.

A one-sided diagram, either the cause side or the effect side, is appropriate when the input or output of a situation is predetermined and unalterable. A study featuring the cause side is called an ***Ishikawa diagram*** after its originator, Kaoru Ishikawa of Tokyo University, who devised this form of analysis in 1953 for quality-control studies. In this version the quality goal is known, and the contributors to quality are sought. A similar condition exists when, for instance, the desired production output is known and the question is whether a current machine or a new one would achieve that output at lower cost. As shown in Figure 10.4, the cost categories are identified but no specific cost figures are given for individual machines. These categories reveal what data must be collected to compare the defender with the challengers in a replacement study.

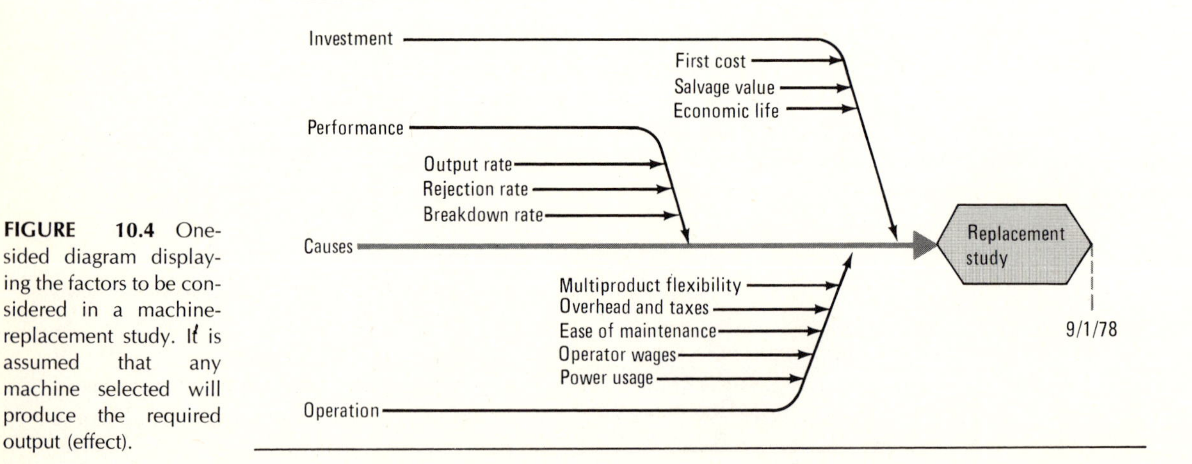


**FIGURE 10.4** One-sided diagram displaying the factors to be considered in a machine-replacement study. It is assumed that any machine selected will produce the required output (effect).

It is sometimes easier to visualize a process by considering factors in an accustomed sequence. Clusters of arrows in a C&E diagram for a material-handling study could group related factors in a sequence that follows the physical flow of material: ***purchase→ receive→store→ship*** on the cause side, and ***inspect→display→sell*** on the effect side. Other sequential scales could be time, organizational hierarchy, level of technological development, etc. Any routine that stimulates a more complete diagram increases the chance that the needed data will be available for a thorough evaluation. Occasionally, the diagraming operation exposes a new alternative or divulges a solution that might otherwise have been overlooked, as described in Example 10.2.

**Example 10.2** **A Search for Alternatives**

The State Department of Vocational Training sent an industrial engineer to a local sheltered workshop to help overcome a production problem. The shop serves as a rehabilitation center for handicapped workers. It provides training and work for the handicapped in assembly operations and in the production of wood products such as pallets, packing cases, wishing wells, and picnic tables. The work is purposely designed to emphasize hand labor, and consequently the processes employ minimal mechanization.

A quality-control problem occurred when the center began work on a contract to produce berry boxes. The customer threatened to cancel the contract if sturdier boxes were not produced within a month. The first analysis of the problem identified the causes and effects shown in Figure 10.5. An obvious solution appeared to be to add an inspection station for outgoing boxes to ensure the use of good wood and adequate fastening. Then it was noticed that attention was being focused strictly on effects, not causes. Maybe it would be better to have the workers inspect as they produced (note 1 at the top of the diagram).

As a result of this internal-inspection idea, attention was given to the need for better physical facilities, brighter lights, lower work tables for wheelchair workers, jigs for more accurate stapling, etc. These and other considerations led to the addition of a new effect, *delays* (shown in Figure 10.5 by note 2 and dashed arrows). To minimize delays, a suggestion was made to improve the motion patterns of workers. Finally, the subject of training was considered, and it was then recognized that training was currently directed to general skills, but little effort was made to tell individual workers what was expected of them and what particularly should be watched in satisfying each production contract.

Thus, the berry-box problem was solved simply by informing the workers what was needed and how to do it with existing facilities. The answer is indicated by the connected and circled factors in the lower portion of the diagram. No investment funds were needed. Perhaps the answer may appear patently obvious, but how many times have sophisticated solutions been installed where a simple change could have produced the same results at lower cost and less loss of dignity?

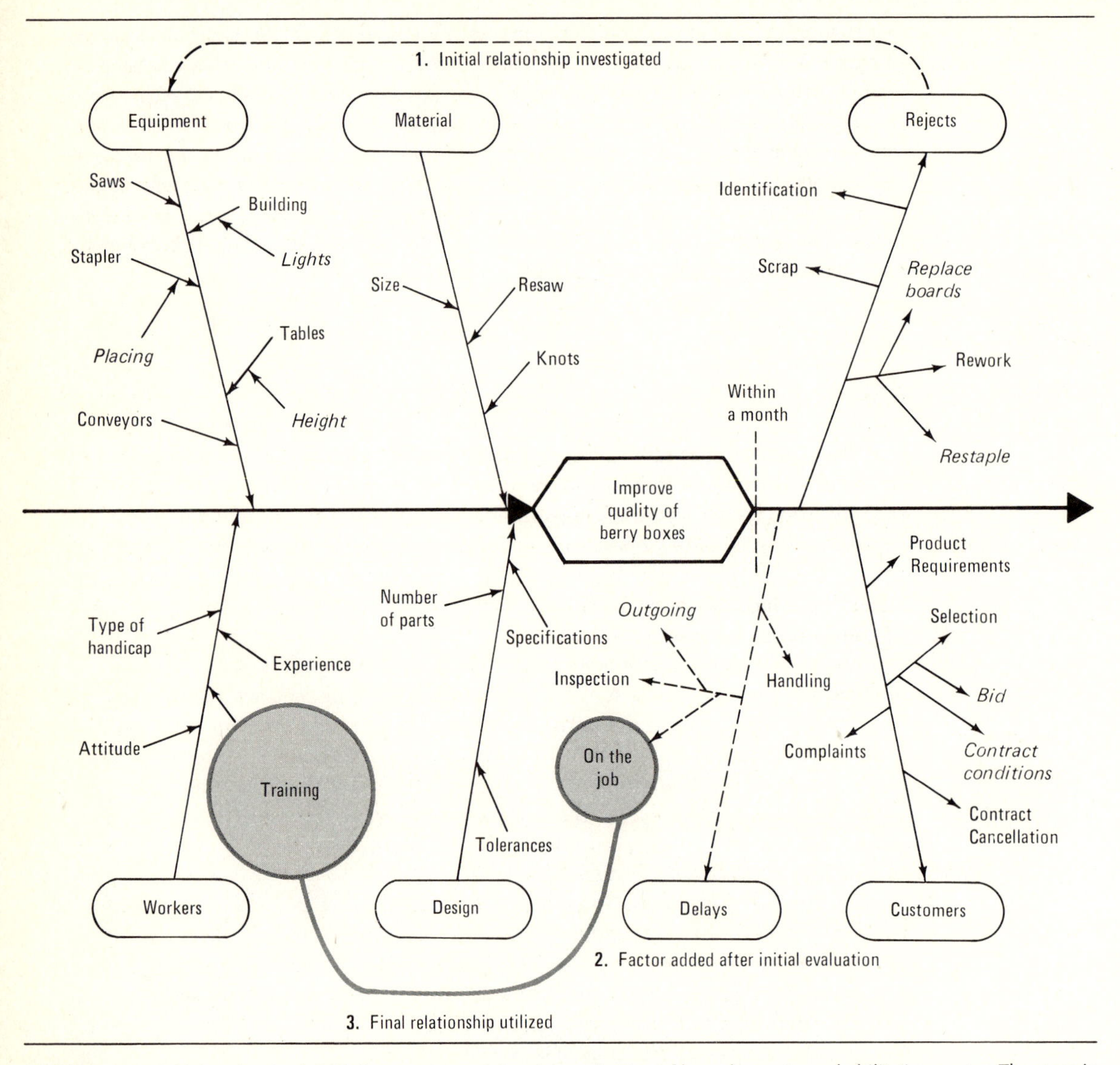


**FIGURE 10.5** Initial and revised C&E diagram representing the production of berry boxes in a rehabilitation center. The steps in the solution are numbered in the order of their development.

## ESTIMATING TECHNIQUES

A new project may be identical in design and mission to one just completed, but the cash flows for the two will probably differ considerably. This variation from one study to the next is characteristic of economic analyses. Unlike physical laws which consistently follow an orderly cause-and-effect relationship, economic laws depend on the behavior of people, and they are erratic. However, past behavior is still a respectable clue to future behavior.

Engineers are well prepared to judge the future performance of materials and machines. Past performance can be extrapolated, and modified if necessary, to predict performance on the next project. The monetary values for the future conditions are less reliably predicted. Strikes, shortages, competition, inflation, and other factors that affect unit costs cannot always be anticipated. Therefore, estimating starts with the calculation of design amounts based on known physical relationships and concludes with the assignment of comparatively less known monetary measures for the design conditions.

## Sources of Cost Data

Figure 10.6 shows the traditional cost and price structure for a manufactured product. Most of the categories are also appropriate for service functions. The principal variation occurs in the relative sizes of the cost categories; manufacturing usually has a higher material-to-labor ratio than service. Public services would be represented by the model by eliminating the top two tiers. Many of the components within the categories were discussed in Chapter 3. Additional aspects are presented in the next chapter from the cost-accounting viewpoint.

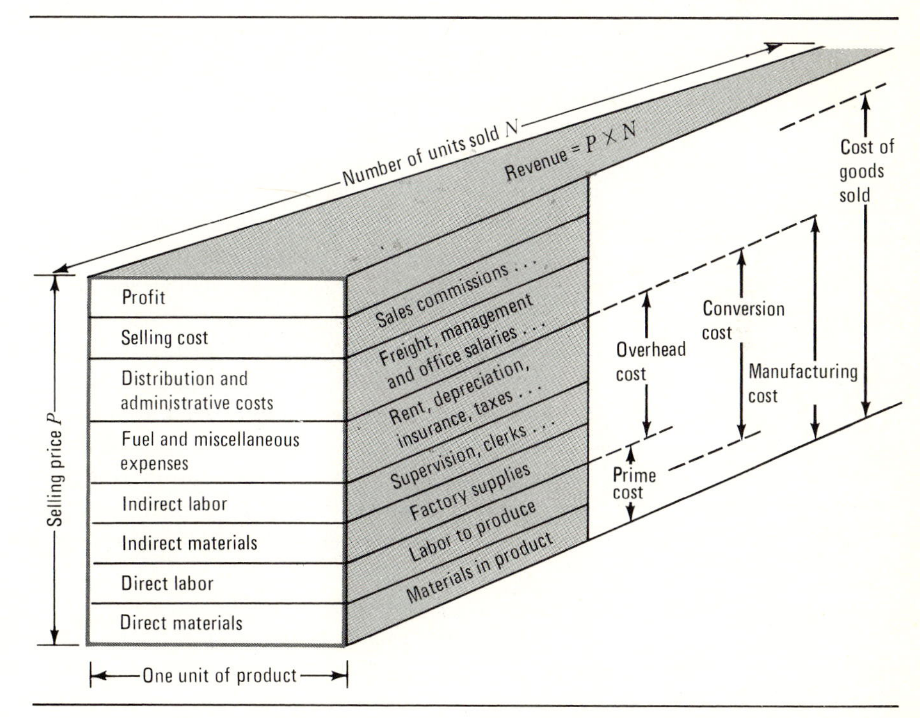


**FIGURE 10.6** Composition of costs traditionally used in accounting for the price of a manufactured product.

The quantities and types of materials required can be determined quite accurately from past records and engineering design documents (e.g., the bill of materials and parts list that accompany blueprints). Current prices are also readily assessable. Difficulties creep in as the price quotes are pushed further into the future. Experts from the purchasing department in a large firm can assist with predictions through their experience with commodity futures and knowledge of the firm's policies toward speculating (buying in large quantities to take advantage of price fluctuations, or buying contracts for future deliveries at prices set now, called *hedging*). A vital part of any estimating process is to be aware of the organization's economic policies and practices.

When purchasing experts are not available, many types of ***cost indexes*** can furnish information. A cost index provides a comparison of cost or price changes from year to year for a fixed quantity of goods or services at particular locations. For building construction alone, there are a dozen major United States indexes compiled for different parts of the country and different types of construction. The Bureau of Labor Statistics is the primary source of national data.

***Labor costs*** are a function of skill level, labor supply, and time required. The saying that "time is money" is especially true in estimating the cost of work. Standards for the amount of output per labor hour have been developed for many classes of work. The standards are based on a "normal" pace for an activity, with allowances included for personal relief and unusually severe working conditions. Standard times combined with expected wage rates provide a reasonable estimate of labor costs for repetitive jobs. Labor costs for specialized work can be predicted from bid estimates or quotes by professionals and agencies offering the service. A project involving a new type of work will likely have higher labor costs at first, while experience is gained. (A method of anticipating the change in output as a function of learning is described in Extension 10.1.)

It should be remembered that the cost of labor includes more than direct wages. Fringe benefits may amount to more than half the base wage in accounting for accident, health, and unemployment insurance; vacation and retirement pay; and special agreements such as guaranteed annual wages or sabbatical leaves. Extra costs due to relocation allowances, accidents, and sickness should also be considered.

***Maintenance costs*** are the ordinary costs required for the upkeep of property and the restoration required when assets are damaged but not replaced. Items under maintenance include the costs of inspecting and locating trouble areas, replacement of minor parts, power, labor, materials, and minor changes in or rearrangements of existing facilities for more efficient use. Maintenance costs tend to increase with the age of an asset because more upkeep is required later in life and the trend of wages and material prices is upward.

***Property taxes and insurance*** are usually expressed as a percentage of first cost in economic comparisons. Although the value of property decreases with age, taxes and insurance seldom show a corresponding decrease. Therefore, a constant annual charge is a realistic appraisal of future expense.

The ***first cost*** of acquiring a major asset may rise well above expectation, especially if the asset originates from a new design or untried process. Cost "overruns" are legendary in government projects, and they haunt the private sector too. After a design is set, direct acquisition costs are readily determined from manufacturers' quotes or competitive bidding. When errors occur, they are typically on the low side, as a result of incomplete listings of desired features and neglecting less obvious costs such as the following:

Materials (freight, sales tax, storage costs, damage)

Installation (extra costs for special arrangements and unconventional designs)

Interest, taxes, salaries, and insurance during a design or construction phase

Change orders during construction (installation costs for additions to or deletions from the original plans)

Investigation, exploratory, and legal fees

Promotional costs

Engineering and associated fees

Debugging and start-up costs

Conservatism

Most construction projects include a "contingency cost" category to account for undefined costs that will assuredly crop up. Although such expenses are ticklish to tally, they can drain capital as thoroughly as design changes for physical assets.

Current levels of costs used in estimating investments may not be applicable to future conditions. Generally this is not a serious problem, but two situations bear watching. One is when an old alternative is resurrected for reconsideration. Old cost estimates may be outdated by new methods or different price levels. The other situation is when current levels reflect abnormal or temporary prices. If prices are adjusted for first costs, similar adjustments should be investigated for other costs.

First costs can immediately eliminate some alternatives. Insufficient capital is a genuine reason to turn down an investment proposal even though it has a handsome rate of return. An investment that would be wise for a firm with adequate capital could be a futile or even disastrous course of action for a firm with limited finances and big ambitions.

*Overhead cost* is by definition that portion of the cost which cannot be clearly associated with particular operations or products and must be prorated among all the cost units on some arbitrary basis. The reason for this catchall category is the prohibitive expense of assigning and charging to each product a specific proportion of such costs as wages of supervisors, factory heat and light, janitorial services, secretarial help, and incidental supplies.

Several methods are used to allocate the composite overhead expense to a product or operation. The versions differ among companies because of differences in the nature of production; it is not uncommon to find different overhead rates used within one firm. Overhead costs may be applied on the basis of

1. Direct labor, direct materials, or prime costs
2. Machine hours or direct labor time
3. Fixed and variable cost classifications (for instance, overhead such as indirect material and power costs would be treated as varying with output, whereas property taxes, depreciation, and indirect labor would be considered as fixed costs)

Once a base is selected, annual overhead as determined from accounting records is divided by the annual cost or usage time of the base category. The resulting ratio is multiplied by the base cost for the alternative being evaluated. For instance, if overhead is charged on the basis of direct labor,

$$\text{Direct labor ratio} = \frac{\text{total annual overhead cost}}{\text{total annual direct labor}}$$

and the overhead charge for one unit of product $X$ is

$$\frac{\text{Overhead charge}}{\text{Unit of product } X} = \text{direct labor ratio} \times \frac{\text{direct labor cost}}{\text{unit of product } X}$$

which is included as a cost in determining the selling price of product $X$. By the same reasoning, a machine used in the production of product $X$ could be assigned an overhead burden based on the operator's wages:

$$\frac{\text{Annual overhead charges}}{\text{Machine } A \text{ producing } X} = \text{direct labor rate} \times \frac{\text{annual wages of operator}}{\text{machine } A \text{ producing } X}$$

An important condition to recognize in economic comparisons is that ***overhead costs are associated with a certain level of output.*** This can be a critical factor in a comparison such as the purchase of either a $200,000 numerically controlled milling machine or a standard $20,000 general-purpose milling machine. If the labor rate of operators of both machines is $9 per hour and the burden rate is based on a direct labor ratio of 300 percent, the machine-hour costs of both machines would be $27 per hour. This figure has to be an incorrect machine-hour costing because the investment in the numerically controlled machine is 10 times that in the standard model. Fortunately, overhead costs often have an identical effect on several alternatives. That is, the same value for overhead costs would apply for different alternatives being compared, making their inclusion redundant in the comparison.

---

**Economy Exercise 10-2** An equipment sales representative claims that the installation of a new machine will allow equal quality and quantity of output with one fewer worker. Using the firm's required rate of return (15 percent) and direct labor ratio (0.6), the representative substantiates the claim with the following figures:

| | |
|---|---|
| *Annual savings for the elimination of one worker* | |
| Direct labor: 1 worker × $10,000/worker-year | $10,000 |
| Overhead: $10,000 × 0.60 | 6,000 |
| Total saving | $16,000 |
| | |
| *Annual cost of new machine ($P$ = $20,000, $S$ = 0, $N$ = 10)* | |
| Capital recovery: ($20,000 − 0)($A/P$, 15, 10) = $20,000(0.19925) | $ 3,985 |
| Operation | 9,000 |
| Maintenance | 600 |
| Taxes and insurance | 400 |
| Total cost | $13,985 |
| *Net saving* | $2,015 |

Does the sales representative's evaluation appear valid?

---

## Sources of Income Data

In an industrial setting, revenue is the money received from customers for the services or products sold to them. There are many patterns of revenue flow. A retail store has an essentially continuous influx of revenue during working hours. Plumbers are often paid after each service call. Utility services are paid for by the month. Farmers usually get their money after a crop is harvested. A homebuilder has to wait until a house is sold before receiving revenue. The timing of the revenues associated with an alternative may have a significant bearing on its acceptability. One of the main reasons that new businesses often fail is the time lag between incurred first costs and the establishment of an expected level of revenue.

Revenue is somewhat harder to estimate than costs for many industrial projects. If a new investment is to serve the same purpose as an existing asset, historical data provide a reliable estimate of future revenue. When the investment is destined to satisfy a new function, revenue estimates are less certain. What looks like a sure bet on the drawing board may end as a miserable flop in the market, where it is exposed to the buying whims of the public.

Occasionally, a precise measure of revenue contribution is impossible. Sums spent on customer goodwill and improved employer-employee relationships are at best extremely difficult to measure in terms of revenue increments. This situation often results in the setting aside of a certain sum for "public relations" or investment in intangible returns. Then the sum is divided among projects rated according to a nonmonetary scale of attractiveness as described in Chapter 5.

Funds for government or public activities result from various types of taxes, charges for specific services, and borrowing. Everyone is familiar with income taxes and charges for services such as mail delivery. Both are revenue sources secured from the public for benefits expected from governing agencies. Because of their mandatory nature and long history, they can be estimated quite accurately. And as a supplement, when the perceived demand for public expenditures exceeds public revenue, governments at all levels tend to engage in deficit financing. The money so borrowed is a pledge that future members of society will pay for projects undertaken by the present society. (The economics of public projects is examined in Chapter 13.) Although the base for public revenues is broad and solid, the portion justified for a specific public project is as difficult to determine as with industrial projects.

Privately owned public utility companies occupy a position midway between government and industry. This position results from the great amount of invested capital required to provide public services such as electricity, water, telephone communications, and railroad transportation. Only by developing a high usage factor can an acceptable return on invested capital be obtained from low service rates. In order to assure both reasonable returns and reasonable rates, an exclusive geographical franchise is allotted to a utility company. Coupled with the grant of a monopolistic position are regulations controlling rates and standards of service. A regulatory body seeks to set a quality of service which satisfies the customers while permitting rates that allow the utility to earn an acceptable rate of return. Because of this control, revenue can be quite accurately forecast.

## FORECASTING

Estimating is more of a science than an art when conditions are stable; then standard costs and catalog prices are satisfactory data. But nothing is static for long. Changes in the economic environment may be as gradual as wind erosion in the desert or as sudden as a tornado. Economic estimators are particularly sensitive to a sudden shift, for it may mean the difference between fortune and fiasco. The prediction of a shift is a work of art.

The function of forecasting is to predict future events by the best means possible. The simplest form of forecasting is a mental voyage into the future to guess what might occur. Experienced estimators can do amazingly well, but "seat of the pants" forecasting promises a rude jolt to the inexperienced. More disciplined forecasting methods utilize statistical analysis and other formal procedures to draw conclusions about future events from past events. Regardless of the methods used (see Figure 10.7), the basic issues to predict for economic comparisons are the amount and timing of resource requirements and the associated cash-flow streams.

**FIGURE 10.7** There is more than one way to tell or seek a fortune.

Forecasting is predominately concerned with future economic conditions. ***Business forecasting*** comprises the prediction of market demands, conversion costs, availability of labor, and the like. Nearly every major engineering decision is affected by business forecasts because engineering projects inevitably are launched to satisfy future demands. Both the pace and the magnitude of engineering expenditures are linked to economic conditions, and business activity is the barometer of these conditions.

Prediction of new engineering developments and scientific advances, assumed to be independent of business activity, is called *technical forecasting*. It is concerned with activities that are largely under the control of an organization or an association sharing common interests. For example, technical forecasts have been developed for advances in nuclear-power technology and laser developments. These forecasts could be used by engineering economists to estimate the economic life and usage of assets in investment proposals for nuclear reactors and laser applications.

Attempts to seek a competitive edge from a glimpse into the future are nothing new. Rulers in the Middle Ages had their favorite crystal gazers and fortune tellers. Today astrologers and palm readers have believers, including some business people and engineers. Other organizational leaders rely on market surveys, product questionnaires, and quantitatively trained forecasting experts to anticipate levels of future activity. Yet no forecasting technique is infallible. Even the overwhelming amount of timely data about current conditions and advanced computer analyses now available cannot ensure accurate forecasts. Just ask any weather forecaster.

There are many ways to develop predictions. For convenience in getting an overview of the methods utilized, they can be grouped into the following three categories:

*Subjective methods,* based on opinions
*Historical methods,* based on past performance
*Causal methods,* based on underlying causes of events

Since more precise predictions are customarily more expensive to develop, it would be financial foolishness to allocate the same estimating time and cost to small and large investment evaluations. Similarly, only limited forecasting effort need be devoted to comparisons of alternatives that are identically affected by future conditions, and many comparisons do fit this category. An indication of the relationship between forecasting methods and costs is given in Figure 10.8.

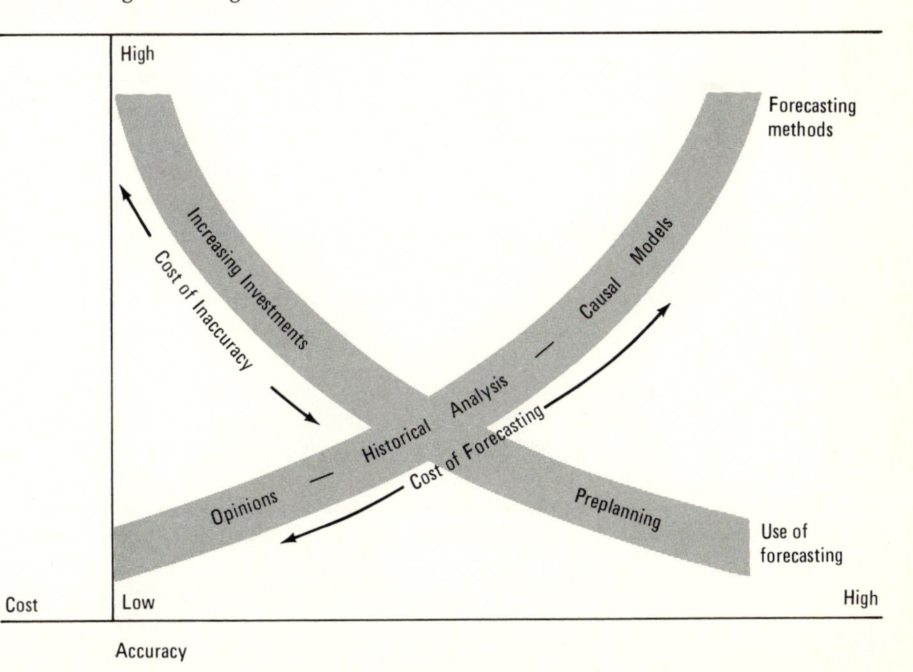


**FIGURE 10.8** The method of forecasting must be justified by its economic feasibility. More sophisticated techniques should produce more accurate predictions. Since the total cost includes both the expense of making a forecast and the cost of damage resulting from inaccurate forecasts, more funds should be allowed for predictions upon which expensive investments will be judged.

## FORECASTS BASED ON OPINIONS

When an estimator does not have access to a staff of forecasting specialists, subjective predictions may be the most feasible forecasting method. A prime source is the opinion of the forecaster. This judgment has a better chance for accuracy if it is grounded in facts. An abundance of free information is available to aid estimators. General data about national economic health, pricing indexes, consumer spending trends, etc., are offered in magazines, newspapers, and publications from trade associations and government agencies. Even specific data and item-by-item forecasts about particular industries are available from various publishers and consultants.

The danger in basing forecasts on opinions is that individuals may let their personal feelings take command over what they know are facts. As an example of well meant but biased advice, consider the case of a young entrepreneur who went to an unfamiliar area some years ago to invest in a coin-operated car wash. He sought advice on the investment and was put in touch with a person who had considerable experience in the automotive field; he had been a mechanic, an automotive service manager, and the operator of a taxi company. The entrepreneur asked him, "What do you think of the possibilities for success of coin-operated car washes in this area?" Now this person had a particular aversion to any mechanical device for cleaning the finish of a car. He simply did not believe that it was good for the finish to use pressure hoses, mechanical scrubbers, spray waxes, etc. He did believe in hand washing with sponge and chamois and in hand waxing and polishing. As a result, his answer was, "Those coin-op car laundries are a flash in the pan. They'll never have any success around here."

It so happened that the facts indicated the opposite. Sales of new cars, personal incomes, traffic counts, and population growth curves all would have said yes to our entrepreneur's question. That area was one in which salt was used on the roads in winter. This fact alone leads many people to wash their cars frequently in coin-operated washes. Now there are several hundred of these car-wash units in the area. The sad part of the tale is that all these coin-ops are now operated by other people. The prediction our entrepreneur got was based upon subjective rather than objective data.

Personal opinions are not necessarily bad sources of advice. In many instances they are the only source of advice. Even when more objective sources can be called upon, opinions may offer valuable insights or provide extra assurance about a prediction. Many companies routinely seek opinions from people who buy or sell their products. Five formal methods of obtaining subjective information are described below.

1 *Consumer survey* A person who has actually bought a product is the most logical source of information about that product's sales appeal. Questions to the buyers are often part of the product's guarantee. Follow-up questionnaires are frequently mailed to the purchasers of a large item, such as an automobile. Sometimes surveys are made of potential rather than actual consumers. Replies from this audience have to be interpreted carefully because consumers' tastes change very rapidly, and what a customer intends or hopes to do in the future may be far removed from what he or she actually does.
2 *Opinions of sales representatives and distributors* An experienced sales force is in a position to observe both the actions of suppliers and the behavior of consumers. They

can give warnings about changes in buying trends and the activities of competitors. Sales engineers are particularly well qualified to suggest design changes to improve a product's acceptance. The optimism or pessimism of individuals in the sales force can be balanced by averaging the predictions made by several sales representatives and sales managers.

3 *Executive views* Many forecasts used by executives are made by executives. The effect of individual biases is reduced by generating forecasts as a group effort. The mixture of interests and experience that ensures a good cross section of estimates makes a consensus difficult to obtain.

4 *Market trials* The most elaborate means of gathering consumer opinions is to sell a new product on a trial basis. The trial is designed to be a carefully controlled experiment. A small market area is selected to represent, as realistically as possible, the whole market that the product will compete in. Care is taken to avoid obstructing the competition and to be sure the results are timely. Since the cost of these experiments is high, market trials are limited to significant product promotions.

5 *Delphi method* A systematic routine for combining opinions into a reasoned consensus has the prophetic name *Delphi.* The technique was developed at the Rand Corporation and has gained fame in technical forecasting of future scientific developments. The procedure followed is to solicit and collate opinions about a certain subject from experts, and feed back digested appraisals to narrow the differences among opinions until a near agreement is obtained.

A carefully prepared questionnaire is delivered to a panel composed of experts from professional specialties that pertain to the forecasting problem. The survey can be conducted in a group meeting or by mail. Each questionnaire solicits written opinions about specific topics and requests supporting reasons for the opinions. These reasons are summarized by the Delphi moderator to assure anonymity of responses and then returned for consideration by the whole panel. The process is continued until the exchanged arguments and transfer of knowledge forges a consensus prediction. The end product may be a time-scaled "map" charting the nature of future technological developments. Advocates of the Delphi method claim the anonymity of written responses preserves the desirable features of a committee of specialists while reducing the "bandwagon" and dominant-personality effects that unduly sway group opinions.

## FORECASTS BASED ON HISTORICAL RECORDS

Basing forecasts on historical evidence relieves most of the uneasiness connected with relying on personal opinions. Historical data are simply facts. It is up to the forecaster to make interpretations from them. A forecaster can, of course, introduce personal bias into the interpretation, but this bias is open to exposure because the facts from which the forecast was drawn are still available for reexamination. In addition to the factual attractiveness of historical data, two adages bear out the value of past performance in estimating future performance: "History repeats itself," and "We learn from experience."

### Line Fitting

The most direct way to observe historical patterns is by plotting data on a time scale, a natural approach for engineering-minded analysts. A quarterly sales record for 10 years is

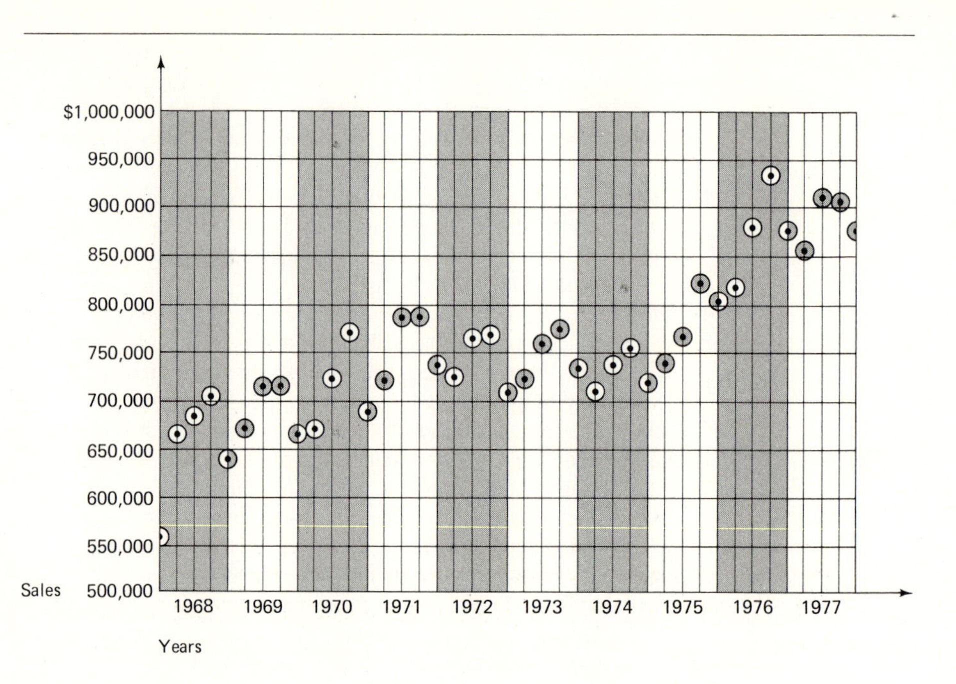


**FIGURE 10.9** Ten-year record of sales, where each dot represents quarterly sales.

plotted in Figure 10.9. The purpose of graphing is to expose a recurring shape or pattern that gives a clue as to where the next point or series of points should fall. The assumption that existing patterns will continue into the future is more likely to be correct over a short than a long estimating horizon, unless the patterns are unusually stable.

The scattered sales-level dots from Figure 10.9 are repeated in Figure 10.10 with the addition of a *trend line*. This straight line is constructed to show the *long-term trend* in sales. It can be sketched freehand or fitted mathematically by use of regression techniques. By either method, roughly half the dots should be above the line, and half below. The best possible fit exists when the sum of the squared distances from each dot to the line is smaller than that sum for any other line that might be drawn. A routine to quantify this condition is appropriately called the *least-squares method* and is explained in Example 10.3. By this method, the formula for the trend line shown is

$$\text{Sales} = \$625{,}000 + \$26{,}500X$$

where $X$ is the number of years in the future from the end of the base year 1967 for which an estimate is sought. For instance, the sales level indicated by the trend line for year 1977 ($X = 1977 - 1967 = 10$) is

$$\begin{aligned}\text{Sales}_{1977} &= 625{,}000 + (26{,}500)(10)\\ &= 625{,}000 + 265{,}000 = \$890{,}000\end{aligned}$$

Similarly, the forecast for 1978 would be

$$\text{Sales}_{1978} = 625{,}000 + (26{,}500)(11) = \$916{,}500$$

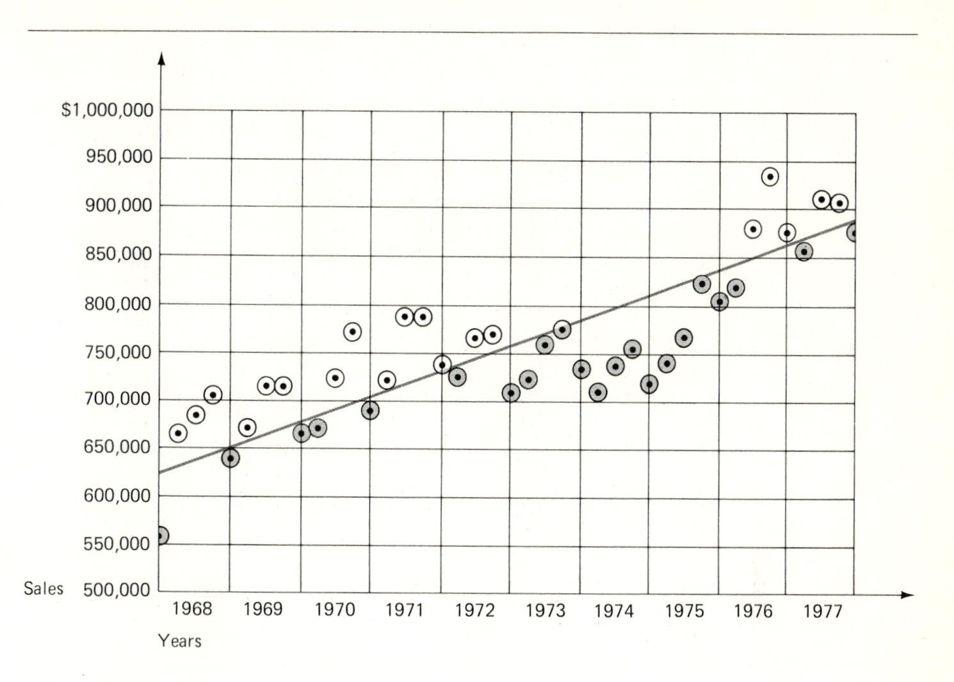


**FIGURE 10.10** Long-term trend line imposed on the sales data from Figure 10.9. The trend line shows the general direction of sales during the 1968–77 decade.

## ➧ Example 10.3 Line Fitting with the Least-Squares Method

Whenever plotted data points appear to follow a straight line, the least-squares method can be used to determine the line of best fit. This line is the one that comes closest to touching all the data points. Calculations provide the equation of the line for which the sum of the squares of the vertical distances between the actual values and the line is at a minimum. A further property of the line is that the sum of the same vertical distances equals zero.

A straight line is defined by the equation $Y = a + bX$. For our purposes, $Y$ is a forecast value at a point in time, and $X$ is measured in increments such as years from a base point. The objective is to determine $a$, which is the value of $Y$ at the base point, and $b$, which is the slope of the line.

Two equations are employed to determine $a$ and $b$. The first is obtained by multiplying the straight-line equation by the coefficient of $a$ and then summing the terms. With the coefficient of $a$ equal to 1 and $N$ as the number of data points, the equation becomes

$$\Sigma Y = Na + b\Sigma X$$

The second equation is developed in a similar manner. The coefficient of $b$ is $X$. After multiplying each term by $X$ and summing all the terms,

$$\Sigma XY = a\Sigma X + b\Sigma X^2$$

The two equations thus obtained are called *normal equations*.

The four sums required to solve the normal equations ($\Sigma Y$, $\Sigma X$, $\Sigma XY$, and $\Sigma X^2$) are obtained from a tabular approach. The calculations can be simplified by carefully selecting the base point. Because $X$ equals the number of periods from the base point, selecting a midpoint in the time series as the base makes $\Sigma X$ equal to zero. The smaller numbers resulting from a centered base point also make other required products and sums easier to handle. After the four sums are obtained, they are substituted in the normal equations, and the values of $a$ and $b$ are calculated. Then these values are substituted into the straight-line equation to complete the simple regression forecasting model:

$Y_F = a + bX$ *where* $Y_F$ = the forecast value $X$ periods in the future

To illustrate simple linear regression, the linear relationship between an independent and dependent variable, the least-squares method will be applied to the data in Figure 10.11. A forecasting equation is sought to relate demand to time.

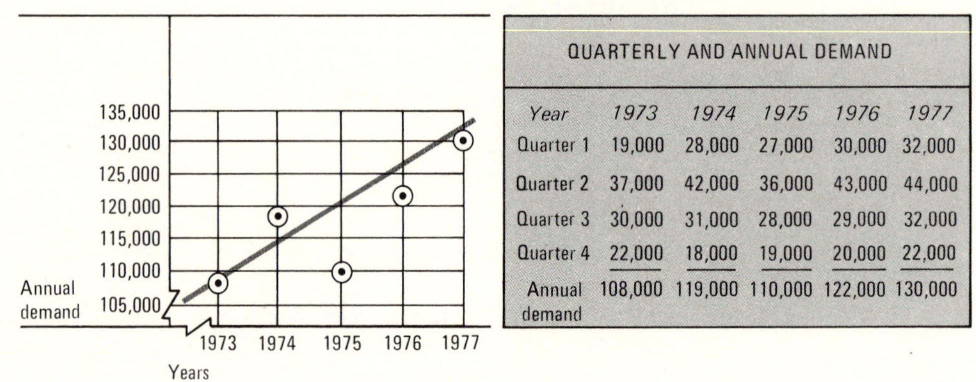


QUARTERLY AND ANNUAL DEMAND

| Year | 1973 | 1974 | 1975 | 1976 | 1977 |
|---|---|---|---|---|---|
| Quarter 1 | 19,000 | 28,000 | 27,000 | 30,000 | 32,000 |
| Quarter 2 | 37,000 | 42,000 | 36,000 | 43,000 | 44,000 |
| Quarter 3 | 30,000 | 31,000 | 28,000 | 29,000 | 32,000 |
| Quarter 4 | 22,000 | 18,000 | 19,000 | 20,000 | 22,000 |
| Annual demand | 108,000 | 119,000 | 110,000 | 122,000 | 130,000 |

**FIGURE 10.11** Sample data for forecasting examples.

A straight, sloping line appears to be a reasonable fit for the data in Figure 10.11. To make the calculations easier, a centered base point is used; 1975 is the midpoint of the historical data and the date selected for $X = 0$. A tabular format with $Y$ as demand in 1000-unit increments is set up to obtain the sums required to solve the normal equations:

| *Year* | $Y$ | $X$ | $X^2$ | $XY$ | |
|---|---|---|---|---|---|
| 1973 | 108 | −2 | 4 | −216 | |
| 1974 | 119 | −1 | 1 | −119 | |
| 1975 | 110 | 0 | 0 | 0 | ←*Base point* |
| 1976 | 122 | +1 | 1 | 122 | |
| 1977 | 130 | +2 | 4 | 260 | |
| *Sums* | 589 | 0 | 10 | 47 | |

The sums are substituted into the normal equations, with $N = 5$, as

$$589 = 5a + b(0)$$
$$47 = a(0) + b(10)$$

and solved to yield $a = 589/5 = 117.8$ or 117,800 units, and $b = 47/10 = 4.7$ or 4700 units.

A forecasting equation is developed by substituting values of $a$ and $b$ into the straight-line equation:

$$Y_F = 117{,}800 + 4700X$$

The forecast for 1978, which is 3 years away from the 1975 base point for $X$, is given by substituting 3 for $X$ in the forecasting equation:

$$\text{Demand}_{1978} = 117{,}800 + 4700(3) = 131{,}900 \text{ units}$$

Historical data may also reveal *cyclic patterns* within the long-term trend. Two types of wavelike variations are noticeable in Figure 10.12. The large wave shows a business cycle of some sort. It may represent repeating fluctuations in local economic conditions, repeating changes in styles or consumer tastes, recurring fads and buying habits, or a combination of these and other factors.

The other cyclic pattern is attributable to seasonal variation. The insert in Figure 10.12 displays the data collected in respective quarters of a year. It is evident that higher sales can be expected in the middle two quarters of each year. Many products that depend on a

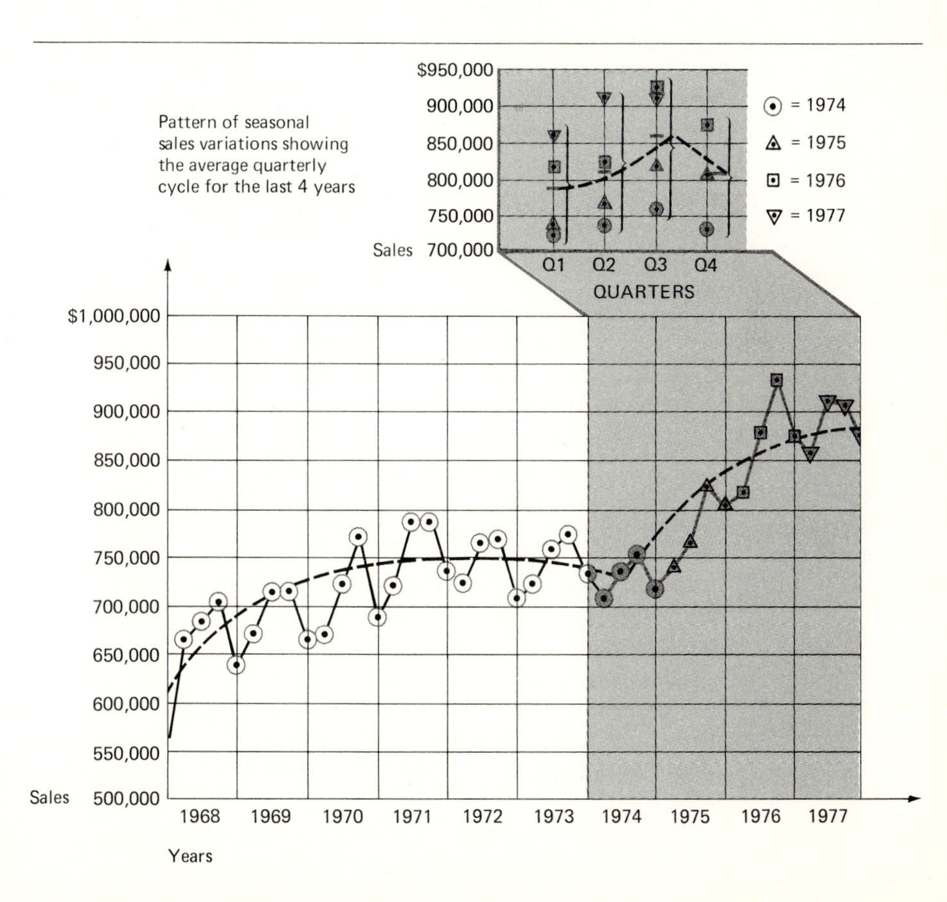


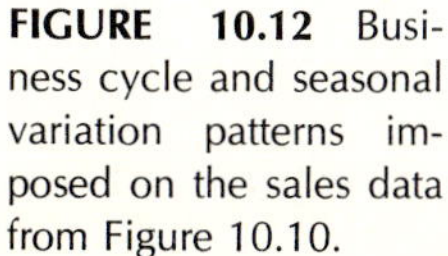
**FIGURE 10.12** Business cycle and seasonal variation patterns imposed on the sales data from Figure 10.10.

particular season (ski equipment, swimming wear, gardening supplies, holiday treats, etc.) exhibit a repetitive sales pattern each year.

The purpose of distilling different patterns from a collection of data is to improve the forecast. A long-term trend gives a valid prediction of the general region in which future sales should fall. (This trend is modified by cyclic influences when they appear.) The longer-term cycles are difficult to identify, but seasonal cycles are quite apparent and predictable. The forecast for 1978 from the given sales data would probably be just slightly higher than the current (1977) sales, with the highest demand concentrated in the middle two quarters. A forecast for 1979 and 1980 would likely predict a sales pattern about the same as 1978, owing to the long-run cyclic influence. Less confidence can be placed in the forecasts as they are pushed further into the future, because new factors could enter to disrupt the traditional patterns.

**Economy Exercise 10-3** A rough indication is needed to determine whether additional trucks will be required to transport raw materials during the coming year. If the preliminary indication is positive, an economic study will be made to see how many and what kinds of trucks should be added to the company fleet.

Annual demands for the past 5 years have been plotted, and free-hand trend lines drawn as shown in Figure 10.13(a). The "eyeballed" trend line appears to start at 60,000 units and increase by 12,500 units per year.

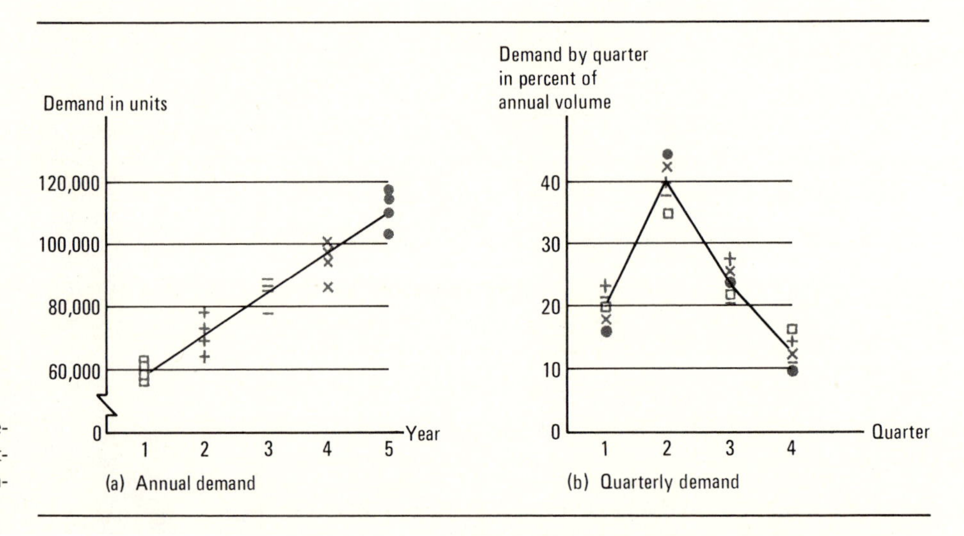


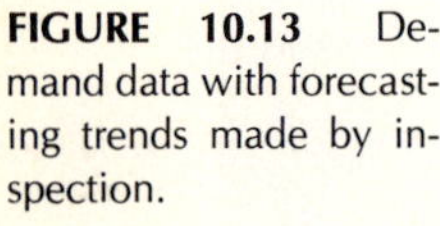
**FIGURE 10.13** Demand data with forecasting trends made by inspection.

A breakdown of annual demand by percentage per quarter is shown in Figure 10.13(b). A simple forecast by quarters would result from an arithmetic average of all the data. This average value would be of little use for the trucking problem because fleet size is a function of the *maximum* transportation needs any time during the year. A visual inspection is made to reveal that second-quarter demand accounts for about 41 percent of the year's total and is therefore the critical figure.

Statistics on national demand over the past 30 years suggest that the current business cycle is on an upswing. Therefore, a judgment value of +8 percent, or 1.08 times the trend-line forecast, is selected to represent the effect of this cycle.

Based on these unsophisticated forecasting methods, what forecast seems reasonable for the maximum demand that will occur during the next year?

## Exponential Smoothing

A line-fit forecast gives equal weight to all the historical data included in the regression model. It may be advisable to give more weight to recent occurrences. This is accomplished in the exponential-smoothing model by selecting a weighting factor $\alpha$ that gives more or less prominence to recent happenings. The exponential-smoothing formula is very convenient to apply since it has only three inputs:

**1** LD = latest demand, the most recent data on actual accomplishments of the subject being forecast. LD for a sales forecast would be the sales made during the current period.

**2** PF = previous forecast, the forecast produced for the current period by applying the exponential-smoothing formula. The forecast made with the LD for the current period will be the PF for the next forecasting period.

**3** $\alpha$ = alpha, the smoothing constant, with a value between 0.0 and 1.0. The value selected for $\alpha$ determines how much emphasis is put on the most recent data.

The above factors are combined as shown below to yield a forecast NF for the next period:

$$\text{Next forecast} = \text{NF} = \alpha(\text{LD}) + (1 - \alpha)(\text{PF})$$

This expression can be rewritten to show that the new forecast is simply the previous forecast corrected by adding a percentage $\alpha$ reflecting how much the old forecast missed the actual demand:

$$\text{NF} = \text{PF} + \alpha(\text{LD} - \text{PF})$$

As an example, let alpha have a value of 0.4, sales for June be 4200 units, and the forecast made in May for the sales expected in June be 4500 units. Applying the two versions of the exponential-smoothing formula shows

$$\begin{aligned} \text{NF}_{\text{July}} &= \alpha(\text{LD}_{\text{June}}) + (1 - \alpha)(\text{PF}_{\text{June}}) \\ &= 0.4(4200) + (1 - 0.4)(4500) \\ &= 1680 + 2700 = 4380 \text{ units} \end{aligned}$$

$$\begin{aligned} \text{and, } \text{NF}_{\text{July}} &= \text{PF}_{\text{June}} + \alpha(\text{LD}_{\text{June}} - \text{PF}_{\text{June}}) \\ &= 4500 + 0.4(4200 - 4500) \\ &= 4500 - 120 = 4380 \text{ units} \end{aligned}$$

It is apparent from the above equations that the value used for alpha significantly affects the forecast. If alpha is set equal to zero, the original forecast never changes, no matter how much the actual demand varies from it. At the other extreme, when $\alpha = 1.0$, the next forecast always equals the last demand experienced. A projection that the next period will be the same as the last period is extensively used for short-term scheduling and is known as *persistence forecasting* (in sporting circles it is called "sticking with a winner"). Between the extremes, $\alpha$ values closer to zero produce more stable predictions that may not detect current trends, and values closer to 1.0 closely track actual demands with forecasts that may fluctuate erratically.

The best way to choose alpha is to apply different values to historical data to see which one would have provided the most useful forecasts in the past. A simulation exercise to evaluate alpha values is shown in Figure 10.14. As expected, forecasts made with higher alphas are more responsive to current demand and consequently follow the actual demand pattern more closely. Other formulas using more involved smoothing functions can yield even better tracking,* but the simplicity of record-keeping, the easy computations, and the often commendable accuracy of the basic exponential-smoothing formula make it attractive for tactical planning.

---

**Economy Exercise 10-4** Develop a forecast by the exponential-smoothing method for the demand in 1978 based on the sample data in Figure 10.11. Let $\alpha = 0.3$. To develop the PF value, start the process with a forecast for 1975 based on the demand in 1973 as PF, and the demand in 1974 as LD:

$$NF_{1975} = 108 + 0.3(119 - 108) = 111.3 \quad \text{or} \quad 111{,}300 \text{ units}$$

and,

$$NF_{1976} = 111.3 + 0.3(110 - 111.3) = 110.9 \quad \text{or} \quad 110{,}900 \text{ units}$$

Continue the procedure to obtain $NF_{1978}$. Compare this forecast with the one obtained using the least-squares regression model in Example 10.3.

---

## Moving Average

Another popular forecasting method predicts the demand for the next period from the average demand experienced during several recent periods. This method is appropriately called the *moving average*. It is quite similar to exponential smoothing in that both methods "smooth" or "average out" the fluctuations in past demands to produce a forecast. The main difference between the two methods is that the choice of alpha in exponential smoothing determines how sensitive this forecast is to the most recent demands, while a moving-average forecast gives equal weight to each demand period included in the forecast.

A 3-month moving-average forecast for the next month would simply be the average demand experienced for the last 3 months. For example, the 3-month moving-average forecast for June would be calculated as

$$NF_{June} = \frac{\text{March demand} + \text{April demand} + \text{May demand}}{3}$$

In the same pattern, the forecast for July would result from averaging the demands for June,

*More elaborate versions are incorporated in the Box-Jenkins technique; see G. E. P. Box and G. M. Jenkins, *Time Series Analysis, Forecasting, and Control,* Holden-Day, San Francisco, 1970.

FORECASTS DEVELOPED FROM IDENTICAL DATA WITH ALPHA VALUES OF 0.2, 0.4, AND 0.6

| Date | Actual Demand | Forecast at $\alpha = 0.2$ | Absolute error | Forecast at $\alpha = 0.4$ | Absolute error | Forecast at $\alpha = 0.6$ | Absolute error |
|---|---|---|---|---|---|---|---|
| Oct. 8 | 180 | | | | | | |
| 15 | 164 | | | | | | |
| 22 | 162 | 177 | 15 | 174 | 12 | 170 | 8 |
| 29 | 151 | 174 | 23 | 169 | 18 | 165 | 14 |
| Nov. 5 | 143 | 169 | 26 | 162 | 19 | 157 | 14 |
| 12 | 158 | 164 | 8 | 154 | 10 | 148 | 10 |
| 19 | 154 | 163 | 9 | 156 | 2 | 154 | 0 |
| 26 | 160 | 161 | 1 | 158 | 7 | 154 | 6 |
| Dec. 3 | 172 | 161 | 11 | 159 | 16 | 158 | 14 |
| 10 | 178 | 163 | 15 | 164 | 16 | 166 | 12 |
| 17 | 188 | 166 | 22 | 170 | 18 | 173 | 15 |
| 24 | 174 | 170 | 4 | 177 | 4 | 182 | 8 |
| 31 | 172 | 171 | 1 | 175 | 4 | 177 | 5 |
| Jan. 7 | | 171 | — | 174 | — | 174 | — |
| Average error: | | | 12.3 | | 11.5 | | 9.6 |

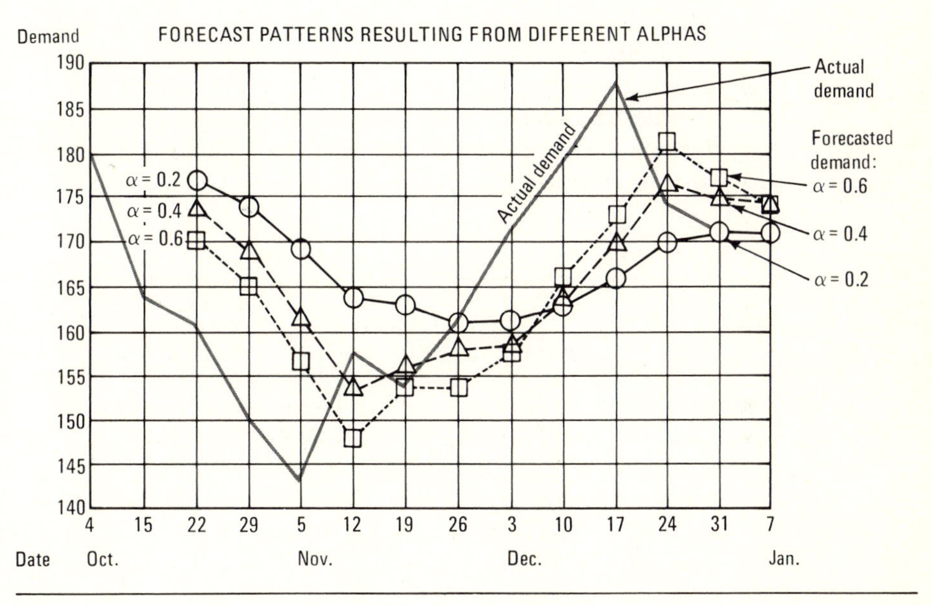


**FIGURE 10.14** Comparison of forecasts provided by the exponential smoothing method using different alphas. To start the forecasting process, the first prediction is approximated by letting it equal the first demand. Thus, the forecast for Oct. 22, when $\alpha = 0.2$, was calculated as $NF_{Oct.\ 22} = 0.2(164) + 0.8(180) = 177$. Subsequent forecasts then include the estimate for the actual demand in each weekly period. The graph demonstrates how the lag between forecast and actual demand fluctuations is decreased by larger alpha values.

May, and April. The most recent block of actual demands is used in each successive forecast.

The choice of how many periods to include in the moving average involves the same considerations used to select an alpha value for exponential smoothing. A forecast based on very few periods is more sensitive to the latest events, as is an alpha value near 1.0. An average derived from data taken over several periods will better reduce fluctuations caused by random events, but it may be too stable to detect current trends. As was the case for exponential smoothing, the best way to decide how much smoothing is desirable is to experiment with past demands, trying different spans for the moving average and selecting the one that would have predicted previous demands most accurately.

A *seasonal index* referenced to a moving average improves the forecast unless the demand pattern is relatively constant. An index value is calculated by dividing the actual demand by the centered moving average for that period. A more reliable index is obtained by averaging several index values for common time periods. The forecast is thereby the product of the most recent centered moving average for a period and the index value for that period. The procedure for developing and using a seasonal index is demonstrated in Example 10.4.

**Example 10.4**

**Moving-Average Forecast Applied to the Quarterly Demand for the Sample Data in Figure 10.11**

With the demand by quarters given in Figure 10.11, a four-quarter moving average will be used to forecast the demand for the next two quarters, $Q\,1_{1978}$ and $Q\,2_{1978}$. The four-period average is chosen because it smooths out seasonal variations over a full year.

The first moving average is one-fourth of the total sales for 1973; it represents a point in time between the end of the second quarter and the start of the third quarter. The second moving average is the sum of the last three quarters of 1973 and the first quarter of 1974 divided by 4. This value is associated with the end of $Q\,3_{1973}$ and the start of $Q\,4_{1973}$. An average of these two numbers gives a moving average

| *Year* | *Quarter* | *Sales 1000s of Units* | *Four-Period Moving Average* | *Centered Moving Average* | *Seasonal Index* |
|---|---|---|---|---|---|
| 1973 | Q1 | 19.0 | | | |
| | Q2 | 37.0 | | | |
| | | | 27.0 | | |
| | Q3 | 30.0 | | 28.1 | 1.07 |
| | | | 29.2 | | |
| | Q4 | 22.0 | | 29.8 | 0.74 |
| | | | 30.5 | | |
| 1974 | Q1 | 28.0 | | 30.6 | 0.91 |
| | | | 30.7 | | |
| | Q2 | 42.0 | | 30.2 | 1.39 |
| | | | 29.7 | | |
| | Q3 | 31.0 | | 29.6 | 1.04 |
| | | | 29.5 | | |
| | Q4 | 18.0 | | 28.7 | 0.63 |
| | | | 28.0 | | |
| 1975 | Q1 | 27.0 | | 27.6 | 0.98 |
| | | | 27.3 | | |
| | Q2 | 36.0 | | 27.4 | 1.32 |
| | | | 27.5 | | |
| | Q3 | 28.0 | | 27.9 | 1.00 |
| | | | 28.3 | | |
| | Q4 | 19.0 | | 29.2 | 0.66 |
| | | | 30.0 | | |
| 1976 | Q1 | 30.0 | | 30.1 | 1.00 |
| | | | 30.3 | | |
| | Q2 | 43.0 | | 30.4 | 1.42 |
| | | | 30.5 | | |
| | Q3 | 29.0 | | 30.7 | 0.94 |
| | | | 31.0 | | |
| | Q4 | 20.0 | | 31.1 | 0.64 |
| | | | 31.2 | | |
| 1977 | Q1 | 32.0 | | 31.6 | 1.01 |
| | | | 32.0 | | |
| | Q2 | 44.0 | | 32.2 | 1.37 |
| | | | 32.5 | | |
| | Q3 | 32.0 | | | |
| | Q4 | 22.0 | | | |

**TABLE 10.1** Computation of four-period moving averages, centered moving averages, and quarterly seasonal index for the sample data from Figure 10.11.

*centered* at $Q\,3_{1973}$. This procedure for finding the midpoint value for all quarters is continued to obtain the four-period and centered moving averages in Table 10.1.

The last column in Table 10.1 is the seasonal index for each quarter. It is obtained by dividing the actual sales for a quarter by the centered moving average for that quarter. A better estimate of a quarter's index is obtained by averaging all the values available (see Table 10.2).

**TABLE 10.2** Calculation of an adjusted seasonal index.

| *Year* | *Q1* | *Q2* | *Q3* | *Q4* |
|---|---|---|---|---|
| 1973 | | | 1.07 | 0.74 |
| 1974 | 0.91 | 1.39 | 1.04 | 0.63 |
| 1975 | 0.98 | 1.32 | 1.00 | 0.66 |
| 1976 | 1.00 | 1.42 | 0.94 | 0.64 |
| 1977 | 1.01 | 1.37 | | |
| Totals | 3.90 | 5.50 | 4.05 | 2.67 |
| Average seasonal index | 0.975 | 1.375 | 1.0125 | 0.6675 |
| *Adjusted seasonal index* | *0.97* | *1.37* | *1.00* | *0.66* |

Before the average seasonal index is applied, two checks should be made:

1 The average of the periodic indexes should total to 1.0. In the example, the average is

$$\frac{0.9750 + 1.3750 + 1.0125 + 0.6675}{4} = \frac{4.03}{4} = 1.0075$$

Therefore, the indexes must be adjusted or else the quarterly forecasts will exceed the implied annual forecast by 0.75 percent.

2 Attention should be given to any obvious trends in a quarterly index. In Table 10.2, $Q\,1$ appears to be increasing and $Q\,3$ has a distinct downward trend.

The adjusted index reflects the above considerations. A fraction (4.00/4.03) of each average index was taken, and the results were rounded off with respect to the trend in $Q\,3$. Thus, calculations provide the forecasting framework, but finishing touches are supplied by judgment.

The final step is to make the forecast. It results from taking the product of the most recent centered moving average and its respective seasonal index. Forecasts for the first two quarters of 1978 are

$$Q\,1_{1978} = 31.6 \times 0.97 = 30.7 \quad \text{or} \quad 30{,}700 \text{ units}$$
$$Q\,2_{1978} = 32.2 \times 1.37 = 44.1 \quad \text{or} \quad 44{,}100 \text{ units}$$

## Extensions of Historical-Basis Forecasting Methods

Only the perimeter has been explored in this survey of methods of forecasting from historical data. Simple regression, a model relating a dependent to an independent variable, can yield fits of many shapes besides straight lines. Curvilinear regression based on the equation

$Y_F = aX^b$ could be used to develop certain curve-shaped fits. Hyperbolic and polynomial fits are also possible. Multiple regression, either linear or nonlinear, is used when the forecast depends on the relationship of several variables.

An indication of how much faith can be put on a forecast results from calculating statistical control limits for regression lines. Such calculations can also provide prediction limits. For instance, a statement could be made that there are only 5 chances in 100 that the accumulated demand will exceed a certain value at a certain time in the future.

The main value of regression models as compared with weighted-average techniques (exponential smoothing and moving average) is that the regression line can be extended several periods beyond the present. This advantage is gained at the expense of extra computational effort. But this burden can be shifted to computers; canned programs that handle huge regression models are widely available.

## FORECASTS BASED ON CAUSE*

Very reliable predictions are possible if the causes for future demand can be identified and measured. The causal approach to forecasting has produced sophisticated models that attempt to include all the relevant causal relationships; in the most elaborate models there may be hundreds of quantified relationships. But causal models can also be quite simple. Perhaps knowledge of a single cause is sufficient for predicting, as in the case of a college bookstore, which orders books based on knowledge of the number of students admitted to the school for next year. The manager of the bookstore could apply a factor based on past history to the number admitted, to forecast the demand by courses in the forthcoming terms.

*Leading indicators* are frequently used in business forecasting. These indicators are statistics compiled and published by many sources: *The Wall Street Journal, Business Week,* United States government reports on almost everything, local government reports on economic activity, university publications, etc. If an indicator can be found that consistently predicts what demand to expect in, say, 6 months, current activities can be planned accordingly. It is even more comfortable to have two or three leading indicators pointing in the same direction. The problem is to find the accurate indicators.

### Correlation Model

A search for a leading indicator starts with assumptions as to the causes of demand. A house builder could hypothesize that housing demand depends on population growth, prices for houses, personal incomes, demolition of old homes, housing starts by competitors, and availability of financing. After data about the possible indicators have been collected, correlation studies are conducted to see if they are indeed related to housing demand. Ideally, an indicator will be identified that has a pattern similar to the historical demand, but that leads it by several periods. For example, it might be discovered that housing demand is correlated with interest rates: As rates climb, demand declines; but the change

*Adapted from J. L. Riggs, *Production Systems,* 2d ed., Wiley, New York, 1976, chap. 3.

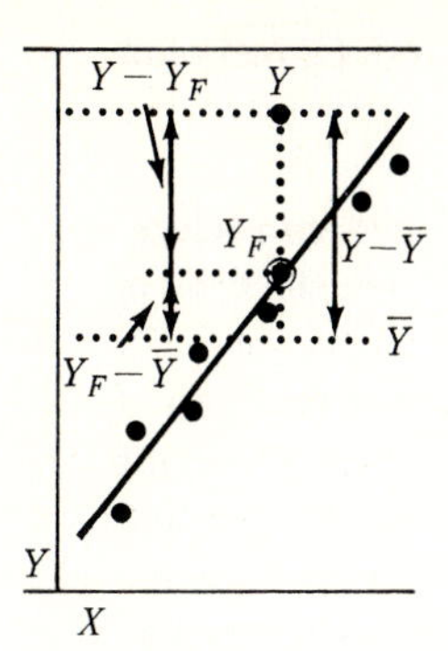


FIGURE 10.15 Deviations of a data point.

in demand follows a change in interest rates by 6 months. In this case, a builder could observe today what is happening to interest rates and use the observation as a leading indicator to predict with some confidence what the demand for houses will be in 6 months.

A correlation study examines the degree of relationship between variables. The variables are represented in mathematical expressions, such as the forecasting formula developed by the least-squares method in Example 10.3. Because not all the data points coincide with the fitted line, part of the relationship among variables remains unexplained. This dispersion of data points about regression lines is characterized by three sums-of-squares:

$$\Sigma(Y - \overline{Y})^2 = \Sigma(Y_F - \overline{Y})^2 + \Sigma(Y - Y_F)^2$$

| Total variation | Explained variation | Unexplained variation |
|---|---|---|

1. *Total variation* The deviation $Y - \overline{Y}$, as shown in Figure 10.15, is the vertical distance between a data point and the mean of all observations, $\overline{Y} = \Sigma Y/N$. This term is a measure of the total variation of the dependent variable (in simple regression) and is divided into two parts: the explained and unexplained variations.
2. *Explained variation* The variation explained by the regression line is represented by $Y_F - \overline{Y}$. In Figure 10.15, this deviation appears as the vertical distance between the line of regression and a horizontal line at $\overline{Y}$.
3. *Unexplained variation* The sum-of-squares of the unexplained variation expresses the error between the forecast and actual values, $Y - Y_F$. Each deviation appears as the vertical distance from a data point $Y$ to the regression line.

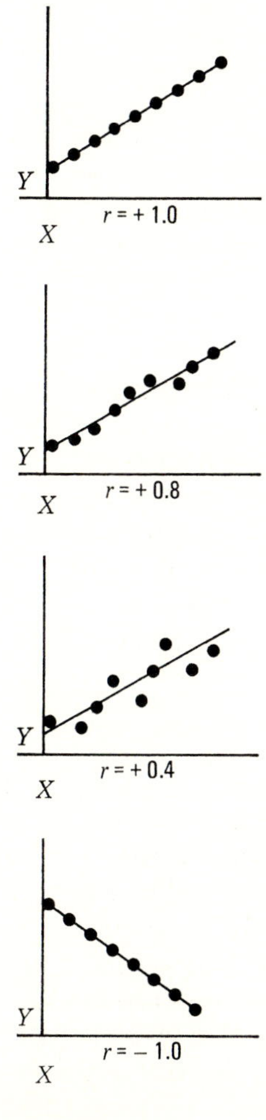


FIGURE 10.16 Correlation patterns.

The ratio of the sum-of-squares of the unexplained variation to the sum-of-squares of the total variation, $\Sigma(Y - Y_F)^2/\Sigma(Y - \overline{Y})^2$, measures the proportion of the total variation that is not explained by the regression. Therefore,

$$1 - \frac{\Sigma(Y - Y_F)^2}{\Sigma(Y - \overline{Y})^2}$$

measures the proportion of the total variation explained by the regression line, and the square root of this expression is the ***coefficient of correlation***, $r$:

$$r = \sqrt{1 - \frac{\Sigma(Y - Y_F)^2}{\Sigma(Y - \overline{Y})^2}}$$

The value under the radical can never be greater than 1 nor less than 0. However, because the radical has both positive and negative roots, the value of $r$ is between +1 and −1. The plus or minus is indicative only of the slope of the regression line as depicted in Figure 10.16. When $r = +1$, all the data points fall on an upward-sloping regression line. When $r$ is between +1 and 0, the regression line still slopes upward, but data points fall on either side of the line. The closer they cluster around the line, the closer $r$ approaches 1.

When sufficient data are available, $r$ is calculated from the sum-of-squares expression or, more directly, from the formula

$$r = \frac{N\Sigma XY - (\Sigma X)(\Sigma Y)}{\sqrt{N\Sigma X^2 - (\Sigma X)^2}\sqrt{N\Sigma Y^2 - (\Sigma Y)^2}}$$

**Example 10.5** **Development of a Leading-Indicator Forecast by Regression and Correlation Models**

The Slam-Bang Company manufactures automatic nail drivers as modern replacements for the conventional hammer. The company sensibly reasoned that the sales of nail drivers should be related to the total amount spent on building construction. If a relationship does exist, published government and construction-industry forecasts of anticipated building levels can be used as a sales indicator. First a check was made to see that the building-level forecasts were relatively accurate. Next the national figures were broken down to conform to the Slam-Bang marketing areas. Then the records of monthly building volume and nail-driver sales for corresponding months were collected. The resulting data are tabulated in Table 10.3, where the construction volume is in \$100-million units and product sales are in \$10,000 units. The column headings correspond to the values needed for the calculation of $r$ and a linear-regression equation.

From this reasonably large sample, the coefficient of correlation was calculated as

$$r = \frac{N\Sigma XY - (\Sigma X)(\Sigma Y)}{\sqrt{N\Sigma X^2 - (\Sigma X)^2}\sqrt{N\Sigma Y^2 - (\Sigma Y)^2}}$$

$$= \frac{20(519.5) - (50.4)(202.4)}{\sqrt{20(130.5) - (50.4)^2}\sqrt{20(2076.62) - (202.4)^2}}$$

$$= \frac{10{,}390 - 10{,}201}{\sqrt{2610 - 2540.16}\sqrt{41{,}532.4 - 40{,}965.7}}$$

| Nail-Driver Sales $\times 10^{-4}$, $Y$ | Construction Volume $\times 10^{-8}$, $X$ | $Y^2$ | $X^2$ | $XY$ |
|---|---|---|---|---|
| 7.1 | 1.8 | 50.51 | 3.24 | 12.78 |
| 9.9 | 2.3 | 89.01 | 5.29 | 22.78 |
| 9.0 | 1.9 | 81.00 | 3.61 | 17.10 |
| 10.4 | 2.6 | 108.16 | 6.76 | 27.04 |
| 11.1 | 3.1 | 123.21 | 9.61 | 34.41 |
| 10.9 | 2.8 | 118.81 | 7.84 | 30.52 |
| 10.5 | 2.9 | 110.25 | 8.41 | 30.45 |
| 9.8 | 2.4 | 96.04 | 5.76 | 23.52 |
| 11.1 | 2.8 | 123.21 | 7.84 | 31.08 |
| 10.2 | 2.5 | 104.04 | 6.25 | 25.50 |
| 9.7 | 2.3 | 94.09 | 5.29 | 22.31 |
| 10.9 | 2.8 | 118.81 | 7.84 | 30.52 |
| 8.8 | 2.1 | 77.44 | 4.41 | 18.48 |
| 8.6 | 1.9 | 73.98 | 3.61 | 16.34 |
| 12.3 | 3.2 | 151.29 | 10.24 | 39.36 |
| 11.4 | 3.0 | 129.96 | 9.00 | 34.20 |
| 11.2 | 2.8 | 125.44 | 7.84 | 31.36 |
| 10.2 | 2.6 | 104.04 | 6.76 | 26.52 |
| 10.7 | 2.7 | 114.49 | 7.29 | 28.89 |
| 8.6 | 1.9 | 73.96 | 3.61 | 16.34 |
| 202.4 | 50.4 | 2076.62 | 130.50 | 519.50 |

**TABLE 10.3** Correlation data for sales of Slam-Bang nail drivers and construction volume. The last three columns are the computations required to calculate $r$ and $Y_F$.

$$= \frac{189}{(8.36)(23.81)} = 0.95$$

Although the correlation is not exact, it is definitely worthy of consideration for prediction purposes. With the sums developed in Table 10.3, the least-squares method is applied to obtain a forecasting equation. The normal equations are

$$\Sigma Y = Na + b\Sigma X \qquad 202.4 = 20a + b(50.4)$$
$$\Sigma XY = A\Sigma X + b\Sigma X^2 \qquad 519.5 = a(50.4) + b(130.5)$$

They are solved for $a$ and $b$ to obtain

$a = 3.37$ or \$33,700 $\quad b = 2.68$ or \$26,800

The forecasting equation is then

$$Y_F = \$33{,}700 + \$26{,}800X$$

Thus, a projected construction volume of \$275,000,000 for the next month would suggest a nail-driver sales volume of

$$Y_F = \$33{,}700 + \$26{,}800(2.75) = \$107{,}400$$

### Econometric Models

Econometrics is a discipline concerned with the measurement and definition of economic systems. It utilizes statistical and programming methods to develop models based on the quantitative aspects of system behavior. The models vary from the simple regression methods previously described to very elegant and rich representations of large systems involving hundreds of variables.

The more sophisticated econometric models are based on causal relationships. They have been developed for different engineering systems, most notably for transportation. The initial effort in developing a causal model is considerable, and after the causal forecast is accepted, the model must be continually monitored to confirm that the modeled relationships are still valid. Causes of demand can change, and the cause of the causes can change.

Causal models are hardly ever constructed just to generate better data for specific economic comparisons. When the models are already available, they should certainly be utilized in the comparisons, but a decision would have to be of "make or break" proportions to provoke the development of an original model. However, causal forecasting is more apt than historical forecasting to predict a turning point in demand, a point at which growth switches to decay or the reverse, and even a hint of a turning point is vital information to the estimator. Reputations and fortunes are made or lost by alternatives selected in anticipation of a shift from the existing trend.

## SENSITIVITY ANALYSIS

Regardless of the care devoted to data collection, there is always a distinct possibility that the data will be misleading. Estimates and forecasts may be inaccurate. Data might be accurate but descriptive of a different situation. When the data are in doubt, as is the usual case, an experienced analyst considers the consequences of using faulty "facts." The sensitivity of a decision is investigated by inserting a range of estimates for critical factors in the comparison model to determine how much variation is permissible for a given outcome. Thus, a *sensitivity analysis measures the relative magnitude of change in one or more elements of an economic comparison that will reverse a decision among alternatives.*

## Sensitivity of a Single Proposal

Assume a decision is to be made about a business opportunity based on the following estimates and tentative before-tax analysis:

| *Economic Factors* | *PW for a 10-Year Study Period at I = 13%* |
|---|---|
| Annual receipts ($35,000) | +$189,917 |
| First cost ($170,000) | −170,000 |
| Salvage value ($20,000) | +5,892 |
| Annual disbursements ($3000) | −16,279 |
| *Net PW* | +$ 9,530 |

The first cost is the most reliably known value in the problem, owing to its immediacy. The other factors could vary considerably over the 10-year period, owing to unforeseeable deviations from anticipated conditions. Even the study period may be inappropriate if the asset's useful life is shorter than 10 years (or longer than 10 years), or if the function it serves does not continue to yield the stated receipts for the full study period. And the 13 percent rate of return might be questioned. Should it be higher to compensate for inflation and the risk of losing invested capital? Or should it be lower?

Such questions are collectively examined by constructing a sensitivity graph as shown in Figure 10.17. Curves are generated by substituting various values for one factor in the PW formula,

$$\text{PW} = -(\text{first cost}) + (\text{salvage value})(P/F, i, n) + (\text{receipts} - \text{disbursements})(P/A, i, N)$$

while holding the values of all other factors constant. The abscissa of the graph is the percentage deviation from original values. This dimensionless scale puts all the factors in the same perspective to allow direct comparison of curve shapes. A steeper rising or falling curve indicates greater sensitivity of the proposal's worth to that factor.

The sensitivity graph reveals that deviations of up to 50 percent of the original estimates for salvage value and annual disbursements will not affect the acceptance of the proposal. The other factors—minimum acceptable rate of return, number of years the proposal will stay in effect, and amount of annual receipts—could switch the verdict to the do-nothing alternative if they deviate by only 10 percent from the original estimates. This condition

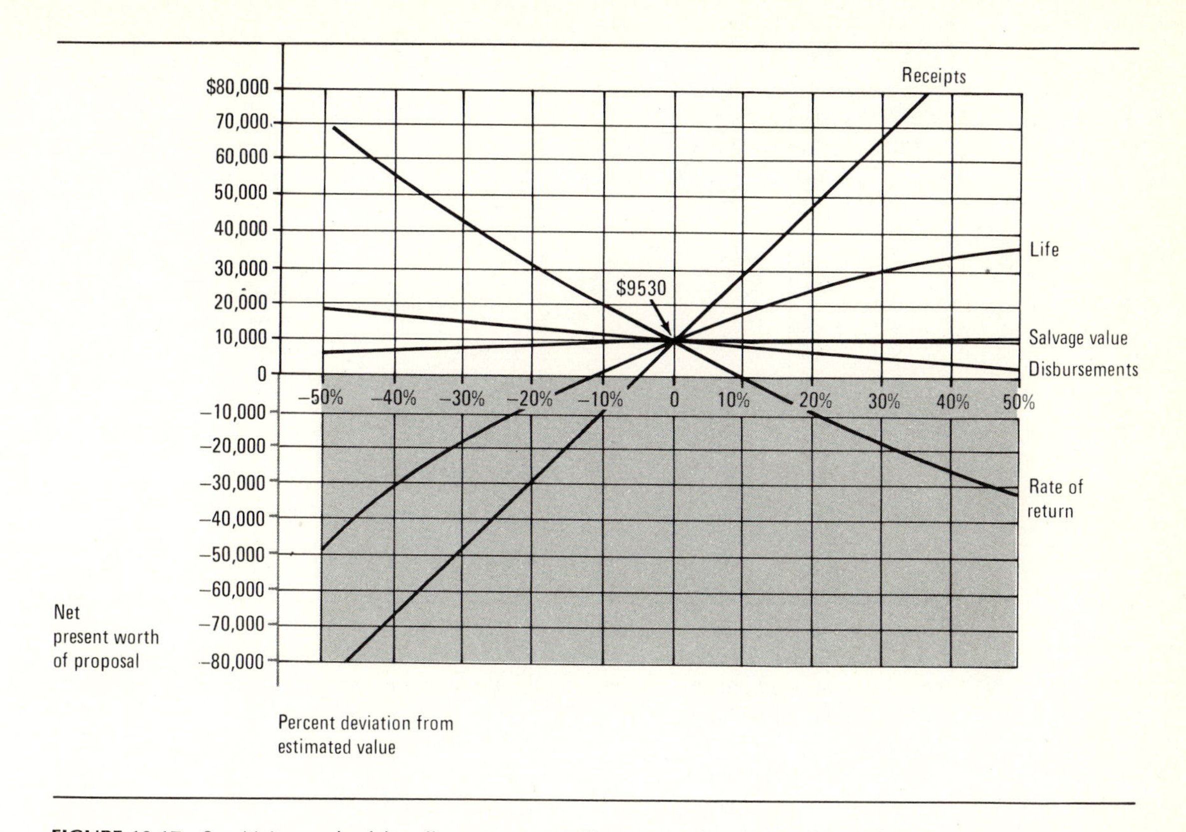


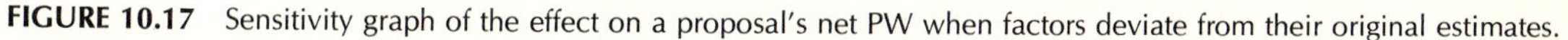

**FIGURE 10.17** Sensitivity graph of the effect on a proposal's net PW when factors deviate from their original estimates.

suggests that extra care be given to forecasts of future business conditions that could affect income flow or the continued demand for the asset purchased.

The life of the asset and the size of receipts are not controlled by the analyst, as is the interest rate used in the comparison. The timing and amount of the cash flows are functions of the operating environment. Since these are the two most sensitive factors in the evaluation of the proposal, the limiting combinations for accepting or rejecting the proposal are graphed as an ***isoquant*** in Figure 10.18. The isoquant forms an ***indifference line*** which indicates the combinations of the proposal's duration and size of receipts that make the present worth of the proposal neither positive nor negative (indifference condition) when the other factors are unchanged. Thus, a reduction in the life of the asset from 10 to 8 years must be accompanied by an increase in annual receipts for 8 years of at least \$36,859 − \$35,000 = \$1859 for the proposal to be minimally acceptable at $i$ = 13 percent:

$$\text{PW} \doteq 0 = -\$170{,}000 + \$20{,}000(P/F,\ 13,\ 8) + (\$36{,}859 - \$3000)(P/A,\ 13,\ 8)$$

$$0 = -\$170{,}000 + \$20{,}000(0.37616) + \$33{,}859(4.7987)$$

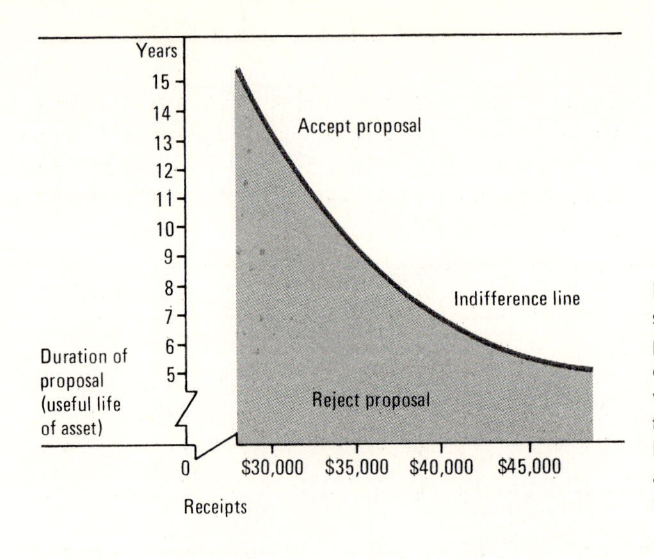


**FIGURE 10.18** An isoquant showing combinations of the proposal's life and annual receipts at which the present worth is zero. Any combination that falls above the indifference line indicates the proposal is acceptable when other factors remain constant.

---

**Economy Exercise 10-5** A would-be engineer spent so much time seeing movies while she was enrolled in engineering that she never passed the differential-equations course. After she dropped out, she inherited $100,000 on her twenty-first birthday (perhaps that is why she did not bother to study much). She greeted this inheritance as a chance to satisfy her passion for flicks by leasing a soon-to-be-vacant supermarket and converting it into a CinemaCenter comprising four small theaters in the one building. Based upon what she remembered from the engineering economics course she took before leaving school, she prepared the following cost estimates:

| | |
|---|---|
| Renovation cost | $80,000 |
| Contingency fund | $10,000 |
| Estimated life of renovations | 9 years |
| Salvage value | 0 |
| Annual operating cost (365 days/year) | $62,000 |
| Annual lease expense | $42,000 |
| Other annualized expenses | $16,000 |
| Desired annual profit | $35,000 |

The CinemaCenter will have 280 deluxe seats in the four projection areas and an elaborate lobby. Income per moviegoer should average $3.35, including net profit from refreshments purchased.

The most questionable estimate involved in this economic evaluation is the number of people who will attend the shows. She has been told that she will be lucky to have a 50 percent utilization rate. What percentage of capacity is necessary to break even (assuming receipts accumulate to year-end totals and the local cost of capital is 12 percent)? What influencing factors should be considered in forecasting utilization rates?

## Sensitivity of Alternatives

An alternative ($A2$) has been developed to accomplish the same mission as the proposal ($A1$) described by the sensitivity relationships in Figure 10.17. Both alternatives are expected to have the same revenue, but proposal $A2$ has a lower first cost. However, the annual disbursements for $A2$ will increase significantly each year as shown by the following cash-flow estimates:

| *Factor* | *Alternative 1* | *Alternative 2* |
|---|---|---|
| Annual receipts | $ 35,000 | $ 35,000 |
| First cost | 170,000 | 116,400 |
| Salvage value (year 10) | 20,000 | 0 |
| Annual disbursements | 3,000 | $3000 the first year and increasing by $2500 each year |

On the basis of the given estimates and a study period of 10 years with $i = 13$ percent, both alternatives have about the same PW:

$$\text{PW}(A1) = \$9530$$

$$\begin{aligned}\text{PW}(A2) &= -\$116{,}400 + [\$35{,}000 - (\$3000 + \$2500)(A/G, 13, 10)](P/A, 13, 10) \\ &= -\$116{,}400 + [\$32{,}000 - \$2500(3.5161)](5.4262) \\ &= \$9540\end{aligned}$$

Therefore, a preference for one or the other rests on the interpretation given to the annual cash flows. Since both alternatives are assumed to produce the same receipts, and disbursements can usually be estimated quite accurately, the most questionable feature remaining is the study period. A sensitivity graph for the alternatives' possible useful lives is given in Figure 10.19.

The graph indicates that $A2$ is preferred to $A1$ when the likelihood is that the proposals

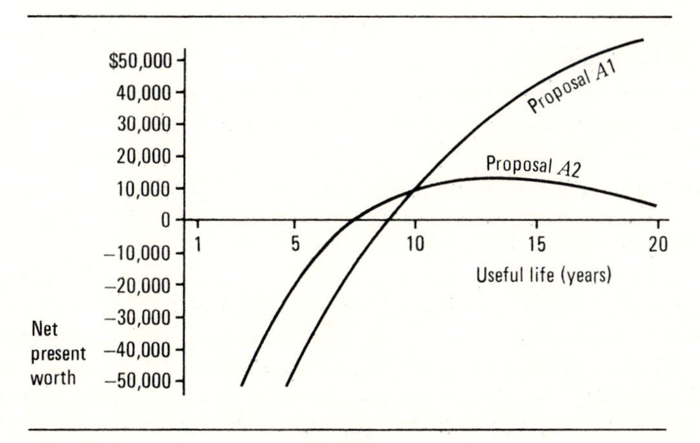


**FIGURE 10.19** Sensitivity of two proposals to deviations from their estimated life.

will have lives of less than 10 years. At $N = 7$, neither proposal is profitable. Proposal $A1$ offers much larger gains should the life extend beyond 10 years. Such observations typically create more questions than solutions. If the questions contribute to a more carefully thought out solution, the sensitivity analysis has served its function.

## SUMMARY

Data are the raw materials for economic comparisons. Without reliable data, a comparison is a futile exercise.

*Cause-and-effect diagrams* are pictorial representations of problems. They provide a systematic procedure for investigating a problem to determine what data are needed for a solution. A standard C&E diagram has cause arrows directed from the left toward a problem identification symbol, and effect arrows branching to the right from the symbol. Modifications include one-sided and sequentially clustered diagrams.

Cash-flow estimates can be generated or secured from a variety of sources: competitive bids, price quotes, accounting records, standard times and costs, cost indexes, etc. Two estimates to watch closely are first cost and overhead cost.

Forecasting methods are used to predict future business conditions and technical developments. *Subjective* methods, which rely on personal opinions, are comparatively inexpensive to obtain but may be biased. Historical data are the bases for *line-fitting* and *smoothing techniques* that extrapolate past trends to predict the future. The *least-squares method* fits a straight line to data to form a simple regression model. The choice of alpha in *exponential smoothing* and the number of periods used in a *moving average* determine how sensitive the forecasts are to recent data. *Causal models* attempt to predict future happenings from changes in the causes of events. A *leading indicator* is a statistic that behaves the same but precedes the behavior of the data sought. A *correlation* model examines the closeness of the relationship.

*Sensitivity analysis* measures the amount of change in one or more elements needed to reverse the decision in an economic comparison. A sensitivity graph displays the effect on PW of changes in each element when all other elements are held constant.

## Discussion of Economy Exercises

**EE 10-1** Possible cures for the C&E diagraming ailments include:

**1** Make the drawing less crowded, and think about how factors are interrelated. Avoid carelessness such as forgetting to note the boundary limit.

**2** Obtain additional expertise to determine if there are other factors involved in the problem.

**3** Have the drawing reviewed to see if the construction has been logical so far. Then get more opinions, or start thinking in terms of other factors that deserve attention. There should always be at least two branch arrows at the same level. Don't be concerned about just one aspect of the problem.

**4** Avoid stereotyped thinking that limits the analysis to binary considerations. Question each factor to see what has been overlooked. Symmetry is rewarded in some art designs, but not in C&E diagraming.

**EE 10-2** Is the sales representative's evaluation valid? Maybe. If the estimated life of the machine and its operating costs are accurate, there is still a question of whether overhead costs will be reduced by $6000. A one-worker reduction in the work force would probably have little effect on total overhead costs. Since the main purpose of an overhead ratio is to allocate indirect costs to products, it is not an exact measure of indirect wages. Therefore, the quoted savings in overhead should be investigated.

**EE 10-3** "Unsophisticated" is not a synonym for worthless. The "eyeball" approach delivers a reasonably close forecast at a minimum cost. Since a "guesstimate" does not depend on a formal procedure, it can put more weight on recent data or other factors that the forecaster feels are especially pertinent. From the given data, maximum demand should occur during the second quarter. The forecast results from multiplying the trend value by the cyclic correction and the quarterly percentage:

$$\text{Demand}_{\text{2d quarter}} = [60{,}000 + 12{,}500(5)](1.08)(0.41) = 54{,}243$$

**EE 10-4** The sequence of calculations leading to the 1978 forecast is

$$NF_{1977} = 110.9 + 0.3(122 - 110.9) = 114.2 \quad \text{or} \quad 114{,}200 \text{ units}$$

$$NF_{1978} = 114.2 + 0.3(130 - 114.2) = 118.9 \quad \text{or} \quad 118{,}900 \text{ units}$$

Compared with the simple regression model in which $NF_{1978} = 131{,}900$ units, the exponential-smoothing model is more pessimistic. This is due to the selection of 0.3 for the value of $\alpha$, which puts more weight on the older data. Changing $\alpha$ to 0.8 increases the forecast for 1978 to 128,000.

**EE 10-5** The percentage of seats occupied by paying customers $X$ needed to break even when the desired profit is not included as a cost is found from

$$0 = (\$3.35)(280)(365)(X) - (\$80{,}000 + \$10{,}000)(A/P, 12, 9) - \$62{,}000 - \$42{,}000 - \$16{,}000$$

$$= \$342{,}370X - \$90{,}000(0.18768) - \$120{,}000$$

$$\text{and, } X = \frac{\$16{,}891 + \$120{,}000}{\$342{,}370} = 0.4 \quad \text{or} \quad 40\%$$

If the estimated 50 percent attendance figure is realistic, the CinemaCenter appears to have a decent chance of surviving, so long as no profit is expected. To achieve the $35,000-per-year profit goal, the utilization rate must exceed 50 percent. Factors to consider in forecasting the chances of bettering 50 percent include the number and success of competitors, demographic data on the number of people in age groups most likely to attend the type of movies to be shown, and the average wage and spending habits of the local population.

## PROBLEMS

**10.1** The State Highway Department is planning to construct several highway rest stops. The purpose of such stops is to improve highway safety by encouraging motorists to get out of their cars and relax for a few minutes during long trips. The locations and appearance of the rest stops largely determine how much they are used. Develop a C&E diagram to show what data are required to compare alternative sites and designs for rest stops. Some of the main causes are ***location, natural features, construction,*** and ***design***; some main effects are ***utilization, upkeep,*** and ***safety***.

**10.2** Many innovative ideas have been proposed to solve the problems of intracity transportation. They range from moving sidewalks to individualized jitney service. A poll to determine what consumers believe are the most important characteristics of a carrier-based city transportation service revealed the following attributes:

**1** Arrive and leave on schedule
**2** Have a seat for the whole trip
**3** No transfers during trip
**4** Low fares, comparable with the cost of using a car
**5** Shelters at pickup points
**6** Available at all hours
**7** Use direct routes to shorten travel time
**8** Easy way to pay fares
**9** Space provided for packages
**10** Adjustable seats provided
**11** Clean vehicle interiors
**12** Adjustable air, light, and sound
**13** Coffee, refreshments, and reading material on board
**14** Stylish vehicle exterior

Develop a C&E diagram that contains the key features by which any intracity transportation proposal could be evaluated. Include the units of measurements by which each entry in the diagram would be measured.

**10.3** A cost index is a dimensionless number that relates a certain type of cost at a given time to that cost during a reference year. That is,

$$C_p = C_r \frac{I_p}{I_r}$$

*where* $C_p$ = present cost, dollars
$C_r$ = cost during reference period
$I_p$ = present value of the cost index
$I_r$ = index value during reference period

**10.3a** Cost estimates are being developed to justify the construction of a warehouse. A structure of similar lay-up design was built 6 years ago at a unit price of \$14.40 per square foot (\$155 per square meter) of wall when the cost index was 116. The index now is 170. What would the construction cost be today for a warehouse with 60,000 square feet (5574 square meters) of wall?

*(\$1,266,207)*

**10.3b** According to present plans, the warehouse construction will not get underway until 3 years from now. If the cost index continues to increase at its present rate, what construction bid can be expected for the warehouse?

**10.4** Persistence predictions may seem too naïve to be classified as a forecasting method, but they can be surprisingly successful under some conditions. Comment on the following observations:

**10.4a** In sporting events the usual forecast is for the current champion to win again.

**10.4b** A decision not to make a decision is a persistence prediction.

**10.5** What approximate mathematical relationship between the $\alpha$ used in exponential smoothing and the number of periods used in a moving average makes both methods equally responsive to changes in historical data?

**10.6** Sales figures for two products first marketed 6 months ago are shown:

| *Month* | *Product 1* | *Product 2* |
|---|---|---|
| January | $110,000 | $ 54,000 |
| February | 102,000 | 63,000 |
| March | 95,000 | 80,000 |
| April | 85,000 | 98,000 |
| May | 78,000 | 112,000 |
| June | 70,000 | 133,000 |

**10.6a** What July forecast for each product is obtained by using a 6-month moving average? How do you explain these forecasts with respect to the dissimilar sales patterns?
**10.6b** What forecast would you make for each product for July?

**10.7** The least-squares method can be applied to fit a curve to data when the forecasting equation is $Y_F = ab^X$, in which $Y$ changes at a constant rate $b$ each period for $X$ periods. When this equation is translated from the exponential form to its logarithmic form,

$$\log Y = \log a + X \log b$$

normal equations can be set up as

$$\Sigma \log Y = N \log a + \Sigma X(\log b)$$
$$\Sigma(X \log Y) = \Sigma X(\log a) + \Sigma X^2(\log b)$$

to allow the tabular approach of the least-squares method in determining the values of $a$ and $b$. When the base point is selected to make $\Sigma X = 0$, the solution reduces to

$$\log a = \frac{\Sigma \log Y}{N} \quad \text{and} \quad \log b = \frac{\Sigma(X \log Y)}{\Sigma X^2}$$

Set up a table with column headings for $Y$, $X$, $X^2$, $\log Y$, and $X \log Y$, and use the annual-demand data from Figure 10.11 to determine a curve-fitting forecasting formula. Compare the forecast for 1978 with the straight-line forecast.

[$Y_F = 117{,}500(1.0404)^X$; $Y_{1978} = 137{,}700$ *units*]

**10.8** Given the following data:

| *Year* | 1 | 2 | 3 | 4 | 5 | 6 | 7 | 8 |
|---|---|---|---|---|---|---|---|---|
| *Demand* | 90 | 100 | 107 | 113 | 123 | 136 | 144 | 155 |

**10.8a** Plot the data, and establish a forecast for year 9 by observation.
**10.8b** Use the least-squares method to develop a forecasting formula. What is the forecast for year 9?
**10.8c** What is the forecast for year 9 by the exponential-smoothing method when $\alpha =$ 0.25?

**10.9** Quarterly unit demands for a product are given:

| *Year* | *Winter* | *Spring* | *Summer* | *Fall* |
|---|---|---|---|---|
| 1 | 81 | 64 | 73 | 83 |
| 2 | 80 | 70 | 84 | 74 |
| 3 | 86 | 59 | 71 | 73 |
| 4 | 98 | 72 | 74 | 64 |
| 5 | 106 | 68 | 75 | 60 |

**10.9a** Using a four-period moving average, determine a seasonal adjusted index, and establish a forecast for each quarter of next year.

**10.9b** Use line-fitting methods to determine a forecast for each period of next year.

**10.10** Assume you have the franchise to sell refreshments for theatrical performances given at the college playhouse. Besides the usual soft drinks and popcorn, you offer homemade tarts. Since the tarts have to be baked within a day of when they are to be sold and the market for stale tarts is quite limited, predicting sales is very important. The sales record for the first 10 performances is tabulated below. Each performance was a sellout, and all future performances are expected to play to capacity crowds.

| *Date* | Jan. 5 | 6 | 12 | 13 | 26 | 27 | Feb. 2 | 3 | 16 | 17 |
|---|---|---|---|---|---|---|---|---|---|---|
| *Tarts sold* | 120 | 120 | 120 | 144 | 191 | 171 | 172 | 186 | 153 | 145 |

In reviewing the sales, you recall that you sold all the tarts you baked for the first four performances. Then you increased production for the remaining performances to 16 dozen (192) and started having leftovers. For the last two performances you barely covered costs because so many tarts remained unsold. Now the question is how many to bake for the next performance.

**10.10a** What forecasts for tart demand would have been made for the last four performances by the exponential-smoothing formula with $\alpha = 0.4$? What is the forecast for the next performance?

*(NF = 160)*

**10.10b** Repeat Problem 10.10*a* using a 3-month moving average.

*(NF = 161)*

**10.10c** What are the forecast tart sales for the next performance when both alpha and the number of moving-average periods are doubled ($\alpha = 0.8$ and $N = 6$)?

$[NF(\alpha = 0.8) = 148;\ NF(N = 6) = 170]$

**10.10d** What forecast do you recommend? What additional information might improve your prediction?

**10.11** A patented new product was introduced 9 years ago. Sales since its introduction are shown:

| *Years since introduction* | 1 | 2 | 3 | 4 | 5 | 6 | 7 | 8 | 9 |
|---|---|---|---|---|---|---|---|---|---|
| *Annual unit sales* | 2023 | 2102 | 2009 | 2768 | 3291 | 3881 | 4622 | 5494 | 5557 |

The original plant that produces the product is now operating at 100 percent capacity. A second plant, to be built immediately, must have the capacity to meet the additional demand expected for the next 5 years. What should be the design capacity of the new plant?

**10.12** Suggest possible leading indicators of readily available data for the following products and services:

**10.12a** Industrial production in a developing country

**10.12b** Enrollment in a private business school specializing in data-processing training

**10.12c** The number of convicts for whom space must be provided in penitentiaries in less populous states

**10.12d** Demand for prepared baby foods

**10.13** Private-aircraft sales in three-state marketing area and the number of students enrolling each year to take flying lessons are shown in the table.

| *Year* | *Aircraft Sales* | *Student Starts* |
|---|---|---|
| 1 | 300 | 4100 |
| 2 | 400 | 4600 |
| 3 | 350 | 4800 |
| 4 | 450 | 4300 |
| 5 | 500 | 5800 |
| 6 | 400 | 5600 |
| 7 | 600 | 6000 |
| 8 | 550 | 6300 |
| 9 | 600 | 6800 |
| 10 | 650 | 7200 |

**10.13a** Using just the past history of aircraft sales, estimate the number of sales to expect in year 11 and year 12.

**10.13b** Calculate the coefficient of correlation to determine if student starts are a leading indicator for aircraft sales. Assume the sales pattern lags the starts pattern by 2 years.

**10.13c** Develop a forecasting formula for sales based on student starts. Develop the equation

$$\text{Sales}_{\text{year } t} = a + b(\text{starts}_{\text{year } t-2})$$

and use it to find the expected sales in years 11 and 12 ($t = 11, 12$) from the student starts in years 9 and 10 ($t - 2$).

**10.14** A proposal is described by the following estimates: $P = \$20{,}000$; $S = 0$; $N = 5$; and net annual receipts = \$7000. A rate of return of 20 percent is desired on such proposals. Construct a sensitivity graph of the life, annual receipts, and rate of return for deviations over a range of ±20 percent. To which element is the decision most sensitive?

**10.15** Tennis enthusiasts in a small town want indoor courts to allow them to play tennis during bad weather. Someone suggested that they band together to form a corporation to build and operate a profit-making tennis facility. Two courts with a small lounge could be built for \$125,000, not including land. The facility would have a 10-year life and a salvage value of \$20,000. Annual operating expenses would be \$23,000.

**10.15a** At a charge of \$7 per hour for playing time on a court, how many hours would the courts have to be rented each year for the investors to break even on construction and operating costs, assuming a rate of return of 10 percent?

**10.15b** Develop a graph of the utilization rate required to yield rates of return between 10 and 25 percent. Assume the facility will be open 14 hours per day for 320 days each year. Would you recommend the investment? Why?

**10.16** Three designs to perform the same function have the cost patterns indicated:

*Design 1* Initial cost is $12,000, and annual expenses are uniform at $4000 per year.

*Design 2* A low initial cost of $5000 is possible because refinements will be made while the facility is in operation. The first year's operating and refinement cost will be $7500, and it will decrease by $500 each year.

*Design 3* Initial cost is $15,000, with annual disbursements that start at $2000 and increase by $1000 each year.

The principal uncertainty in the evaluation is how long the function will be needed for which the designs were developed. All the designs could perform the function for 12 years, if the function lasts that long, and none of the designs would have any salvage value at any time. Determine the range of life over which different designs would be preferred when the minimum attractive rate of return is 11 percent.

**10.17** An aluminum company has to decide whether to install a new type of "air float" conveyor for extra thin aluminum sheets or to retain their conventional conveyors. A pilot test reveals that the float conveyor moves the sheets faster and reduces damage, but maintenance costs to keep it operating properly may be high. The amount of savings expected from the new design depends on the quantity of aluminum sheeting produced in the future and the reliability of the new equipment. The engineering department has provided the following estimates:

| | |
|---|---|
| First cost | $180,000 |
| Economic life | 4 years |
| Annual maintenance expense | $40,000 |
| Annual savings | $100,000 |

There is no realizable salvage value because the cost of removing the equipment would about equal the scrap value. The company uses a minimum attractive rate of return of 12 percent.

**10.17a** Because there is still some doubt about the effectiveness of the air float conveyor design, some of the analysts feel a higher rate of return should be required for the project. Make an isoquant of the maximum first cost that could be incurred to earn rates of return between 5 and 25 percent.

**10.17b** Assuming the life, maintenance cost, and savings may vary as much as 50 percent on both sides of the given estimate, develop a graph of the effects of individual variations on the net present worth of the project. Do you recommend that the new conveyor system be installed? Why?

## EXTENSION

***10.1 Learning Curves*** Some types of repetitive operations exhibit decreasing cost per unit as the production run continues. The declining costs are attributable to experience gained by individual operators and by engineering-management support teams. Studies have shown that about 15 percent of the total time or effort reduction comes from improved operator performance, 35 percent from industrial-engineering cost-reduction activities, and the rest from redesigns or process improvements by product engineering. The total effect of the experience is called *learning*.

Learning curves were first identified by the airframe industry in the 1940s. Since then they have been found applicable in many high-cost, low-volume industries, such as shipbuilding, computers and electronics, machine tools, and building or plant construction. In cost estimating, the existence of learning can have a major effect. Bids or price quotes should be lower on successive orders of major items that benefit from learning. Make-or-buy decisions can be affected by the learning rate in the manufacturing process.

An 80 percent learning curve is a fairly representative figure for many large-scale repetitive projects. This learning rate means that there will be a 20 percent reduction in worker-hours per unit between *doubled* units. That is, if it took 100,000 hours to produce unit 1, it would take 100,000(100 percent − 20 percent) = 100,000(0.8) = 80,000 hours to produce unit 2. Then unit 4 would take 80,000(0.8) = 64,000 hours; unit 8 (double unit 4) would take 64,000(0.8) = 51,200 hours; and so on. This diminishing marginal-improvement pattern reveals the characteristics of a process subject to learning:

1 The time to complete each unit is less than the time to complete the previous unit.
2 Unit times decrease at a decreasing rate.

Note that learning rates usually refer to operational times rather than costs. These times are normally proportional to direct labor costs, but other production costs such as materials and indirect labor are likely paced by other factors.

The relationship between direct labor hours and the number of units produced is

$$Y_N = KN^X$$

*where* $Y_N$ = effort per production unit to produce the $N$th unit, as in direct labor hours/unit

$K$ = effort to produce the first unit, in dimensions compatible to $Y_N$

$N$ = unit number: 1, 2, etc.

$X$ = improvement function = [log (learning rate)]/log 2, as in (log 0.8)/0.3010

Table 10.4 shows values of the improvement function $x$ as it is related to various learning rates. For the product that took 100,000 hours for the first unit, the time to produce the eighth unit when the learning rate is 0.8 is calculated as

$$Y_8 = 100{,}000(8)^{\log 0.8/\log 2} = 100{,}000(8)^{-0.322}$$
$$= 51{,}200 \text{ hours}$$

| *Learning Rate* | *Improvement Function* |
|---|---|
| 1.0 (no learning) | 0 |
| 0.95 | −0.075 |
| 0.90 | −0.152 |
| 0.85 | −0.234 |
| 0.80 | −0.322 |
| 0.75 | −0.415 |
| 0.70 | −0.515 |
| 0.65 | −0.621 |
| 0.60 | −0.737 |
| 0.55 | −0.861 |
| 0.50 | −1.000 |

**TABLE 10.4** Learning rates and associated improvement-function ($x$) values.

Individual and cumulative times to produce 10 units at this improvement function are given in Table 10.5.

**TABLE 10.5** Unit, cumulative, and cumulative average time or effort required to produce 10 units of a product when the first unit took 100,000 hours and the learning rate is 80 percent (improvement function = −0.322).

| *Unit Number, N* | *Unit Production Effort, $Y_N$* | *Cumulative Production Effort* | *Cumulative Average Production Effort* |
|---|---|---|---|
| 1 | 100,000 = $K$ | 100,000 | 100,000 |
| 2 | 80,000 | 180,000 | 90,000 |
| 3 | 70,210 | 250,210 | 83,403 |
| 4 | 64,000 | 314,210 | 78,553 |
| 5 | 59,560 | 373,770 | 74,754 |
| 6 | 56,170 | 429,940 | 71,657 |
| 7 | 53,450 | 483,390 | 69,056 |
| 8 | 51,200 | 534,590 | 66,824 |
| 9 | 49,290 | 583,880 | 64,876 |
| 10 | 47,650 | 631,530 | 63,153 |

When it appears that learning is taking place in a production run, the rate can be calculated from any two unit times, $Y_i$ and $Y_j$. Suppose the only data available about a process are that the fifth unit took 59,560 hours to produce and the ninth unit took 49,290 hours. Then $Y_5 = K(5)^x$ is divided by $Y_9 = K(9)^x$ to obtain

$$\frac{Y_i}{Y_j} = \frac{K(N_i)^x}{K(N_j)^x} \quad \text{or} \quad \frac{59{,}560}{49{,}290} = \left(\frac{5}{9}\right)^x$$

The logs of both sides of the equation are taken as

$$\log \frac{Y_i}{Y_j} = x \log \frac{N_i}{N_j}$$

from which

$$x = \frac{\log Y_i - \log Y_j}{\log N_i - \log N_j} = \frac{\log 59{,}560 - \log 49{,}290}{\log 5 - \log 9}$$

$$= \frac{4.7750 - 4.6928}{0.6990 - 0.9542} = \frac{0.0822}{-0.2552} = -0.322$$

The learning rate is then determined from

$$\log \text{(learning rate)} = x \log 2$$
$$= -0.322(0.301) = 9.9031 - 10$$

where the antilog of both sides reveals the learning rate is 80 percent, as expected from the values in Table 10.5.

**QUESTIONS**

**10.1a** A condominium complex is to be composed of five identical housing units. Direct labor costs for the first unit are $531,000. It is believed that costs for the remaining units will follow a 90 percent learning curve. What is the estimated total direct labor cost for the condominium complex?

*($2,304,009)*

**10.1b** A contractor is supplying 20 custom rigs to your organization. The ninth rig required 59,800 hours of direct labor, and the eighteenth rig required 50,850 hours. Work is now under way on the twentieth rig, and you find that you will need one additional rig beyond the contracted 20. How many direct labor hours should be anticipated for the construction of the twenty-first rig? (*Hint*: Determine $x$, then $K$, then $Y_{21}$.)

($x = -0.234$; $K = 100{,}000$; $Y_{21} = 49{,}046$)

**10.1c** Units produced under proposal $A$ are expected to have a 90 percent learning rate with the first unit costing $12,000. The first unit produced under proposal $B$ will cost $18,000, but production is expected to have an 80 percent learning rate. What would be the cost for the fifteenth unit produced by each proposal? Assume costs are for direct labor only.

[$Y(A)_{15} = \$7951$; $Y(B)_{15} = \$7426$]

# CHAPTER 11

# DEPRECIATION AND TAX EFFECTS

If, indeed, "the name of the game is business and you keep score with money," then the rules are set by government, the scorekeepers are accountants, and the referee is the tax collector. An engineering economist acts as a playmaker.

Engineering economists do not have to become accountants to carry out economic evaluations, but they should be aware of the accounting function and how it can contribute to better analyses. An accounting system records financial data. These data are often raw materials for estimates of future costs, as described in the preceding chapter. Accounting records are also scorecards that show what actually happened to a proposal after it was accepted and augmented. Postaudits that compare accounting records of actual receipts and disbursements with estimated results disclose any weaknesses in the estimating procedures.

Provisions for recovering capital invested in income-producing assets are made by charging depreciation against current income. These depreciation charges and the regular operating expenses are deductible from gross income in determining taxable income. The type of depreciation method used for tax purposes affects the timing and amount of tax payments, which in turn affect the after-tax worth of proposals. Many special provisions in tax regulations influence investments. After-tax comparisons of alternatives occasionally change the preference indicated by before-tax evaluations, but the main purpose of an after-tax analysis is to determine the actual amount of capital available after the government has taken its toll. This amount is the real score of the business game.

The links between accounting, depreciation, and taxes are examined in this chapter.

# ACCOUNTING

An accountant and an engineering economist may work with the same financial data, but they do so for different purposes. The engineering economist deals with the future to evaluate proposed courses of action. Each proposal is analyzed with respect to its future cash flow and potential for earning a certain rate of return.

The accountant deals with transactions that have already occurred to determine what *was* the return on capital. The accountant does not have an entry for the cost of capital, unless there is an actual disbursement to be made, such as interest due on a loan. The cost of money is implied from the net profit of all operations combined. The views of the accountant and engineering economist coincide when a postaudit is conducted to see how closely actual figures came to figures estimated during the evaluation of a proposal; then both are looking at what happened and how it affected the economic health of the organization.

## Balance Sheet

The fundamental accounting equation is

$$\textit{Assets} - \textit{liabilities} = \textit{net worth}$$

Periodically, a *balance sheet* is prepared to show the monetary values of the elements in the accounting equation. This statement is accurate only for the date given on the balance sheet, because specific amounts vary from day to day. It is a measure of financial health in the same way a report of an annual medical checkup is a measure of an individual's health. Both statements might be different if prepared on a different day, because both are summaries of all contributing factors to the day of preparation. These summary effects make the statements analytically valuable as readings on current conditions, and, when compared to previous statements, as an indicator of what to expect in the future. (Extension 11.1 illustrates the potential of analyzing financial ratios.)

An abbreviated balance sheet is shown in Figure 11.1. The left side tabulates the assets that total to an amount equal to the sum of the liabilities and net worth on the right side. The *assets* fall into three categories that represent the resources of the organization:

*Current assets* Those that will become cash within 1 year (materials, credit due, etc.)

*Fixed assets* Those that cannot be readily converted to cash (mortgages, vehicles, buildings, etc.)

*Other assets* Intangible or difficult to classify (prepaid insurance, patents, etc.)

*Liabilities* are obligations of known amounts owed to creditors. They are usually listed in order of due dates; *current liabilities* are payable within 1 year, and *long-term liabilities* are not due for over a year. The *net worth* (of stockholders' equity) represents the current worth of investments in the organization. *Retained earnings* are profits from past operations which belong to the stockholders and have been retained within the organization for reinvestment.

Balance Sheet
June 30, 19xx

| ASSETS | | LIABILITIES | |
|---|---|---|---|
| *Current assets* | | *Current liabilities* | |
| Cash | $ 5,400 | Accounts payable | $35,200 |
| Accounts receivable | 9,000 | Bank loans | 9,000 |
| Raw-material inventory | 11,000 | Accrued taxes | 4,800 |
| Work in progress | 10,000 | *Long-term liabilities* | |
| Finished products | 16,600 | Bonds (due in 12 years). | 15,000 |
| *Fixed assets* | | | $64,000 |
| Land | 37,000 | NET WORTH | |
| Buildings and equipment. | 83,000 | Capital stock | $100,000 |
| *Other assets* | | Retained earnings | 24,000 |
| Prepaid services | 16,000 | | $124,000 |
| | $188,000 | | $188,000 |

**FIGURE 11.1** Balance sheet showing an organization's major accounts as of a specific date.

Fixed assets are usually shown at their current value as determined by the difference between what was paid for the assets minus accumulated depreciation. If the buildings and equipment in Figure 11.1 had originally cost $100,000 and had since been depreciated by $17,000, the entry in the balance sheet could have taken the form:

| *Fixed assets* | | |
|---|---|---|
| Land | | $37,000 |
| Buildings and equipment | $100,000 | |
| Less: reserve for depreciation | 17,000 | |
| | | $83,000 |

There are many other variations and different headings for accounts in balance sheets. Since all the accounts are summaries, supplementary accounting records are necessary to observe the cash flow.

---

**Economy Exercise 11-1** Sam Sales is an engineering sales representative for an air-conditioning manufacturer. He is paid a commission on the amount of commercial air-conditioning equipment he sells. As he traveled within his assigned territory calling on prospective customers, he became fascinated by the variety of novelties displayed near the cash registers at restaurants and stores.

One day Sam saw a person filling up one of the novelty displays and started talking to him. Before they were done talking, Sam agreed to purchase the novelty distributorship. The seller wanted to go to a different climate, and Sam figured he could combine the novelty business with his other travels.

The distributorship was priced at $23,000 and included an exclusive franchise for Freaky postcards, $14,200 in inventory stored in a warehouse where rent was prepaid for 6 months at $200 per month, and accounts receivable of $3100 from ongoing sales. Sam agreed to take over the accounts payable of $1900 and accrued taxes of $300 from the seller.

Sam met the $23,000 selling price by putting up $5000 from his savings, borrowing $8000 on a 2-year bank loan, and getting the former owner to take an unsecured note promising to pay the remainder over 3 years. Then Sam withdrew $10,000 more from his savings for *working capital* to handle day-to-day cash transactions for his venture, now called "Sales' Gimmicks." Construct a balance sheet for the new company as of the current date, April 1, 19x0.

## Income Statement

A statement of the earnings of an organization over a stated period of time is called an *income statement* or a *profit-and-loss statement.* The net profit or loss it discloses represents the difference between revenue from the sale of goods or services and the sum of all the operating expenses, including taxes. The net profit may be held as retained earnings or distributed to the owners. An abbreviated income statement is shown in Figure 11.2. Its development precedes the balance sheet because the bottom-line figure, net income after tax, is an entry in the balance sheet under net worth. Both the income statement and the balance sheet are needed to understand an organization's financial position, and then only the more prominent features are revealed.

Income Statement
Year Ended June 30, 19xx

| | | |
|---|---|---|
| *Income* | | |
| Product sales | $247,500 | |
| Fees for services | 16,500 | $264,000 |
| *Expenses* | | |
| Cost of goods sold | $119,000 | |
| Administrative | 82,000 | |
| Depreciation | 9,200 | |
| Utilities | 6,000 | |
| Other (insurance, rent, etc.) | 7,800 | $224,000 |
| Income before income tax | | $40,000 |
| Federal income tax (at 40%) | | 16,000 |
| *Net income after tax* (to retained earnings) | | $24,000 |

**FIGURE 11.2** Income statement for the same organization during the same time period represented by the balance sheet in Figure 11.1.

## Cost Accounting

The major accounts displayed in the balance sheet and income statement are composed of many smaller accounts, each of which is, in turn, composed of more detailed items. For instance, the cost of goods sold in the income statement includes direct labor, direct material, and factory overhead. More detailing breaks down the costs of direct labor by craft and work assignment, and the cost of material by type, use, and timing. This type of information is in the accounting system and recorded because the summarizing balance sheets and income statements are built up from the elemental cost data. The collection of these costs for producing a product or service is called *cost accounting*.

The relationship of contributing costs to the selling price of an item was given in Figure 10.6. The cost of goods sold category is particularly useful to an engineering economist because it contains data that can be extracted and interpreted to estimate costs for proposals that alter the production process. The cost data about the current process are applicable to the do-nothing or defending alternative, and *predictive* cost accounting estimates the cost of future operations of the same nature from past performance.

---

**Economy Exercise 11-2** After Sam Sales bought the novelty business, he laid out a year-long operating plan in the form of a monthly budget. He knew that his postcards and other novelties would enjoy their highest sales during the summer. He also realized that he would not have time to service all his outlets during the busy season while continuing his primary job in engineering sales, and he was not nearly confident enough in his Gimmicks Company to rely on it alone for a living.

He could hire helpers part time in the summer as route servicers, but he was worried about his cash-flow position. With only $10,000 for working capital, he could not afford to build up his inventory levels too high or he would run out of cash to pay the bills. The results of juggling amounts between cost categories and the time periods are shown in the monthly budget accounts in Table 11.1. The plan anticipates a decline in the postcard inventory during the winter to stock up on novelties. During the summer the monthly receipts would pay for the additional help, and the cash balance should not drop below $4000.

Develop an income statement for the year by assuming that the actual receipts and disbursements came out exactly as budgeted in Table 11.1 for the first year's operation of Sales' Gimmicks. The tax rate is 20 percent on the $18,500 income, and the after-tax total is held as retained earnings.

---

SALES' GIMMICKS COMPANY MONTHLY BUDGET

| | *Jan.* | *Feb.* | *March* | *April* | *May* | *June* |
|---|---|---|---|---|---|---|
| *Sales* | | | | | | |
| Postcards | $6000 | $5,000 | $6,000 | $6,000 | $ 8,000 | $9,000 |
| Other novelties | | 500 | 1,000 | 1,500 | 2,000 | 3,000 |
| Cash received | $6000 | $5,500 | $7,000 | $7,500 | $10,000 | $12,000 |
| *Disbursements* | | | | | | |
| Postcards | 1000 | | 2,000 | 4,000 | 6,000 | 12,000 |
| Other novelties | 3000 | 1,000 | 1,000 | 500 | 500 | 1,000 |
| Route servicers' salaries | 600 | 600 | 600 | 600 | 1,400 | 2,100 |
| Travel expense | 1000 | 1,000 | 1,000 | 1,000 | 1,000 | 1,300 |
| Office salaries | 450 | 450 | 450 | 450 | 450 | 700 |
| Office expense | 100 | 100 | 100 | 150 | 150 | 200 |
| Other expense | 180 | 180 | 180 | 220 | 220 | 280 |
| Loan repayment | | | | 1,000 | | |
| Cash disbursements | $6330 | $3,330 | $5,330 | $7,920 | $9,720 | $17,580 |
| Gain (or loss) from operations | (330) | 2,170 | 1,670 | (420) | 280 | (5,580) |
| Cash balance ($10,000) | $9670 | $11,840 | $13,510 | $13,090 | $13,370 | $ 7,790 |

**TABLE 11.1** Monthly budget for cash receipts and disbursements for Sales' Gimmicks Company. Loan expense includes $2200 in interest and $9000 toward repayment. Before-tax income is $28,500 − $10,000 = $18,500.

| | *July* | *Aug.* | *Sept.* | *Oct.* | *Nov.* | *Dec.* |
|---|---|---|---|---|---|---|
| *Sales* | | | | | | |
| Postcards | $10,000 | $13,000 | $11,000 | $7,000 | $5,000 | $ 6,000 |
| Other novelties | 4,000 | 5,500 | 4,500 | 2,000 | 2,000 | 4,000 |
| Cash received | $14,000 | $18,500 | $15,500 | $9,000 | $7,000 | $10,000 |
| *Disbursements* | | | | | | |
| Postcards | 6,000 | 4,000 | 2,000 | 2,000 | | 1,000 |
| Other novelties | 2,000 | 1,000 | 1,000 | 1,000 | 1,000 | 200 |
| Route servicers' salaries | 2,100 | 2,100 | 1,950 | 650 | 650 | 650 |
| Travel expense | 1,300 | 1,300 | 1,100 | 600 | 600 | 800 |
| Office salaries | 700 | 700 | 700 | 450 | 450 | 1,050 |
| Office expense | 200 | 200 | 200 | 200 | 200 | 300 |
| Other expense | 640 | 340 | 280 | 220 | 1,040 | 220 |
| Loan repayment | 4,700 | | | 1,000 | | 4,500 |
| Cash disbursements | $17,640 | $ 9,640 | $ 7,230 | $ 6,120 | $ 3,940 | $ 8,720 |
| Gain (or loss) from operations | (3,640) | 8,860 | 8,270 | 2,880 | 3,060 | 1,280 |
| Cash balance ($10,000) | $ 4,150 | $13,010 | $21,280 | $24,160 | $27,220 | $28,500 |

# DEPRECIATION

Depreciation means a decrease in worth. Most assets are worth less as they get older. Newly purchased production assets have the advantage of latest technical improvements and operate with less chance of breakdown or need for repairs. Except for possible antique value, production equipment gradually becomes less valuable through wear. This lessening in value is recognized in accounting practices as an expense of operating. Instead of charging the full purchase price of a new asset as a one-time expense, the outlay is spread over the life of the asset in the accounting records. This concept of amortization may seem to disagree with the actual cash flow for a particular transaction, but for all transactions taken collectively it provides a realistic representation of capital consumption in profit-and-loss statements.

## Reasons for Depreciation Accounting

The costs of production can be divided into *direct* money costs—wages, salaries, raw materials, etc.—and *imputed* costs—payments not contractually required for production during a short accounting period. In financial accounting, depreciation is an imputed cost. The principal objectives for charging a depreciation cost can be summarized as (1) to recover capital invested in production assets, (2) to accurately determine imputed costs of production for cost records, and (3) to include the cost of depreciation in operating expenses for tax purposes.

To emphasize the importance of depreciation, consider the plight of an individual who used $8500 in savings and borrowed $10,000 to buy a used tractor, truck, and hauling rig to start a landscaping service. Business was good. A gross income of $20,000 per year allowed annual operating and loan-repayment expenses of $9000 to be met. After 5 years of comfortable living with good wages, the loan was repaid but the equipment was worn out. Its salvage value after the 5 years of use was only $500. When it was time to purchase replacement equipment, the original $8500 was no longer available to use as a down payment. The investment had not been recovered from the earnings it afforded.

From the vantage point of hindsight, it is obvious that the landscaper should have made provisions to set aside $8500 − $500 = $8000 during the life of the equipment. This would have shown the actual earnings during the period, and allowed capital consumption to be shown as an expense on tax reports.

## Depreciation Accounts

In making depreciation studies it is convenient to visualize a charge for depreciation as being a series of payments made to a specific fund for the replacement of the asset being studied. While this notion is quite reasonable in concept, it is seldom followed in industrial practice. A bookkeeping account shows the annual charge for depreciation; the charge is used for tax purposes, but it appears in accounts as "other assets," such as working capital. The physical form of the "depreciation fund" could be stocks of raw materials or finished products.

An exception to the practice of keeping accounts of depreciation funds by book entries is the occasionally used external sinking fund. In this depreciation method a separate reserve is established by payments invested outside the company. The interest earned by the fund is the interest paid by the organization that holds the investments. The annual payments are determined by the sinking-fund formula $(A/F, i, N)$, where $F$ is the amount needed to replace the asset, $N$ is the economic life, and $i$ is the interest received on the invested funds. A firm that is having internal financial troubles may be required by courts to temporarily set up such a sinking fund for its depreciation reserve. Otherwise, a sinking-fund depreciation schedule is seldom used in private industry.

For depreciation allowances kept within an organization, a charge is made each year against the cost of operations and is credited to the asset's reserve for depreciation. The original cost of the asset minus the accumulated depreciation is called the ***book value.*** Land is one of the few assets for which no reserve is needed, because land values normally remain constant or appreciate. Therefore, any portion of an investment representing land is deducted from the original cost when making depreciation calculations.

## Causes of Declining Value

Being aware of the potential causes of decreasing worth may assist in determining the most appropriate depreciation schedule for an asset.

1 *Physical depreciation* The everyday wear and tear of operation gradually lessens the physical ability of an asset to perform its intended function. A good maintenance program retards the rate of decline, but it seldom maintains the precision expected from a new machine. In addition to normal wear, accidental physical damage can also impair ability.

2 *Functional depreciation* Demands made on an asset may increase beyond its capacity to produce. A central heating plant unable to meet the increased heat demands of a new building addition no longer serves its intended function. At the other extreme, the demand for services may cease to exist, as with a machine which produces a product no longer in demand.
3 *Technological depreciation* Newly developed means of accomplishing a function may make the present means uneconomical. Steam locomotives lost value rapidly as railroads turned to diesel power. Current product styling, new materials, improved safety, and better quality at lower cost from new developments make old designs obsolete.
4 *Depletion* Consumption of an exhaustible natural resource to produce products or services is termed *depletion*. Removal of oil, timber, rock, or minerals from a site decreases the value of the holding. This decrease is compensated for by a proportionate reduction in earnings derived from the resource. Theoretically, the depletion charge per unit of the resources removed is

$$\frac{\text{Present value of resource}}{\text{Remaining units of resource}} = \text{depletion rate (\$/unit)}$$

In practice, the depletion rate is largely set by the percentage of a year's income allowed for a depletion allowance by the Internal Revenue Service. Allowances for depletion vary with the type of resource and with the most recent legislation. Highest allowances are theoretically allowed for the resources which require the greatest expenditures for discovery and development.
5 *Monetary depreciation* A change in price levels is a subtle but troublesome cause of decreases in the value of depreciation reserves. Customary accounting practices relate depreciation to the original price of an asset, not to its replacement. If prices rise during the life of an asset, as in the case of recent high inflation rates, a comparable replacement becomes more expensive. This means that the capital recovered will be insufficient to provide an adequate substitute for the worn-out asset. It also suggests that the selling price of the product being produced by the asset does not accurately reflect the cost of production. Because the depreciation is actually happening to the invested capital representing the asset instead of to the asset itself, monetary depreciation is very difficult to accommodate. It cannot be charged as an operating expense for tax purposes.

## DEPRECIATION METHODS

Of the many depreciation methods available, four will be examined in the following pages: *straight line, sum of digits, declining balance,* and *sinking fund*. These methods are based strictly on time. That is, an asset used every day has the same depreciation charge as one used only once per year. Some advocate that depreciation should be based on the amount of use as well as the economic life. It is possible to combine a usage factor with a time-based method such as straight-line depreciation to reflect the rate of decreasing value with use. This approach is acceptable for tax purposes as long as it can definitely be proved that the useful life is a function of the rate of production.

Each depreciation method has unique features which appeal to different management philosophies. A method by which the bulk of the money invested is recovered early in the life of an asset is a popular conservative view. An early write-off guards against sudden changes which could make the equipment less valuable and shifts some taxes toward later years. Methods in which the annual charge is constant simplify the accounting procedure. In general, the desirable features of a depreciation method are that it (1) recovers the capital invested in an asset, (2) maintains a book value close to the actual value of the asset throughout its life, (3) is easy to apply, and (4) is acceptable to the Internal Revenue Service.

Broad patterns of capital recovery for the four methods to be considered are shown in Figure 11.3. The curves are based on the sample data in Table 11.2 and indicate the book value any time during the life of the asset. The steeper the curve, the more rapid the depreciation. The sinking-fund method has the slowest rate of capital recovery. If the interest rate used for sinking-fund calculations is zero, $(A/F, 0, N)$, the sinking-fund curve coincides with the values for the straight-line method. Both the sum-of-digits and declining-balance methods recover a large share of the initial investment early in the depreciable life. In the first half of an asset's economic life, about 70 percent of the depreciation cost is written off by the sum-of-digits method, and 80 percent is written off by the double-declining-balance method.

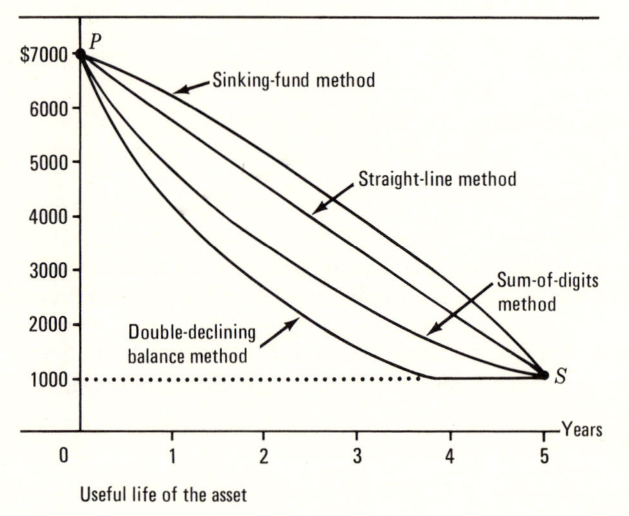


**FIGURE 11.3** Book values of the asset described in Table 11.2 when different depreciation methods are utilized.

All the methods will be illustrated by reference to the problem information shown in Table 11.2.

**TABLE 11.2** Data and problem statement for Examples 11.1 to 11.4.

Trucks purchased by a delivery company cost $7000 each. Past records indicate the trucks should have a useful life of 5 years. They can be sold for an average of $1000 each after 5 years of use. The company currently receives 7 percent interest on invested funds. Determine:

**a** The depreciation charge during year 1
**b** The depreciation charge during year 2
**c** The depreciation reserve accumulated by the end of year 3
**d** The book value at the end of year 3

The symbols used in the development of the formulas are

$P$ = purchase price (present worth at time zero) of asset
$S$ = salvage value or future value at end of asset's useful life
$N$ = useful life of asset
$n$ = number of years of depreciation or use from time of purchase
DC = annual charge for depreciation
BV = book value shown on accounting records

## Straight-Line Method

Straight-line depreciation is the simplest to apply and the most widely used of the depreciation methods. The annual depreciation is constant. The book value is the difference between the purchase price and the product of the number of years of use times the annual depreciation charge:

$$\text{DC} = \frac{P - S}{N}$$

$$\text{BV(end of year } n) = P - \frac{n}{N}(P - S)$$

**▶ Example 11.1** **Straight-Line Depreciation Applied to Basic Data**

**▶ Solution 11.1**

**a and b** Since the annual depreciation cost is constant, the charges for both the first and second year are

$$\text{DC} = \frac{P - S}{N} = \frac{\$7000 - \$1000}{5} = \$1200/\text{year}$$

**c** The depreciation reserve at the end of the third year is the sum of the annual depreciation charges for the first 3 years and is equal to $3 \times \$1200 = \$3600$.

**d**
$$\text{BV}(3) = \$7000 - \frac{3}{5}(\$7000 - \$1000)$$
$$= \$7000 - 0.6(\$6000) = \$3400$$

or, with the book value considered as the difference between the purchase price and the amount accumulated in the depreciation reserve,

$$\text{BV}(3) = \$7000 - 3(\$1200) = \$3400$$

## Sum-of-Digits Method

The sum-of-digits method provides a larger depreciation charge during the early years of ownership than in the later years. The name is taken from the calculation procedure. The annual charge is the ratio of the digit representing the remaining years of life $(N - n + 1)$ to the sum of the digits for the entire life ($1 + 2 + 3 + \cdots + N$) multiplied by the initial price minus the salvage value ($P - S$). Thus, the annual charge decreases each year from a maximum the first year:

$$\text{DC} = \frac{N - n + 1}{1 + 2 + 3 + \cdots + N}(P - S)$$

$$= \frac{2(N - n + 1)}{N(N + 1)}(P - S)$$

$$\text{BV}(n) = \frac{2[1 + 2 + \cdots + (N - n)]}{N(N + 1)}(P - S) + S$$

**Example 11.2** **Sum-of-Digits Depreciation Applied to the Basic Data**

**Solution 11.2**

**a** The sum of digits for the 5-year useful life is

$$1 + 2 + 3 + 4 + 5 = 15$$

or,

$$\frac{N(N + 1)}{2} = \frac{5(5 + 1)}{2} = 15$$

which is the denominator of the formula for

$$\text{DC}(1) = \frac{N - n + 1}{15}(P - S) = \frac{5 - 1 + 1}{15}(\$7000 - \$1000)$$

$$= \frac{5}{15}\$6000 = \$2000$$

**b** After the first year, only 4 years remain in the useful life. Therefore, with $N - n + 1 = 5 - 2 + 1 = 4$,

$$\text{DC}(2) = \frac{4}{15}\$6000 = \$1600$$

**c** The ratio for calculating the depreciation reserve has a numerator equal to the sum of digits representing the years during which the reserve was built up:

$$\text{Depreciation reserve at end of year 3} = \frac{5 + 4 + 3}{15}\$6000 = \$4800$$

**d** $\text{BV}(3) = P - \text{depreciation reserve at end of year 3}$
$= \$7000 - \$4800 = \$2200$

or, by formula,

$$\text{BV}(3) = \frac{2[1 + 2 + \cdots + (N - n)]}{N(N + 1)}(P - S) + S$$

$$= \frac{2[1 + (5 - 3)]}{5(5 + 1)} \$6000 + \$1000$$

$$= \frac{6}{30} \$6000 + \$1000 = \$1200 + \$1000 = \$2200$$

## Declining-Balance Method

The declining-balance method is another means of amortizing an asset at an accelerated rate early in its life, with corresponding lower annual charges near the end of service. An important point with this method is that the salvage value must be greater than zero. A depreciation rate is calculated from the expression

$$\text{Depreciation rate} = 1 - \left(\frac{S}{P}\right)^{1/N}$$

which requires a positive value for $S$ in order to be realistic. This constant rate is applied to the book value for each depreciation period. Since the undepreciated balance decreases each year, the depreciation charge also decreases, and

$$\text{BV}(n) = P(1 - \text{depreciation rate})^n$$

$$= P\left\{1 - \left[1 - \left(\frac{S}{P}\right)^{1/N}\right]\right\}^n$$

$$= P\left(\frac{S}{P}\right)^{n/N}$$

$$\text{DC}(n) = \text{BV}(n - 1)\left(1 - \sqrt[N]{\frac{S}{P}}\right)$$

A much more widely used version of the declining-balance method, allowed by the income-tax code since 1954, is based on a depreciation rate which does not depend on the $S/P$ ratio. Under certain circumstances a rate is allowed that is twice as great as would be proper under the straight-line method. Under other circumstances, the rate is limited to 1.5 or 1.25 times that of the straight-line method.

When the maximum rate is used,

$$\text{Depreciation rate}_{\text{max}} = \frac{200\%}{N}$$

it is called the ***double-declining-balance method*** of depreciation. It has the same characteristics as the declining-balance method.

**Example 11.3** **Double-Declining-Balance Depreciation Applied to the Basic Data**

**Solution 11.3** a Given that depreciation $\text{rate}_{\text{max}} = 200\%/5 = 40\%$ or 0.4,

$$\text{DC}(1) = P(0.4) = \$7000(0.4) = \$2800$$

**b** $DC(2) = BV(1)(0.4) = (\$7000 - \$2800)(0.4)$
$= \$4200(0.4) = \$1680$

**c** The depreciation reserve at the end of year 3 is the sum

$$DC(1) + DC(2) + BV(2)(0.4) = \$2800 + \$1680 + \$2520(0.4) = \$4480 + \$1008 = \$5488$$

**d** $BV(3) = P$ − depreciation reserve $= \$7000 - \$5488 = \$1512$

It is important to note that the book value at the end of year 5 would be, by formula,

$$BV(5) = P(1 - \text{depreciation rate})^5 = \$7000(0.6)^5 = \$544$$

which is less than the asset will likely bring in salvage value. Tax regulations do not allow property to be depreciated below its reasonable salvage value. Therefore, the allowable depreciation charge for the fourth year could be $\$1512 - \$1000 = \$512$, which makes $BV(4) = \$1000$, and maintains the estimated salvage value when no further depreciation is charged.

## Sinking-Fund Method

The sinking-fund concept has already been introduced on page 204 and by compound-interest calculations. The sinking-fund model is used by some government agencies as a realistic representation of public projects that lose value slowly during the first years and more rapidly during later years. The annual depreciation charge is constant *if earned interest is not included.* The amount of accumulated depreciation reserve is equal to the future worth of the series of uniform depreciation-charge payments at a given date. Then the book value is the difference between the purchase price and the depreciation reserve. Book values according to the sinking-fund method are greater than they would be by straight-line depreciation. Increasing the interest rate increases the difference.

**Example 11.4** **Sinking-Fund Depreciation Applied to the Basic Data**

**Solution 11.4** **a and b** The constant annual charge for depreciation is

$$DC = (P - S)(A/F, i, N) = (\$7000 - \$1000)(A/F, 7, 5) = \$6000(0.17389) = \$1043$$

However, it should be recognized that when a sinking fund is accumulated within a firm, funds must be allocated to account for the interest earned by the depreciation reserve. Thus, during the second year the capital recovery includes the depreciation charge (\$1043) plus the interest earned by the first year's depreciation charge ($\$1043 \times 0.07 = \$73$).

**c** The depreciation reserve at the end of year 3 is

$$(P - S)(A/F, i, N)(F/A, i, n) = \$6000(A/F, 7, 5)(F/A, 7, 3) = \$1043(F/A, 7, 3) = \$1043(3.2148) = \$3553$$

**d** The book value at the end of year 3 is

$$P - \text{accumulated depreciation} = \$7000 - \$3553 = \$3447$$

## Equivalence of Depreciation Methods When Interest Charges Are Included

The time-value equivalence of different methods is an interesting aspect of depreciation accounting. Interest charges are not formally included in the straight-line, sum-of-digits, or declining-balance method. Even in sinking-fund calculations the interest factor does not account for the return that would be received if the funds were not invested in the depreciable asset. *If the return on invested capital is included, the different patterns share a common, equivalent annual cost.* The importance of this condition is that it allows one basic formula to represent any depreciation method when alternatives are being compared.

It is possible to derive the equivalent formula from the equations for each depreciation method. Instead of individual derivations, the equivalence is demonstrated by example in Table 11.3. Two depreciation methods are compared according to the present worth of annual capital recovery plus a return on the unrecovered balance of the investment. The return may be thought of as the "profit" that would result if the undepreciated portion of the funds for an asset were invested at a given rate of return.

| Year, N | BV at Beginning of Year N (1) | Return on Unrecovered Capital at 7% Interest [0.07 × (1)] (2) | Depreciation Charge in Year N (3) | Capital Recovery plus Return [(2) + (3)] (4) | (P/F, 7, N) (5) | PW of Payments [(4) × (5)] (6) |
|---|---|---|---|---|---|---|
| | | STRAIGHT-LINE DEPRECIATION METHOD | | | | |
| 1 | $7000 | $490 | $1200 | $1690 | 0.93458 | $1580 |
| 2 | 5800 | 406 | 1200 | 1606 | 0.87344 | 1403 |
| 3 | 4600 | 322 | 1200 | 1522 | 0.81630 | 1242 |
| 4 | 3400 | 238 | 1200 | 1438 | 0.76290 | 1097 |
| 5 | 2200 | 154 | 1200 | 1354 | 0.71399 | 965 |
| | | | *Total PW of payments plus interest on unrecovered balance* | | | $6287 |
| | | SUM-OF-DIGITS DEPRECIATION METHOD | | | | |
| 1 | $7000 | $490 | $2000 | $2490 | 0.93458 | $2327 |
| 2 | 5000 | 350 | 1600 | 1950 | 0.87344 | 1703 |
| 3 | 3400 | 238 | 1200 | 1438 | 0.81630 | 1174 |
| 4 | 2200 | 154 | 800 | 954 | 0.76290 | 728 |
| 5 | 1400 | 98 | 400 | 498 | 0.71399 | 355 |
| | | | *Total PW of payments plus interest on unrecovered balance* | | | $6287 |

**TABLE 11.3** Capital recovery plus return at 7 percent interest for straight-line and sum-of-digits depreciation.

The calculations in Table 11.3 are based on the problem data in Table 11.2. They demonstrate that the total present worth of payments plus a theoretical interest charge on the unrecovered balance total to the same amount for the two depreciation methods. This total, $6287, is obtained by applying the capital-recovery factor to determine the annual cost of ownership as

$(P - S)(A/P, 7, 5) + S(0.07) = (\$7000 - \$1000)(0.24389) + \$1000(0.07) = \$1533.34$

from which the total present worth of all payments is

$$\text{Present worth of five capital-recovery payments} = \$1533(P/A, 7, 5) = \$1533(4.1001) = \$6287$$

The significance of the agreement in PW of the totals in Table 11.3 and that determined by the standard discounted cash-flow model is that it proves discounted cash-flow comparisons provide equivalent results regardless of the depreciation method utilized in bookkeeping. The only exception is the after-tax evaluation discussed in the next section.

## DEPRECIATION AND TAXES

Depreciation is treated for tax purposes as a deductible expense of doing business. The taxation rules governing the deductions are elaborate and complex. Only very general observations are included here, in relation to selecting a depreciation method and using it to compute taxes.*

### Depreciable Property

The property on which depreciation may be claimed must contribute to income produced in a trade or business and have a useful life of more than 1 year. Tangible property may be depreciated to the extent that it is subject to wear and tear, to decay or decline from natural causes, to exhaustion, and to obsolescence. Buildings may be depreciated; land may not.

Intangible assets with a limited period of usefulness can be depreciated. These include patents, copyrights, franchises, and trademarks. The burden of proving the value of the intangibles for tax purposes rests with the owner. Straight-line depreciation is the only method allowed for intangible property.

### Useful Life and Salvage Value

The useful life of an asset for depreciation purposes is not necessarily the economic life, but is the period over which the asset may reasonably be expected to be useful in the production of income. The useful life depends on an asset's use, age when acquired, repair policy, and other factors. It may be modified as a result of obsolescence and other causes, apart from physical wear and tear, that actually diminish the value of the property or shorten its life.

Samples of useful life recommended under the Class Life Asset Depreciation Range system (ADR system) are given in Table 11.4. Under the ADR system, the allowance for depreciation of property is based upon the cost without deduction for salvage value. However, the allowed depreciation for a tax year may not exceed the cost less the sum of the depreciation reserve plus the salvage value. When, as occasionally happens, the depre-

*Based on *Publication 534* (revised October 1974), Department of the Treasury, Internal Revenue Service.

| *Description of Depreciable Assets* | ASSET DEPRECIATION RANGE, YEARS *Lower Limit* | *Guideline Period* | *Upper Limit* |
|---|---|---|---|
| *Transportation* | | | |
| Automobiles, taxis | 2.5 | 3 | 3.5 |
| Buses | 7 | 9 | 11 |
| General-purpose trucks: Light | 3 | 4 | 5 |
| Heavy | 5 | 6 | 7 |
| Air transport | 5 | 6 | 7 |
| *Petroleum* | | | |
| Exploration and drilling assets | 11 | 14 | 17 |
| Refining and marketing assets | 13 | 16 | 19 |
| *Manufacturing* | | | |
| Sugar and sugar products | 14.5 | 18 | 21.5 |
| Tobacco and tobacco products | 12 | 15 | 18 |
| Knitwear and knit products | 7 | 9 | 11 |
| Lumber, wood products, and furniture | 8 | 10 | 12 |
| Paper and paperboard | 13 | 16 | 19 |
| Chemicals and allied products | 9 | 11 | 13 |
| Cement | 16 | 20 | 24 |
| Fabricated metal products | 9.5 | 12 | 14.5 |
| Electrical equipment | 9.5 | 12 | 14.5 |
| Aerospace products | 6.5 | 8 | 9.5 |
| *Communication* | | | |
| Telephone: Central-office buildings | 36 | 45 | 54 |
| Distribution poles, cables, etc. | 28 | 35 | 42 |
| Radio and television broadcasting | 5 | 6 | 7 |
| Satellite space segment property | 6.5 | 8 | 9.5 |
| *Electric utility* | | | |
| Hydraulic plant | 40 | 50 | 60 |
| Nuclear plant | 16 | 20 | 24 |
| *Services* | | | |
| Office furniture and equipment | 8 | 10 | 12 |
| Computers and peripheral equipment | 5 | 6 | 7 |
| Data handling—typewriters, copiers, etc. | 5 | 6 | 7 |
| Recreation—bowling alleys, theaters, etc. | 8 | 10 | 12 |

**TABLE 11.4** Range of useful life allowed for the depreciation of selected classes of assets under the ADR system of the Internal Revenue Service, *Publication 534,* revised October 1974.

ciation reserve more than compensates for the loss in value of the asset, the excess is recognized as a gain in the tax year and is treated as ordinary income, or part of it may sometimes be considered as capital gain.

## Choice of Depreciation Method

According to IRS *Publication 534,* "Any reasonable method that is consistently applied may be used in computing depreciation. The three methods most generally used are: (1) straight line; (2) declining balance; and (3) sum of the years digits." The total tax paid under any of the methods is the same over the life of the asset for a given set of tax rates, income, and expenses. But the timing of the tax payments is different. The more rapid capital recovery by accelerated depreciation methods allows a higher rate of return on invested capital because the present worth of higher early after-tax returns is greater than the same total returns when a larger proportion comes later in the life. The time value of after-tax returns is illustrated in Table 11.5.

The table is based on an annual income of $500 after deducting operating expenses, but before deducting depreciation and income taxes, resulting from ownership of an asset that cost $1500 when new and has a useful life of 5 years with no salvage value. The effective tax rate is 40 percent. As is apparent in column 6, the total tax paid is the same whether depreciation is calculated by the straight-line or sum-of-digits method. However, as shown in column 8, the total present worth of the series of tax payments is less for the accelerated capital-recovery pattern of the sum-of-digits method. The difference when money is worth 10 percent is $304 − $274 = $30; the tax savings amount to about 10 percent for the faster write-off.

| *End of Year N (1)* | *Income before Taxes and Depreciation (2)* | *Annual Depreciation Charge (3)* | *Income less Depreciation [(2) − (3)] (4)* | *Income-Tax Rate (5)* | *Income Tax (4) × (5) (6)* | *Present-Worth Factor (P/F, 10, N) (7)* | *PW of Income Tax [(6) × (7)] (8)* |
|---|---|---|---|---|---|---|---|
| | | | STRAIGHT-LINE DEPRECIATION METHOD | | | | |
| 1 | $ 500 | $ 300 | $ 200 | 0.4 | $ 80 | 0.90909 | $ 73 |
| 2 | 500 | 300 | 200 | 0.4 | 80 | 0.82645 | 66 |
| 3 | 500 | 300 | 200 | 0.4 | 80 | 0.75132 | 60 |
| 4 | 500 | 300 | 200 | 0.4 | 80 | 0.68302 | 55 |
| 5 | 500 | 300 | 200 | 0.4 | 80 | 0.62092 | 50 |
| *Total* | $2500 | $1500 | $1000 | | $400 | | $304 |
| | | | SUM-OF-DIGITS DEPRECIATION METHOD | | | | |
| 1 | $ 500 | $ 500 | $ 0 | 0.4 | $ 0 | 0.90909 | $ 0 |
| 2 | 500 | 400 | 100 | 0.4 | 40 | 0.82645 | 33 |
| 3 | 500 | 300 | 200 | 0.4 | 80 | 0.75132 | 60 |
| 4 | 500 | 200 | 300 | 0.4 | 120 | 0.68302 | 82 |
| 5 | 500 | 100 | 400 | 0.4 | 160 | 0.62092 | 99 |
| *Total* | $2500 | $1500 | $1000 | | $400 | | $274 |

**TABLE 11.5** Comparison of taxes resulting from depreciating an asset ($P$ = $1500; $N = 5$; $S = 0$) with the straight-line and sum-of-digits methods when income is $500 per year. Both depreciation methods lead to the same total tax payments over the life of the asset, but the present worth of the taxes is less for the sum-of-digits method.

**Economy Exercise 11-3** Utilize the double-declining-balance method to depreciate the asset described in Table 11.5. Assume the income is still $500 per year and that you can "carry forward" any unused expenses from one year to the next. Compare the present worth of the taxes when the tax rate is 0.4 with those in Table 11.5.

## TAXES

To tax and to please, no more than to love and be wise, is not given to men *(Edmund Burke in a speech, "On American Taxation," in 1774).*

The art of taxation consists in so plucking the goose as to obtain the largest possible amount of feathers with the smallest possible amount of hissing *(attributed to Jean Baptiste Colbert, 1665).*

When I catch myself resenting not being immortal, I pull myself up short by asking whether I should really like the prospect of having to make out an annual income-tax return for an infinite number of years ahead *(Arnold J. Toynbee, in* Saturday Review, *1969).*

Death and Taxes are inevitable *(Richard Halliburton).*

## Types of Taxes

Federal, state, and sometimes city or county taxes are imposed on income, property, and/or transactions. The transfer of wealth through the taxing mechanism is a major concern of governments, and the payment of those taxes is a major concern of income producers, both corporate and individual. The principal types and their relevance to engineering economic studies are described below.

1. *Property taxes* are charged by local governments on land, buildings, machinery and equipment, inventory, etc. The amount of the tax is a function of the appraised value of the assets and the tax rate. Property taxes are usually not a significant factor in an engineering economics study because of their small magnitude compared to income taxes and their similar effect on competing proposals.
2. *Excise taxes,* imposed on the production of certain products such as tobacco and alcohol, rarely affect economic comparisons. Other taxes that are not normally relevant, but may become so in specific situations, are *sales tax* on retail products, *user's tax, value-added tax, unemployment tax,* and *social security contributions.*
3. *Income taxes* are levied on personal and corporate income at increasingly higher rates for higher incomes. They are based on net income after deductions allowed for permissible "expenses." The tax effects of different types of expenses on the cash flow of proposals have significant influence on their acceptability. The rest of this chapter is devoted to the examination of income-tax effects.

## Changing Taxes

The federal government controls the monetary and fiscal policy of the nation to influence the level of economic activity. *Monetary policy* influences the availability and cost of credit, and *fiscal policy* deals with government receipts and expenditures. Taxation is the key instrument in fiscal policy. The principal methods for altering government receipts are (1) changing the tax rate, (2) changing the depreciation requirements, and (3) allowing tax credits.

Tax rates imposed on incomes may be raised to dampen the level of economic activity when rapid expansion threatens inflationary consequences. In theory, the reduction in disposable incomes reduces the purchasing power of individuals and thereby decreases demand for goods and services. An associated reduction in after-tax profits by corporations reduces the funds available for new investments and discourages expansion. The reverse, a tax-rate cut, theoretically encourages purchasing and expansion when the fiscal policy attempts to stimulate a depressed economy characterized by high unemployment.

In the last decade, both tax increases and tax decreases have been legislated. In 1968, Congress raised taxes in the form of a *surcharge* (a percentage added to the amount of taxes computed by an existing method): 10 percent in 1969, and 5 percent in 1970. More comprehensive tax changes were made in 1971. Then tax rates were reduced by varying amounts, and rebates were given in 1974. Engineering economic studies would ideally be based on the tax rates in effect during the lives of the assets being evaluated, but this is an unrealistic expectation, so current or "typical" rates are utilized. The rates utilized in this chapter are representative but not necessarily currently correct.

Changes in tax laws to allow accelerated depreciation have already been described. The fiscal-policy implications of a more rapid write-off of invested capital are faster cash flows which have a stimulating effect on new investing. The most sweeping changes in depreciation requirements occurred in 1954, 1962, and 1970, the latter two featuring reductions in assets' useful lives.

A special provision to encourage corporate investments was first enacted in 1962. Since then, versions of an *investment tax credit* have been canceled and reinstated at least twice. This credit concept allows businesses to deduct a specified percentage of the dollar amount of new investments as a *credit* against their income taxes. The investments must be in specified types of assets to be eligible, and the size of the credit depends on the useful life of the assets, as illustrated by the representative percentages tabulated below.

| *Investment in an Asset with a Useful Life of* | *Allows a Tax Credit on the Amount of the Investment up to* |
|---|---|
| 8 years or more | 7% |
| 6 to 8 years | 2/3 of 7% or 0.04667 |
| 4 to 6 years | 1/3 of 7% or 0.02333 |
| Less than 4 years | 0% |

Thus, a firm that otherwise would have a \$50,000 tax bill could utilize a tax credit, on the purchase of a \$100,000 asset with a 10-year useful life, of $\$100,000 \times 0.07 = \$7000$. The adjusted tax liability would then be $\$50,000 - \$7000 = \$43,000$ for the year. Other provisions that affect investment tax credits include a maximum allowance for any year and the opportunity to carry the credit forward or backward a limited number of years to apply against tax liabilities in those years.

**Economy Exercise 11-4** An asset with a useful life of 7 years that qualifies for an investment tax credit is purchased for \$40,000. It is expected to have a salvage value of \$5000.

**a** If the tax liability without the investment tax credit is \$20,000 for the year the asset was purchased, what tax payment is due?

**b** If straight-line depreciation is used, what depreciation charge should have been included in calculating the original tax liability?

# CORPORATE INCOME TAXES

Income taxes are due from corporations and businesses whenever revenue exceeds allowable tax deductions. Revenue includes sales to customers of goods and services, dividends received on stocks, interest from loans and securities, rents, royalties, and other gains from ownership of capital or property. Deductions embrace a wide range of expenses incurred in the production of revenue: wages, salaries, rents, repairs, interest, taxes, materials, employee benefits, advertising, etc. Also deductible, sometimes under special provisions, are losses from fire and theft, contributions, depreciation and depletion, bond interest, research and development expenditures, outlays to satisfy legislated objectives such as pollution control, etc. The difference between the revenue and deductions is taxable income. In general,

*Taxable income* = *gross income* − *expenses* − *interest on debt* − *depreciation*

and

*Corporate income tax* = *taxable income* × *effective tax rate*

## Effective Income-Tax Rate

The first $25,000 of taxable income is subject to a federal tax of 22 percent, and all income over $25,000 is taxed at 48 percent (nominally composed of the base 22 percent corporate rate plus a 26 percent surtax). The result is an average tax rate that increases progressively up to about a $1 million taxable income, as shown in Table 11.6, after which it is virtually constant at 48 percent.

**TABLE 11.6** Representative marginal tax rates (tax on each additional taxable dollar) and average tax rates (total tax divided by the upper limit of taxable income in each increment) for corporations.

| *Corporate Taxable Income* | *Marginal Tax Rate, %* | *Average Tax Rate, %* |
|---|---|---|
| $0–$25,000 | 22 | 22 |
| 25,001–50,000 | 48 | 35 |
| 50,001–100,000 | 48 | 41.50 |
| 100,001–200,000 | 48 | 44.75 |
| 200,001–500,000 | 48 | 46.70 |
| 500,001–1,000,000 | 48 | 47.35 |
| 1,000,001–5,000,000 | 48 | 47.87 |
| 5,000,001–10,000,000 | 48 | 47.99 |

Most states and some cities also impose a corporate income tax. The federal and state taxes can be combined into an *effective income-tax rate* that represents the total corporate tax liability. This one rate for all tax obligations is convenient to use in economic studies. It is based on the conditions expected during the study period. Recognizing that state income taxes are always deductible from federal taxes but that the opposite is rarely true, we have

Effective tax rate = state rate + (1 − state rate)(average federal rate)

Capital Gains Profit 50% Tax Free
50% Ordinary Inc

For example, if a firm expects a taxable income of $100,000 during a study period, and state taxes are 6 percent on this amount, then

$$\begin{aligned}
\text{State tax due} &= \$100{,}000 \times 0.06 = \$6000 \\
\text{Taxable income for federal tax} &= \$100{,}000 - \$6000 = \$94{,}000 \\
\text{Federal tax due} &= \$25{,}000(0.22) + (\$94{,}000 - \$25{,}000)(0.48) \\
&= \$38{,}620 \\
\text{Total tax due} &= \$38{,}620 + \$6000 = \$44{,}620 \\
\text{Effective tax rate} &= \$44{,}620/\$100{,}000 = 0.4462 \quad \text{or} \quad 44.62\%
\end{aligned}$$

Using the effective-tax formula with an average federal tax rate interpolated from Table 11.6 gives

$$\begin{aligned}
\text{Effective tax rate} &= 0.06 + (1.00 - 0.06)(0.41) \\
&= 0.4454 \quad \text{or} \quad 44.54\%
\end{aligned}$$

which varies from the other calculation by the approximation of the federal rate. A typical value for the effective corporate income-tax rate used in engineering economic analyses is 50 percent.

---

**Economy Exercise 11-5** What is the effective income-tax rate when the state rate is 5 percent, the federal rate is 48 percent, and the federal income tax is deductible in computing the state income tax?

---

## Some Significant Features of Tax Regulations*

An effective income-tax rate is appropriate for making economic comparisons but should not be utilized for computing annual tax returns without a very careful examination. There are many special provisions in the income-tax regulations that should be evaluated in preparing tax returns. Some of the considerations are described below. These are only selected features and are not intended to be definitive. A collection of all the tax regulations that affect corporations and interpretations of their significance would occupy a book several times the size of this one.

*Capital gains and losses* occur when assets, such as investments in securities, are bought and sold. When such assets are held for not more than 6 months, the gains or losses are considered *short term*, and when held over 6 months are *long term*. Net short-term gains are added to the firm's ordinary income. Net long-term gains are taxed at a maximum rate of 30 percent. For instance, if a corporation buys stock in another corporation, holds it more than 6 months, and sells it at a profit, the difference between the purchase and selling prices is subject to a maximum tax of 30 percent, regardless of the rate being paid on ordinary income.

Interest paid for funds borrowed to conduct a corporation's business is a deductible expense, but dividends paid on its own stock are not deductible. This differential treatment of dividends and interest payments has an important effect on the sources of funds for a corporation (see page 382).

*See Extensions 11.1 and 11.2 for additional data and discussion.

A special surtax is imposed on excessive income accumulation by a corporation. The purpose of the surtax is to penalize stockholders who increase retained earnings in a corporation they control to avoid personal income taxes. This temptation occurs because individual tax rates are as high as 70 percent of taxable income, whereas the maximum corporate federal income-tax rate is 48 percent.

A company is allowed to use a loss incurred in one year to offset profit in another year, or years. An operating loss can have a *carry-back* up to 3 years and a *carry-forward* up to 5 years. The loss must be carried back to the earliest year, the remainder applied to the next earliest year, and so on. For instance, a loss in 1978 may be used to reduce taxable income in 1975, 1976, 1977, 1979, 1980, 1981, 1982, and 1983—always in this sequence. If the 1978 loss just equaled the 1975 gain, the carry-back would allow recovery of all the taxes paid in 1975. The purpose of permitting this loss averaging is to avoid penalizing businesses that have widely fluctuating revenues and expenses.

## AFTER-TAX ECONOMIC COMPARISONS

All the economic analyses in previous chapters were based on before-tax cash flows. In many situations, before-tax analyses provide adequate solutions. When the alternatives being compared are to satisfy a required function and are affected identically by taxes, the before-tax comparison yields the proper preference. Evaluations of public projects rarely include tax effects and are conducted as before-tax analyses (see Chapter 13).

Tax effects occasionally cause the preference to switch among alternatives between before- and after-tax evaluations. The principal causes are differences in depreciation schedules, deductions for interest payments, and special tax regulations such as applicable investment credits. Often, the net return after taxes, the amount actually available, is the main interest. This is usable capital, not subject to downward revision when taxes are withdrawn.

A simple adjustment of the rate of return calculated without regard for taxes gives a reasonable approximation to the after-tax rate of return:

$$RR_{\text{after-tax}} \doteq RR_{\text{before-tax}}(1 - \text{effective income-tax rate})$$

Thus, the after-tax rate of return resulting from a before-tax RR of 15 percent and an effective income-tax rate of 40 percent would be

$$RR_{\text{after-tax}} \doteq 15\%(1 - 0.4) \doteq 9\%$$

This figure is exact when the assets involved are nondepreciable, financing is by equity capital alone, and no special tax provisions are applicable.

### After-Tax Cash Flow

A tabular approach is convenient for modifying the before-tax cash flow to show the effects of taxes. It is normally sufficient to assume tax payments occur at the end of each period, just as the other cash flows are assumed to occur collectively at the year's end. The number of entries in the table depends on the number of tax considerations involved. The most common are depreciation and interest deductions. Table headings based on these tax effects are shown below.

| End of Year (1) | Before-Tax Cash Flow (2) | Depreciation Charges (3) | Loan and Interest (4) | Taxable Income [(2) − (3) − (4)] (5) | Taxes [(5) × tax rate] (6) | After-Tax Cash Flow [(2) − (6)] (7) |
|---|---|---|---|---|---|---|

## After-Tax Comparison of Proposals

An after-tax evaluation can be made using any of the comparison methods: AW, PW, or RR. All the precautions discussed for before-tax comparisons are applicable to after-tax analyses. Once the tax effects on cash flows have been determined, the computational procedures and interpretation of results are the same.

**Example 11.5** **After-Tax Evaluation of a Depreciable Asset**

The budget includes $45,000 for the purchase of a new testing machine requested by the maintenance department. All the major investments in the budget are being checked to see if they meet the new required rate of return that has been raised to 12 percent *after taxes*.

The testing machine will have a useful life of 5 years with no salvage value. During the 5 years, it is estimated to save $23,000 per year in maintenance costs while annual operating costs are $7300. It will be depreciated by the sum-of-digits method. The firm has an effective income-tax rate of 42 percent. Does the proposal to buy the testing machine satisfy the firm's new minimum acceptable rate of return?

**Solution 11.5**

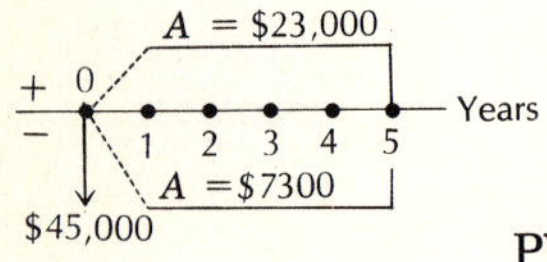


A quick check by the after-tax RR approximation can be made to see if the proposal is promising. Based on the cash-flow diagram, where savings are considered to be income

$$PW = -\$45{,}000 + (\$23{,}000 - \$7300)(P/A,\ i,\ 5) \stackrel{?}{=} 0$$

At $i$ = 22 percent,

$$PW = -\$45{,}000 + \$15{,}700(2.8701) = \$61$$

Using 22 percent as the before-tax rate of return,

$$RR_{\text{after-tax}} \doteq 22\%\,(1 - 0.42) \doteq 12.8\%$$

which is just above the required minimum.

After-tax computations are based on the values in Table 11.7. The depreciation charge for year 1 using sum-of-digits depreciation is

$$DC(1) = \frac{5}{15}(\$45{,}000) = \$15{,}000$$

and it decreases by (1/15)($45,000) = $3000 per year.

TABLE 11.7 Tabulated after-tax cash flow based on sum-of-digits depreciation and an effective income-tax rate of 42 percent.

| End of Year, N | Before-Tax Cash Flow | Depreciation Charges | Taxable Income | Taxes (at 42%) | After-Tax Cash Flow |
|---|---|---|---|---|---|
| 0 | −$45,000 | | | | −$45,000 |
| 1 | 15,700 | −$15,000 | $ 700 | $ 294 | 15,406 |
| 2 | 15,700 | −12,000 | 3700 | 1554 | 14,146 |
| 3 | 15,700 | −9,000 | 6700 | 2814 | 12,886 |
| 4 | 15,700 | −6,000 | 9700 | 4074 | 11,626 |
| 5 | 15,700 | −3,000 | 12,700 | 5334 | 10,366 |

Recognizing that the after-tax cash flow is a uniform gradient decreasing from \$15,406 by \$1260 each year, the present-worth formula for calculating the rate of return is

$$\text{PW} = -\$45{,}000 + [\$15{,}406 - \$1260(A/G, i, 5)](P/A, i, 5) \stackrel{?}{=} 0$$

At $i = 14$ percent,

$$\text{PW} = -\$45{,}000 + [\$15{,}406 - \$1260(1.7398)](3.4330) = \$363$$

At $i = 15$ percent,

$$\text{PW} = -\$45{,}000 + [\$15{,}406 - \$1260(1.7227)](3.3521) = -\$634$$

So,

$$\text{RR}_{\text{after-tax}} = 14\% + 1\% \frac{363}{\$363 + \$634} = 14.4\%$$

The proposal apparently meets the after-tax rate-of-return criterion. Note that the after-tax RR is about 65 percent of the before-tax RR and exceeds the after-tax approximation based on the before-tax RR.

**Example 11.6** **After-Tax Evaluation When the Asset Is Eligible for an Investment Income-Tax Credit**

Further investigation of the testing-machine proposal described in Example 11.5 reveals that its purchase allows an investment income-tax credit of 2⅓ percent on the purchase price. However, the tax regulations specify that it must then be depreciated by the straight-line method. What is the revised after-tax rate of return?

**Solution 11.6** If the investment tax credit of (0.0233)(\$45,000) = \$1049 is applied against the taxes due in the first year, the after-tax cash flow in Table 11.8 for year 1 is \$2814 − \$1049 = \$1765. The other change from Table 11.7 is the timing of taxes caused by the constant straight-line depreciation charge of (1/5)(\$45,000) = \$9000 per year.

$$\text{PW} = -\$45{,}000 + [\$13{,}935 + \$12{,}886(P/A, i, 4)](P/F, i, 1) \stackrel{?}{=} 0$$

At $i = 14$ percent,

$$\text{PW} = -\$45{,}000 + [\$13{,}935 + \$12{,}886(2.9137)](0.87719) = \$159$$

TABLE 11.8 After-tax cash flow when an investment tax credit is applicable.

| End of Year, N | Before-Tax Cash Flow | Depreciation Charges | Taxable Income | Taxes (at 42%) | After-Tax Cash Flow |
|---|---|---|---|---|---|
| 0 | −$45,000 | | | | −$45,000 |
| 1 | 15,700 | −$9000 | $6700 | $1765 | 13,935 |
| 2 | 15,700 | −9000 | 6700 | 2814 | 12,886 |
| 3 | 15,700 | −9000 | 6700 | 2814 | 12,886 |
| 4 | 15,700 | −9000 | 6700 | 2814 | 12,886 |
| 5 | 15,700 | −9000 | 6700 | 2814 | 12,886 |

At $i$ = 15 percent,

$$\text{PW} = -\$45{,}000 + [\$13{,}935 + \$12{,}886(2.8549)](0.86957) = -\$893$$

and

$$\text{RR}_{\text{after-tax}} = 14\% + 1\% \frac{\$159}{\$159 + \$893} = 14.2\%$$

Compared with that in Example 11.5, the after-tax rate of return is lower even with an investment tax credit included because straight-line depreciation does not allow as high deductions in the early years.

**Example 11.7** **After-Tax Evaluation When Investment Capital Is Borrowed**

An alternative to the testing machine favored in Example 11.5 can be purchased for $67,000. It will have a useful life of 5 years, $12,000 salvage value, and net savings of $22,000 per year. Compare it to the machine in Example 11.5, using sum-of-digits depreciation. The additional $67,000 − $45,000 = $22,000 required for the purchase must be borrowed at 9 percent annual interest with the principal due at the end of the fifth year.

**Solution 11.7** Table 11.9 contains the figures employed in calculating the after-tax cash flow for the more expensive testing machine. The middle column indicates the amount of the loan and the annual interest payments, which are tax deductible. The extra row at the bottom indicates the salvage value received and the loan repayment at the end of year 5. Note that the last column in the table is essentially a cash-flow diagram in tabular form.

TABLE 11.9 After-tax cash flow involving both borrowed and equity funds.

| End of Year, N | Before-Tax Cash Flow | Depreciation Charges | Loan and Interest | Taxable Income | Taxes (at 42%) | After-Tax Cash Flow |
|---|---|---|---|---|---|---|
| 0 | −$67,000 | | $22,000 | | | −$45,000 |
| 1 | 22,000 | −$20,000 | −1,980 | $ 20 | $ 8 | 20,012* |
| 2 | 22,000 | −16,000 | −1,980 | 4,020 | 1688 | 18,332 |
| 3 | 22,000 | −12,000 | −1,980 | 8,020 | 3368 | 16,652 |
| 4 | 22,000 | −8,000 | −1,980 | 12,020 | 5048 | 14,972 |
| 5 | 22,000 | −4,000 | −1,980 | 16,020 | 6728 | 13,292 |
| 5 | 7,000 | | −22,000 | | | −15,000 |

*After-tax cash flow is obtained by subtracting interest payments ($22,000 × 0.09 = $1980) and taxes based on taxable income from the before-tax cash flow. For year 1, after-tax cash flow = $22,000 − $1980 − $8 = $20,012.

Recognizing that the annual net returns decrease by the constant amount of $1680 each year, a gradient factor is included in the present-worth formula as

$$PW = -\$45{,}000 + [\$20{,}012 - \$1680(A/G, i, 5)](P/A, i, 5) - \$15{,}000(P/F, i, 5) \stackrel{?}{=} 0$$

At $i = 20$ percent,

$$PW = -\$45{,}000 + [\$20{,}012 - \$1680(1.6405)](2.9906) - \$15{,}000(0.40188) = \$578$$

At $i = 25$ percent,

$$PW = -\$45{,}000 + [\$20{,}012 - \$1680(1.5631)](2.6893) - \$15{,}000(0.32768) = -\$3070$$

and $$RR_{\text{after-tax}} = 20\% + 5\% \frac{\$578}{\$578 + \$3159} = 20.8\%$$

The new proposal is obviously superior to the proposal in Example 11.5.

## After-Tax Replacement Study

Before- and after-tax replacement evaluations are conducted in the same way once the tax effects have been imposed on the cash-flow patterns. There are, however, a couple of points that occasionally cause confusion. One is the use of the book value as the worth of the presently owned asset in a replacement study.

The *book value* is the asset's purchase price less its accumulated depreciation charges. This value is the result of accounting practices and is unlikely to be the same as the asset's value in an open market. The difference between the book value and the present realizable value is the *sunk cost.* As depicted in Figure 11.4, the actual amount obtainable for a

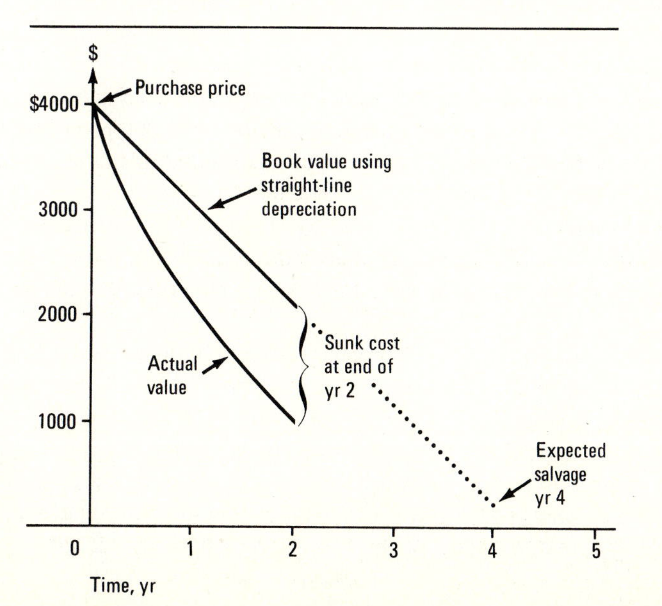


**FIGURE 11.4** Sunk cost as a function of book value and actual price for an asset sold 2 years before the end of its expected useful life.

defender in a replacement study is less than its value shown by the accounting records. If the asset is replaced, the sunk cost is a loss for the year, charged against profit. However, part of the loss is absorbed by a reduction in taxes for that year. For instance, let the book value of the asset in Figure 11.4 be $2100 and the market value be $1000, to produce a sunk cost of $2100 − $1000 = $1100. Taxable income for the year would be reduced by $1100, which "saves" $1100 × 0.40 = $440 in taxes when the effective income-tax rate is 40 percent. Thus, the effective loss is only $1100 − $440 = $660.

The mental static generated by sunk costs is hard to comprehend unless there is personal involvement. A large sunk cost can be detrimental to the current year's profit picture and may cause censure of the current managers, even if they had nothing to do with the original purchase and the depreciation schedule that caused it. There may also be a feeling that, "We've got so much invested in it that we can't afford to sell it for what it would bring." The weakness of this argument is that the owners should be looking ahead to determine the course of action that will minimize *future* costs, not the course that will make past expenditures look good.

Another possible error in replacement studies is to add, to the market value, the cost of an overhaul or major repair made on a defender. The overhaul or repair is a current expenditure and deductible as such from current income. If this expense were instead added to the market value of an asset to increase its capital cost, it would be spread over the remaining years of life of the asset. Not only is the realizable value likely incorrect when overhaul and repair expenses are added on, but a tax loss is caused by not charging the full expenses immediately.

**Example 11.8** **After-Tax Comparison of a Defender and Challenger**

The present machine has a book value of $11,000, but the net realizable value is $8000 because technological advances have made its use less attractive. If an overhaul is made for $5000, its operating costs will decrease from the current $19,000 to $16,000 per year for the remaining 5 years of its useful life. It will then have no salvage value.

A new machine of advanced design to perform the same function as the presently owned machine will have operating expenses of $12,000 per year. The new machine is priced at $36,000 and is expected to have a useful life of 5 years with a $6000 salvage value.

If both machines are depreciated by the straight-line method based on current values, which has the lowest after-tax annual cost when the effective income-tax rate is 50 percent and the required after-tax rate of return is 8 percent?

**Solution 11.8** The after-tax cash flows of the defender and challenger are given in Table 11.10. The $5000 overhaul expense is a single entry in year 1 of column 2 for the defender. The difference between the $11,000 book value and the $8000 realizable value of the defender is a loss shown as a deductible charge at year 0 in column 4. Other figures in column 4 are the repeated sums of the operating expenses and depreciation charges for years 1 through 5. These amounts are deductible from net income and reduce taxes at the 50 percent effective tax rate to produce the tax

TABLE 11.10 Factors leading to after-tax cash-flow data for a defender and challenger in a replacement study.

| *End of Year, N (1)* | *Before-Tax Cash Flow (2)* | *Depreciation Charges (3)* | *Deductible Charges (4)* | *Tax Savings (5)* | *After-Tax Cash Flow (6)* |
|---|---|---|---|---|---|
| | | DEFENDER | | | |
| 0 | −$8,000 | | −$3,000 | $1500 | −$9500 |
| 1 | −5,000 | | −5,000 | 2500 | −2500 |
| 1-5 | −16,000 | −$1600 | −17,600 | 8800 | −7200 |
| | | CHALLENGER | | | |
| 0 | −$36,000 | | | | −$36,000 |
| 1-5 | −12,000 | −$6000 | −$18,000 | $9000 | −3,000 |
| 5 | 6,000 | | | | 6,000 |

savings given in column 5. These savings are then subtracted from the before-tax cash flows in column 2 to give the after-tax cash flows in column 6. Based on column 6, with the negative values treated as positive costs, the annual costs of the defender and challenger are

$$\begin{aligned} AC_{\text{after-tax}}(\text{defender}) &= \$9500(A/P,\ 8,\ 5) + \$2500(P/F,\ 8,\ 1)(A/P,\ 8,\ 5) + \$7200 \\ &= \$9500(0.25046) + \$2500(0.92593)(0.25046) + \$7200 \\ &= \$10{,}159 \end{aligned}$$

$$\begin{aligned} AC_{\text{after-tax}}(\text{challenger}) &= \$36{,}000(A/P,\ 8,\ 5) + \$3000 - \$6000(A/F,\ 8,\ 5) \\ &= \$36{,}000(0.25046) + \$3000 - \$6000(0.17046) = \$10{,}994 \end{aligned}$$

**Economy Exercise 11-6** What would the annual cost have been for the defender in Example 11.8 if the overhaul expense had incorrectly been added to the market price so that the before-tax cash flow at year 0 was $8000 + $5000 = $13,000?

## SUMMARY

Accounting records indicate the financial condition of an organization, record data from which estimates for future activities can be obtained, and reveal how closely implemented proposals met expectations. An *income statement* shows the sources of revenue and expenses over a period of time. Net earnings from the income statement is an input to the *balance sheet,* where accounts are tabulated for a given date to give values for the basic accounting equation: *assets minus liabilities equal net worth.*

*Depreciation charges* are made in accounting records to recover capital spent for production assets and to substantiate deductions from current income for tax calculations. The three most commonly used depreciation methods are *straight line, sum of digits,* and *declining balance.* The latter two are accelerated methods which allow higher depreciation charges earlier in the write-off period. Total taxes are the same under all three methods, but accelerated charges decrease the time value of taxes. Rigid rules for depreciation accounting for tax purposes are set by the Internal Revenue Service.

Taxes are a major factor in any profit-seeking venture. *Property, sales, excise,* and/or *income taxes* affect net returns for both individuals and corporations. Types and amounts of taxes vary as governments pursue their *fiscal policies*. In the past few years, *surcharges* and *investment income-tax credits* have been enacted, rescinded, and reinstated. Other features change too, such as provisions for *capital gains and losses*, options to *carry amounts forward* and *backward*, and *depletion allowances*.

An *effective income-tax rate* that represents total corporate tax liability can be developed for after-tax economic evaluations. Special tax provisions and charges for depreciation and/or interest are applied to the before-tax cash flow to determine *taxable income*. Then the taxable income is multiplied by the effective tax rate, and the resulting product is subtracted from the before-tax cash flow to obtain the *after-tax cash flow* for the year. A tabular format is convenient for these calculations and provides information equivalent to a cash-flow diagram. Once the after-tax data are tabulated, economic comparisons are conducted as described in Chapters 7 through 9.

## Discussion of Economy Exercises

**EE 11-1** Sam Sales' beginning balance sheet is shown in Figure 11.5.

SALES' GIMMICKS
Balance Sheet
April 1, 19x0

| ASSETS | | LIABILITIES | |
|---|---|---|---|
| Cash | $10,000 | Accounts payable | $ 1,900 |
| Accounts receivable | 3,100 | Accrued taxes | 300 |
| Inventory | 14,200 | Bank loan | 8,000 |
| Prepaid rent | 1,200 | Note due seller | 10,000 |
| Franchise | 4,500* | NET WORTH | |
| Total assets | $33,000 | Sam Sales' equity | $12,800† |
| | | Total liability and equity | $33,000 |

FIGURE 11.5

*The value of the franchise is the amount included in the purchase price for existing customer goodwill and the exclusive territorial rights to distribute Freaky postcards.

†Sam's equity is derived from the $15,000 withdrawn from savings minus the amount of accounts and taxes payable ($1900 + $300).

**EE 11-2** The income statement for the first year's operation of Sales' Gimmicks Company, based on the budget shown in Table 11.1, is in Figure 11.6. The income and expense categories summarize the expected effects of operations during the 1-year period analyzed. Sam did pretty well; the return on his initial investment was about 100 percent in 1 year.

**EE 11-3** The depreciation pattern and tax calculations for the asset described in Table 11.5 are shown in Table 11.11. Double-declining-balance depreciation is applied to an asset with an initial cost of $1500, a useful life of 5 years, and no salvage value.

**FIGURE 11.6**

| SALES' GIMMICKS COMPANY<br>Year Ending April 1, 19x1 | | |
|---|---|---|
| *Income* | | |
| Net Sales: Postcards | $92,000 | |
| Other gimmicks | 30,000 | |
| Total net sales | | $122,000 |
| Cost of goods sold: Postcards | 40,000 | |
| Other gimmicks | 13,200 | |
| Total cost of goods sold | | 53,200 |
| Gross profit on sales | | $68,800 |
| *Expenses* | | |
| Route servicers' salaries | $14,000 | |
| Travel expenses | 12,000 | |
| Office salaries and expenses | 9,100 | |
| Other expenses (storage, pilferage, etc.) | 4,000 | |
| Loan repayment ($2200 interest + $9000) | 11,200 | |
| Total expenses | | $50,300 |
| Net operating profit | | $18,500 |
| Income tax | | 3,700 |
| Net profit (to retained earnings) | | $14,800 |

**TABLE 11.11**

| *End of Year, N* | *Income before Taxes and Depreciation* | *Annual Depreciation Charges* | *Income less Depreciation* | *Income-Tax Rate* | *Income Tax* | *Present-Worth Factor (P/F, 10, N)* | *PW of Income Tax* |
|---|---|---|---|---|---|---|---|
| 1 | $ 500 | $600 ($500)* | $ 0 | 0.4 | $ 0 | 0.90909 | $ 0 |
| 2 | 500 | 360 (460)* | 40 | 0.4 | 16 | 0.82645 | 13 |
| 3 | 500 | $216 | 284 | 0.4 | 114 | 0.75132 | 86 |
| 4 | 500 | 130 | 370 | 0.4 | 148 | 0.68302 | 101 |
| 5 | 500 | 194† | 306 | 0.4 | 122 | 0.62092 | 76 |
| *Totals* | $2500 | $1500 | $1000 | | $400 | | $276 |

*For purposes of comparison, the depreciation expense that exceeds income in a given year is carried forward and deducted from profit the following year. The excess in year 1 ($600 − $500 = $100) is added to the depreciation charge in year 2 ($360 + $100 = $460) to give a net income for year 2 of $500 − $460 = $40.

†The double-declining-balance method always leaves a positive salvage value based on the initial cost and useful life. In the last year of the life of the asset shown, the depreciation charge is adjusted to make the depreciation reserve equal to $P - S$ ($1500). In the table, the amount required to make the total depreciation charges accumulate to $1500 in year 5 is $1500 − $600 − $360 − $216 − $130 = $194, which is $116 more than the amount calculated by the double-declining-balance formula.

As indicated in the last column of Table 11.11, there is very little difference in the present worth of taxes between the sum-of-digits method shown in Table 11.5 and the double-declining-balance method as used. However, the total PW could have been reduced by switching from the double-declining-balance method to straight-line depreciation in year 4 to recover the remaining capital invested in the asset more rapidly. This practice is allowed by the tax laws and is used to avoid a large depreciation write-off in the last years of a double-declining-balance depreciation schedule. The time to switch is a function of the book value, salvage value, and remaining years of useful life of the asset. The revised schedule is supposed to come out very close to the actual salvage value.

**EE 11-4**

**a** The investment income-tax credit is based on the purchase price and does not include consideration of the salvage value. The 7-year life allows a tax credit in the first year of two-thirds of 7 percent of the investment: tax liability = \$20,000 − (0.04667)(\$40,000) = \$18,133.

**b** The depreciation charge for year 1 is (\$40,000 − \$5000)/7 = \$5000; it is not related to the investment tax credit but is a deductible expense.

**EE 11-5** Let the federal income tax be $X$ (in dollars). Then the income subject to state tax is $T - X$, where $T$ represents taxable income. The state income tax, at a 5 percent tax rate, is

$$0.05(T - X) = 0.05T - 0.05X$$

The income subject to federal tax is

$$T - 0.05T + 0.05X = 0.95T + 0.05X$$

The federal income tax is

$$\begin{aligned} X &= 0.48(0.95T + 0.05X) \\ X - 0.024X &= 0.456T \\ X &= 0.4672T \quad \text{or} \quad 46.72\% \text{ of taxable income} \end{aligned}$$

Then state tax $= 0.05\ (T - 0.4672T) = 0.0266T$ or 2.66% of taxable income

Therefore,

$$\begin{aligned} \text{Effective income-tax rate} &= 0.0266 + 0.4672 \\ &= 0.4938 \quad \text{or} \quad 49.38\% \end{aligned}$$

If the federal tax had not been deductible from state taxes, the formula on page 355 is appropriate:

$$\begin{aligned} \text{Effective income-tax rate} &= 0.05 + (1.0 - 0.05)(0.48) \\ &= 0.05 + 0.456 = 0.506 \quad \text{or} \quad 50.5\% \end{aligned}$$

**EE 11-6** Assuming the tax savings due to a sunk cost would still be claimed on the difference between book and market values (\$11,000 − \$8000 = \$3000), the worth of the challenger at year 0 when the overhaul expense is added to the market value is \$8000 + \$5000 + 0.5(\$3000) = \$14,500, but the annual depreciation charge is \$13,000/5 = \$2600. This charge added to the \$16,000 annual operating expense gives yearly tax savings of 0.5(\$16,000 + \$2600) = \$9300. Then,

$$\begin{aligned} AC_{\text{after-tax}} &= \$14{,}500(A/P,\ 8,\ 5) + \$16{,}000 - \$9300 \\ &= \$14{,}500(0.25046) + \$6700 = \$10{,}332 \end{aligned}$$

which is slightly greater than the correct solution given in Example 11.8. The only time it would be appropriate to add repair expenses to the market price is when the repairs raise the market value an amount equal to their cost.

## PROBLEMS

**11.1** What use can be made of general-accounting and cost-accounting records by an engineering economist? What precautions should be exercised in this usage?

**11.2** Compare the features that are desirable in a depreciation method with the causes of declining value of an asset. What does this comparison suggest about the accounting concept of depreciation versus the considerations involved in an economic analysis?

The following basic data about the In-Out Universal Corporation are to be utilized in Problems 11.3–11.5 and Extension 11.2. The two tables show the condition of I-O-U on December 31 and the transactions that occurred during the prior 6 months.

**TABLE 11.12**

State of Accounts for In-Out Universal on December 31, 19xx

| *Account* | *Balance* |
|---|---|
| Accounts payable | $ 50,000 |
| Accounts receivable | 80,000 |
| Accrued taxes | 20,000 |
| Buildings (net value) | 205,000 |
| Cash on hand | 40,000 |
| Dividends payable | 20,000 |
| Equipment (net value) | 180,000 |
| Finished goods | 30,000 |
| Land value | 115,000 |
| Long-term mortgages | 390,000 |
| Material inventory | 40,000 |
| Notes payable | 60,000 |
| Stockholders' equity | 150,000 |

Transactions for In-Out Universal during the 6 Months prior to December 31, 19xx

| *Category* | *Amount* |
|---|---|
| Administration expenses | $ 30,000 |
| Depreciation charged | 30,000 |
| Direct labor costs | 70,000 |
| Factory overhead charged | 35,000 |
| Finished goods (July 1, 19xx) | 75,000 |
| Interest and debt payments | 40,000 |
| Inventory, materials (July 1, 19xx) | 15,000 |
| Materials purchased (July 1 to December 31) | 115,000 |
| Sales: Cash | 200,000 |
| Credit | 200,000 |

**11.3** Construct a balance sheet from the accounts given in Table 11.12. What are the dollar values of the terms in the basic accounting equation?

Assetts - Liabilites = Net Worth *($690,000 − $540,000 = $150,000)*

**11.4** Prepare a cost-of-goods-sold schedule from the amounts given in Table 11.12. Include materials, labor, and factory overhead. What was the cost of goods sold in the period from July 1 to December 31?

*($240,000)*

**11.5** Prepare an income statement for the 6-month period from the data in Table 11.12. What is the net profit if the accrued-taxes account represents the taxes owed for the period?

*($40,000)*

**11.6** Why is the sinking-fund method of depreciation used regularly by some government agencies but seldom by private industry? How does a court order requiring a company in financial difficulties to set up a temporary sinking-fund depreciation schedule protect the shareholders of the company?

**11.7** An asset cost $400 when purchased 4 years ago. A scrap value of $50 was expected at the end of its 7-year useful life. Determine the depreciation charge during the coming year and the asset's current book value by:

**11.7a** Straight-line depreciation

[$DC(5) = \$50$; $BV(4) = \$200$]

**11.7b** Sum-of-digits depreciation

[$DC(5) = \$37.50$; $BV(4) = \$125$]

**11.7c** Declining-balance depreciation (using the salvage value to determine the depreciation rate)

[$DC(5) = \$31.40$; $BV(4) = \$122$]

**11.7d** Sinking-fund depreciation (6 percent)

[$DC(5) = \$52.62$, *including interest*; $BV(4) = \$218$]

**11.8** A production machine with a first cost of $2000 is expected to last 4 years, when it will be worth $200. Tabulate the book value at the end of each year by the declining-balance and double-declining-balance methods of depreciation.

**11.9** A sales representative made an agreement to purchase a building lot by making six annual payments of $900 each. She was then transferred to a different territory before she could build a home. When she returned 2 years later, she found she could buy an equivalent lot for $3000 because land values had decreased during her absence. She now feels that she will lose the $1800 she has already paid if she drops her contract to buy the equivalent lot. Assuming that she will suffer no penalty for reneging on the original contract, discuss the "sunk cost" concept as it applies to the situation. What would you advise her to do? Does an assumption that she can invest her savings at 7 percent have any bearing on the situation?

**11.10** An asset has an initial cost of $65,000 and an estimated salvage value of $5000 after 12 years.

**11.10a** What depreciation rate for the declining-balance method would produce a book value at the end of year 12 that comes closest to equaling the estimated salvage value?

*(Maximum allowable = 0.1667)*

**11.10b** What amount is accumulated in a depreciation reserve after 5 years if straight-line depreciation is used?

*($25,000)*

**11.10c** What would the depreciation charge be in the eighth year by the sinking-fund method if $i = 4$ percent?

*(Assuming internal funding: DC = $5253)*

**11.10d** If the declining-balance depreciation rate is 10 percent, what would the book value be after 3 years?

*($47,385)*

**11.10e** If the asset had to be sold after 7 years of use for $7000, what would be the sunk cost, by sum-of-digits depreciation?

*($9,538)*

**11.11** A machine has a present book value of $400. It cost $1000 new, 3 years ago. It can now be sold for $100. A new machine to replace it will cost $1200 and has an expected useful life of 4 years. Neither machine has any salvage value. Conduct a preliminary analysis of the replacement question by assuming straight-line depreciation, 0 percent interest rate, and no taxes, to determine:

**11.11a** The sunk cost of the old machine

**11.11b** The depreciation rate for the new machine in dollars per year

**11.11c** The depreciation rate for the old machine as it will be used in a comparison for replacement (dollars per year)

**11.11d** The rate of return for the replacement machine if it could save $200 per year

**11.12** Three corporations have the following data for 1 year of operation:

| | *Macro, Inc.* | *Meso, Ltd.* | *Micro Co.* |
|---|---|---|---|
| Sales | $20,000,000 | $2,000,000 | $200,000 |
| Expenses | 15,250,000 | 1,000,000 | 140,000 |
| Other revenue | 750,000 | 250,000 | 50,000 |
| Depreciation | 3,000,000 | 750,000 | 60,000 |

**11.12a** If all three corporations operate in areas where no state taxes are charged, what is their federal income tax?

*($17,500 for Micro)*

**11.12b** If state taxes average 5 percent, and the federal income tax is not an allowable deduction in state tax returns, what is the effective income-tax rate for each corporation?

*(37.6% for Micro)*

**11.13** A corporation has an effective tax rate of 54 percent. Gross revenue for the past year was $8 million. Operating expenses and depreciation accounted for $6 million. Interest on outstanding debts was $1.2 million. What amount is left for dividends and surplus after taxes?

*($368,000)*

**11.14** An individual in the 30 percent income-tax bracket purchased $10,000 worth of tax-exempt municipal bonds which mature in 15 years and pay $200 in interest twice a year. If the buyer remains in the same tax bracket, what effective interest rate would he have to receive on a taxable investment to equal the yield from the bond purchase?

*(5.77%)*

**11.15** A company in the 40 percent income-tax bracket is considering the purchase of equipment which will eliminate a rental expense of $9000 per year. The equipment will cost $30,000 and have a zero salvage value at the end of its 8-year life. Maintenance and operating costs are estimated at $1800 per year. Insurance is 1 percent of the first cost. Straight-line depreciation is used for tax purposes. The company requires a 7 percent after-tax rate of return. Make a recommendation regarding the purchase of the equipment.

**11.16** Alternative $A$ will produce direct cost savings of $2800 per year more than alternative $B$. However, alternative $B$ offers a more impressive layout than $A$. If the income-tax rate for the corporation is 38 percent, what is the implied worth of the prestige value of alternative $B$, assuming it is selected in preference to alternative $A$? The depreciation schedules for the alternatives are identical.

*($1736)*

*The following basic data pertain to a proposal that is analyzed according to the conditions and objectives of Problems 11.17 through 11.25:*

> A proposed investment in a depreciable income-producing asset is expected to produce annual net savings of $13,500. The asset has a first cost of $56,000 and an estimated useful life of 7 years with no salvage value. The effective income-tax rate of the organization is 45 percent. Unless otherwise stated, it is assumed all cash flows occur at the end of a year, equity funds are used for the purchase, and any deductions beyond the $13,500 net savings can be applied against other income. In effect, accelerated depreciation can produce an apparent savings greater than $13,500 in the early years.

**11.17** What is the before-tax rate of return for the asset described in the basic data? Determine the approximate after-tax RR from the before-tax figure.

*(15.1% and 8.3%)*

**11.18** Compute the after-tax rate of return when straight-line depreciation is charged on the 7-year useful life of the asset described in the basic data.

*(8.7%)*

**11.19** Compute the after-tax rate of return when sum-of-digits depreciation is charged on the asset in the basic data.

*(9.6%)*

**11.20** Compute the after-tax rate of return when the double-declining-balance method is used and an extra depreciation charge is made in the last year to reduce the book value to zero.

*(9.5%)*

**11.21** Would a tax saving result from switching to the straight-line method some time during the depreciation schedule in Problem 11.20 to avoid the year-7 extra large write-off? If so, when should the switch be made? What rule can you devise to indicate when a switch should be made?

*(RR after switch = 9.3%)*

**11.22** An investment income-tax credit is allowed on the asset's purchase price and is taken on the first year's income, but the asset must then be depreciated by the straight-line method over a 10-year period. If the useful life is still only 7 years and the operating savings cease at the end of the seventh year, what is the after-tax RR?

*(9.0%)*

**11.23** The asset qualifies for a special depreciation schedule reserved for certain national-security projects. It can be written off by straight-line depreciation over a 5-year period, although the operational savings will continue through the seventh year. What is the after-tax RR for this special arrangement?

**11.24** Half the purchase price will come from equity funds, and half from capital borrowed at 7 percent. The borrowed capital is to be repaid at the end of the seventh year, but interest is due on the principal at the end of each year. If sum-of-digits depreciation is used, what will be the after-tax rate of return?

**11.25** If the purchase of the asset could be considered an operating expense instead of a depreciable investment, what would be the after-tax rate of return? Assume there is sufficient income to charge the entire amount in the first year. The savings from the purchase of the asset will still continue for 7 years.

**11.26** Rework Problem 9.8 when half the purchase price is for the land value, and the land appreciates by 50 percent during the 10-year study period. This increase in value is subject to a 30 percent long-term capital-gains tax payable at the end of the tenth year. Sum-of-digits depreciation is used by the company, and the effective income-tax rate is 50 percent.

**11.27** Determine the after-tax equivalent annual costs for reinforcing the bridge and for the culvert-and-fill approach described in Example 7.8. The minimum acceptable after-tax rate of return is 6 percent, and straight-line depreciation is used. The effective income-tax rate is 35 percent.

**11.28** Rework Problem 7.19 to determine the after-tax annual costs for machines $A$ and $B$ when the required after-tax rate of return is 8 percent. Sum-of-digits depreciation is to be applied to the challenger, and straight-line depreciation will continue to be used for the defender. The corporate effective income-tax rate is 50 percent.

**11.29** Compare the after-tax present worths of the two types of boat landings in Problem 8.11. Assume the useful life for tax purposes is 20 years for both alternatives, and depreciation is charged on a straight-line basis. The required rate of return remains the same, and the effective income-tax rate is 30 percent.

**11.30** Assume the value of the land in Problem 9.16 holds at $20,000 during the study period, and the building can be depreciated by the double-declining-balance method over 10 years. The purchaser's effective income-tax rate is 35 percent. What after-tax rate of return can be expected from the purchase?

**11.31** An overhead crane was installed in a warehouse 14 years ago at a cost of $38,000. Straight-line depreciation shows a current book value of $14,200 based on a 20-year useful life. Operating costs, including labor and maintenance, are $16,500 per year.

Two small lift trucks could handle the same capacity as the crane with a one-third reduction in operating costs because the factory no longer utilizes the heavy parts formerly handled by the crane. However, 2000 square feet (185.8 square meters) of storage space valued at \$2 per square foot per year (\$21.52 per square meter per year) would be lost in providing lanes for the lift trucks. The trucks can be purchased for \$4800 apiece. They have a useful life of 10 years and an expected salvage value of one-tenth their purchase price.

If the trucks are purchased, the crane can be dismantled and sold for a net gain of \$7800. Capital is charged at 6 percent before taxes or 12 percent after taxes, since the effective income-tax rate for the organization is 50 percent.

**11.31a** What will be the sunk cost of the crane if the change is made?

**11.31b** What was the original estimate for the salvage value of the crane?

**11.31c** What are the annual costs of the defender and the challenger in a before-tax analysis?

**11.31d** If all depreciation is charged on a straight-line schedule, what are the after-tax equivalent annual costs of the defender and challenger?

**11.32** An investment proposal is expected to have the following characteristics:

| | *Year 1* | *Year 2* | *Year 3* |
|---|---|---|---|
| Gross income provided | \$17,000 | \$22,000 | \$19,000 |
| Investments needed | 16,000 | 14,000 | 0 |
| Operating expenses | 6,000 | 8,000 | 11,000 |
| Depreciation charges | 10,000 | 10,000 | 10,000 |

Investments occur during the year and allow the indicated earnings, but both the income and investments are considered to have been made at the end of each year for analysis purposes. In any year in which deductions exceed income, the tax benefits can be considered as additional income because the organization can apply the deductions against income earned from other operations.

**11.32a** Determine the rate of return before taxes.

**11.32b** If the effective income-tax rate is 42 percent, what is the approximate after-tax rate of return?

**11.32c** What is the actual after-tax rate of return?

**11.33** A consulting engineer in acoustical design has purchased and equipped a mobile "sound lab" at a cost of \$38,000. She borrowed \$20,000 on a 3-year contract at 7 percent interest compounded annually, with the loan to be repaid in three equal installments (including interest) due at the end of each of the next 3 years. Her average gross income over the next 4 years is expected to be \$75,000 per year. Her average annual expenses for the period, in addition to sound-lab expenses, are expected to be \$30,000. The effective income-tax rate is 36 percent. The sound lab can be written off for tax purposes in 4 years, and the salvage value at the end of its useful life will be \$8000.

**11.33a** What is the present worth of the engineer's annual income for the next 4 years at $i = 10$ percent when the sound lab is depreciated by the straight-line method?

**11.33b** How much will the present worth of the engineer's income increase for the 4 years if double-declining-balance depreciation is used ($i = 10$ percent)?

**11.33c** The engineer is considering withdrawing money from a savings account that now earns 7 percent annually, to pay the \$20,000 owed on the sound lab instead of borrowing that amount. If she does use equity funds for the entire purchase, and straight-line depreciation is charged, what is the PW of her income at $i = 10$ percent for the next 4 years?

## EXTENSIONS

***11.1 Analysis of Financial Statements*** A sound financial position is the bedrock upon which successful enterprises are built. Accounting practices measure the financial conditions. Accounting statements are the reports of the measurement. Interpretation of the reported figures is the responsibility of the analyst. Engineering economists may become involved in financial-statement analyses through their professional responsibilities to evaluate dealings with other companies, such as a subcontractor's capacity to meet contractual commitments or a supplier's ability to continue follow-up services, and in personal investing to evaluate the operating performance of a company.

The relationship of accounts to the return on investment is broached in Figure 11.7. The left leg of the progression shows the accounts that comprise *total investment,* where *permanent investment* represents land, buildings, equipment, etc. *Sales* divided by total investment is the *turnover,* or number of times per year the capital is used; it suggests how hard the investment is being worked. This turnover times the earnings as a percent of sales gives the *return on investment.* For earnings on sales of 3 percent, as determined by subtracting the value of sales from the cost of sales and then dividing this amount by total sales, and for a turnover of 4 times per year, the annual return on investment is 12 percent.

Certain figures from accounting records have been found to be good barometric readings of a firm's future performance. A half dozen of the more prominent ratios

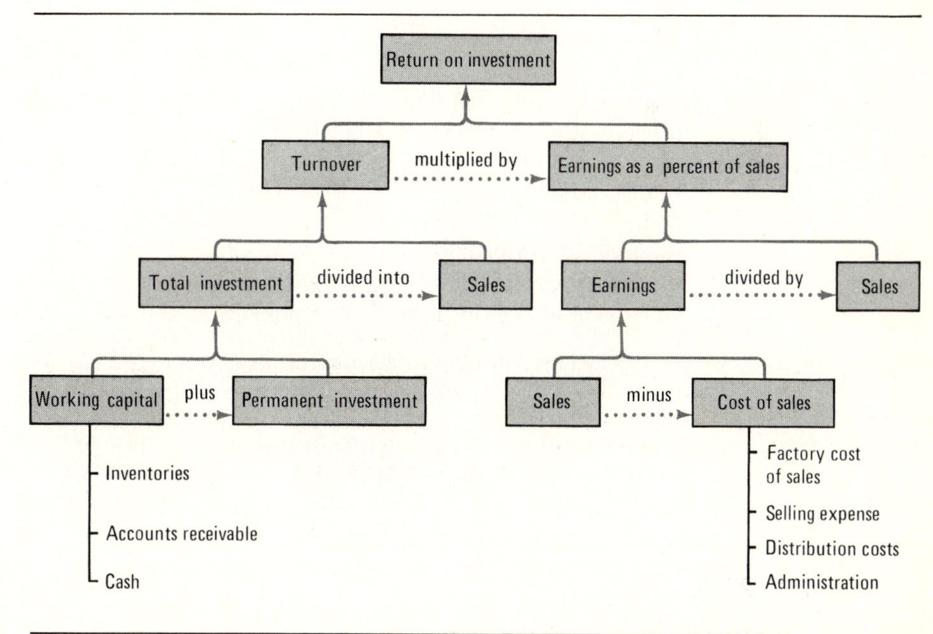


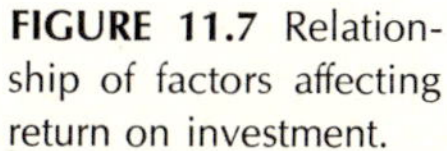
**FIGURE 11.7** Relationship of factors affecting return on investment.

employed to analyze financial conditions are described below and are applied to the data in Figures 11.1 and 11.2.

1 The *current ratio* gives an indication of a company's liquidity—its ability to meet current financial obligations. The current ratio applied to the balance-sheet data given in Figure 11.1 is

$$\text{Current ratio} = \frac{\text{current assets}}{\text{current liabilities}} = \frac{\$52{,}000}{\$49{,}000} = 1.06$$

which is below the level of 2:1 or 3:1 that is generally considered satisfactory. A lower ratio suggests the company may have a cash-flow problem in meeting current obligations.

2 The *acid-test ratio* is a more rigorous evaluation of liquidity made by subtracting inventories and other prepaid expenses from current assets to obtain *quick assets*—assets readily convertible into dollars. With a 1:1 ratio deemed reasonable, the firm represented in Figure 11.1 appears to have a fairly serious liquidity problem:

$$\text{Acid-test ratio} = \frac{\text{current assets} - \text{inventories}}{\text{current liabilities}}$$

$$= \frac{\$52{,}000 - \$37{,}600}{\$49{,}000} = 0.29$$

3 The *equity ratio* indicates the dependency of a company upon its creditors for working capital:

$$\text{Equity ratio} = \frac{\text{total capital}}{\text{total liabilities and capital}}$$

Based on the data from Figure 11.1 and assuming retained earnings are actually owned by the stockholders rather than the corporation,

$$\text{Equity ratio} = \frac{\$124{,}000}{\$188{,}000} = \frac{\$100{,}000 + \$24{,}000}{\$64{,}000 + \$124{,}000} = 0.66$$

which is below the generally preferred lower limit of 0.75.

4 The *operating ratio* shows the percentage of every dollar received from net sales which is needed to meet the cost of production and operations:

$$\text{Operating ratio} = \frac{\text{total expenses}}{\text{net sales}} = \frac{\$224{,}000}{\$247{,}500} = 0.905$$

A lower ratio is preferred by management for obvious reasons. The relatively high ratio for the firm represented in the income statement of Figure 11.2 suggests cost-reduction or sales-promotion efforts should be considered.

5 *Inventory turnover* is a ratio used to evaluate the passage of goods through a firm's operations:

$$\text{Inventory turnover ratio} = \frac{\text{net sales}}{\text{inventory}} = \frac{\$247{,}500}{\$37{,}600} = 6.6$$

where the total inventory given in the balance sheet is assumed to be the average level for the year. With typical turnover ratios of 5 to 10, the given firm appears to have a reasonable balance of inventory on hand. Desirable ratios vary widely for industries with different operating requirements.

**6** A *net-profit ratio* compares the profit after taxes with sales, total assets, or stockholders' equity and is usually expressed as a percentage:

$$\text{Net-profit ratio} = \frac{\text{profit after taxes}}{\text{sales}}$$

$$\text{or} \quad \frac{\text{profit after taxes}}{\text{total assets}}$$

$$\text{or} \quad \frac{\text{profit after taxes}}{\text{stockholders' equity}}$$

$$\text{Net-profit ratio} = \frac{\$16{,}000}{\$247{,}000} = 6.5\%$$

$$\text{or} \quad \frac{\$16{,}000}{\$188{,}000} = 8.5\%$$

$$\text{or} \quad \frac{\$16{,}000}{\$124{,}000} = 12.9\%$$

With sales in the denominator, the ratio indicates the volume of business needed to earn a certain profit. When the denominator is total assets or equity, the ratio is indicative of the return on investment. The sample company appears to be providing adequate but not handsome returns.

Financial ratios are much more meaningful when they are compared with industrial averages and past ratios. Industrial averages are published by several sources to indicate norms for specific industries. A graph that plots changes in ratios with respect to time is an indicator of management effectiveness, especially when compared with an equivalent graph of industry norms. However, the ratios offer only general guidelines. A deeper investigation is needed if it is important to fully understand a company's operations. For instance, a major research effort could absorb funds that would otherwise be displayed as profit. The resulting low net-profit ratio would make the firm appear weak, while in reality it might have developed a strong position for future profits.

## QUESTIONS

**11.1a** Consolidated balance sheets and income statements for two companies engaged in similar operations are presented in Figures 11.8 and 11.9. Compare the two companies as potential investments, or with respect to their financial health as potential suppliers to your company. Discuss the meaning of each ratio in terms of managerial effectiveness and the financial positions of the companies.

Income Statement
Year Ended December 31, 19xx

| | *Salt Co.* | *Pepper Co.* |
|---|---|---|
| Net sales | $220,000 | $126,800 |
| Cost of goods sold | 108,000 | 55,000 |
| Gross profit on sales | $112,000 | $ 71,800 |
| *Operating expenses* | | |
| Maintenance and repairs | $ 6,000 | $ 4,000 |
| Depreciation | 15,200 | 12,000 |
| Bad-debt expense | 3,600 | 3,200 |
| Selling expense | 37,000 | 32,000 |
| Administrative and general | 12,600 | 11,000 |
| Total operating expense | $74,000 | $62,200 |
| Net operating profit | $37,600 | $9,600 |
| Other income | 3,000 | 1,600 |
| Other expense | 17,000 | 8,000 |
| Net income | $23,600 | $3,200 |

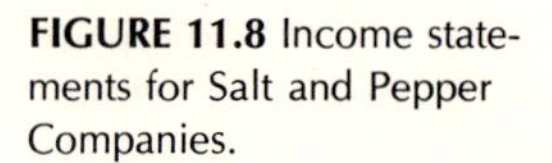

**FIGURE 11.8** Income statements for Salt and Pepper Companies.

**11.1b** Utilize the basic data in Table 11.12 to determine the values of the ratios listed below for the In-Out Universal Corporation. Industry norms are also listed. Compare the calculated ratios to the industry norms in appraising the economic health of I-O-U. Assume the income-statement values are doubled for a full year's operation, and that 10 workers are employed.

| *Ratio* | *Industry Norms* |
|---|---|
| Current | 2.0 |
| Acid test | 1.0 |
| Equity | 0.8 |
| Net profit/equity | 15% |
| Net profit/worker | $2000 |
| Accounts-receivable turnover: $\left(\frac{\text{credit sales}}{\text{average accounts receivable}}\right)$ | 8.0 |

*11.2* ***Taxes: Who Pays How Much for What?*** It has long been argued whether taxes should be based on the amount of service received or the ability to pay. Such arguments will likely continue as long as taxes are assessed, which means they will go on for a long time because taxes have a history of being levied indefinitely and some people will surely be dissatisfied with the levies.

A consolidated listing of federal income and expenditures for 1975 is shown in Table 11.13. It indicates a deficit of $43.6 billion, a condition which also occurred in 14 of the 15 previous years. The bulk of the income is derived from individual and corporate income taxes, plus social-security payments by individuals and organizations. Among the millions who benefited from expenditures in 1975 were people

Balance Sheet
December 31, 19xx

| | *Salt Co.* | *Pepper Co.* |
|---|---|---|
| *Current assets* | | |
| Cash | $ 60,000 | $ 30,000 |
| Marketable securities | 12,000 | 16,000 |
| Accounts receivable | 19,000 | 25,400 |
| Inventories | 80,000 | 48,000 |
| Total current assets | $171,000 | $119,400 |
| *Fixed assets* | | |
| Investments | $ 32,000 | $ 40,000 |
| Land, buildings, equipment (net) | 346,000 | 251,200 |
| Total fixed assets | $378,000 | $291,200 |
| *Prepaid expenses* | 10,400 | 3,000 |
| *Total assets* | $559,400 | $413,600 |
| *Current liabilities* | | |
| Notes payable | $ 7,000 | $16,000 |
| Accounts payable | 32,000 | 31,200 |
| Other current liabilities | 58,000 | 48,000 |
| Total current liabilities | $97,000 | $95,200 |
| *Fixed liabilities* | | |
| Mortgage payable (due in 10 years) | $111,600 | $180,000 |
| *Total liabilities* | $208,600 | $275,200 |
| *Net worth* | | |
| Preferred stock (6% cumulative) | $ 84,000 | $ 40,000 |
| Common stock | 120,000 | 68,000 |
| Appropiated surplus | 48,000 | 21,600 |
| Unappropriated surplus | 98,800 | 8,800 |
| Total net worth | $350,800 | $138,400 |
| *Total liabilities and net worth* | $559,400 | $413,600 |

**FIGURE 11.9** Balance sheets for Salt and Pepper Companies.

| *Income* | *Billions of Dollars* | *Expenditures* | *Billions of Dollars* |
|---|---|---|---|
| Taxes on individuals | $122.4 | National defense | $ 86.6 |
| Taxes on corporations | 40.6 | Interest on debt | 32.7 |
| Social-security taxes | 73.7 | Social-security benefits | 63.6 |
| Excise taxes | 16.6 | Medicare and health | 27.6 |
| Estate and gift taxes | 4.6 | Public assistance, education, and unemployment compensation | 56.3 |
| Unemployment-insurance taxes | 6.8 | Aid to veterans and civil-service retirement | 23.6 |
| All other revenue | 16.3 | All other | 34.2 |
| | $281.0 | | $324.6 |

**TABLE 11.13** Federal income and expenditures in the United States for the year ending June 30, 1975.

collecting social security (32.1 million), school lunches (26.6 million), medicaid (23.2 million), food stamps (18.4 million), medicare (13.2 million), and unemployment compensation (11.5 million). When the many additional federally subsidized programs (housing, aid to businesses, farm price supports, college tuition loans, etc.) are included, nearly everyone in the nation is affected in one way or another.

The rates of the graduated individual income tax climb to 70 percent on taxable incomes above $100,000 for single persons, and $200,000 for married couples. But the actual amount of taxes paid tends to be much lower, owing to various "tax breaks," as is indicated in Table 11.14 by the average tax paid in 1973 according to adjusted gross income levels.

**TABLE 11.14** Personal income taxes paid on adjusted gross incomes in 1973. (*Source:* Internal Revenue Service.)

| *Adjusted Gross Income* | *Average Income* | *Average Tax* | *Average Tax Rate, %* |
|---|---|---|---|
| Under $5000 | $ 3,571 | $ 188 | 5.3 |
| 5000–10,000 | 7,469 | 674 | 9.0 |
| 10,001–15,000 | 12,470 | 1,350 | 10.8 |
| 15,001–25,000 | 18,879 | 2,473 | 13.1 |
| 25,001–50,000 | 33,018 | 5,676 | 17.2 |
| 50,001–100,000 | 69,945 | 17,448 | 24.9 |
| 100,001–200,000 | 146,002 | 44,864 | 30.7 |
| 200,001–500,000 | 344,012 | 113,322 | 33.1 |
| 500,001–1,000,000 | 920,229 | 305,568 | 33.2 |

A similar condition exists for corporate tax returns. In 1972, three out of four corporations paid federal tax at the 22 percent rate because their incomes were less than $25,000. However, 90 percent of all corporate income is earned by big firms subject to the 48 percent income-tax rate. But no industry group paid an average tax rate as high as 40 percent on 1972 income, as shown by Table 11.15.

**TABLE 11.15** Federal taxes paid as a percentage of net income in 1972 for selected industry groups.

| *Corporate Group* | *Average Tax on Net Income, %* | *Corporate Group* | *Average Tax on Net Income, %* |
|---|---|---|---|
| Motor vehicles | 39.9 | Retail, wholesale trade | 30.1 |
| Food processing | 36.0 | Construction | 26.8 |
| Chemicals | 34.8 | Banking | 16.7 |
| Electrical equipment | 34.2 | Petroleum refining | 9.9 |
| Utilities, transportation | 33.6 | Oil and gas production | 3.8 |

**QUESTIONS**

**11.2a** Citizens' groups and legislators perennially debate the merits of various tax-saving provisions. Several of the more prominent ones are briefly described below. Discuss the factors for and against each of them.

1 *Substitution of taxes paid in foreign countries for taxes owed in the United States* Foreign taxes are traded one-for-one on domestic taxes instead of counting as a business expense; designed to encourage industry to expand and modernize.

2 *Capital gains* Gains on securities or property held over 6 months are taxed at a

maximum rate of 25 percent; intended to provide investment funds for businesses.

3 *Tax-exempt securities* Interest income on bonds of state and local governments is exempt from federal income tax; makes borrowing less expensive for local public projects.

4 *Investment tax credits* Tax credits cut $3 billion from the tax liabilities of corporations in 1972, and the allowance was increased from 7 percent to 10 percent in 1975; designed to modernize production capabilities and to encourage expenditures for plant and equipment to increase the level of business activity during slack periods.

5 *Depletion allowances* Certain extractive industries may deduct a specified percentage of their sales in determining income (22 percent for the petroleum industry, up to one-half its taxable income), which may accumulate a total deduction greater than the value of the depleted resource; intended to stimulate highly risky oil exploration and the development of natural resources.

**11.2b** Some people feel that all corporate taxes should be eliminated because the taxes are simply passed on to consumers as a "regressive" levy, and corporate profits are taxed twice—once on corporate earnings and again on the dividends declared which are counted as personal income to the receiving individuals. Others say that bigger corporations would be taxed more and at a "progressive" rate, that all the earnings are not returned to stockholders and some thereby escape taxation, and that these corporate retained earnings contribute to the effectiveness of the capital-gains loophole.

Discuss the above arguments and others concerning the appropriate tax rate for corporations. What are "regressive" and "progressive" taxes? How do they enter the argument?

# CHAPTER 12

# INDUSTRIAL PRACTICES

Every year millions of Americans consider buying a new car. Ten million or so decide to do it. In magnitude, the purchase of an automobile lies between expenditures for everyday necessities, such as food, and the largest investment an average family makes, purchasing a home. Factors involved in the car-purchase decision include:

1. Where the money will come from to pay for it. A loan? Savings?
2. When it should be purchased. Will prices rise? Will income rise faster or slower?
3. How much strain the payments and upkeep will put on the family budget. Payments for how long? Upkeep costs?
4. Which car is the best one to buy. Price? Resale value? Maintenance expenses? Operating costs? Taxes?
5. Whether a new car is more important than other things the money could buy. Vacations? Furniture? Clothing?

The same practical questions face industrial enterprises. Firms are confronted with continuing expenses necessary to keep operating, intermediate-sized costs to improve operations, and occasional major investments to change the mode or direction of operations. A firm's economic decisions are equivalent to family considerations of expenditures for food, a car, and a home. The industrial counterparts of the factors involved in the family's car-purchase decision include:

1 *Sources of funds for investments* Retained earnings? Debt? Selling stock? Tax effects of each option?
2 *Inflation and price instability* Effect of general increases in price? Gain or loss from specific cost changes?
3 *Payback period* How long before an investment becomes profitable?
4 *Decision-making practices* Who prepares and who reviews estimates? Proper form for an economic analysis? Who decides?
5 *Capital budgeting* Which proposals get funded when there are more good ones than there is money to support them all?

These considerations form the subject matter of this chapter—the practices followed by industrial organizations in meeting their financial needs and evaluating investments.

## SOURCES OF FUNDS

Securing capital is a prerequisite to implementing an acceptable proposal. Ideas for profitable investments are generated and data are collected to support or refute the investment proposals on the assumption that funds will become available to carry out the best ones. These funds can be generated internally or acquired from external sources.

The amount of *internal funds* available for investment during a given time period is the positive difference between cash receipts and all cash disbursements for that period. The disbursements include operating expenses, interest charges, taxes, and dividends, but *not* depreciation charges. The flow of funds is shown in Figure 12.1. The influx of *external funds* is mainly from *debt* (bonds, mortgages, short-term notes, etc.) and *equity* (issues of common or preferred stock and other forms of partial ownership).

One of the best ways to cultivate funding sources is to have profitable operations or activities that strongly promise future profits. Even then there may be a scarcity of capital owing to inflated prices for new investments and the worldwide tightening of capital supplies. Mr. F. W. Searby, a director of McKinsey & Company, described the plight of not-so-profitable capital-intensive companies as follows:

> In the last 20 years, U.S. industrial companies in basic industries have substantially increased their debt in absolute terms and, perhaps what is more important, in relation to their equity. (By one recent measure, the debt of U.S. manufacturing companies has risen as a percentage of equity from 21% to 44%.) Capital requirements as measured by assets to sales are up; return on capital is down; the financing burden as measured by the dividend yield demanded by the market is up; and the total bite on profits taken by interest payments is also up.*

*From F. W. Searby, "Return to Return on Investment," *Harvard Business Review*, March–April 1975.

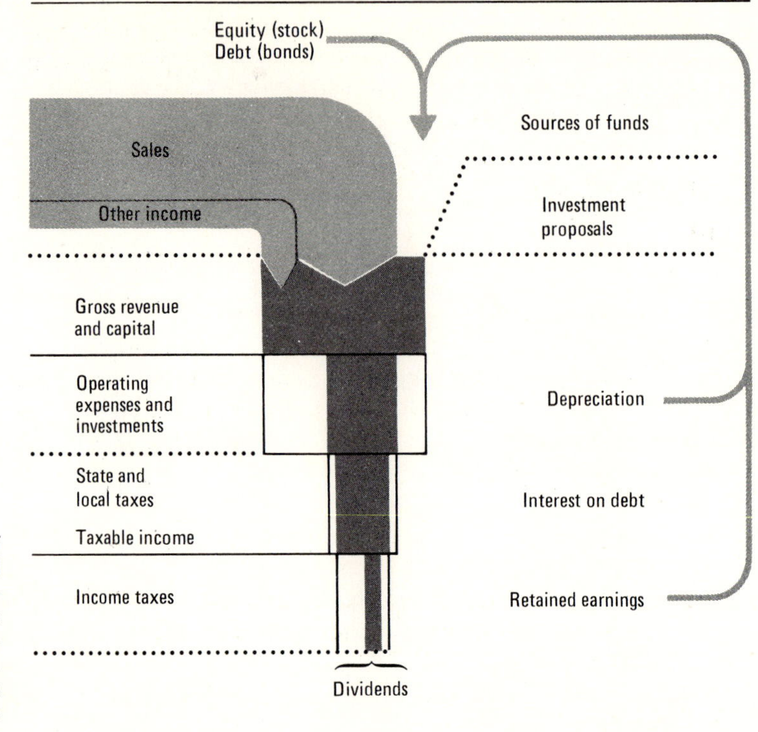


**FIGURE 12.1** Flow of funds into, through, and out of an industrial organization. New funds are derived from borrowing or selling equity. Sources of internal funds are depreciation charges and retained earnings.

## STRATEGIC CONSIDERATIONS CONCERNING FUNDING SOURCES

The management of an industrial enterprise works for the owners. In smaller firms the managers may be the owners. Either in the role of an owner or representing owners, management has the responsibility of conducting operations in ways that financially benefit the equity holders. Owners receive gains from successful corporate operations in two forms: dividend payments and increases in the value of their stock. Each of the four main sources of funds shown in Figure 12.1 has its own effect on returns to equity holders.

*Depreciation charges* are the least controversial source of funds for new investments. How a firm allocates the depreciation reserve shown on accounting records is not specified by legal requirements. However, there is an implied obligation that depreciation charges be applied to purchases that maintain the production capabilities of the firm. Investments in replacements for worn assets and new processes to update production functions are customary practices.

*Retained earnings* come from after-tax income that could be distributed to shareholders as dividends. Expenditures of retained earnings are judged by the effect they have on the market price of stock. Stock prices typically reflect current earnings per share and the growth potential of a firm. If the stock market is confident that a firm is building up to a more profitable position, the stock price can rise even when the earnings are modest and no dividends are declared over a prolonged period.

An indication of the after-tax cost of equity financing is

$$\text{Capitalization rate} = \frac{\text{earnings per share}}{\text{market price per share}}$$

where

$$\text{Earnings per share} = \frac{\text{annual dividends} + \text{retained earnings}}{\text{number of shares of stock}}$$

and the market price is the amount bid for a share of stock. The capitalization rate is the "value" of earnings to an individual stockholder before taxes on personal income are imposed. For instance, if corporate after-tax earnings are \$5 per share and the stock is selling for \$50 per share, then

$$\text{Capitalization rate} = \frac{\$5}{\$50} = 0.10 \quad \text{or} \quad 10\%$$

This suggests that cash kept for internal operations rather than distribution to stockholders should have a reinvestment rate of return of at least 10 percent.

*Equity financing* results from selling part ownership in a firm. When additional shares of stock are issued and sold, and profits do not increase, there is a dilution of earnings that decreases the capitalization rate.

*Debt financing* does not dilute ownership, and it offers certain tax advantages, but the incurred indebtedness can be an onerous drain on finances during periods of low economic activity. Example 12.1 illustrates some of the considerations involved in an equity-versus-debt financing decision.

**Example 12.1** **A Decision to Borrow Funds or to Sell Equity**

Two brothers own and operate a vegetable cannery valued at \$300,000. They have an opportunity to buy a firm which now supplies most of their warehousing and distribution services. Only an insignificant amount of the to-be-acquired firm's assets is depreciable because the warehouses and equipment are leased. An annual saving of \$28,000, exclusive of financing costs, should result from the acquisition. There are insufficient equity funds to meet the \$100,000 purchase price.

The needed funds can be secured from a bond issue or by selling 25 percent control of the firm. The 20-year bonds would carry 8 percent annual interest. Which source of funds would provide the greater return to the brothers on their equity when their effective income-tax rate is 40 percent?

**Solution 12.1** The bond expense would amount to an annual interest charge of \$100,000(0.08) = \$8000 and a commitment to have \$100,000 available in 20 years to retire the debt. If the brothers' tax rate is assumed to remain unchanged, the annual after-tax return from the acquisition would be

$$\begin{aligned}\text{Net annual return} &= \text{before-tax returns} - \text{interest payments} - \text{taxes}\\ &= \$28{,}000 - \$100{,}000(0.08) - (\$28{,}000 - \$8000)(0.4)\\ &= \$28{,}000 - \$8000 - \$8000 = \$12{,}000\end{aligned}$$

which provides a

$$\text{Return on equity} = \frac{\$12{,}000}{\$300{,}000} = 0.04 \quad \text{or} \quad 4\%$$

when the acquisition is made from funds obtained by borrowing.

If they sell one-quarter interest in their organization for \$100,000 and invest that amount in the firm, the brothers' equity is still (\$100,000 + \$300,000)(0.75) = \$300,000. The annual \$28,000 saving now comes from ownership funds for which no interest is owed and no tax deductions are allowed:

$$\text{Net annual return} = \$28{,}000 - \$28{,}000(0.4) = \$16{,}800$$

The brothers' share is then \$16,800(0.75) = \$12,600 which provides a

$$\text{Return on equity} = \frac{\$12{,}600}{\$300{,}000} = 0.042 \quad \text{or} \quad 4.2\%$$

Both return-on-equity figures require interpretation. The return provided by borrowing \$100,000 does not include provisions for paying off the debt. A sinking fund that draws interest at the same rate as the loan would require annual payments from the revenue earned by the investment of

$$\text{Annual redemption reserve} = \$100{,}000(A/F,\ 8,\ 20) = \$2185$$

which reduces the net return to \$12,000 − \$2185 = \$9815. This figure is less than the net return for the brothers when they sell one-quarter ownership, but it represents *only* the return from the investment in the distribution service. The brothers would also have to share the profit from their regular cannery operations with the new owners.

If it is assumed the brothers had an after-tax return on investment of 15 percent before acquiring the distribution service, the new return on equity when \$100,000 is borrowed would be

$$\text{Return on equity by debt financing} = \frac{\$300{,}000(0.15) + \$9815}{\$300{,}000} = 0.183 \quad \text{or} \quad 18.3\%$$

The preacquisition profit due the brothers is reduced by one-fourth through the sale of equity, because the new owners share the total returns. The resulting proportion of the brothers' share of total returns after the acquisition would be

$$\text{Return on equity after selling 25\% ownership} = \frac{\$45{,}000 - \$45{,}000(0.25) + \$12{,}600}{\$300{,}000} = 0.155 \quad \text{or} \quad 15.5\%$$

which is slightly better than the preacquisition return on equity but less than that afforded by debt financing. In addition, the brothers would have to share their management prerogatives with the new owners. The effect on managerial decision making is not a major consideration in selling a small block of shares in a large corporation, but the sale of a controlling interest or even a significant minority interest can create an uneasy leadership situation. Therefore, the brothers would probably be wise to employ debt financing.

## Financial Leverage

The ratio of total debt to total assets is called the *leverage factor*. A firm having assets of $50 million and a debt of $20 million has a leverage factor of 0.4. Financial leverage, as measured by the leverage factor, is akin to the *operating leverage* described in Economy Exercise 3-2. The degree of operating leverage measures how sensitive operating levels are to changes in sales volume, and the *degree of financial leverage* suggests how sensitive the return on investment is to changes in earnings. (See Problem 12.4 for a quantitative measurement of the degree of financial leverage.)

The effects of different leverage factors on the percentage return on equity when earnings vary as a function of economic conditions are portrayed in Table 12.1. Three states of financial leverage are shown for total assets of $1 million with equity proportions of $1 million, $600,000, and $300,000. When all the assets are composed of equity holdings, the after-tax rate of return on investment is equal to half the before-tax rate of return, with the effective income-tax rate at 50 percent. Under unfavorable business conditions causing low sales, the percentage return on equity drops as the leverage factor increases. The trend is

| FINANCIAL LEVERAGE (IN $1000 UNITS) | LEVERAGE FACTOR = 0: EQUITY = 1000 AND DEBT = 0 | | | LEVERAGE FACTOR = 0.4: EQUITY = 600 AND DEBT = 400 | | | LEVERAGE FACTOR = 0.7: EQUITY = 300 AND DEBT = 700 | | |
|---|---|---|---|---|---|---|---|---|---|
| *Economic Conditions* | *Low Sales* | *Average Sales* | *High Sales* | *Low Sales* | *Average Sales* | *High Sales* | *Low Sales* | *Average Sales* | *High Sales* |
| RR before taxes and interest | 2% | 8% | 14% | 2% | 8% | 14% | 2% | 8% | 14% |
| Net earnings before interest and taxes | $20 | $80 | $140 | $20 | $80 | $140 | $20 | $80 | $140 |
| Interest on debt at $i$ = 8% | 0 | 0 | 0 | $32 | $32 | $32 | $56 | $56 | $56 |
| *Taxable income* | $20 | $80 | $140 | −$12 | $48 | $108 | −$36 | $24 | $84 |
| Taxes at effective tax rate of 50%* | $10 | $40 | $70 | 0 | $24 | $54 | 0 | $12 | $42 |
| *Available for equity returns* | $10 | $40 | $70 | −$12 | $24 | $54 | −$36 | $12 | $42 |
| Percentage return on equity | 1% | 4% | 7% | −2% | 4% | 9% | −12% | 4% | 14% |

*Assumes losses are carried back and result in tax credits.

**TABLE 12.1** Equity holders' returns under different leverage factors and economic conditions when assets total $1 million (shown in units of $1000).

reversed when favorable business conditions allow high sales. At average sales, when the before-tax rate of return equals the interest rate for borrowed funds, all three leverage positions result in the same after-tax return on equity investments. In general, *whenever favorable business conditions allow a before-tax rate of return greater than the interest rate on debt, the higher the leverage factor, the higher the after-tax percentage return on equity.*

A highly leveraged firm produces very attractive return-on-investment figures during periods of prosperity, yet it can quickly become a target for management criticism or a victim of bankruptcy during a recession. Lenders are often a determinant of the degree of leverage a firm can attain. The credit standing of a borrower is reduced by excessive borrowing, and lenders attempt to protect their loans by limiting the leverage to certain norms. The norms vary among industries with respect to the stability of incomes, and within an industry according to the confidence lenders have in a firm's management; service industries and public utilities typically have very high debt-to-asset ratios (0.6+) and manufacturing has relatively low leverage (0.3 to 0.4). A thriving business can expand faster with the infusion of debt funding, and a sick business may recover more rapidly with the aid of borrowed funds, but a firm weathers brief spells of adversity better when it is less leveraged and not obligated to pay off large debts.

## Tactical Considerations concerning Funding Sources

After an organization has established its strategic leverage policy, there is still a question of long-term versus short-term debt. This question is most critical to working-capital management—the administration of a firm's short-term assets, such as cash and accounts receivable, to meet day-by-day financial obligations. The question also arises in investment decisions. The answer depends on the cost of loans and the nature of operations.

Short-term (under 1 year) loans conventionally carry a lower interest rate than long-term loans. The actual percentages vary from year to year, depending mostly on the national monetary policy. Typical values might be 6 percent for a 90-day note, 8 percent for a 3-year loan, and 10 percent for a 20-year bond. In taking advantage of the lower interest rates allowed for short-term debt, a firm incurs the risk of being unable to repay the loan as it matures. Temporary revenue deficiencies could prevent a firm from refunding its debts on time and cause it to negotiate renewals from its creditors. The creditors may perceive the difficulties as a decrease in credit worthiness and raise the interest rates or even refuse new loans. Long-term (over 5 years) loans reduce the risk of failure to refund by decreasing the number of times debts mature.

A conservative tactic of relying completely on long-term debt is compared with a more aggressive short-term debt tactic in Table 12.2. Both approaches are based on an equity holding of $500,000, a leverage factor of 0.5, and earnings before tax and interest of 14 percent. The question of relying completely on long-term or short-term credit is narrowed to a choice between a 2 percent loss on the percentage of equity return and taking the associated risk of loan-due dates falling in periods of low earnings.

The allowable degree of leverage and long- or short-term credit policy may also influence an economic comparison of proposals. Particularly in a small firm, where the use of credit

**TABLE 12.2** Effect of debt maturity on the percentage return on equity when assets are $1 million and earnings before tax and interest are 14 percent. The conservative tactic, holding only long-term loans, yields a lower return on investment than the riskier, aggressive tactic of using all short-term credit.

| | *Conservative Tactic* | *Aggressive Tactic* |
|---|---|---|
| Equity | $ 500,000 | $ 500,000 |
| Debt | 500,000 | 500,000 |
| Total debt and equity | $1,000,000 | $1,000,000 |
| Short-term debt, 6% | 0 | 500,000 |
| Long-term debt, 10% | 500,000 | 0 |
| Earnings before interest and taxes | 140,000 | 140,000 |
| Less interest on debt | 50,000 | 30,000 |
| *Taxable income* | $ 90,000 | $ 110,000 |
| Less taxes at 50% rate | 45,000 | 55,000 |
| *Available for returns on equity* | $ 45,000 | $ 55,000 |
| Percentage return on equity | 9% | 11% |

is decided more by operating requirements than a strategic financial policy, the evaluation of individual proposals is based on the current level of equity or debt funding available and the interest rate required for loans with different maturities. The following example demonstrates the difference in the cost of an asset acquired with debt financing by loans of contrasting durations.

**Example 12.2** **Comparison of Costs for Purchasing an Asset by Different Financing Plans**

A construction company must obtain a piece of heavy equipment to work on a long-term project. The initial cost is $100,000, and it will have no salvage value at the end of its 10-year useful life. The company uses double-declining-balance depreciation with a switch to straight-line depreciation and a minimum after-tax rate of return of 10 percent. Bank loans of 3 years at an annual interest rate of 8 percent and of 8 years at 9 percent are being considered for financing the purchase. Both loans require 10 percent down payment. The construction company has an effective tax rate of 40 percent.

Which financing plan would be the most economical?

**Solution 12.2**

The discounted cash flows for the purchase of the construction equipment by both funding plans are shown in Table 12.3. The longer loan offers a lower total discounted cash flow than the shorter loan, even though the 8-year loan has a higher interest rate. A casual observation that interest charges on the longer loan exceed those on the shorter loan by $40,084 − $14,769 = $25,315 might leave the impression that the 3-year financing plan is superior. The tax savings generated by the additional stretched-out interest and loan charges make the longer financing plan preferable.

| | | THREE-YEAR LOAN AT 8% | | | | EIGHT-YEAR LOAN AT 9% | | | |
|---|---|---|---|---|---|---|---|---|---|
| *End of Year, N* | *Annual Depreciation Charge (1)* | *Annual Loan Payment (2)* | *Interest Charge (3)* | *Tax Saving (4)* | *PW of Cash Flow (5)* | *Annual Loan Payment (2')* | *Interest Charge (3')* | *Tax Saving (4')* | *PW of Cash Flow (5')* |
| 0 | | | | | −$10,000 | | | | −$10,000 |
| 1 | $ 20,000 | −$ 34,923 | $ 7,200 | $10,880 | −21,857 | −$ 16,261 | $ 8,100 | $11,240 | −4,606 |
| 2 | 16,000 | −34,923 | 4,982 | 8,393 | −21,926 | −16,261 | 7,366 | 9,346 | −5,820 |
| 3 | 12,800 | −34,923 | 2,587 | 6,155 | −21,614 | −16,261 | 6,565 | 7,746 | −6,575 |
| 4 | 10,240 | | | 4,096 | 2,798 | −16,261 | 5,692 | 6,373 | −7,005 |
| 5 | 8,192 | | | 3,277 | 2,035 | −16,261 | 4,742 | 5,174 | −7,206 |
| 6 | 6,554 | | | 2,622 | 1,480 | −16,261 | 3,705 | 4,104 | −7,249 |
| 7 | 6,554 | | | 2,622 | 1,346 | −16,261 | 2,575 | 3,651 | −6,898 |
| 8 | 6,554 | | | 2,622 | 1,223 | −16,261 | 1,343 | 3,159 | −6,576 |
| 9 | 6,554 | | | 2,622 | 1,112 | | | 2,622 | 1,112 |
| 10 | 6,552 | | | 2,621 | 1,011 | | | 2,621 | 1,011 |
| *Totals* | $100,000 | −$104,769 | $14,769 | $45,910 | −$64,392 | −$130,008 | $40,084 | $50,793 | −$59,812 |

**TABLE 12.3** Calculation of the present worth of alternative financing plans. An asset with $P = \$100{,}000$, $S = 0$, and $N = 10$ is depreciated (1) by the double-declining-balance method for the first 5 years and by the straight-line method for the last 5 years. Annual payments (2) repay a $90,000 loan plus interest charges (3). Tax savings (4) result from multiplying the deductible charges for interest plus depreciation [(1) + (3)] by the effective income tax rate, 40 percent. The cash flow at time zero is the $10,000 down payment, and the cash flow for successive years is the difference between loan payments and tax savings [(4) − (2)]. The present worth of this flow (5) is determined by multiplying each year's cash flow by the appropriate present-worth factor, $(P/F, 10, N)$.

**Economy Exercise 12-1** Two other options are available for acquiring the heavy construction equipment described in Example 12.2. It can be an outright cash purchase with equity funds or it can be leased. The lease would have an annual rental charge of $20,000 per year with an option to buy the equipment at the end of the eighth year for $5000. Assuming the option is exercised, the equipment would be depreciated by the straight-line method for the remaining 2 years of its useful life.

Compare these two alternatives with the 8-year financing plan in Example 12.2. The previous tax rate of 40 percent, minimum rate of return of 10 percent, and depreciation schedule are applicable to the cash-purchase and lease alternatives.

## INFLATION AND PRICE INSTABILITY

The complexities of forecasting were cited in Chapter 10, and the troublesome aspect of monetary depreciation was mentioned in Chapter 11. These difficulties are manifested by inflation and unstable prices. ***Inflation*** is a rising general level of prices, and ***price instability*** describes a condition in which prices for certain items do not change proportionately to others. Even during periods of acute inflation, some specific prices remain relatively constant, and some actually decline. ***Deflation,*** which has not been a consideration in recent years, is the opposite of inflation.

In 1974 and 1975, inflation became a national (and worldwide) concern as the general price level rose at a rate over 10 percent, higher by several percentage points than the rates that people were accustomed to. Those living on fixed incomes were hurt the most as the buying power of a dollar declined rapidly. Debtors benefited at the expense of creditors, because the buying power of the money used to pay off a debt was less than the buying power of the money when borrowed. Inflation can also affect certain aspects of economic analyses.

## Responsiveness of Receipts and Disbursements

A savings bond bought for $75 in 1972 to yield $100 in 1978 is ***unresponsive*** to price changes. It is a fixed contract that takes no cognizance of the reality that $100 in 1978 could not buy the same amount of goods as $75 in 1972. The investor's real wealth, as represented by the bond, declined over the 6 years. A ***responsive*** investment has receipts or appreciation that keep up with general price levels.

Most before-tax economic comparisons are insensitive to inflation. For a comparison to be valid, it must be based on equivalent outcomes. That is, the alternatives must satisfy the same objective criteria. They may do so by radically different means, but the means are related by the similarity of the functions expected. These similarities normally cause the cash flows to be equally responsive to inflation. Therefore, estimates of future cash flows expressed in ***current dollars*** will have essentially the same ***future dollar*** amounts.

Consider the two proposals shown below, in which the before-tax cash flows are stated in ***current dollars***—what the dollars will buy today.

| | COST | RECEIPTS IN CURRENT DOLLARS | | | | |
|---|---|---|---|---|---|---|
| | *Year 0* | *Year 1* | *Year 2* | *Year 3* | *Year 4* | *PW at 12%* |
| Proposal *A* | −$10,000 | $4000 | $4000 | $4000 | $4000 | |
| Proposal *B* | −14,000 | 5500 | 5500 | 5500 | 5500 | |
| *Net difference* | −$4,000 | $1500 | $1500 | $1500 | $1500 | $556 |

The net difference might also have been estimated in ***future dollars*** to show what receipts would have to be in years ahead to equal today's purchasing power. If inflation during the next 4 years is expected to be 6 percent per year, a consistent estimator would place the net difference as

| | COST | RECEIPTS IN FUTURE DOLLARS | | | |
|---|---|---|---|---|---|
| | *Year 0* | *Year 1* | *Year 2* | *Year 3* | *Year 4* |
| Net difference between proposal *A* and proposal *B* | −$4000 | $1590 | $1685 | $1787 | $1894 |

The procedure for evaluating proposals stated in future dollars is to convert them to current dollars. Doing so to the future dollars above, by applying the $(P/F, 6\%, N)$ factor, naturally converts them to the previously given current dollars. By either approach the present worth of the additional investment in proposal $B$ is \$556.

The same present worth is obtained by determining a combined discount rate that represents both the minimum required rate of return and the inflation rate. Let $i$ = rate of return and $i_f$ = inflation rate; then the combined rate is $(1 + i)(1 + i_f) - 1$. For the rates from the previous examples, where $i = 12$ percent and $i_f = 6$ percent,

$$\text{Combined interest-inflation rate} = (1.12)(1.06) - 1 = 1.1872 - 1 = 0.1872 \quad \text{or} \quad 18.72\%$$

Applying this rate to the given net difference expressed in future dollars produces the same present worth as previously calculated:

**TABLE 12.4** Combined interest-inflation rate applied to a future-dollar cash flow to obtain the before-tax present worth.

| *Year, N* | *Cash Flow in Future Dollars* | *(P/F, 18.72, N)* | *Present Worth* |
|---|---|---|---|
| 0 | −\$4000 | 1.00000 | −\$4000 |
| 1 | 1590 | 0.84237 | 1339 |
| 2 | 1685 | 0.70964 | 1195 |
| 3 | 1787 | 0.59786 | 1068 |
| 4 | 1894 | 0.50372 | 954 |
| | | | \$ 556 |

After-tax evaluations incorporating future-dollar estimates cannot be handled so conveniently. The problem lies with interest and depreciation charges that are necessarily based on current dollars which impinge on future-dollar cash flows; the charges in current dollars are deducted from future dollars to obtain the amount of taxable income. Because interest and depreciation are not responsive to inflation like other cash flows, the taxable income tends to be higher and the rate of return is consequently lower than the same calculations based strictly on current dollars. This is illustrated in Problem 12.14.

A companion problem occurs when the responsiveness varies among cash flows in a comparison. Then the receipts or disbursements for one alternative increase at a different inflationary rate than those for other alternatives. Several routines have been suggested to overcome these difficulties, but none has been awarded recognition as the best way. Fortunately, most alternatives in engineering economic comparisons are affected so nearly the same by inflation that it is seldom a deciding factor.

## Comparisons Based on a Range of Estimates That Reflect Unstable Prices

Forecasting future inflation rates is exceedingly difficult. Expert economic forecasters have missed badly in recent years. An industrial practice that avoids the need to forecast a percentage rate for general price-level advances is to estimate a range of likely cash flows for each alternative and discount each flow as if it were sure to happen. Then the ranges for alternatives are compared to select the preferred course of action. Judgment enters the comparison when one alternative performs better for a condition represented by one segment of the range, and another alternative is more attractive for a different segment. The contest is resolved formally by weighting one future more likely than another (see Chapter 14), or informally by considering intangibles and the management stance toward future cash commitments—the overall effect on the organization of unstable prices and the options open for responding to price changes.

Three estimates are typically utilized in the range:

1 *Objective estimate* The most likely cash flow and the one that would be used if only a single estimate were made
2 *More-favorable estimate* An optimistic appraisal based on an advantageous interpretation of future events
3 *Less-favorable estimate* A pessimistic assessment of the future that adversely affects the cash flow

Neither the more- nor less-favorable estimate is based on an extreme—the best or worst that could conceivably happen. They assess the outcomes of border conditions that are reasonably likely to occur. The three-estimate approach is also applicable to uncertainties besides price; analyses based on ranges of possible levels of activity, such as utilization rates and output quantities, or levels of performance may be more useful than single-estimate evaluations.

Item-by-item estimates are more credible than collective estimates. Additional work is required to develop estimates for smaller cash-flow categories, but the resulting summations are more likely to be accurate than block estimates, and the effort expended makes the analyst better informed about the situation. The cost of extra calculations and estimating is usually insignificant compared with the value of even a small improvement in the decision process. Therefore, when price instability casts doubt on future cash flows, or operating levels cannot be anticipated with any assurance, an evaluation encompassing a range of itemized estimates is a sound practice.

**Example 12.3 Range of Estimates That Reflect Unstable Future Prices for a Training Proposal**

A proposal has been made to introduce a training program to improve a production process which relies primarily on manual operations. New motions and fixtures for individual operators are expected to lower manufacturing costs from their present level by about $50,000 per year. It is difficult to predict the benefits of the training proposal because the size of the savings depends on the scale of operations and prices, which are functions of the marketplace and general economic conditions.

A range of estimates for possible future conditions is shown. Should the training program be adopted?

| *Items Estimated* | *Less-Favorable Estimate* | *Objective Estimate* | *More-Favorable Estimate* |
|---|---|---|---|
| Additional units produced annually | 60,000 | 75,000 | 100,000 |
| Price/unit | $2 | $3 | $3.50 |
| Annual income | $120,000 | $225,000 | $350,000 |
| Duration of income | 5 years | 6 years | 7 years |

| *Items Estimated* | *Less-Favorable Estimate* | *Objective Estimate* | *More-Favorable Estimate* |
|---|---|---|---|
| Training cost/year | \$45,000 | \$35,000 | \$30,000 |
| Required period of intensive training | 2 years | 2 years | 1 year |
| Operating expenses of new process | \$90,000 | \$160,000 | \$275,000 |
| Investment in consumable supplies | \$30,000 | \$30,000 | \$30,000 |

**Solution 12.3** A before-tax analysis using a 15 percent required rate of return is conducted by calculating the present worth of the cash flows for the three possible conditions: less favorable (LF), objective estimate (OE), and more favorable (MF).

$$\begin{aligned}\text{PW(LF)} &= (\$120{,}000 - \$90{,}000)(P/A, 15, 5) - \$45{,}000(P/A, 15, 2) - \$30{,}000\\ &= \$30{,}000(3.3521) - \$45{,}000(1.6257) - \$30{,}000\\ &= \$100{,}563 - \$73{,}157 - \$30{,}000 = -\$2594\end{aligned}$$

$$\begin{aligned}\text{PW(OE)} &= (\$225{,}000 - \$160{,}000)(P/A, 15, 6) - \$35{,}000(P/A, 15, 2) - \$30{,}000\\ &= \$65{,}000(3.7844) - \$35{,}000(1.6257) - \$30{,}000\\ &= \$245{,}986 - \$56{,}900 - \$30{,}000 = \$159{,}086\end{aligned}$$

$$\begin{aligned}\text{PW(MF)} &= (\$350{,}000 - \$275{,}000)(P/A, 15, 7) - \$30{,}000(P/A, 15, 1) - \$30{,}000\\ &= \$75{,}000(4.1604) - \$30{,}000(0.8696) - \$30{,}000\\ &= \$312{,}030 - \$26{,}087 - \$30{,}000 = \$255{,}943\end{aligned}$$

Although there is a chance the training program will result in a small loss if things go unfortunately, the opportunity for a very large gain under more favorable conditions makes the proposal very attractive. The three-phase analysis admits the chance of low returns and shows how good returns can be if things turn out favorably.

## THE PAYBACK CRITERION

One way to guard against the consequences of unexpected price (and cost) increases is to require proposals to return their original investment from the savings they generate in a very short period of time. A rough measure of this period is obtained by the *payback* or *payout* method of investment evaluation:

$$\text{Payback method} = \frac{\text{required investment}}{\text{annual receipts} - \text{annual disbursements}}$$

$$\text{or} \qquad = \frac{\text{first cost}}{\text{net annual savings}}$$

Data utilized in applying the formula are usually direct, *not discounted,* cash-flow amounts, and no salvage values are included. The resulting figure tells how long before the amount invested is recovered in actual dollars.

Claims such as "This investment will pay for itself in 18 months" are commonly heard in industry, and they indicate anxiety about the elapsed time before a proposed investment begins to show a profit. The payback period is an extremely popular investment criterion in this country and throughout the world (see Extension 12.2). Polls consistently reveal that the payback method is used more than any other comparison method by United States industry to rate investments, particularly proposals from operating units for relatively small capital expenditures to improve operations. In a 1964 survey* of 108 large corporations, over half used the payback method alone or in combination with other methods. Just 14 percent used discounted-cash-flow methods exclusively.

In actual practice the simple payback formula is sometimes modified to recognize capital recovery through depreciation charges and to include some discounted values. As will be seen in the next section of this chapter, the payback concept is often incorporated into discounted cash-flow analyses as an auxiliary criterion upon which proposals are judged. Still, the simple payback formula is widely used without elaborations, though it yields ratings that may lead to incorrect conclusions unless carefully interpreted. Its deficiencies arise from failing to give recognition to savings that occur after the payback period has passed and ignoring the time value of money.

For instance, as an extreme illustration of payback-period deception, an investment of \$1000 in an asset with a life of 1 year and an associated net return of \$1000 would yield a

$$\text{Payback period} = \frac{\$1000}{\$1000/\text{year}} = 1 \text{ year}$$

Another investment of \$1000 promises to return \$250 per year during its economic life of 5 years and yields

$$\text{Payback period} = \frac{\$1000}{\$250/\text{year}} = 4 \text{ years}$$

Favoring the alternative with the shortest payback period would rate the first alternative best, yet this alternative actually earns nothing: \$1000 − \$1000 = 0. Meanwhile, the spurned second alternative would have provided an annual return of 8 percent: $P/A = \$1000/\$250 = 4 = (P/A, 8, 5)$.

If the results of payback calculations are questionable, why are they used? There are at least two apparent reasons. One is simplicity. Since both depreciation and interest effects are usually ignored, calculations are quick and simple, and the results are intuitively logical. The other reason stems from a preoccupation with the flexibility of capital. If the money spent on an improvement is recovered rapidly, the funds can be allocated again to other desired projects. This concept tends to engender a false sense of security with reasoning such as, "If the project can quickly pay for itself, it must be good" or, "Only the best projects can meet our short-payback-period requirement."

While the payback-period criterion is not always appropriate, it does address the problem of working-capital management by attempting to protect a firm's liquidity position. During times of constricted income, when a firm may have trouble meeting operating expenses and have very limited capacity for funding new investments, an extremely short

*According to G. A. Christy, *Capital Budgeting: Current Practices and Their Efficiency*, Research Study 21, University of Oregon, 1966.

(as low as 6 months) payback period assures only quick-profit projects will be endorsed. In such exceptional situations, cash-availability considerations may be equal to or more important than total earnings. As an auxiliary criterion, requiring a short payback period guards against the chance of losses due to new technological developments.

**Example 12.4** **Payback-Period Comparisons for Alternatives with Different Lives**

The supervisor of a small machine shop has received three suggestions for reducing production costs. Suggestion $A$ is for new jigs and fixtures; $B$ is to rebuild an existing machine to improve its performance; and $C$ is a new machine to replace some manual labor. The following estimates have been made for the three alternative investments:

| | $A$ | $B$ | $C$ |
|---|---|---|---|
| First cost | \$1800 | \$2350 | \$4200 |
| Economic life, years | 3 | 4 | 8 |
| Net annual saving | \$645 | \$840 | \$1100 |
| Payback period, years | 2.8 | 2.8 | 3.8 |

The supervisor selects alternative $B$, explaining that, because of limited capital for investments, shorter payback periods are preferable. With alternatives $A$ and $B$ having the same payback period, $B$ is favored because the annual savings are greater than for $A$. What are the fallacies in this reasoning?

**Solution 12.4** Based on just the payback-period criterion, consistent attention to the condition of limited capital would dictate a preference for alternative $A$ over $B$ because $A$ requires less investment capital. Based on the rate of return on each investment, $C$ is the most attractive, as attested to by

$$\text{RR}(A)\text{:} \quad (P/A, i, 3) = 2.8 \quad \text{or} \quad i = 3.5\%$$
$$\text{RR}(B)\text{:} \quad (P/A, i, 4) = 2.8 \quad \text{or} \quad i = 15\%$$
$$\text{RR}(C)\text{:} \quad (P/A, i, 8) = 3.8 \quad \text{or} \quad i = 20\%$$

The flaws in the reasoning stem from a strict reliance on the payback criterion. Too short a period can obviously block acceptance of some high-return alternatives. If successive investments in shorter-lived alternatives produce a smaller rate of return than investment in one longer-lived alternative, flexibility is purchased by the losses in total earnings. In the example, successive utilization of alternatives $A$ and $B$ limits the maximum investment level to \$2350 during 7 years, but it also limits the rate of return to less than half that achievable by accepting alternative $C$.

## DECISION-MAKING PRACTICES FOR INDUSTRIAL INVESTMENTS

The procedures followed to arrive at a decision to accept or reject an investment proposal naturally differ among industrial organizations. A decision-making process gets more structured as the size of an organization increases. Larger firms specify channels for successive screenings of a proposal and provide preprinted forms for economic analyses. Evaluation methods vary according to the type of expenditures requested and the special requirements established by the firm, but most of the larger organizations do use some form of discounting.

One of the most elaborate analysis procedures is the MAPI (Machinery and Allied Products Institute) method. It has been refined over several years to provide a complete package of formulas, worksheets, and charts applicable to most economic evaluations. MAPI is basically a rate-of-return test that compares a defender with a challenger and with an improved challenger. Obsolescence is cleverly handled by factors for "accumulated inferiority" of a present machine with respect to improvements expected in future machines. Three variations of the basic MAPI formula represent different patterns of inferiority accumulation.

The standard-projection MAPI formula with sum-of-digits tax depreciation is shown below.*

$$C = \frac{n(Q^n - w^n)(Q - 1)^2 - (1 - b)P\left[(Q^n - 1) - n(Q - 1)\right]}{nQ^n(Q - 1) - (Q^n - 1)} - (Q - 1)$$

with $P = w^n\left[1 - w + py + \frac{(1 - P)z}{1 - b}\right]$

and $Q = 1 + i - bpy$

*where* $C$ = next year's capital consumption as a ratio of $P$
$n$ = useful life
$w$ = ratio of salvage value in year $n + 1$ to salvage value in year $n$
$w^n$ = salvage value expressed as a decimal fraction of original cost
$p$ = fixed ratio of debt to equity funding = 25%
$y$ = fixed interest rate on debt = 3%
$b$ = fixed income-tax rate = 50%
$z$ = fixed rate of return on equity after taxes = 10%
$i$ = fixed interest rate for capitalizing future sums = 8.25%

Prepared charts allow the formula to be solved graphically. When the assumptions built into the MAPI model fit a firm's operations and objectives, it offers a convenient and tested method for analyzing investment alternatives. Other, less rigid, approaches are described next.

## Request for Expenditure

Proposals for investments normally originate with operating personnel or staff analysts. They observe deficiencies in the present system and conceive ways to overcome them. Improvements may be directed toward *cost reduction* or *profit expansion,* and may or may not require commitments of capital. For instance, a training program for cost reduction

*From G. Terborgh, ***Business Investment Policy,*** Machinery and Allied Products Institute, Washington, D.C., 1958.

is "expensed," or classified as an operating cost, when no durable assets are required to conduct it. Similarly, a change in packaging styles designed to increase sales could be an expense if only cosmetic alterations are involved, or a capital cost when new equipment is needed to produce redesigned containers. A list of proposed objectives is shown in Table 12.5.

**TABLE 12.5** Objectives of investment proposals for industrial operations. (OSHA is the Occupational Safety and Health Act, and EPA is the Environmental Protection Agency.)

| *Categories of Proposed Expenditures* | |
|---|---|
| 1 Cost reduction | 7 Process modernization |
| 2 Profit improvement | 8 Replacement of equipment or materials |
| 3 Additional capacity, new plant | 9 Energy conservation |
| 4 Redesign or new product | 10 OSHA conformance |
| 5 Product improvement | 11 EPA conformance |
| 6 Plant modernization | 12 Metrication |

After preliminary data have been collected to confirm that an idea for cost reduction or profit expansion is worth pursuing, a proposal for expenditure is typically turned over to a staff engineer or analyst. If the initial assessment of the proposal is encouraging, firmer cost and demand figures are generated. These figures are based on estimates from engineering design, operations, purchasing, marketing, and other departments that would be involved in the proposed project in some way. Checklists of easily overlooked costs are often provided as reminders. An example is shown in Table 12.6.

A form to show the advantage of a proposed replacement or cost-reduction plan is given in Figure 12.2. In addition to the comparison of operating costs required on the form, the present and proposed facilities or equipment are described by model number, anticipated usage, net value or installed cost, and estimates of useful life and salvage value.

**TABLE 12.6** Checklist of cost categories to be accounted for in the preparation of a request for expenditure.

COST AND OPERATING SPECIFICATIONS FOR EXPENDITURE REQUESTS

1 *Land* Acquisition cost, legal and escrow fees, title search, etc.
2 *Buildings* Architectural design costs, contractor construction billings, transfer costs, waste treatment, power, fire control, heating and ventilating, elevators, loading docks, etc.
3 *Grounds* Grading and drainage, sidewalks, parking lots, art, fencing, utility tunnels, lighting, sprinklers, landscaping and landscape-architect fees, etc.
4 *Machinery and equipment* Invoice prices, taxes, freight, installation costs, utility connections, consultant fees, building modifications, material-handling systems, spare parts, down time during replacement, maintenance costs, output rates over time, insurance, utility usage rates and other operating costs, etc.
5 *Dies, molds, fixtures* Invoice price, freight, internal and external labor costs, modifications, associated materials, etc.
6 *Materials and supplies* Type, price and discounts, delivery time and reliability of deliveries, build-up-to-capacity costs, storage, equipment modifications, handling requirements, protection, associated equipment required, inspection costs, etc.
7 *Labor* Number, wages, skills, training costs, tools needed, affirmative action, overtime, and typical employment costs as indicated below.

| *Standard Wage* | *Incentives* | *Hourly Additives* | *Vacation, Shift Premiums, Allowances* | *Fringe Benefits* | *Employment Cost* |
|---|---|---|---|---|---|
| \$3.240 | \$0.661 | \$1.565 | \$1.640 | \$2.00 | \$9.106 |

## Investment Analysis

When no capital outlays are required, a request for expenditure (RFE) is based on just operating costs of the type shown in Figure 12.2, plus estimates of expanded profits, if appropriate. Some firms use only the payback criterion for relatively small investments, say up to $300; if the cost data meet the before-tax payback criterion, the RFE is transmitted to the screening group that decides which requests are funded.

Larger investments are usually subjected to a discounted cash-flow analysis. Again the type of analysis and what is included vary among organizations. For instance, some analyses use continuously compounded interest, and firms that process large quantities of raw materials may require a "working-capital cost" to be charged against any investment that increases the amount of inventory held.

Three methods of analyzing investments in use by large corporations in different industries are shown in Figures 12.3 to 12.6. They are designed to assure the uniformity of RFE preparations and to provide sufficient information for deciding which requests are most deserving. In essence, the forms are programmed routines for conducting the after-tax computations described in Chapter 11; only minor variations are added to customize the procedure.

Figure 12.3 is a cash-flow worksheet that utilizes the data collected from the "operating advantage" form shown in Figure 12.2. Irregular expenses (column 4) and depreciation or depletion charges (column 5) are subtracted from the annual savings in column 3 to determine the taxable profit (column 6). Columns 7, 8, and 9 are the tax effects and lead to the after-tax cash flow in column 10. When the amount invested is entered as a negative value in the "zero point" row, the cumulative total in column 11 reveals the payback period—the number of years before the cumulative cash flow switches from negative to positive. This payback is the flow in actual dollars rather than time-value dollars.

A convenient worksheet for calculating the rate of return, called the *profitability index*, is shown in Figure 12.4. The after-tax cash flow as developed in the last column of Figure 12.3 is the input for the "actual amount" column in the "operating benefits" section of Figure 12.4. Receipts and disbursements that occur irregularly before the nominal implementation date, the zero point, are listed in the negative-year rows; the regular cash flow is listed below the zero-point row. Similarly, regular and irregular investments are listed in the "capital costs" section. Then values in the "actual amount" column (PW at 0 percent interest rate) are multiplied in trial 2 by the present-amount factors for 10 percent. As long as the $A/B$ ratio at the bottom of the trial column is less than 1.0, another trial is needed. The present-amount factors for each trial are multiplied by the "actual amount" figures to obtain trial values. The given interest factors are based on mid-period continuous compounding, and consequently are different than the factors given in Appendix B.

The profitability index is determined from the sums of the discounted capital cost $A$ and the operating benefits $B$. $A/B$ ratios are calculated for each trial interest rate up to a rate which produces a ratio greater than 1.0. The profitability index lies between the interest rate which produced the last $A/B$ ratio less than 1.0 and the next larger one. The actual index percentage is determined graphically by plotting the $A/B$ ratios on the "interpolation chart." The profitability index is indicated by the point where the line connecting the plotted $A/B$ ratios crosses the vertical 1.0 line on the interpolation chart. The return on

OPERATING COST COMPARISON: REPLACEMENT OR COST REDUCTION

| | PRESENT | PROPOSED | (PRESENT – PROPOSED) DIFFERENCE | EFFECT ON VOLUME IN UNITS | |
|---|---|---|---|---|---|
| Direct Labor ______ @ ______ (Hrs. / Rate) | ______ | ______ | ______ | | |
| Indirect Labor ______ @ ______ (Hrs. / Rate) | ______ | ______ | ______ | INCREASE | DECREASE |
| Other Labor ______ @ ______ (Hrs. / Rate) | ______ | ______ | ______ | ______ | ______ |
| Subtotal Labor ______ | ______ | ______ | ______ | ______ | ______ |
| Benefits @ ______ % of Labor | ______ | ______ | ______ | OTHER | |
| | ______ | ______ | ______ | | |
| Maintenance Costs ______ | ______ | ______ | ______ | | |
| Tooling Costs ______ | ______ | ______ | ______ | | |
| Materials & Supplies ______ | ______ | ______ | ______ | | |
| Down Time Cost ______ | ______ | ______ | ______ | Attach Supporting Documents | |
| Utilities (power, air, etc.) ______ | ______ | ______ | ______ | ______ | |
| Floor Space (in feet) ______ | ______ | ______ | ______ | | |
| Subcontracting Costs ______ | ______ | ______ | ______ | COST BENEFITS IN SUBSEQUENT YEARS. | |
| Inventory (incr/decr) ______ | ______ | ______ | ______ | YR 2 $ ______ | |
| Safety ______ | ______ | ______ | ______ | YR 3 ______ | |
| Start-up Costs ______ | ______ | ______ | ______ | YR 4 ______ | |
| Training Costs ______ | ______ | ______ | ______ | YR 5 ______ | |
| Software Costs ______ | ______ | ______ | ______ | YR 6 ______ | |
| Other ______ | ______ | ______ | ______ | YR 7 ______ | |
| ______ | ______ | ______ | ______ | | |
| ______ | ______ | ______ | ______ | | |
| ______ | ______ | ______ | ______ | | |
| TOTALS | $ ______ | $ ______ | ______ | Prepared by | Date |
| Cost Benefits of Project—First Year (Present Minus Proposed) | | | $ ______ | | |

**FIGURE 12.2** Cost-analysis worksheet for a replacement or cost-reduction proposal.

CASH FLOW FROM OPERATIONS

| 1 | 2 | 3 | 4 | 5 | 6 | 7 | 8 | 9 | 10 | 11 |
|---|---|---|---|---|---|---|---|---|---|---|
| | | | | | | Taxes Paid | | | Cash Flow | |
| Calendar Year | Year No. | Operating Advantage Form xxx | Expense Portion of Project | Tax Depreciation & Depletion | Taxable Profit Col 3 Less 4 & 5 | Normal Tax | Tax Credits | Total Col 7 Less 8 | Annual Col 3 Less 9 | Cumulative |
| | -3 | | | | | | | | | |
| | -2 | | | | | | | | | |
| | -1 | | | | | | | | | |
| Zero Point | | | | | | | | | | |
| | 1 | | | | | | | | | |
| | 2 | | | | | | | | | |
| | 3 | | | | | | | | | |
| | 4 | | | | | | | | | |
| | 5 | | | | | | | | | |
| | 6 | | | | | | | | | |
| | 7 | | | | | | | | | |
| | 8 | | | | | | | | | |
| | 9 | | | | | | | | | |
| | 10 | | | | | | | | | |
| | 11 | | | | | | | | | |
| | 12 | | | | | | | | | |

Date: ________________, 19 ____ Division: ________________ RFE No. ________

Prepared by: ________________ Plant: ________________ Payback: ________

**FIGURE 12.3** Worksheet for calculating the after-tax cash flow and payback period before discounting.

CALCULATION OF PROFITABILITY INDEX (INTEREST RATE)

**Capital Costs**

| Year: Actual | Year: Relative to Zero | Trial No. 1, 0% Interest Rate: Actual Amount | Trial No. 2, 10% Interest Rate: Factor | Trial No. 2: Present Worth | Trial No. 3, 25% Interest Rate: Factor | Trial No. 3: Present Worth | Trial No. 4, 40% Interest Rate: Factor | Trial No. 4: Present Worth |
|---|---|---|---|---|---|---|---|---|
| Cash Outflow | | | | | | | | |
| | -3 | | 1.285 | | 1.873 | | 2.736 | |
| | -2 | | 1.162 | | 1.459 | | 1.834 | |
| | -1 | | 1.052 | | 1.136 | | 1.230 | |
| Zero Point | | | 1.000 | | 1.000 | | 1.000 | |
| | 1 | | .952 | | .885 | | .824 | |
| | 2 | | .861 | | .689 | | .553 | |
| | 3 | | .779 | | .537 | | .370 | |
| | 4 | | .705 | | .418 | | .248 | |
| | 5 | | .638 | | .326 | | .166 | |
| Totals A | | | | | | | | |

**Operating Benefits**

| Year: Actual | Year: Relative to Zero | Trial No. 1, 0% Interest Rate: Actual Amount | Trial No. 2, 10% Interest Rate: Factor | Trial No. 2: Present Worth | Trial No. 3, 25% Interest Rate: Factor | Trial No. 3: Present Worth | Trial No. 4, 40% Interest Rate: Factor | Trial No. 4: Present Worth |
|---|---|---|---|---|---|---|---|---|
| Cash Inflow | | | | | | | | |
| | -3 | | 1.285 | | 1.873 | | 2.736 | |
| | -2 | | 1.162 | | 1.459 | | 1.834 | |
| | -1 | | 1.052 | | 1.136 | | 1.230 | |
| Zero Point | | | 1.000 | | 1.000 | | 1.000 | |
| | 1 | | .952 | | .885 | | .824 | |
| | 2 | | .861 | | .689 | | .553 | |
| | 3 | | .779 | | .537 | | .370 | |
| | 4 | | .705 | | .418 | | .248 | |
| | 5 | | .638 | | .326 | | .166 | |
| | 6 | | .577 | | .254 | | .112 | |
| | 7 | | .522 | | .197 | | .075 | |
| | 8 | | .473 | | .154 | | .050 | |
| | 9 | | .428 | | .119 | | .034 | |
| | 10 | | .387 | | .093 | | .023 | |
| | 11 | | .350 | | .073 | | .015 | |
| | 12 | | .317 | | .057 | | .010 | |
| Totals B | | | | | | | | |
| Ratio A/B | | | | | | | | |

Interpolation Chart

Profitability Index (%): 0, 10, 20, 30, 40, 50

Ratio (A/B): 0, 0.5, 1.0, 1.5, 2.0

PROFITABILITY INDEX = ____________

RFE No. ____________ Date: ____________

Plant: ____________ Div.: ____________

**FIGURE 12.4** Worksheet for determining the profitability index of an investment proposal. The present worth of capital costs and operating benefits are summed to obtain *A/B* ratios at different interest rates. The *A/B* ratios are then graphed to indicate the profitability-index percentage.

| RETURN ON INVESTMENT (ROI) | | | | | | | | | | | | | | | |
|---|---|---|---|---|---|---|---|---|---|---|---|---|---|---|---|
| Year | 0 | 1 | 2 | 3 | 4 | 5 | 6 | 7 | 8 | 9 | 10 | 11 | 12 | 13 | 14 |
| A. Annual Savings | | $3.45 | 3.62 | 3.80 | 3.99 | 4.19 | 4.40 | 4.62 | 4.85 | 5.10 | 5.35 | 5.62 | 5.90 | 6.20 | 6.51 |
| B. S. O. D. Factor | | .1333 | .1238 | .1143 | .1048 | .0952 | .0857 | .0762 | .0667 | .0571 | .0476 | .0381 | .0286 | .0190 | .0095 |
| C. Depreciation Cost (C = Investment X B) | | $ (.96) | (.89) | (.82) | (.75) | (.68) | (.61) | (.55) | (.48) | (.41) | (.34) | (.27) | (.21) | (.14) | (.07) |
| D. Savings before Tax (D = A – C) | | $2.49 | 2.73 | 2.98 | 3.24 | 3.51 | 3.79 | 4.07 | 4.37 | 4.69 | 5.01 | 5.35 | 5.69 | 6.06 | 6.44 |
| E. Tax @ 52% (E = .52 X D) | | $1.30 | 1.42 | 1.55 | 1.68 | 1.82 | 1.96 | 2.12 | 2.27 | 2.44 | 2.60 | 2.78 | 2.96 | 3.15 | 3.35 |
| F. Savings after Tax (F = D – E) | | $1.19 | 1.31 | 1.43 | 1.55 | 1.68 | 1.82 | 1.97 | 2.10 | 2.25 | 2.40 | 2.57 | 2.73 | 2.91 | 3.09 |
| G. Depreciation Credit (G = –C) | | .96 | .89 | .82 | .75 | .68 | .61 | .55 | .48 | .41 | .34 | .27 | .21 | .14 | .07 |
| H. Annual Return (H = F + G) | | $2.16 | 2.20 | 2.25 | 2.30 | 2.36 | 2.43 | 2.51 | 2.58 | 2.66 | 2.74 | 2.84 | 2.94 | 3.05 | 3.16 |
| I. Cumulative Return | | $2.16 | 4.36 | 6.61 | 8.91 | 11.27 | 13.70 | 16.21 | 18.79 | 21.45 | 24.19 | 27.03 | 29.97 | 33.02 | 36.18 |
| J. Investment | $(7.168) | | | | | | | | | | | | | | |

Calculations

| | 0 | 1 | 2 | 3 | 4 | 5 | 6 | 7 | 8 | 9 | 10 | 11 | 12 | 13 | 14 | |
|---|---|---|---|---|---|---|---|---|---|---|---|---|---|---|---|---|
| Discount Factor—35% | 1.000 | .864 | .640 | .474 | 3.51 | .260 | .193 | .142 | .106 | .078 | .058 | .043 | .032 | .024 | .017 | Total |
| Present Value—35% | $(7.168) | $1.87 | 1.41 | 1.07 | .81 | .61 | .47 | .36 | .27 | .21 | .16 | .12 | .09 | .07 | .05 | .40 |
| Discount Factor—40% | 1.000 | .849 | .607 | .433 | .309 | .221 | .158 | .113 | .081 | .057 | .041 | .030 | .021 | .015 | .010 | Total |
| Present Value—40% | $(7.168) | $1.83 | 1.34 | .97 | .71 | .52 | .38 | .28 | .21 | .15 | .11 | .09 | .06 | .05 | .03 | (.44) |

Payout - - - - - - - - - - - - - - - - $= 3 + \frac{.558}{2.30} =$ 3.24 Years

Return on Investment (ROI) $= 35\% + \frac{.40}{.84} \times 5\% =$ 37.4%

Payout = 3.24 years  ROI = 37.4%

Prepared by: DR. J. KEARNS  Date: 5/29/79

Checked by: J. LOWRY  Date: 5/31/79

**FIGURE 12.5** Calculation procedure for determining the after-tax return on investment and payout period (figures in thousands of dollars).

| STEP NUMBER 1 | | |
|---|---|---|
| INVESTMENT | | COST |
| ACQUISITION PRICE | (a) | 10,500 |
| FREIGHT | (b) | 500 |
| INSTALLATION | (c) | |
| | (d) | |
| TOTAL CAPITAL | (x) | 11,000 |

| | | |
|---|---|---|
| DEPRECIABLE LIFE | five YRS | |
| RESIDUAL VALUE | 10 % | $ 1000 |

STEP NUMBER 4

INTERPOLATION CHART
(PLOT X/Y RATIO FOR EACH TRIAL)

DISCOUNT RATE %: 0, 5, 10, 15, 20, 25, 30, 35, 40, 45, 50, 55

TRIAL #4 · TRIAL #3 · TRIAL #2 · TRIAL #1

18%

X/Y RATIO: 0, 0.5, 1.0, 1.5, 2.0, 2.5, 3.0, 3.5
(ENTER DISC. RATE % INTERSECTING AT 1.0 IN STEP 6 RIGHT)

2ND PART OF STEP NUMBER 5

PAYOUT PLOT
PLOT FLOWBACK % FOR EACH YEAR

CUMULATIVE PERCENT: 20, 40, 60, 80, 100, 120

YEARS TO PAY OUT: 0, 1, 2, 3, 4, 5
(INTERSECTION AT 100% REPRESENTS PAYBACK PERIOD IN YEARS-ENTER IN STEP 6)

| PROJ. DESCRIPTION |
|---|
| |

| STEP NUMBER 6 |
|---|
| RATE OF RETURN |
| 18% |
| (FROM STEP 4) INTERSECTION OF CURVE WITH 1.0 X/Y RATIO LINE |
| PAYBACK |
| 3 YRS. |
| (FROM STEP 5) INTERSECTION OF CURVE WITH 100% LINE |

| STEP NUMBER 2 | | | | | | STEP NUMBER 3 | | | | | | |
|---|---|---|---|---|---|---|---|---|---|---|---|---|
| | | | | | | TRIAL 1 % Disc. Rate | TRIAL 2 10% Disc. Rate | | TRIAL 3 25% Disc. Rate | | TRIAL 4 40% Disc. Rate | |
| | COST BENEFITS | (LESS) DEPRE-CIA-TION | NET COST BENEFITS | AFTER PROFIT SHARING | AFTER-TAX EARNINGS | AFTER-TAX CASH FLOW | FAC-TOR | CASH VALUE | FAC-TOR | CASH VALUE | FAC-TOR | CASH VALUE |
| YEAR | (P) | (D) | (P-D) = C | (Cx.65)=E | (Ex.50)=A | (A+D)=B | f1 | B(f1) | f2 | B(f2) | f3 | B(f3) |
| 1st | 5000 | 2000 | 3000 | 1950 | 975 | 2975 | .91 | 2707 | .80 | 2380 | .71 | |
| 2nd | 7000 | 2000 | 5000 | 3250 | 1625 | 3625 | .83 | 3009 | .64 | 2320 | .51 | |
| 3rd | 9000 | 2000 | 7000 | 4550 | 2275 | 4275 | .75 | 3206 | .51 | 2180 | .36 | |
| 4th | 8000 | 2000 | 6000 | 3900 | 1950 | 3950 | .68 | 2686 | .41 | 1620 | .26 | |
| 5th | 6000 | 2000 | 4000 | 2600 | 1300 | 3300 | .62 | 2046 | .33 | 1089 | .19 | |
| 6th | | | | | | | .56 | | .26 | | .13 | |
| 7th | | | | | | | .51 | | .21 | | .10 | |
| | | | | TOTALS | Y1 | 18,125 | Y2 | 13,654 | Y3 | 9588 | Y4 | |
| | | | | RATIOS | X/Y | 0.61 | X/Y2 | 0.81 | X/Y3 | 1.15 | X/Y4 | |

11,000

| FIRST PART OF STEP # 5 | | |
|---|---|---|
| CUMULATIVE CASH FLOW BACK | | |
| AMOUNT | | |
| (CUM B) | CUM B / X | YEAR |
| 2975 | .27 | 1st |
| 6600 | .60 | 2nd |
| 10,875 | .99 | 3rd |
| 14,825 | 1.35 | 4th |
| | | 5th |

| | |
|---|---|
| REF. NO. 132 | |
| PREPARED BY J LOWRY | DATE 5/3/79 |

**FIGURE 12.6** Completed worksheet for determination of the rate of return on invested capital and the payback period for a profit-sharing organization.

investment is the percentage obtained by reading across from that intersection to the vertical scale on the left.

Another approach to investment analysis ensues from the use of written instructions rather than preprinted forms. This approach tends to be similar to the after-tax RR computations described in Chapter 11. As shown in Figure 12.5, a typical analysis starts with the estimation of gross savings before taxes $A$. On long-term projects, projections of these savings are responsive to expected future price trends. Depreciation charges $C$ are calculated with percentage tables based on the organization's depreciation policy; the sum-of-digits method $B$ is employed in the sample calculations. Savings less depreciation costs $D$ are multiplied by the tax rate $E$ to obtain after-tax savings $F$. Savings from depreciation $G$ added to the after-tax earnings yields the annual returns $H$, which are accumulated in $I$ to facilitate payback calculations. Then the investments ($7.168 at time zero in the example) are entered in $J$ to provide data for the return-on-investment (ROI) calculations shown below the table.

**Example 12.5** **Investment Analysis for a Profit-sharing Company**

A proposed investment will have an initial outlay, including freight charges, of $11,000. It is expected to produce cost savings of $5000, $7000, $9000, $8000, and $6000 over the next 5 years. At the end of the fifth year it will have a salvage value of $1000.

The company has a profit-sharing plan by which 35 percent of before-tax earnings are distributed among employees. This proportion of earnings is thus an operating expense to the company.

A "capital investment evaluation" worksheet is used to analyze proposals. It is based on straight-line depreciation, a 50 percent income-tax rate, and interest rates for discrete compounding periods. The proposal is to be evaluated by the six-step procedure given in Figure 12.6.

**Solution 12.5** In "step 1" of Figure 12.6, the purchase price and associated costs are recorded. The $11,000 total is the amount $X$ used in "step 3."

Cost savings, lease payments avoided, and other net benefits are entered in column $P$ at "step 2"; nonrecurring expenses are deducted in the year they occur. Depreciation in column $D$ results from

$$\text{Annual depreciation} = \frac{\text{total capital cost } (X) - \text{residual value}}{\text{depreciable life}}$$

This charge is deducted from amounts in column $P$ to obtain net cost benefits $C$. The portion of these benefits available to the company for capital recovery is 65 percent $E$. After-tax earnings $A$ result from applying a 50 percent effective income-tax rate to the company's profit share. The after-tax cash flow $B$ is earnings $A$ plus the depreciation charges $D$.

In "step 3" the after-tax yearly values $B$ are multiplied by single-payment present-worth interest factors to obtain the equivalent cash flow at different interest rates. The sums of the discounted values ($Y_1, Y_2, Y_3$) are the present worths of the net benefits and are the denominators for $X/Y$ ratios representing each interest rate. The $X/Y$ ratio is simply

$$\frac{X}{Y} = \frac{\text{investment at time zero}}{\text{PW of net benefits}} \quad \text{at } i = ?$$

The rate of return is determined graphically in "step 4." At the point where the line connecting $X/Y$ ratios crosses the vertical line at $X/Y = 1.0$, the value of $i$ that makes $X = Y$ is identified.

Similarly, for the graph in "step 5," the cumulative values of the after-tax cash flow at $i = 0$ $(B)$ are divided by the investment $X$ to obtain ratios to plot. The payback period is interpolated from the crossing point of the line connecting the plotted ratios and the horizontal 100 percent line.

Resulting interpolated values for the return on investment, 18 percent, and the payback period, 3 years, are recorded in "step 6."

## Letter of Justification

Most industrial organizations require submission of an explanatory letter along with the economic analysis of a proposed investment. A typical summary sheet for a request for expenditure is shown in Figure 12.7. The key figures from the analysis are displayed, and space is provided for approvals from the staff engineers and managers responsible for budgeting.

The accompanying written justification explains the reasons for the RFE. It describes the origin of the problem or the opportunity for which the RFE is submitted, and how the proposal will be implemented if approved. Intangible factors not included in the analysis are discussed. Supplementary data on related effects of the proposal are also supplied if pertinent to the decision.

## Postaudits

Following the progress of a design from the drawing board through construction to on-the-job performance is a natural progression to engineers. They take pride in their designs, welcome the opportunity to make modifications to improve performance, and realize they will be able to make future designs better by knowing what was previously successful and what failed. Follow-ups on capital investments tend to be neither as extensive nor as rigorous. Engineering economists would be well served by the same monitoring instincts displayed in other engineering pursuits.

The purpose of auditing investments is not to punish those who approved the proposals any more than quality-control inspections are intended to penalize a production process. The aim is to improve future analyses. How else can analysts learn whether their estimated cash flows are realistic? Many factors influence estimated receipts and disbursements; some are easily overlooked unless brought to attention by postaudits.

Cost-reduction proposals tend to be more accurately estimated than profit-expansion proposals because costs are more controllable than sales volume and market prices.* Yet

*According to J. L. Bower, *Managing the Resource Allocation Process*, Harvard, Cambridge, Mass., 1970.

LETTER OF JUSTIFICATION RFE No.

Description of Request:

Budget Spending 19 ______ $ ____________ Budget Total 19 ______ $ ____________

| | Total Investment | 19 ____ | 19 ____ | 19 ____ | 19 ____ | 19 ____ |
|---|---|---|---|---|---|---|
| Investment: Capital | | | | | | |
| Expense | | | | | | |
| Total Authorized Working Capital | | | | | | |
| Future Obligations | | | | | | |
| Total Justified | | | | | | |
| Benefit: Sales | | | | | | |
| Operating Profit after Tax | | | | | | |
| Cash Flow from Operations | | | | | | |
| % Return on Capital Employed after Tax | | | | | | |

| Type of RFE | PI | Payout |
|---|---|---|
| | | |

Most Critical Assumptions

Prepared by ____________ Sponsored by ____________

| Staff Review | | | Approvals | |
|---|---|---|---|---|
| Date | Dept. | Signature | Signature | Date |
| | | | | |
| | | | | |
| | | | | |
| | | | | |
| | | | | |

**FIGURE 12.7** Summary page for a letter of justification to support a Request For Expenditure.

the savings generated by implementing a lower-cost process or by replacing a higher-cost asset cannot be checked because the performance of the replaced assets cannot be determined for the latest operating conditions. A good indication is provided, however, by the comparison of estimated operating costs to the actual costs of the replacement.

A major advantage of a systematized postaudit program is the availability of the most recent cost data for categories not itemized in cost-accounting records. Maintenance costs for various types of machines, operating efficiency during training periods, labor costs for specific activities such as installation or set-up activities, and other data utilized in preparing requests for expenditure would be available from postaudits. The more complete computer-based management information systems that are spreading throughout industry promise to make such information conveniently accessible.

## CAPITAL BUDGETING

It could be said that the central theme of this book is *capital budgeting–the process of analyzing and determining optimum capital expenditures.* Successful capital budgeting is vital for the long-run prosperity of any industrial organization. Funding decisions affect all departments within a firm, and mistakes in forecasting or fulfilling asset requirements for any type of operation can damage the others. Both immediate and long-term needs have to be considered. And these needs have to be evaluated in conjunction with questionable future developments in unique markets and in the economy as a whole. The combined effects of internal pressures and external uncertainties make the capital-budgeting decision at once the most important and difficult kind of decision in financial management.

Many of the considerations involved in capital budgeting have been discussed in earlier chapters. Some critical concepts are referenced below to refresh the discussion.

1. *Supply and demand relationships* Capital is a resource that generally obeys the "laws" of supply and demand (Chapter 2).
2. *Analysis of intangibles and mixed-rating comparisons* Intangible factors are just as pervasive and influential in strategic capital-budgeting decisions as in tactical operating decisions (Chapter 5).
3. *Recognizing alternatives as dependent or independent* Relationships among requests for expenditure must be defined before capital allocations can be logically judged (Chapter 8).
4. *Cost of capital and minimum attractive rate of return* The cost of money to a firm sets a lower limit on the rationing of capital to proposed investments (Chapters 8 and 9).
5. *Management of capital* The flow of current and future funds resulting from capital acquisition, depreciation, and operating policies underlies capital-allocation decisions (Chapters 10 to 12).

Other economic comparison considerations and techniques from previous chapters contribute to more accurate and rational proposal inputs to the capital-budgeting process, but they have less direct bearing than the listed concepts on the actual allocation procedure.

The basic role of an engineering economist in capital budgeting is to uncover, examine, and prepare analyses of profitable ways to invest capital. The prepared analyses are acted upon by screening groups that vary in size and composition for different organizations; successive screenings by departmental groups may be forwarded to a board of top-level managers, all proposals may be put through a committee-type screening mechanism before a final allocation by select executives, or some other management device may be designed to guarantee a *careful, objective scrutiny* of requested expenditures. Major outlays for strategic objectives usually await approval of the board of directors or top management.

## Capital Inventory

It is almost certain that in any ambitious firm there will be more requests for expenditures than there is capital to satisfy them. An abundance of RFEs is a healthy sign that operating personnel, engineers, and managers are actively pursuing improvements. A firm might increase the amount of capital available in a budget period by external financing or by retaining a larger than normal share of earnings, but considerations of debt obligations and stockholders' interests limit the amount procurable. Unless unusually attractive investment opportunities become apparent, the amount of capital available is essentially set by the long-term financial policies of the firm. Asset expansion is necessarily related to future sales, because a decision to invest in a fixed asset is made in anticipation of future returns from sales, and future sales are the basis of capital formulation.

The *capital-inventory* approach for selecting which proposals to fund matches the cost of capital to the returns expected from investments. As portrayed in Figure 12.8, the shaded blocks form a ladder of investment proposals promising various rates of return; $RR_1$ is a smooth curve representing the blocks. Curves $RR_2$ and $RR_3$ represent other configurations of investment opportunities that might become available during a budgeting period. The cost of each additional dollar acquired for purposes of making capital expenditures is given by the bold MCC (*marginal cost of capital*) curve. This curve is relatively flat up to an amount which exhausts the normal sources of capital; beyond this point, it rises sharply as more expensive sources are tapped.

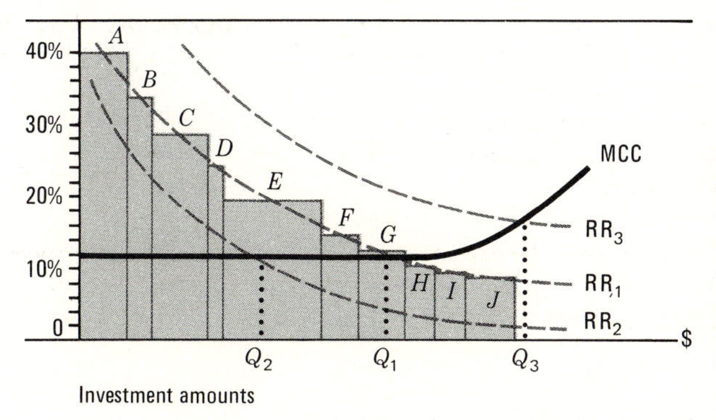


**FIGURE 12.8** Capital-inventory graph showing the marginal cost of capital (MCC) and three investment-proposal schedules ($RR_1$, $RR_2$, $RR_3$). The smooth-curve schedules represent discrete investment projects as indicated by the lettered ladder of proposals represented by $RR_1$.

If all proposals that promise a rate of return greater than the marginal cost of capital are funded in schedule $RR_1$, capital allocations would be made to proposals $A$ through $G$. These investments total $Q_1$. The same criterion would indicate investments in the amount $Q_2$ for the $RR_2$ schedule; $Q_2$ is smaller than $Q_1$ because the investment opportunities of schedule $RR_2$ are not so lucrative as those of $RR_1$. A wealth of proposals with high rates of return might induce the firm to secure additional capital to allow the $Q_3$ level of investment for the $RR_3$ schedule. Any proposals that have a rate of return smaller than the marginal cost of capital are rejected by the investment-inventory criterion.

## Capital Rationing

The practice of placing a fixed limit, or limited range, on total investments rations capital to the best proposals that can be funded within the budget limits. The floor for capital rationing is the cost of capital; that is, if the available investment capital exceeds the requested expenditures, no request would be granted that yields a rate of return smaller than the cost of capital. Such requests should not even be made if economic evaluations are conducted properly, because the minimum rate of return for an acceptable alternative is always equal at least to the cost of capital.

Figure 12.9 shows a ***cutoff rate*** that conforms to a ceiling on capital. The point on the ***cutoff index*** is set by the intersection of a line from the amount of capital available for investments $Q_1$ and the $RR_1$ curve which represents proposals laddered according to the cutoff index. For the condition shown, the cutoff point limits acceptable proposals to a number less than would be admissible using a capital inventory. Rationing capital to amount $Q_1$ rather than allowing all investments $Q_2$ which exceed the marginal cost of capital (MCC) increases the ***average*** rate of return on accepted investments. However, the firm may be missing an opportunity to increase its ***total*** profit by not funding all proposals that earn more than the cost of acquiring the capital for funding.

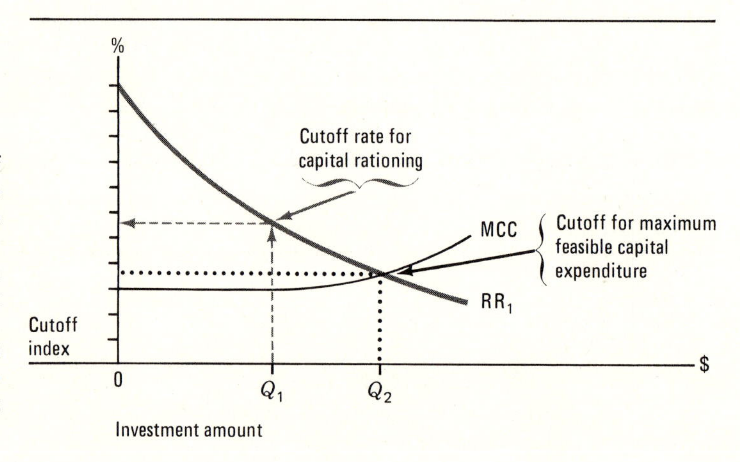


**FIGURE 12.9** Cutoff rate established by the point where the investment-opportunity schedule ($RR_1$) crosses a projection of the amount of capital available $Q_1$. Capital rationing limits acceptable proposals to a higher cutoff index than allowed by $Q_2$.

The cutoff rate for a budget ceiling becomes the ***minimum acceptable rate of return*** for the budgeting period when the cutoff index is expressed as a percentage. The laddering of proposals could also be accomplished by rating proposals according to

$$\text{Present-worth index} = \frac{\text{PW(receipts} - \text{disbursements)}}{\text{PW(investments)}}$$

$$\text{or} \quad \frac{\text{PW(net savings)}}{\text{first cost}}$$

when the entire investment occurs at time zero. Ranking proposals and accepting only the best ones until all the available capital is allocated works reasonably well, but care should be exercised to apply the concept adroitly and without bias.

**Example 12.6**

**Capital Rationing Using a Cutoff Rate of Return and a Present-Worth Cutoff Index**

Six independent alternatives are listed in Table 12.7. None of the proposals is expected to have any realizable salvage value. If the marginal cost of capital is 12 percent and the amount of investment capital is limited to $100,000, which requests for expenditure should be granted?

**TABLE 12.7** Cash flows for independent alternatives considered for investment.

| *Proposal* | *Request for Expenditure* | *Investment* | *Net Savings* | *Useful Life, Years* |
|---|---|---|---|---|
| *A* | Remodel loading dock | $30,000 | $12,600 | 9 |
| *B* | Modernize office | 35,000 | 12,250 | 4 |
| *C* | Purchase shredder | 10,000 | 4,450 | 4 |
| *D* | Install conveyor | 25,000 | 10,025 | 4 |
| *E* | Purchase press | 30,000 | 9,700 | 4 |
| *F* | Construct storage shed | 45,000 | 11,150 | 9 |

**Solution 12.6**

The rate of return for proposal $A$ is calculated as

$$\text{PW}(A) = -\$30{,}000 + \$12{,}600(P/A, i, 9) \stackrel{?}{=} 0$$

At RR = 40 percent,

$$\text{PW}(A) = -\$30{,}000 + \$12{,}600(2.3790) = -\$25$$

or RR $\doteq$ 40 percent. Rates of return for the other proposals are shown below.

| *Proposal* | A | B | C | D | E | F |
|---|---|---|---|---|---|---|
| *RR* | 40% | 15% | 28% | 22% | 11%* | 20% |

*Unacceptable because the rate of return is smaller than the cost of capital.

Since present-worth calculations are based on a given interest rate, the minimum cost of capital, 12 percent, is assumed reasonable for budgeting and is used to determine the present-worth index of proposal $A$ as

$$\text{Present-worth index}(A) = \frac{\text{PW(net savings)}}{\text{investment}}$$

$$= \frac{\$12{,}600(P/A, 12, 9)}{\$30{,}000} = \frac{\$12{,}600(5.3282)}{\$30{,}000}$$

$$= 2.24$$

The remaining present-worth indexes are given below.

| *Proposal* | *A* | *B* | *C* | *D* | *E* | *F* |
|---|---|---|---|---|---|---|
| *PW index* | 2.24 | 1.06 | 1.35 | 1.22 | 0.98* | 1.32 |

*Unacceptable because the index is smaller than 1.0.

The proposals, ranked according to their index values and cumulative investment totals, are tabulated to determine the cutoff level as shown below.

| | *Ranked Proposals* | *Rate of Return, %* | *Investment* | *Cumulative Investment* | *Ranked Proposals* | *PW Index* | *Investment* | *Cumulative Investment* |
|---|---|---|---|---|---|---|---|---|
| | *A* | 40 | $30,000 | $ 30,000 | *A* | 2.24 | $30,000 | $ 30,000 |
| | *C* | 28 | 10,000 | 40,000 | *C* | 1.35 | 10,000 | 40,000 |
| *Cutoff level* | *D* | 22 | 25,000 | 65,000 | *F* | 1.32 | 45,000 | 85,000 |
| | *F* | 20 | 45,000 | 110,000 | *D* | 1.22 | 25,000 | 110,000 |
| | *B* | 15 | 35,000 | 145,000 | *B* | 1.06 | 35,000 | 145,000 |
| | *E* | 11 | 30,000 | Not acceptable | *E* | 0.98 | 30,000 | Not acceptable |

The rankings produced by the rate-of-return and present-worth calculations are different. The reasons for the difference are assumptions embodied in the two methods. The rate-of-return method measures the *rate* of excess income accumulation over outgo under the reinvestment assumption discussed in Chapter 9. A single *amount* by which income exceeds outgo is measured by the present-worth method based on an explicit discounting rate. In addition, the present-worth index does not fully account for differences in the lives of the assets; it is assumed that the returned excess from a shorter-lived asset draws 0 percent interest for the period from the end of its life to the end of the life of a longer-lived asset.

For the proposals in Example 12.6, and often in other capital-budgeting situations, the two methods lead to the same selections. It might appear that the cutoff rate is 22 percent because the inclusion of proposal $F$ would push the total investment over the $100,000 limit. The combination of proposals $A$, $C$, and $D$ uses only $65,000 of the available capital. Assuming the remaining $35,000 is invested in proposal $B$, which still has a return greater than the marginal cost of capital, the overall rate of return is

$$\text{RR}(A + C + D + B) = \frac{\$30{,}000(0.4) + \$10{,}000(0.28) + \$25{,}000(0.22) + \$35{,}000(0.15)}{\$100{,}000}$$

$$= 0.256 \quad \text{or} \quad 25.6\%$$

A different and better combination is to fund proposals $A$, $F$, and $D$. This combination gives

$$\text{RR}(A + F + D) = \frac{\$30{,}000(0.4) + \$45{,}000(0.2) + \$25{,}000(0.22)}{\$100{,}000}$$

$$= 0.265 \quad \text{or} \quad 26.5\%$$

The most profitable combination is also identified by the net present worth:

$$\text{PW}(A + C + D + B) = -\$100{,}000 + \$12{,}600(P/A,\ 12,\ 9) + (\$4450 + \$10{,}025 + \$12{,}250)(P/A,\ 12,\ 4)$$

$$= \$48{,}302$$

$$\text{PW}(A + F + D) = -\$100{,}000 + (\$12{,}600 + \$11{,}150)(P/A,\ 12,\ 9) + \$10{,}025(P/A,\ 12,\ 4)$$

$$= \$56{,}993$$

---

**Economy Exercise 12-2** Another possible cutoff index is the net difference between the present worths of income (or savings) and expenditures. Apply this index to obtain a ranking for the six proposals in Example 12.6.

**a** Which proposals should be funded when capital is limited to $100,000?
**b** Which proposals should be funded when the capital budget is only $75,000? Compare the combinations indicated by the net-present-worth method and the rate-of-return method for this budget level.

---

## Qualifications for Capital Budgeting

The discussion of capital inventory and rationing was simplified by ignoring several considerations that frustrate actual applications:

1. Determining a cutoff rate, or an index cutoff point, does not automatically identify the best combination of proposals to take advantage of the limited capital, as evidenced in Example 12.6. *Linear-programming* models have been developed to assist the rationing exercise, but each model has its own assumptions that limit general applicability.
2. *Independent* projects are not always *completely* independent. Sometimes a proposal from one department is made on the assumption that a proposal from another department will be funded. A proposal with a very high rate of return might be made for an addition to an activity that is only marginally profitable; the highly profitable addition could be voided if the marginal activity is modified or eliminated. An *interdependency* could exist as an informal policy of the capital budgeters to try to spread investments along geographical or organizational lines.
3. All the proposals in Example 12.6 were treated as if they had identical *risk.* Rarely do different courses of action have the same probability of success. Riskier ventures typically are expected to yield a higher rate of return. Adding a percentage to the required rate of return to compensate for risk is a reasonable practice, but its conservative flavor thwarts the objective laddering implicit in an investment-opportunity schedule. Ways to analyze risk and to select preferred alternatives subject to risk are presented in Chapter 14.
4. The capital-budgeting process normally includes no formal provisions for evaluating

the effects of *intangibles,* mainly because there is no completely satisfactory way to analyze and include intangible factors in decision making. Still, since intangibles do influence decisions, and rightfully so, they should be exposed and examined. The methods given in Chapter 5 for quantifying intangibles and including them in the decision process encourage careful examinations, at a minimum, and certainly add objectivity to the analysis.

**5** Mandatory projects are effectively exempt from capital rationing when failure to fund them can shut down operations. Government regulations, such as required pollution controls, are the cause of many *mandatory investments.* These investments may have a negative rate of return, but they still top the list of requests for expenditure. Other projects that are not compulsory but are rated urgent may be ranked by management edict above proposals with higher rates of return.

One large corporation divides its investment-opportunity schedule into distinct categories: New projects must have a rate of return greater than 25 percent to be even considered for funding; improvements or modifications to existing projects have a floor at 17 percent; and investments that increase labor utilization are accepted if they produce a 10 percent return. These minimum required percentages reflect the risk involved and management views about the most effective use of capital. An amount of capital is allocated for each category and is rationed to proposals within the category according to the rankings. Projects considered mandatory, such as safety improvements, are funded from the top of the capital pool, irrespective of their return on investment.

**6** Subtle forms of persuasion may intimidate or prejudice the workings of the decision-making process. *Politic* considerations surround most decisions, and capital-allocation evaluations are not exempt. Perhaps it is important to reward the originator of an RFE by accepting his or her most recent proposal, after several previous ones were rejected, even if it has a modest RR; the encouragement may spur more activities of the type desired. Successful proposal sponsors gain stature, and the types of proposals currently accepted tend to guide the submission of future RFEs. Susceptibility to the "squeaky wheel" principle that directs oil to the noisiest wheel may bias judgment, yet the aggressiveness exhibited in seeking attention (and change) may be in the best interest of the firm. A manager's relationship to peers and subordinates can be affected by success in pushing proposals through, but should personal relationships be a factor in determining the merits of proposals? Though the questions are many and the answers are not easy, competent capital budgeting is at stake.

## SUMMARY

The "moment of truth" in engineering economics is the final decision to fund or reject a proposed investment.

Investment funds are acquired *internally* from depreciation reserves and retained earnings, and *externally* by borrowing capital or selling equity. The reinvestment rate of retained earnings should exceed the firm's capitalization ratio (per-share earnings divided by the stock price) for fairness to stockholders. Tax deductions allowed for interest payments enhance the attractiveness of debt financing. A higher *leverage factor* (ratio of debt to total assets) generally produces higher after-tax returns on equity when business conditions are favorable, but it lowers returns when conditions are unfavorable. An aggressive tactic of relying on short-term debt keeps financing charges lower by incurring greater risk of running out of working capital during periods of constricted cash flow.

Increases in future prices can be included in an economic analysis by estimating cash flow in *future dollars* responsive to general price trends, or by a *combined interest-inflation rate,* $(1 + i)(1 + i_f) - 1$, where $i_f$ is the inflation rate. A *range* of estimates, developed by forecasting cash flows for detailed income and cost categories under more- or less-favorable future conditions, is also used to evaluate alternatives when prices are unstable.

The *payback method* (first cost divided by net annual savings) indicates how much time will elapse before the amount invested is recovered from the net earnings it generates. In its simplest form, the recovery period is based on actual dollar flows; in more sophisticated applications the flow is discounted. The payback method is popular because it is simple to apply and stresses the turnover of capital, but it may show a preference for an alternative that is inferior by a discounted cash-flow comparison.

Many industrial organizations utilize customized economic evaluation procedures, often complemented by worksheet forms. The MAPI method is a respected procedure. A ***request for expenditure*** (RFE) includes supporting analysis documents and a letter of justification which explains the proposed investment. ***Postauditing*** is a systemized follow-up on approved RFEs to appraise the accuracy of previous estimates in order to improve future estimations.

Proposals to be funded are selected by a *capital-budgeting* process. The *capital-inventory* approach matches the rates of return from different investments against the cost of acquiring capital; all proposals with returns greater than capital costs are acceptable. In *capital rationing* a limit is set on the amount of investment capital, and the best proposals are funded until the supply of capital is exhausted. Objective capital budgeting is difficult, owing to the need to consider numerous combinations of proposals affecting all parts of the organization, dependencies among RFEs, various degrees of riskiness, intangible aspects, mandatory or "urgency" ratings for certain investments, and politic factors.

## Discussion of Economy Exercises

**EE 12-1** Of the three methods of acquiring the $100,000 construction equipment, the least expensive is still the 8-year loan plan. As shown in Table 12.8, a cash purchase using internal funds has an after-tax present worth of $72,588. The actual disbursements for the cash purchase are less than for the other plans, but tax savings are less too; in effect, the government shares part of the interest cost through its taxing rules.

An 8-year lease has the highest total disbursements and the highest tax savings, but the total discounted cost is greater than that of the loan plan, largely because the tax savings are spread over the length of the lease to reflect the lease payments as operating costs instead of being concentrated in the early years as in an accelerated depreciation schedule.

**EE 12-2** The net present worth of proposal $A$ at the minimum discount rate of 12 percent is

$$\text{PW}(A) = -\$30{,}000 + \$12{,}600(P/A, 12, 9) = \$37{,}135$$

| | CASH PURCHASE | | | LEASE-BUY OPTION | | | |
|---|---|---|---|---|---|---|---|
| *End of Year* | *Annual Depreciation Charge (1)* | *Tax Saving (2)* | *PW of Cash Flow (3)* | *Annual Rental Charge (4)* | *Depreciation Charge (5)* | *Tax Saving (6)* | *PW of Cash Flow (7)* |
| 0 | | | −$100,000 | | | | 0 |
| 1 | $ 20,000 | $ 8,000 | 7,273 | −$ 20,000 | | $ 8,000 | −$10,909 |
| 2 | 16,000 | 6,400 | 5,289 | −20,000 | | 8,000 | −9,917 |
| 3 | 12,800 | 5,120 | 3,847 | −20,000 | | 8,000 | −9,016 |
| 4 | 10,240 | 4,096 | 2,798 | −20,000 | | 8,000 | −8,196 |
| 5 | 8,192 | 3,277 | 2,035 | −20,000 | | 8,000 | −7,451 |
| 6 | 6,554 | 2,622 | 1,480 | −20,000 | | 8,000 | −6,774 |
| 7 | 6,554 | 2,622 | 1,346 | −20,000 | | 8,000 | −6,158 |
| 8 | 6,554 | 2,622 | 1,223 | −20,000 | | 8,000 | |
| | | | | (5,000)* | | | −7,931 |
| 9 | 6,554 | 2,622 | 1,112 | | $2500 | 1,000 | 424 |
| 10 | 6,552 | 2,621 | 1,011 | | 2500 | 1,000 | 386 |
| *Totals* | $100,000 | $40,002 | −$ 72,588 | $165,000 | $5000 | $66,000 | −$65,542 |

*$5000 cost at the end of year 8 to exercise the buy option for the equipment.

**TABLE 12.8** Comparison of acquiring the 10-year use of a $100,000 asset with no salvage value by cash purchase and by lease with an option to buy for $5000 after 8 years. Tax savings (2) for the cash purchase result from multiplying the annual depreciation charge (1) by the 40 percent tax rate; the total discounted cash flow (3) is the sum of the present worths of total tax savings plus the purchase price. Tax savings (6) for the lease-buy plan result from the 40 percent tax rate multiplied by the annual lease cost (4) or depreciation charges (5), and the present worth of the cash flow (7) is the sum of the net outlays [(6) − (4)] discounted at 10 percent.

The remaining PWs are calculated similarly to yield the amounts by which the proposals are ranked.

| *Ranked Proposals* | *Net PW* | *Net Savings* | *Investment* | *Cumulative Investment* |
|---|---|---|---|---|
| *A* | $37,135 | $12,600 | $30,000 | $30,000 |
| *F* | 14,409 | 11,150 | 45,000 | 75,000 |
| *D* | 5,449 | 10,025 | 25,000 | 100,000 |
| *C* | 3,516 | 4,450 | 10,000 | 110,000 |
| *B* | 2,207 | 12,250 | 35,000 | 145,000 |
| *E* | −538 | 9,700 | 30,000 | Not acceptable |

**a** The ranked proposals suggest the same combination of investments as identified in Example 12.6: Fund proposals *A*, *F*, and *D*.

**b** When available capital is limited to $75,000, the rankings by net present worth suggest allocations to proposals *A* and *F*, which yield an average rate of return based on calculations in Example 12.6 of

$$\text{RR}(A + F) = \frac{\$30{,}000(0.4) + \$45{,}000(0.2)}{\$75{,}000} = 0.28 \quad \text{or} \quad 28\%$$

But a more profitable combination includes proposals *A*, *C*, and *D*, which consume

$65,000 of the $75,000 available. Assuming the remaining $10,000 can always be invested to earn at least the cost of capital,

$$\text{RR}(A + C + D + \$10{,}000 \text{ invested at } 12\%) = \frac{\$30{,}000(0.4) + \$10{,}000(0.28) + \$25{,}000(0.22) + \$10{,}000(0.12)}{\$75{,}000}$$

$$= 0.287 \quad \text{or} \quad 28.7\%$$

Rating by the net present worth emphasizes the discounted amount of total benefits. Unless corrected for unequal lives among proposals, the higher ratings naturally go to the proposals which yield returns for a longer period of time, even when these returns are at a lower rate than is obtainable with shorter-lived projects. Some analysts feel the total return from long-lived assets is a legitimate comparison criterion, especially for public projects (discussed in the next chapter).

## PROBLEMS

**12.1** The owners of a prosperous partnership have profits of $40,000 they wish to invest outside their firm. One partner wants to invest in a venture which will yield a rate of return of 9 percent. The other partner wants to put $20,000 in bonds and the remainder in a venture for which the returns are difficult to estimate. The returns from the bonds (4 percent per year) are nontaxable, but other returns will be taxed at the partnership's tax rate of 35 percent. Assuming that the risks involved are approximately the same, what rate of return must the other half of the bond venture earn in order to make the returns from the two alternatives equivalent?

*(11.8%)*

**12.2** A cooperative which pays a 50 percent tax rate on gross profit made two investments which have a total future value of $300,000 in 10 years. One of the investments (taxable) has a present worth of $50,500 and will yield before-tax returns of 16 percent compounded semiannually. The other investment is in nontaxable municipal bonds. One hundred bonds, each with a face value of $1000 and semiannual interest periods for 10 years, were purchased at a price to yield the same after-tax return as the other investment. What interest was stated on the bonds if it is assumed that dividends can be reinvested at the rate of return for taxable investments?

*(3%/period)*

**12.3** The owner of a drive-in restaurant believes there is a good potential for another drive-in at a nearby town. She has $40,000 of owned capital with which to expand the business. Using this amount she can open a new operation which should yield a gross income before taxes of $8000. She feels that a conservative study period of 10 years is reasonable, and she will use straight-line depreciation for tax purposes. The salvage value is expected to be $10,000.

After checking with a bank, she finds that she can borrow $20,000 at 6 percent interest per year on the unpaid balance. The loan extends for 10 years and would be repaid in 10 equal annual installments.

If her effective income-tax rate is 40 percent for the period, which means of financing is the most attractive?

**12.4** The degree of operating leverage at point $Q$ was defined in Economy Exercise 3-2 as

$Q(P - V)/Q(P - V) - F$. At a given level of earnings before interest and taxes,

$$\text{Degree of financial leverage} = \frac{Q(P - V) - F}{Q(P - V) - F - I}$$

*where* $I$ = annual interest payments on borrowed capital

**12.4a** Derive the expression for the degree of financial leverage from the relationship that the degree of financial leverage is the percentage change in earnings per share of stock (let $N$ = number of shares outstanding) divided by the percentage change in earnings before interest and taxes.

**12.4b** Assume a company produces 2000 units per year which sell for $1000 apiece. The variable cost per unit is $400, and the fixed costs are $400,000. What is the degree of operating leverage? What does this figure mean?

*(1.50)*

**12.4c** If the interest paid annually on long-term debt is $250,000, what is the degree of financial leverage? What does this figure mean in terms of an increase in earnings per share as a function of the increase in earnings before interest and taxes?

*(1.45)*

**12.5** A new manager has been hired by All-Sports Warehouse, Inc., a regional retail center specializing in skiing and camping equipment. Total sales for next year are expected to remain relatively constant at $1 million and be concentrated in the early summer and late fall months. The continuing policy is to have a leverage factor of 0.5 based on fixed assets of $500,000. Earnings are expected to be 15 percent on the total sales before interest and taxes are paid. The effective income-tax rate is 50 percent.

The new manager has to decide what level of current assets to maintain during the coming year. He can adopt (1) an aggressive tactic of requiring current assets of only 45 percent of prospective sales, (2) the present policy of 50 percent of sales as current assets, or (3) a conservative tactic which would put the current asset level at 60 percent of sales.

**12.5a** What return on equity will result from each of the working-capital tactics when the interest on long- and short-term debt is the same annual 6 percent?

*[(1) = 12.8%; (2) = 12%; (3) = 10.6%]*

**12.5b** What return on equity can be expected from the three tactics when short-term debt is at 6 percent and long-term debt has an interest rate of 10 percent?

**12.5c** Discuss considerations the new manager should keep in mind as the decision is made. (***Hint:*** Consider seasonality of sales and riskiness of assumptions.)

**12.6** A corporation plans to buy out a supplier and needs $2 million in new capital to do so. The proportion of equity to debt financing will be 40:60 for the acquisition. The current cost of equity capital is 14 percent after taxes, and it is 12 percent before taxes for debt financing. The effective income-tax rate is 55 percent.

**12.6a** What is the minimum amount of before-tax earnings necessary per year from the purchase to justify raising the required capital?

*($392,000)*

**12.6b** What minimum after-tax earnings should be expected?

*($180,000)*

**12.7** A firm needs $300,000 to install secondary waste-recovery equipment. Since this is a mandatory investment to meet federal pollution-control regulations, the firm can use internal funds and forgo other investment opportunities or borrow the entire amount. Internal funds are expected to earn 12 percent per year after taxes, and the cost of borrowing money on a 15-year loan (the useful life of the equipment) is 9 percent.

The investment has a positive net flow after taxes, owing to deductions for depreciation of $10,000 per year for 15 years when the equipment is purchased from equity funds. The income-tax rate of the firm is 50 percent.

**12.7a** Compare the present worths of full debt and equity financing when the required rate of return is 12 percent. Assume the debt is repaid in 15 equal installments of $47,000 each, of which $17,000 is the constant annual interest charge, and depreciation is by the straight-line method.

**12.7b** How much would the secondary waste-recovery equipment have to increase annual income in order to earn a 15 percent rate of return when funded 50 percent from equity and 50 percent from borrowed funds? Assume the given equity after-tax cash flow and the given loan-installment payments are each reduced by half in the 50:50 financing plan.

**12.8** A consulting engineer has a 10-year contract to monitor waste discharges from several mills located on a small bay. Since the contract also calls for a hydrographic study of the area, a boat will be needed. A properly equipped craft can be purchased for $20,000. At the end of the 10-year study it can likely be sold for 10 percent of its purchase price. Moorage and maintenance costs will be $400 per year. It will be depreciated by straight-line charges. The engineer's effective income-tax rate is 40 percent.

A loan for the entire amount, requiring annual payments of $2600, can be obtained. Interest charges during each year of the loan are shown below.

| *Year* | 1 | 2 | 3 | 4 | 5 | 6 | 7 | 8 | 9 | 10 |
|---|---|---|---|---|---|---|---|---|---|---|
| *Interest* | $1000 | $920 | $840 | $760 | $680 | $560 | $460 | $360 | $260 | $160 |

It would also be possible to lease an adequate boat from a marina on the bay. A 10-year lease that allows use of a boat whenever needed, if a 2-day warning is given, will cost $3000 per year. Maintenance and moorage fees are included in the lease cost. Operating expenses will be the same under both plans.

**12.8a** What are the present worths of the costs of the two plans for the use of a boat when the discount rate is 3 percent?

[*PW(buy) = $15,838; PW(lease) = $15,354*]

**12.8b** What other factors might be considered in the comparison of the two plans? Would a higher required rate of return, say 10 percent after taxes, affect the decision? Why?

**12.9** An asset has a purchase price of $50,000, a 10 percent salvage value, and a useful life of 5 years. It is to be depreciated by the sum-of-digits method. Operating costs will be the same under any of the financing plans to be considered. A discount rate of 10 percent is to be used in the comparisons. The effective income-tax rate is 52 percent.

**12.9a** What is the present worth of ownership if the asset is purchased with equity funds?

**12.9b** What would the annual lease cost have to be to make leasing as attractive as cash purchase?

**12.9c** A 7 percent, 5-year loan can be obtained if a down payment of $25,000 is made. What is the present worth of costs under this plan?

**12.10** Net cash flow from the purchase of an asset for $1000 is expected to be responsive to inflation. The inflation rate is forecast to be 5 percent for the next 3 years. Based on this forecast, the expected cash flow in future dollars is shown below.

| *Year* | 0 | 1 | 2 | 3 |
|---|---|---|---|---|
| *Cash flow* | −$1000 | $400 | $600 | $500 |

**12.10a** What is the combined interest-inflation rate if the organization expects a 10 percent return on investments?

*(15.5%)*

**12.10b** Using this interest rate, calculate the present worth.

*($121)*

**12.10c** What cash flow estimates in current dollars would produce the same present worth when discounted at 10 percent?
**12.10d** Do you feel it is a better practice to estimate future cash flow in "then-current" future dollars or in "now-current" constant dollars? Why?

**12.11** There is considerable doubt about the need for and performance to be obtained from a new process developed by the R&D department. A decision about launching a small-scale pilot project is being evaluated. Three estimates of possible outcomes of the pilot project are given below.

| | *Objective Estimate* | *Less-Favorable Estimate* | *More-Favorable Estimate* |
|---|---|---|---|
| Income/year | $200,000 | $150,000 | $250,000 |
| Expenses/year | $80,000 | $80,000 | $90,000 |
| Start-up cost | $300,000 | $350,000 | $300,000 |
| Life of project | 3 years | 1 year | 4 years |
| Salvage value | $100,000 | $50,000 | $100,000 |

**12.11a** What is the present worth of each possible future when the minimum attractive rate of return is 15 percent?
**12.11b** What other considerations could affect the decision to launch the pilot project? Should the money already invested in research and development be a consideration? Why?
**12.11c** Compare the range-of-estimates method of evaluation with sensitivity analysis.

**12.12** Use the payback criterion to select a preferred investment proposal from the five cash-flow patterns given in Problem 8.22.

*(B at 2.4 years)*

**12.13** Six independent investment proposals have the same payback period based on the after-tax cash flows shown below.

| | YEAR | | | | | |
|---|---|---|---|---|---|---|
| *Proposal* | *0* | *1* | 2 | 3 | 4 | 5 |
| 1 | −$1000 | $200 | $300 | $500 | $300 | $200 |
| 2 | −900 | 300 | 300 | 300 | 300 | 300 |

| | YEAR | | | | | |
|---|---|---|---|---|---|---|
| *Proposal* | *0* | *1* | *2* | *3* | *4* | *5* |
| 3 | −800 | 0 | 400 | 400 | 800 | 0 |
| 4 | −700 | 0 | 0 | 700 | 400 | 400 |
| 5 | −600 | 300 | 200 | 100 | 300 | 500 |
| 6 | −500 | 300 | 100 | 100 | 200 | 0 |

**12.13a** Which alternative would be preferred when the cash flows are discounted to their present worths at a discount rate of 8 percent?
**12.13b** Comment on the cash-flow patterns that are most likely to cause the payback criterion to disagree with a preference indicated by the discounted cash-flow criterion.
**12.13c** What cash-flow pattern is most likely to make the payback and discounted cash-flow comparisons agree?
**12.13d** Which alternatives would be selected by a capital-inventory approach when the cost of capital is 10 percent?
**12.13e** If the available capital is limited to $2000 and any unused funds can be invested at 10 percent, what combination of investments should be selected? What average rate of return results from this combination?

**12.14*** The relationship of interest and depreciation charges, which are predicated on current values, to future dollar receipts tends to cause the rate of return on investments to be overstated during periods of inflation. As explained by George Terborgh, "The reason is not far to seek: tax deductions for depreciation, for 'basis' in computing terminal gains, and for interest on borrowed capital are ***unresponsive*** to inflation."

**TABLE 12.9** Cash flow of an asset purchased for $6194 and partially financed by a $1843 loan at 5 percent. Future dollars responsive to 3 percent inflation result from current dollars multiplied by $(1.03)^N$. Straight-line depreciation charges and interest apply equally to current- and future-dollar cash flows.

| *End of Year, N* | *Net Receipts in Current Dollars (1)* | *Depreciation and Interest Charges (2)* | *Net Receipts in Future Dollars (3)* |
|---|---|---|---|
| 0 | −$4351 | $ 0 | −$4351 |
| 1 | 1000 | 711 | 1030 |
| 2 | 1000 | 702 | 1061 |
| 3 | 1000 | 693 | 1093 |
| 4 | 1000 | 684 | 1126 |
| 5 | 1000 | 675 | 1159 |
| 6 | 1000 | 665 | 1194 |
| 7 | 1000 | 656 | 1230 |
| 8 | 1000 | 647 | 1267 |
| 9 | 1000 | 638 | 1305 |
| 10 | 1000 | 629 | 1344 |

To illustrate the effect of inflation on rate-of-return evaluations, consider the asset described in Table 12.9. Its cost is $6194, and it is expected to produce receipts of $1000 (column 1) in present buying power for 10 years. There is no salvage value. Depreciation and interest charges are based on 30 percent financing at 5 percent interest and straight-line

*Adapted from G. Terborgh, ***Effects of Anticipated Inflation on Investment Analysis,*** Machinery & Allied Products Institute, Washington, D.C.

depreciation (column 2). These charges remain the same regardless of the effect of inflation upon receipts. With an inflation rate of 3 percent, the current-dollar receipts in column 1 would grow to the values in column 3 in $N$ years.

**12.14a** Calculate the after-tax rate of return for the asset's purchase based on the current-dollar cash flow (columns 1 and 2). The effective income-tax rate is 50 percent. *(RR = 14.2%)*

**12.14b** Calculate the after-tax rate of return based on future-dollar cash flow when the tax rate is 50 percent.

**12.14c** Explain the erosion of the real after-tax return in your own words. How could it affect engineering economic analyses?

**12.15** A request for expenditure for a paint-drying system is being developed. The investment in new assets includes $36,000 for the oven, $23,200 for ventilation and controls, and $14,000 for the filter assembly. The system is to be depreciated collectively in 10 years with no salvage value, but it will be kept in operation for 12 years.

The major reasons for a forced drying system are (1) to comply with new federal health standards, (2) to lower costs by reducing worker-hours and rework time, (3) to dry trucks and trailers at the same rate as the painting and shot-blast operations, and (4) to reduce stenciling costs and to be able to utilize outside storage which will be needed for the higher production runs expected in the future. The last reason is the least substantial, but painting will be a bottleneck for any other expansion plans unless the fast-dry system is installed.

Direct savings from the installed system will result from an approximate annual 9860 hours reduction in operator time costed at $6.55 per hour, and $35,000 annually from reduced material expense. Other savings after the first year should amount to $5166. Extra expenses the first year will reduce the operating advantage to $24,759 for that year. No increase in working capital is required, and there are no future obligations involved in adopting the new drying system.

The company has an effective income-tax rate of 51 percent. A tax credit of 10 percent is allowed on the first $50,000 for investments of the type anticipated. Depreciation will be deducted from earnings *after* the first year and will continue for 10 years; thus, the first and last year of the 12-year service life of the paint-drying system will have no depreciation charges. The depreciation schedule is shown below.

| *Year* | 1 | 2 | 3 | 4 | 5 | 6 | 7 | 8 | 9 | 10 | 11 | 12 |
|---|---|---|---|---|---|---|---|---|---|---|---|---|
| *Depreciation charge* | 0 | $11,566 | $9736 | $8198 | $6881 | $6734 | $6734 | $6661 | $6661 | $6661 | $6661 | 0 |

**12.15a** Use the worksheet form in Figure 12.3 to calculate the after-tax cash flow. (*Hint:* Entries in the worksheet for the first 2 years are shown below.)

| *Year* | *(3)* | *(4)* | *(5)* | *(6)* | *(7)* | *(8)* | *(9)* | *(10)* | *(11)* |
|---|---|---|---|---|---|---|---|---|---|
| 1 | $24,759 | 0 | $ 0 | $24,745 | $12,627 | $5000 | $ 7,627 | $17,132 | −$56,068 |
| 2 | 73,249 | 0 | 11,566 | 61,683 | 31,458 | 0 | 31,458 | 41,791 | −14,277 |

**12.15b** What is the payback period indicated on the worksheet? How does this compare with the simple before-tax payback period? Which approach is more realistic? Why? *(2.35 years; 1.66 years)*

**12.15c** Use the worksheet form in Figure 12.4 to calculate the profitability index.

*Hint:* Trial values are determined by consecutive multiplications as shown below for year 1, where $17,132 is the year-1 after-tax cash flow:

Trial 1 × factor = trial 2 × factor = trial 3 × factor = trial 4
$17,132 × 0.952 = $16,310 × 0.885 = $15,162 × 0.824 = $14,117

*(PI = 45%)*

**12.15d** Use the information in the problem description and the values determined in Problems 12.15*a* to 12.15*c* to prepare the summary page for a letter of justification as shown in Figure 12.7. Fill in only the cost data for the first 2 years. Interpret the problem information to develop a "description of request" and "most critical assumption."

**12.16** A cost-reduction proposal to equip a locomotive with remote radio control is being developed. Because the radio-control equipment, which costs $28,000, can be considered an operating expense, no depreciable assets are involved.

Two operators are now employed to shuttle scrap-weighing cars between the scrap yard and the steel-making furnace. By equipping the locomotive with remote radio-control equipment, the need for one operator each shift is eliminated. The locomotive is utilized 4000 hours per year. Operators are paid $9 per hour including fringe benefits. Installation costs for the controls are expected to be $2000. There have been significant improvements in remote-control systems since the company had its unpleasant experience with a radio-controlled locomotive 10 years ago.

Complete the form in Figure 12.7 to justify the request for expenditure.

**12.17** Measurement of space consumed by aisles in a company-owned warehouse reveals that there are 200 feet (61 meters) of 12-foot-wide (3.66-meter-wide) aisles and 750 feet (228.6 meters) of 10-foot (3.05-meter) aisles. The 10-foot (3.05-meter) aisles are used because a 6000-pound (2722-kilogram) capacity fork truck has to operate in them to handle some heavy equipment. All other loads run under 4000 pounds (1814 kilograms). If the heavy items are concentrated in one storage area, the 6000-pound (2722-kilogram) truck will not need to operate in all aisles. The layout could be arranged as follows:

| | *Width* | *Length* |
|---|---|---|
| Central aisle for 6000-lb (2722-kg) truck access | 12 ft (3.66 m) | 30 ft (9.14 m) |
| Heavy storage area | 10 ft (3.05 m) | 75 ft (22.9 m) |
| Other central aisles | 10 ft (3.05 m) | 175 ft (53.3 m) |
| All other aisles cut to 7 ft (2.13 m) | 7 ft (2.13 m) | 670 ft (204.2 m) |

The 7-foot (2.13-meter) aisles are possible only if three new 4000-pound (1814-kilogram) narrow-aisle fork trucks are purchased at a cost of $5000 apiece. They will have a useful life of 4 years and no salvage value. The replaced 6000-pound (2722-kilogram) fork trucks will be kept for standby and scavenger purposes. Operating costs will increase by $1750 per year with the new trucks.

Rented storage space can be reduced by the amount of square footage saved in the new arrangement. Savings will be at the rate of $5 per square foot per year ($53.82 per square meter per year).

**12.17a** Utilize the format shown in Figure 12.5 to calculate the after-tax cash flow. What are the payback period and rate of return?
**12.17b** Utilize the profit-sharing worksheet to determine the after-tax cash flow. What are the payback period and rate of return (see Figure 12.6)?

**12.17c** Of what value is the payback-period calculation as used in Problems 12.17*a* and 12.17*b*? What is the purpose of an "auxiliary criterion?"

**12.17d** Prepare a summary page for a letter of justification to support the aisle arrangement and fork-truck purchase (see Figure 12.7).

**12.17e** Assume the savings in warehouse rental space are responsive to a 6 percent inflation rate. Use the format of Figure 12.5 to calculate the after-tax rate of return. Compare the RR obtained with the answer to Problem 12.17*a*, and comment on the comparison.

**12.17f** To analyze the problem more completely, let the worth of storage space vary between −10 percent and +20 percent of the given values. Operating costs could double under unfavorable conditions. The fork trucks might last as long as 6 years or only for 3 years.

Calculate the before-tax rate of return under favorable, given, and unfavorable conditions. Make a recommendation based on the company's minimum acceptable before-tax rate of return of 20 percent.

**12.18** Ten requests for expenditure have been received. All are for investments to reduce operating costs, and all have about the same risk. The assets involved have a useful life of 4 years and no salvage value. Investment amounts and the uniform cash flows are shown in Table 12.10.

**12.18a** If the alternatives are independent and the cost of capital is 13 percent, how much should be invested?

**12.18b** If the alternatives are independent and capital is limited to $100,000, which proposals should be accepted? What is the cutoff rate? What is the average rate of return?

**12.18c** Answer the questions in Problem 12.18*b* when the only change is that proposal 4*A* is mandatory.

| *RFE* | *Investment* | *After-Tax Cash Flow* |
|---|---|---|
| 1*A* | $10,000 | $ 3,087 |
| 1*B* | 20,000 | 7,434 |
| 1*C* | 30,000 | 10,722 |
| 1*D* | 40,000 | 14,012 |
| 2*A* | 10,000 | 4,159 |
| 2*B* | 20,000 | 8,020 |
| 2*C* | 30,000 | 11,151 |
| 3*A* | 10,000 | 3,863 |
| 3*B* | 20,000 | 6,864 |
| 4*A* | 30,000 | 9,877 |

**TABLE 12.10** After-tax cash flows expected to result from capital expenditures for the listed proposals. All alternatives have the same useful life and risk.

**12.18d** Use the present-worth index to answer Problem 12.18*a*.

**12.18e** Let the numbers in the RFE identifications represent four different departments in the organization. Determine the answers to the questions in Problem 12.18*b* if it is company policy to accept at least one proposal from every department.

**12.18f** Assume the policy in Problem 12.18*e* is changed to the capital-inventory approach in which departments 1 and 3 have a 15 percent limit on cost of capital and the other two departments have a 20 percent minimum. What is the average rate of return resulting from this policy?

**12.18g** Let the numbers in the RFE identifications represent four categories in which the lettered alternatives ($A$, $B$, $C$, and $D$) are mutually exclusive. Which proposal should be selected from which categories if capital is limited to $70,000? What is the average rate of return? (*Hint:* Calculate incremental rates of return.)

**12.18h** What combination of proposals would yield the highest average rate of return for the condition in Problem 12.18*g* if the capital is limited to $50,000? Any unallocated capital can earn 13 percent. Does a lower limit placed on capital always raise the average rate of return of accepted proposals?

**12.18i** Again assuming lettered alternatives within numbered categories are mutually exclusive, what average rate of return results when only $40,000 is available in the capital budget?

## EXTENSIONS

***12.1 Investment Involvements: A Success Story**** For several years the A. H. Robins Company had contemplated building a plant to produce one of its critical raw materials, methocarbomal. Impetus was given to the idea in 1973–74 when shortages and soaring prices hit methocarbomal supplies and caused production closedowns at Robins. Two management teams were set up, one to investigate the feasibility of building a plant, and the other buying a plant that already produced the needed chemical.

The "build" team selected an engineering firm to develop design criteria. A 1-inch-thick, $16,000 report from the consultants stated a plant with the required capacity would cost $6 million, plus 30 percent or minus 20 percent. The plant could produce for $3.5 million the annual demand for which the company was currently paying $6,050,000. This saving suggested a before-tax return on investment of better than 40 percent and a payback period of less than 3 years.

Meanwhile, the "buy" team identified over 100 existing plants that met Robins' needs. One of the best was Hexagon Laboratories Inc., which was for sale and already a big supplier of methocarbomal to Robins. It was turned down because its location in the Bronx, New York, allowed no room for expansion and because of its unionization. A plant in Wisconsin was rated the best buy.

A keen rivalry developed between the "build" and "buy" teams. Personal reputations and egos became tied to the outcome and, as it drew nearer, doubt arose about the whole project. A vice-president, one of the discoverers of methocarbomal in 1958, expressed concern that the company might not have the know-how and experience to run such a chemical plant efficiently. The hesitation was supported by a study commissioned by Chase Manhattan Bank which argued that the venture was not wise because of the cyclical nature of the chemical business. It suggested that prudent hedging, buying large quantities when prices were low, would provide a reliable and reasonably priced supply of raw materials. The recession, coupled with high inflation in 1974–75, also dampened enthusiasm for a large commitment of capital.

After 10 months of study, a decision session was scheduled in March 1975 to select

*Adapted from V. F. Zonana, "How A. H. Robins Co. Made a Decision to Build a Chemical Facility," *The Wall Street Journal*, Oct. 22, 1975.

either the build or buy option. Worries about supply interruptions overrode doubts about operating inexperience, and the company's ability to finance the project with equity funds reduced the recession's impact. A decision was made to build a new plant near the company's headquarters in Virginia.

As the search for a building site narrowed, a new possibility arose. Hexagon Laboratories, the Bronx firm previously considered for purchase, had been acquired by a big pharmaceutical and chemical corporation based in West Germany. This corporation learned of Robins' intent to produce its own methocarbomal and hurriedly started negotiations to keep the Robins business for Hexagon. An offer was made by Hexagon to build a plant in Virginia based on a long-term sales agreement with Robins.

But Robins wanted its own captive supplier. Eventually they agreed to 50 percent joint ownership of a new plant. Most of the output would be sold to Robins; profits would be divided equally; and Robins would have an option to buy the whole works after a period of time. At the outset of the investment study there was no glimpse of the actual outcome, but advantages and disadvantages of the build and buy options were handy criteria for weighing the acceptability of the joint venture. Both sides won.

**QUESTIONS**

**12.1a** Discuss the merits of creating tense competition in developing alternative investment proposals.

**12.1b** Consider the factors involved in the decision, at various stages, that appear to be intangible or at least difficult to define and quantify. Discuss these factors in terms of a sensitivity analysis. How would you present arguments in favor of an option that depended on such factors?

*12.2* ***A World of Investment Criteria*** Industrial organizations in every nation face similar problems in screening out the most advantageous investments to improve production. Discounted cash-flow analysis is the most celebrated screening device in this country and is widely used elsewhere. The payback period also has international acceptance, apparently more so than the theoretically sounder discounting methods. In developing countries where capital is severely limited, the emphasis on capital turnover in the payback-period criterion is more appreciated. In centralized socialist societies, the payback criterion in various versions is broadly applied because it conforms nicely to ideological beliefs concerning the social harm of interest charges.

Paul Samuelson observes in his ninth edition of *Economics* (McGraw-Hill, New York, 1973):

> Recent studies show that the social engineers of the Soviet Union are anxious not to be denounced as capitalistic apologists; yet they need some form of interest rate (or "discount factor," or "payoff period") for making efficient investment calculations. As a result, about a dozen different accounting methods are in vogue there for introducing a thinly disguised interest-rate concept into Soviet planning procedures.

This observation is supported by Dr. Kljusev in his writings:

> The methods for estimating the efficiency of the investment which are characteristic of the circumstances of capitalist production cannot automatically be used in a socialist economy because the stimulus and interests of the investment are different.*

Dr. Kljusev also notes that the efficiency criterion (known too as the "lucrativity index")

$$E = \frac{\text{difference in annual cost between two alternatives}}{\text{difference in investment between two alternatives}}$$

or its reciprocal (the "period of return") is the most widely used economic evaluation method in the USSR and other socialist countries. But it still is considered inadequate because it is viable only for the relative economic efficiencies of alternative ways to produce the same article; it does not account for the apportionment of new revenue between continued investment to broaden production and amounts made available for personal and social expenditures.

Several extensions of basic economic comparison models have been forwarded by writers from different countries. The suggestions generally reflect particular communal concerns for strategic or tactical effects of capital expenditures, as indicated by the samples below.

1. Special attention should be given to investments involving convertible currency: Encourage projects that conserve or build foreign-exchange credits.
2. A weighting method should be used to evaluate the territorial impact of investments: Direct industrial development toward underdeveloped areas.
3. The time between appropriation and realization should be considered in comparing investments: Proposals that promise quick results should be rewarded.
4. Consider the total returns expected from a completely developed program rather than piecemeal returns in deciding which alternatives are most beneficial: Variations due to budgetary constraints and general economic conditions may mask the total effect of a program.
5. Develop ways to quantitatively include the social effects as well as the economic effects: Supplement economic models with a quantitative mechanism to express societal worth.
6. Stress considerations such as maintainability of equipment, flexibility of use, and variability of inputs when evaluating tactical investments: Detailed anticipation of future supply and demand conditions should influence present choices among long-lived assets, possibly causing a sacrifice of immediate profits in expectation of higher later benefits.

**QUESTIONS**

**12.2a** Discuss the suggested extensions in terms of their applicability to engineering economic studies. What current trends in this country (e.g., consumerism, recycl-

*N. Kljusev, *About Criteria and Methods of Valuation of Economic Effectiveness of Investment,* Ekonomiski Institut of Skopje, Yugoslavia.

ing, environmental protection) are likely to affect future investment decisions, and how might their influence be shown quantitatively in economic analyses?
**12.2b** Comment on the quote below by Raymond Mayer. Do you agree with the statements? Why? Does this perspective account for the popularity of the payback-period criterion?

> Theoretical methods of capital investment analysis, although sound and well established, have proven to be unacceptable to many industrial managers. The typical business man agrees that the relevant factors in the evaluation of an investment proposal are: the dollar investment, the cost of money, the alternative's life and salvage value, and its operating costs and revenues. But he also knows that accurate estimates of their values are not usually possible. Given this risk of error, it follows that he cannot appreciate the need for tedious calculations involving obscure compound interest factors. Even more important, he feels that the use of these interest factors gives the calculated results an aura of precision which simply is not there due to the approximate nature of the estimates.*

*R. Mayer, "Capital Investment Analysis Another Way," *Industrial Engineering,* July 1970.

# CHAPTER 13

# ANALYSIS OF PUBLIC PROJECTS

The public and industrial sectors of the United States economy interact as mutually supportive, but occasionally contradictory, forces. Government functions to facilitate the operation of a free-enterprise system may thwart or make difficult certain actions deemed advantageous by certain industrial subsystems. Despite claims of "government is best which governs least," the private sector simply does not have the means to provide all the necessary social goods and services. Government activities regulate and support the legal framework of the marketplace, establish and collect taxes for the redistribution of income, and allocate resources to programs and projects believed beneficial to society. Ideally, these activities will mirror the disposition of the citizenry with reasonable accuracy, while not completely satisfying the loudly vocal minorities calling either for no governing or for governmental solutions to all the perceived ills of society. The diversity of public-sector activities designed to serve so many interests complicates the evaluation of their effectiveness.

Comparison methods discussed in previous chapters are suitable for the evaluation of government activities. Some considerations, such as taxes and payoff periods, are less significant in the public sector, and other considerations such as the secondary effects of benefits are more prominent. In this chapter, attention is directed toward the evaluation of specific projects rather than the general economic functions of government. Thus, the role of government to provide national defense, education, and dams is not questioned; subjects of evaluation are explicit investment proposals to implement these government functions.

Evaluations take the form of *benefit-cost* (sometimes called *cost-benefit*) analyses. Operationally, the technique is similar to the present-worth index. Conceptually, an analysis considers the worthiness of shifting resources from the private sector to the public sector, and to what extent a public project should be pursued when its benefits exceed its costs. Many factors besides a benefit-cost analysis influence the final decision about funding public projects, and weaknesses of benefit-cost analyses as traditionally conducted have cast doubts on the resulting ratings, but the analytical procedures still invite credibility and objectivity in appraisals of public expenditures.

## BENEFIT-COST ANALYSIS

The U.S. Department of Defense, in DOD *Instruction 7041.3* (1972), defines benefit-cost analysis as

> An analytical approach to solving problems of choice. It requires the definition of objectives, identification of alternative ways of achieving each objective, the identification, for each objective, of that alternative which yields the required level of benefits at the lowest cost. This same analytical process is often referred to as cost-effectiveness analysis when the benefits or outputs of the alternatives cannot be quantified in terms of dollars.

The need for formal evaluation of public expenditures can be traced back to the 1844 writings of J. Dupuit in France.* The Rivers and Harbor Act of 1902 in this country stipulated that a board of engineers report on the merits of river and harbor projects of the Army Corps of Engineers. The reports were to include the amount of commerce benefited with respect to the estimated cost. A later act required a statement of local benefits to facilitate sharing of project costs with local interests which would benefit from the project. Government participation in public projects was extended by the Flood Control Act of 1936 which justified improvements to waterways for flood control *"if the benefits to whomsoever they may accrue are in excess of the estimated costs."* In the 1940s this principle was expanded to justify other projects or programs for social welfare, and it is now an accepted measure of desirability of projects at the federal, state, and local levels of government.

### Which Comparisons to Make

In comparing benefit (B) to cost (C), several different perspectives are reasonable. Consider the simplified data in Table 13.1 that describe the alternatives for a small flood-control project. The current average annual damage from flooding is $200,000. Three feasible options are available to reduce the damages; each larger investment of public funds provides greater protection.

*J. Dupuit, "On the Measurement of Utility of Public Works," *International Economic Papers*, vol. 2.

**TABLE 13.1** Annual costs and benefits from different levels of investment in a small flood-control project.

| *Alternative* | *Equivalent Annual Cost of Project* | *Average Annual Flood Damage* | *Annual Benefit* |
|---|---|---|---|
| *A*: No flood control | $ 0 | $200,000 | $ 0 |
| *B*: Construct levees | 40,000 | 130,000 | 70,000 |
| *C*: Small reservoir | 120,000 | 40,000 | 160,000 |
| *D*: Large reservoir | 160,000 | 10,000 | 190,000 |

The three projects and the do-nothing alternative are mutually exclusive. Data for selecting the most attractive alternative are shown below, where the figures are in thousands of dollars.

| | | | TOTAL | | INCREMENTAL | | | |
|---|---|---|---|---|---|---|---|---|
| *Alternative* | *Annual Benefit, B* | *Annual Cost, C* | *B/C* | *B − C* | Δ*B* | Δ*C* | Δ*B*/Δ*C* | Δ*B* − Δ*C* |
| *A* | $ 0 | $ 0 | 0 | $ 0 | | | | |
| *B* | 70 | 40 | 1.75 | 30 | $70 | $40 | 1.75 | $30 |
| *C* | 160 | 120 | 1.33 | 40 | 90 | 80 | 1.125 | 10 |
| *D* | 190 | 160 | 1.19 | 30 | 30 | 40 | 0.75 | −10 |

The following criteria indicate different plausible preferences among the alternatives:

1. Minimum investment: Choose *alternative A*. If funds are severely limited, this may be the only possible choice.
2. Maximum benefit: Choose *alternative D*. Flooding would occur only during extremely wet seasons.
3. Aspiration level: Depends on the threshold set for cost or benefit. If, for instance, the aspiration level is to reduce flood damage by 75 percent, *alternative C* should be chosen because it meets the aspiration with a lower cost than alternative *D*. Similarly, an annual-cost threshold of $100,000 indicates a preference for *alternative B*.
4. Maximum advantage of benefits over cost (B − C): Choose *alternative C*.
5. Highest benefit-to-cost ratio (B/C): Choose *alternative B*.
6. Largest investment that has a benefit-to-cost ratio greater than 1.0: Choose *alternative D*.
7. Maximum incremental advantage of benefit over cost (ΔB − ΔC): Choose *alternative B*.
8. Maximum incremental benefit-to-cost ratio (ΔB/ΔC): Choose *alternative B*.
9. Largest investment that has an incremental B/C ratio greater than 1.0: Choose *alternative C*.

A liberal interpretation of the wording of the 1936 Flood Control Act suggests that the highest investment in which benefits exceed costs is the most desirable. A more realistic interpretation is that the act sets a lower limit on acceptability, as to both total project amount and the incremental amounts in multilevel alternatives. By the latter interpretation, alternative *D* is rejected because of its negative ΔB − ΔC component and its ΔB/ΔC ratio less than 1.0. Then alternatives *B* and *C* are both acceptable, so other considerations should enter into the final choice. These considerations include the availability of investment funds, capital-rationing criteria, and the special features of social merit and economic objectives that affect most public projects.

### Which Measurements to Compare

A benefit-cost comparison directly rejects a project when the extra benefits are less than the costs and, by implication, when the extra benefits from a project are less than the extra benefits obtainable for alternative uses of the funds in the private sector of the economy. It is questionable whether society as a whole, present and future members included, is rewarded by public projects that have markedly lower returns than are currently earned in private commerce.

Reallocation of resources from the private sector, comprised mostly of tax collections, to a public project usually affects the distribution of wealth. Suppose the taxes collected to finance the flood-control project were drawn disproportionately from low-income taxpayers, and the land to be protected by the levee or dams were owned by wealthy farmers. If equality of income distribution is a government policy, the project would receive a low recommendation regardless of its B/C ratio. Similarly, the project could be questioned if the flood plain were inhabited by homeowners who knew the risk of flooding when they bought the building lots at lower prices than flood-free land prices, but are now pushing for flood protection at the expense of the general public.

Most government projects and social programs entail *spillovers* or *externalities*—benefits or costs that affect other parties besides those immediately involved in an activity. Depending on how rigorously spillovers are pursued, they can add up to tremendous amounts. The main problem is to quantify the effects. In the flood-control case, the cost of taxes lost from land occupied by a reservoir or levee could be calculated accurately, but how would the recreational benefits from the reservoir's lake or the disbenefits from lost open space be measured? Determining which spillovers are appropriately included and then quantifying them is an onerous but necessary task for conscientious benefit-cost analysts.

---

**Economy Exercise 13-1** Consider a freeway project that proposes a new four-lane highway leading from the downtown core of a city to a major arterial bypass in the suburbs. What factors besides the construction requirements could be considered in the total project cost?

---

## BENEFIT-COST COMPARISONS

The mechanics of benefit-cost comparisons are straightforward and simple—deceptively so. The basic comparison formulas are

$$B/C = \frac{\text{present worth of benefits}}{\text{present worth of costs}}$$

$$\text{or} \quad \frac{\text{equivalent annual benefits}}{\text{equivalent annual costs}}$$

and

Present value of net benefit (B − C) = PW(benefits) − PW(costs)

where, according to ***Circular No. A-94*** (1972) from the Office of Management and Budget, Executive Office of the President, the present worth of benefits is the discounted, ***constant*** dollar value of goods and services expected to result from a project or program for each of the years it is in effect. Estimates may reflect changes in the ***relative*** prices, when there is a reasonable basis for estimating such changes, but should not include any forecast change in the general price level during the planning period.

The present worth of costs is the discounted annual value in ***constant*** dollars of resources, goods, and services required to establish and carry out a project or program. All economic costs, including acquisition, possession, and operating costs, must be included whether or not actually paid by the government. Such costs, ***not*** generally involving a direct payment by the government, include imputed market values of public property and state and local property taxes forgone.

***Circular No. A-94*** recommends the use of a 10 percent discount rate that "represents an estimate of the average rate of return on private investment, before taxes and after inflation." Exceptions to the general 10 percent rate recommendation are admissible. The circular also states that "the evaluation should ***not*** use different discount rates to reflect the relative uncertainty of alternatives." Instead, a ***sensitivity analysis*** is suggested in which expected benefit and cost are supplemented with estimates of minimum and maximum values. Then the present value of those possible benefits and costs should be calculated, and the relative likelihood of each discussed.

**Example 13.1** **Benefit-Cost Calculations**

**Data for a Youth Corps Training Program are shown below. The expected duration of the program is 6 years. Both benefits and costs are uncertain. Analyze the desirability of the program from the given data.**

| | BENEFIT | | | COST | | | | PRESENT WORTH | | | | | |
|---|---|---|---|---|---|---|---|---|---|---|---|---|---|
| | | | | | | | | EXPECTED | | MINIMUM | | MAXIMUM | |
| *Year* | *Min.* | *Exp.* | *Max.* | *Min.* | *Exp.* | *Max.* | *(P/F, 10, N)* | *B* | *C* | *B* | *C* | *B* | *C* |
| 1 | 0 | 0 | 5 | 10 | 15 | 25 | 0.90909 | 0 | 13.6 | 0 | 9.1 | 4.5 | 22.7 |
| 2 | 5 | 10 | 15 | 10 | 10 | 15 | 0.82645 | 8.3 | 8.3 | 4.1 | 8.3 | 12.4 | 12.4 |
| 3 | 15 | 20 | 25 | 5 | 5 | 10 | 0.75132 | 15.0 | 3.8 | 11.3 | 3.8 | 18.8 | 7.5 |
| 4 | 20 | 30 | 30 | 4 | 5 | 8 | 0.68302 | 20.5 | 3.4 | 13.7 | 2.7 | 20.5 | 5.5 |
| 5 | 10 | 20 | 25 | 3 | 5 | 6 | 0.62092 | 12.4 | 3.1 | 6.2 | 1.9 | 15.5 | 3.7 |
| 6 | 5 | 15 | 20 | 2 | 5 | 5 | 0.56448 | 8.5 | 2.8 | 2.8 | 1.1 | 11.3 | 2.8 |
| *Totals* | | | | | | | | 64.7 | 35.0 | 38.1 | 26.9 | 83.0 | 54.6 |

**Solution 13.1** The *expected* benefit-to-cost ratio based on the expected values is

$$B/C = \frac{64.7}{35.0} = 1.85 \quad \text{and} \quad B - C = 29.7$$

which makes the program acceptable. As is apparent in the table, the B/C ratios under all three conditions are greater than 1.0. However, if maximum cost occurred while the benefit was at a minimum, then

$$B/C(\text{worst condition}) = \frac{38.1}{54.6} = 0.70 \quad \text{and} \quad B - C = -16.5$$

The worst combination might be excused from consideration if supporting evidence could be collected to show that the chance is very remote that the highest cost would accompany the lowest benefit. (See Chapter 14 for ways to evaluate risk.)

**Economy Exercise 13-2** A proposal has been made to modify certain navigational aids that will decrease the cost of operation by $10,000 per year for the next 30 years. A loss in benefit during the same period amounts to $1000 per year. Conduct a benefit-cost evaluation with a discount rate of 10 percent.

## Incremental B/C Comparisons

A preference for $X2$ is indicated by a comparison of the two present-worth levels given below for project $X$ according to the incremental differences for benefits and costs:

| *Alternative* | *B* | *C* | *B/C* | *B − C* | $\Delta B$ | $\Delta C$ | $\Delta B/\Delta C$ | $\Delta B - \Delta C$ |
|---|---|---|---|---|---|---|---|---|
| $X1$ | 4 | 2 | 2 | 2 | 3 | 2 | 1.5 | 1 |
| $X2$ | 7 | 4 | 1.75 | 3 | | | | |

Although $X1$ has a higher B/C than $X2$, the incremental ratio for the extra benefit and cost exceeds 1.0, making $X2$ an acceptable alternative. If it can be assumed that sufficient capital is available to fund either alternative, $X2$ is selected because it provides greater benefits. However, when the total costs for projects exceed the resources allocated to the proposing agency, capital-budgeting procedures are required and intangible effects may influence the selection.

**Example 13.2** **Incremental Benefit-to-Cost Evaluation**

A number of small earthen dams are contemplated for the headwaters of a drainage system. Four tributaries originate in a national forest and flow together to form a river which passes through private lands. Each year there is some flooding, and every few years a major inundation occurs. Construction of one or more dams will ease the threat of high water. Dams on all the tributaries would largely eliminate the chance of a major flood.

In addition to the damage to private lands, floods also ruin fire and logging roads in the forest. Other benefits from the dams include the value of the impounded water for fire protection and recreational use. The following benefit and cost estimates have been developed for the only topologically feasible combinations of dams:

| *Dam Sites* | *Construction Costs* | *Annual Maintenance and Operation* | *Annual Flood Benefits* | *Annual Fire Benefits* | *Annual Recreation Benefits* |
|---|---|---|---|---|---|
| 1 | $1,200,000 | $20,000 | $200,000 | $20,000 | $30,000 |
| 1 and 2 | 1,500,000 | 35,000 | 190,000 | 40,000 | 30,000 |
| 1, 2, and 3 | 2,700,000 | 50,000 | 280,000 | 60,000 | 60,000 |
| 1, 2, 3, and 4 | 3,500,000 | 60,000 | 300,000 | 70,000 | 70,000 |

A 40-year life and no salvage value is assumed for earthwork dams. An interest rate of 4 percent is deemed appropriate for the investment. This rate reflects the low risk involved and is in line with the historical interest rate for bonds issued by the federal agency to finance public projects.

Based on B/C ratios, which of the four alternatives should be selected?

**Solution 13.2** A B/C ratio based on equivalent annual values for each alternative is calculated from

$$\text{B/C ratio} = \frac{\text{annual flood and fire savings + recreation benefits}}{\text{equivalent annual construction costs + maintenance}}$$

where

Equivalent annual construction cost = initial construction cost × $(A/P, 4, 40)$

The incremental B/C ratio is determined from the additional benefits returned by an increment of cost above the last acceptable alternative (B/C ratio > 1.0).

| *Dam Sites* | *Annual Benefits* | *Annual Cost* | INCREMENTS OF *Benefit* | INCREMENTS OF *Cost* | *Total B/C Ratio* | *Incremental B/C Ratio* |
|---|---|---|---|---|---|---|
| 1 | $250,000 | $ 80,630 | | | 3.10 | |
| 1 and 2 | 260,000 | 110,770 | $ 10,000 | $30,140 | 2.34 | 0.32 |
| 1, 2, and 3 | 400,000 | 186,400 | 140,000 | 75,630 | 2.14 | 1.42 |
| 1, 2, 3, and 4 | 440,000 | 236,820 | 40,000 | 50,420 | 1.85 | 0.79 |

For an accurate evaluation, the requirement that annual benefits must equal annual costs should be applied to each separable increment of project costs. Site 1 is compared to the alternative "no action" and yields a total B/C ratio = incremental B/C ratio = 3.10, which qualifies it as an acceptable alternative. The next increment that meets the B/C-ratio standard is the location of dams at sites 1, 2, and 3 compared with a dam at site 1:

$$\text{B/C(3)} = \frac{\$10{,}000 + \$140{,}000}{\$30{,}140 + \$75{,}630} = 1.42$$

Adding another dam site fails to produce an acceptable B/C ratio:

$$B/C(4) = \frac{\$40,000}{\$50,420} = 0.79$$

Therefore, preference is indicated for the third alternative, dams at sites 1, 2, and 3.

Without an incremental analysis, the last alternative (four dams) might have been selected, because it does possess a B/C ratio greater than 1.0 and offers the greatest total benefits. Another mistake would be to eliminate all other alternatives because the dam at site 1 has a larger benefit-to-cost ratio than the rest of the options. The reasoning errors behind such conclusions are the same as those examined for incremental rates of return (Chapter 9). The conclusion to accept the three-dam alternative based on the given data would also result from a rate-of-return or present-worth evaluation.

It is interesting to note the *sensitivity* of the selection to changes in the data. Using a required interest rate of 7 percent instead of 4 percent would change the choice to the first alternative (one dam) because all the added benefit and cost increments would produce B/C ratios smaller than unity. Including only the flood-control benefits would also make site 1 the only alternative with an acceptable B/C ratio. The number of spillover benefits to include in an analysis and the monetary rating given to less tangible benefits can significantly influence decisions.

**Economy Exercise 13-3*** The present worth of benefits and costs for two mutually exclusive "base" proposals are shown below, along with data for three supplementary projects which can be combined with either base to yield additional benefits. The supplementary projects are not mutually exclusive. Which combination of projects is preferred when resources are limited to $400,000?

| BASIC-PROJECT PROPOSALS | | | | SUPPLEMENTARY PROJECTS | | |
|---|---|---|---|---|---|---|
| *Project* | *B* | *C* | *B/C* | *Project* | *B* | *C* |
| *P*1 | $300,000 | $150,000 | 2.0 | *S*1 | $75,000 | $50,000 |
| *P*2 | 450,000 | 250,000 | 1.8 | *S*2 | 140,000 | 100,000 |
| | | | | *S*3 | 300,000 | 150,000 |

## Irregularities in B/C Comparisons

No investment criterion seems to be able to escape all theoretical objections. The benefit-cost criterion has been subjected to considerable criticism based on its oversimplification of complex inputs, susceptibility to misinterpretation, and potential for misuse. Some of the more prominent objections are discussed below.

*Adapted from G. A. Fleisher, ***Benefit-Cost Analysis: An Introductory Exposition***, Tech. Report 72-1, Department of Industrial and Systems Engineering, University of Southern California.

## 1 INSENSITIVITY TO PRIORITIES

A B/C ratio includes the cost and benefit of a particular project but no indication of how valuable the benefit is, as compared with other projects, or the relative amount of resources involved.* As an extreme example, two projects could have the same B/C ratio and have costs of $1 million and $1 billion. The billion-dollar project might have long-range consequences for improvement of the environment in a sparsely populated region while the million-dollar project would serve a critical inner-city need. Eventually both projects might serve the same number of people, but the smaller project would have far greater immediate benefits for the taxpayers who are funding it. A third $100,000 project, with the same B/C as the other two, could be to preserve a site that will be destroyed if not subsidized immediately. Assuming there are not enough resources available to undertake all three projects, the B/C ratios are of little help in setting priorities.

A related objection is that the B/C standard assumes benefits are equally valuable to the rich and poor. Thus, a project to push a new freeway through a densely populated low-cost-housing neighborhood would use money partially drawn from the poor to force them to relocate to make room for an expressway that would mainly benefit wealthier suburban commuters.

The argument against the above criticisms is that the B/C criterion was not designed to rank projects; it sets only a minimum level of acceptability and does not pretend to identify the source of investment funds. Its use as a guide to decision makers is poignantly described in a compendium of papers submitted to the Subcommittee on Priorities and Economy in Government of the United States Congress:

> . . . If benefit-cost analysis is to be implemented and used to its fullest potential, renewed efforts must be made by policy makers in both the executive and legislative branches of government. The economics profession has made significant advances in the level of sophistication of their analyses which should aid this task, but one thing is clear—benefit-cost analysis does not make decisions.
>
> Analysis can provide an important and helpful tool for making decisions, but it is no more than a tool. Problems involving social policy and value judgments must be considered and weighed in conjunction with the results of benefit-cost analysis and the final decision made by the human policymaker.†

## 2 VARIATIONS DUE TO FORMULA INPUTS

The numerator of the B/C ratio is usually taken to mean the net benefit—the sum of all benefits minus all disbenefits. If the disbenefits were treated as costs in the denominator, the ratio would obviously change. For instance, if the present worth of benefit is 10, disbenefit is 3, and cost is 5, then

$$\text{B/C} = \frac{10 - 3}{5} = 1.40$$

*Discussed in A. Maass, "Benefit-Cost Analysis: Its Relevance to Public Decisions," *Quarterly Journal of Economics*, vol. LXXX, May 1966.

†*Benefit-Cost Analysis of Federal Programs*, Government Printing Office, Washington, D.C., 1973.

But if the disbenefit were counted as a cost, then

$$B/C = \frac{10}{5 + 3} = 1.25$$

This variation presents no difficulty when it is remembered that the purpose of the B/C > 1 criterion is simply to distinguish between acceptable and unacceptable alternatives. There is no way in which shifting of disbenefits from the numerator to the denominator can cause a B/C ratio to change from greater than 1.0 to less than 1.0. Equivalently, the new present value (benefit minus disbenefit minus cost) must be the same regardless of the order of subtractions. Nor will an incremental analysis indicate any difference in the absolute acceptability of alternatives owing to the location of disbenefits in the B/C ratio.

Two alternatives are evaluated below, first by subtracting disbenefits from benefits and then by adding them to costs. In both cases, though the value of the ratios differ, the extra increment of investment is found to be acceptable according to the $\Delta B/\Delta C > 1$ standard.

| | PRESENT WORTH OF BENEFIT, DISBENEFIT, AND COST | | | | DISBENEFITS SUBTRACTED FROM BENEFITS | | | | DISBENEFITS ADDED TO COSTS | | | |
|---|---|---|---|---|---|---|---|---|---|---|---|---|
| *Alternative* | *B* | *D* | *C* | *B−D−C* | *B/C* | $\Delta B$ | $\Delta C$ | $\Delta B/\Delta C$ | *B/C* | $\Delta B$ | $\Delta C$ | $\Delta B/\Delta C$ |
| *X* | 10 | 3 | 5 | 10−3−5=2 | 7/5=1.4 | 7−9=2 | 6−5=1 | 2/1=2 | 10/8=1.25 | 13−10=3 | 10−8=2 | 3/2=1.5 |
| *Y* | 13 | 4 | 6 | 13−4−6=3 | 9/6=1.5 | | | | 13/10=1.3 | | | |

The relative sizes of B/C ratios provide an indication of investment productivity that may be misleading. Consider a project *P*1, in which an investment of 10 and annual costs of 95 produce yearly benefits of 100 for 20 years: At $i = 10$ percent,

$$B/C(P1) = \frac{100(P/A, 10, 20)}{10 + 95(P/A, 10, 20)} = \frac{100(8.5135)}{10 + 95(8.5135)}$$

$$= \frac{851}{10 + 809} = 1.04$$

Compare this project to *P*2, which also has an initial investment of 10, no annual costs, and a yearly benefit of 1.25 for 20 years: At $i = 10$ percent,

$$B/C(P2) = \frac{1.25(8.5135)}{10} = \frac{106}{10} = 1.06$$

Both projects are barely acceptable, and *P*2 has a slightly higher B/C ratio, but *P*1 will "pay back" the initial investment in 10/(100 − 95) = 2 years, while *P*2 requires 10/1.25 = 8 years before the sum of annual benefits equals the investment. Stated another way,

$$RR(P1): \quad (100 - 95)(P/A, i, 20) - 10 = 0 \quad \text{at } i = 50\%$$
$$RR(P2): \quad 1.25(P/A, i, 20) - 10 = 0 \quad \text{at } i = 10.9\%$$

The rate of return of *P*1 far exceeds that of *P*2, although their B/C ratios were nearly the same.

This apparent conflict is caused by the inability of the B/C formula to distinguish between cash-flow patterns; it tends to underrate the quality of a project with high annual costs. This same characteristic was discussed in Chapter 9, where ranking reversals were noted for alternatives evaluated by PW and RR comparisons when the discount rate was raised to a certain level. In the example above, a discount rate of 11 percent would make $P2$ an unacceptable alternative:

$$\text{B/C}(P2) = \frac{1.25(P/A, 11, 20)}{10} = \frac{1.25(7.9633)}{10} = \frac{9.95}{10} = 0.995$$

whereas $P1$ remains acceptable at the 11 percent rate:

$$\text{B/C}(P1) = \frac{100(P/A, 11, 20)}{10 + 95(P/A, 11, 20)} = \frac{796}{10 + 757} = 1.038$$

**Economy Exercise 13-4** Benefit-cost ratios for two mutually exclusive alternatives have been calculated with disbenefits assigned to both numerator and denominator as shown:

| | PRESENT WORTH | | | | |
|---|---|---|---|---|---|
| *Alternative* | *B* | *D* | *C* | *B/C = (B − D)/C* | *B/C = B/(C + D)* |
| *A*1 | 150 | 0 | 100 | (150 − 0 )/100 = 1.5 | 150/(100 + 0) = 1.5 |
| *A*2 | 200 | 20 | 120 | (200 − 20)/120 = 1.5 | 200/(120 + 20) = 1.43 |

What reply would you make to a claim that $A1$ is economically superior because its B/C ratio is equal to or better than that obtained from $A2$, regardless of the way disbenefits are handled?

# THE DISCOUNT-RATE QUESTION

The appropriate interest rate to use in evaluating public investments is a voluminously debated issue. Although numerous suggestions have been forwarded, none has been awarded complete acceptance. About the only near consensus is that discount rates established for early benefit-cost analyses in the depressed 1930s (around 2½ to 3¼ percent) are not valid today. Studies made of the Corps of Engineers' projects reveal that 80 percent of the projects authorized in 1962 would have been rejected at a discount rate of 8 percent, and California's big Feather River project would be uneconomical at a 6 percent rate.* The effect of the discounting rate on benefit-cost ratios for projects with different lives is shown in Figure 13.1. The five curves represent B/C ratios at different discount rates for an initial investment of $1 that returns a total of $2 over $N$ years. Therefore, at a 0 percent discount rate and $N = 10$,

*See J. A. Stockfish, "The Interest Rate Applicable to Government Projects," in H. H. Hinricks and G. M. Taylor, eds., ***Program Budgeting and Benefit-Cost Analysis***, Goodyear, Pacific Palisades, Calif., 1969.

$$B/C = \frac{(2/10)(P/A,\ 0,\ 10)}{1} = \frac{0.2(10)}{1} = 2$$

but at $i$ = 10 percent for 10 years,

$$B/C = \frac{(2/10)(P/A,\ 10,\ 10)}{1} = \frac{0.2(6.1445)}{1} = \frac{1.23}{1} = 1.23$$

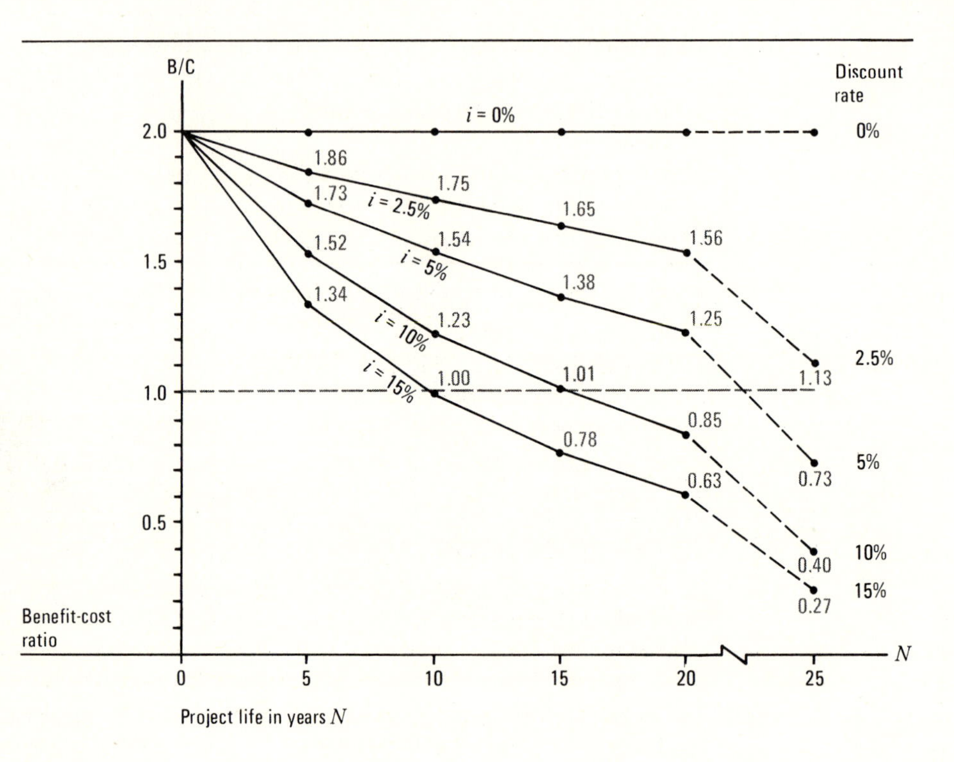


**FIGURE 13.1** Graph of benefit-cost ratios at five discount rates for different project lives of a proposal to invest $1 in order to receive $2 in equal installments prorated over the project's life.

## Range of Discount Rates

The government borrowing rate is an obvious measure of financing costs for public projects. It is akin to the *cost-of-capital* concept by which a minimum attractive rate of return is determined for investments in the private sector. The cost of capital to the government can also be taken as a minimum discount rate for public projects. In recent years the interest rate for government borrowing has varied between 5 percent and 9 percent, depending on the level (federal, state, or local), tax-exemption privileges offered, current economic conditions, and length of the loan. Since government bonds are generally considered to be riskfree, any public project should promise a minimum return at least equal to the bond rate. Otherwise, the resources that could be used to fund projects should instead be applied to debt repayment.

The rates of return expected in the private sector are higher than the interest rates on government bonds. This is a reasonable condition because a firm would invest in riskfree government bonds if it did not expect larger returns from industrial investments. (See page 274 for industrial rates of return.) In the opinion of W. A. Baumal,* "The correct discount rate [for public projects] is the percentage rate of return that the resources would otherwise provide in the public sector." His view recognizes the opportunity cost to general welfare that occurs when resources are used to produce benefits smaller than could have been obtained if the resources were applied elsewhere; specifically, a public project that barely exceeds B/C = 1 when evaluated at a low discount rate would, if carried out, produce fewer benefits than could have been obtained by leaving the amount invested in the private sector.

An argument against charging higher rates for public projects is that certain socially desirable programs would never meet higher evaluation standards but are nonetheless worthwhile. This line of reasoning emphasizes the responsibility of the present population to assure adequate living conditions for future generations. Since most people are more concerned with satisfying their current needs and wants than with saving for the future, proponents of low discount rates say only government action can enforce investments for future benefits and this transfer of resources necessarily yields a lower return than resources used for current consumption.

Counterarguments question the legitimacy of allowing a few policymakers to decide what is best for general welfare, especially when their allocations appear to contradict public preferences shown in a free market. It is generally agreed that no generation has the right to completely consume resources that cannot be replaced by its successors (for example, erosion of soil, extinction of a wildlife species, destruction of natural phenomena). The dispute is with subsidizing future generations by resource expenditures acceptable only if evaluations are based on artificially low discount rates; such special assistance is not warranted, critics say, because the future is also served by public and private investments that earn returns comparable to current resource commitments. From a historical perspective that indicates each generation has been wealthier than its predecessor, it appears imprudent to justify "wasteful" resource allocations on the basis of protecting future beneficiaries.

## Social Discount Rate

The rate of interest used in evaluating public projects is often referred to as the *social discount rate*. It reflects the cost of capital obtained from the private sector and the opportunity cost of resources applied to public projects rather than private investments. One approach to the calculation of a social discount rate based on the cost of government borrowing is given below;† it is a summation of contributing cost components that provides a composite percentage figure.

*W. A. Baumal, "On the Appropriate Discount Rate for the Evaluation of Public Projects," in ibid.

†Adapted from E. B. Staats, "Survey of Use by Federal Agencies of the Discounting Technique in Evaluating Future Programs," in ibid.

SAMPLE CALCULATION OF A SOCIAL DISCOUNT RATE BASED ON THE COST OF GOVERNMENT BORROWING

| *Cost Components* | *Rate, %* |
|---|---|
| **1** Approximate interest rate for long-term, low-risk borrowing | 6.0 |
| **2** Corporate taxes forgone by the government, assuming the average corporate return is 13% before taxes, the fraction of money borrowed by the government that would have gone into corporate investment is 65%, and the marginal corporate tax rate is 40%: (0.13)(0.65)(0.4) = | 3.4 |
| **3** Personal taxes forgone by the government, assuming the 35% not put into corporate investment earns 10% for individual taxpayers and the marginal tax rate is 30%: (0.10)(0.35)(0.3) = | 1.0 |
| **4** Taxes forgone from dividends and bond interest generated by the 65% of government-borrowed dollars that would have gone into corporate investment: | |
| **a** Assuming individuals are taxed at 30%, dividend payments are 40% of corporate profits, and 60% of corporate earnings are after-tax profits: (0.3)(0.4)(0.6)(0.13)(0.65) = 0.6% | |
| **b** Assuming interest on corporate bonds is 7% and the individual marginal tax rate is still 30%: (0.07)(0.30)(0.65) = 1.4% | |
| assuming an equal weight for taxes forgone from bond interest and dividend payments: (0.6% + 1.4%)/2 = | 1.0 |
| **5** Income taxes collected on interest payments from government borrowing are a negative cost of capital, assuming the borrowing rate is 6% [from (1) above] and the composite tax rate for corporate and individual bondholders is 35%: (0.06)(0.35) = | −2.1 |
| *Social discount rate* | 9.3% |

An alternative aggregate method for determining the discount rate based on the cost of government borrowing assumes a composite personal and corporate tax rate of 50 percent and a taxable return of 12 percent on any money not borrowed by the government:

| | |
|---|---|
| Interest cost of borrowing | 6.0% |
| Taxes forgone = (0.50)(0.12) | 6.0% |
| Less taxes on bond interest: (0.06)(0.35) | −2.1% |
| *Social discount rate* | 9.9% |

The above calculations do not include any consideration of ***risk***. It is well known in the financial market that government bonds are perceived to have low risk which is rewarded by the lower interest rate needed to attract capital. The inference for public projects is that the government is in a better position than a private entrepreneur to cope with risk; governments undertake so many projects that their operations are essentially immune to risk, on the average. However, critics point out that there have been some obvious and major failures in government-funded projects and proposals. Therefore, like investments in the private sector, the social discount rate should include a risk factor. A surcharge of ½ to 1 percent has been suggested, but this across-the-board addition fails to account for differences in the nature of risk for various classes of projects.

The recommended 10 percent discount rate in DOD *Circular No. 7041.3* is evidence that the continuing dialogue about discount rates has brought them more in line with rates of return utilized in the private sector.

## BENEFIT AND COST DATA

According to the theorems of economics, a person is indifferent to the choice between having an additional unit of any commodity and paying the equilibrium price of that commodity. Many benefits and costs applicable to public projects can be determined from the going market price of commodities. However, there is a class of ***public goods*** that are available to all if they are available to a single individual. Public parks, drinking water, and clean air are examples of public goods. The value an individual attaches to such goods is elusive because the benefits may be taken for granted and the individual feels that they will be available whether or not he or she contributes an amount equal to the value received. Also, in many cases there is no marketplace in which to measure the value of a public good, even if it could be assumed everyone enjoys a public good in the same amount and would willingly pay for it.

### Consumers' Surplus

The concept of ***consumers' surplus*** has been utilized to explain the difference between what consumers actually pay for a public good and what they might be willing to pay. As shown in Figure 13.2, the demand curve indicates the value (expressed in dollars) placed by consumers on various quantities of a public good: The willingness to pay is a function of quantity. The shaded area is the ***consumers' surplus,*** representing the surplus value a consumer receives over the amount paid for at price $P$. This surplus is difficult to measure because demand curves are difficult to develop, but some measure of the value of goods above price $P$ is desirable in determining the worth of certain kinds of benefits derived from public projects.

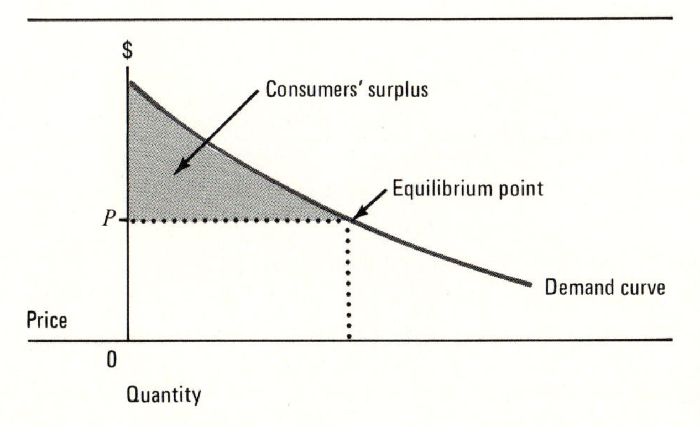


**FIGURE 13.2** Consumers' surplus as the extra amount over the market price $P$ that consumers would be willing to pay for quantity $Q$.

Assume electricity is to be supplied to a region where it was never before available. A hydroelectric project produces power priced at $\$P$ per kilowatthour to residents of the region. Benefits from the project might be based on the market value of the power sold, $P \times Q$, when the residents buy $Q$ kilowatthours. Then, true benefits are understated, because the market value is based on the price of the *last* unit of power purchased. In general, a consumer would willingly pay an amount significantly greater than the market price for the first kilowatthour furnished, a little less for the second kilowatthour, and so on. This bonus to the consumer represents an extra lump sum the consumer would theoretically be willing to pay for the opportunity to buy electric power at price $P$. The correct measure of the benefit provided by the project should include the surplus in addition to the actual dollar value of power sold.

A consumers'-surplus analysis should take into account several conceptual questions concerning appropriateness to a given project. Exact quantitative measures may be difficult to justify. However, one frequently encountered case is easily handled. When a project provides a quantity of public good at a lower price than previously possible, the increase in usage is the difference between the previous and present quantities consumed. In Figure 13.3, usage has increased from $Q_1$ to $Q_2$ as a function of a price decrease from $P_1$ to $P_2$ realized from a new public project. The new dollar amount of consumption is $P_2Q_2$, and the increase from the previous level is $P_2Q_2 - P_1Q_1$. But the benefit from the new price should also include the incremental change in the consumers' surplus. If the demand curve can be assumed linear over the range of $Q_1$ to $Q_2$, then the change in consumers' surplus is calculated as

$$\Delta(\text{consumers' surplus}) = \frac{(P_1 - P_2)(Q_1 + Q_2)}{2}$$

which is added to the dollar value of increased consumption to provide a measure of total benefit.

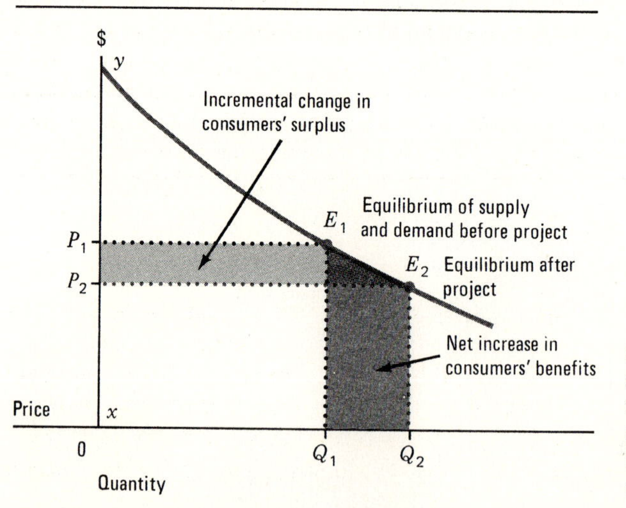


**FIGURE 13.3** Change in consumers' surplus caused by a new project that decreases the price from $P_1$ to $P_2$ with a resulting increase in quantity consumed from $Q_1$ to $Q_2$.

**Economy Exercise 13-5** Let the conditions in Figure 13.3 represent a road-construction project. The original unimproved road provided benefits to users that are shown in Figure 13.3 by the enclosed area $xyE_1Q_1$. Paving, grading, and straightening can improve the value of the road until its benefits are represented in the figure by the area $xyE_2Q_2$. Because the new road completely replaced the existing one, the total benefit claimed is the entire area $xyE_2Q_2$. Is this the correct benefit for the road-improvement project?

## Spillover Benefits and Costs

Spillovers (or externalities) exist when there is a divergence between private and social benefit or between private and social cost. The causes and effects of spillovers from environmental pollution were described in Extension 8.2. An example of spillover benefits is the increase in land values brought about by the construction of a new highway or park. Spillovers are easily overlooked and are sometimes tricky to quantify, owing to filtering effects, but engineering economists should be conscious of spillover considerations for public projects because they constitute a significant source of benefits and costs. Cognizance is necessary to avoid B/C ratios bloated by exaggerated spillover benefits. Similarly, spillover costs deserve close inspection since their magnitude may be exaggerated by damage reports from injured parties. It is reasonable to view any spillover claim with some suspicion, if only to trace its logic.

Enumeration of spillovers may suggest the most effective direction for public expenditures. For example, in the case of production plants that are dumping pollutants, a tax on the effluent which is a measure of the external costs of pollution can make private costs identical with social costs. Or a subsidy could be granted to the offending plants to assist them in reducing the undesirable discharges. Both courses of action are probably better alternatives than public ownership of the production facilities. In other cases the spillovers may be so extensive that public ownership is preferable, as in a multipurpose river-basin project. It would be more practical for the government to plan, undertake, and operate the entire river-basin development than it would be to create a system of market incentives capable of directing the combined resources of private investors toward the desired social objectives.

**Economy Exercise 13-6** There are public schools and private schools. From consideration of spillover benefits, why is the public investment in education logical?

## Project Costs

Costs for all resources required to achieve the stated objectives of a project should be included in the analysis. Each alternative in multilevel proposals should be self-contained in terms of cost, to avoid double counting or overlapping expenses. Costs which have already been incurred at the time of the analysis, ***sunk costs,*** should not be included. The ways to collect estimates described in Chapter 10 are as applicable to public as private projects. A few cost considerations of particular significance to benefit-cost analyses are:

1 Imputed costs of existing assets employed on a project should be included when there are alternative uses for the assets. For instance, land or facilities desired for use in another project besides the one being analyzed would be treated in both projects as costs based on fair market values. Neglecting such costs would be similar to a contrac-

tor building houses on inherited land and then basing the selling price of the homes on just the construction costs because the land was free.

2 Preliminary costs of investigation and technical services required to get a project started are part of the project's budget. Once the project is under way, interest charges during the construction period before earnings begin are applicable expenditures. Management costs, whether incurred by the sponsoring agency or contributed by the beneficiaries, are part of the total project cost.

3 Spillover costs comprise all significant adverse effects caused by the construction and operation of a project and are expressed in terms of market prices whether or not actual outlays for compensation are made. For instance, an irrigation project could reduce the quality of water downstream from the area to be serviced; the additional treatment facilities necessary to restore the quality of the water or provisions for water from another source would be costs to the irrigation project.

## Project Benefits

Public projects are typically conceived to provide a certain benefit or family of benefits. Many of the difficulties of quantifying these benefits have already been discussed. The nature of some social benefits often precludes their measurement by direct market pricing, because there is no market in the private sector which offers the benefits.

To illustrate the valuation of benefits directly attributable to gains that are valued by marketplace statistics, consider a proposal to provide irrigation to 10,000 acres of farmland.* Water supplied by irrigation is an input to the production of food and fiber. Beneficial effects from irrigation are based upon their total value to agricultural producers and are measured as the increase in net farm income with and without a plan for providing irrigation water. Changes in net farm income may be estimated by analyzing changes in gross farm income and expenses for each separate enterprise or by using representative farm budgets. A sample compilation is shown in Table 13.2.

**TABLE 13.2** Summary of components contributing to benefits from an irrigation project.

| *Component* | *Without Project* | *With Project* | *Difference Due to Project* |
|---|---|---|---|
| Area in acres | 10,000 | 10,000 | |
| Number of farms | 20 | 50 | 30 |
| Farm investment | $1,500,000 | $5,000,000 | $3,500,000 |
| Farm receipts | $490,000 | $2,500,000 | $2,010,000 |
| Farm interest expense | ($67,500) | ($225,000) | ($1,560,000) |
| Other farm expenses | ($372,500) | ($1,775,000) | |
| *Net farm income* | $50,000 | $500,000 | $450,000 |
| Disbenefit for alternative earnings forgone by new farmers: 30 new farms × $2000 | | | $60,000 |
| *Irrigation benefit* | | | $390,000 |

Data used to estimate farm investment, receipts, and expenses are drawn from representative practices, prices, and budgets experienced during comparable situations. The dis-

*Adapted from *Guidelines for Implementing Principles and Standards for Multiobjective Planning of Water Resources*, U.S. Department of the Interior, 1972.

benefit for the opportunity cost of wages that farm workers would receive if they did not operate their own farms is based on census data for farm-work earnings compiled by the Department of Agriculture.

Statistics for estimating many types of benefits are not as directly applicable as in the irrigation project. It is then necessary to derive a benefit value by research that combines data from several sources. Recreational benefits are a good example of derived values. Factors affecting recreational values include the number and distribution of the population in the project area; socioeconomic characteristics including disposable income, occupation, education, age, and mobility of the relevant population; and the population's leisure-time and recreational habits as indicated by trends in hunting and fishing license sales, sales of recreational equipment, and total demand. Such factors suggest the potential participation rate but do not put dollar amounts on the value of the participation.

To facilitate valuation, two types of recreation days are recognized, a ***recreation day*** being defined as a unit of use consisting of a visit of one individual to a recreation development during a reasonable portion of a 24-hour period:

| *Type of Recreation Day* | *Value of 1 Day's Use* |
|---|---|
| *General* Type of activity attracting the majority of outdoor recreationists—requires the development of convenient access and maintenance of adequate facilities (for example, swimming, picnicking, nature studies, tent and trailer camping, canoeing) | \$0.75–\$2.25 |
| *Specialized* Type of outdoor activity for which opportunities are limited, intensity of use is low, and which often involve a large personal expense by the user (for example, cold-water fishing, big-game hunting, wilderness pack trips) | \$3.00–\$9.00 |

Indications of the values that users put on such experiences can be obtained from charges for similar activities in the private sector, time and cost of travel required to partake in an activity, cost of alternative uses of the facilities, and questionnaires about how much a user would be willing to pay for various experiences. A suggested composite value for a visitor day to any area is simply the gross national product per day divided by the total population.

Some attributes of a public project are ***intangible***. In the 1940s, aesthetic values were considered of immeasurable worth, so any cost that made them available to the public was acceptable. This line of reasoning meant every project that promised aesthetic values automatically had a B/C ratio of at least 1.0. An equivalent conclusion results from placing an infinite value on irreplaceable assets such as wilderness areas.

A more systematic approach to the evaluation of intangible characteristics defines categories of attributes and employs a subjective rating system to rank the quality of different projects. Formal environmental quality-evaluation procedures have been developed to compare intangible benefits and disbenefits for alternative water- and land-use plans. (Rating concepts for intangibles are discussed in Chapter 5.) The procedure starts with the identification of the environmental resources involved in a proposed project. Resources include open spaces, stream systems, beaches, wilderness areas, historical and cultural features, and ecological systems. These resources are then evaluated according to the following criteria:

1 *Quantity* Specific environmental features are enumerated and measured in units such as acres of wilderness; miles of white water; area of lakes; and number of waterfalls, animals, scenic attractions, and significant sites.
2 *Quality* The desirability of an environmental feature is rated by assigning a number from 1 to 10 after comparing the features with known or projected conditions at other locations.
3 *Human influence* Subjective 1-to-10 ratings are made to indicate the degree to which people would use the resource identified; the degree to which it is protected for continued use; and the degree that it contributes to education, scientific knowledge, and human enjoyment.
4 *Uniqueness* A 1-to-10 rating measures the frequency of occurrence of a specific resource in the project relative to its occurrence elsewhere.
5 *Significance* Irreversible damage to a resource is rated on a 1-to-10 scale that indicates the magnitude of adverse effects on the environment. The rating reflects the scarcity of supply of a resource and the capability of its returning to its original or natural state after the proposed project is implemented.

The ratings for intangible benefits are typically reported as an appendix to the monetary comparisons. A narrative may accompany the presentation of intangible considerations to explain how the ratings were developed, to discuss opposing views, and to provide supplementary information.

## BENEFIT-COST APPLICATIONS

The methodology of benefit-cost analysis is applied to government projects in many nations. Its applicability as a public-investment criterion in most sociopolitical environments is discussed in ***Guidelines for Project Evaluation,**** a publication of the United Nations Industrial Development Organization:

> The main reason for doing social benefit-cost analysis in project choice is to subject project choice to a consistent set of general objectives of national policy. . . . When one project is chosen rather than another, the choice has consequences for employment, output, consumption, savings, foreign exchange earning, income distribution and other things of relevance to national objectives. The purpose of social benefit-cost analysis is to see whether these consequences taken together are desirable in the light of objectives of national planning. . . .
>
> It is worth remembering that social benefit-cost analysis is not a technique but an approach. It provides a rational framework for project choice using national objectives and values. Projects are judged in terms of their precise impact on the economy, and this impact is evaluated by using parameters reflecting national goals, social objectives and global facts. This is relevant not merely for evaluation of given projects but also for formulating new ones and implementing the chosen projects. There are always a very large number of elements one can vary in formulating new projects. Which of these are important and worth pursuing will depend on the guidance provided by social benefit-

*P. Dasgupta, A. Sen, and S. Marglin, ***Guidelines for Project Evaluation,*** United Nations, 1972.

cost analysis. Similarly, in implementation there are always small choices between variants that are not fully specified in the plan and are left to the implementors, and it helps to have a clear picture of the social benefits and costs of these choices.

## Criticisms

Many of the charges of misuse of benefit-cost analyses are the result of specific projects that have gone awry. The well-publicized cases of dams that failed to stop flooding as promised and project overruns that boosted costs well above anticipated benefits could be considered as evidence of the need for more exacting benefit-cost analysis, rather than of deficiencies in the B/C criterion.

In addition to the theoretical critiques of the applicable discount rate and the fairness of benefit measurements, criticism has been leveled at the way the analyses are conducted. It is claimed that unrealistically high values have been assigned to intangibles to compensate for low monetary benefits in projects that would be unacceptable when evaluated only on quantifiable data. Indirect or intangible costs seldom seem to get the same recognition as nonmonetary benefits. Costs have been underestimated, say critics, because the local impact of major federal projects was not anticipated; labor and material prices went up because of increased local demand caused by the project, to the detriment of both project expenses and consumers in the community.

Projects are sometimes undertaken without the support of the residents in the area. Planners are thus forced into the position of telling residents what is good for them, which requires a selling campaign that may create costly delays or added community-relations costs. Part of the problem could be alleviated by exposing a project to a public vote. However, difficulties of informing the voters about the issues involved and the remoteness of many projects from their base of funding (the taxpayers) limit the workability of decisions by vote. A choice between a football stadium and an art museum in a city would probably arouse enough interest by those affected to assure a careful appraisal and representative turnout of voters. Yet an equal expenditure to develop a wildlife refuge might inspire less interest, even if all the affected voters could be given a chance to vote. On all questions of national objectives, there will likely be a few opponents, and many of these will be in the geographical locality where a project is to be carried out. Local protests can naturally be expected from people whose lives are disrupted by a project designed to serve the general welfare. It is the duty of policymakers to give consideration and adequate compensation to the local interests while supporting regional and national social interests.

## Examples

Three benefit-cost analyses are described in the following pages. The applications are digested to point out certain highlights; the actual studies and techniques employed were much more exhaustive. Example 13.3 illustrates a feasibility study that disqualified public expenditures in solving a local problem. The sensitivity of inputs to B/C calculations is demonstrated in Example 13.4. A postaudit to determine the actual benefit and cost of a program is featured in Example 13.5.

## Example 13.3 Feasibility Investigation of a Water-Resource Project*

"The Corps of Engineers was directed by the Congress of the United States to make a study of Marys River Basin. The purpose of the study was to determine what could be done to reduce flood damages and to conserve, use, or develop the basin's water resource."

Six alternatives were investigated: (1) floodproofing of individual structures, (2) building levees, (3) improving river channels, (4) instituting a system of landowner-constructed dams, (5) implementing a system of small tributary reservoirs, and (6) erecting multipurpose reservoirs. The costs of floodproofing, levees, and channel improvements were found to be much greater than the benefits, mainly because of the lack of secondary benefits. Landowner-constructed dams built with government financial support would have some localized benefits, but would require an unreasonable amount of land for the storage obtained, and problems of balancing outflows made the alternative impractical.

The last two alternatives also proved to be economically infeasible, as shown in Figure 13.4. Flood-control benefits were based on historical flood damages updated to 1974 prices, estimates of flood-stage reductions by operation of the projects investigated, and projections of economic growth in the area. Recreation benefits were based on user-day projections. The cost of the cheapest alternative water supply was used in the analysis as a measure of the water-supply benefits that could reasonably be expected. Data provided by the U.S. Fish and Wildlife Service showed that fish and wildlife benefits resulting from any of the project alternatives would be negligible. Costs included construction and, as appropriate, land; relocation of roads, railroads, and utilities; fish-passage facilities; wildlife-migration features; recreation developments; design costs; and construction supervision costs.

## Example 13.4 Critical Examination of a B/C Analysis†

The Chicod Creek project consists of land-treatment measures and channel improvement. In 1971 an original proposal for the project was reevaluated to include additional expenditures for fish and wildlife mitigation, modification and addition of laterals, inclusion of sediment-control features, and the updating of installation costs. The original study in 1965 had a benefit-cost ratio of 2.1 and included the following annual benefits:

| | |
|---|---|
| Flood-damage reduction | $46,666 |
| More intensive use | 7,044 |
| Improved efficiency | 21,100 |
| *Total benefit cited in the 1965 study* | $74,810 |

*Adapted from *Summary Report on Marys River Basin Water Resource Study*, United States Corps of Engineers, 1975.

†Adapted from R. K. Davis, B. J. Ingle, and W. J. Gillen, "An Economic Perspective on the Small Watershed Program," one of a compendium of papers submitted to the Subcommittee on Priorities and Economy in Government, Joint Economic Committee, Congress of the United States, 1973.

| | Noon Dam | Wren Dam | Tumtum Dam | Tributary Dam System |
|---|---|---|---|---|
| *Costs, total* | | | | |
| Construction | $51,352,000 | $52,944,000 | $37,024,000 | $19,090,000 |
| Investment* | 57,386,000 | 59,165,000 | 41,374,000 | 21,333,000 |
| *Costs, annual* | | | | |
| Interest and amortization† | 3,380,000 | 3,485,000 | 2,437,000 | 1,257,000 |
| Operation, maintenance, and replacement | 210,000 | 180,000 | 170,000 | 135,000 |
| *Total average annual cost* | $ 3,590,000 | $ 3,665,000 | $ 2,607,000 | $ 1,392,000 |
| *Benefits, average annual* | | | | |
| Flood control | $ 945,400 | $483,100 | $258,500 | $126,300 |
| Recreation | 573,000 | 345,000 | 401,000 | 163,000 |
| Irrigation water supply | 85,000 | 85,000 | 85,000 | 85,000 |
| Municipal and industrial water supply | 90,600 | 63,400 | 8,800 | 0 |
| Fish and wildlife enhancement | 0 | 0 | 0 | 0 |
| *Total average annual benefits* | $1,694,000 | $976,500 | $753,300 | $374,300 |
| *Benefit-cost ratio* | 0.47:1 | 0.27:1 | 0.29:1 | 0.27:1 |

*Construction cost plus interest during construction.
†The annual cost equivalent to the total cost spread over a 100-year period computed at 5 7/8 percent interest rate.

**FIGURE 13.4** Benefit-cost analyses for alternative reservoir projects.

The interest rate used for discounting costs in the original study was 3 1/8 percent. The current rate reported by the Water Resources Council is 5 3/8 percent. The revised costs in 1971 evaluated at the two discount rates and the corresponding B/C ratios are shown below.

| | ANNUAL COSTS | | | |
|---|---|---|---|---|
| *Discount rate, %* | *Structural* | *O and M* | *Total* | *B/C* |
| 3 1/8 | $38,655 | $19,100 | $57,765 | 1.30 |
| 5 3/8 | 55,040 | 19,100 | 74,140 | 1.01 |

While the project does not fail the benefit-cost test, at the revised interest rate it clearly becomes marginal. A more realistic cost for the land acquired would make the project uneconomic. An appropriate rate of interest for the land adjustment cost is determined as

| | |
|---|---|
| Ratio of annual income to market value | 5.6% |
| Plus rate of capital appreciation | 5.0% |
| Less rate of inflation | 1.6% |
| *Discount rate* | 9.0% |

which changes the project cost to

| | *Annual cost* |
|---|---|
| Structural costs (less land) at 5 3/8% | $45,152 |
| Land cost at 9% | 15,559 |
| Operation and maintenance | 19,100 |
| *Total cost* | $79,100 |

and leads to

$$\text{B/C} = \frac{\$74{,}810}{\$79{,}100} = 0.95$$

"The project fails the benefit-cost analysis even before environmental costs and excessive agricultural benefits are deducted."

**Example 13.5** **Follow-up Study to Evaluate Benefit versus Cost for a Human-Resource Program***

The Roswell Employment Training Center was established in 1967 under contract with the Bureau of Indian Affairs. The Center was designed as a residential employment training program for Indian families, solo parents, and single adults. Emphasis is placed on training and counseling intended to aid trainees in adjusting to typical work situations and to living off the reservation.

Using earnings differentials to measure benefits from human-resource investment programs is a standard method of evaluation. The major resource costs are staff services, where salaries are taken as the measure of their value in an alternative use; other direct costs such as supplies; and the opportunity costs of training—the earnings forgone as a result of undertaking training.

In June 1970, a follow-up study was conducted by mail of all trainees who had entered Roswell Center between March 1968 and February 1970. The calculation of a benefit-cost ratio using the data from this survey is shown below. Of the group of 170 returning adequate survey forms, 70 were currently employed at an average hourly wage of $2.25 after training, whereas 82 members of the same group were employed before training was given. Because the average wage rate after training was significantly higher than the average wage before training, there was still a positive earnings differential associated with training, equal to $573 per year per trainee:

$$\text{Additional annual earnings/trainee} = \frac{(\$4680 \times 70) - (\$2808 \times 82)}{170} = \$573$$

*Adapted from B. F. Davis, "Using Benefit-Cost Analysis to Assess a Human Resource Investment Program," in ibid.

*Benefit of program* PW of extra annual earnings over 41 years (the number of remaining years in the labor force for trainees whose average age was 24) at a 6% discount rate $8674

*Cost of program* Contracted cost for an average length of training of 8 months at $658 per month per trainee = $5264. Forgone earnings (number of trainees working before training times their monthly earnings times the months of training, divided by the total number of trainees) = [(82)($2808/12)(8)]/170 = $902. Then, PW of total cost = $5264 + $902 $6166

$$B/C = \frac{\$8674}{\$6166} = 1.41$$

The benefit-cost analysis described above is about the simplest possible for a human resource investment program. Earnings before and after training are compared, differentials projected, at a constant amount, over the years ex-trainees can be expected to remain in the labor force. The present value of these amounts are then compared to direct program costs plus the opportunity costs of trainees to estimate benefit-cost ratios. Every step of the analysis can be challenged on theoretical and empirical grounds. Pretraining earnings may be understated; perhaps earnings differentials should be projected at growing, rather than constant, amounts in future years; the lack of complete followup data may bias the estimated earnings differentials; a control group should really be established; maybe 6 percent is not the "right" discount to use; an estimate of the value of the physical facilities in an alternative should really be included as a cost—all this and probably more could be charged. To focus on these niceties of analysis would be to miss an important aspect of program analysis. Analysis should not only ask the question, "Is this a good program?" or "Has the benefit-cost ratio been appropriately and accurately estimated and is it greater than unity?" but should provide program managers with the information they need to change particular aspects of their program in an effort to improve performance.*

## SUMMARY

*Benefit-cost analysis* is a well-rooted method of evaluating public projects. The basic measure of acceptability is a benefit-to-cost ratio greater than 1.0 or, equivalently, a positive net difference between benefit and cost. Incremental B/C ratios should also exceed unity for added increments of investment. $B/C > 1$ is a standard of minimum approval but is not an effective ranking criterion. Discounting procedures, computational practices, and capital-rationing considerations are essentially the same in the private and public sectors.

The appropriate interest rate for the evaluation of public projects is a much-discussed issue. Social projects are particularly sensitive to the rate of discount applied, because most major expenditures occur early while the benefits extend many periods into the future; thus, a high discount rate tends to reduce the relative proportion of benefit to cost. Low discount rates are recommended by those who feel it is government's social responsibility

*Ibid.

to undertake projects that do not necessarily provide immediate returns comparable with those earned in the private sector. Counterarguments claim that resources are wasted when transferred from private uses, where returns are high, to low-yield public investments. One measure of the ***social discount rate*** results from computing the total cost of government borrowing.

Social projects often provide ***public goods***—facilities or services available to all if available to one. The actual amount paid for a public good may not be a true measure of its worth. A ***consumers' surplus*** exists when a good is worth more to the user than the price paid for it. ***Spillover benefit*** or ***cost*** occurs when an activity affects third parties not directly involved in a project or program. These special properties of social involvement complicate the evaluation of benefits and costs.

Many project benefits and most project costs can be determined directly, as in industrial studies, and are subject to the same cautions for economic completeness. Owing to the nature of some benefits, monetary values must be derived from a combination of contributing factors. Intangible qualities that defy monetary assignments can be systematically, yet still subjectively, rated and appended to benefit-cost computations to provide more complete information about a project. It is then the responsibility of policymakers to decide which projects to fund of those that pass the B/C hurdle.

## Discussion of Economy Exercises

**EE 13-1** Benefit-cost analyses for some major highway-construction projects consume hundreds of pages. A few of the considerations include:

1. Air pollution from traffic flow
2. Expenditures for traffic control
3. Accident frequency and severity, past and forecast
4. Parking expectations in affected areas
5. Time and cost consequences of congestion, before and after
6. Shift of business competition
7. Accessibility of outlying jobs to central-city poor
8. Changes in property values due to accessibility
9. Relocation of dwellings in the highway path; associated disruption of neighborhoods
10. Visual and auditory impact
11. Temporary effects of construction: wages, price of materials, living costs, employment, etc.

**EE 13-2**

$$B/C = \frac{-\$1000(P/A, 10, 30)}{-\$10{,}000(P/A, 10, 30)} = 0.1$$

The criterion of B/C > 1.0 for acceptability is not applicable when both the benefit and cost of a project are negative. The ratio of savings to costs is obviously 10:1 in favor of the proposal. Thus, when both the numerator and denominator of a B/C ratio are negative, a ratio of less than 1.0 indicates acceptability. Further, a project with a positive B and a negative C is automatically accepted.

**EE 13-3** Since $P1$ has a higher benefit-cost ratio than $P2$, it is a logical base for combinations. Only those combinations that approach but do not exceed the $400,000 limit are considered because the objective is to maximize benefits with the resources available.

| *Combination* | *Benefit* | *Cost* | *B − C* | *B/C* |
|---|---|---|---|---|
| $P1 + S1 + S2$ | $515,000 | $300,000 | $215,000 | 1.72 |
| $P1 + S3$ | 600,000 | 300,000 | 300,000 | 2.0 |
| $P1 + S1 + S3$ | 675,000 | 350,000 | 325,000 | 1.93 |
| $P1 + S2 + S3$ | 740,000 | 400,000 | 340,000 | 1.85 |

Of these combinations, the one that spends all the available resources, $P1 + S2 + S3$, provides the greatest benefits ($740,000), yields the highest net benefits ($340,000), has a B/C ratio (1.85) greater than 1.0, and gives an incremental benefit-cost ratio exceeding unity,

$$\Delta B/\Delta C = \frac{\$740,000 - \$675,000}{\$400,000 - \$350,000} = \frac{\$65,000}{\$50,000} = 1.3$$

is the most satisfying.

However, the best possible combination is one that utilizes the other base, $P2$. The combination of $P2 + S3$ has a total benefit of \$450,000 + \$300,000 = \$750,000 and a total cost of \$250,000 + \$150,000 = \$400,000 to yield

$$(B - C)(P2 + S3) = \$750,000 - \$400,000 = \$350,000$$

and $(B/C)(P2 + S3) = \dfrac{\$750,000}{\$400,000} = 1.875$

This combination might be overlooked by the preselection error of eliminating $P2$ from consideration because its benefit-cost ratio is less than the B/C for $P1$. It is also interesting to note that in this case the B/C ratio provides an accurate ranking of top contenders because both have the same total cost.

**EE 13-4** The net benefit of $A2$ is 200 − 20 − 120 = 60, versus 150 − 100 = 50 for $A1$. Thus, if any preference is shown by the benefit-cost criterion, it is bestowed on $A2$. An incremental analysis conducted either way,

| | DISBENEFIT IN NUMERATOR | | | | DISBENEFIT IN DENOMINATOR | | |
|---|---|---|---|---|---|---|---|
| | *B* | *C* | *B/C* | | *B* | *C* | *B/C* |
| $A1$ | 150 | 100 | 30/20 = 1.5 | $A1$ | 150 | 100 | 50/40 = 1.25 |
| $A2$ | 180 | 120 | | $A2$ | 200 | 140 | |

confirms that $A2$ is an acceptable project because its incremental B/C ratio is consistently greater than 1.0.

**EE 13-5** The claimed benefit overstates the value added by the road-improvement project. The original benefit $xyE_1Q_1$ produced by the old road exists regardless of improvements.

The new benefit from the project that can be counted is represented by the area $Q_1E_1E_2Q_2$, which can be calculated as

$$\text{Added benefit} = P_2Q_2 - P_1Q_1 + \frac{(P_1 - P_2)(Q_1 + Q_2)}{2}$$

$$= P_2Q_2 - P_1Q_1 + \frac{1}{2}(P_1Q_1 + P_1Q_2 - P_2Q_1 - P_2Q_2)$$

$$= \frac{1}{2}(P_2Q_2 - P_2Q_1 + P_1Q_2 - P_1Q_1)$$

$$= \frac{1}{2}(P_2 + P_1)(Q_2 - Q_1)$$

**EE 13-6** A primary benefit of education is the personal development of talents by students; this is a private benefit. However, the social benefits of education are enormous. The economy as a whole benefits from a more versatile and productive labor force. Spillover benefits extend to reductions in the costs of crime prevention, welfare programs, and many other areas in which the lack of education raises the cost of correction to society. It would be very difficult to coordinate and monitor a nationwide network of private schools to maintain standards and to direct educational efforts toward desired social objectives.

## PROBLEMS

**13.1** A 2.2-mile (3.54-kilometer) stretch of highway is known locally as "Fog Hollow." In an attempt to reduce accidents, the shoulders along Fog Hollow could be widened at a cost of $103,000 per mile ($64,000 per kilometer), and large electric warning signs could be installed at both ends of the dangerous stretch. The signs would be controlled by a computer hooked to sensing devices to display recommended speeds for various fog conditions. The fog-warning signs could be installed for $165,000 and would have annual maintenance costs of $7000 during their 10-year life.

If the average accident cost in Fog Hollow amounts to $2385, how many accidents per year would have to be avoided by the project to make it acceptable for a 20-year study period? The discount rate used for such projects is 6 percent.

*(21)*

**13.2** A public good is represented by a linear demand schedule ranging from equilibrium at $1 for the first unit available to zero when 1 million units are available. The current level of availability and consumption is 300,000 units.

**13.2a** What is the dollar volume of current consumption?

*($210,000)*

**13.2b** What is the dollar volume of the current consumers' surplus?

*($45,000)*

**13.2c** If a public project can decrease the cost to 40 cents per unit, what additional total benefit does it produce?

*($135,000)*

**13.2d** What would the consumers' surplus be if the project were carried out?

*($180,000)*

**13.2e** What is the dollar volume of consumption after the project is completed?

*($240,000)*

**13.2f** Calculate by formula the increase in consumers' surplus resulting from the project. Check the figure obtained with respect to the answers to the previous problems.

*($135,000)*

**13.2g** Would it be possible for a demand schedule to be such that total consumption would decrease from a previous level after implementation of a project that allowed the quantity consumed to increase as a function of a decreased price? Illustrate your answer. Comment on the change in consumers' surplus.

**13.3** Two projects from each of four departments have been submitted for evaluation. The projects from each department are mutually exclusive.

| *Department* | *Project* | *Benefit* | *Cost* |
|---|---|---|---|
| A | A1 | $100,000 | $ 70,000 |
| | A2 | 112,000 | 80,000 |
| B | B1 | 70,000 | 55,000 |
| | B2 | 76,000 | 60,000 |
| C | C1 | 160,000 | 100,000 |
| | C2 | 184,000 | 120,000 |
| D | D1 | 110,000 | 75,000 |
| | D2 | 122,000 | 85,000 |

**13.3a** Which projects should be funded according to the B/C criterion if one project must be selected from each department?

*(A2, B2, C2, and D2)*

**13.3b** Which projects should be funded if only $300,000 is available?

*(A1, B1, C1, and D1)*

**13.4** A state-sponsored "forest protective association" is evaluating alternative routes for a new road into a formerly inaccessible region. Different routes for the road provide different benefits, as indicated in the following table:

| *Route* | *Construction Cost* | *Annual Saving in Fire Damage* | *Recreational Benefits* | *Timber Access* | *Annual Maintenance Cost* |
|---|---|---|---|---|---|
| A | $185,000 | $5000 | $3000 | $ 500 | $1500 |
| B | 220,000 | 5000 | 6500 | 500 | 2500 |
| C | 310,000 | 7000 | 6000 | 2800 | 3000 |

The roads are assumed to have an economic life of 50 years, and the interest rate normally required is 3 percent per year.

**13.4a** According to a B/C comparison, which route should be selected?

$[B/C(B) = 1.09;\ \Delta B/C(B) = 1.48]$

**13.4b** Would the choice be changed if the interest rate were doubled?

*(None acceptable)*

**13.4c** Would the choice be changed if annual maintenance costs were not included?
[*B/C(C)* = *1.31*; $\Delta$*B/C(C)* = *1.09*]

**13.5** The State Highway Department has collected data for a proposal to construct an overpass and extra lanes at a busy, dangerous highway intersection. Estimated land-acquisition, demolition, and construction costs are:

| | |
|---|---|
| Land and demolition | $410,000 |
| Additional lanes [0.6 mile at $1 million/mile (0.966 kilometer at $621,400/kilometer)] | $600,000 |
| Overpass construction | $2,740,000 |

and annual maintenance is expected to average $105,000 per year during the 30-year life of the project.

Improved safety is the primary benefit expected from the project. A road count revealed that 12,000 vehicles per day pass through the intersection. An average of 2.4 fatalities per year have occurred at the present usage rate. Usage is expected to rise to 15,000 vehicles per day by the time improvements are completed, and accidents would be expected to rise commensurately. For every fatal accident there are 19 nonfatal accidents and 160 property-damage accidents. Settlements for fatal, nonfatal, and property accidents are calculated, respectively, at $65,000, $5000, and $1000. The improved intersection is expected to reduce all accidents by 90 percent.

Expenses associated with patrolling, maintaining the traffic signals, and directing traffic at the intersection are currently $31,000 per year; these would be eliminated by the project. Twenty-one percent of the vehicles using the intersection now must stop, with an average wait of 0.8 minute. The average cost of a stop for a vehicle's operation is estimated at 6 cents, and waiting time is assumed to be worth $2.40 per hour. When the overpass is in operation, all stops will be avoided, but 40 percent of the vehicles will have an added travel distance of 0.2 mile (0.322 kilometers). A vehicle operating cost of 18 cents per mile (11.2 cents per kilometer) is assigned as an average value.

The overpass will reduce pollution. This reduction is valued at $50,000 per year. Annual taxes forgone from property condemned for the expansion amount to $8000. The discount rate for the study is 8 percent.

**13.5a** What is the benefit-cost ratio based only on the primary benefit of improved safety and just the building costs?

**13.5b** What is the B/C based on secondary benefits and the building cost?

**13.5c** What is the total B/C? What other factors might be considered in the project?

**13.6** A dam on Roily River is being evaluated. A 50-year life and a 9 percent discount rate are to be used. Estimates have been obtained for the following benefits and costs:

| | |
|---|---|
| Flood losses prevented in the Roily River area | $ 900,000/year |
| Flood losses reduced in downstream rivers | 800,000/year |
| Increases in property values along the Roily River (present worth of values) | 1,000,000 |
| Income from electric power produced | 3,200,000/year |
| Construction of dam and access roads | 25,000,000 |
| Cost of powerhouse and transmission facilities | 10,000,000 |
| Interest costs during construction | 2,000,000 |
| Operating and maintenance costs | 100,000/year |

Assume construction costs occur at time zero but the benefits do not begin until the start of the fifth year and interest charges during production are prorated evenly over the 4-year period.

**13.6a** What is the benefit-cost ratio?

**13.6b** Assume opponents of the dam are questioning the study on the basis that no costs were included for the destruction of a popular recreation area that would be inundated by the dam's reservoir. The land is all federally owned, but a very popular park that features mineral-water springs will be inundated. What value would have to be placed on this park and associated recreational areas to make the project unacceptable? Could a reasonable case be developed to abandon the project? What other benefits and disbenefits might be considered?

**13.7** In recent years many cities have constructed large municipal stadiums. The sales pitch to get voters' approval for funding often appealed to civic pride and implied that the stadium would be self-supporting as well as an attraction for visitors whose purchases would increase revenues for many merchants. After completion, some stadium projects failed to meet advertised expectations and caused difficulties that the voters had not anticipated.

Disregarding misjudgments of construction costs, which far exceeded original estimates in many cases, and inaccuracies which overestimated the income that would be received from promotions booked into a stadium, what disbenefits and spillover costs could have been logically anticipated and accounted for in a thorough benefit-cost analysis? List 10 considerations, and discuss how values could be obtained for the factors involved.

**13.8** Assume that a modest-size city park with which you are familiar is the end product of a public project. That is, a project produced the present park based on an accepted benefit-cost analysis. Further assume that the land became available for the park because a school on the site burned down, and since there was sufficient classroom space available in nearby schools, the city decided not to rebuild the facility. Instead, it decided the land would be used for other city purposes or sold to add money to the city treasury. One of the proposals for the land was to build the park that now exists.

**13.8a** Based on the above scenario, list the benefit and cost categories that should have been included in the evaluation of the park project.

**13.8b** Insofar as possible, quantify the benefit and cost categories in terms of current values to complete the benefit-cost analysis. Use a 30-year life and a discount rate of 7 percent. Discuss the intangible benefits involved.

**13.9** Two alternative routes for a new expressway are being evaluated. One follows the valley along a river, and the other takes a short cut through a range of hills.

The river route has a length of 30 miles (48.28 kilometers) and a first cost of $7,125,000. Its annual cost of maintenance will be $2500 per mile ($1553 per kilometer), and a major overhaul will be required every 10 years at a cost of $750,000.

The hilly route will be 7 miles (11.27 kilometers) shorter than the river route, but it will cost $9.5 million. Annual maintenance costs will be $4000 per mile ($2486 per kilometer), and the major overhaul and surfacing every 10 years will cost $575,000.

Traffic on either expressway is predicted to average 6000 vehicles per day, one-fourth of which will be commercial traffic. Anticipated average speed on either route should be 50 miles per hour (80.5 kilometers per hour). Time is valued at $7.00 per hour for commercial traffic and $2.50 per hour for other vehicles. The average operating costs for commercial and noncommercial traffic are, respectively, 35 cents and 10 cents per mile (21.8 cents and

6.2 cents per kilometer) for the river route and 40 cents and 11 cents (24.8 cents and 6.8 cents per kilometer) on the steeper hilly route.

Compare the alternative routes according to an incremental benefit-cost ratio based on a 30-year life and a discount rate of 7 percent.

**13.10** When the owner of an old mansion in a residential neighborhood died, she left a will deeding 4 acres (16,187 square meters) to the city if the city agreed to use it for a park and maintain the ornate gardens of the site for 40 years. If the city chooses not to accept the offer, the land will be sold to a developer at a price of $260,000. The city uses its municipal-bond rate of 7 percent for evaluating recreational projects.

Modifying the garden to make it into a suitable park would require an investment of $110,000. Annual maintenance, service, and policing would cost $21,000. Forty adjacent homes would increase in value by $2000 each if the land were used as a park. Property taxes in the area are 2.4 percent of the assessed value.

The park would be used mostly for passive recreation such as walking, picnicking, and resting. No team sports would be allowed, and playground equipment for children would be minimal. It is estimated that 30,000 visitors per year would spend an average of 1 hour in the park.

**13.10a** What value per hour would visitors to the park have to place on their visits to have the benefit equal cost?

**13.10b** Is the value per visit calculated in Problem 13.10*a* a reasonable figure? Compare it with the cost of roughly equivalent experiences which people pay for.

**13.10c** What action would you recommend that the city take?

**13.11** A proposal to develop a new type of mine-shaft borer is being evaluated by a government agency. The project, if approved, will be carried out as a joint venture with a private company which will share the research-and-development investment with the government. The new borer is an adaptation of a tunnel-boring machine made by the cooperating company to perform as a coal-mining machine. Benefits and costs are described below.

*Direct benefits* (Advantage over conventional shaft-sinking techniques) The conventional cost for sinking a 1000-foot (304.8-meter) shaft, which takes an average of 18 months, is $1.7 million per shaft. The cost for the new borer to sink a 1000-foot (304.8-meter) shaft is:

First cost prorated over a life of 15 shafts = $137,000/shaft
Operation (cutter and maintenance costs) = $152,000/shaft
Labor (crew × wages × 6 months to sink shaft) = $420,000/shaft

*Time benefits* Because a borer will complete a shaft in 6 months instead of the conventional 18 months, production profits are available 1 year sooner. Assuming a typical shaft produces 250,000 tons (226,795,000 kilograms) of coal, average profit is $2.80 per ton, and financing costs are 10 percent for the coal-mining companies, a company will benefit both from obtaining profits sooner and from reduced financing charges because the investment in a mine is recovered 1 year sooner when a new borer is used.

*Spillover benefits* Savings expected from other mining operations besides coal mining which can utilize the advanced technology available from the borer are $3.9 million per year.

*Derived benefits* Additional benefits from improved safety are likely but are not included in the analysis.

*Costs* The government share of the 3-year development costs are, respectively, $100,000,

$500,000, and $500,000 for years 1, 2, and 3. An equal amount will come from the private sector for deployment and demonstration activities.

Twenty-seven new borers are expected to be built and sold. Five will be in operation in year 4, 14 in year 5, 24 in year 6, and 27 thereafter. Each borer is expected to sink two shafts per year. Using a 10-year study period and a discount rate of 10 percent, what is the B/C ratio for the borer project? State all assumptions required in your analysis.

**13.12** As one part of a model program to demonstrate ways to rehabilitate a decaying portion of a city, a proposal has been made to provide swimming facilities for residents of the model area. This can be accomplished by constructing swimming pools in small parks or by busing to a nearby lake. Pertinent data are listed below.

1. Population in the 5-square-mile (12.95-square-kilometer) area that might take part in the swimming program is 35,000.
2. Average daily usage for a pool could vary from 1 percent to 5 percent of the population living within 1½ miles (2.41 kilometers) of a pool.
3. Minimum standard for the surface area of water per swimmer is 20 square feet (1.86 square meters).
4. A standard swimming pool has 7000 square feet (650.3 square meters) of water surface and requires 2 acres (8094 square meters) of land (part of the land serves also as a small park). Construction costs including equipment would be $200,000. Operating costs, including lifeguard wages, would be $35,000 during the swimming season. The life of a pool is 15 years.
5. The swimming season lasts 100 days each year.
6. Land prices in the area are $80,000 per acre ($19.77 per square meter). Taxes forgone on land acquired by the government would average $1800 per acre per year ($0.445 per square meter per year). Demolition and relocation expenses would amount to $70,000 per acre ($17.30 per square meter).
7. School buses could be used during the swimming season at a cost of $105 per bus per day including drivers' wages.
8. Four round trips could be made each day per bus during the swimming season to carry swimmers from the model area to a nearby lake. A bus can carry 70 people. The same number of people are expected to participate in the swimming program if either the pools are constructed or busing is provided.
9. Extra lifeguards and operating expenses at t e city park on the lake to handle an influx of additional swimmers would be $200 per day. Sufficient space is available at the lake to accommodate any foreseeable usage.
10. A "recreation leader" would be hired for each 100 people expected to take part in the busing arrangement. A recreation leader would cost the program $25 per day, including fringe benefits. Four administrators would receive $4000 each during the season if busing is used.
11. A 50-cent admission fee is charged for pools in other parts of the city, and the same charge is reasonable for bus fare for a round trip to the lake.

Analyze the swimming program based on a 10 percent discount rate.

**13.12a** What is the equivalent annual cost per swimmer per day under each plan? Assume no fees are paid for swimming. State your assumptions.

**13.12b** Assuming a 50-cent fee per day is charged each swimmer, what amount of additional benefits for the swimming pools would have to be counted to allow them to pass the B/C > 1 criterion? Discuss the benefits in terms of consumers' surplus, spillover effects, and derived or intangible benefits.

**13.12c** Conduct a sensitivity analysis. Is there any combination of readily quantifiable factors that can provide a B/C > 1 for the swimming pools? State your assumptions.

**13.12d** Based on your calculations above, what decision would you make if you were the government policymaker who had to decide whether a summer swimming program should be part of the model-city rehabilitation demonstration? Write a short summary supporting your conclusions.

# EXTENSION

### *13.1 Recreation, Conservation, and Money*

(The following case study is designed as a group project. It has been used several times in industrial and university classes, usually as a competitive exercise between teams representing the three competing companies in the case. Depending on the interpretation given to the data, the additional spillover benefits and costs developed, and the presentation techniques employed, each of the three companies has "won" on different occasions. All teams have access to the same hard data upon which to base their arguments; thereafter it is the generation of supplementary values and their substantiation that decides the issue.)

Your advice as an economic analyst is sought for a multifaceted decision involving several groups with contrasting objectives. The situation centers on how a particular tract of land owned by the government should be developed. If the land did not have special recreational appeal, it would be managed under the normal policies followed by the Forest Service. Because so many influential groups became involved, the Forest Supervisor solicited each group to make a proposal that they believed would use the natural resources most effectively. The supervisor stressed that each group should quantify their proposed course of action as to expected costs and benefits.

**BACKGROUND INFORMATION**

Three years ago the Forest Service announced it would accept bids for the timber on approximately 2600 acres around Jewel Lake. This tract includes one of the few remaining stands in the area of old-growth Douglas fir. It also contains classically beautiful Jewel Lake and 180-foot Mac Falls.

The timber-harvesting plan called for a clear cut of about 1500 acres to yield approximately 45 million board feet of Douglas fir and 15 million board feet of hemlock. Selective cutting was planned for another 800 acres to provide corridors around Jewel Lake, along Quick Creek, and other sites which might be developed as recreational areas at a later date. The yield from selective logging was expected to average 22,000 board feet per acre. The remaining 300 acres, comprising particularly rugged terrain with sparse growth, would not be logged. The stated mix of Douglas fir and hemlock is expected to yield sawn lumber with an average value (selling price) of $285 per thousand board feet. The minimum net stumpage value acceptable to the Forest Service for the stated mix (including road improvements) was set at $125 per thousand board feet.

The advertised bids included conditions for constructing a two-lane paved road from Forest Highway 3264 to Jewel Lake, 6.3 miles (Route *A* in Figure 13.5). In addition, three spur roads (single-lane, unpaved) totaling 3.2 miles would be re-

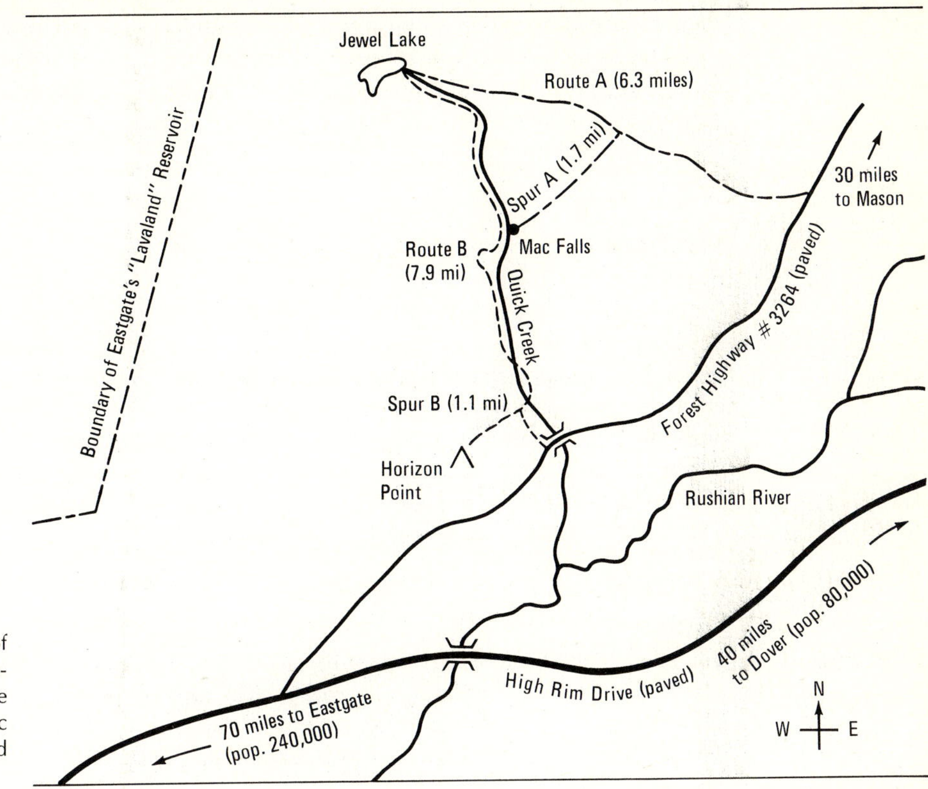


**FIGURE 13.5** Map of the Quick Creek Drainage showing the relative positions of geographic features and proposed roads.

quired to get timber out. The locations of and specifications for all roads were provided. The estimated cost of constructing Route *A* would be $1.1 million, and the spur roads would be $15,000 per mile. It was further stipulated that the timber must be removed within one logging season after the roads are completed.

As soon as the advertisement for bids was out, responses came from three large lumber companies near Eastgate and an international log exporter. Then pressures began to build up on all sides. A senator from the state said the log exporter should be disqualified because it was exporting local jobs with the logs. Two state representatives and the governor accused the Forest Service of favoring the big national lumber companies by including such a short time limit for the logging operation to be completed; they wanted it extended to allow smaller mills located in Mason to bid. No single mill in Mason, a town of 3000 with a high unemployment rate, had the necessary capacity. Also, unlike the big mills near Eastgate, the Mason mills had no company-owned lands to harvest, and operations had been curtailed for several months, owing to recently limited sales from federal lands in that area.

The most pressure came from conservation groups. Initially led by statements from the Siskiyou Club, other ecology and sports groups soon took up the cry to save Jewel Lake and Mac Falls. They condemned individually or in concert the practice of clear cutting, destruction of one of the few remaining tracts of virgin timber, putting roads into what should be a primitive area, possible damage to the Eastgate water supply, and putting private interest for more logs above public interest in recreation. A student group at the university in Eastgate started a campaign to save Jewel Lake. They obtained an injunction to halt the sale.

**DIPLOMACY**

Rather than let the situation be decided in court, or develop into a media battle of "good guys versus bad guys" with the government on the side of the bad guys, a new approach was initiated. The Forest Service asked all the interested parties to form teams that could develop a proposal and had the capability of carrying out the objectives. Forest Service personnel and cruise data would be made available to assist each proposal group. Very few restrictions were placed on the request for proposals, but the Forest Service retained the right to select or reject all or portions of any plans. It agreed to state in writing the reasons behind the final decision.

Six months later, four proposals were received.

**PROPOSAL 1**

The student-inspired ecology group proposed that the area be declared a primitive area. They also outlined a plan to construct a parking lot where Forest Highway 3264 crossed Quick Creek, with an improved trail that would lead to Mac Falls and Jewel Lake. Primitive camping facilities would be constructed at both the falls and the lake.

This proposal was rejected by the Forest Supervisor. Among the points included in the written reply were reasons why the area did not meet the requirements of a primitive area, that improvement plans were incompatible with the primitive designation, the waste of a prime forest stand, the lack of money for proposed construction and maintenance, the availability of other primitive areas in the region, and the current Forest Service policy of encouraging multiple use of its lands.

No organized efforts appear likely to be made to override the rejection.

**PROPOSAL 2**

Perhaps as a result of collusion, only one mill from Eastgate submitted a proposal. The Pacific Mill, second largest in Eastgate, but the one with the least company-owned timber land, has a capacity of 75 million board feet per year. It is located 61 miles from the Jewel Lake sale. Key points of the proposal are listed below.

*Bid* $97.80 per thousand board feet, regardless of specie.

*Logging plan* Clear cut 1500 acres, and selectively log 800 acres.

*Road plan* Construct two-lane paved and improved roads for Route *A* and Spur *A* (estimated cost: $175,000 per mile), and 3.2 miles of unimproved logging spurs (estimated cost $15,000 per mile). The unimproved spurs are not needed for fire protection.

*Recreation facilities*

| JEWEL LAKE | | MAC FALLS | |
|---|---|---|---|
| 20 tent campsites | $22,000 | Fenced viewing area and parking lot (20 cars) | $25,000 |
| 2 restrooms (double) | 12,000 | 10 picnic tables | 3,000 |
| Water system | 8,000 | 1 restroom (double) | 6,000 |
| 1 cabin-storage and pier | 20,000 | | |
| Landscaping | 6,000 | Landscaping and water system | 17,000 |
| | $68,000 | | $51,000 |

Accompanying the proposal was an explanation of the plan:

1 All construction and logging would be completed within 2 years, at which time the Forest Service would be paid the $97.80 per thousand board feet. The area would not be open for recreational use during the 2-year period.
2 A consultant's report stated that the recreational value of the area was more for local residents than for tourists. Mac Falls is high but not very spectacular because of the limited water flow. The view from the top of the falls, reached by Spur *A*, is more impressive than the view from the bottom; $17,000 was allocated for fencing to protect viewers. Jewel Lake was considered too small for motorboats and for extensive camping facilities. Therefore, only tent camping and rowboats would be allowed. A two-bedroom cabin for one or two Forest Service personnel was included.

**PROPOSAL 3**

Three locally owned mills in Mason with a combined annual capacity of 14 million board feet joined to submit a proposal. They were assisted by a coordinator from the Governor's office. The basic elements of the plan are listed below.

1 The longer and more expensive Route *B* is planned because it provides access to Horizon Point and the length of Quick Creek. Construction costs for two-lane paved roads along the indicated routes are expected to average $205,000 per mile. The 5.2-mile stretch from Forest Highway 3264 to Mac Falls will be completed in 2 years, and the remainder to Jewel Lake will be open at the end of year 5. Spur *B* will be ready by the end of the fourth year. Associated recreational facilities will be ready for use as each section of the road is opened.
2 A picnic area and campground will be built at the foot of Mac Falls. It will have 15 trailer (or tent) sites ($60,000), five picnic sites ($1000), two double restrooms ($8000), 1.1 miles of improved trail ($1000), electric and water utilities ($3000), and landscaping ($5000). Owing to the terrain, it is believed that camping facilities are better suited to the Mac Falls area than Jewel Lake.
3 Horizon Point affords a panoramic view of several mountain peaks. Facilities at the point will include eight picnic sites ($3000), parking area and trails ($2500), and two single (dry) restrooms ($500).

4 State recreational consultants suggest that Jewel Lake be limited to day use to protect the area. To encourage picnic use, three group shelters, a swimming area, open play fields, and a canoe rental facility will be included. A cabin for the caretaker and a storage building for canoes are provided:

| | | | |
|---|---|---|---|
| 3 group shelters | $18,000 | Caretaker's house | $14,000 |
| 35 picnic sites | 7,000 | Canoe storage | 8,000 |
| 3 double restrooms | 14,000 | 2 piers and beach | 3,000 |
| Parking lot and trails | 9,000 | Canoes | 6,000 |
| Complete utilities | 12,000 | Landscaping | 14,000 |

5 Roads and facilities will be maintained during the period of the contract. Revenue from canoe rentals is expected to pay for caretaker work in all three recreational areas.
6 Seven years are to be allowed for the logging. It will conform to the acreages, yields, and types of cut in the original bid advertisement.
7 Five equal annual payments of $1.7 million will be made to the Forest Service, beginning at the end of the third year.

**PROPOSAL 4**

The log exporter combined with a national resort corporation to submit a "consortium" proposal for harvesting the timber and building a resort on Jewel Lake. To avoid the criticism of exporting jobs, an offer was made to provide up to 14 million board feet per year for 2 years to the Mason mills at a delivered-to-the-mill price of $200 per thousand board feet. The key points of the proposal are as follows:

*Roads* Construct 6.3 miles of two-lane paved road ($175,000 per mile) at Route *A* with an additional 12 miles of logging spurs ($15,000 per mile).

*Facilities* Construct a campground along Forest Highway 3264 where it crosses Quick Creek. It will include 20 improved trailer sites ($70,000), two double restrooms ($7000), 10 picnic tables ($3000), utilities and parking ($13,000), and landscaping ($18,000). No facilities will be built at Mac Falls or Horizon Point, but a network of paved trails will be built along Quick Creek from the campground to Jewel Lake and to Horizon Point (16 miles at $6000 per mile). Facilities at Jewel Lake, besides a resort, will include 15 picnic sites, restrooms, and parking, at a cost of $26,000.

*Resort* A commercial development comprised of a resort, cabins, boating and swimming facilities, gift shop, and restaurant, will be built at a cost of $1.2 million. It is expected to gross $1.5 million per year with a net profit before taxes of 14 percent of gross.

*Payment* In accordance with the original bid advertisement, $100 per thousand board feet will be paid at the end of 2 years for the timber. In addition, during a 15-year lease commencing in 2 years, 10 percent of the net receipts of the resort will be paid in lieu of taxes to the Forest Service, paved roads and trails will be maintained at the resort's expense, and all improvements will revert to government ownership at the termination of the lease (expected future value is $300,000).

The "consortium" had consultant reports that suggested that the trail network would be better suited to recreational interests than more roads and developed sites. However, 4 miles of spur roads must be maintained by the Forest Service to provide fire protection. All roads, recreational facilities, and the resort will be finished by the end of year 3.

**LAST-MINUTE INFORMATION**

Accompanying the four proposals were several telegrams and other communications. Most were unsolicited, but a few were in reply to inquiries about the proposals. Summaries are given below.

1 Many letters from outdoor enthusiasts supported the plan to make the entire Jewel Lake tract into a primitive area. The repetitious wording of the letters suggests a form letter had been provided for a concerted write-in campaign. A surprising number wrote in favor of the plan to have extensive hiking trails and a scenic viewing area at Horizon Point. These comments were apparently generated after newspaper articles appeared that described the proposals. The large number of letters indicates strong public interest.
2 The Department of Human Resources estimates that about 100 families will be added to the welfare rolls if mill activity is curtailed by 50 percent at Mason. Over a 6-month period, unemployment at Mason could cost the state $250,000, not including ripple effects.
3 An engineering report favors Route *A* over Route *B* with respect to environmental damage to the drainage. However, the report also states that careful construction along the stream bed would make the environmental effects similar, but the low-level route would be much more expensive initially, and upkeep would likely be higher in future years.
4 The State Tourist Bureau sent a strong endorsement of the resort construction. It was estimated that out-of-state visitors would spend $150,000 per year in the state while going to and from the resort.
5 Statistics from the Port of Eastgate, 52 miles from Jewel Lake, indicate that the average price for old growth delivered to the pier is $236 per thousand board feet, and that log shipments have been increasing each year.
6 A Forest Service economist calculated that the original bid conditions would have produced a present-worth net return to the government of $8,164,490. In addition, 6.3 miles of paved roads would have been provided from the contract. It was suggested that the total returns from the original conditions be a base point for comparing the present worth of the other proposals to harvest the same stand.

**COMPARISON OF PROPOSALS**

In analyzing the alternative uses of the Jewel Lake area, several assumptions have been made about costs and utilization:

1 The Pacific Mill and the consortium are expected to have hauling costs of $1.50 per thousand board feet per mile. The Mason mills will likely spend $2.00 per thousand board feet per mile for log hauling. Hauling distance for the consortium is 52 miles.
2 Logging costs for all companies should be about $16 per thousand board feet for clear cutting and double that for selective logging. Mill costs will average $30 per thousand board feet for Pacific and $38 per thousand board feet for the Mason mills.
3 During the 100-day tourist season, each picnic site will serve 3.7 people per day on the average, and the camping sites will be full 95 percent of the time with an average occupancy of 4.2 people per day.
4 If the area is developed to serve recreation purposes according to any of the three proposals, $45,000 per year will be spent on stocking Quick Creek and Jewel Lake; annual road maintenance will be $400 per mile; and caretaker-supervision costs will be $10,000 per season.
5 The road network should save about $3000 annually in fire loss. If any logging spurs are maintained for Forest Service use, maintenance costs will average $1200 per mile per year (4 miles of logging roads would have to be maintained for proposal 4 to provide fire protection).
6 Since the proposals are not answers to advertised bids, the Forest Service can bargain with the proposers to change some aspects of their plans if the expected profit seems unreasonably large.
7 The values used by government agencies in determining public benefits for recreational facilities are considered applicable: $1 per outing for each member of a picnic party, $2 per night for each camper, $5 for each person fishing. It is estimated that 30 percent of the people who use any of the recreational facilities in the Jewel Lake tract will fish.
8 The interest rate to be used in the analysis is 9 percent.

# SECTION FOUR

# RECOGNITION OF RISK AND UNCERTAINTY

Over 100 years ago a delightful and prophetic paper appeared in the *Journal of the Statistical Society.** It was written by William Farr, Esq., M.D., D.C.L., F.R.S., and bore the inclusive title "On the Valuation of Railways, Telegraphs, Water Companies, Canals, and other Commercial Concerns, with Prospective, Deferred, Increasing, Decreasing, or Terminating Profits." The following excerpts from that treatise reveal the century-old roots of engineering economic evaluation concepts:

> The value of things depends to some extent on their utility. . . . Goods are in general good things: they give life, health, and strength; they yield enjoyment to the highest faculties as well as the lowest wants of human nature. . . . The value of a thing bears no definite or constant relation to its excellence in an aesthetic sense, or to its high place in philosophy. . . . Thus it is found in every case value expressed in money is measured by the mind, and that its price is practically fixed by the concurrence of the seller and the buyer. . . .
>
> As the same thing differs in value at different distances in space, so its value—expressed in money—differs in time. It may be worth 1,000£ if paid for in ready

*Volume XXXIX, London, September 1876.

money, but if payable in a year's time it may be worth 905£: due in fourteen year's time it may only be worth 505£ . . .

If there is risk, that is valued by the doctrine of probabilities; thus, if in a lottery there are one thousand prizes of 1£ and forty blanks, the value of one thousand and forty tickets of 1£ will only be 1,000£. The same result, the same depreciation of value, may be caused by risk as is caused by remoteness of payment; and as both risks and profits vary, and their combination produces corresponding effects on the values involved, several of these combinations are conveniently included in the rates of interest.

The value of annuities at constant risk diminishes with the factor of risk. Thus let an annuity of 4£ in perpetuity be equivalent to 4 percent interest on 100£, then it will be worth twenty-five years' purchase; but if there is a constant risk equivalent to 1/105, it will be worth only twenty years' purchase. The risk rate enters into the dividends of all commercial undertakings; so rate of dividends includes the ordinary profit of capital at no risk, and the insurance premium to cover such normal risk as the concern is exposed to.

The views expressed by Dr. Farr are forerunners of modern considerations for investment decisions. His approaches to the utility of money and risk analysis have been refined over the years, but they are still not completely developed. As long as there are doubtful data involved in an investment decision, outcomes will be unsure, and no analysis technique can guarantee the most profitable solution for a given transaction. The purpose of recognizing risk is to make economic evaluations as complete as possible, thereby better understanding the opportunities, and to develop a policy for consistent decisions that in the long run will maximize returns.

# CHAPTER 14

# ECONOMIC ANALYSES RECOGNIZING RISK

Economic analyses in previous chapters were based on the assumption that complete information was available and that any uncertainty connected with a comparison could be tolerated. Thus, a salvage value assigned today is expected to be valid, say, 15 years hence. A positive cash flow of $1000 per year in a comparison model presupposes that exactly $1000 will become available each period on schedule. Unfortunately, real-world conditions do not always follow the models developed to represent them.

Variability is a recognized factor in most engineering and management activities. People are expected to possess individual skills and temperaments, and to behave impetuously when exposed to some situations. The properties of materials vary over time. Seemingly identical machines exhibit diverse operating characteristics. Environmental factors are never constant, and economic conditions change irregularly. But recognizing variability is much easier than including its consequences in economic comparisons.

Risk analysis recognizes the existence of variability and accordingly provides ways to select a preferred alternative. Consideration of inflation, sensitivity analysis, and comparisons based on a range of estimates were modifications of assumed certainty suggested earlier to deal with unpredictable future events. In this chapter probability theory and associated management-science tools are introduced. They are applied to discounted cash flows and minimum-cost models. Such analyses contribute to a more complete economic evaluation when there are ***significant*** risks involved that can be represented by the assignment of ***meaningful*** probabilities.

## INCLUDING RISK IN ECONOMIC ANALYSES

The problem-solving and decision-making processes depicted in the first two figures in this book set the framework for decisions under risk. Every decision of any consequence has overtones of risk. Sometimes the risk is so remote that it can be disregarded as a factor. Even when risk is recognized, it may have to be ignored because of insufficient data for evaluation or because the evaluation would require too much time or money. Such secondary attention to risk factors is neither laziness nor stupidity; it is often a necessity. If every decision were subject to a searching appraisal of risk, management functions would grind along at an intolerably slow pace.

### Definition of the Problem

The investigation leading to a problem definition should give strong hints as to the appropriateness of including risk considerations in the evaluation. Objectives that can be satisfied almost immediately are not as subject to chance variations that intrude over extended time periods. Narrow objectives bounded by limited means of accomplishment effectively exclude risk by confining the problem and solution to known quantities. Risk enters into the decision situation when current activities are projected into a distant future, and the resulting activities are subject to conditional influences.

### Collection of Data

Collecting data is a more demanding task when risk effects are to be considered. The first obstacle is the identification of possible alternatives. Both the obvious and the subtle warrant consideration. Creative effort is commonly considered a virtue in developing alternatives and is subordinated to statistical and accounting efforts in assessing outcomes. This is a mistaken belief. Both phases benefit from creativity, and both rely on documentation. Moreover, the knowledge acquired in one phase is bound to contribute to the other phases.

In comparing a problem to a tree (Figure 14.1), the main branches might be alternative solutions, and the outcomes the secondary branches and twigs. Assessment of these outcomes includes (1) the identification of future states or conditions, (2) the prediction of the probability of each state, and (3) the determination of returns associated with each state. Future states can be anticipated, but they cannot be controlled. Noncontrollable conditions may include weather, economic or technological developments, political legislation, world affairs, whims of buyers, and so forth. A key question is to decide which states are relevant to the problem objectives.

To continue the problem-tree analogy illustrated in Figure 14.1, the roots of the tree or

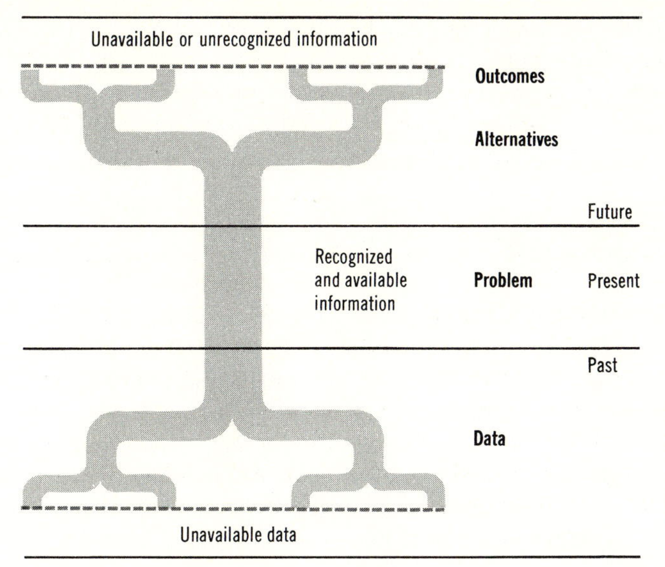


**FIGURE 14.1** Problem tree linking data sources to alternative outcomes. The problem results from a current situation which has roots in the past and outcomes in the future.

problem lie in the past and constitute the source material. Some information about the outcomes is obtained by investigating the roots. Other information is available from the structure of the limbs, or alternatives. The least promising alternatives can be pruned out, thereby reducing the effort required for outcome assessments to reasonable limits.

A shortage of pertinent information is a common handicap in risk evaluation. This condition is depicted in Figure 14.1 by the cutoff lines bounding the problem tree. Lack of information may be due to a scarcity of historical data or the inability to decipher past trends in terms of future outcomes. Failure to identify promising alternatives can be traced to unavailability of data or to simply having overlooked a course of action that should have been apparent. Estimating distant outcomes is at best precarious with good information; without applicable data it becomes more heroic than rational.

## Formulating the Model

A natural reaction to risk is to adopt a conservative stance. Conservatism in economic evaluations often takes the form of overly pessimistic estimates of future cash flows or requiring an abnormally high rate of return (or an exceptionally short payback period) when there is a significant likelihood that an investment might fail. Both approaches are informal attempts to ration capital to proposals that have better chances of profitability. The weakness of such informal weighting is the subjectivity of the assigned risk penalties. It would be difficult to explain to someone why, for instance, a 5 percent boost in the minimum required rate of return is realistic for one proposal and not another. For the 5 percent penalty to be meaningful, it should be associated with specific receipts or expenditures that are doubtful. If this is done, only slightly more definitive data are required to utilize a formal model of risk.

The decision-making literature offers many models of risk to choose from. Taken one at a time, each appears logical. Taken as a group, they may blend into a maze of diverse solution paths, difficult to decipher for a particular application. The models presented in this chapter are limited to future states defined by discrete probabilities. They are sufficient for most practical problems.

### Evaluation

The selection of the most advantageous alternative can be compared with a simple weight balance. Each alternative is weighed. Rewards are balanced against outlays. Opportunity costs are compared to operating costs. The alternative that best meets the decision criteria on the economic balance is chosen.

The inclusion of risk factors does not automatically increase the accuracy of a study. Probabilities representing risk appear on both sides of the economic balance. Sometimes they can decidedly swing the balance to one side. The important point is to realize that some evaluations are trivial or misleading without risk considerations, while others will not benefit from the additional effort required to include risk. The ability to distinguish between the two comes from confidence nourished by familiarity and practice.

Including risk in an analysis can be likened to adding a small flashlight to a key ring; it can be a novelty and a conversational gambit, or it can be used for illumination.

---

**Economy Exercise 14-1** The Owlglass Company is considering the manufacture of two mutually exclusive types of sunglass clip-on attachments. One clip-on design is specifically devised for Owl frames only and would not fit the eyeglass frames produced by other manufacturers. The other type is adaptable to competitors' frames but would not fit as securely or be color-coordinated with Owl frames. The investment required to get into the production of either design is $100,000. Estimates of net annual receipts for a 4-year market period are $43,000 for Owl-only clip-ons, and $50,000 for general-purpose clip-ons.

Because the "captive" market for clip-ons designed just to fit and complement Owl frames appears to have less risk, management decides that the general-purpose clip-on proposal should be evaluated at a risk-adjusted rate of return 8 percent greater than the regular 10 percent required rate. What effect does this decision have on the economic analysis?

---

## PROBABILITY ESTIMATES

A formal evaluation of risk is feasible when the likelihood of possible futures can be estimated, and associated outcomes from alternative courses of action can be identified. The first step is to determine categories of future states that affect the alternatives being compared. If the questionable aspect is expected market conditions, the states might be low sales, average sales, and high sales, defined by specific ranges such as annual sales below $50,000, $50,000 to $125,000, and over $125,000. Finer categories for expected sales could be very low, low, average, high, and very high, with corresponding dollar ranges declared. The breadth of categories appropriate for risk analysis depends mostly on how much the outcomes of alternatives vary between states; when cash flows do not differ much over a wide spectrum of possible future operating conditions, a smaller number of states adequately defines the problem. After the future states are identified and bounded, cash-flow outcomes can be estimated, using the methods of Chapter 10, by assuming each state, in turn, is sure to occur.

The next step in risk analysis is to determine the probability that each state will actually occur. The sources may be objective or subjective. Objective evidence of probability is usually in the form of historical documentation or common experience. It is readily accepted that a coin has two sides and a die has six. A reasonable person would not question the assignment of a probability of 0.50 to the occurrence of a head on the flip of a coin or a probability of 1/6 for an ace from the roll of a die (assuming that both the coin and die were fair or unbiased and that they were tossed or rolled in a fashion that assured an equal chance for all possible outcomes; equivalently, probability data for economic studies are assumed to be impartial and intrinsic). This type of information is ***prior*** (also known as *a priori*) knowledge.

In practice, prior knowledge of probabilities is seldom possible. Most of the time it is necessary to look at past records of events and use this empirical knowledge as a basis for current probabilities. However, vigilance is required to determine the relationship of historical records to the present action; seemingly unrelated events could influence each other. On the other hand, the records of one machine can sometimes be used to predict the performance of a similar machine. Experimental and other measurements that provide *after the fact* (*a posteriori*) probabilities are necessarily approximate, but they can still be the basis for practical applications.

Subjective probability estimates are derived from opinions based on general experience and knowledge that pertain to the situation under consideration. It has been academically disputed whether "guesstimates" of likelihood qualify as legitimate probability relationships, but the practice is well established for decision-making applications. Any subjective estimate suffers some from the bias and ignorance of the estimator, since no one predicts perfectly, but even purely intuitive valuations often have surprising roots in fact when their lineage is probed.

## PROBABILITY CONCEPTS

Betting in poker is a decision under risk. By keeping track of the cards played and relating them to the known distribution of card values in a complete deck, a betting policy can be developed to theoretically maximize returns. However, it takes considerable discipline to abide by the policy consistently and to do it long enough for the laws of probability to have effect. Somewhat the same problems affect economic decisions under risk. Each event has a single outcome and an associated probability of occurrence. It can be discouraging when an undesirable outcome occurs that has odds against it of 99 to 1, but in the long run probability theory promises that, if the odds are accurate, the temporary disappointment will be replaced by satisfaction when the same situation is replicated a number of times. Several basic properties of probability are collected in the following short survey.

## Independent Events

Events can be either statistically independent or dependent. Statistical *dependence* means that the probability of an outcome is dependent on or influenced by the occurrence of some other event, whereas an *independent* event is not affected by the occurrence of any other event.

The probabilities of mutually exclusive independent events can be added. The probability of drawing a queen from an honest deck of cards is the sum of the probabilities of all four queens in the deck. In functional notation this would appear as

$$P(Q) = P(Q_S) + P(Q_H) + P(Q_D) + P(Q_C)$$

where $P$ is probability, $Q$ means queen, and the subscripts stand for the four suits in the deck. And the numerical values are

$$P = \frac{1}{52} + \frac{1}{52} + \frac{1}{52} + \frac{1}{52} = \frac{1}{13} = 1/13 \qquad \text{or} \qquad 0.077$$

Intuitively, the probability of not drawing a queen is 1 minus the probability of drawing a queen:

$$P(Q) = 1 - P(Q) = 1 - \frac{1}{13} = \frac{12}{13} \qquad \text{or} \qquad 0.923$$

because the outcomes are mutually exclusive and collectively exhaustive (add up to 1.0). Mutually exclusive sets are also additive. The probability that a card will be either a spade or a heart is

$$P(S + H) = P(S) + P(H) = \frac{13}{52} + \frac{13}{52} = \frac{26}{52} \qquad \text{or} \qquad 0.5$$

**Example 14.1** **Additive Probabilities**

The output of a machine has been classified into three grades: superior (A), passing (B), and failing (C). The items in each class from an output of 1000 items are 214 in A, 692 in B, and 94 in C. If the run from which this sample was taken is considered typical, the probability that the machine will turn out each grade of product is

$P(\text{A}) = 214/1000 = 0.214$
$P(\text{B}) = 692/1000 = 0.692$
$P(\text{C}) = 94/1000 = 0.094$

What is the probability of making *at least* a passable product?

**Solution 14.1** The probability of producing a passable product includes the set of superior and passing grades.

$$P(\text{A} + \text{B}) = P(\text{A}) + P(\text{B}) = 0.214 + 0.692 = 0.906$$

or $P(\text{A} + \text{B}) = P(\not{\text{C}}) = 1 - P(\text{C}) = 1 - 0.094 = 0.906$

The probability that two or more independent events will occur together or in succession is the product of all the individual probabilities. In terms of the deck of cards, the probability of drawing the queen of hearts twice in a row (provided the queen is reinserted in the deck and the deck is reshuffled) is

$$P(Q_H Q_H) = P(Q_H)P(Q_H) = 1/52 \times 1/52 = 1/2704$$

The same reasoning applies to drawing any predesignated cards in a predesignated order or to drawing the same predesignated cards simultaneously from more than one deck.

A probability tree provides a pictorial representation of sequential events. In the probability tree of Figure 14.2 a new deck is used for each draw, and each suit is considered a set. The probability of drawing three hearts in a row is represented by the double line in the probability tree. By formula, this action is equal to

$$P(H_1 H_2 H_3) = P(H_1)P(H_2)P(H_3) = 0.25 \times 0.25 \times 0.25 = 0.015625$$

The same probability is evident for the sequence of drawing two hearts in a row and then drawing any predesignated suit on the third draw.

The probability of not drawing a heart in two consecutive draws is

$$P(\not{H})^2 = [1 - P(H)]^2 = (1 - 0.25)^2 = (0.75)^2 = 0.5625$$

It is shown in the tree by the sum of the probabilities in shaded circles which end the paths containing no hearts. Then the probability of at least one heart in two draws is the sum of the probabilities in the light circles and is equal to 0.4375.

**Example 14.2** **Multiplication of Probabilities**

A series of samples from the output of a machine reveals that 4 items out of every 100 sampled are defective. Thus, $P = 0.04$. For an order of 5 items taken directly from the machine without preliminary inspection, what is the probability that the order will be filled without containing a defective item?

**Solution 14.2** The probability of no defects, $P(0)$, in an order of five is

$$P(0) = (0.04)^0(0.96)^5 = (0.96)^5 = 0.82$$

Further, the probability that the order has not more than one defective is

$$P(1) = 0.82 + 5(0.04)^1(0.96)^4 = 0.82 + 0.17 = 0.99$$

## Dependent Events

An event is termed ***statistically dependent*** when its outcome is affected by the occurrence of another event. Dependency is illustrated by two boxes, labeled $X$ and $Y$, containing black and white balls. Box $X$ contains three white and two black balls, while box $Y$ holds one white and four black balls. This situation is depicted in Figure 14.3.

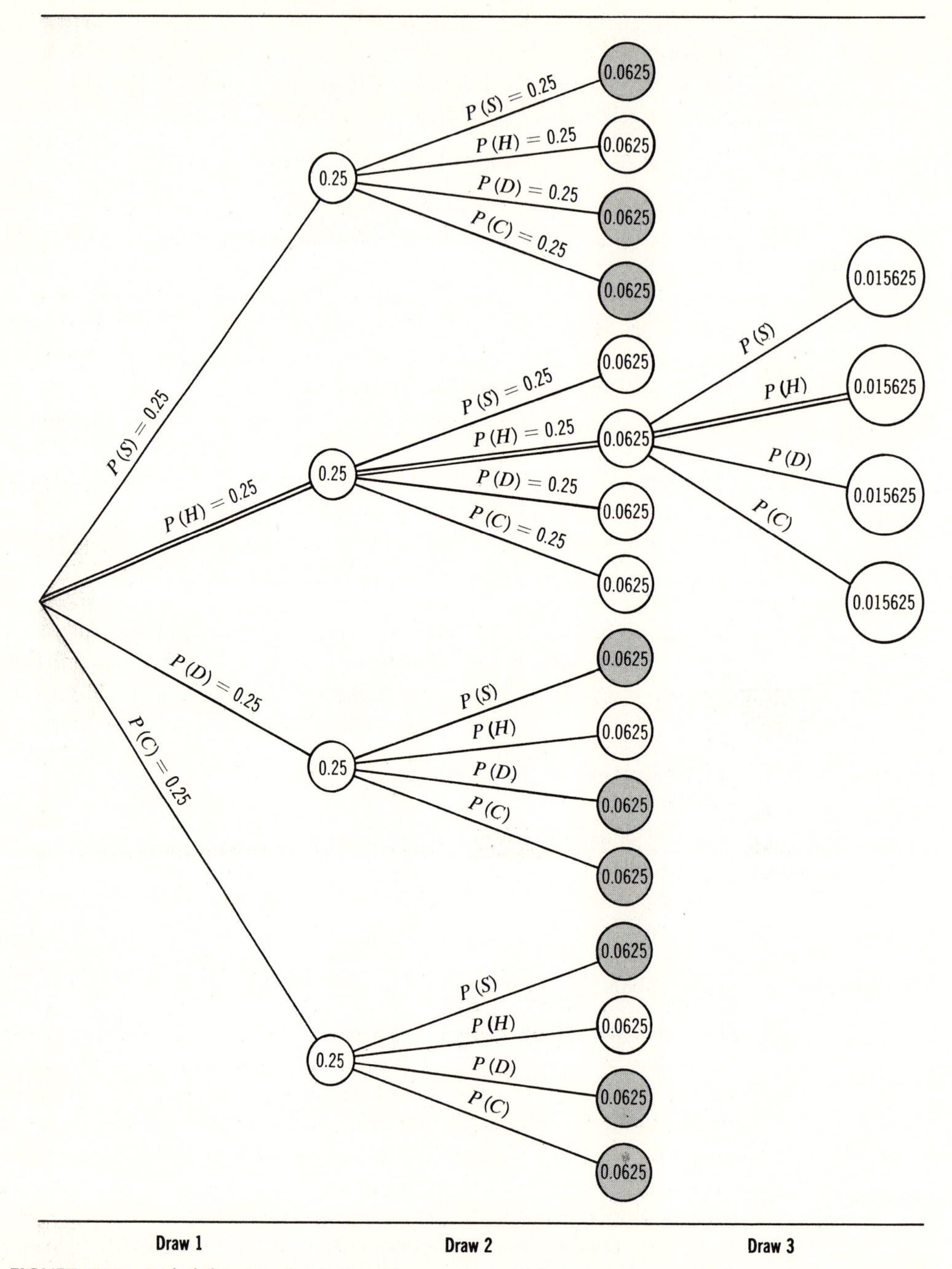


**FIGURE 14.2** Probability tree for independent events which registers the suit of a card drawn from a complete deck. The sum of the independent probabilities in each column of circles is 1.0 and represents all possible outcomes. By extension of the tree to include three draws (the incomplete column on the right), the probability of at least two hearts ($H$) in three draws is $10 \times 0.015625 = 0.156$.

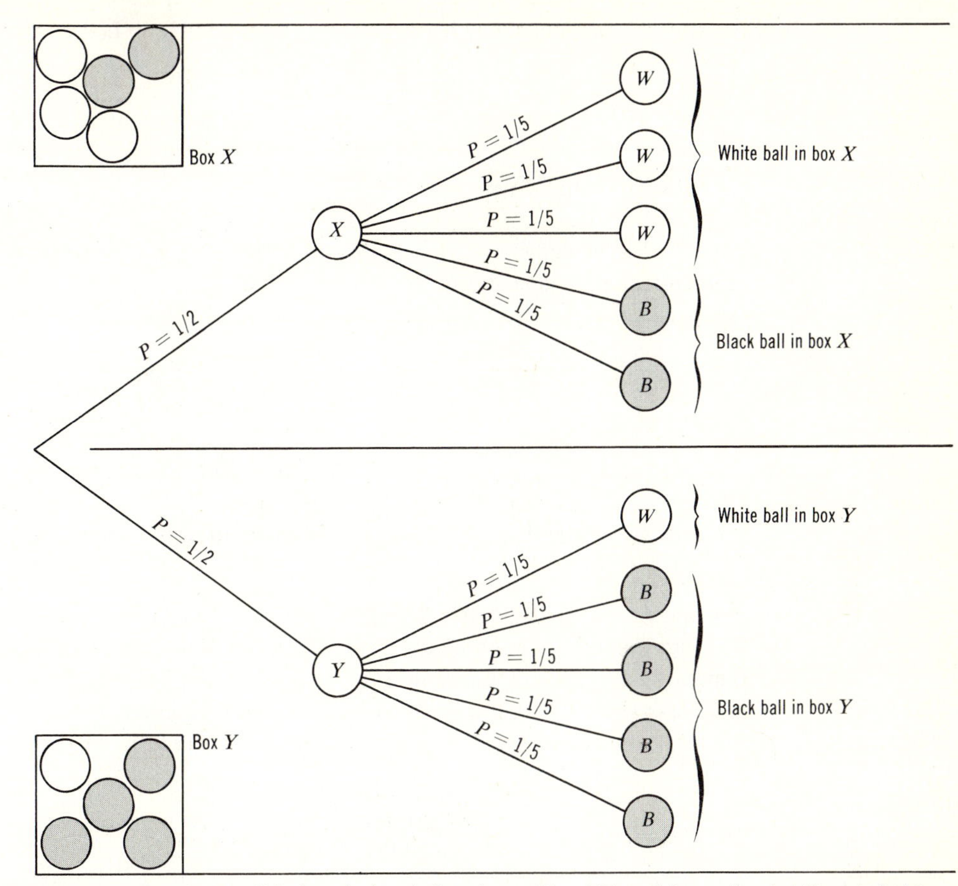


**FIGURE 14.3** Population of black and white balls in boxes $X$ and $Y$. Each burst of probability-labeled lines represents additional available information.

The *marginal probability* of drawing a white ball in this situation is 0.4 [the sum of the probabilities of individual white balls, $P(W) = 1/2 \times 1/5 = 1/10$]. Even though two events may be related, a marginal probability refers to just one of the dependent events. From the probability tree in Figure 14.3, it is clear that the probability of drawing a white ball is affected by the box from which it is drawn. Therefore the two events (drawing from one of the boxes and drawing a white ball) are related, but the marginal probability of drawing a white ball is still 0.4, because there are 10 balls with an equal probability of selection on the initial draw, and 4 of them are white.

What is the probability that a black ball will be drawn from box $Y$? This *conditional probability* is expressed symbolically as $P(B|Y)$, where the vertical line is read "given." From inspection of the probability tree, there are four chances in five that a black ball will be selected from box $Y$. Expressed as an equation, the conditional probability that a black ball will be drawn from box $Y$ is equal to the probability that a black ball will occur in box

$Y$ divided by the probability that the draw will be made from box $Y$. Thus

$$P(B|Y) = \frac{P(BY)}{P(Y)} = \frac{0.4}{0.5} = 0.8$$

**Example 14.3** **Conditional Probability**

The number of defective and acceptable items received in a shipment from two different companies is shown in the table below.

| | *Company C1* | *Company C2* | *Total* |
|---|---|---|---|
| Defective | 500 | 1000 | 1,500 |
| Acceptable | 9500 | 4000 | 13,500 |
| | 10,000 | 5000 | 15,000 |

What is the probability of receiving a defective item from company $C1$?

**Solution 14.3** The marginal probability of a defective item in the entire shipment is 1500/15,000 = 0.10, but the conditional probability of a defective item, given that it was supplied by company $C1$, is 0.05. This result could be obtained logically by dividing the number of defective items $D$ in the portion of the shipment supplied by company $C1$ (500) by the total number received from company $C1$ (10,000). It could also be calculated by formula:

$$P(D|C1) = \frac{P(DC1)}{P(C1)} = \frac{500/15{,}000}{10{,}000/15{,}000} = \frac{1}{20}$$

The *joint probability* that two dependent events will occur is given by the general equation

$$P(AB) = P(A|B) \times P(B) \quad \text{or} \quad P(B|A) \times P(A)$$

It is readily apparent that this is a restatement of the formula used to calculate conditional probabilities. In the illustration of black and white balls in boxes $X$ and $Y$, the joint probability that a ball will be black and will come from box $Y$ is

$$P(BY) = P(B|Y) \times P(Y) = 0.8 \times 0.5 = 0.4$$

Similar applications to other joint probabilities for this situation can be verified by referring to Figure 14.3.

**Example 14.4** For several years, records were kept of the number of absentees in two offices of a firm. The average number of employees absent from work on each day of the work-week at each office are summarized in the following table:

| | *M* | *T* | *W* | *Th* | *F* | *Total* |
|---|---|---|---|---|---|---|
| Denver office | 40 | 28 | 29 | 32 | 44 | 173 |
| Miami office | 26 | 21 | 20 | 22 | 28 | 117 |
| | 66 | 49 | 49 | 54 | 72 | 290 |

Knowing only that an employee was absent one day, how is the absence related to a day of the week (Friday) and the place of employment (Denver)?

**Solution 14.4** The probability that the employee was absent on Friday is $P(F) = 72/290 = 0.248$. The probability that the employee worked in Denver is $P(D) = 173/290 = 0.6$. The probability that an employee who works in Denver was absent on Friday is $P(F|D) = 44/173 = 0.255$.

The probability that an employee was absent on Friday *and* worked in Denver is

$$P(FD) = P(F|D) \times P(D) = 0.255 \times 0.6 = 0.15$$

or $P(DF) = P(D|F) \times P(F) = 44/72 \times 72/290 = 0.15$

## Bayesian Analysis

The opportunity to update and refine probability forecasts by taking advantage of additional information is a powerful analytic tool. Original objective or subjective probability assignments are developed from current knowledge to anticipate possible future outcomes. As time passes, we often have access to new information about the events we are predicting. The concept of revising prior probability estimates to reflect new data is attributed to Thomas Bayes. His basic formula, $P(A|B) = P(AB)/P(B)$, was described in connection with the calculation of conditional probability. The routine followed in finding posterior probabilities is based on this basic formula. The sequence of operations follows this pattern: The equality

$$P(AB) = P(B|A) \times P(A)$$

can be rewritten as

$$P(BA) = P(A|B) \times P(B)$$

which provides the equality

$$P(B|A) \times P(A) = P(A|B) \times P(B)$$

which can be converted to

$$P(B|A) = \frac{P(A|B) \times P(B)}{P(A)}$$

and the marginal probability of $A$ is

$$P(A) = P(A|B_1) \times P(B_1) + \cdots + P(A|B_x) \times P(B_x) + \cdots + P(A|B_n) \times P(B_n)$$
$$= \sum_x P(A|B_x) \times P(B_x)$$

The correct use of this routine allows an event $B$ to be reevaluated when new information concerning the outcome of $A$ becomes available.

The illustration in Figure 14.3, black and white balls in boxes $X$ and $Y$, will be adapted to illustrate a Bayesian analysis. Assume that the content of each box is known but the identity of the boxes is unknown. This situation is equivalent to knowing the outcome of two possible futures without being certain which future will occur.

In this case we want to identify which box is $X$ and which is $Y$. Since the two boxes are indistinguishable, there is an equal opportunity that either could be nominated as $X$ or $Y$.

**EVENT 1**

The first event is a random selection of one of the boxes. The prior probability of this event is represented by the probability tree of Figure 14.4.

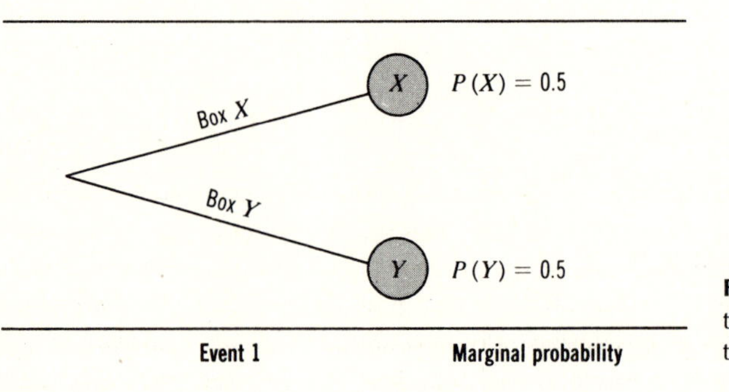


**FIGURE 14.4** First event in the Bayesian analysis illustration—box selection.

**EVENT 2**

Next a ball is drawn from whichever box was picked in event 1. We will assume that it is a black ball. Since we know the proportion of black to white balls in each box, we can calculate the probability of drawing a black ball, given that it came from a designated box. In box $X$ three of the five balls are white, so the conditional probability for the top branch of an expanded probability tree is 0.6. The likelihood of drawing a ball of a given color from a given box is shown by the joint probability for each branch of the tree. The complete tabulation of joint probabilities (collectively exhaustive) totals to 1.0, but we are interested primarily in the probabilities relating to black balls, because our first draw was black. The sum of the probabilities pertaining to black balls is the marginal probability of drawing a black ball and is shown in the last column of Figure 14.5.

The posterior probability of identifying the boxes, based on the additional information derived from the draw, is calculated by Bayes basic formula. Box $Y$ is arbitrarily used in the formula to give

$$P(Y|B) = \frac{P(YB)}{P(B)} = \frac{0.4}{0.6} = 0.67$$

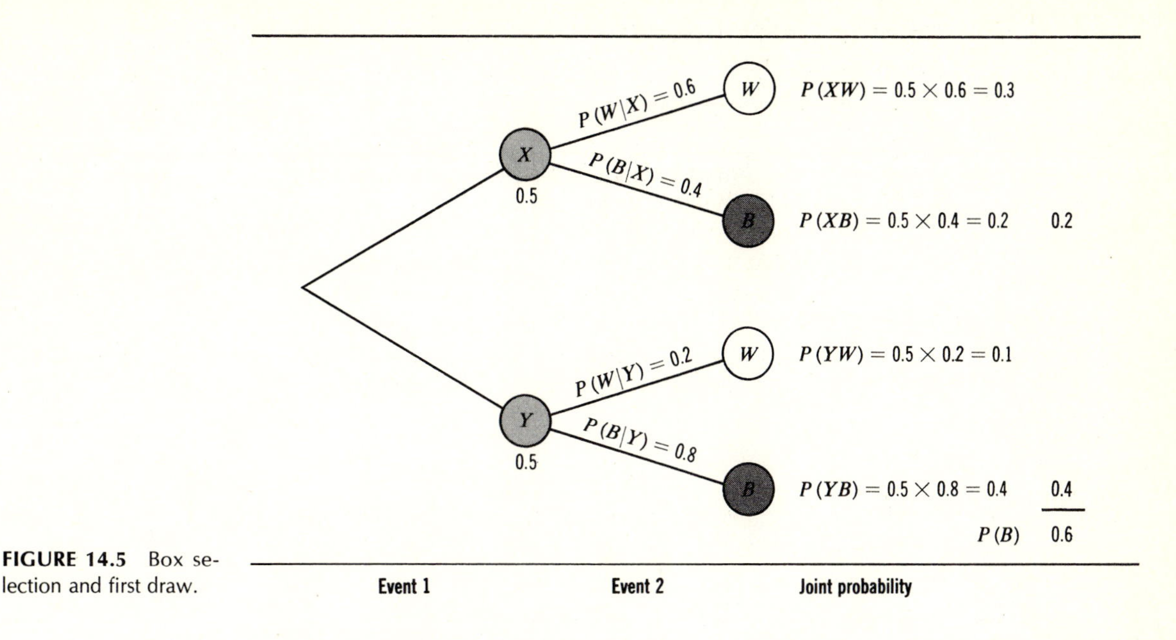


**FIGURE 14.5** Box selection and first draw.

Thus the added data have allowed us to revise our probability estimate from 0.5 to 0.67 that the selected box is indeed box $Y$.

## EVENT 3

Now assume that another ball is drawn from the same box and is black. Conditional probabilities for this draw are calculated on the basis of the four balls remaining in the box after the first draw. Then the joint probabilities are determined for two successive black draws from either box. These procedures are depicted in the third section of the probability tree in Figure 14.6.

A further revision of the likelihood that the chosen box is box $Y$ now becomes

$$P(Y|B_1B_2) = \frac{P(B_1B_2Y)}{P(B)} = \frac{0.3}{0.35} = 0.857$$

Suppose the second draw had revealed a white rather than a black ball. This possibility is represented by the dotted lines in the probability tree. A format for revising the probability that the draws ($B_1$ and $W_2$) were from box $Y$ is given in the following table:

| *Event 1* | *Event 2* $= B_1$ | *Event 3* $= W_2$ | *P(E1E2E3)* |
|---|---|---|---|
| $P(X) = 0.5$ | $P(B_1|X) = 0.4$ | $P(W_2|B_1X) = 0.75$ | $0.5 \times 0.4 \times 0.75 = 0.15$ |
| $P(Y) = 0.5$ | $P(B_1|Y) = 0.8$ | $P(W_2|B_1Y) = 0.25$ | $0.5 \times 0.8 \times 0.25 = 0.10$ |
| | | | 0.25 |

$$P(Y|B_1W_2) = \frac{0.10}{0.25} = 0.40$$

We would intuitively suspect that a reversed order of the draws ($W_1$ and $B_2$) would not alter our revised probability. This suspicion is confirmed by the following values:

| *Event 1* | *Event 2* $= W_1$ | *Event 3* $= B_2$ | *P(E1E2E3)* |
|---|---|---|---|
| $P(X) = 0.5$ | $P(W_1 \mid X) = 0.6$ | $P(B_2 \mid W_1X) = 0.5$ | $0.5 \times 0.6 \times 0.5 = 0.15$ |
| $P(Y) = 0.5$ | $P(W_1 \mid Y) = 0.2$ | $P(B_2 \mid W_1Y) = 1.0$ | $0.5 \times 0.2 \times 1.0 = 0.10$ |
| | | | 0.25 |

$$P(Y \mid W_1B_2) = \frac{0.10}{0.25} = 0.40$$

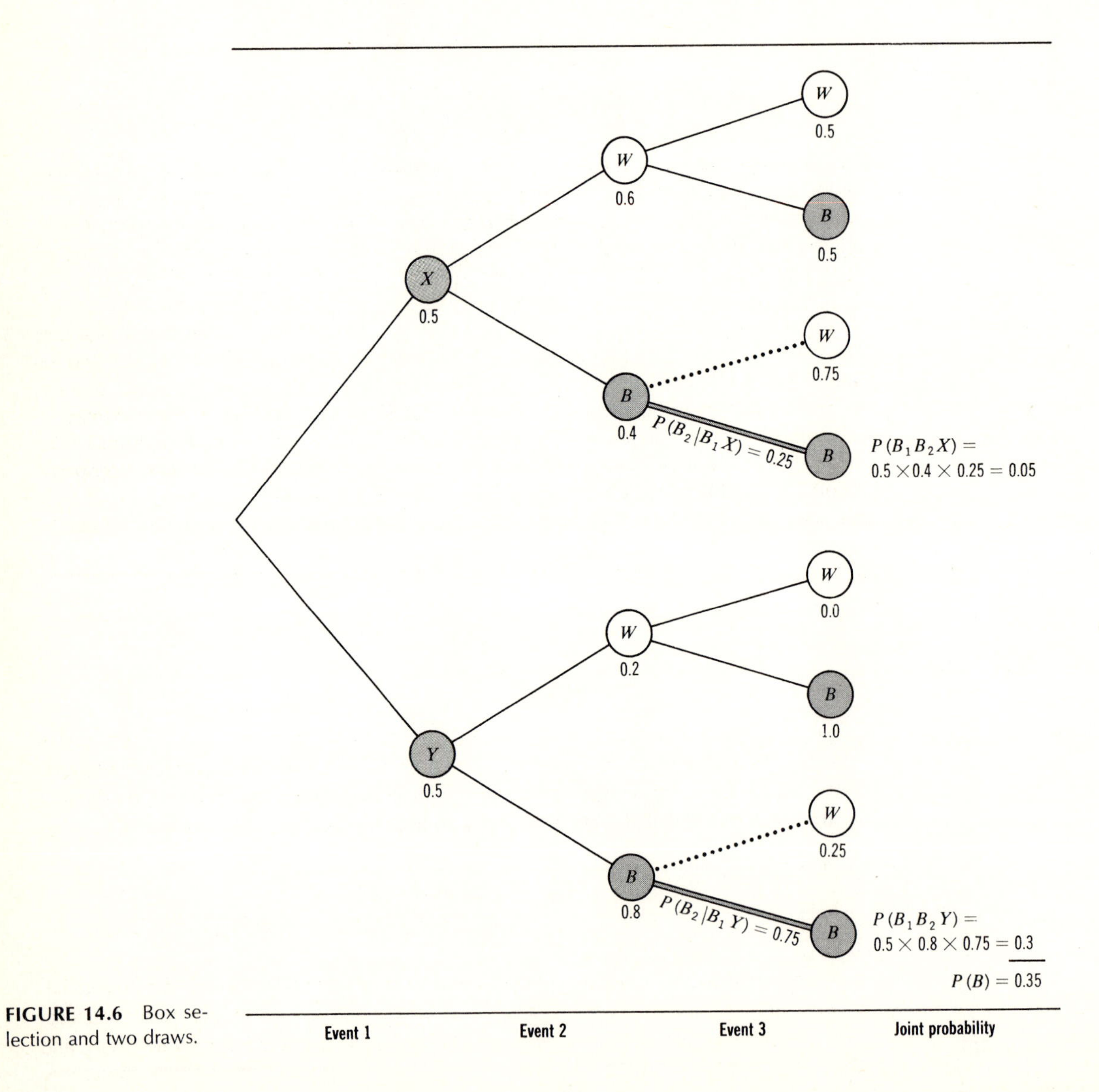


**FIGURE 14.6** Box selection and two draws.

**EVENT 4**

As a final possibility, again assume that the first two balls drawn were black, and now a third black ball is picked from the same box. We have obviously been drawing from box $Y$, because box $X$ originally contained only two black balls. Although formal calculations are unnecessary in this case, the conclusion is verified easily by recognizing that

$$P(\text{3 black-ball draws} \mid \text{box } X) = 0.4 \times 0.25 \times 0.0 = 0.0$$

and

$$P(\text{3 black-ball draws} \mid \text{box } Y) = 0.8 \times 0.75 \times 0.67 = 0.4$$

which makes the marginal probability of three successive black draws $0.0 + 0.4 = 0.4$, so

$$P(\text{box } Y \mid \text{3 black-ball draws}) = \frac{0.4 \times 0.5}{0.4 \times 0.5} = 1.0$$

---

**Economy Exercise 14-2** A complaint was received by company $C$ from one of its wholesalers that a certain key part on appliances recently delivered by company $C$ had often failed to operate properly. The company immediately stopped shipment on the remaining appliances made in the same production run as those supplied to the wholesaler. An investigation of similar malfunctions revealed that 70 percent of the time the trouble was caused by poor assembly, and 30 percent of the time by inferior materials. It was also determined that the probability of failure from incorrectly assembled parts is 0.40, and the probability of failure from faulty materials is 0.90. If the material is bad, the whole part has to be replaced, but if the trouble was caused by improper assembling, the part may be adjusted to perform adequately. The only means of determining the cause of trouble is by destructive testing. A decision was made to test enough parts to determine with a probability of 0.95 that the trouble stemmed from one cause or the other.

If the first five parts tested all failed because of poor materials, what conclusion can be drawn?

---

# EXPECTED VALUE

*Expected value* is a standard measure for economic comparisons involving risk. It incorporates the effect of risk on potential outcomes by means of a weighted average. Outcomes are weighted according to their probability of occurrence, and the sum of the products of all outcomes multiplied by their respective probabilities is the expected value, EV:

$$\text{EV}(i) = \sum_j P_j O_{ij}$$

where $P_j$ is the independent probability of future $j$, $\Sigma P_j = 1.0$, and $O_{ij}$ is the outcome of alternative $i$ for future $j$.

As a common example of the expected-value approach, consider the coin-flipping game: A coin is flipped, and if it comes up *heads* you win the coin, but when *tails* show you lose. Let the coin be a "fair" dime; then the probability of a head or a tail is, of course, 0.5, and the expected value of winnings from the game is

$$\text{EV(flipping dimes)} = P(H)(\$0.10) + P(T)(-\$0.10) = 0.5(\$0.10) + 0.5(-\$0.10) = 0$$

The given probabilities of future states are collectively exhaustive (both heads and tails are accounted for, and $\Sigma P = 0.5 + 0.5 = 1.0$); an outcome is associated with each future state (the gain or loss of a dime); and the expected value is the long-term expectation from repeated performances under the same conditions.

A profit-seeking individual or organization prefers a positive expected value, when gains are plus and losses are minus. In comparing alternatives, the one with the highest EV is preferred, other things being equal. For instance, a variant of the coin-flipping game might be to pay $1 for the chance to flip two coins. The payoff could be to win $2 for two heads and $1 for two tails, and to lose $1 when the coins split head and tail. The expected-value calculation,

$$\begin{aligned}\text{EV} &= P(H_1H_2)(\$2 - \$1) + P(H_1T_2)(-\$1) + P(T_1H_2)(-\$1) + P(T_1T_2)(\$1 - \$1)\\ &= 0.25(\$1) + 0.25(-\$1) + 0.25(-\$1) + 0.25(0) = -\$0.25\end{aligned}$$

shows that this is not a very good game to get into because the *average* loss per play is 25 cents. But the *actual* outcome for any given gamble is a gain of a dollar, a loss of a dollar, or no change.

The appropriateness of the expected-value model for a once-in-a-lifetime decision is questionable. Other decision criteria such as those presented in Chapter 15 would enter the verdict. However, truly unique decisions do not occur too often in actual practice; alternatives vary individually as to amounts, possible futures, and probabilities, but their evaluation is based on a consistent long-term objective of profit maximization or cost minimization. Since most industries and governments are long-lived, and new investment projects are being continually initiated, expected value is a rational measure for most comparisons that recognize risk.

## Payoff Tables

A format for organizing and displaying outcomes of alternative courses of action is called a *payoff table*. Each row in the table represents an alternative with its outcomes arranged in columns according to respective future states. In the payoff table shown, the present worths of two new products, *A* and *B*, are shown for three states of sales success during a 3-year marketing period.

| | STATE OF MARKET ACCEPTANCE | | |
|---|---|---|---|
| *Alternative* | *Rejection* | *Average* | *Domination* |
| Product *A* | −$50,000 | $200,000 | $500,000 |
| Product *B* | −200,000 | 100,000 | 1,000,000 |

The initial cost of developing product *A* is $50,000, and this amount would be lost if the product were rejected by consumers. If product *A* received average acceptance, the expected gain would be \$200,000, and if it dominated the market, the payoff would be \$500,000. Product $B$ would cost four times as much as product $A$ to put into production, but if it became a best seller it would double $A$'s profit. Because production costs for $B$ are higher, an average demand would result in only half the payoff expected from $A$.

Some conclusions might be drawn from just the payoffs included in the table. For instance, a loss of \$200,000 could be considered disastrous to the company, while a loss of \$50,000 would at least be tolerable. With such a severe penalty for failure, alternative $B$ would practically be eliminated regardless of the potentially large payoff. However, even more meaningful observations can be made by including the relative likelihood of each outcome. The payoff table below has been modified to incorporate probability factors.

| | OUTCOME AND RISK | | |
|---|---|---|---|
| *Product* | $P(R) = 0.1$ | $P(A) = 0.6$ | $P(D) = 0.3$ |
| A | −\$50,000 | \$200,000 | \$ 500,000 |
| B | −200,000 | 100,000 | 1,000,000 |

It might have been necessary to assign probabilities to each outcome of each alternative, but it is assumed that products $A$ and $B$ are similar enough to possess the same consumption pattern. In the example both products have a probability of 0.1 of rejection, 0.6 of normal demand, and 0.3 of booming acceptance. If all the possible futures are included, the sum of probabilities will equal 1.0.

**Example 14.5**

### Expected Value of Mutually Exclusive Incremental Alternatives

Owing to the lack of other suitable land, a mill is forced to construct its new settling ponds for waste disposal along the bank of a river. The stream has no dams to control its flow. Flood-level records for the last 70 years reveal that the river rose above the minimum required height of the settling-pond walls 42 times. Therefore, the economic analysis must include not only construction costs but also the risk of flood damage. Building the walls higher increases the initial investment but lowers the threat of flooding. Engineering calculations for the cost of higher retaining walls in 5-foot increments and the expected damage from floods of different crests are shown in the table below.

| *River Levels above Minimum Wall Height in Range, Feet* | *Years the River Crests in Given Range above Wall Height F* | *Probability of River Cresting in Range above Height F* | *Flood Damage when River Crests above Walls* | *Cost to Construct Walls to Height F* |
|---|---|---|---|---|
| 0–5 | at $F = 0$: 14 | 14/70 = 0.20 | \$ 70,000 | \$135,000 |
| 5–10 | at $F = 5$: 14 | 14/70 = 0.20 | 105,000 | 200,000 |
| 10–15 | at $F = 10$: 8 | 8/70 = 0.11 | 150,000 | 280,000 |
| 15–20 | at $F = 15$: 6 | 6/70 = 0.09 | 200,000 | 370,000 |
| Over 20 | at $F = 20$: 0 | 0/70 = 0.0 | 0 | 450,000 |

The settling ponds are expected to be needed for 10 years before the process can be improved sufficiently to eliminate the need for special waste treatment. A rate of return of 10 percent is used by the mill to evaluate federally mandated investments. To what height should the retaining walls be built to minimize annual costs?

**Solution 14.5** The initial cost of the settling ponds is translated to an equivalent annual cost by the capital-recovery factor. For the minimum wall height ($F = 0$), the equivalent annual investment cost is

$$\text{AC(investment at } F = 0) = \$135{,}000(A/P,\ 10,\ 10) = \$135{,}000(0.16275) = \$21{,}971$$

Equivalent annual investment costs for successively higher retaining walls are calculated similarly.

Data arranged for the calculation of expected values of flood damage for alternative wall heights are shown in Table 14.1. The outcomes in the payoff table indicate damages expected from floods exceeding a given wall height by amounts $F$. Thus, for the minimum wall height, flood waters will cause damage 3 years in every 5 years, on the average, to produce an expected value of

$$\begin{aligned} \text{EV}(F = 0) &= \$70{,}000(0.20) + \$105{,}000(0.20) + \$150{,}000(0.11) + \$200{,}000(0.09) \\ &= \$14{,}000 + \$21{,}000 + \$16{,}500 + \$18{,}000 \\ &= \$69{,}500 \end{aligned}$$

which, when added to the equivalent annual investment cost, makes a total expected cost for the minimum-height settling-pond walls of $\$21{,}971 + \$69{,}500 = \$91{,}471$. Expected values for other wall heights are calculated in the same manner, as indicated in the shaded area of Table 14.1. Note that the probabilities do not add up to 1.0 because river levels below $F = 0$ (probability $= 0.40$) have no bearing on the flood-damage expectations.

| *Wall Height above Minimum, Feet* | PROBABILITY OF RIVER CRESTING ABOVE WALL HEIGHT IN RANGE, FEET | | | | *Expected Value* | *Equivalent Annual Investment* | *Total Expected Annual Cost* |
|---|---|---|---|---|---|---|---|
| | *0–5 P = 0.20* | *5–20 P = 0.20* | *10–15 P = 0.11* | *15–20 P = 0.09* | | | |
| 0 | \$70,000 | \$105,000 | \$150,000 | \$200,000 | \$69,500 | \$21,971 | \$91,471 |
| 5 | 0 | 70,000 | 105,000 | 150,000 | 39,050 | 32,550 | 71,600 |
| 10 | 0 | 0 | 70,000 | 105,000 | 17,150 | 45,570 | 62,720 |
| 15 | 0 | 0 | 0 | 70,000 | 6,300 | 60,218 | 66,518 |
| 20 | 0 | 0 | 0 | 0 | 0 | 73,238 | 73,238 |

Flood damage from incremental river levels above alternative settling-pond wall heights, $F$

**TABLE 14.1** Payoff table (shaded) for five alternative wall heights for settling ponds. Total expected annual costs are the sums of expected flood damages and equivalent annual investment costs.

The settling-pond retaining-wall height that minimizes total annual cost is 10 feet above the minimum level.

---

**Economy Exercise 14-3** A process line that will continue to be needed for 3 years has annual costs of $310,000, which are expected to remain constant. A novel redesign for the line has been suggested. The new approach will cost $150,000 to install and has a 50 percent chance of cutting annual operating costs to $210,000. However, there is a probability of 0.25 that annual costs for the redesigned line will increase, from $210,000 for the first year, by $20,000 or $75,000 in each of the next 2 years. If a 12 percent rate of return is required, should the new design be installed?

---

## Investment-Risk Profiles

The range-of-estimates approach (Chapter 12) can be modified to reflect risk. Probabilities of occurrence are assigned to each future state in the range. Then all possible combinations of outcomes are collected, and their dollar amounts and joint probabilities are calculated. The result is an ***investment-risk profile*** that reveals the likelihood for an investment to realize various net present-worth returns.

Assume an investment proposal for income expansion requires an initial outlay of $200,000. The most likely outcome is for after-tax returns to amount to $100,000 per year for 4 years. There is no salvage value, and the minimum attractive rate of return is 9 percent. Under the assumption of certainty,

$$\begin{aligned} \text{PW} &= -\$200{,}000 + \$100{,}000(P/A,\ 9,\ 4) \\ &= -\$200{,}000 + \$100{,}000(3.2396) = \$123{,}960 \end{aligned}$$

A risk analysis of the same proposal recognizes that the most likely cash flow has just a probability of 0.5 of occurrence. A pessimistic appraisal of the future concedes returns might amount to only $50,000 per year. Under the most optimistic assessment, returns could total $125,000 per year. The probabilities of pessimistic and optimistic returns materializing are 0.3 and 0.2, respectively. The duration of returns is also questionable; probabilities for 2, 3, 4, and 5 years are, respectively, 0.2, 0.2, 0.5, and 0.1. These data are summarized on the decision-probability tree shown in Figure 14.7.

The joint probability for each outcome is the product of the independent probabilities representing each factor involved in the outcome. These probabilities are entered in the nodes of the probability-decision tree, and the factors are labeled on the lines connecting the nodes. For example, the joint probability of returns of $50,000 per year for 2 years (top of the column in Figure 14.7) is 0.06. It results from multiplying the probability of the definite investment (1.0) by the probabilities of $50,000 returns (0.3) and a 2-year period (0.2).

Present worths of the outcomes are calculated by the usual procedures. The value listed at the top of the "PW of outcome" column, based on a required rate of return of 9 percent, is computed as

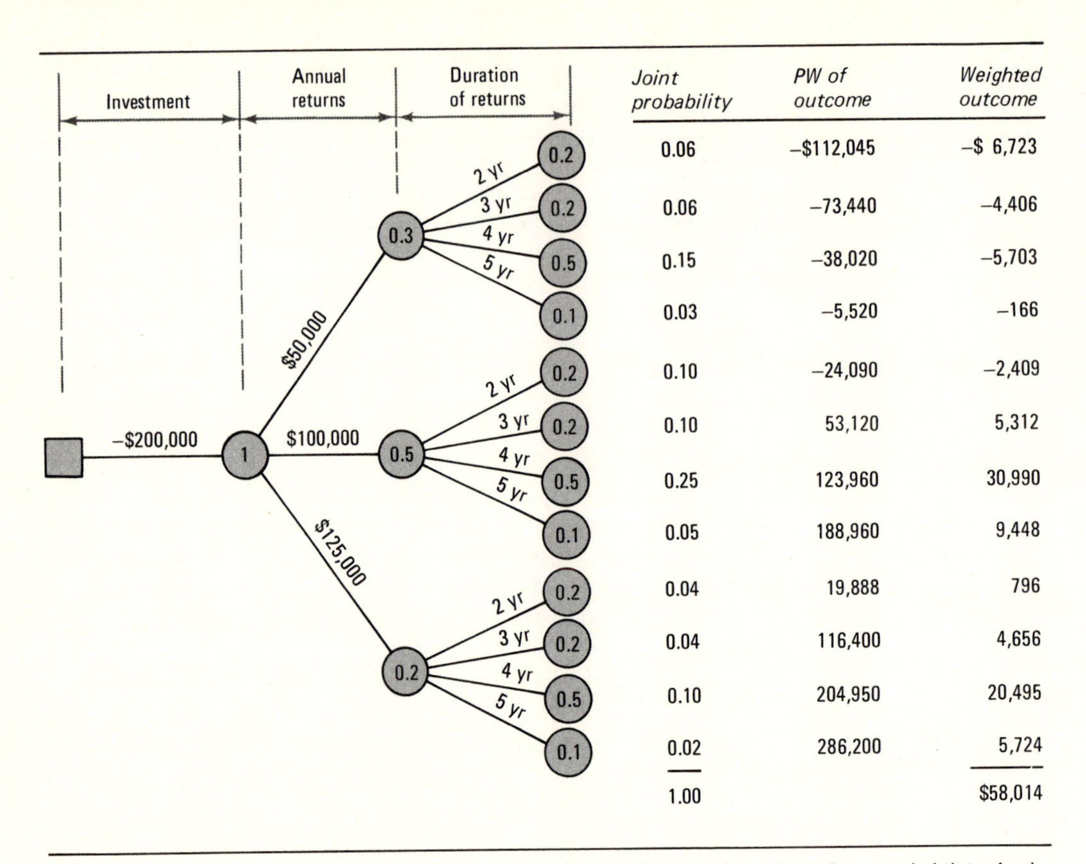


**FIGURE 14.7** Decision-tree format for an investment-risk profile analysis. Node entries indicate probabilities for the economic factors noted on lines leading to the nodes. Each burst represents one economic condition subject to risk.

$$\begin{aligned} \text{PW}(P = 0.06) &= -\$200{,}000 + \$50{,}000(P/A, 9, 2) \\ &= -\$200{,}000 + \$50{,}000(1.7591) = -\$112{,}045 \end{aligned}$$

The weighted value of this present worth is the top entry of the last column on the right in Figure 14.7, and is the product

$$\text{Joint probability}(i) \times \text{PW(outcome } i) = 0.06 \times -\$112{,}045 = -\$6{,}723$$

The remaining outcomes are computed similarly. The sum of these values is the expected value of the proposal, EV = $58,014. For the given probabilities of occurrence, the expected value is much less than the present worth based on the most likely outcomes. Although the probability estimates are quite likely subject to some error, the exercise of calculating an investment profile contributes to a more complete analysis and a better appreciation of the factors involved in the proposal.

**Economy Exercise 14-4** The most likely cash flow for a cost-reduction proposal is for an investment of $4000 to produce after-tax present-worth savings of $1200 per year for 5 years. The net present worth is

$$\text{PW(most likely)} = -\$4000 + \$1200(5) = \$2000$$

Upon further investigation it appears that the initial investment has probabilities of 0.4 of being as high as \$5000, and 0.6 of being \$3000. The after-tax present worths of annual savings could amount to \$2000, \$1200, or \$800 with respective probabilities of 0.2, 0.3, and 0.5. Determine the investment-risk profile and the expected value of the proposal.

## INVESTMENT-RISK DECISION CRITERIA

An investment-risk profile highlights more than just the expected value. When the outcomes and related probabilities are graphed, a proposal's prospects are clearly displayed for capital-budgeting discussions. The data from Figure 14.7 are graphed in Figure 14.8. Lines connect the outcome-risk points to better define the cumulative probability distribution of net present values. The connected points make it quite evident that the proposal has a probability greater than 0.4 of showing a loss.

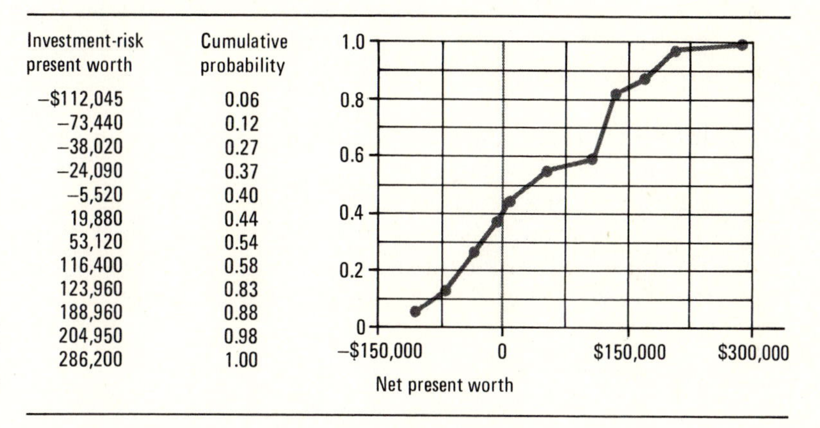

| Investment-risk present worth | Cumulative probability |
|---|---|
| −\$112,045 | 0.06 |
| −73,440 | 0.12 |
| −38,020 | 0.27 |
| −24,090 | 0.37 |
| −5,520 | 0.40 |
| 19,880 | 0.44 |
| 53,120 | 0.54 |
| 116,400 | 0.58 |
| 123,960 | 0.83 |
| 188,960 | 0.88 |
| 204,950 | 0.98 |
| 286,200 | 1.00 |



**FIGURE 14.8** Tabulated data and graph of an investment-risk profile.

### Acceptable-Investment Diagram

Another graphic display to assist in the evaluation of alternatives subject to risk is called an *acceptable-investment diagram* (AID). Its format is a horizontal axis representing rates of return and a vertical axis scaled to show the probability an investment will surpass a given rate of return. Criteria for an acceptable investment are blocked off in the chart by setting limits for:

1. The required probability that an investment's rate of return exceeds a minimum percentage (loss coefficient)
2. The desired probability that an investment's rate of return will exceed an attractive level (payoff coefficient)
3. A line connecting the two coefficients, called an *aspiration level*

Any investment-risk profile that does not intrude on the reject area defined by the listed limits is considered an acceptable investment; its risk-return potential is greater than the minimum risk-return requirements.

**Example 14.6**

## Application of an AID

An investment of $1000 for 1 year has the possible after-tax returns shown. The probability of each outcome is estimated, and the rate of return is apparent.

| *Net Return (outcome)* | *Rate of Return, %* | *Probability of Outcome* | *Probability Investment's RR Will Exceed Rate of Return* |
|---|---|---|---|
| −$900 | −10 | 0.05 | 0.95 |
| 1050 | 5 | 0.15 | 0.80 |
| 1150 | 15 | 0.40 | 0.40 |
| 1300 | 30 | 0.30 | 0.10 |
| 1500 | 50 | 0.10 | 0.00 |

The investors seek proposals that provide a probability of 0.95 that they will not lose more than 5 percent, and a 0.30 likelihood that their rates of return will be greater than 15 percent. Use an AID to determine the acceptability of the $1000 investment proposal.

**Solution 14.6**

The reject area for the acceptable-investment diagram in Figure 14.9 results from lines connecting the loss coefficient horizontally to the ordinate, the loss coefficient diagonally to the payoff coefficient, and the payoff coefficient to the abscissa. The sloping line is the implied aspiration level relating desired returns to allowable risk. The investment's risk-return potential is represented by lines connecting the rates of return associated with their probabilities of occurrence.

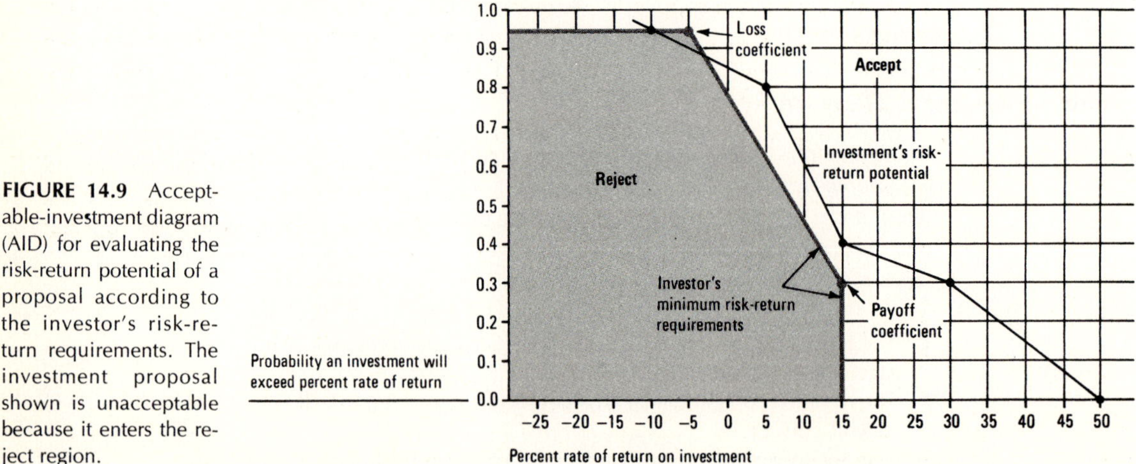


**FIGURE 14.9** Acceptable-investment diagram (AID) for evaluating the risk-return potential of a proposal according to the investor's risk-return requirements. The investment proposal shown is unacceptable because it enters the reject region.

Because part of the risk-return curve crosses the reject area, the investment is not acceptable according to the given criteria. Investments passing these criteria can be further evaluated on other merits such as timing of returns, capital availability, and the decision criteria discussed next.

## Auxiliary Decision Criteria

Once an investment-risk profile is developed, there are several ways to appraise its desirability. The basic criterion, as already observed, is ***maximization of expected returns,*** which is determined by expected-value calculations. Other criteria include the ***most probable future*** and the ***aspiration level.*** These criteria are illustrated by reference to Table 14.2.

**TABLE 14.2** Probability of returns from three equal-size, equal-life investments.

| | POSSIBLE NET PRESENT WORTHS OF PROPOSALS | | | | | |
|---|---|---|---|---|---|---|
| *Alternative* | *–$1000* | *0* | *$1000* | *$2000* | *$3000* | *$4000* |
| *A* | 0 | 0.11 | 0.26 | 0.22 | 0.02 | 0.39 |
| *B* | 0.29 | 0.18 | 0.07 | 0 | 0 | 0.46 |
| *C* | 0.14 | 0.10 | 0.11 | 0.37 | 0.28 | 0 |

According to the most-probable-future criterion, the investment that has the greatest return for the most probable future is preferred. The most probable futures for the alternatives in Table 14.2 are

| *Most Probable Future* | *PW at Most Probable Future* |
|---|---|
| Alternative $B: P = 0.46$ | $4000 |
| Alternative $A: P = 0.39$ | 4000 |
| Alternative $C: P = 0.37$ | 2000 |

They indicate a preference for $B$, because its maximum return, though the same as that of alternative $A$, has a greater probability of occurrence. This criterion thus assumes the future with the highest probability is ***certain*** to occur, and the alternative with the highest return for its "certain" future is best.

An ***aspiration level*** is based on a minimum amount that will satisfy a decision maker. It is closely related to utility theory, which is explored in Extension 15.2. The rationale behind an aspiration level is to achieve assurance that the investment will "at least return a respectable amount" or "surely not lose money." For the sample data from Table 14.2 and an aspiration level of $2000, we have

| *Alternative* | *Probability of Return of $2000 or More* |
|---|---|
| *C* | $0.37 + 0.28 = 0.65$ |
| *A* | $0.22 + 0.02 + 0.39 = 0.63$ |
| *B* | $0.46 = 0.46$ |

Alternative $C$ is preferred because it has a higher cumulative probability of providing \$2000 and more. If the aspiration level had been to "at least make some money," alternative $A$ would be selected because it has a probability of $0.26 + 0.22 + 0.02 + 0.39 = 0.89$ of having returns greater than zero, which is larger than that of any other alternative.

Since alternative $A$ would also be selected by the maximum-expected-return criterion [because EV($A$) = \$2320, EV($B$) = \$1620, and EV($C$) = \$1550], any of the alternatives in Table 14.2 could receive the nod of approval by one criterion or another. This should cause no dismay; it merely demonstrates the importance of a decision maker's perspective, the theme of Chapter 15.

## DISCOUNTED DECISION TREES

Discounted cash flows, probabilities, and expected values are combined in a tree format to generate a graphic *discounted decision tree.* Its unique feature is its capacity to display future decision points—times in the future when a decision maker can appraise actual outcomes from earlier decisions to decide if a previously determined course of action should be modified to cope with current conditions.

### Successive Decisions

A discounted decision tree shows decisions separated by time intervals and susceptible to external influencing factors. Branches radiate from an initial decision point to indicate the primary alternatives. Each main branch is divided to show foreseeable outcomes associated with possible future events. Then the events are rated with respect to their probable occurrence. When gains can be maximized by introducing new alternatives at a future date, a second decision point is established. A succession of decision points can extend to the limit of forecasting ability. The time value of monetary outcomes is effected by discounting the outcomes to a common point in time.

The layout of a discounted decision tree follows the extensive form in which decision points are represented by squares, and circles symbolize outcomes. Dotted links represent courses of action, and solid links show possible consequences of these actions. Alternatives are evaluated on an equivalent basis by discounting receipts and disbursements of future occurrences back to a decision point. The decision criterion is the expected value of alternatives at each decision point.

### Formulation of a Discounted Decision Tree

A warehousing problem of a small novelty manufacturing company will serve to illustrate a decision tree for successive decisions. The company is relatively new and has captured a limited segment of the novelty market. It must have additional storage space to meet customer demands and to allow more flexible production scheduling. A primary decision has been made to secure additional inventory storage.

An initial investigation has revealed the availability of only one suitable rental warehouse, and it is available only if leased for 10 years. The warehouse has more space than is immediately required, but the company feels that some of the space could be sublet if desired. Estimates solicited from building contractors confirm that the construction of a new warehouse of equivalent size would amount to more than the $23,000 per year lease cost.

Another alternative is to build a small warehouse now and enlarge it if future business activity warrants expansion. The owners feel that in 3 years they will know whether the company's growth will support the addition. In order to evaluate the alternatives, estimates were made of possible business patterns and the likelihood of each. Their optimistic forecasts are shown in Table 14.3.

**TABLE 14.3** Growth patterns.

| *Growth Pattern* | *Probability* |
|---|---|
| No increase in activity for 10 years | 0.15 |
| No increase for 3 years, but an expanded growth rate during the next 7 years | 0.15 |
| Increasing growth for the next 3 years, but no increase during the following 7 years | 0.14 |
| Increasing activity for the full 10 years | 0.56 |

The owners place the probability for increased growth during the next 3 years at 0.56 + 0.14 = 0.70. If the growth materializes, they can use more room than is available in the anticipated small warehouse. Therefore, if they initially decide to build, they will have to

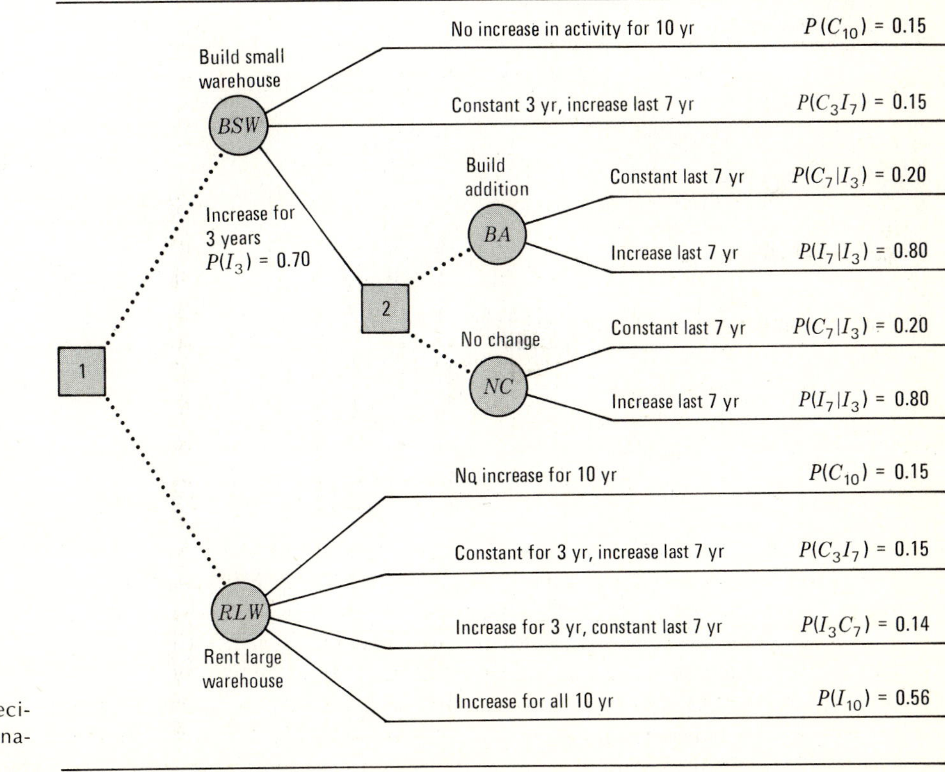


**FIGURE 14.10** Decision tree with alternatives and forecasts.

make a decision in 3 years about whether to add to the small warehouse or find other means to obtain extra storage space. At that time the conditional probability that the company will continue to grow is 0.56/0.70 = 0.80. The spectrum of forecasts and alternatives can be summarized in a decision tree as shown in Figure 14.10.

## OUTCOMES

The outcomes for the warehouse proposals are rated according to expected costs. Initial building costs for a small warehouse should be accurate, but the estimated price for an addition is less firm because of possible changes in building conditions at the time of construction. Yearly rental fees for the leased warehouse are a fixed amount. Other annual costs are less certain. Savings, net positive cash flows, could result if the entire capacity is not required for the company's inventory and the extra portion is rented. Conversely, additional costs are incurred when a lack of storage space forces production runs below the economical lot size or causes out-of-stock costs in supplying customers. Estimates of net annual costs for the outcomes of each alternative are tabulated in the decision tree of Figure 14.11.

The cash-flow estimates shown in the figure include the first costs associated with each

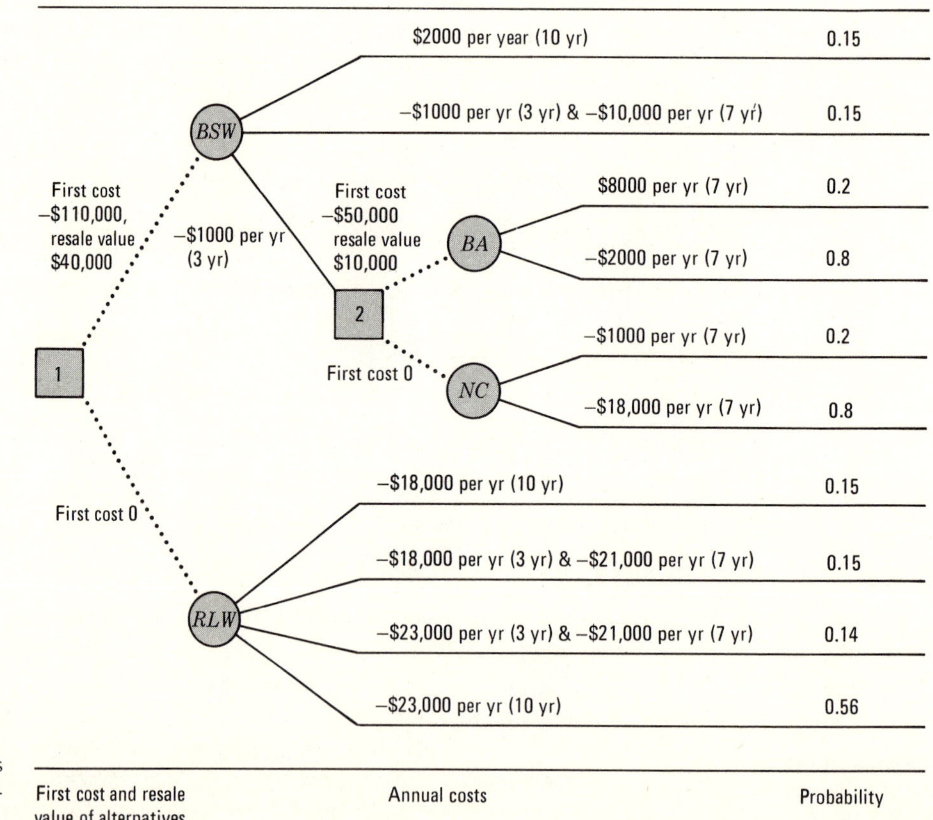


**FIGURE 14.11** Costs associated with warehouse alternatives.

alternative and the annual costs related to the outcomes. Building a small warehouse, ***BSW***, has a first cost of $110,000 and a resale value of $40,000 after 10 years; renting a large warehouse, ***RLW***, has no initial cost. The second decision point 3 years in the future is a choice between building an addition, ***BA***, at a cost of $50,000 with a $10,000 resale value, and making no change, ***NC***. Net annual returns comprise repairs, taxes, insurance, leasing expense, and opportunity costs, as well as the savings from subletting when possible. The lease for the rental warehouse states an annual charge of $23,000, which includes taxes, insurance, and repairs. If there is no increase in the company's activities, this annual charge can be defrayed by $5000 from subletting extra space. Less rental income is anticipated when the company needs part or all of the space to handle its own increasing activity. The two periods of returns, 3 and 7 years, correspond to the original growth patterns indicated in Figure 14.10.

## EVALUATION

Two types of calculations are involved in evaluating alternatives. The present values of receipts and expenditures are determined for each outcome and are then weighted according to their probability of occurrence. This procedure amounts to finding the expected value of the present worth of the outcomes. These expected values are compared at a decision point to select the most advantageous alternative.

Comparisons are made in a reverse chronological order. That is, the most distant decision point from time zero is evaluated first. The selected alternative from the first decision then becomes an input to the next decision. The backward pass through successive points is continued until the primary decision is resolved.

For the warehouse example, the discounting procedure begins at decision point 2. At this point in time, 3 years away from the primary decision, the company must decide whether to build an addition to the plant or make no change. The outcomes of each of these two alternatives depend on the level of business activity during the last 7 years of the study period. We shall assume that the company uses an interest rate of 12 percent. The present worths of the four outcomes at decision point 2 are calculated as shown.

$$\begin{aligned}\text{PW}(BA{:}C_7) &= -\$50{,}000 + \$10{,}000(P/F,\ 12,\ 7) + \$8000(P/A,\ 12,\ 7)\\ &= -\$50{,}000 + \$10{,}000(0.45235) + \$8000(4.5637)\\ &= -\$8{,}966\end{aligned}$$

$$\begin{aligned}\text{PW}(BA{:}I_7) &= -\$50{,}000 + \$10{,}000(P/F,\ 12,\ 7) + -\$2000(P/A,\ 12,\ 7)\\ &= -\$54{,}603\end{aligned}$$

$$\text{PW}(NC{:}C_7) = -\$1000(P/A,\ 12,\ 7) = -\$1000(4.5637) = -\$4{,}564$$

$$\text{PW}(NC{:}I_7) = -\$18{,}000(P/A,\ 12,\ 7) = -\$82{,}142$$

These present worths at decision point 2 are entered in a payoff table as shown in Figure 14.12, where the calculated expected values indicate a preference for the alternative to build an addition, ***BA***.

Decision point 1 also has two alternatives. Since we have already considered decision point 2, we shall continue on this branch of the decision tree. The initial cost of building is $110,000, with an expected resale value of $40,000 after 10 years. Three outcomes may occur from this course of action. Company activity could increase or remain constant for

| | COMPANY GROWTH PATTERNS | | |
|---|---|---|---|
| *Alternative* | *Constant (0.2)* | *Increase (0.8)* | *Expected Value (costs)* |
| *BA* | −$8966 | −$54,603 | −$45,476 |
| *NC* | −4564 | −82,142 | −66,626 |

**FIGURE 14.12** Payoff table for decision point 2.

the entire 10 years, or it could increase the first 3 years and then level off or continue to increase the remaining 7 years. The conditional decision made at point 2 was based on the possible outcomes of the last 7 years, ***given*** that activity increased during the first 3 years. This decision becomes an outcome of the primary alternative to build a warehouse. A comparison, assuming end-of-year returns and a 12 percent interest factor, is shown below.

$$\begin{aligned}\text{PW}(BSW{:}C_{10}) &= -\$110{,}000 + \$40{,}000(P/F,\ 12,\ 10) + \$2000(P/A,\ 12,\ 10)\\ &= -\$110{,}000 + \$40{,}000(0.32197) + \$2000(5.6502)\\ &= -\$85{,}821\end{aligned}$$

$$\begin{aligned}\text{PW}(BSW{:}C_3I_7) &= -\$110{,}000 + \$40{,}000(P/F,\ 12,\ 10) + -\$1000(P/A,\ 12,\ 3) +\\ &\qquad -\$10{,}000(P/A,\ 12,\ 7)(P/F,\ 12,\ 3)\\ &= -\$110{,}000 + \$40{,}000(0.32197) + -\$1000(2.4018) +\\ &\qquad -\$10{,}000(4.5637)(0.71178) = -\$132{,}007\end{aligned}$$

$$\begin{aligned}\text{PW}(BSW{:}I_3) &= -\$110{,}000 + \$40{,}000(P/A,\ 12,\ 7) + -\$1000(P/A,\ 12,\ 3) +\\ &\qquad -\$45{,}476(P/F,\ 12,\ 3) = -\$131{,}892\end{aligned}$$

Equivalent calculations applied to the other main fork of the decision tree give the present worths of annual costs associated with leasing a large warehouse.

$$\text{PW}(RLW{:}C_{10}) = -\$18{,}000(P/A,\ 12,\ 10) = -\$18{,}000(5.6502) = -\$101{,}704$$

$$\begin{aligned}\text{PW}(RLW{:}C_3I_7) &= -\$18{,}000(P/A,\ 12,\ 3) + -\$21{,}000(P/A,\ 12,\ 7)(P/F,\ 12,\ 3)\\ &= -\$18{,}000(2.4018) + -\$21{,}000(4.5637)(0.71178)\\ &= -\$111{,}448\end{aligned}$$

$$\begin{aligned}\text{PW}(RLW{:}I_3C_7) &= -\$23{,}000(P/A,\ 12,\ 3) + -\$21{,}000(P/A,\ 12,\ 7)(P/F,\ 12,\ 3)\\ &= -\$123{,}457\end{aligned}$$

$$\text{PW}(RLW{:}I_{10}) = -\$23{,}000(P/A,\ 12,\ 10) = -\$129{,}955$$

Grouping the outcomes from the two alternatives into a payoff table yields the values in Figure 14.13.

| | COMPANY GROWTH PATTERNS | | | | |
|---|---|---|---|---|---|
| *Alternatives* | *Constant 10 Years (0.15)* | *Constant 3 Years, Increase 7 Years (0.15)* | *Increase 3 Years, Constant 7 Years (0.14)* | *Increase 10 Years (0.56)* | *Expected Value (cost)* |
| *BSW* | −$ 85,821 | −$132,007 | −$131,892 | −$131,892 | −$124,998 |
| *RLW* | −101,704 | −111,448 | −123,457 | −129,955 | −122,032 |

**FIGURE 14.13** Payoff table for decision point 1.

From this analysis it is apparent that the most economical alternative is to sign the lease for renting the large warehouse. The margin of difference is \$124,998 − \$122,032 = \$2966.

## Value of Additional Information

The novelty company now has a quantitative base for making its decision. It may be satisfied with the information and proceed to sign the lease. It is also possible that some segment of management may not agree on the indicated course of action. There may be a disagreement as to estimated growth patterns or cost figures. Assuming that the need for some kind of storage facility is unanimously recognized, there is still another alternative available: The decision can be postponed. This alternative would be procrastination unless it were coupled with a firm desire to obtain more information about the problem. Such information could be provided by further intensive investigation by company personnel or by a study conducted by an independent agency.

The first step in considering the advisability of further research is to determine the worth of additional information. Securing new data costs money, whether the investigation is conducted by company or outside investigators. This additional investment should be exceeded by the expected value of added profits or reduced costs realized from the information.

With *perfect* information the company would *know* which alternative would cost the least. Figure 14.13 revealed that the best alternative would be to build a small warehouse without an addition if it were *known* that the company would experience a constant growth pattern for 10 years. The cost of building (\$85,821) is \$15,883 less than leasing for this growth pattern. In a similar fashion the best alternative for each state assumed *known* can be determined. The expected value of such perfect foresight is calculated as shown in Figure 14.14.

Four future states were identified in the warehouse illustration. It was observed that a small warehouse should be built for a condition of constant growth for 10 years. The best alternative for 3 years of constant growth followed by 7 years of increasing activity is to lease. Figure 14.13 shows that the present worth of renting expense is \$111,448 versus building costs of \$132,007. The other two states are not so apparent from previous calculations.

For the state of accelerated growth for 3 years followed by 7 years of level activity, the cost for no change in the capacity of the small warehouse was determined to be \$4564. Discounting this value to the present and including the initial cost, resale, and annual costs for the first 3 years give

$$\text{PW}(BSW{:}I_3C_7) = -\$110{,}000 + \$40{,}000(P/F,\ 12,\ 10) + -\$1000(P/A,\ 12,\ 3) + -\$4564(P/F,\ 12,\ 3)$$

$$= -\$102{,}772$$

which reveals a cost of building that is less than the lease expense of \$123,457. A similar computation for the building cost with 10 years of increasing activity is

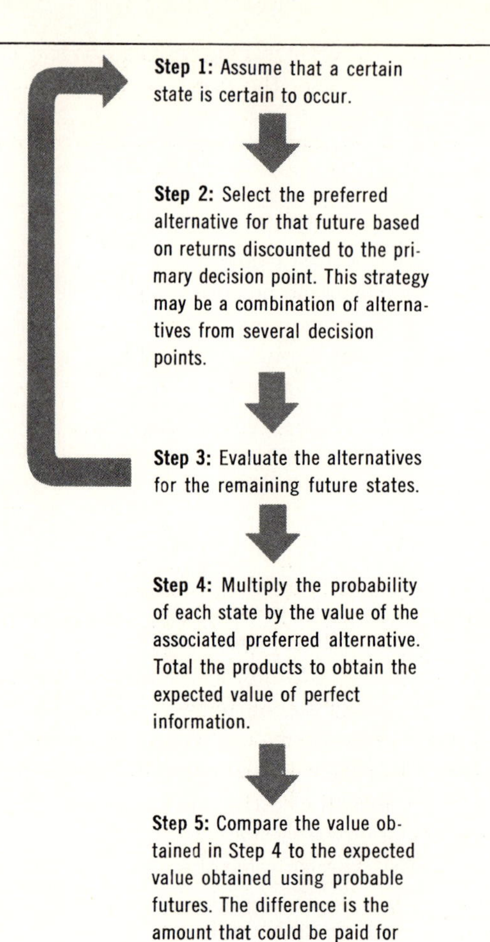


**FIGURE 14.14** Flow chart for determining the value of perfect information.

$$\text{PW}(BSW{:}I_3I_7) = -\$110{,}000 + \$40{,}000(P/F,\ 12,\ 10) + -\$1000(P/A,\ 12,\ 3) + -\$54{,}603(P/F,\ 12,\ 3)$$
$$= -\$138{,}388$$

which exceeds the leasing cost of $129,955 for the same future state.

The expected value from the preferred alternatives for each state evaluated separately is shown in Table 14.4.

Comparison of the expected value of perfect information with the expected value of the best alternative obtained earlier (Figure 14.12) indicates that the novelty company could afford to pay up to $122,032 − $116,753 = $5279 for a perfect forecast. Securing perfect information is at best a wild hope, but these figures do suggest that significant savings could result from further study of the uncertainties in expected growth patterns.

| *State* | *Preferred Alternative* | *Cost* | *Proba-bility* | *Product* |
|---|---|---|---|---|
| Constant 10 years | *BSW* | −$85,821 | 0.15 | −$12,873 |
| Constant 3 years, increase 7 years | *RLW* | −111,448 | 0.15 | −16,717 |
| Increase 3 years, constant 7 years | *BSW-NC* | −102,772 | 0.14 | −14,388 |
| Increase 10 years | *RLW* | −129,955 | 0.56 | −72,775 |
| | | | Expected value | −$116,753 |

**TABLE 14.4** Expected value of perfect information.

**Economy Exercise 14-5** An ambitious entrepreneur has signed a regionally well known rock band to give a concert on the Fourth of July. The concert can be staged in a conveniently located pasture or in a school auditorium. If the weather is sunny, a concert in the pasture could be very profitable, but rain would ruin attendance and cause a loss due to the high fixed costs of preparing the site. Attendance at the auditorium would be largely unaffected by weather, but the maximum capacity is far less than the pasture. Weather records indicate that the odds against rain on July 4 are 9 to 1. Using the probabilities given and the estimated returns shown for concerts held at the two locations, calculate the *expected value* and the amount of money that could be paid for *perfect information* about the weather, if it could be purchased.

| | NET RETURNS IF THERE IS | |
|---|---|---|
| *Alternative* | *Rain* | *No Rain* |
| Auditorium | $2400 | $3,000 |
| Pasture | −2700 | 9,000 |

## RISKS IN INVENTORY POLICY

The analysis methods for determining minimum-cost operating levels in Chapter 4 are based on assumptions of certainty. Inputs to the basic formula employed in the analyses, $C_{\min} = AX' + B/X' + K$, are assumed to be known and unchanged in future states. Such assumptions allow direct evaluation of the basic properties of the problem at the sacrifice of some realism. The appropriateness of developing and applying more rigorous minimum-cost models is subject to the same benefits and drawbacks discussed for developing more sophisticated discounted cash-flow models. Sample applications that include risk in the analysis of inventory policies are presented on the following pages.

## Inventory Policy for Perishable Products

Some firms handle merchandise which has negligible utility if it is not sold almost immediately. Products in this category include newspapers, printed programs for special events, fresh produce, and other perishable commodities. Such items commonly have a high markup. The large difference between wholesale cost and retail price is due to the risk a vendor faces in stocking the item. The vendor faces obsolescence costs on one hand and opportunity costs on the other.

In most cases a retailer cannot forecast exactly what the demand will be on a given day. The loss incurred from any unsold items left after their limited selling period must be balanced against the loss in profit caused by unfilled orders. With records of past sales, and believing that the pattern will not change appreciably in the future, the retailer can calculate an order size which will permit a maximum profit over an extended period of time.

Consider the case of a vendor who has an exclusive franchise to sell programs at a municipal coliseum. During the basketball season he orders programs in increments of 500. Using his records from the previous two seasons, he has developed the sales forecast shown in Table 14.5. It will be used to determine an ordering policy for typical games. Special attractions would require a separate forecast, because they have a different pattern from regular games.

| *Sales per Game* | *Number of Games Sold* | *Probability of Programs Being Sold* |
|---|---|---|
| 2500 | 4 | 0.10 |
| 3000 | 6 | 0.15 |
| 3500 | 10 | 0.25 |
| 4000 | 16 | 0.40 |
| 4500 | 4 | 0.10 |
| | 40 | 1.0 |

**TABLE 14.5** Program sales.

The vendor purchases the programs for $0.15 each and sells them for $0.50. From his markup of $0.35 he must pay the commissions and fixed expenses. As a simplifying measure, we shall call the $0.35 markup his profit.

His forecast does not tell him how many programs will be sold at a regular game, but it does tell him, for instance, that there is a 50 percent chance he will sell at least 3500 copies. It also indicates his profit outlook. If he orders 4000 programs but sells only 3500, his profit will be $3500 \times \$0.35 - 500 \times \$0.15 = \$1150$. In an extreme case of misjudgment, he could order 4500 and sell 2500. This calamity would result in the minimum profit of $2500 \times \$0.35 - 2000 \times \$0.15 = \$575$. A maximum profit of $1575 would result from ordering and selling 4500 programs.

Between $575 and $1575 is a range of possible profits. These are conditional profits. That is, each is the profit that will result from a possible demand, given that a certain number of programs have been ordered. The range of possible profits is determined by the range of possible sales. There is no need to extend the range, because all past performances are included. All the conditional profits are tabulated below.

| | POSSIBLE DEMAND | | | | |
|---|---|---|---|---|---|
| *Possible Order Supply* | *2500* | *3000* | *3500* | *4000* | *4500* |
| 2500 | $875 | $875 | $875 | $875 | $875 |
| 3000 | 800 | 1050 | 1050 | 1050 | 1050 |
| 3500 | 725 | 975 | 1225 | 1225 | 1225 |
| 4000 | 650 | 900 | 1150 | 1400 | 1400 |
| 4500 | 575 | 825 | 1075 | 1325 | 1575 |

The vendor's profit status can be further refined by analyzing his sales history in combination with his conditional profits. This combination takes the form of a payoff table. The future states are the possible sales levels. Associated with each state is the probability of its occurrence, derived from sales records. The alternatives are the various order sizes, and the outcomes are the conditional profits.

| | STATE OF DEMAND | | | | | |
|---|---|---|---|---|---|---|
| *Alternative (order size)* | *2500 (0.10)* | *3000 (0.15)* | *3500 (0.25)* | *4000 (0.40)* | *4500 (0.10)* | *Expected Profit* |
| 2500 | $875 | $875 | $875 | $875 | $875 | $875.00 |
| 3000 | 800 | 1050 | 1050 | 1050 | 1050 | 1025.00 |
| 3500 | 725 | 925 | 1225 | 1225 | 1225 | 1137.50 |
| 4000 | 650 | 900 | 1150 | 1400 | 1400 | 1187.50 |
| 4500 | 575 | 825 | 1075 | 1325 | 1575 | 1137.50 |

**FIGURE 14.15** Expected profit from program sales.

The vendor can expect to make the most profit by ordering 4000 programs each time. By following this policy he will average $1187.50 profit for each game, provided the actual demand follows the pattern of past sales. He is not guaranteed this amount for a given game, but any other quantity will leave him with less *average* profit.

After studying the calculations reported in Figure 14.15, the vendor might ponder the possibility of improving his operation. Perhaps he could use weather forecasts to anticipate the size of the crowd at a game. Maybe he could devise a program which would retain its appeal for postgame sales. The amount he could afford to pay to develop improvements is the difference between the expected profit calculated in Figure 14.15 and the profit he could make if he possessed perfect information. Table 14.6 shows that the maximum

| *Program Sales* | *Preferred Alternative* | *Profit* | *Probability of Demand* | *Expected Profit from Perfect Data* |
|---|---|---|---|---|
| 2500 | 2500 | $875 | 0.10 | $87.50 |
| 3000 | 3000 | 1050 | 0.15 | 157.50 |
| 3500 | 3500 | 1225 | 0.25 | 306.25 |
| 4000 | 4000 | 1400 | 0.40 | 560.00 |
| 4500 | 4500 | 1575 | 0.10 | 157.50 |
| *Maximum profit* | | | | $1268.75 |

**TABLE 14.6** Maximum profit with perfect information.

possible profit is \$1268.75. Therefore, he could afford \$1268.75 − \$1187.50 = \$81.25 *per game* to develop means to avoid accumulating obsolete programs.

## Safety Stock to Counteract Risk

The classic sawtooth pattern of inventory usage given in Figure 4.8 rests on assumptions that the demand rate is constant and replenishments are received when ordered. Many practical realities frustrate these assumptions. Causes of disruptions to the regular flow of inventory are shown in Figure 14.16. Receiving an order a week later than anticipated may cause an interruption in production or a loss of sales. An unexpected increase in demand may cause a stockout of merchandise or supplies. Such risks are relieved by carrying a reserve inventory, or ***safety stock.***

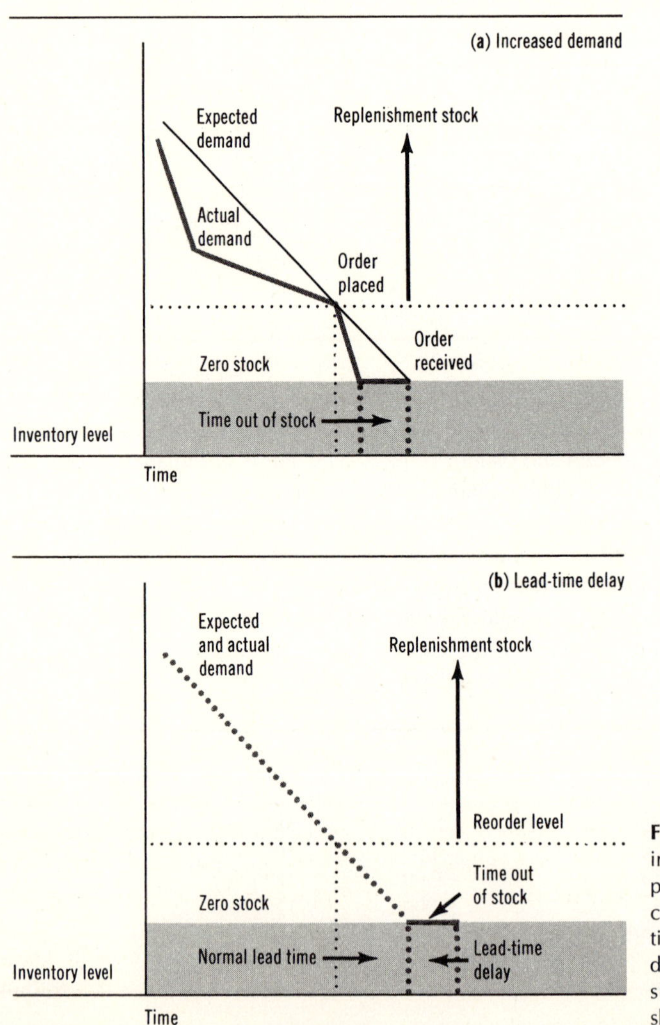


**FIGURE 14.16** Irregularities in the classic inventory-usage pattern (see Figure 4.8) that contribute to stockouts. The time out of stock would be reduced by holding a reserve supply represented by the shaded area.

Safety stock acts as a buffer between anticipated and actual inventory levels. In Figure 14.16, safety stock would appear as a plus (+) inventory in the shaded section to lower the zero stock level, serving thereby to eliminate or reduce the out-of-stock time. Reducing stockouts reduces opportunity costs, but it increases holding costs. A very large safety stock could completely eliminate the chance of stockouts, and, correspondingly, it would substantially raise insurance, damage, interest, and other holding costs. The problem is to determine a safety-stock level which will provide adequate protection from stockouts at reasonable holding costs.

## Minimum-Cost Inventory Model for Variable Lead Times

Placing orders for replenishments well in advance of the time stock levels are expected to drop to zero protects against stockouts by creating a safety stock. One method of determining a reasonable ordering policy when lead times vary is to modify the basic EOQ formula to include *opportunity costs,* the loss in revenue (or increase in costs) from not having stock available when needed. To illustrate the problem of irregular replenishment periods, consider a manufacturing plant which uses special chemicals in its finishing department. The chemicals must be stored in a controlled environment. Because the chemicals deteriorate with age, the producer must make a new batch for each delivery. This leads to a considerable variation between the time an order is placed and the time at which it is received. The distribution of lead times is shown in Table 14.7. The minimum delivery time is 7 days. Cumulating the lead times shows there is a probability of 0.5 that a delivery will take at least 10 days.

| *Lead Time, Days* | *Relative Frequency* |
|---|---|
| 6 | 0.00 |
| 7 | 0.04 |
| 8 | 0.08 |
| 9 | 0.38 |
| 10 | 0.24 |
| 11 | 0.12 |
| 12 | 0.09 |
| 13 | 0.03 |
| 14 | 0.02 |
| | 1.00 |

**TABLE 14.7** Probabilities of different lead times.

Many factors in the problem are known with relative certainty. Continuous production entails an annual demand for 1150 cylinders used at a constant rate through the year. Each cylinder costs $40, in addition to the order and inspection costs of $55 per order for a typical order size. Holding costs are 25 percent of the value of the average inventory in storage.

Disregarding stockout costs temporarily, an order size based on known conditions for costs and demand can be calculated by the economic lot-size formula developed for conditions of assumed certainty:

$$Q = \sqrt{\frac{2OD}{H}} = \sqrt{\frac{2 \times 55 \times 1150}{10}} = 113 \text{ cylinders}$$

*where* $O$ = order costs = \$55/order
$D$ = annual usage rate = 1150 cylinders
$H$ = holding costs = 0.25 × \$40 = \$10/cylinder-year

Since $Q$ and $D$ are known, the number of orders per year is

$$\text{Orders/year} = \frac{D}{Q} = \frac{1150}{113} \doteq 10.2$$

Assuming 230 working days per year, the time between orders is

$$\text{Order interval} = \frac{230}{10.2} \doteq 23 \text{ days}$$

and the daily usage rate is

$$D_{\text{daily}} = \frac{1150}{230} = 5 \text{ cylinders/day}$$

Now the effect of the lead-time distribution on the basic inventory plan can be determined. For a fixed lead time of 7 days the company would place an order whenever the inventory fell to 7 × 5 = 35 cylinders. Under conditions of certainty there would be no chance of late delivery and therefore no danger of running out of chemicals. However, from the table of lead times it is apparent that a 7-day lead time occurs with a probability of only 0.04. This indicates that the chances of running out of stock for at least 1 day are 96 out of 100 if an order is not placed until the inventory level reaches 35. Ordering when the stock level is higher than 35 reduces the chance of a stockout, but adds the costs of holding a safety stock. Both holding and opportunity costs must be included in the economic-lot-size calculations to achieve a minimum-cost inventory policy.

Figure 14.17(a) shows the effect of each ordering alternative (amount of lead time pro-

| | LEAD TIME REQUIRED, WEEKS | | | | | | | |
|---|---|---|---|---|---|---|---|---|
| *Lead Time Provided* | *7* | *8* | *9* | *10* | *11* | *12* | *13* | *14* |
| | Number of cylinders not supplied | | | | | | | |
| 7 | | −5 | −10 | −15 | −20 | −25 | −30 | −35 |
| 8 | 5 | | −5 | −10 | −15 | −20 | −25 | −30 |
| 9 | 10 | 5 | | −5 | −10 | −15 | −20 | −25 |
| 10 | 15 | 10 | 5 | | −5 | −10 | −15 | −20 |
| 11 | 20 | 15 | 10 | 5 | | −5 | −10 | −15 |
| 12 | 25 | 20 | 15 | 10 | 5 | | −5 | −10 |
| 13 | 30 | 25 | 20 | 15 | 10 | 5 | | −5 |
| 14 | 35 | 30 | 25 | 20 | 15 | 10 | 5 | |
| | Number of cylinders stored | | | | | | | |

**FIGURE 14.17*a*** Excess or deficient cylinder supply.

vided). The upper portion indicates the shortage that will occur for each day of delay from the lead time provided. The lower portion shows the incremental size of safety stock which will be accumulated by providing more lead time than is required. If orders are based on a 7-day lead time, there will never be old stock on hand when a new order arrives. At the other extreme, an order policy based on a lead time of 14 days assures the company that it will never run out of stock, although as many as 35 cylinders could be in storage when a new supply is delivered.

The opportunity cost of running out of stock is estimated to be $40 per cylinder. This cost accrues from the disruption of production. The manufacturing plant would sustain a loss of 5 × $40 = $200 per day for late deliveries based on a daily usage of five cylinders. By multiplying the stock deficiencies in Figure 14.17(a) by $40 and the excess stock by the $10-per-cylinder holding costs, the payoff table of Figure 14.17(b) is generated.

| | LEAD TIME REQUIRED, WEEKS | | | | | | | | EXPECTED VALUE | |
|---|---|---|---|---|---|---|---|---|---|---|
| *Lead Time Provided* | *7 (0.04)* | *8 (0.08)* | *9 (0.38)* | *10 (0.24)* | *11 (0.12)* | *12 (0.09)* | *13 (0.03)* | *14 (0.02)* | *HC* | *OC* |
| 7 | $ 0 | $200 | $400 | $600 | $800 | $1000 | $1200 | $1400 | $ 0 | $562 |
| 8 | 50 | 0 | 200 | 400 | 600 | 800 | 1000 | 1200 | 2 | 370 |
| 9 | 100 | 50 | 0 | 200 | 400 | 600 | 800 | 1000 | 8 | 194 |
| 10 | 150 | 100 | 50 | 0 | 200 | 400 | 600 | 800 | 33 | 94 |
| 11 | 200 | 150 | 100 | 50 | 0 | 200 | 400 | 600 | 70 | 42 |
| 12 | 250 | 200 | 150 | 100 | 50 | 0 | 200 | 400 | 113 | 14 |
| 13 | 300 | 250 | 200 | 150 | 100 | 50 | 0 | 200 | 161 | 4 |
| 14 | 350 | 300 | 250 | 200 | 150 | 100 | 50 | 0 | 210 | 0 |

**FIGURE 14.17*b*** Expected costs of lead-time alternatives.

In Figure 14.17(b) the alternatives are the lead times provided, and the future states are lead times that could occur. The outcomes result from two types of costs: holding costs and opportunity costs. These two costs affect the ordering policy in different ways. Opportunity costs occur only when an order takes longer to arrive than the lead time that was allowed. Since this cost is associated with the order interval, it can be treated like an ordering expense or a set-up cost. Holding costs are annual expenses and reflect the extra inventory or safety stock that is held in storage.

The expected-value column on the right of the table has two divisions. One applies to holding costs for an alternative, and the other to the opportunity costs. Both are calculated from pertinent outcome costs. When the lead time provided coincides with the actual lead time required, the outcome is a zero cost. Obviously, one zero outcome must occur for each alternative. The costs to the left of the zero outcome arise from holding costs, while those to the right are opportunity costs. Each type is multiplied by the likelihood of its occurrence, and the sum of the products is entered in the appropriate expected-value column. Thus, a 7-day lead time shows no holding costs, and a 14-day lead time has no opportunity costs. For a lead-time alternative of 11 days,

$$\text{EV}(HC_{11}) = \$200 \times 0.04 + \$150 \times 0.08 + \$100 \times 0.38 + \$50 \times 0.24$$
$$= \$8 + \$12 + \$38 + \$12 = \$70$$

*where HC* = holding costs of safety stock

and

$$\text{EV}(OC_{11}) = \$200 \times 0.09 + \$400 \times 0.03 + \$600 \times 0.02$$
$$= \$18 + \$12 + \$12 = \$42$$

*where OC* = opportunity costs

| *Lead Time Provided* | *Economic Lot Size* |
|---|---|
| 7 | 377 |
| 8 | 314 |
| 9 | 239 |
| 10 | 186 |
| 11 | 150 |
| 12 | 126 |
| 13 | 117 |
| 14 | 113 |

The expected values for the other alternatives are obtained in a similar manner.

Each lead-time alternative provides a different economic lot size because of different opportunity costs. By adding the expected value of the opportunity costs to the other costs, we can apply the same economic-lot-size formula used previously. For a 7-day lead time,

$$Q_7 = \sqrt{\frac{2(O + OC)D}{H}} = \sqrt{\frac{2(55 + 562)1150}{10}} = 377$$

All the lot sizes are shown in the accompanying table.

The final step is to determine which lot size will allow the lowest total cost. If the purchase price of cylinders is not subject to quantity discounts, price breaks will not affect the ordering policy. Then the total cost becomes the sum of annual ordering costs and holding costs. The ordering costs include both the cost of placing an order and the opportunity cost. Holding costs are composed of the storage expense due to the order size, $H$, and the additional storage expense incurred from providing a safety margin for delivery delays, $HC$. This latter cost is the expected value of the holding cost for each lead-time alternative in Figure 14.17(b). Total costs are calculated from the formula

$$C_{LT} = \frac{(O + OC)D}{Q} + \frac{HQ}{2} + HC$$

This yields the total cost for a 7-day lead time as

$$C_7 = \frac{(55 + 562)(1150)}{377} + \frac{10 \times 377}{2} + 0 = \$3767$$

| *Lead Time* | *Total Cost* |
|---|---|
| 7 | $3767 |
| 8 | 3130 |
| 9 | 2402 |
| 10 | 1884 |
| 11 | 1564 |
| 12 | 1373 |
| 13 | 1326 |
| 14 | 1335 |

and for an 8-day lead time as

$$C_8 = \frac{(55 + 370)(1150)}{314} + \frac{10 \times 314}{2} + 2 = \$3130$$

From the complete tabulation of costs shown in the total-cost table, it is clear that the most economical policy is to allow a 13-day lead time. An order should be placed whenever the stock level declines to 13 × 5 = 80 cylinders. By following this policy, the company should expect to run out of chemicals twice in every 100 order periods.

Opportunity costs are considered more difficult to estimate than holding costs. It is interesting to observe the ***sensitivity*** of the lead-time calculations to different opportunity-cost estimates. Total annual costs for the cylinder-inventory problem with four different opportunity costs are shown in Table 14.8. All other factors in the problem are unchanged. It is apparent that estimates of the same general magnitude will not change the results significantly. Consequently, there should not be too much concern if opportunity costs cannot be estimated with the desired degree of accuracy.

| *Opportunity Cost* | *Lead Time* | *Total Cost* |
|---|---|---|
| $10 | 12 | $1272 |
| 20 | 13 | 1305 |
| 30 | 13 | 1314 |
| 40 | 13 | 1326 |

**TABLE 14.8** Sensitivity to opportunity costs.

## Consideration of Variable Demand Rates

Variations in demand or usage rates produce situations similar to those produced by variable lead times. If the lead time is fixed, the risk of a stockout is dependent on the actual demand exceeding the anticipated demand. Holding a safety stock reduces the risk. The problem then becomes one of selecting a safety-stock size which balances the expense of holding the safety stock against the cost of a stock shortage.

Consider a maintenance section which receives deliveries of spare parts from a distributing firm. Deliveries are made regularly a week after an order is placed. The ordering policy has been to reorder class 2 parts whenever the inventory level dips to 200 parts. A series of stockouts has indicated that the policy should be reviewed. It was decided to retain the same lot size, but a safety stock could be carried for certain parts. Part no. 323 has the usage rate depicted by the histogram in Figure 14.18. It shows the relative frequency of demand for a 1-week period. Annual holding costs for the part are $2, and the stock is replenished about six times a year. Opportunity costs, estimated to be $1 per week, are the result of special ordering procedures and inconveniences caused by the part's being required but unavailable.

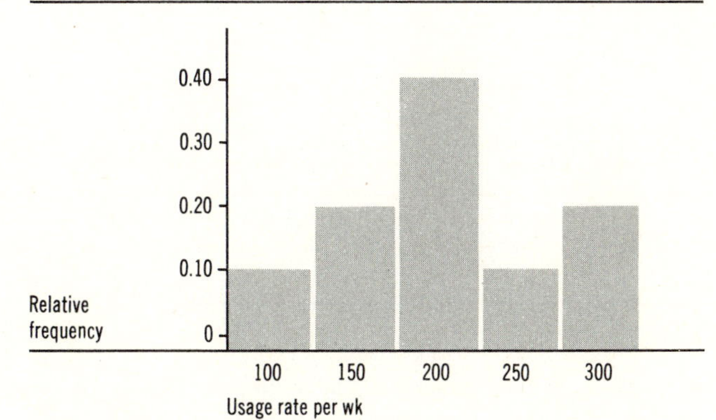


**FIGURE 14.18** Histogram of demand for Part no. 323.

Before seeking a solution to this inventory situation, let us inspect the conditions carefully. Here we have a fixed order size and a fixed lead time. A stockout can occur only if the demand increases during the lead time. The solution space has been limited to the investigation of an appropriate safety stock for certain parts. These conditions considerably simplify the calculations. Without a fixed order size, we would follow the same general approach to a solution as that used for variable lead times. A zero safety stock would be associated with a 200-part reorder point. This is equivalent to decreasing the reorder point, and it would be a reasonable alternative if storage space is limited or if negligible inconvenience results from stockouts. In turn, altering the reorder point would affect the order quantity by increasing costs associated with the order interval. But with the fixed order interval, only the effect of adding a buffer stock to the accepted reorder level needs to be considered.

A reorder level of 200 parts will lead to a stockout 30 percent of the time. Holding an extra 50 parts would limit running out of parts to a probability of 0.20, and a 100-part cushion would completely eliminate the chances of a stockout if the demand pattern remains unchanged. The expected value of the opportunity costs for each of these alternatives is shown in Figure 14.19. Each outcome is the product of the cost of a shortage and the number of times a shortage would occur in a year.

**FIGURE 14.19** Expected value of opportunity costs.

| | | | SAFETY STOCK REQUIRED | | |
|---|---|---|---|---|---|
| *Parts Provided* | *Reorder Point* | *Safety Stock* | *50 (0.10)* | *100 (0.20)* | *Expected Value OC* |
| 200 | 200 | 0 | 50 × \$1 × 6 = \$300 | 100 × \$1 × 6 = \$600 | \$150 |
| 250 | 250 | 50 | 0 | 50 × \$1 × 6 = \$300 | 60 |
| 300 | 300 | 100 | 0 | 0 | 0 |

The total cost of each of the alternatives is the sum of the opportunity costs and the holding costs. No additional holding costs are incurred for a zero safety stock. For each safety-stock increment of 50 parts, the annual holding cost is 50 × \$2 = \$100. From the totals in Table 14.9, it appears that the original ordering policy was sound and the reorder point should remain at 200 for Part no. 323.

**TABLE 14.9** Total-cost table for Part no. 323.

| *Safety Stock Provided* | *Opportunity Costs* | *Holding Costs* | *Total Cost* |
|---|---|---|---|
| 0 | \$150 | \$ 0 | \$150 |
| 50 | 60 | 100 | 160 |
| 100 | 0 | 200 | 200 |

Far more sophisticated models are available for including risk in the evaluation of inventory policies. For instance, the case in which both demand and lead times vary complicates the issue considerably and requires a more extensive analysis. ***Simulation*** is often used for such situations because it can closely duplicate actual conditions without the development of a specific optimization model. (See Extension 14.1 where simulation is applied to inventory analysis.) At the other extreme from elegant mathematical formulations are rules of thumb and SOPs that have evolved from long experience with inventory problems. How deeply to pursue reality in risk analysis is akin to limiting routine responses to operational problems as described by Figure 1.3; the depth depends on the cost of operating errors and the prospect of eliminating errors by the analysis.

## RISKS IN MAINTENANCE-REPLACEMENT POLICY

Maintenance costs were influencing factors in long-term replacement decisions made under assumed certainty, and they are equally significant in short-term replacement decisions which include risk. The short-term effect of wear often follows a consistent pattern. After the pattern has been identified, a user has the alternative of dealing with each item as it fails or following a policy of collective remedial action before failure. Preventive maintenance is a well-known example of the latter alternative.

Maintenance policies are extremely varied in practice. They depend on the characteristics of the items, available maintenance facilities, and relevant costs. We shall consider a basic model which may be adapted to fit the circumstances of a particular situation. It includes the factors common to most situations: probable life pattern, alternative courses of preventive action, and outcome costs for the alternatives. The objective is to find the interval at which preventive action is most suitable, and then compare this with the alternative of remedial action to decide which is more economical.

| *Months after Maintenance* | *Probability of Failure* |
|---|---|
| 1 | 0.2 |
| 2 | 0.1 |
| 3 | 0.1 |
| 4 | 0.2 |
| 5 | 0.4 |

To illustrate a maintenance policy, assume that a factory has 30 similar machines, which exhibit the probability distribution of failures shown in the accompanying table. The cost of remedial action after a breakdown averages $100, and the cost of providing preventive maintenance is $30 per machine.

The total cost of a preventive-maintenance program is the sum of servicing expense for all the machines each maintenance period (30 machines × $30 per machine = $900) and the cost of breakdowns occurring between services. For a *monthly* preventive-maintenance policy, PM1, the cost is $900 plus $100 for each breakdown expected in the first month after servicing. This amounts to

PM1 = $900 + $100 × 30 × 0.20 = $1500/month

A *bimonthly* policy must again include the $900 basic group-servicing cost and the cost of individual breakdowns. In the first month of the period, 30 × 0.20 = 6 machines are expected to break down. During the second month, 30 × 0.10 = 3 machines serviced at the regular period are likely to break down, and 20 percent of the individual breakdowns treated during the first month will *again* fail. These calculations are depicted in Figure 14.20, where a modified expected-value table is utilized.

| | INDIVIDUAL FAILURES DURING MONTH | | EXPECTED VALUE | |
|---|---|---|---|---|
| *PM Periods* | *1 (0.2)* | *2 (0.1)* | *Individual* | *Cumulative* |
| 1 | 30 | 0 | 6 | 6 |
| 2 | 6 | 30 | 4.2 | 10.2 |

**FIGURE 14.20** Expected failures for two PM periods.

The first alternative in the table represents a monthly preventive-maintenance policy. The expected value of this alternative is six machines treated each month, which agrees with previous calculations. The second line of outcomes depicts the failures of the second month. The total number of expected breakdowns in 2 months is the cumulative value for both months (6 + 4.2 = 10.2 machines). The cost of a bimonthly policy is

PM2 = $900 + $100 × 10.2 = $1920

which makes the cost per month equal to $1920/2 = $960.

A policy of servicing all the machines every 3 months would lead to the expected failure record shown in Figure 14.21. Now a pattern can be observed. Some of the original group of 30 machines continue to break down each month, and some of those repaired fail again individually. The repairs made each month start a new cycle which must follow the failure distribution. Thus, the expected number of breakdowns from one month is always the first outcome (0.20 probability of failure) for the next month. Similarly, the expected value 2 months ago is the second outcome for the current month, and the cumulative value is the total number of expected breakdowns for a cycle of so many months. A 3-month cycle would have a total cost of

PM3 = $900 + $100 × 14.64 = $2364

or a monthly cost of $2364/3 = $788

**FIGURE 14.21** Expected failures for three PM periods.

| | INDIVIDUAL FAILURES DURING MONTH | | | EXPECTED VALUE | |
|---|---|---|---|---|---|
| *PM Periods* | *1 (0.20)* | *2 (0.10)* | *3 (0.10)* | *Individual* | *Cumulative* |
| 1 | 30 | 0 | 0 | 6 | 6 |
| 2 | 6 | 30 | 0 | 4.2 | 10.2 |
| 3 | 4.2 | 6 | 30 | 4.44 | 14.64 |

Figure 14.21 can be converted to costs and expanded to include all the necessary calculations for the cost of every preventive-maintenance alternative. The outcomes are converted from units of machines to dollar values by multiplying the number of machines by the individual remedial cost ($100). Columns are added to account for the cyclic group PM costs, the sum of individual and group costs, and the prorated monthly costs. The completed table in Figure 14.22 displays the same recurring pattern of outcomes described previously.

The last step is to determine the costs associated with a policy of performing no preventive maintenance; machines are serviced whenever they break down. The expected period between breakdowns is calculated from the original failure distribution. The expected period is

1 month × 0.2 + 2 months × 0.1 + 3 months × 0.1 + 4 months × 0.2 + 5 months × 0.4 = 3.5 months between breakdowns

| | INDIVIDUAL FAILURE COSTS DURING MONTH | | | | | EXPECTED VALUE | | | | |
|---|---|---|---|---|---|---|---|---|---|---|
| *PM Period* | *1 (0.20)* | *2 (0.10)* | *3 (0.10)* | *4 (0.20)* | *5 (0.40)* | *Individual* | *Cumulative* | *PM Cost* | *Total Cost* | *Monthly Cost* |
| 1 | $3000 | $ 0 | $ 0 | $ 0 | $ 0 | $ 600.00 | $ 600 | $900 | $1500 | $1500 |
| 2 | 600 | 3000 | 0 | 0 | 0 | 420.00 | 1020 | 900 | 1920 | 960 |
| 3 | 420 | 600 | 3000 | 0 | 0 | 444.00 | 1464 | 900 | 2364 | 788 |
| 4 | 444 | 420 | 600 | 3000 | 0 | 791.00 | 2255 | 900 | 3155 | 789 |
| 5 | 791 | 444 | 420 | 600 | 3000 | 1564.60 | 3819 | 900 | 4719 | 944 |

**FIGURE 14.22** Expected cost of preventive-maintenance alternatives.

Then, with the cost of servicing individual breakdowns pegged at $100 per breakdown, a remedial-action policy costs

$$\frac{30 \text{ machines} \times \$100/\text{machine service}}{3.5 \text{ months/service}} = \$857/\text{month}$$

A comparison of the $857 monthly cost for the remedial-action alternative with the minimum monthly cost of a preventive-maintenance policy from Figure 14.22 ($788) indicates that the latter alternative is preferable. Both the 3- and 4-month preventive-maintenance periods show a lower expected cost than that for dealing with machines only after they fail.

This approach can be applied to a variety of situations. The items being evaluated could fail completely, like electric light bulbs or electronic tubes. Failures may represent personnel who are no longer available owing to transfers or retirement, and replacement could be the recruiting or training policy. There are also numerous modifications which fit special maintenance and repair situations.

## INVESTMENTS TO AVOID RISK

An investment made to avoid the consequences of risk may provide self-protection or insurance. An outlay which effectively reduces the cause of risk or the penalty associated with it is self-protection. The purchase of insurance transfers the burden of risk to another party. In either case the objective is to minimize the effect of a disaster.

### Penalty Estimates: Intangible Considerations

The cost of avoiding a disaster can be very high. The price is most commonly accepted in projects dealing with public safety. When human lives are involved, designers invariably strive to include an ample safety factor. The question is, "How large is ample?" Before we can answer this question we must know the value of human life. So many intangible considerations surround the value that we would probably receive a different estimate from each person questioned. A crude yardstick can be developed from previous safety efforts. If

$100,000 has been spent in a given area to reduce the probability of a fatal accident from 0.001 to 0.0001, the consequent value implied for a life is

$$\frac{\$100{,}000}{0.001 - 0.0001} = \$111{,}111{,}111$$

Although this method is not too precise in that it recognizes no other contributory factors, it does provide a reference level for difficult or intangible considerations.

The same approach can assist in the often difficult estimate of opportunity costs. An idea of the magnitude of these costs can frequently be obtained by investigating a present level of investment committed to risk reduction. For example, if a standby generator costs $300 per year to maintain and is used on the average of once a year, the value placed on an electrical outage is

$$\frac{\$300}{1/365} = \$109{,}500$$

Then $109,500 could be used as an estimate of opportunity costs for evaluation of other investments to reduce risks from a similar source.

## Self-Protection

An adequate investment can essentially eliminate risk in unique cases. Ski resorts can make provisions for artificial snow in anticipation of a dry winter. Investments are made in massive irrigation projects to eliminate the danger of drought. On a smaller scale, the selection by a city engineer of a storm drain sized to handle the maximum expected runoff is an attempt to nullify risk.

A more typical response in avoiding risk is a compromise. Even if the cause or penalty associated with a risk can conceivably be eliminated, the cost may be prohibitive. Yet the effects of risk often can be limited to a tolerable level by a reasonable expenditure. We dealt with this compromise when we considered safety-stock levels for an inventory policy. Similarly, the city engineer could select a storm-drain size which balances the extra costs of larger-diameter sewers against damages that could occur in exceptional storms.

## Quality Control

A quality-control program balances the cost of passing defective items against the cost of an inspection procedure capable of detecting inferior quality. (This concept was introduced in Chapter 4 as the cost of vigilance versus error.) Consider the case of a company which receives a contract to build 3000 new type precision instruments. The terms of the contract strongly suggest an inspection system to avoid the risk of supplying defective instruments; the penalty clause states damages of $300 per faulty unit.

Inspection equipment which would limit undetected faulty units to 0.5 percent could be purchased for $19,000. It would have no value after termination of the 3-year contract. A trained operator for the testing equipment would have to be paid $7000 per year, and operating costs would amount to $1000 annually. Assuming a desired rate of return of 10 percent, the annual cost of inspection would be

| | |
|---|---|
| Testing machine: \$19,000($A/P$, 10, 3) = \$19,000(0.40212) | \$ 7,640 |
| Operator | 7,000 |
| Operation | 1,000 |
| *Total cost of inspection* | \$15,640 |

In addition to the inspection costs, there is still a penalty for the few defective instruments which pass undetected and adjustment or reworking costs for the detected faulty instruments:

$$\text{Annual cost of penalty} = \frac{3000 \text{ instruments}}{3 \text{ years}} 0.005 \times \$300/\text{instrument} = \$1500$$

$$\text{Annual adjustment, reworking, and scrap costs} = X - 0.005 \times 1000 \text{ instruments/year} \times C_{av}/\text{instrument}$$

*where* $X$ = percentage of defective instruments produced
$C_{av}$ = average cost of reworking a defective instrument to enable it to pass inspection

Then the total cost of the inspection program is the sum of inspection, penalty, and reworking costs. Setting the cost of reworking at $C_{av}$ = \$50 per instrument, we have

$$\begin{aligned} \text{Total cost} &= \$15{,}640 + \$1500 + (X - 0.005) \times 1000 \times \$50 \\ &= \$17{,}140 + (X - 0.005) \times \$50{,}000 \end{aligned}$$

By equating the cost of the inspection program to the penalty cost of no inspections, we get

$$\begin{aligned} \text{Cost of vigilance} &= \text{cost of error} \\ \$17{,}140 + (X - 0.005) \times \$50{,}000 &= X \times 1000 \times \$300 \\ X &= 6.8\% \end{aligned}$$

$X$ represents the percentage of defective units produced at which the two costs are equal. If the production capabilities can limit the percentage of defective instruments to 6.8 percent or less, the investment in this inspection program is unwarranted from strictly a cost viewpoint. However, reputation and other intangible considerations should influence the decision. It would also be wise to explore other inspection programs which are not so precise in detecting errors, but are less expensive investments.

## Insurance

The dichotomy of risk costs is again apparent in the use of an insurance program which transfers the burden of risk. Paying a premium to an insurer is an investment made to avoid the consequences of a specific disaster. The amount of the premium should reflect the potential magnitude of the disaster in proportion to the probability of its occurrence (plus administrative costs and profit). For a disaster which could cause damages of \$100,000 per year with a probability of 0.001, it would be reasonable to expect an annual premium cost of at least \$100,000 × 0.001 = \$100.

The reason insurance programs are so prevalent is that they spread the cost of protection over a period of time and transfer the risk of a financial calamity to a group better prepared to meet the payment. Just because the chance of a disaster is one in a thousand, it cannot be stated in exactly what year it will occur or even that it could not occur 2 years in a row. Premiums prorate the disaster shock to the pooled assets of the absorbing insurance company.

A program of self-insurance is followed by some companies for specific risks. Such a policy presumes sufficient resources to cover potential damages and is usually limited to minor risks. Accident liability is a popular area for self-insurance.

Investments may be made to reduce premiums for a group insurance policy or to reduce the probability of risk for a self-insurance policy. Annual investments in a plant safety program are aimed at reducing the probability of accidents. They should result in less damage payment for the self-insured or lower premiums for group insurance.

A feasible amount to spend for a risk-reduction investment can be estimated from insurance rates. Companies which sell insurance normally have a great deal of experience to draw on in setting rates. These rates may be used as estimates of risk for related situations as described in the following example.

The owners are considering an automatic sprinkling system for a warehouse. The company insuring the warehouse will reduce the annual \$1200 fire-insurance premium by one-third if the system is installed. The owners estimate that opportunity costs incurred as a result of a serious fire, such as relocation expenses, disruption of deliveries, and loss of reputation, would amount to one-half the value of the warehouse. Using the insurance company's evaluation for the extent of risk reduction, the expected value of annual savings in opportunity costs from the installation would be $\$1200 \times 1/2 \times 1/3 = \$200$.

If annual taxes and maintenance costs for the sprinklers are \$100, the life of the proposed system is 20 years, and capital is worth 15 percent, a feasible price for the sprinkler system could be determined from the following equality:

$$P(A/P, 15, 20) = \text{premium reduction} + \text{opportunity-cost reduction} - \text{annual taxes and maintenance}$$

$$P = \frac{\$1200 \times 1/3 + \$200 - \$100}{0.15976} = \$3130$$

## SUMMARY

At least some risk is inherent in any economic decision. Inputs and outputs for short-term investment alternatives are usually subject to less variability than long-term investment proposals. Risk analysis is appropriate when *significant* outcome variations are likely for different future states and *meaningful* probabilities can be assigned to those states.

Probability principles govern risk analyses. Both objective and subjective probability estimates are utilized in economic evaluations. Future states are usually defined by independent probabilities. The widely used *expected-value* measure of preference is a weighted average of outcomes—the sum of the products of outcomes from independent states multiplied by their associated probabilities of occurrence. ***Investment-risk profiles*** and ***acceptable-investment diagrams*** are graphic aids that assist in the evaluation of expected values. Other criteria such as the ***most probable future*** and ***aspiration level*** may influence the final selection of the preferred alternative.

Conditional probabilities are the basis for a ***discounted-decision-tree*** analysis. A graphic tree format displays the effects of successive decisions occurring at timed intervals during a study period. Computations begin with expected values calculated for the most distant decision and roll back to the present by accepting the preferred alternative at each distant decision as a certain outcome for the next closer decision. A completed backward pass discloses the present worth of each immediate course of action.

Additional information about a decision is used in ***Bayesian analysis*** to revise the probability of a future event. The ***value of perfect information*** for estimating occurrences of future independent states reveals how much could be spent to obtain better forecasts.

Minimum-cost models for short-term operating decisions may include risk considerations. Variations in ***lead times*** and ***demand rates*** affect inventory practices. The most economical ***safety-stock*** size is determined by calculating the expected value of opportunity costs and holding costs for different ordering alternatives. Similarly, the lowest-cost maintenance-replacement tactic can be determined from the expected value of individual and group servicing alternatives.

Investments to avoid risk involve self-protection (outlays to reduce the cause or penalty of risk) and insurance (transfer of risk to another party). There is no way to completely eliminate risk from operations, but risk-conscious investment policies can mitigate the hazards of unfavorable outcomes.

## Discussion of Economy Exercises

**EE 14-1** The present worths of the two proposals are calculated using a 10 percent discount factor for the clip-ons that fit only Owl frames $O$, and 18 percent for general-purpose clip-ons $G$:

$$\begin{aligned}\text{PW}(O) &= -\$100{,}000 + \$43{,}000(P/A,\ 10,\ 4)\\ &= -\$100{,}000 + \$43{,}000(3.1698) = \$36{,}301\end{aligned}$$

$$\begin{aligned}\text{PW}(G) &= -\$100{,}000 + \$50{,}000(P/A,\ 18,\ 4)\\ &= -\$100{,}000 + \$50{,}000(2.6952) = \$34{,}760\end{aligned}$$

The risk-adjusted discount rate switches the preference from general-purpose to Owl-only clip-ons. An engineering economist should question whether the 8 percent addition to the required rate of return fairly represents the perceived riskiness.

**EE 14-2** Based on the analysis on page 516, enough parts have been tested to determine that the probability is 0.96 that the trouble is caused by faulty materials. Since the company planned to take action when it was 95 percent sure of the cause, the five tests provide adequate reason to begin replacement of the defective parts.

| *Cause* | *P(E1)* | *P(failure \| cause)* | *P(5 failures \| cause)* | *P(5 failures)* |
|---|---|---|---|---|
| Material | 0.3 | 0.90 | $(0.9)^5 = 0.59$ | 0.177 |
| Assembly | 0.7 | 0.40 | $(0.4)^5 = 0.01$ | 0.007 |
| | | | | 0.184 |

$$P(\text{poor material} \mid 5 \text{ failures}) = \frac{0.177}{0.184} = 0.96$$

**EE 14-3** Annual costs for each future state of the redesigned process line are

$$\begin{aligned}\text{AC(at } P = 0.5) &= \$150{,}000(A/P,\ 12,\ 3) + \$210{,}000 \\ &= \$150{,}000(0.41635) + \$210{,}000 = \$272{,}453\end{aligned}$$

$$\begin{aligned}\text{AC(at } P = 0.25) &= \$150{,}000(A/P,\ 12,\ 3) + \$210{,}000 + \$20{,}000(A/G,\ 12,\ 3) \\ &= \$272{,}453 + \$20{,}000(0.9245) = \$290{,}943\end{aligned}$$

$$\begin{aligned}\text{AC(at } P = 0.25) &= \$150{,}000(A/P,\ 12,\ 3) + \$210{,}000 + \$50{,}000(A/G,\ 12,\ 3) \\ &= \$272{,}453 + \$75{,}000(0.9245) = \$341{,}790\end{aligned}$$

These costs are inserted in a payoff table, as shown below, where the calculated expected values indicate that the redesign should be attempted even when there is one chance in four of increasing the process-line costs.

| | FUTURE ANNUAL COSTS | | | |
|---|---|---|---|---|
| *Alternative* | *P(0.5)* | *P(0.25)* | *P(0.25)* | *Expected Value* |
| No change | \$310,000 | \$310,000 | \$310,000 | \$310,000 |
| Redesign | 272,453 | 290,943 | 341,790 | 284,410 |

**EE 14-4** If $I$ and $S$ represent investment and savings levels, respectively, then the pattern of possible future outcomes is shown in the table. There is a 20 percent chance the proposal will lose money, but the expected value is \$2000.

| *Possible Futures* | *Net After-Tax Present Worth* | *Joint Probability* | *Weighted Outcome* |
|---|---|---|---|
| $S(P = 0.2)I(P = 0.6)$ | \$2000(5) − \$3000 = \$7000 | (0.2)(0.6) = 0.12 | \$ 840 |
| $S(P = 0.3)I(P = 0.6)$ | 1200(5) − 3000 = 3000 | (0.3)(0.6) = 0.18 | 540 |
| $S(P = 0.5)I(P = 0.6)$ | 800(5) − 3000 = 1000 | (0.5)(0.6) = 0.30 | 300 |
| $S(P = 0.2)I(P = 0.4)$ | 2000(5) − 5000 = 5000 | (0.2)(0.4) = 0.08 | 400 |
| $S(P = 0.3)I(P = 0.4)$ | 1200(5) − 5000 = 1000 | (0.3)(0.4) = 0.12 | 120 |
| $S(P = 0.5)I(P = 0.4)$ | 800(5) − 5000 = −1000 | (0.5)(0.4) = 0.20 | −200 |
| | | 1.00 | \$2000 |

**EE 14-5** The payoff table indicates that holding the concert in the pasture has the greater expected value.

| | *Rain* (*P* = 0.1) | *No Rain* (*P* = 0.9) | *Expected Value* |
|---|---|---|---|
| *Auditorium* | $2400 | $3000 | $2940 |
| *Pasture* | −2700 | 9000 | 7830 ←*Preferred* |

The shaded outcomes in the payoff table show the highest possible returns for each future state. If the entrepreneur ***knew*** that it was going to rain on the Fourth of July, the concert would surely be scheduled in the auditorium to reap a $2400 profit instead of a loss. Equivalently, assurance of dry weather would lead to a profit of $9000 from the concert held in the pasture. Since it is dry 9 out of 10 July Fourths, the expected value of perfect information is

$$\$2400(0.10) + \$9000(0.9) = \$8340$$

and the amount that could be paid for it is $8340 − $7830 = $510, the difference between the returns from a commitment to hold the concert in the pasture and the option to schedule it at the site best suited to the weather.

## PROBLEMS

**14.1** Given a fair coin,

**14.1a** What is the probability of at least one head in three tosses?

*(0.875)*

**14.1b** What is the probability of flipping a tail, a head, and a head in that order?

*(0.125)*

**14.1c** What is the probability of flipping at most two heads in three tosses?

*(0.875)*

**14.2** In the example used in the chapter, the probability of drawing a queen was 1/13, and the probability of a heart was 1/4. The probability of drawing a queen *or* a heart is called a ***union,*** because the two sets overlap; they are not mutually exclusive. Therefore, when the probabilities of the two sets are added the portion that overlaps must be subtracted. By this means we arrive at the formula for the probability of drawing a queen or a heart,

$$P(Q + H) = P(Q) + P(H) - P(QH) = 1/13 + 1/4 - 1/52 = 4/13$$

**14.2a** Show the union by a sketch of the overlapping heart and queen sets.

$[P(H + Q) = 4/13]$

**14.2b** Let three overlapping sets be $A$, $B$, and $C$. Sketch the sets, and determine a formula for $P(A + B + C)$.

**14.3a** Given the probability that 1 out of every 1000 tire valves is defective, what is the probability that two of the four tires on a car will have defective valves?

$(5.98 \times 10^{-6})$

**14.3b** In super-safety tires which have a separate inner tire, the valves for the inner tire are defective 1 time in 500. What is the probability that a tire will have both valves defective? (See Problem 14.3*a*.)

*($2 \times 10^{-6}$)*

**14.4** When a machine is properly adjusted, it will produce an acceptable product 9 times in 10. When it is out of adjustment, the probability of an acceptable product is 0.4. The probability of the machine's being adjusted properly is 0.95.

**14.4a** If the first part tested after an adjustment is not acceptable, what is the probability that the machine was correctly adjusted?

*(0.76)*

**14.4b** If the first two parts were acceptable, what is the probability of a correctly adjusted machine?

*(0.99)*

**14.5** A shipment of parts contains 20 items, 8 of which are defective. Two of the items are randomly selected from the shipment and inspected.

**14.5a** What is the probability that the first one selected is good?
**14.5b** What is the probability that both are good?
**14.5c** What is the probability that one is good and one is bad?

**14.6** A data processing firm mails 1000 bimonthly newsletters to present or potential clients. In one issue the firm announced a new service and asked that interested parties write for more information. The firm believed that one of every two replies would come from one of its present customers. From past experience, it is estimated that the probability of a reply from noncustomers is 0.40. On the assumption that the mailing list includes the names of 300 present customers, how many replies can be expected?

*(560 replies)*

**14.7** New types of concrete mixes are tested in a laboratory by batching four test cylinders. The probability that a trial batch will yield the specified strength is 0.90 if the mix is properly prepared and tested. Occasionally, about once every 20 times, the trial batch will be improperly handled or the ingredients inaccurately measured. The probability that a poorly prepared mix will yield the specified strength is 0.20. If only one cylinder in a trial batch of four meets the specified strength, what is the probability that the mix was correctly prepared?

*(0.145)*

**14.8** A Louisiana oil operator owns a $5-million oil rig. It costs $75,000 to pull the drills to safety and batten down the rig in anticipation of a bad storm. An uninsured average loss of $400,000 results from a bad storm when no precautionary measures are taken. A weather-forecasting service provides an assessment of the probability of a severe storm. Four out of five times that a severe storm is predicted with a probability of 1.0, it does occur. Only 1 severe storm in 100 arrives unpredicted. Should the rig owner pull the drills when the forecasting service predicts a storm at 1.0?

**14.9** A manufacturer has three inspection plans: *A*, *B*, and *C*. The chance that a faulty unit will pass undetected is 2 percent in plan *A*, 5 percent in plan *B*, and 10 percent in plan *C*. The respective inspection costs per unit are $0.35, $0.10, and $0.01. A defective unit going undetected causes opportunity costs of $3.00. The manufacturing process averages 12 percent defectives.

**14.9a** Which inspection plan should be used?
**14.9b** Compare the plan selected in Problem 14.9*a* with a policy of no inspection.

**14.10** Ninety percent of the fruit received at a cannery comes from local growers. The fruit from local sources averages 80 percent grade 1 and 20 percent grade 2. The fruit obtained from other sources averages 40 percent grade 1 and 60 percent grade 2. The markings on a shipment of bins full of fruit were lost. One bin was sampled, and from five pieces of fruit inspected, four were of grade 1. What is the probability that the bin came from a local grower?

*(0.98)*

**14.11** How would you explain the reason for using the expected-value measure to evaluate a unique engineering design project that is unlikely ever to be repeated again in exactly the same form?

**14.12** A logging company must decide the most advantageous duration for a paving project. The beginning date of the project has been definitely set. A critical-path analysis has shown that three project durations are feasible. If the paving is completed in 4 months, the basic project cost will be $80,000. A 5-month duration will allow construction savings of $20,000, and it will cost an extra $40,000 over the basic cost to crash the project to 3 months. However, transportation expenses can be cut by $10,000 over the 4-month schedule if the paving is done in 3 months, and an extra transportation expense of $15,000 will be incurred for an extension of the paving time to 5 months.

Since the project must be completed during a period of expected foul weather, the extra expense due to possible weather conditions should also be considered. Weather records indicate that the probabilities for mild rain, heavy rain, and wind and rain are, respectively, 0.3, 0.5, and 0.2. The costs that must be included for these conditions are given in the following table:

| *Weather conditions* | *3 Months* | *4 Months* | *5 Months* |
|---|---|---|---|
| Mild rain | $10,000 | $15,000 | $ 5,000 |
| Heavy rain | 10,000 | 40,000 | 60,000 |
| Wind and rain | 15,000 | 55,000 | 65,000 |

Which duration has the lowest expected total cost?

**14.13** There are several methods available to discover defective welds. A company has investigated two methods. Method 1 costs $0.50 per inspection and detects defects 80 percent of the time. Method 2 costs $2.00 per test, but it always detects a defective weld. When a defective weld goes undetected, the estimated cost to the company is $30.00 for replacement and other incidental costs. The probability of a defective weld is 0.05. Using the expected-value criterion, determine whether method 1 or 2 should be used, or whether the company is better off with no inspection procedure.

*(Method 1: $0.80)*

**14.14** An investment is being considered that requires $1 million and commits the money for 10 years. During that period it is equally likely that the annual returns from the investment will be $100,000, $150,000, and $200,000. The probability is 0.75 that the salvage value will be $300,000, but there is one chance in four that it will be zero. A minimum rate of return of 10 percent is expected.

**14.14a** Construct an investment-risk profile for the proposal on a chart in which the horizontal axis registers the net PW and the vertical axis is a probability scale ranging from 0 to 1. Draw the curve to show the probability of returns equal to or less than the scaled PWs.

**14.14b** How could the investment-risk profile contribute to the economic evaluation of the million-dollar investment?

**14.15** Three mutually exclusive alternatives are described by the data below, which show the probabilities of earning four rates of return.

| | RATE OF RETURN | | | |
|---|---|---|---|---|
| *Alternative* | *−5%* | *0* | *10%* | *20%* |
| *A* | 0.3 | 0.1 | 0.2 | 0.4 |
| *B* | 0.0 | 0.3 | 0.5 | 0.2 |
| *C* | 0.15 | 0.15 | 0.4 | 0.3 |

**14.15a** Which alternative would be selected using the most-probable-future criterion? Why?

*(A)*

**14.15b** Which alternative would be selected using the expected-value criterion? What is the EV?

*(C, 9.25%)*

**14.15c** Which alternative would be selected if the decision maker had an aspiration level of 10 percent? Why?

*(C)*

**14.15d** Construct an acceptable-investment diagram with a loss coefficient of −5 percent at $P = 0.7$ and a payoff coefficient of 10 percent at $P = 0.3$. Which alternative would be eliminated by the AID?

*(B)*

**14.16** A payoff table (in thousands of dollars) is given below for three investments of equal size and duration.

| *Alternative* | *Boom* *(P = 0.3)* | *So-so* *(P = 0.5)* | *Bust* *(P = 0.2)* |
|---|---|---|---|
| A | 1000 | 200 | −500 |
| B | 300 | 400 | 0 |
| C | 400 | 600 | −300 |

**14.16a** Which alternative would you select? Why?
**14.16b** How much could be paid for perfect information?

**14.17** A cost-saving modification to an existing process is being evaluated. The savings will affect products I and II. The rate of return earned by the investment in the modification depends on how much the process is utilized, which depends on the market conditions for future sales of the two products. Three future states have been identified: ***good***, with RR = 20 percent; ***average***, with RR = 10 percent; and ***bad***, with RR = −5 percent. Since there are two products involved, the maximum possible rate of return under good conditions is

20 percent + 20 percent = 40 percent. The probability of each future for both products is given below.

| | FUTURE | | |
|---|---|---|---|
| *Product* | *Good* | *Average* | *Bad* |
| I | 0.20 | 0.70 | 0.10 |
| II | 0.40 | 0.30 | 0.30 |

**14.17a** What is the expected value of the cost-saving modification?

*(20%)*

**14.17b** Draw an investment profile for the proposal on an acceptable investment diagram. The ordinate is the probability the investment will exceed the percentage return, and the horizontal axis is the rate of return. The investment criteria are to limit losses to 5 chances in 100 of a −10 percent return and to be 90 percent sure that the rate of return is at least 20 percent. Should the proposal be accepted? Why?

**14.18** For the decision situation depicted in the decision tree below, where capital is valued at 8 percent:

**14.18a** What is the expected profit at decision point 2?

[*EV(A1)* = *$4438*]

**14.18b** Which alternative should be selected at decision point 1, and what is its expected value?

[*EV(B)* = *$1835*]

**14.18c** How much could be paid for perfect information?

*($0)*

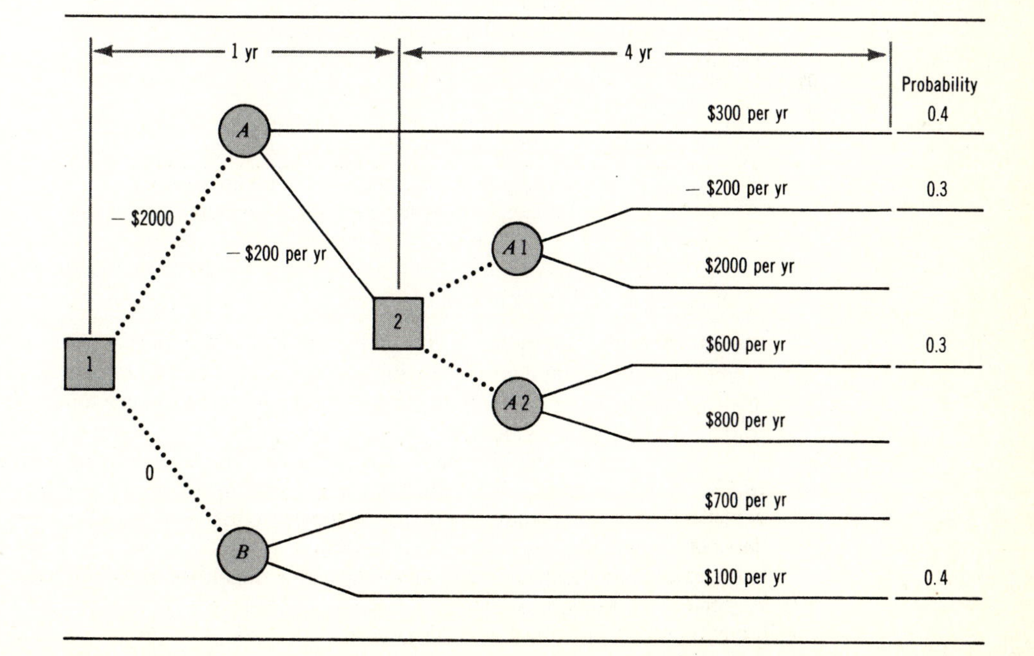

**14.19** Plans are being developed for the construction of a new school. The city engineer now feels that the probability of growth ($G$) in the school area is 0.6 as opposed to a stable ($S$) census probability of 0.4. Two alternative designs are being considered. One is to build a medium-size school ($M$) with provisions for adding ($A$) onto it if needed, and the other is to construct a large ($L$) facility with the possibility of leasing part of the space to city and county departments if the classroom space is not required.

The study period for the school question is 15 years. It is believed that after 5 years the growth pattern will be evident. If the population is stable for 5 years, there is still a 50 percent chance it will remain stable for the rest of the 15-year period; there is no chance the population will decrease. Given growth during the first 5 years, the probability of continued growth is 0.8, with a corresponding probability of 0.2 for a stable census during the next 10 years.

Additional data are as follows:

| *Estimated construction costs* | |
|---|---|
| Medium-size school to accommodate stable census | $1,000,000 |
| Addition to medium-size school to accommodate growth | 800,000 |
| Large-size school to accommodate a growing census | 1,500,000 |
| Remodeling of large school to provide rental space if all the classrooms are not needed | 100,000 |
| *Annual costs and income* | |
| Maintenance: Medium-size school | $40,000/year |
| Large or enlarged school | $70,000/year |
| Revenue expected from rental space in a large school if all the capacity is not needed for classrooms | $20,000/year |
| Busing and overcrowding costs if the school is not large enough to accommodate the population after the first 5 years | $100,000/year |

The outcomes for the various options are labeled on the partially completed discounted decision tree given on page 523. Nodes are identified by the symbols above, and $N$ means no change. The interest rate for the study is 7 percent.

**14.19a** Construct payoff tables for each decision point 2, and determine the preferred alternatives.

**14.19b** Develop a payoff table for decision point 1, and determine the preferred course of action for school construction.

**14.20** A small foundry has had trouble with its old arc furnace. This furnace has been completely depreciated for accounting purposes, but it could currently be sold for $6000. The immediate alternatives are to overhaul and modify the old machine or to buy a current model which has many desirable features that could not be incorporated in the modification of the old machine. The plans are complicated by the general opinion in the industry that a breakthrough could be made in furnace technology in the near future.

The best estimate the foundry owners can make is that there is a 40 percent chance that a radically improved furnace will be available in about 3 years. If it is developed, the probability that it will make present models noncompetitive is 0.90, and that it will be only a minor improvement is 0.10.

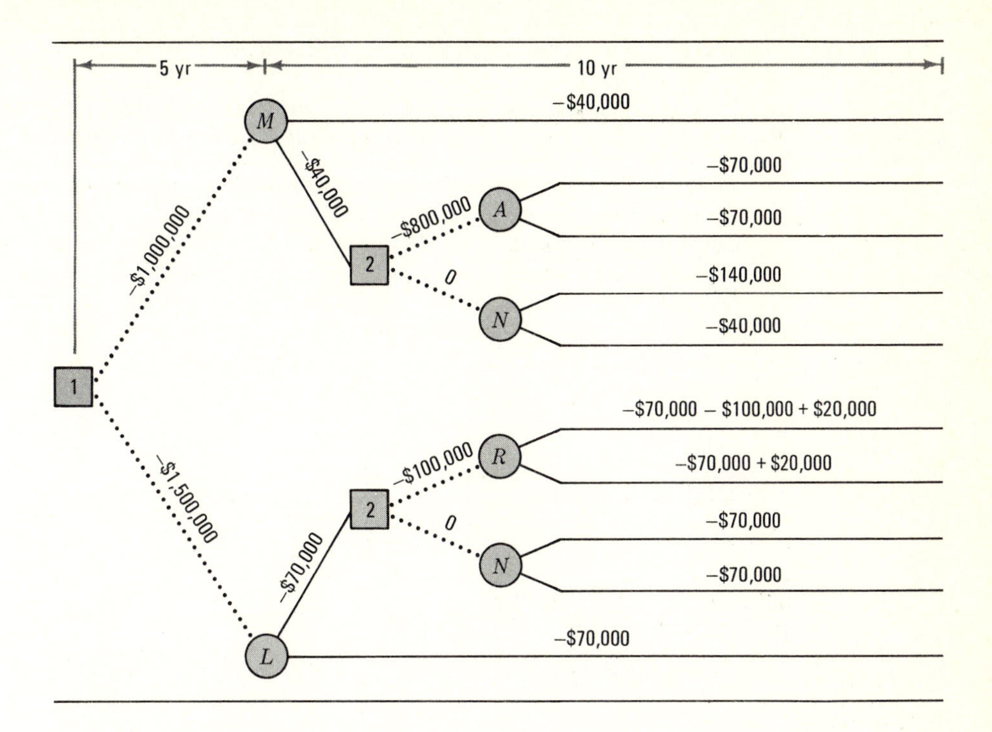


The cost of modifying the old machine is $8000, and the cost of a new, current-model machine is $25,000. Expected savings and resale values are given in the accompanying table, based on the following three possible future states:

$S1$ = no technological breakthrough
$S2$ = furnace developed which provides significant savings
$S3$ = furnace developed which provides minor savings

| | BUY NEW | | MODIFY | |
|---|---|---|---|---|
| *Possible Outcomes* | *Savings per Year* | *Resale at 8 Years* | *Savings per Year* | *Resale at 8 Years* |
| $S1$ | $6000 | $8000 | $2000 | $4000 |
| $S2$ | 2000 | 2000 | 1000 | 2000 |
| $S3$ | 3000 | 4000 | 1000 | 3000 |

The table is based on a study period and life of 8 years for both furnaces. The sharp decreases in savings and salvage in states 2 and 3 occur because the development of a radically different or even improved furnace would probably cut into the foundry's demand and its general competitive position.

Another alternative exists for the foundry. If the new type of furnace is developed in 3 years, the modified furnace could be sold at that time for $9000 and the new one purchased for an estimated $45,000. This new furnace would provide a saving of $13,000 per year with a probability of 0.90, and $8000 per year with a probability of 0.10. It will be worth $20,000 or $15,000 after 5 years with respective probabilities of 0.90 and 0.10.

If a new machine is purchased now, it will be used for 8 years regardless of new developments.

Using a discounted decision tree, determine whether the old furnace should be modified or a new, current model should be purchased. Interest is 10 percent.

**14.21** A firm has produced a new product which was unusually successful, and in order to meet the unexpectedly high demand it will be necessary to add additional production facilities. The troubling question is whether the high demand will continue, increase, or decrease. Plan *A* provides a permanent capacity increase and will be more profitable if the demand continues to increase. Plan *B* is a stopgap measure which can be converted to permanent capacity by a supplementary investment *B'* after 3 years, when the demand pattern is better known. For a steady or lower demand, plan *B* is more profitable than plan *A*. The estimated future outcomes for an 8-year study period are as indicated in the table.

| *First 3 Years* | *Last 5 Years* | *Probability* |
|---|---|---|
| High | High | 0.40 |
| High | Low | 0.20 |
| Low | High | 0.30 |
| Low | Low | 0.10 |

Initial cost estimates are

| | |
|---|---|
| Plan *A* | $100,000 |
| Plan *B* | 70,000 |
| Plan *B'* | 45,000 |

*B'*, the supplementary investment in plan *B*, will take place after the demand is known for the first 3 years.

Annual income estimates are as follows:

Plan *A* with a high demand will yield a cash flow of $40,000 per year.

Plan *A* with a low demand will yield $5000 per year.

Plan *B* with a high demand will yield $30,000 per year in the first 3 years and $20,000 annually in the last 5 years.

Plan *B* combined with *B'* will yield $40,000 per year with high demand.

Plan *B* combined with *B'* will yield $10,000 per year with low demand.

Plan *B* with a low demand will yield $30,000 per year.

With interest at an annual rate of 8 percent, determine which plan or combination of plans appears most attractive.

**14.22** The owners of the novelty company described in the chapter disagree as to the solution of their storage problem. There is a minority feeling that more research should be given to the question. One member of the minority group is anxious to do such a study and estimates that it would cost about $5000 and take approximately 6 months to complete. The company could limp through this period with existing inventory facilities. Although the other owners have faith in the person who would make the study, they feel that she is overconservative. Because of this attitude they estimate (1) that if company activity in the next 6 months is very strong, the probability that the study will indicate a continuation of increasing growth will be 0.60, and (2) that if the next 6 months' activity is relatively constant, the chance of a forecast for increasing growth will be 0.10. On the basis of these estimates an engineer in the company is asked to calculate:

**14.22a** The probability that the study will indicate increasing growth

[$P(I_s) = 0.45$]

**14.22b** If the study indicates increasing growth, the probability of continued increase and the chance of leveling off

[$P(I_a|I_s) = 0.933$; $P(C_a|I_s) = 0.067$]

**14.22c** If the study indicates no increase, the probability of increase and the chance of level activity

[$P(I_a|C_s) = 0.509$; $P(C_a|C_s) = 0.491$]

Carry out the engineer's assignment, and show the results in a decision-tree format without costs. (*Hint:* You must use the original estimates disclosed in the chapter example in conjunction with those in the problem. The decision tree has three primary alternatives.)

**14.23** Engineering Service, a large consulting firm, is considering the acquisition of a computer to lower its project accounting and control costs. It can rent a large computer for $180,000 per year on a noncancellable but renewable 3-year lease. The other alternative is to buy a smaller computer at a cost of $200,000.

There is a probability of 0.70 for a high service demand in the next 3 years. If the demand is large the first 3 years, the probability that it will continue large is 0.60. Expected annual net savings during a period of high demand are $300,000 from a large computer and $170,000 from a smaller model. If demand is low, the large computer will permit a saving of $140,000 per year, and the smaller computer's net annual saving will be $110,000. The probability of a continuously low demand for the 6-year study period is 0.27.

After 3 years the lease on the large computer could be terminated and a smaller, used computer could be purchased for $120,000. Either purchased computer would have a negligible salvage value at the end of the 6-year period. Also, after 3 years the smaller computer could be sold for $120,000, and a large one could be leased for the remaining 3 years for $200,000 per year.

Assume that all receipts and disbursements are end-of-year payments and the acceptable interest rate is 8 percent before taxes. Using a decision tree, determine the most attractive alternative for Engineering Services.

| *Delivery Lead Time, Weeks* | *Probability* |
|---|---|
| 2 | 0.1 |
| 3 | 0.2 |
| 4 | 0.3 |
| 5 | 0.3 |
| 6 | 0.1 |

Data for Problem 14.24

**14.24** An automobile dealer has kept a record of the lead time for delivery of new cars from the factory. He estimates that the opportunity cost of lost sales averages $60 for each car not on display when it might have been. His sales are relatively uniform at seven per week. Ordering costs average $20 per order, and holding costs are $120 per year for each car in stock. What ordering policy should the dealer use?

| *Demand during Lead Time* | *Probability* |
|---|---|
| 60 | 0.05 |
| 65 | 0.05 |
| 70 | 0.15 |
| 75 | 0.40 |
| 80 | 0.20 |
| 85 | 0.10 |
| 90 | 0.05 |

Data for Problem 14.25

**14.25** Compute the reorder point for the following situation:

| | |
|---|---|
| Average lead time | 5 days |
| Holding costs/unit/year | $2 |
| Optimum order interval | 2 months |
| Cost of stockouts/order interval | $20 |
| Average demand rate | 15 units/day |

| *Demand* | *Probability* |
|---|---|
| 6 | 0.10 |
| 5 | 0.20 |
| 4 | 0.50 |
| 3 | 0.20 |

**14.26** A small plant uses wood chips as a raw material for one of its products. Chips are delivered in units of railway cars. The average demand is four cars per month, but the demand varies according to the accompanying table.

Since special unloading equipment is rented when a delivery arrives, it is desirable to have large orders. Space is available to store any size of order, but stored inventory is subject to weather damage. The costs associated with wood-chip inventory are

| | |
|---|---|
| Holding costs | $60/year for each unit |
| Order costs | $300/order (includes rental of unloading equipment) |
| Opportunity costs | $80/unit (includes altering production schedules) |
| Lead time | 1 month (4 weeks) |

What safety stock should be carried?

**14.27** The Able Company places an order for 300 gallons of paint (the EOQ when opportunity costs are not considered) 25 times each year. The average lead time is 7 calendar days. The usage rate is 20 gallons per day, 60 percent of the time. The probabilities of 15-, 25-, and 30-gallon usage are, respectively, 0.10, 0.20, and 0.10. If the order costs are $30 per order and the cost of being out of stock is $20 per gallon, what safety stock should be carried? Develop an expected-cost table, and determine the total cost for each safety-stock alternative.

| *Lead Time, Weeks* | *Probability* |
|---|---|
| 1 | 0.1 |
| 2 | 0.1 |
| 3 | 0.3 |
| 4 | 0.5 |

**14.28** Because of a shortage of wood chips, the lead time for the small plant's orders (Problem 14.26) can no longer be considered fixed. The estimated future lead times are described by the distribution shown in the accompanying table. The costs and demand rate from Problem 14.26 are still applicable. Determine the order size and the inventory level at which this size order should be placed. The problem can be solved by simulation (see Extension 14.1, which introduces simulation methods) or by direct calculations considering all possible futures.

*(Q = 23 at a stock level of 4)*

| *Tube Failure during Week* | *Probability of Failure* |
|---|---|
| 1 | 0.3 |
| 2 | 0.1 |
| 3 | 0.1 |
| 4 | 0.2 |
| 5 | 0.3 |

**14.29** It has been suggested to a data-processing firm that it adopt a policy of periodically replacing all the tubes in certain pieces of equipment. A given type of tube is known to have the mortality distribution shown in the table. There are approximately 1000 tubes of this type in all the combined equipment. The cost of replacing the tubes on an individual basis is estimated to be $1.00 per tube, and the cost of a group-replacement policy averages $0.30 per tube. Compare the costs of preventive versus remedial replacement.

**14.30** A vending machine operator has machines in 40 locations. There is an equal probability each day during a 10-day period that the machines in one location will be emptied. After 10 days all the machines will be empty. The cost to individually replenish the machines at one location (travel and working time) is $18. The loss in profit from idle machines in one location is $10 per day. Replenishments may be made individually as requested when the machines in one location are empty, or they may be made all at one time with a total cost to service the 40 locations of $250. What is the lowest-cost replenishment policy?

*(Replenish every 4 days with group replenishment cost = $122.50)*

**14.31** A resident engineer on a construction project has been troubled with parts received from a supplier. Eight percent of the parts have been defective. The scrap and reworking costs for each item average $6. At least 2000 parts will be used over the next 2 years. Two methods have been suggested to reduce the risk of defective parts:

**1** Visual inspection by part-time labor would reduce the risk by half. The labor cost would be $250 per year.

**2** Only 0.5 percent defective parts would be accepted if gages were purchased for $90 and a worker were trained to use them. The training cost would be $30, and an annual wage for the inspection time would total $350.

Which of the alternatives should be adopted if a 10 percent rate of return is considered acceptable?

| *Number of Machines Out of Order at One Time* | *Probability* |
|---|---|
| 0 | 0.5 |
| 1 | 0.2 |
| 2 | 0.1 |
| 3 | 0.1 |
| 4 | 0.1 |

**14.32** An assembly line has 30 identical machines. The pattern of breakdowns is shown in the accompanying table. Breakdowns can usually be fixed in a short period, but the disruption of the production line creates considerable expense. One way to eliminate this disruption is to provide standby machines. The daily cost of keeping a standby machine is estimated to be $12. The cost of an out-of-order machine is $150 per day. How many standby machines should be provided?

*(4 machines)*

**14.33** A private golf course lies in a hollow beside a river. Every other year the river rises to a level that prohibits play. The loss of revenue and damage to the course average $8000 each time the river rises to this level. A new clubhouse has been constructed on higher ground, but flood data indicate a 1/20 chance the river will reach a level which will inundate the facility. The owners estimate that a flood of this magnitude would cause $40,000 damage to the clubhouse.

**14.33a** How much could the course owners afford to pay for a levee that would protect them from the biennial damage? The required rate of return is 8 percent, and the economic life of the levee is expected to be 30 years.

**14.33b** What would be a reasonable annual premium for a full-coverage flood-insurance policy on the clubhouse?

**14.33c** Compare the premium in Problem 14.33*b* with a self-insurance policy the owners could follow of laying aside 1/20 × $40,000 = $2000 per year in anticipation of a major flood.

## EXTENSIONS

***14.1 Simulation*** To simulate, in a general sense, means to feign or to assume the appearance of something without being the real thing. In an engineering economics sense, simulation is used to feign a real system in order to observe and learn from the behavior of the replica. This departure from reality has several advantages over observing the real system. It is usually easier, not so disrupting, far less expensive, and may be more illuminating by confining attention to characteristics of particular interest.

The use of simulation techniques is increasing rapidly, owing to the wider availability of computers along with the greater sophistication of today's analysts. Computer simulation is an effective way to treat complex economic relationships without suffering the penalties of real trial-and-error experiences. Some problems do not yield conveniently to ready-made solution methods and models, as in inventory problems in which both lead times and demand rates are variable. In such cases, when no brief, handy models are available, problems can be "run" to see the numerical effect of different alternatives rather than solved by analysis. However, a run does not guarantee an optimum solution for the given data as does an optimization model.

*Monte Carlo* is the colorful name given to a simulation technique by which random numbers are generated to select events from a probability distribution of occurrences. The name is derived from possible random-number generators: a flipped coin, a tossed die, a cut of a deck of cards, or even a roulette wheel. However, the most-used generator is a random-number table, as displayed in Figure 14.23. The groups of numbers follow no pattern or special order; they are randomly distributed. The main concern a user should have is to avoid imposing a pattern by repeatedly using the same set in a consistent order. The figures can be read in any manner desired—by rows or columns, diagonally, up or down, etc.

| | | | | | | |
|---|---|---|---|---|---|---|
| 32867 | 53017 | 22661 | 39610 | 03796 | 43693 | 18752 |
| 43111 | 28325 | 82319 | 65589 | 66048 | 04944 | 61691 |
| 38947 | 60207 | 70667 | 39843 | 60607 | 63948 | 49197 |
| 71684 | 74859 | 76501 | 93456 | 95714 | 87291 | 19436 |
| 15606 | 13543 | 09621 | 68301 | 69817 | 39143 | 64893 |
| 82244 | 67549 | 76491 | 09761 | 74484 | 91307 | 64222 |
| 55847 | 56155 | 42878 | 23708 | 97999 | 40131 | 52360 |
| 94095 | 95970 | 07826 | 25991 | 37584 | 56966 | 68623 |
| 11751 | 69469 | 25521 | 44097 | 07511 | 88976 | 30122 |
| 69902 | 08995 | 27821 | 11758 | 64989 | 61902 | 32131 |

**FIGURE 14.23** Random numbers.

For an example of simulation,* the pie charts in Figure 14.24 may be assumed to represent the relative frequency of lead times and usage rates for a commodity. The relative size of the slices in the "pies" corresponds to the chance occurrence of each increment of lead time or demand. Thus, possible lead times are 8, 9, 10, and 11 days, occurring with respective relative frequencies of 0.20, 0.20, 0.30, and 0.30.

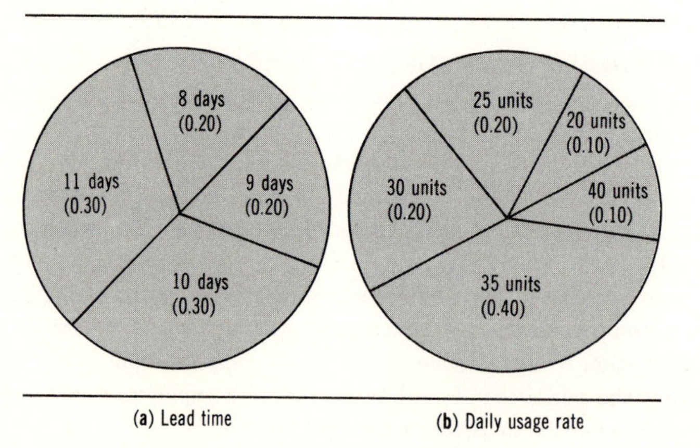


**FIGURE 14.24** Relative frequency of lead times and usage rates.

The simulation of stock movement is conducted by randomly selecting both a lead

*See Extension 14.2 for another example of a simulation application.

time and a usage rate. This could be done by mounting spinners on the two pies. A spin on the lead-time pie might show 9 days, and the associated spin on the usage pie could point to 30 units. The two spins are combined to indicate one level of total demand that could occur after an order has been placed.

A random-number table would commonly be used in place of spinners. In a group of 10 numbers, any digit from 0 to 9 is equally likely to occur. By letting each digit be equal to 10 percent, we can assign digits to each increment of lead time and usage according to its likelihood of occurrence. An arbitrary assignment might be

| LEAD TIME | | USAGE RATE PER DAY | |
|---|---|---|---|
| *Days* | *Digits* | *Units* | *Digits* |
| 8 | 0 and 1 | 20 | 0 |
| 9 | 2 and 3 | 25 | 1 and 2 |
| 10 | 4, 5, and 6 | 30 | 3 and 4 |
| 11 | 7, 8, and 9 | 35 | 5, 6, 7, and 8 |
| | | 40 | 9 |

Then each pair of numbers in the random-number table could represent a lead-time duration and the number of units used per day during that duration. Using the first two numbers in the first column of Figure 14.23, the inventory pattern would take the shape shown in Table 14.10. The reorder level is based on an average lead time of 10 days and an average usage rate of 30 units per day. A typical order is placed when the stock on hand falls to $10 \times 30 = 300$ units.

If the simulation procedure started in Table 14.10 is continued, a distribution can be determined for expected stock levels at the time of delivery. From this distribution an inventory policy can be developed which establishes a minimum-cost balance between holding and opportunity costs. The computations would be similar to those followed for a variable-lead-time inventory policy. A more direct tactic would be to set a tolerable limit for stockouts per year and hold a safety stock which conforms to this limit.

| *Random Number* | *Lead Time (1)* | *Usage Rate (2)* | *Demand during Lead Time [(1) × (2)] (3)* | *Average Reorder Level (4)* | *Stock on Hand when Order Arrives [(4) − (3)] (5)* |
|---|---|---|---|---|---|
| 32 | 9 | 25 | 225 | 300 | +75 |
| 43 | 10 | 30 | 300 | 300 | 0 |
| 38 | 9 | 35 | 315 | 300 | −15 |
| 71 | 11 | 25 | 275 | 300 | +25 |
| 15 | 8 | 35 | 280 | 300 | +20 |

**TABLE 14.10** Simulation of five reordering periods.

It is logical that the greater the number of trials, the more closely the simulation will correspond to the actual inventory pattern. Computers are almost always used for simulation because actual distributions usually cover a far greater range than those used in the example, and a relatively large number of trials is required to give reliable information. A flowchart for a computerized simulation of investment returns is shown in Figure 14.25.

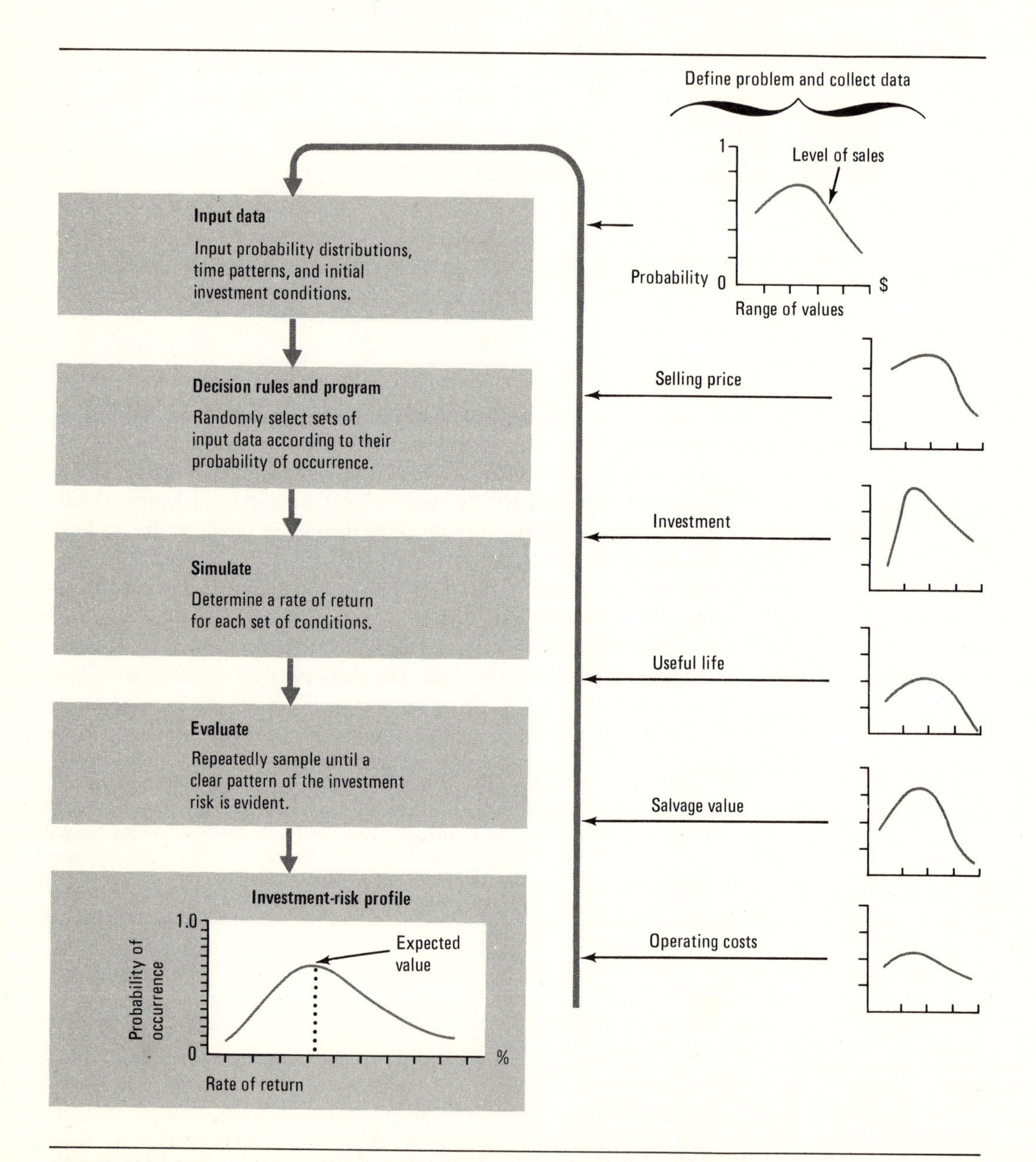


**FIGURE 14.25** Flowchart for simulation procedures to evaluate an investment subject to risk.

**QUESTIONS**

**14.1a** A machinery supply and service company advertises that orders received by 8:00 A.M. will be delivered that day or the customer will only have to pay half price for the order. The amount of an average order is $124. The number of orders received before the deadline varies according to the following pattern:

| *Orders per day* | 16 | 17 | 18 | 19 | 20 | 21 | 22 | 23 |
|---|---|---|---|---|---|---|---|---|
| *Probability* | 0.05 | 0.10 | 0.10 | 0.20 | 0.25 | 0.15 | 0.10 | 0.05 |

Each order is delivered by van. The fixed cost of a van is $10 for a normal 8-hour working day. Van variable costs depend on the length of time required to deliver an order. The daily average order-time distribution and the associated variable costs are as shown:

| *Average Hours/Order during 1 Day* | *Variable Cost* | *Probability* |
|---|---|---|
| ½ | $ 3.50 | 0.40 |
| 1 | 7.00 | 0.30 |
| 1½ | 10.50 | 0.20 |
| 2 | 14.00 | 0.10 |

How many vans should the company operate?

**14.1b** Simulate a waiting-time situation by using two dice to represent an arrival distribution. The total number of points on a toss of the two dice represents the time between arrivals. Let a single die represent the service times. Each time an arrival occurs, throw a die to determine how long that arrival spends in the service station. [See below for an explanation of waiting-time (queuing theory) situations.]

**1** What is the average arrival rate?
**2** What is the average service time?
**3** What is the average length of the queue?
**4** What is the average waiting time in the queue?
**5** Compare the answers to (3) and (4) above with the answers you get by using a Poisson arrival distribution and an exponential service rate. Comment on your results.
**6** Reverse the distribution for arrival and service rates described in the problem statement. How many service stations are required to limit the average waiting time to that determined in (4)?

***14.2 Queuing Theory*** Whenever waiting lines form in an industrial system there is reason to question the situation. A lineup means congestion. The cost of congestion may be directly observable, as trucks waiting to be unloaded, or the costs may be more subtle, as when a potential customer leaves a line that has formed at a sales counter. But there is also a cost associated with relieving congestion. Providing more services—more loading crews or sales clerks—eliminates congestion at the risk of creating excess service capacity. Surplus capacity merely transfers idle time to the service facility. The objective must be to minimize the sum of congestion and service costs.

Two methods are available to reduce total costs: schedule and control the flow of arrivals into the system and/or provide the correct service capacity. The preferred approach depends on whether the input flow or service facilities (or both) can be altered suitably. Such problems are analyzed according to mathematical relationships collected under the title of *queuing theory.*

In order to utilize queuing theory to evaluate a waiting-time situation, the characteristics of the system must be known. The input to the system is defined by an *arrival rate A* in units per time period. The units may be human or inanimate. They arrive in a pattern described by a probability distribution of time between arrivals. After arriving, they are subject to an order in which they are served, such as "first come, first served." They receive service at a designated *number of servicing stations N*. Each station has the capacity for a certain *number of services per unit time S,* and a distribution of service times.

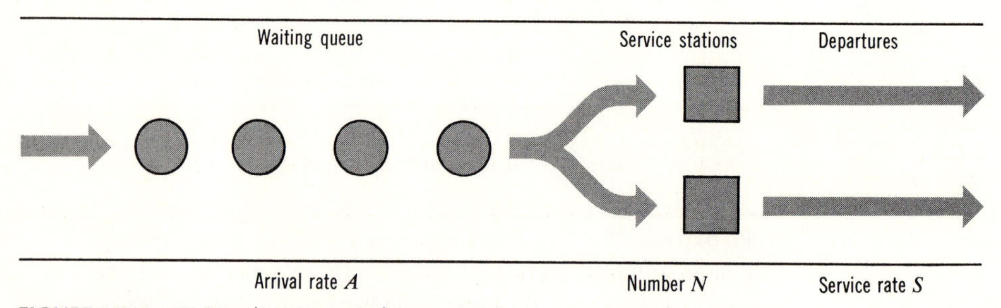


**FIGURE 14.26** Waiting-line (queue) characteristics.

## DISTRIBUTION OF ARRIVALS AND SERVICE TIMES

Constant arrival and service rates are the easiest to evaluate but are rare in practice. An automatic machine feeding arrivals to another automatic machine exemplifies a constant pattern if no breakdowns occur. In such cases the second machine will process the products without delay as long as its service rate $S2$ is equal to or shorter than the arrival rate from the first machine, $A1$. A waiting line forms when $S2$ is greater than $A1$. In this "explosive" case the waiting line increases indefinitely with time.

Random arrival and service rates are much more common than a uniform pattern. With random rates it is possible for new arrivals to be forced to wait even though the service time is actually smaller than the average time between arrivals. This condition exists because arrivals may be bunched together. In the period between bunches the service facility can reduce or eliminate the queue, but the temporary waiting time may still be a source of significant costs.

## SIMULATED QUEUES

Simulation methods (discussed in Extension 14.1) are appropriate means of evaluating the effect of waiting time. This general approach can be used regardless of the distribution of arrivals or service times. The techniques involved are the same as those described for inventory simulation. In a waiting-time application we have the objective of determining a service policy which will minimize the total cost of idleness within the system.

Machine breakdowns in a factory could be considered as arrivals to a repair (servicing) facility. Breakdowns are randomly distributed, with an average arrival rate of 40 minutes. The cost of an idle machine awaiting repair is estimated at $12 per hour. A mechanic hired at a wage of $5 per hour can repair machines according to the distribution of service times shown in Table 14.11. The service rate for two mechanics working together is also shown. Should one or two mechanics be employed?

**TABLE 14.11** Distribution of service times.

| ONE MECHANIC AT $5/HOUR | | TWO MECHANICS AT $10/HOUR | |
|---|---|---|---|
| *Time, Minutes* | *Probability* | *Time, Minutes* | *Probability* |
| 10 | 0.10 | 5 | 0.10 |
| 20 | 0.10 | 10 | 0.10 |
| 30 | 0.30 | 15 | 0.20 |
| 40 | 0.30 | 20 | 0.20 |
| 50 | 0.20 | 30 | 0.30 |
| | | 35 | 0.10 |
| 34 minutes average | | 21 minutes average | |

One way of simulating random arrivals with random numbers is to select one digit to represent a breakdown. This digit will average one appearance in a group of 10 random numbers. Then, for this example, each group of 10 digits represents 40 minutes of factory time. The number of times the selected digit appears in each group of 10 numbers indicates how many breakdowns occurred in that 40-minute period. The number of breakdowns will vary from zero in a period to the remotely possible maximum of 10.

A more definitive arrival simulation could be used if more information were available. A detailed probability distribution of periods between breakdowns would specify the breakdown pattern more precisely. In this example we are assuming that at most 10 breakdowns could possibly occur in a 40-minute period, and we must arbitrarily set the pattern within the period for multiple breakdowns. We shall assume that:

If two occur, one is at the beginning of minute 21 and the other is at the end of the period.

If three occur, one will be at the start of the period, one at the end, and one at the beginning of minute 21.

If more occur, they will follow a similar periodic arrangement.

Service-time simulation is accomplished by assigning digits in proportion to the frequency of each repair time. For the distribution of service times with one mechanic, we could let digit 0 = 10 minutes, 1 = 20 minutes, 2, 3, and 4 = 30 minutes, 5, 6, and 7 = 40 minutes, and 8 and 9 = 50 minutes.

With digit 0 = an arrival, a table of waiting times for a single mechanic could appear as shown in Table 14.12. The random numbers are from Figure 14.23, columns 4 and 5.

| ARRIVALS | | | | SERVICE | | | | |
|---|---|---|---|---|---|---|---|---|
| *Work Period, Minutes* | *Random No.* | *Arrival No.* | *Breakdown Time* | *Random No.* | *Service Period* | *Service Begins* | *Service Ends* | *Queue (machine waiting time)* |
| 0–40 | 3961003796 | 1 | 20 | 1 | 20 | 20 | 40 | 0 |
| | | 2 | 40 | 8 | 50 | 40 | 90 | 0 |
| 40–80 | 6658966048 | 3 | 80 | 7 | 40 | 90 | 130 | 90 − 80 = 10 |
| 80–120 | 3984360607 | 4 | 100 | 5 | 40 | 130 | 170 | 130 − 100 = 30 |
| | | 5 | 120 | 2 | 30 | 170 | 200 | 170 − 120 = 50 |
| 120–160 | 9345695714 | ... | ... | ... | ... | ... | ... | ... |
| 160–200 | 6830169817 | 6 | 200 | 6 | 40 | 200 | 240 | 0 |
| ..... | ..... | ..... | ..... | ..... | ..... | ..... | ..... | ..... |

**TABLE 14.12** Waiting-time simulation for one mechanic.

The tabulations in Table 14.12 are just a beginning. Many more samples would be required before legitimate conclusions could be drawn, but the same procedures would be followed. For each arrival (breakdown) a service time is simulated. The difference between the time a breakdown occurs and the time a service mechanic can begin repairs is the waiting time per machine. The total waiting time divided by the number of machines repaired is the average waiting time per machine. In an 8-hour day (480 minutes) the average number of breakdowns is

$$\frac{480 \text{ min/day}}{40 \text{ min/machine}} = 12 \text{ machines/day}$$

and the total daily cost of waiting time plus servicing is calculated from

Repair costs per day = number of mechanics × daily wage +
average number of breakdowns per day ×
average waiting time (hours) per breakdown × hourly cost of idle machines

For the simulated data from Table 14.12, this amounts to

$$\text{Repair costs} = 1(8 \times \$5) + 12\,\frac{10 + 30 + 50}{60 \times 6}\,\$12$$
$$= \$40 + \$36 = \$76$$

The same simulation procedure is followed for each alternative waiting-time policy. For the example situation we would next determine the total repair costs with a crew of two. Repair costs for each alternative are then compared to decide which policy is most suitable.

## MATHEMATICAL QUEUING MODELS

A more direct calculation method is available for some types of waiting-line (queuing) problems. We can use formulas developed for special cases without going into the underlying statistical foundations if we recognize the restrictions on their applications. Basic assumptions of the mathematical queuing models are as follows:

1. Arrivals follow a Poisson distribution. A histogram of this distribution displays a range of values from zero to infinitely large, with higher relative frequencies occurring toward the lower end of the scale. Many practical situations are well approximated by a Poisson arrival rate.
2. Arrivals are not affected by the length of the queue or any previous experience with the waiting line. This means a long queue will not discourage arrivals, and disgruntled arrivals will not leave the queue after spending time in a waiting line.
3. Arrivals are served on a "first come, first served" basis. No special priority or emergency provisions are included.
4. Service times follow an exponential distribution. This distribution, which shows continually decreasing relative frequencies from a zero value, is a good model of service time associated with Poisson arrivals. Theoretically, an infinitely long waiting line could develop, but we shall consider only reasonable relationships between arrival and servicing rates.

**Single Service Facility** When the above assumptions are valid conditions for an actual situation, we can apply established equations to evaluate the system. For example, assume that logging trucks arriving at a lumber mill follow a Poisson distribution of arrival times. An average of three trucks per hour arrive at the log pond for unloading. The trucks are unloaded at an exponential rate averaging four per hour.

With just the arrival $A$ and servicing rates $S$ we can investigate several aspects of the system. The equation for the average number of trucks in a waiting line is

$$\text{Length of queue} = \frac{A^2}{S(S-A)}$$

and for the example,

$$\frac{(3)^2}{4(4-3)} = \frac{9}{4} = 2.25 \text{ trucks in the waiting line}$$

To find the average waiting time for the trucks in the queue (including trucks which are served immediately), we can use the equation

$$\text{Waiting time in queue} = \frac{A}{S(S-A)}$$

which yields

$$\frac{3}{4(4-3)} = \frac{3}{4} = 0.75 \text{ hour in line per truck}$$

When service time is included with the waiting time, the formula for total unloading time is

$$\text{Waiting time} = \frac{1}{S - A}$$

and for the logging trucks the average time spent at the unloading dock is

$$\frac{1}{4 - 3} = \frac{1}{1} = 1 \text{ hour for unloading}$$

The proportion of arrival times a trucker can expect to find the service facility busy is $A/S$, or 3/4 of the trucker's trips to the mill. Conversely, the proportion of time the unloading facility is idle is $1 - A/S$, or 1/4 of the time.

The lumber mill is considering the replacement of its present log dump with a new unloading facility which is expected to provide an exponential service rate of six trucks per hour. The initial cost will be $200,000, with no salvage value at the end of its 10-year life. The cost of an idle truck is $20 per hour and is composed of overtime pay and contract carrier charges for additional log deliveries to meet mill requirements. With money valued at 10 percent and no salvage expected from the present facility, should the investment be made?

Annual costs for lost transportation time are based on a 40-hour week with 50 working weeks per year ($40 \times 50 = 2000$ hours per year). Then the annual waiting-time cost for the *present* facility is

$$\begin{aligned}\text{Unloading cost per year} &= A \times W \times \text{opportunity cost per hr} \times \text{hours/year} \\ &= 3 \times 1 \times \$20 \times 2000 = \$120{,}000\end{aligned}$$

The servicing costs of the *new* facility will be

$$\text{Unloading cost per year} = 3\frac{1}{6 - 3}\$20 \times 2000 = \$40{,}000$$

Expected annual savings for the new unloader are, then $120,000 − $40,000 = $80,000.

The new facility has no extra operating costs compared to the present method. Judging from the expected savings, we would surmise that the new investment would be profitable even without converting the initial cost to annual costs for comparative purposes. This suspicion is confirmed by

$$\begin{aligned}\text{Annual cost of new unloader} &= \text{initial cost} \times (A/P, 10, 10) \\ &= \$200{,}000 \times 0.16275 \\ &= \$32{,}550\end{aligned}$$

**ALTERNATIVES**

Although the lumber company may be satisfied with its plan to replace the present log-dump facility, it should also give consideration to other means of improving its unloading capabilities. General alternatives and specific considerations include:

*Changing the arrival rate* An appointment system or a priority rating could be initiated. Trucks could be dispatched from logging areas according to a schedule, but this arrangement would have to be coordinated with the woods crew. Incentives can occasionally be instigated, such as cheaper rates at certain hours to alter the arrival distribution.

*Changing service conditions* Service rates could be altered by adding stations or by redesigning existing facilities. Special arrangements for periods of congestion can often be provided. The logging company should consider adding a smaller, less expensive unloader to the present facilities. This would allow the use of two service stations on an as-required basis.

*Changing queue discipline* Arrivals requiring a short service time could be served first, or arrivals could be assigned to the most appropriate service station when multiple stations are available. Trucks with the longest hauls could be given service priority in order to reduce overtime resulting from adverse starting times for round-trip cycles.

Formulas are available for a number of these alternatives, but care must be exercised to assure that the assumptions made in developing the formulas pertain to the actual system. It is just as necessary to be sure the data used in simulation are descriptive of the real-life situation. Whenever models are used, it is advisable to give particular attention to the most sensitive characteristics. In queuing models the total cost is highly influenced by small changes in arrival and service rates.

**QUESTIONS**

**14.2a** An automobile dealer has used cars which need reconditioning before they can be sold. They have a Poisson arrival rate which averages four per week. The dealer is going to hire one of two people to clean up the cars. The first is capable of reconditioning cars at the rate of seven per week, but requires a wage of $100 per week. The alternative is to hire a less efficient worker, for $65 per week, who is capable of reconditioning only five cars per week. The dealer estimates the cost of holding a car off the used-car lot at $45 per week. Which worker should be hired? Assume the cars are reconditioned at an exponential rate.

**14.2b** Referring to the waiting-time example in this extension, suppose the lumber company could equip its trucks with special equipment which would make all the unloading times constant rather than exponentially distributed. The average unloading time is 10 minutes per truck, while all the other conditions remain the same. The special equipment has a life of 10 years with no salvage value. What is the maximum amount that could be spent on refitting the trucks to make this alternative comparable to providing a new unloader? [*Hint*: Waiting time $= A/2S(S-A)$ applies to a constant service time.]

**14.2c** Machines break down in a factory at an average rate of four per day. Breakdowns occur in a manner which closely follows a Poisson time distribution. An idle machine costs the company $30 per day. The current method of servicing follows an exponential repair rate averaging six per day. The cost of this facility is $60 per day. By purchasing diagnostic equipment for $12,000, the capacity of this facility will be doubled while still following an exponential repair-time distribution. The diagnostic equipment will last 10 years, and the operating costs will be $35 per day. A minimum rate of return of 12 percent is expected on all new investments. There are 250 working days per year. Should the equipment be purchased?

# CHAPTER 15

# PERSPECTIVES FOR DECISION MAKING

Discussions of decision processes usually start with the premise that the decision maker is ***rational***. Beyond the flippant assertion that "You are rational if you agree with me," rationality implies a reasoned response. In economic decisions, reasoning involves the specification of options, evaluation of their economic outcomes, and selection of the preferred option based on outcome criteria such as profit maximization and cost minimization. Difficulties hamper the decision process when there is insufficient information to evaluate the outcomes. Then the choice of a preferred alternative depends less on the objective properties of the outcomes than on the subjective perspective of the decision maker. Two opposing choices may be completely rational according to personal reference frames based on different objectives and attitudes.

Insufficient information is an undesirable but often unavoidable circumstance. Under the condition of ***uncertainty***, different potential outcomes can be recognized with the realization that their probability of occurrence cannot be legitimately projected. This admission is a step into the unknown beyond conditions of ***risk*** under which probabilities were estimated to indicate a solution. The step is a big one because it leaves behind the comfortable security of established preference criteria. Neat, precise solutions are replaced by judgments which reflect the disposition of individual decision makers.

The concepts presented in this chapter should contribute to a fuller appreciation of problem conditions and a better understanding of decision attributes. Conditions of uncertainty are categorized as either *noncompetitive* or *competitive*. The former are characterized by future states in which one outcome will presumedly occur and its likelihood is not influenced by the decision maker's choice; the decision environment is *neutral*. A competitive environment presupposes an intelligent opponent capable of selecting a future state for our options that maximizes the opponent's returns at our expense, while concurrently we choose a course of action that maximizes our returns with respect to the opponent's anticipated actions.

Competitive and noncompetitive viewpoints are theoretical constructs that attempt to expose the underlying rationale of decision making. They reveal effects of pessimistic and optimistic attitudes, adventurism and conservative tactics, possibilities of collusion or mutual destruction, and differences in the perceived value of returns, but they do not provide definitive decision rules. It is still the responsibility of the decision maker to decide how to decide.

## NONCOMPETITIVE DECISIONS

A decision situation labeled *noncompetitive* may appear trivial or unimportant. If so, the impression is deceptive. The term "noncompetitive" refers to the treatment of outcomes for any alternative. Each course of action has several possible outcomes. We do not know the probability associated with each outcome, but we do know that a probability exists and that it will not change as a result of our choice of a certain alternative. There is no active opposition controlling the likelihood of outcomes. Thus, our objectives may be competitive in the financial sense, but the competition is neutral or isolated from any cause-and-effect reactions.

Nature is a good example of a neutral opponent. Human beings are continuously in competition with nature. They suffer the inconvenience of carrying an unbrella to guard against rain or altering plans to "beat the weather." They purchase insurance to soften the loss from storms and launch tremendous research efforts to discover the "secrets" of nature. In each case nature is the opponent. Yet nature makes no willful effort to thwart our actions. Carrying an umbrella neither increases nor decreases the chance of rain. The probability of storms is not a function of the amount of insurance carried. Even though researchers may swear nature tries to frustrate their experiments, there is no evidence to support their claims. Nature is neutral.

Nature pervades social and economic situations as well as physical problems. All of us are under the same handicap in determining probabilities for future states. Weather forecasting seeks to foretell coming meteorological states, and public-opinion polls try to forecast social behavior; neither is completely reliable. Probing or spying is unproductive, because nature fails to react to the stimulus of our effort. Even records of past performance, such as stock-market or weather charts, often give false clues because only a portion of the influencing factors are identified. As a result, when unidentified factors exert unknown influences on future states, we turn to decision criteria attuned to our personal philosophies.

# COMPARISON OF DECISIONS UNDER CERTAINTY, RISK, AND UNCERTAINTY

A problem under uncertainty is treated initially like a problem under risk. Different alternative solutions are developed. The possible futures or states of nature are identified as completely as feasible. Then outcomes are estimated for each alternative under the assumption that each future state is certain to occur. The resulting figures are entered in a payoff matrix. Similarities and differences between certainty, risk, and uncertainty decisions are illustrated by the typical formats in Figure 15.1.

| | $P(1.0)$ |
|---|---|
| $A$ | $O_i$ |
| $B$ | $O_i$ |

**(a)** *CERTAINTY.* SOLUTION: GREATEST PROFIT

| | $P(0.7)$ | $P(0.3)$ |
|---|---|---|
| $A$ | $O_{ij}$ | $O_{ij}$ |
| $B$ | $O_{ij}$ | $O_{ij}$ |

**(b)** *RISK.* SOLUTION: HIGHEST EXPECTED VALUE

STATE OF NATURE

| | $N1$ | $N2$ |
|---|---|---|
| $A$ | $O_{ij}$ | $O_{ij}$ |
| $B$ | $O_{ij}$ | $O_{ij}$ |

**(c)** *UNCERTAINTY.* SOLUTION: PREFERRED STRATEGY

**FIGURE 15.1** Problem formats for conditions of certainty, risk, and uncertainty.

## Payoff Matrix

A payoff matrix under uncertainty closely resembles that used for problems under risk. There can be any number of alternative courses of action. It is important to identify all the pertinent states of nature ($N1, N2, \ldots, Nn$). Omitting states of nature is equivalent to neglecting states of risk, but it is easier to detect an omission under risk, because the associated probabilities would sum to less than 1. Sometimes related states of nature are lumped together because their effect on alternatives is the same. The check for using composite states comes when estimating outcomes $O_{ij}$. If one payoff for each alternative accurately measures the effect of every condition implied by the composite state, the combination is valid.

## Dominance

After the payoff matrix has been developed, the next step is to check for dominance. The problem objective is to select the best-paying alternative. If one alternative produces a greater payoff than another alternative for ***every*** state of nature, then a rational decision maker would never select the lower-paying course of action. The higher-paying alternative is said to *dominate* the lower one.* In Figure 15.2, alternative $A$ dominates alternative $C$.

*When ***all*** payoffs for one alternative are better than those of a second alternative, a condition called ***strict dominance*** exists. However, an alternative is still dominant when some of its payoffs are ***equal to*** and others are ***greater than*** the corresponding payoffs of another alternative.

Consequently, the dominated alternative *C* may be dropped from the matrix. An early check for dominance avoids unnecessary calculations by reducing the size of the problem matrix.

STATE OF NATURE

| | *N1* | *N2* | *N3* | *N4* |
|---|---|---|---|---|
| *A* | 10 | 1 | 3 | 6 |
| *B* | 6 | 2 | 5 | 3 |
| *C* | 7 | 0 | 2 | 4 |

**FIGURE 15.2** Dominance relationship.

# CRITERIA FOR NONCOMPETITIVE DECISIONS UNDER UNCERTAINTY

The most difficult aspect of noncompetitive problems under uncertainty is to decide what type of criteria to use for making a decision. In essence, we must determine the criteria for the criterion. The choice should be consistent with management philosophy. Is the current management outlook optimistic or pessimistic, conservative or adventurous? Certain criteria are compatible only with certain management views. Thus, it is necessary to understand both management policy and the principles of choice before selecting a decision criterion.

## Minimum-Maximum Criterion

A conservative approach to a decision is to look at the worst possible outcome for each alternative and select the course of action which assures the best results for the worst conditions. The underlying viewpoint is that nature is malicious. If things can go wrong, they will. This pessimistic philosophy dictates that attention be focused only on the most damaging outcomes in order to limit the damage as much as possible.

The words "minimax" and "maximin" are derived from the measures taken to identify the limiting loss or the guaranteed gain. A *minimax* decision minimizes the maximum loss. The *maximin* principle is associated with positive payoffs, where it maximizes the minimum gain or profit. For either criterion the smallest payoff (or greatest loss) for each alternative is noted. Then the alternative having the most favorable of the collected worst payoffs is selected.

## Maximax Criterion

A maximax philosophy is one of optimism and adventure. Nature is considered to be benevolent, so greatest gains are highlighted. The principle of choice is to identify the maximum gain possible for each alternative and then choose the course of action with the greatest maximum gain.

## Hurwicz Criterion

A moderate outlook between the extremes of optimism and pessimism is allowed by the Hurwicz criterion. The degree of optimism is established by a coefficient called alpha ($\alpha$), which may take any value between 0 and 1.0, with the following interpretation:

| | | | |
|---|---|---|---|
| Coefficient of optimism ($\alpha$): | 0.0 | | 1.0 |
| Decision maker's philosophy: | Pessimistic | ⟶ | Optimistic |

After deciding the value of $\alpha$ which measures the decision maker's degree of optimism, maximum and minimum gains are identified for each alternative. Then the maximum payoffs are multiplied by $\alpha$, and the minimum payoffs by $1 - \alpha$. The two products for each alternative are added, and the alternative with the largest sum is chosen.

The minimax-maximin and maximax criteria are special cases of the Hurwicz criterion. When $\alpha = 1$ only the maximum payoffs are included in the final alternative selection, because the minimum payoffs have been eliminated by zero multiplication. The opposite is true for $\alpha = 0$, a completely pessimistic outlook. Any value of $\alpha$ other than 1 or 0 is a compromise opinion about the hostility or benevolence of nature.

### ➧ Example 15.1 Different Degrees of Optimism

Two sons and their mother own and operate an import shop in a medium-sized city. They have been successful enough to be in a position to expand their operations. Three courses of action are deemed most desirable: (1) expand their present operations by opening a store in a nearby city; (2) start a catalog business from their present location; or (3) invest their extra money in real estate and rentals. Each alternative will utilize about the same amount of capital and will require equivalent management. They recognize that the returns from each of these investments depend on the national economy (inflation versus recession) and on the local economy (growth versus stagnation). However, they have no consensus of the probabilities of future conditions.

Outcomes for each of the alternatives have been developed for four possible levels of business activity: very high ($VH$), high ($H$), medium ($M$), and low ($L$). The payoffs shown are the estimated percentage returns on invested capital for expanding ($E$), starting a catalog service ($C$), and investing in property ($P$):

STATE OF NATURE

| | $VH$ | $H$ | $M$ | $L$ |
|---|---|---|---|---|
| $E$ | 20 | 12 | 8 | 4 |
| $C$ | 26 | 10 | 4 | −4 |
| $P$ | 10 | 8 | 7 | 5 |

The youngest son is an optimist and a risktaker. The mother is conservative, and the other son is midway between his mother and brother. Which alternative would probably appeal to each member of the family?

**Solution 15.1**

The younger son, being an optimist, would use the maximax criterion. By doing so he would limit his selection area to the *VH* level of business activity, where the largest gains occur. From the possible gains of 20, 26, and 10 percent, he would naturally select the highest, which results from opening a catalog store.

The conservative mother would lean toward the maximin criterion, where she could be assured a minimum gain of 5 percent (the maximum gain under the worst condition of low business activity) by choosing to invest in real estate and rentals.

The other son might use the Hurwicz criterion with a coefficient of optimism $\alpha$ of 0.5. The consequent calculations reveal that his choice would be to expand the import business to the nearby town.

| *Alternative* | *Max* $O_i \times \alpha + min\ O_i(1 - \alpha) = total$ | |
|---|---|---|
| *E* | $20 \times 0.5 + 4(1 - 0.5) = 12$ | ←*Maximum* |
| *C* | $26 \times 0.5 + (-4)(1 - 0.5) = 11$ | |
| *P* | $10 \times 0.5 + 5(1 - 0.5) = 7.5$ | |

The relationship of the three philosophies can be better interpreted by plotting the maximum and minimum gains from each as a function of $\alpha$. As shown in Figure 15.3, the topmost lines indicate the alternative that would be selected for different levels of optimism. At $\alpha = 0$ the maximin criterion is in effect, and property appears to be the most attractive investment. Property continues to be favored until the less pessimistic attitude of $\alpha = 0.09$ is attained. The next switch point occurs at the intersection of the lines representing "expand" and "catalog." Using the equation for the return expected from these two alternatives, the value of $\alpha$ at which a decision

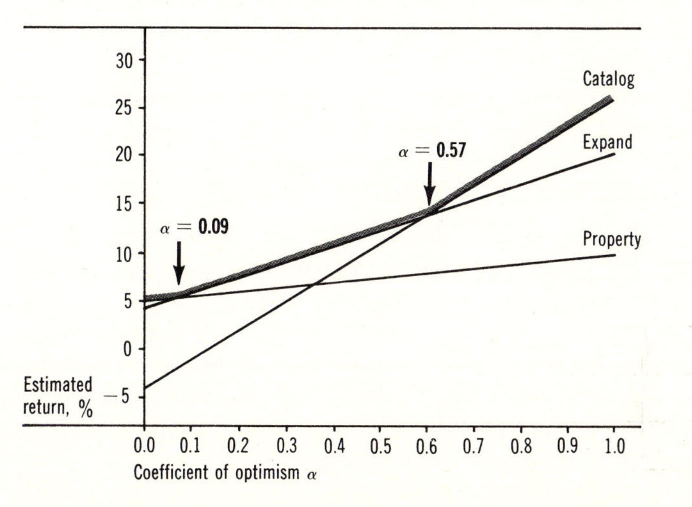


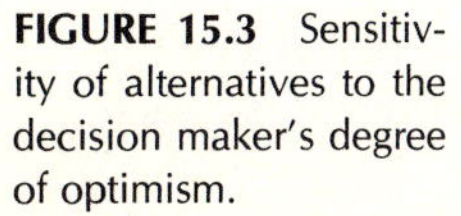
**FIGURE 15.3** Sensitivity of alternatives to the decision maker's degree of optimism.

maker is indifferent to the choice between the two is calculated as

$$20\alpha + 4 \times (1 - \alpha) = 26\alpha + (-4) \times (1 - \alpha)$$
$$16\alpha + 4 = 30\alpha - 4$$
$$\alpha = 0.571$$

Mapping the alternatives gives an indication of their sensitivity with respect to the degree of optimism of the decision maker. The property alternative is quite sensitive because its selection requires a very pessimistic attitude. The remaining alternatives are relatively insensitive because each would be chosen over a considerable range of $\alpha$ values. The apparent range of attitudes which favor each alternative could aid the mother and sons in a search for a compromise solution.

## Minimax-Regret Criterion

Opportunity costs have been used in previous chapters to express the loss incurred by not selecting the best alternative. The minimax-regret criterion is based on similar costs. The opportunity costs are determined for each state of nature by subtracting the largest payoff in each column from all other payoffs in the column. The absolute value of each subtraction is the amount of "regret" that results from not selecting the best alternative for the occurrence of a given state. This procedure converts the original matrix to a regret matrix.

A rational decision maker attempts to minimize regret. By applying the minimax principle, the alternative with the minimum maximum (lowest value of the worst regret for each row) is selected. The minimax-regret procedure applied to the data from Example 15.1 is shown in Figure 15.4. The indicated preference is for the "expand" alternative. In general, the minimax-regret criterion tends toward a conservative viewpoint.

ORIGINAL PAYOFF MATRIX

| | *VH* | *H* | *M* | *L* |
|---|---|---|---|---|
| *E* | 20 | 12 | 8 | 4 |
| *C* | 26 | 10 | 4 | −4 |
| *P* | 10 | 8 | 7 | 5 |

REGRET MATRIX

| | *VH* | *H* | *M* | *L* | *Worst Regret* |
|---|---|---|---|---|---|
| *E* | 6 | 0 | 0 | 1 | 6 |
| *C* | 0 | 2 | 4 | 9 | 9 |
| *P* | 16 | 4 | 1 | 0 | 16 |

**FIGURE 15.4** Original and regret matrix.

## Equal-Likelihood Criterion

When it is possible to assign probabilities to future states, we use the expected-value criterion to select a preferred alternative. An extension of this approach is the basis of the *equal-likelihood* criterion. Under uncertainty we admit that we cannot reasonably estimate outcome probabilities. Therefore, since we have no excuse to believe otherwise, why not treat each outcome as the same? The rationale behind this theory is that there is insufficient reason to believe one state of nature more probable than another, so each should be assigned an equal probability of occurrence.

The equal-likelihood criterion is certainly the simplest to apply. Under the assumption that each future is equally likely to occur, the expected value of an alternative becomes its average outcome. The alternative with the largest average payoff is preferred. As applied to the data from Example 15.1, we have

$$E(E) = 20 \times 1/4 + 12 \times 1/4 + 8 \times 1/4 + 4 \times 1/4 = 11$$

$$E(C) = 26 \times 1/4 + 10 \times 1/4 + 4 \times 1/4 - 4 \times 1/4 = 10$$

$$E(P) = 10 \times 1/4 + 8 \times 1/4 + 7 \times 1/4 + 5 \times 1/4 = 7.5$$

which again leads to the decision to expand.

**Example 15.2** **Application of Decision Criteria**

Given the payoff matrix below, which alternative would be selected under each of the noncompetitive decision criteria? Assume that the coefficient of optimism is 0.375 ($\alpha = 3/8$) for the Hurwicz criterion.

| | *N1* | *N2* | *N3* | *N4* |
|---|---|---|---|---|
| *A* | 2 | 2 | 2 | 2 |
| *B* | 1 | 5 | 1 | 0 |
| *C* | 1 | 4 | 1 | 1 |
| *D* | 1 | 3 | 1 | 4 |
| *E* | 3 | 4 | 3 | 0 |

**Solution 15.2** The procedures followed for each criterion reduce the original matrix to a single column of outcomes from which the desired value is selected. The results and preferences indicated by applying all the criteria are as shown:

| *Alternative* | *Maximin (pessimist)* | *Maximax (optimist)* | *Hurwicz ($\alpha = 3/8$)* | *Minimax Regret* | *Equal Likelihood* |
|---|---|---|---|---|---|
| A | **2** | 2 | 2 | 3 | 2.0 |
| B | 0 | **5** | 15/8 | 4 | 1.75 |
| C | 1 | 4 | **17/8** | 3 | 1.75 |
| D | 1 | 4 | 17/8 | **2** | 2.25 |
| E | 0 | 4 | 12/8 | 4 | **2.5** |

It should not be too surprising that each criterion indicates a different preferred alternative. Each criterion has a slightly distinctive underlying principle. The key is to decide which criterion best fits the decision environment for each specific application.

# EVALUATION OF DECISION CRITERIA

Several different criteria for making noncompetitive decisions under uncertainty have been offered because no one criterion is unanimously preferred. Each one has certain weaknesses. Often one criterion is more intuitively appealing than the others. This appeal seems to vary among individuals and to vary with time or circumstances. As there is no universal preference, a sound recourse is to investigate the criteria limitations in order to select the principle which accommodates a given decision environment.

## Partial-Optimist Principle

The key factor in applying the Hurwicz criterion is the choice of a value for $\alpha$. An arbitrary choice defeats the intent of portraying an individual outlook. A deliberate choice is a forced judgment often based on slim evidence. However, the judgment does allow a measure of added knowledge, even though it is undefined, to be included in the decision. This additional knowledge could be described as "a feel for the problem" or an "educated guess." Until methods are developed to determine $\alpha$ objectively, its value will remain uniquely individual.

Pure optimism and pessimism are special cases of the Hurwicz criterion. Since these are extreme outlooks, they often leave a decision maker uncomfortable in applying them to special situations. For instance, the maximin criterion applied to

| | *N1* | *N2* | *Minimum* |
|---|---|---|---|
| *A* | \$0.01 | \$0.01 | \$0.01 ←Maximin |
| *B* | 0 | \$100 | 0 |

would indicate that alternative $A$ should be selected, yet most people would be inclined to choose $B$. In the same vein, the maximax criterion applied to the matrix below would lead to the choice of $B$; most decision makers would express a decided preference for alternative $A$.

| | *N1* | *N2* | *Maximum* |
|---|---|---|---|
| *A* | \$99 | \$99 | \$99 |
| *B* | 0 | \$100 | \$100 ←Maximax |

By selecting a value for $\alpha$ other than 1 or 0, a more moderate outlook is achieved. However, the criterion still possesses some dissatisfying aspects. According to the criterion, the two alternatives in the following matrix are considered equivalent, because attention is given only to the best and worst outcomes:

| | *N1* | *N2* | *N3* | *N4* | *N5* | *N6* |
|---|---|---|---|---|---|---|
| *A* | $100 | $100 | $100 | $100 | $100 | 0 |
| *B* | 0 | 0 | 0 | 0 | 0 | $100 |

Many people faced with this choice would cast a strong vote in favor of $A$ over $B$.

There is also another difficulty. Some critics object to a decision criterion that changes preferences when a constant is added to all the outcomes in one column. Such a criterion is said to lack the property of "column linearity." To illustrate, the maximin criterion would indicate a preference for alternative $B$ when applied to Figure 15.5(a). The outcomes in both matrices are the same, with the exception of 100 added to both alternatives under $N1$ in Figure 15.5(b). Such an exception could be caused by a discovery that a bonus payoff would result from the occurrence of state $N1$ regardless of which alternative were selected. The disturbing feature is that although the bonus has the same effect on both alternatives, it changes the preference from $B$ to $A$.

| | *N1* | *N2* |
|---|---|---|
| *A* | 0 | 100 |
| *B* | 50 | 50 |

**(a)**

| | *N1* | *N2* |
|---|---|---|
| *A* | 100 | 100 |
| *B* | 150 | 50 |

**(b)**

**FIGURE 15.5** Column linearity.

## Opportunity-Loss Principle

The minimax-regret criterion is plagued by many of the same defects as the Hurwicz criterion. Attention is focused on only the largest opportunity costs, with a resultant disregard for other payoffs.

A further argument against the minimax-regret criterion is that the addition of irrelevant information to the matrix can switch the alternative preferences. From the payoff matrix

| | *N1* | *N2* | *N3* |
|---|---|---|---|
| *A* | 100 | 125 | 25 |
| *B* | 25 | 125 | 75 |

the resulting regret matrix is

| | *N1* | *N2* | *N3* | *Worst Regret* |
|---|---|---|---|---|
| *A* | 0 | 0 | 50 | 50 ←Minimax |
| *B* | 75 | 0 | 0 | 75 |

which leads to the choice of alternative $A$. Now an additional, rather unattractive, course of action is included in the matrix as

| | *N1* | *N2* | *N3* |
|---|---|---|---|
| *A* | 100 | 125 | 25 |
| *B* | 25 | 125 | 75 |
| *C* | 25 | 25 | 125 |

The addition, alternative $C$, changes the regret matrix to

| | *N1* | *N2* | *N3* | *Worst Regret* |
|---|---|---|---|---|
| *A* | 0 | 0 | 100 | 100 |
| *B* | 75 | 0 | 50 | 75 ←Minimax |
| *C* | 75 | 100 | 0 | 100 |

which indicates that alternative $B$ rather than $A$ should be selected. Although the switch caused by irrelevant information may seem disturbing, it can also be argued that the additional alternative is not necessarily "irrelevant," because it can reveal more information about the states of nature, such as the change in the maximum possible payoff for $N3$.

## Average-Outcome Principle

Most of the doubts raised for the other criteria are not applicable to the equal-likelihood criterion. It has column linearity, and includes all the outcomes in an evaluation, and the addition of an unattractive alternative cannot switch an earlier preference. However, it also has one serious defect: The selection is very sensitive to the number of future states identified for a problem.

Consider the problem faced by a company that is bidding on a contract. They can build special low-operating-cost equipment which will be ready if they get the contract, or they can buy higher-operating-cost equipment after they know they have won the bidding. The payoffs, in thousands of dollars, are shown in the accompanying matrix.

| | *No Contract* | *Win Contract* | *Average Outcome* |
|---|---|---|---|
| *Build* | −50 | 125 | 37.5 |
| *Buy* | 0 | 100 | 50.0 ←Maximum |

When the future states are limited to winning or losing the one contract, the equal-likelihood criterion shows a preference for buying the equipment. If other future states, such as the possibilities of winning two more similar contracts, are included, the same criterion indicates a preference for building the equipment.

| | *No Contract* | *Win Contract 1* | *Win Contract 2* | *Win Contract 3* | *Average Outcome* |
|---|---|---|---|---|---|
| *Build* | −50 | 125 | 125 | 125 | 81.25 ←Maximum |
| *Buy* | 0 | 100 | 100 | 100 | 75.0 |

Beginning with an exhaustive list of futures will eliminate the chance of a preference change. If one state is subdivided into substates such that the sum of the probabilities for the substates is equal to the probability of the original state, the problem is also avoided. However, each of these methods presumes some knowledge of the future that contradicts the assumption of uncertainty upon which the criterion is based.

### Application of Principles

None of the criteria is perfect. None can take the place of an accurate forecast. They should be considered guidelines which will help in the interpretation and consideration of possible choices.

It should be noted that most of the reservations about the criteria were more intuitive than deductive. Perhaps the adoption of a certain criterion must also rely to some degree upon intuition, because such insight is often a function of knowledge not yet formalized into distinct views. Nevertheless, the decision maker must understand the characteristics of each principle to be able to select the one which corresponds most closely to the uncertainties of a situation.

## COMPETITIVE DECISIONS

In competitive decision making, two or more decision makers are pitted against each other. Both are considered to be equally informed and intelligent. They are referred to as *players,* and their conflicts are called *games.* The rationale of their competition is the basis of *game theory.**

Some of the terms associated with game theory need a special explanation. The generic term "game" in no way implies that game theory is limited to parlor games or similar entertainment contests. The players could be nations maneuvering on the brink of war, labor and management negotiating to ward off a strike on their own terms, or department store owners seeking a strategy to end a price war with an acceptable compromise.

In a game context, alternative courses of action are ***strategies.*** There is a subtle difference between alternatives in games of skill (akin to conditions of certainty), games of chance (risk), and games of strategy. In the last category the best course of action for a player depends on what that player's adversaries can do. Thus, an optimum strategy in a

*The publication of *Theory of Games and Economic Behavior* by von Neumann and Morgenstern in 1947 laid the framework for and stirred subsequent interest in game theory.

competitive environment may be always to use one alternative or to mix alternatives—that is, to use different alternatives for successive plays. The considerations inherent in determining strategies make game theory a rich source of fundamental ideas for decision making.

## GAME THEORY

In our treatment of game theory we shall limit attention to games involving two players. Allowing more than two players greatly increases the mathematical complication without a corresponding accrual of basic logic. Multiplayer games rely on essentially the same type of inductive and deductive reasoning as two-person games. Actually, the two-player limitation is not as restrictive as a first impression might indicate. Strategy for one side in a conflict is often designed to counter the aims of the most dangerous member of an opposing team. In other cases, a player may view the entire array of business competitors as a single opponent (a viewpoint similar to treating nature as hostile) and thereby reduce the conflict to a two-person contest.

### Payoff Matrix

In two-person games the rows of the payoff matrix contain the outcomes for one player, and the columns show the outcomes for the other player. The outcomes are written in terms of the player on the left, player $A$ in Figure 15.6. The alternatives for player $A$ are $A1$, $A2$, $A3$, and $A4$. Player $B$ also has four possible courses of action, $B1$, $B2$, $B3$, and $B4$. If player $A$ selects alternative $A1$ and player $B$ uses $B1$, the outcome is a loss of 1 for $A$ and the gain of 1 for $B$. At $A1$, $B4$ the outcome is a gain of 5 for $A$ and a corresponding loss of 5 for $B$. Thus, a positive number is always a gain for $A$, and a negative number is a gain for $B$. Because the sum of the payoffs for any choice of alternatives is zero (when $A$ wins, $B$ loses), the game described by Figure 15.6 is called a *zero-sum* game.

PLAYER $B$

| PLAYER $A$ | | $B1$ | $B2$ | $B3$ | $B4$ |
|---|---|---|---|---|---|
| | $A1$ | −1 | −2 | 4 | 5 |
| | $A2$ | −1 | 4 | −3 | −2 |
| | $A3$ | 0 | 3 | 1 | 2 |
| | $A4$ | −2 | −3 | 3 | 1 |

**FIGURE 15.6** Two-person zero-sum game.

Several assumptions are implied for a conflict situation represented by a two-person zero-sum payoff matrix:

1. The conflict is between only two opponents.
2. Each opponent has a finite number of alternatives.
3. Both players know all the alternatives available.
4. Each gain or loss can be quantified to a single number.

**5** Both players know all the outcomes (numbers).
**6** The sum of the payoffs for each outcome is zero.

## Dominance

The first step in analyzing a game matrix is to check for dominance. In previous matrices we had to compare only the rows for dominance. In competitive decisions the columns as well as the rows represent alternatives. Therefore it is necessary to check both horizontally and vertically for dominance. Furthermore, the discovery of one dominant relationship may reveal another dominant condition that was previously indistinguishable.

A sequential dominance relationship can be observed in Figure 15.6. A check of the columns, the alternatives available to player $B$, shows that no one alternative is better for every outcome than any of the other alternatives. For instance, $B3$ is preferred to $B4$ for all outcomes except those in the bottom row, $A4$.

Turning to the alternatives available to player $A$, we see that $A1$ is always preferred to $A4$. Eliminating $A4$ from the matrix, we have

| | *B1* | *B2* | *B3* | *B4* |
|---|---|---|---|---|
| *A1* | −1 | −2 | 4 | 5 |
| *A2* | −1 | 4 | −3 | −2 |
| *A3* | 0 | 3 | 1 | 2 |

Now, from the viewpoint of player $B$, $B3$ is always preferable to $B4$. A further check reveals no additional dominant relationships.

The assumption that allows two-way dominance checks is that both players are intelligent. Thus $B$ would recognize that $A$ would never use $A4$, which means $B$ would never have a use for $B4$. Viewed from either side, the end result would be the 3 × 3 matrix (three alternatives for each player) shown in Figure 15.7.

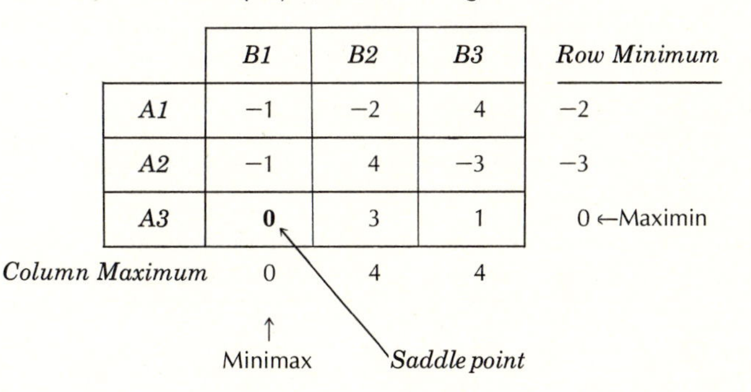

| | *B1* | *B2* | *B3* | *Row Minimum* |
|---|---|---|---|---|
| *A1* | −1 | −2 | 4 | −2 |
| *A2* | −1 | 4 | −3 | −3 |
| *A3* | **0** | 3 | 1 | 0 ←Maximin |
| *Column Maximum* | 0 | 4 | 4 | |
| | ↑ Minimax | *Saddle point* | | |

**FIGURE 15.7** Saddle point for the reduced matrix.

## Saddle Point

When a player uses the same alternative at every play, we say the player is following a *pure strategy*. If it is advantageous for both players to use pure strategy, a saddle point is present. A *saddle point* is identified by an outcome which is *both* the smallest number in its row *and* the largest number in its column. In Figure 15.7 the smallest numerical values for each row are listed to the right of the matrix, and the largest numerical values for the columns are shown below the matrix. The outcome 0 at $A3$, $B1$ satisfies the requirements for a saddle point.

The significance of a saddle point develops from an investigation of the player's motives. Alternatives $A1$ and $A2$ are attractive to $A$ because they allow a potential gain of four units. However, $B$ can be assured that $A$ will lose one unit (−1) by using $B1$ whenever $A$ chooses to utilize $A1$ or $A2$. $A$ would also be attracted to $A3$ because no negative outcomes can occur from this alternative. Again $B$ can thwart gains by $A$ through the use of $B1$. Since $A$ is astute enough to observe the advantage $B1$ affords $B$, $A$ would select the alternative that minimizes $B$'s gain from following the pure $B1$ strategy. This alternative is $A3$. Therefore both players would use pure strategy, with $A$ always employing alternative $A3$ and $B$ always using $B1$. The result is a standoff, where neither side gains an advantage.

The underlying principle of choice for the players is maximin-minimax. $B$ seeks the least of the maximum losses, minimax; $A$ identifies the greatest of the minimum gains, maximin. This conservative viewpoint acknowledges that each player is capable of selecting the strategy which will satisfy the objective of maximizing gain or minimizing loss.

### Value of a Game

The return from playing one game is the amount each player nets from the ensuing outcome. For instance, if one player gets a return of +5 after each has followed a certain course of action, the other player gets −5 in a zero-sum game. The average gain or loss per play, taken over an extended series of plays, is called the *value* of the game.

When the optimal strategy for both players is a pure strategy, the value of the game is the outcome at the saddle point. In Figure 15.7, the value of the game is zero; neither player wins or loses.

When no saddle point exists, the players turn to a policy called *mixed strategy*. This means that different alternatives are used for a fixed proportion of the plays, but the alternative employed for each play is a random choice from those available. The value of the game is the average return resulting from each player's having followed an optimal mixed strategy.

## SOLUTION METHODS FOR MIXED-STRATEGY ZERO-SUM GAMES

Several different methods have been developed to solve zero-sum games. Every two-person zero-sum game with a finite number of alternatives can be transformed into a linear-programming problem and solved accordingly. In addition, there are special methods appropriate to certain conditions. Often these special methods are much less demanding than the general approach. A different solution method is offered in each of the following sections for games of different sizes. Other methods are available from the references at the end of the book.

## 2 × 2 Games

| | *B1* | *B2* |
|---|---|---|
| *A1* | 1 | 5 |
| *A2* | 3 | 2 |

**FIGURE 15.8** 2 × 2 matrix.

A game with only two alternatives for each player readily reveals the wisdom of mixed strategy. Referring to Figure 15.8, we see that all the payoffs are positive. Therefore player *A* is bound to be a winner. The objective of *B* is to limit losses (*A*'s gains) as much as possible. A pure strategy will not do this.

A quick check shows that there is no dominance among the alternatives and no saddle point. The absence of a saddle point means that the players should rely on mixed strategy. Sometimes player *A* will use *A*1 and other times *A*2. Similarly, *B* switches randomly between *B*1 and *B*2. The problem is to determine the proportion of time each player will use each alternative.

There is a very simple method for solving 2 × 2 games. The procedural steps are listed in Figure 15.9 and illustrated by application to the matrix of Figure 15.8.

*Step 1* Obtain the absolute value of the difference in payoff for each row and column.

| | *B1* | *B2* | *Payoff Difference* |
|---|---|---|---|
| *A1* | 1 | 5 | **4** = 5 − 1 |
| *A2* | 3 | 2 | **1** = 3 − 2 |
| *Payoff Difference* | **2** | **3** | |

*Step 2* Add the values obtained in step 1 for the rows and columns. The sum of the column differences should equal the sum of the row differences.

| | *B1* | *B2* | |
|---|---|---|---|
| *A1* | 1 | 5 | 4 |
| *A2* | 3 | 2 | 1 |
| | 2 + | 3 | = **5** |

*Step 3* Form a fraction associated with each row and column by using the values obtained in step 1 as the numerator and the number obtained in step 2 as the denominator.

| | *B1* | *B2* | |
|---|---|---|---|
| *A1* | 1 | 5 | 4/5 |
| *A2* | 3 | 2 | 1/5 |
| | 2/5 | 3/5 | |

*Step 4* Interchange each pair (row and column) of fractions obtained in step 3. The fraction now associated with each alternative is the proportion of plays that that alternative should be used for an optimal mixed strategy.

| | *B1* | *B2* | |
|---|---|---|---|
| *A1* | 1 | 5 | **1/5** |
| *A2* | 3 | 2 | **4/5** |
| | **3/5** | **2/5** | |

**FIGURE 15.9** Solution steps for 2 × 2 games.

The indicated strategy is that $A$ should use $A1$ 20 percent of the time, and $A2$ for 80 percent of the plays. $B$ should play $B1$ 60 percent of the time, and $B2$ the other 40 percent.

The strategies appear intuitively sound when we remember the conservative viewpoint inherent in game theory. $A$ would use $A2$ more often because of its higher minimum gain (two units instead of one). $B$ would rely on $B1$ more than $B2$ because of the lower minimum and maximum losses. The validity of the strategies becomes even more apparent when the value of the game is considered.

The long-run game return is the expected value of all the outcomes to one player. In order to calculate the probability of each outcome we must know how often each player plans to follow a strategy which includes the outcome. This information is provided by the optimal mixed strategy. Figure 15.9 shows the strategies determined for the sample-problem data. The cells of the matrix in Figure 15.10 contain the probable occurrence of each outcome in the original payoff matrix (Figure 15.8). Each cell value is the product of the probabilities for its row and column. The sum of these joint probabilities must equal 1.0. From Figure 15.10,

$$3/25 + 2/25 + 12/25 + 8/25 = 25/25 = 1.0$$

| | *B1* (3/5) | *B2* (2/5) |
|---|---|---|
| *A1* (1/5) | 3/25 | 2/25 |
| *A2* (4/5) | 12/25 | 8/25 |

**FIGURE 15.10** Probability of outcomes.

By multiplying each outcome by the probability of its occurrence and adding them together, we have the expected value of the game. In terms of player $A$, the value of the game is

$$3/25 \times 1 = 3/25 \qquad 2/25 \times 5 = 10/25$$
$$12/25 \times 3 = 36/25 \qquad 8/25 \times 2 = 16/25$$

$$\text{EV}(A) = 3/25 + 10/25 + 36/25 + 16/25 = 65/25 = 2.6 \text{ units}$$

which means $A$ can expect an average payoff of 2.6 units per play over a large number of plays. Since we are dealing with a zero-sum game, we would expect $B$'s loss to be equal to $A$'s gain. This is confirmed by

$$\begin{aligned}\text{EV}(B) &= 3/25 \times (-1) + 2/25 \times (-5) + 12/25 \times (-3) + 8/25 \times (-2)\\ &= (-3/25) + (-10/25) + (-36/25) + (-16/25)\\ &= -2.6 \text{ units}\end{aligned}$$

The best $B$ can do is an average loss of 2.6 units.

If $B$ tried to improve this position by changing strategy, say to a pure $B1$ strategy, $A$ would soon recognize the pattern and switch accordingly to a pure $A2$ strategy. The value of the game would then change to 3, the outcome of $A2$, $B1$. If $A$ continued with the optimal mixed strategy while $B$ switched to a pure $B1$ strategy, the value of the game for $A$ would remain unchanged:

| | *B1* (1.0) |
|---|---|
| *A1* (1/5) | 1/5 |
| *A2* (4/5) | 4/5 |

$$\text{EV}(A) = 1/5 \times 1 + 4/5 \times 3 = 13/5 = 2.6 \text{ units}$$

From this result, we can observe that in 2 × 2 games one player can assure the value of the game by using the optimal strategy, regardless of what the other player does.

**Example 15.3** **2 × 2 Matching Contest**

Two prudent but impoverished gamblers were matching dimes. One player, Alice, won both dimes when two heads or two tails occurred. The other player, Bob, won both dimes when the coins did not match. After a long series of flips, both players were tired, bored, and nonwinners. Then Bob offered a suggestion: "Let's stop flipping coins and just turn up whichever face we want. To make it interesting, you win the dime if we both show heads, I win a nickel if we don't match, and neither of us wins if we both show tails. It's a fair game, because you have one chance of winning a dime while I have two chances of winning a nickel."

Alice stopped to think. She visualized the game in matrix form and recognized it as a two-person zero-sum game. With no saddle point, it had to be mixed strategy. Calculating her optimal strategy, she found that she should turn up a head one-fourth of the time and a tail three-fourths of the time. After calculating Bob's strategy, she found her value of the game to be

$$\begin{aligned} \text{EV(Alice)} &= 1/16 \times \$0.10 + 3/16(-\$0.05) + 3/16(-\$0.05) + 9/16 \times 0 \\ &= -\$0.0125 \end{aligned}$$

| | | BOB | |
|---|---|---|---|
| | | *H* (1/4) | *T* (3/4) |
| ALICE | *H* (1/4) | \$0.10 | −\$0.05 |
| | *T* (3/4) | −\$0.05 | 0 |

After cogitating a bit, she said to Bob: "Your game sounds like fun, but we don't have any nickels to use as payoffs, so let's use only dimes. I'll give you four dimes if we both show heads and one dime for two tails. You give me two dimes if I show a head while you turn up a tail, and three dimes if I have a tail and you have a head. Since both our total payoffs come to 50 cents, the game is just as fair as the one you suggested."

| | | BOB | |
|---|---|---|---|
| | | *H* | *T* |
| ALICE | *H* | −4 | 2 |
| | *T* | 3 | −1 |

Now Bob stopped to think. He mentally pictured the payoff matrix, determined the strategies, and calculated his expected return:

BOB

| | | *H (0.3)* | *T (0.7)* |
|---|---|---|---|
| ALICE | *H (0.4)* | −\$0.40 | \$0.20 |
| | *T (0.6)* | \$0.30 | −\$0.10 |

$$\text{EV(Bob)} = 0.3 \times 0.4 \times \$0.40 + 0.3 \times 0.6(-\$0.30) + 0.7 \times 0.4(-\$0.20) + 0.7 \times 0.6 \times \$0.10 = \$0.048 - \$0.054 - \$0.056 + \$0.042$$

$$= -\$0.02$$

Noting his negative average payoff, he replied: "This takes too much thinking. Let's go back to flipping coins."

## 2 × *n* and *n* × 2 Games

When the number of alternatives increases beyond two, the method used to solve 2 × 2 games is not directly applicable. However, through the use of a graphical technique we can convert a 2 × *n* or *n* × 2 game to a 2 × 2 game. The obvious limitation is that one of the players still has to have only two alternatives.

To introduce the graphical method, we can apply it to the 2 × 2 game already solved in the previous section and shown again in Figure 15.11. The data from the payoff matrix are entered on a graph which shows payoffs on the ordinates and mixed strategy on the abscissa. The vertical scale must accommodate the largest payoff, and the horizontal scale runs from 0 to 1.0.

Figure 15.11 represents player *A*'s interests. Alternative *A* 1 is depicted by the abscissa, and the ordinates are *A*'s payoffs. If *A* used pure *A* 2 (left vertical scale), *A* could expect to win 2 or 3 units, depending on *B*'s strategy. The right vertical scale shows the possible payoffs from a pure *A* 1 strategy. The lines connecting points on the two vertical scales represent *B*'s alternatives. They serve to constrain the payoffs to *A* as indicated by the heavy

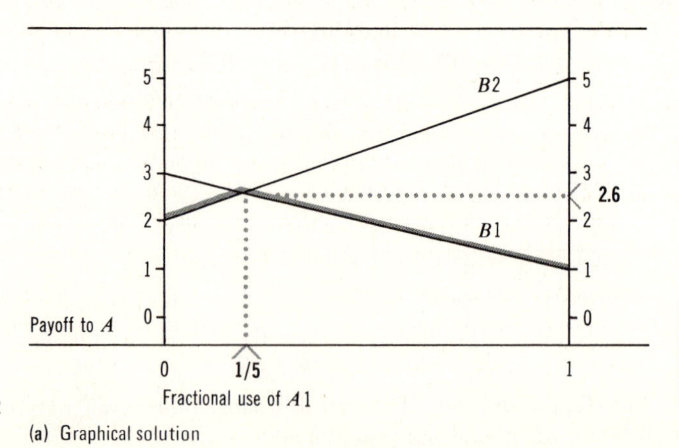


(a) Graphical solution

| | *B*1 | *B*2 |
|---|---|---|
| *A*1 | 1 | 5 |
| *A*2 | 3 | 2 |

(b) Payoff matrix from Fig. 15.8

**FIGURE 15.11** 2 × 2 game solution.

line, the minimum possible gains by $A$. Using the maximin criterion, $A$ selects the strategy which gives the best minimum payoff. This payoff is the value of the game (2.6) and is indicated by the intersection of lines $B1$ and $B2$. The point on the horizontal scale directly below the intersection designates the fraction of the time $A$ should use $A1$.

The graphical method contributes little to the solution of a $2 \times 2$ game; it was applied above only to illustrate the technique. Its value lies in identifying the two limiting alternatives among several possessed by one player. As will be observed later, when one opponent has only two alternatives, the other player can advantageously play only two alternatives, even though several others are available. This condition* allows the game to be reduced to $2 \times 2$ and solved accordingly, once the limiting alternatives are known.

A $3 \times 4$ payoff matrix is shown in Figure 15.12(b). The first step in its solution is to check for dominance and a saddle point. The dominance of $A2$ over $A3$ reduces the matrix to a $2 \times 4$, which allows a graphical approach. The scales of the graph always represent the player with only two alternatives. Letting the abscissa denote $A1$, each of $B$'s alternatives is plotted with $A2$ payoffs (0 use of $A1$) on the left and $A1$ payoffs on the right.

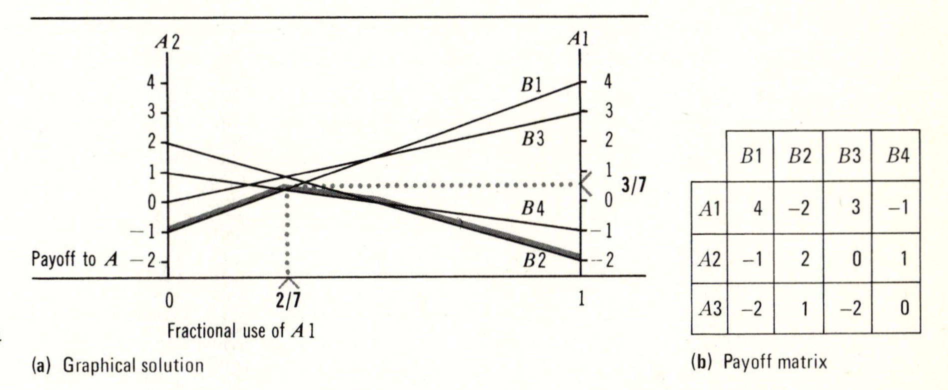


|  | B1 | B2 | B3 | B4 |
|---|---|---|---|---|
| A1 | 4 | −2 | 3 | −1 |
| A2 | −1 | 2 | 0 | 1 |
| A3 | −2 | 1 | −2 | 0 |

**FIGURE 15.12** 2 × 4 game solution. (a) Graphical solution (b) Payoff matrix

After the graph has been completed, there are two ways to determine the mixed strategies and value of the game. The first is to read the value of the game and the strategy for $A$ directly from the graph. If the scale were sufficiently large for accurate readings, the value of the game could be taken from Figure 15.12(a) as 3/7, and the strategy for $A$ as 2/7 $A1$ and 5/7 $A2$. The value of the game is the maximum amount $A$ can gain if $B$ uses the optimal strategy. This strategy is defined by the intersection of $B1$ and $B4$ at the highest point of $A$'s minimum gains. Knowing the value of the game and the payoffs for each of $B$'s limiting alternatives, we can set up the relations

$$x \times 4 + (1 - x)(-1) = 3/7$$
$$x(-1) + (1 - x)1 = 3/7$$

*where* $x$ = fractional use of $B1$
$1 - x$ = fractional use of $B4$

*An exception occurs when one player has an optimal pure strategy. Problem 15.10 illustrates this condition.

| | *B1** | *B4** |
|---|---|---|
| *A1* | 4 | −1 |
| *A2* | −1 | 1 |

*Limiting alternatives are selected graphically.

| | *B1* | *B4* | |
|---|---|---|---|
| *A1* | 4 | −1 | 2/7 |
| *A2* | −1 | 1 | 5/7 |
| | 2/7 | 5/7 | |

**FIGURE 15.13** Solution of a graphically reduced payoff matrix.

Then, solving for $x$,

$$4x + x - 1 = -x - x + 1$$
$$7x = 2$$
$$x = 2/7$$

we find that $B$ should use $B1$ two-sevenths of the time, and $B4$ five-sevenths of the time. $B2$ and $B3$ are never used. The validity of this strategy can be interpreted from the graph and confirmed by calculating the value of the game for any other strategy.

The other way to calculate the optimal mix of strategies is to let the graph just point out the two limiting alternatives and solve the evident 2 × 2 game. This method of solution for the sample problem is shown in Figure 15.13. A short-cut method to calculate the value of the game is given below.

$$\text{EV}(A) = 2/7 \times 4 + 5/7 \times (-1) = 2/7\,(-1) + 5/7 \times 1 = 3/7 \text{ units}$$

The solution of an $n \times 2$ game is analogous to that used for $2 \times n$ games. The payoff scales on the graph represent the player with two alternatives (by our previous notation, this would be player $B$). Then the alternatives are plotted for the other player. Since the payoffs are written in terms of $A$, the player with two alternatives ($B$) seeks to minimize the maximum losses. Therefore, ***the solution plane lies along the topmost set of lines.*** The minimax solution is the lowest point on the plane. The lines intersecting at this point indicate the limiting alternatives for $A$. An $n \times 2$ game is illustrated in the following example.

## Example 15.4 Competitive Selling Efforts

A new restaurant, Anton's, has opened in a city where the main competition is from Blue's restaurant. Anton's management has observed that Blue's promotional effort varies between reduced prices ($B1$) on certain nights and special menus ($B2$) on other nights. To attract customers from Blue's, Anton's can also reduce prices ($A1$), serve special meals ($A2$), or emphasize a takeout service ($A3$).

Both restaurants use the same advertising media to promote one special attraction each week. Because of differences in physical facilities and experience, each restaurant excels in certain specialties. Payoffs, the percentage of gain or loss in net revenue, have been estimated by Anton's management for the different outcomes shown.

| | | BLUE'S | |
|---|---|---|---|
| | | *B1* | *B2* |
| ANTON'S | *A1* | −3% | 6% |
| | *A2* | 2% | −6% |
| | *A3* | −1% | 5% |

What strategy should Anton's follow in scheduling its weekly specials?

**Solution 15.4** Anton's can expect a slight gain of 0.286 percent in revenue by serving special menus 44 percent of the time and featuring takeout meals 56 percent of the time.

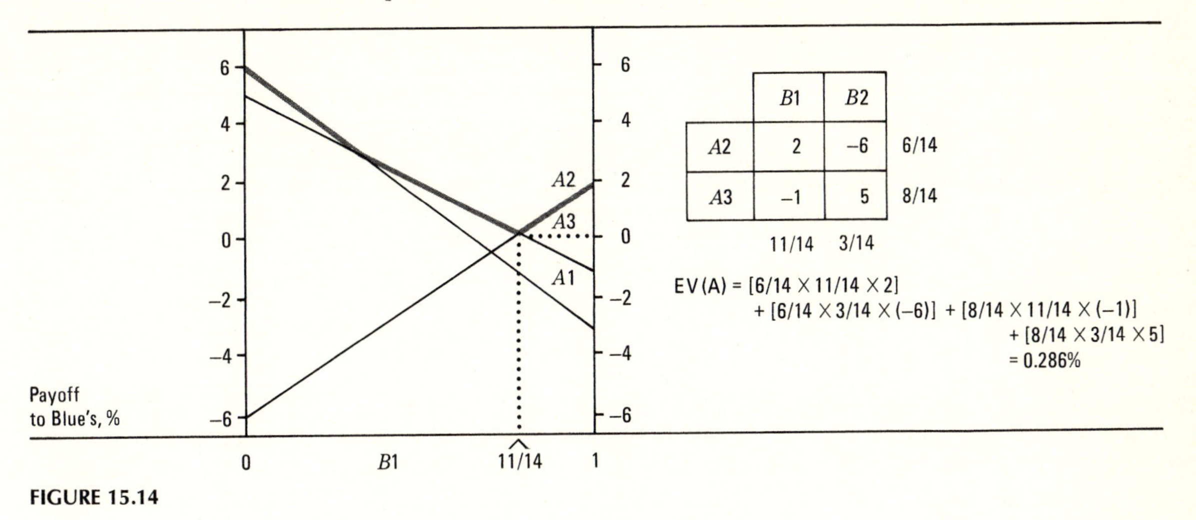


**FIGURE 15.14**

## $n \times n$ Games

When a game matrix has more than two alternatives per player, the first effort should be a check for dominance and a saddle point. If these steps fail to produce a solution or a matrix reduced to two alternatives, an exact solution by linear programming or other methods can be undertaken. These measures can be very tedious. There is also an approximation method which avoids mathematical complications and provides good estimates of strategy and the value of the game.

The approximation method is based on a series of fictitious plays of the game. In this respect it resembles the simulation methods we applied to queuing and inventory problems. The difference between the two methods is that the game returns are not a function of chance. When one player makes a play, the other player reacts by selecting the course of action which minimizes the gain of the opponent. In this way both players act under the assumption that past plays offer the best guide to the future. At each play an optimal pure strategy is selected according to the mixture of outcomes which represent all the opponent's past plays.

Detailed procedures for fictitious play can best be described by a sample application. A normal payoff matrix represents the contest. There can be any number of alternatives for either player. The sample problem shown in Figure 15.15 is a two-person, zero-sum, 3 × 3 game.

The value of the game is estimated by dividing the boldface payoff at each play by the number of that play. In Figure 15.16, the tenth play for $A$ shows a payoff of 22, and the tenth play for $B$ shows 18. Therefore the average payoff at the tenth play is 22/10 = 2.2 for $A$ and 18/10 = 1.8 for $B$. The value of the game lies between the ***highest*** average payoff to

*Step 1* One of player $A$'s alternatives is arbitrarily selected for use in the first play. We shall choose $A1$. The payoffs for $A1$ are then entered directly below the bottom row of the matrix.

| | B1 | B2 | B3 |
|---|---|---|---|
| A1 | 4 | 2 | 0 |
| A2 | 2 | 0 | 3 |
| A3 | 2 | 4 | 1 |
| | 4 | 2 | 0 |

*Step 2* The response of player $B$ to $A1$ is to select and play an alternative which minimizes the gain to $A$. Therefore the smallest payoff, 0, is selected (boldface), to indicate that $B3$ is $B$'s best counter to $A$'s use of $A1$.

| | B1 | B2 | B3 |
|---|---|---|---|
| A1 | 4 | 2 | 0 |
| A2 | 2 | 0 | 3 |
| A3 | 2 | 4 | 1 |
| | 4 | 2 | **0** |

*Step 3* Now $B$'s first play, the use of $B3$, is entered to the right of the matrix. The best pure strategy for $A$ against the use of $B3$ is $A2$. The largest number in the column, 3, indicates this choice.

| | B1 | B2 | B3 | |
|---|---|---|---|---|
| A1 | 4 | 2 | 0 | 0 |
| A2 | 2 | 0 | 3 | **3** |
| A3 | 2 | 4 | 1 | 1 |
| | 4 | 2 | **0** | |

*Step 4* Add the payoff from the alternative selected in step 3 ($A2$) to the last play by $A$ as shown below:

4 2 0 first play
2 0 3 second play
6 2 3 cumulative payoff

Enter the cumulative payoff as the next row. Again select the smallest number, 2.

| | B1 | B2 | B3 | |
|---|---|---|---|---|
| A1 | 4 | 2 | 0 | 0 |
| A2 | 2 | 0 | 3 | **3** |
| A3 | 2 | 4 | 1 | 1 |
| | 4 | 2 | **0** | |
| | 6 | **2** | 3 | |

*Step 5* Add the payoff from the indicated column ($B2$ from step 4) to the last column on the right of the matrix. Enter the payoff sums as a new column, and indicate the largest number.

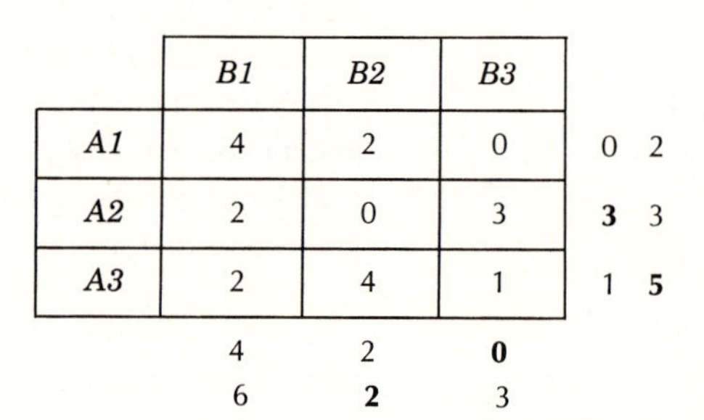

| | B1 | B2 | B3 | | |
|---|---|---|---|---|---|
| A1 | 4 | 2 | 0 | 0 | 2 |
| A2 | 2 | 0 | 3 | **3** | 3 |
| A3 | 2 | 4 | 1 | 1 | **5** |
| | 4 | 2 | **0** | | |
| | 6 | **2** | 3 | | |

*Step 6* For each successive play the boldface number designates the row or column alternative to be added to the cumulative payoff. When a tie for high or low payoffs exists, any consistently applied rule may be used to select the next number. In the sample problem, to break a tie we designate an alternative different from the one last chosen. The number of plays is continued until a desired accuracy is obtained. Figure 15.16 shows 20 iterations for the sample game.

**FIGURE 15.15**

$B$ (maximum loss or minimum gain) and the *lowest* average payoff to $A$. From the sample problem, the value of the game lies between 1.9 (occurring at play 20 for $B$) and 2.0, which occurs several times for $A$.

***The optimal strategy is approximated by the number of times each alternative is used during fictitious play. In our illustration,*** $A$ 2 and $A$ 3 were each used half the time, while $B$ 2 and $B$ 3 were used, respectively, for 7/20 and 13/20 of the plays. The proximity of the exact values to the approximate values is indicated in Figure 15.16.

| | B1 | B2 | B3 | 3.0 | 2.5 | 2.0 | 2.5 | 2.2 | 2.0 | 2.14 | 2.13 | 2.0 | 2.2 | 2.09 | 2.0 | 2.08 | 2.07 | 2.0 | 2.12 | 2.06 | 2.0 | 2.05 | 2.05 | | Av payoff |
|---|---|---|---|---|---|---|---|---|---|---|---|---|---|---|---|---|---|---|---|---|---|---|---|---|---|
| A1 | 4 | 2 | 0 | 0 | 2 | 2 | 4 | 4 | 4 | 4 | 6 | 6 | 8 | 8 | 8 | 8 | 10 | 10 | 12 | 12 | 12 | 12 | 14 | **0** | Response by A |
| A2 | 2 | 0 | 3 | **3** | 3 | **6*** | 6 | 9 | **12*** | **15** | 15 | **18*** | 18 | 21 | **24*** | **27** | 27 | **30*** | 30 | 33 | **36*** | **39** | 39 | **10** | |
| A3 | 2 | 4 | 1 | 1 | **5** | 6 | **10** | **11** | 12 | 13 | **17** | 18 | **22** | **23** | 24 | 25 | **29** | 30 | **34** | **35** | 36 | 34 | **41** | **10** | |
| | | | | 1 | 2 | 3 | 4 | 5 | 6 | 7 | 8 | 9 | 10 | 11 | 12 | 13 | 14 | 15 | 16 | 17 | 18 | 19 | 20 | | Play |

| Av payoff | Response by B | | | Play |
|---|---|---|---|---|
| 0.0 | 4 | 2 | **0** | 1 |
| 1.0 | 6 | **2** | 3 | 2 |
| 1.33 | 8 | 6 | **4** | 3 |
| 1.5 | 10 | **6** | 7 | 4 |
| 1.6 | 12 | 10 | **8** | 5 |
| 1.5 | 14 | 14 | **9** | 6 |
| 1.72 | 16 | 14 | **12** | 7 |
| 1.75 | 18 | **14** | 15 | 8 |
| 1.78 | 20 | 18 | **16** | 9 |
| 1.8 | 22 | **18** | 19 | 10 |
| 1.82 | 24 | 22 | **20** | 11 |
| 1.75 | 26 | 26 | **21** | 12 |
| 1.85 | 28 | 26 | **24** | 13 |
| 1.86 | 30 | **26** | 27 | 14 |
| 1.87 | 32 | 30 | **28** | 15 |
| 1.88 | 34 | **30** | 31 | 16 |
| 1.88 | 36 | 34 | **32** | 17 |
| 1.83 | 38 | 38 | **33** | 18 |
| 1.89 | 40 | 38 | **36** | 19 |
| 1.90 | 42 | **38** | 39 | 20 |
| | **0** | **7** | **13** | |

Results of ficticious play

Value of game between 1.9 and 2.0

Strategy for $A$:

| A1 | A2 | A3 |
|---|---|---|
| 0 | 10/20 | 10/20 |

Strategy for $B$:

| B1 | B2 | B3 |
|---|---|---|
| 0 | 7/20 | 13/20 |

Exact solution

Value of game 2.0

Mixed strategies:

| A1 | A2 | A3 | B1 | B2 | B3 |
|---|---|---|---|---|---|
| 0 | 1/2 | 1/2 | 0 | 1/3 | 2/3 |

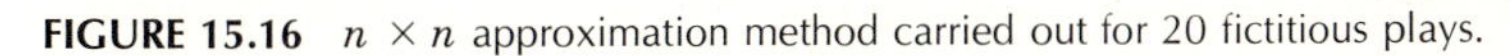

*Ties broken by avoiding the last used alternative

**FIGURE 15.16** $n \times n$ approximation method carried out for 20 fictitious plays.

The approximation method is a powerful tool for very large games. Its iterative nature makes it a natural candidate for machine computations. If there are several optimal solutions to a game, fictitious play will approximate one of them. It can be shown that the approximate solutions will converge on the actual solution with continued playing.

## NONZERO-SUM GAMES

Many, if not most, real-life games are nonzero-sum games. In a conflict between nations neither side completely wins; if anything, both sides in a war are losers, regardless of which side surrenders. On the other hand, in labor-management negotiations both sides may be winners; labor can win an increase in wages, while the company wins increased productivity. Perhaps there is collusion in the game by the two players, labor and management, against an unrepresented but respected third party, the consumers.

Bargaining positions occur in nonzero-sum games because different players place different values on the same outcomes. A conciliation offer by a food processor to farmers might be less than either side wanted, but it could be accepted to avoid consequences both sides fear. In a compromise each side gives up certain alternatives on the basis that the other side will not take advantage of the reduced game.

Unfortunately, there is no generally acceptable solution method for nonzero-sum games. However, the concepts of pure and mixed strategies, decision criteria, matrix representation, and game values provide considerable assistance in analyzing any conflict situation.

**Example 15.5** **Analysis of a Two-Person Nonzero-Sum Game**

Two large shopping centers are competing for the same market. The Alpha company attracts customers by staging contests ($A1$) and by special sales or loss leaders ($A2$). The other firm, Beta, also uses contests ($B1$) and special bargains ($B2$). In addition, Beta has the exclusive franchise in the area for trading stamps ($B3$). Both firms periodically change advertising emphasis in their promotions. During these periods the entire advertising budget is spent exclusively on only one of the promotional alternatives.

The estimated gains and losses for each possible outcome are shown in the accompanying payoff matrix. The payoffs represent the net daily change in revenue resulting from competing promotional activities.

| | | BETA | | |
|---|---|---|---|---|
| | | *B1* | *B2* | *B3* |
| ALPHA | *A1* | −$100 | 0 | $100 |
| | *A2* | $400 | $200 | −$300 |

Alpha is a new, locally owned firm. The company's utility for money is directly proportional to the amount gained or lost. Beta is one store of a large chain which has a conservative management philosophy. The utility each firm places on changes in daily revenue is indicated by the following utility functions (see Extension 15.1 for an explanation of utility theory.)

| | NET CHANGE, DOLLARS | | | | | | | | | | |
|---|---|---|---|---|---|---|---|---|---|---|---|
| | 500 | 400 | 300 | 200 | 100 | 000 | −100 | −200 | −300 | −400 | −500 |
| Alpha (U) | 1.0 | 0.9 | 0.8 | 0.7 | 0.6 | 0.5 | 0.4 | 0.3 | 0.2 | 0.1 | 0.0 |
| Beta (U) | 1.0 | 0.95 | 0.9 | 0.85 | 0.8 | 0.7 | 0.6 | 0.5 | 0.3 | 0.2 | 0.0 |

If the monetary payoffs are converted to utility payoffs, the game matrix takes the values shown, where the first payoff in each cell is the utility of that outcome to Alpha and the second number is Beta's utility payoff for the same outcome. Analyze the positions of the two competitors.

| | *B1* | *B2* | *B3* |
|---|---|---|---|
| *A1* | 4,8 | 5,7 | 6,6 |
| *A2* | 9,2 | 7,5 | 2,9 |

**Solution 15.5**

The usual check for dominance and a saddle point is unrewarding. Considerable insight can be gained by plotting graphs for both the utility payoffs. The resulting 3 × 2 games are then solved in terms of each player's utility function. From the first graph in Figure 15.17 it appears that Alpha can expect an average payoff of 5.11 utiles by using $A$ 1 seven-ninths of the time. The second graph indicates that this strategy for Alpha will allow a value of the game for Beta of 6.667 if Beta uses 1/3 $B$1 and 2/3 $B$3. If Beta decides to use maximum strategy for its own utility payoffs, the value of the game for Alpha would not be changed:

| | *B1* (1/3) | *B3* (2/3) |
|---|---|---|
| *A1* (7/9) | 4 | 6 |
| *A2* (2/9) | 9 | 2 |

$$\begin{aligned} \text{EV}(A) &= 7/27 \times 4 + 14/27 \times 6 + 2/27 \times 9 + 4/27 \times 2 \\ &= 5.11 \end{aligned}$$

Therefore both firms should be quite pleased with their mixed strategy; Alpha receives a slight cash gain ($11.11 per play), and Beta's concern for maintaining the status quo (as implied by its utility function) is satisfied by the high utility payoff.

The competition might include other factors that could lead to collusion, conciliation, or compromise. Perhaps both firms consider contests too expensive and difficult to administer. Alpha might avoid using contests if Beta would agree to use $B$2 70 percent of the time and $B$3 30 percent of the time. This strategy would allow an average utility for Alpha of $0.7 \times 7 + 0.3 \times 2 = 5.5$, and Beta's utility would be $0.7 \times 5 + 0.3 \times 9 = 6.2$.

If Beta wanted to stop staging contests and special bargains ($B$1 and $B$2), it might get Alpha to agree to use only $A$1. This compromise would allow both firms an equal utility payoff, 6. Beta would be giving Alpha an increase in average payoff for the privilege of administering only trading stamps. There are many such bargaining positions that can be established in nonzero-sum games.

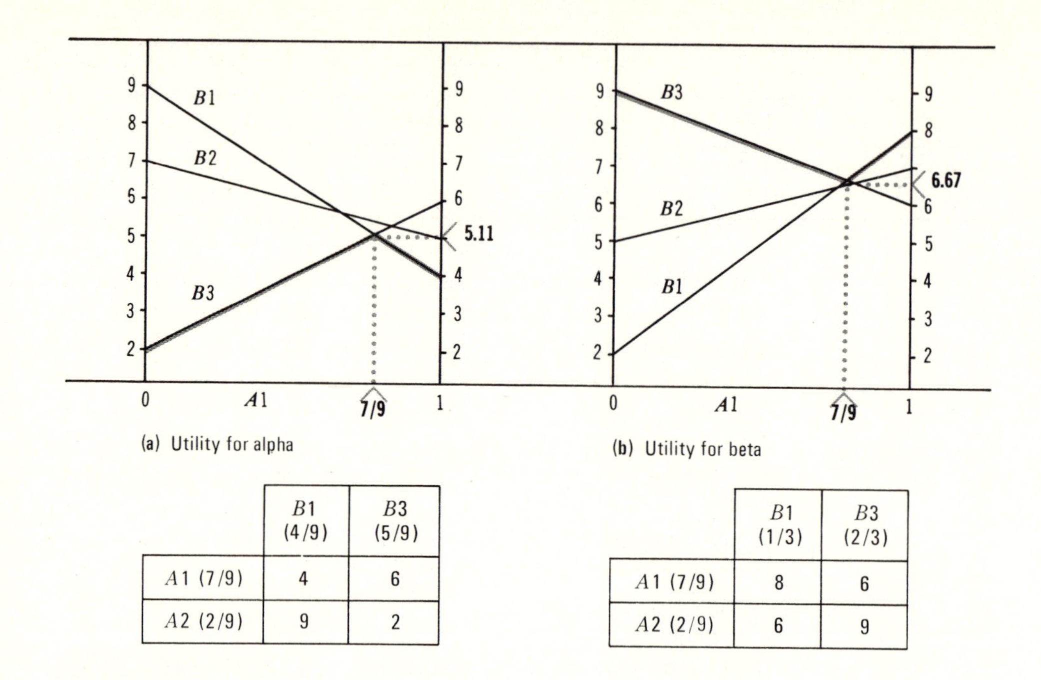


| | B1 (4/9) | B3 (5/9) |
|---|---|---|
| A1 (7/9) | 4 | 6 |
| A2 (2/9) | 9 | 2 |

| | B1 (1/3) | B3 (2/3) |
|---|---|---|
| A1 (7/9) | 8 | 6 |
| A2 (2/9) | 6 | 9 |

**FIGURE 15.17**

## PRACTICALITY OF GAME THEORY

Game theory is not a panacea for decision makers, nor is it a purely theoretical exercise in logic. While few practical industrial applications have been realized, it must be remembered that game theory is still a relatively new approach to decision making. Proponents of game theory believe continued research and development will make it more applicable to realistic problems. Some of the main difficulties preventing more widespread usage are:

1. Assigning meaningful payoffs
2. Solving very large matrices
3. Handling nonzero-sum and multiplayer games
4. Taking into consideration the possibility of collusion, conciliation, irrational players, and conditions which do not conform to the conventional game assumptions

Despite the difficulties, there is much to be gained from just an awareness of the reasoning behind game theory. It forces attention to an opponent's strategy as well as our own. Payoffs have to be quantified, ranked, or otherwise evaluated. The potential of mixed strategy is stressed. It suggests a philosophy for dealing with human conflict, an area in which decision makers have long needed assistance.

## SUMMARY

A decision under ***uncertainty*** is characterized by competence to estimate outcomes for possible future states but inability to estimate the probability of occurrence for those states. As a consequence, criteria for selection of a promising alternative are a function of the personal perspective of the decision maker.

Several *principles of choice* have been developed to systematize *noncompetitive decisions*—choices made in a neutral decision environment. Once the outcomes of alternatives are arranged in a payoff matrix, attention to the best or worst outcomes leads to the *maximin* or *maximax* criterion. An intermediate outlook is defined by a coefficient of optimism in the *Hurwicz* criterion. Opportunity costs assigned to less favored alternatives guide decisions according to the *minimax-regret* criterion. The *equal-likelihood* criterion registers the belief that each state must have an equal chance to occur if there is insufficient reason to rate one more likely than another. Each criterion has certain unattractive properties, but they are still useful in directing attention to the attitudes and objectives of the decision makers.

*Game theory* is the basis for *competitive* decisions under uncertainty. In a *zero-sum, two-person* game, two opponents have equal knowledge of outcomes, and the winnings of one are the losses of the other. An optimal *pure* or *mixed* strategy, selecting a single course of action or randomly mixing actions within given proportions, can be calculated to assure a minimum return, the value of the game. Strategies for multiplayer and nonzero-sum games are complicated to formulate, but future developments in these areas could increase the practicality and acceptance of game-theory analyses for competitive situations.

FIGURE 15.18

## PROBLEMS

**15.1** The profit expected from four alternative courses of action, *A, B, C,* and *D,* under four states of nature, is given on the next page:

| | *1* | *2* | *3* | *4* |
|---|---|---|---|---|
| *A* | 7 | 9 | 5 | 2 |
| *B* | 8 | 1 | 10 | 4 |
| *C* | 6 | 6 | 6 | 6 |
| *D* | 5 | 7 | 9 | 8 |

**15.1a** Which alternative would be selected by applying each of the following criteria: maximax, maximin, regret, and equal likelihood?
**15.1b** Subtract 12 from each number in the matrix, and multiply the resulting difference by 3. Apply the criteria from Problem 15.1*a* to the modified payoffs. What do the results indicate?
**15.1c** Suppose the states of nature are hair colors: brown, black, blonde, and red. A payoff will be made according to the color of hair of the next person that passes. You do not know which state applies to each hair color. Which alternative would you choose if the numbers in the matrix represented $1000 bills? If the numbers in the matrix represented $1 bills? Which criterion do the above choices resemble?
**15.1d** Add 10 to each number in the first column of the original matrix. Apply the same criteria as in Problem 15.1*a*. What do the results indicate?

**15.2** Apply the Hurwicz criterion to the matrix below, and show how different degrees of optimism will affect the selection of a preferred alternative:

| | *V* | *W* | *X* | *Y* | *Z* |
|---|---|---|---|---|---|
| *A* | −8 | 6 | 2 | −5 | 4 |
| *B* | 4 | −2 | −3 | 3 | 2 |
| *C* | 0 | 1 | 2 | 1 | 0 |
| *D* | 6 | 11 | −10 | 4 | −7 |

If each payoff represented years added to or subtracted from your life expectancy, which alternative would you choose?

**15.3** An enterprising young man read about a national Nature Club meeting to be held in a remote desert setting. He visualizes a great potential for the sale of ice-cream products during their meeting in the hot desert. He can carry up to 3000 ice-cream bars in a rented refrigeration truck. The bars will be sold for $.50 each and cost only $.05. Any bars he fails to sell will have no salvage value because of melting and refreezing. His only problem is that he does not know how many will attend the meeting or how many of those that do attend will buy his products. A rough guess is that 750 people will attend, and his expenses, exclusive of the cost of ice-cream bars, would be $125 to make the trip. What criterion would you suggest he use to help decide whether or not to go and how many bars to take? Why? What preference is indicated? Assume that his alternatives are increments of 500 bars.

**15.4** Figure 14.17(b) gave the expected costs of lead-time alternatives for a chemical-inventory problem. The solution to the problem was based on an estimated distribution of possible lead times. Now assume that nothing is known about the expected lead times except that they will vary from 7 to 14 days. Apply the minimax and equal-likelihood criteria to select a safety-stock size. Base your conclusion on the direct-holding and opportunity costs, without regard to economic lot size.

**15.5** The expected rates of return for investment in securities and investment in expanded plant facilities are estimated for two levels of future business activity:

| | *Recession* | *Inflation* |
|---|---|---|
| Securities | 5 | 7 |
| Expansion | 1 | 15 |

**15.5a** The company is undecided as to the likelihood of each of the future conditions. It has been suggested that each future be considered equally likely. Then the probability of each future at which the two alternatives are equivalent can be calculated. These are "indifference probabilities." The decision rule would be to select the alternative which has the highest return for the future state that has the greatest difference between the equal-likelihood probability and the indifference probability. Apply this decision rule, and comment on the results.

**15.5b** Consider the states of business activity as being controlled by an intelligent opponent. What strategy should the company use under this assumption?

**15.6** Determine the optimal strategies and values of the following 2 × 2 two-person zero-sum games:

| | |
|---|---|
| −1 | 6 |
| 7 | 2 |

**15.6a**

| | |
|---|---|
| 4 | 6 |
| −1 | 3 |

**15.6b**

| | |
|---|---|
| −11 | 0 |
| −1 | −4 |

**15.6c** *(EV = −3.14)*

| | |
|---|---|
| 8 | 2 |
| 6 | 1 |

**15.6d** *(EV = 2)*

**15.7** Determine the optimal strategies and values of the following 2 × *n* and *n* × 2 two-person zero-sum games:

| | | | |
|---|---|---|---|
| −3 | 4 | 1 | −3 |
| 5 | 0 | 6 | −5 |

**15.7a** *(EV = −3)*

| | | |
|---|---|---|
| −8 | −1 | −7 |
| −2 | 6 | 9 |

**15.7b**

| | | | |
|---|---|---|---|
| 4 | 0 | 1 | 7 |
| 2 | 8 | 4 | 2 |

**15.7c** *(EV = 2.8)*

| | |
|---|---|
| 7 | 1 |
| 3 | 0 |
| 4 | 8 |

**15.7d**

| | |
|---|---|
| −6 | 9 |
| −10 | −10 |
| 8 | −4 |

**15.7e** *(EV = 1.78)*

| | |
|---|---|
| 7 | 2 |
| 8 | 0 |
| 4 | 7 |

**15.7f**

| | |
|---|---|
| 6 | −4 |
| −1 | 3 |
| 4 | −2 |

**15.7g** *(EV = 1)*

**15.8** Determine the optimal strategies and values of the following $n \times n$ two-person zero-sum games:

| | | | |
|---|---|---|---|
| 8 | 4 | 2 | −1 |
| 9 | 2 | 4 | 3 |
| 4 | −5 | 3 | 0 |
| 3 | −1 | 5 | 2 |
| 7 | 3 | 0 | −3 |

**15.8a** *(EV = 2.33)*

| | | |
|---|---|---|
| 2 | −4 | 3 |
| 5 | 0 | −2 |
| 4 | 1 | 0 |
| −2 | 3 | 1 |

**15.8b**

| | | | | |
|---|---|---|---|---|
| 1 | 6 | 3 | −4 | 0 |
| −2 | 7 | −1 | 1 | 2 |
| 3 | 2 | 5 | −2 | 4 |

**15.8c** *(EV = −0.125)*

| | | | |
|---|---|---|---|
| 4 | −1 | −7 | 2 |
| −2 | 0 | 3 | 4 |
| 1 | 2 | −1 | −6 |

**15.8d**

| | | | | |
|---|---|---|---|---|
| 400 | 100 | 200 | 0 | −500 |
| 0 | −100 | 300 | −200 | −400 |
| 500 | 0 | 300 | 0 | −100 |
| 400 | 200 | −100 | 100 | 100 |
| 200 | 100 | 300 | 100 | −100 |

**15.8e**

| | | |
|---|---|---|
| 1 | −2 | 3 |
| −2 | 3 | −4 |
| 3 | −4 | 5 |

**15.8f** *(EV = 0)*

**15.9** Abner devised a new card game to play with Betty during the long winter months. The new game uses only eight playing cards: the ace, king, queen, and jack of spades and hearts. Each player has four different face cards in one suit. At each play a particular card is selected by each player from the four available. Each player shows the other the selected card at the same time. Then they turn to the matrix below to see how many points were scored.

| | | BETTY | | | |
|---|---|---|---|---|---|
| | | *A* | *K* | *Q* | *J* |
| ABNER | *A* | −2 | 3 | 4 | −1 |
| | *K* | 4 | 1 | −1 | 0 |
| | *Q* | −3 | −2 | 3 | −2 |
| | *J* | 3 | 0 | −3 | −4 |

What strategy should each player use? Did Abner set up the game for his benefit?

*(Betty: Q = 1/6, J = 5/6; Abner: A = 1/6, K = 5/6)*

**15.10** What strategy should $A$ and $B$ use for the game described? What is the value of the game?

| | $B1$ | $B2$ | $B3$ |
|---|---|---|---|
| $A1$ | −$1000 | $3000 | $1000 |
| $A2$ | $4000 | 0 | $1000 |

($A1 = 35\%$; $B3 = 100\%$; $EV = \$1000$)

**15.11** Two ruthless but modern underworld gangs are competing in the same city. Once a month they must make a decision as to where to set up their illegal gambling operations to minimize police interference. Some locations are more profitable than others. The outcomes of the competing alternatives are indicated in the matrix, where the payoffs are in thousands of dollars:

| ANGELS \ BANDITS | $B1$ | $B2$ | $B3$ |
|---|---|---|---|
| $A1$ | −5 | 8 | 6 |
| $A2$ | 4 | −3 | 2 |
| $A3$ | −10 | 7 | −2 |

Suppose the Angel gang were able to place a spy in the Bandit gang. What amount could the Angels afford to pay the spy for perfect information about where the Bandits plan to stage each new operation?

**15.12** A hotel has 100 rooms that rent for an average of $22 per day. The fixed cost per room is $9 per day, and the variable cost is $4 per day. The hotel policy is to accept more reservations than there are rooms available during rush periods. They have found that between 0 and 8 percent of those reserving rooms will fail to claim them. A disgruntled customer who has reserved a room but finds none available is estimated to cost the hotel $50 in loss of future business. Analyze the situation as a competitive and noncompetitive problem. How many rooms should be rented under each condition? Assume that no deposit is required with a reservation.

**15.13** Two very large firms with big government contracts have a serious shortage of technical people. Both firms attempt to fill their personnel needs by recruiting in colleges and by pirating from other firms. Company $B$ has a very desirable location and pays excellent extra benefits. Company $A$ is not so well situated but is known for its enlightened management and superb working conditions. Because of their respective reputations, $B$ appeals to recent college graduates, and $A$ is more attractive to people with industrial experience.

Both companies have several teams scouring the country for workers. In any given locality a team can attempt to (1) recruit or (2) pirate, but not both at one time because of time and personnel limitations. Often, teams from both companies meet in one locality, which sets up the conflict situation shown in the following matrix:

| | *B1* | *B2* |
|---|---|---|
| *A1* | −4,−1 | 0,0 |
| *A2* | −1,0 | −1,−3 |

The payoffs denote extra thousands of dollars beyond an average annual salary required to hire an employee under the conditions of each outcome. The first number in each cell refers to $A$'s payoff, and the second to $B$'s.

**15.13a** Analyze the problem, and suggest independent solutions for each company.

**15.13b** Comment on the possibility of collusion. Which company has the strongest bargaining position? Suggest a possible compromise solution.

**15.14** Two companies compete in a relatively small market with very similar products. Ajax is the larger of the two and sells approximately twice as many units as Besta. Since the products sold by Ajax and Besta serve the same purpose with about equal efficiency, each company attempts to elevate its product through consumer-oriented promotional plans. Three basic promotions are currently in vogue: (1) television and magazine advertising, (2) special packaging with bonus "gifts," and (3) short-term price reductions advertised in stores and local newspapers. Alternatives 2 and 3 tend to differentiate the two brands of product, but alternative 1 tends to increase the general awareness of the buying public to the product type without regard to the manufacturer. Thus, a strong advertising policy by Ajax would also increase the sales of Besta. Since Ajax is larger and works with a bigger budget, it benefits proportionately less from Besta's advertising than Besta does from Ajax's. Other alternatives also show a carry-over effect between the gains registered by each company.

Ajax has developed the payoff estimates for the most commonly used alternatives of the two companies. The values are in percent of net gain or increased increment of profit expected from the following various alternatives during one quarter. The payoffs for Ajax from advertising by television and magazines ($A1$) depend on the extent of advertising conducted; the values for Besta are based on an extrapolation of historical data:

| | | BESTA | | |
|---|---|---|---|---|
| | | *B1* | *B2* | *B3* |
| AJAX | *A1* | $x$,12 | $y$,21 | $z$,18 |
| | *A2* | −6,4 | 26,−12 | −12,−20 |
| | *A3* | 4,0 | 9,9 | −6,0 |

**15.14a** What strategy should Ajax follow to maximize gains if $x$, $y$, and $z$ are respectively, 14, 8, and 12? Discuss the pros and cons of the alternatives open to Ajax.

**15.14b** Based on the given values of $x$, $y$, and $z$, what is the best strategy for Besta?

**15.14c** Ajax recognizes that Besta benefits from an advertising campaign conducted by Ajax, but the reciprocal is not true. Ajax believes that its advertising budget can be manipulated from the level given in Problem 15.14*a* so that for every 2 percent reduction in profit for Ajax, Besta's profit will decrease 3 percent. However, if advertising is increased by Ajax, Besta will gain 3 percent for every 1 percent increase. The maximum and minimum payoffs possible for Ajax are 20 percent and 0. What strategy combined with what advertising policy will be most advantageous for Ajax?

## EXTENSIONS

***15.1 The Engineering Economist's Dilemma*** The classic game-theory fable known as the *prisoner's dilemma* illustrates a decision situation that is not strictly competitive. It highlights considerations of trust and accommodation that are factors in many decisions faced by engineering economists.

The two prisoners in the fable were caught committing a crime and were immediately jailed in separate cells where they could not communicate. A police officer interviews each individually to deliver the same message: "I have enough evidence now to send you to jail for 18 months. If you confess to the crime that usually deserves a 10-year jail term, I'll let you off with a 6-month sentence, providing you will testify against your partner. If you both confess, you will both be sentenced to 3 years."

The payoff matrix, written in terms of lost years for prisoners 1 and 2, is shown in Figure 15.19. Prisoner 1's loss is shown in the lower left of each cell, and the loss for prisoner 2 is at the upper right. Neither prisoner knows what the other will do. Prisoner 1 wonders what prisoner 2 will do, and prisoner 2 thinks about what prisoner 1 is thinking: "Should I confess to play it safe with the possibility of getting a really short term, or should I hold out in hope that my partner will also hold out so we'll both get only 18 months? But if I hold out and my partner doesn't, I'll be stuck for 10 years." The equilibrium outcomes for confessions from both prisoners suggest the generally preferred strategy: confess.

| | | PRISONER 2 | |
|---|---|---|---|
| | | *Don't Confess* | *Confess* |
| PRISONER 1 | *Don't Confess* | −18 months (upper right); −18 months (lower left) | −6 months (upper right); −10 years (lower left) |
| | *Confess* | −10 years (upper right); −6 months (lower left) | −3 years (upper right); −3 years (lower left) |

**FIGURE 15.19** Possible sentences for two prisoners caught jointly committing a crime.

Hopefully no engineering economist will be faced with a decision about confessing to a crime, but most likely somewhat similar circumstances will be faced in advising which actions will be most compatible with an adversary's probable strategy. For example, replace the prisoners with two competing firms supplying the same product. For simplicity, say the options are to price the product either high or low. Each firm would profit most by undercutting the other's price and least by being undercut. Assume too that both would profit more when both charge high than when both charge low. The resulting payoff matrix has exactly the same pattern as that for the prisoner's dilemma. The most reasonable independent course of action would appear to be to set low prices.

But how should the analyst explain this advice to set low prices when there is a possibility of much higher profits from setting a high price? Should a suggestion be made to initiate communication between the competitors to seek a mutually advantageous price? Such counsel could lead the analyst to a practical experience of the prisoner's dilemma if price-fixing charges were proved. Should the company state its strategy ahead of an actual price increase and hope its competitor will "rationally" agree to raise its price too? However, a unilateral price increase might provoke consumer wrath and hurt sales regardless of what the competitor did.

Competitive situations are not exclusively restricted to industry. In the public sector it is often necessary to select from alternatives that will solicit differing degrees of support or opposition. Future states may be sensitive to "opposition" activities such as law suits, budget curtailments, new legislation, adverse publicity, and, in extreme cases, voter approval or rejection. Consider the following situations, and recommend a strategy for player *A*. Discuss the influencing factors that led to your recommendation.

## QUESTIONS

**15.1a** A government agency must provide increased capacity for handling nuclear wastes. The agency can expand and modernize its current noncontroversial facility, or it can build a more efficient facility at a new site. Constructing at a new site would likely cause considerable public discussion and opposition which would increase costs if prolonged. The present worths of net benefits from the two plans, in relation to delays and modifications that might arise from public agitation, are shown below in the lower left of each cell.

PUBLIC RESPONSE

| | *Accept* | *Oppose* |
|---|---|---|
| *A1: Expand at present site* | 5<br>$4 | 3<br>$3 |
| *A2: Develop at a new site* | 1<br>$10 | 10<br>$2 |

The strategies may be opposed by local groups backed by national environmentalist organizations. Payoffs for the opposition are rated on a scale of 1 to 10, with 10 being the most rewarding. The groups would gain most from stopping or delaying a new site development. Their credibility would suffer most from accepting the new site without a protest.

What should the agency do? What factors are implied in the agency outcomes that might be negotiable?

**15.1b** Two companies have nearly identical pulpwood plants. Both have been accused by local citizens' groups of being too slow in adopting air-quality-improvement equipment to reduce discharges of noxious odors and pollutants. Resentment is growing, even though it is realized that the local economy would suffer from a plant closure; excusing the odor as the "smell of money" is no longer a pervasive argument.

The discharge-control equipment is expensive. Both firms have been reluctant to make the investment until forced to. Now they face the prisoner's dilemma from legal suits brought against them. If only one plant admits it should have cleaned up its discharges earlier and will do so now, it will gain public goodwill and avoid stiff fines that will be levied against the plant that continues to resist. If both plants continue to claim they cannot afford the equipment and still remain competitive with plants in other locations, the eventual total outlay will probably be smaller than if both agree to install the needed equipment immediately.

Assume the proportions of gains and losses are the same as the outcomes in the payoff matrix for the prisoner's dilemma (Figure 15.19). What would you advise? Why? What other factors not expressly noted in the payoff matrix might influence the decision?

***15.2 Utility Theory*** The value attached to an outcome for any alternative and a given future can vary among individuals or from one business enterprise to another. Response to risk is a very personal matter. It permeates our daily activities without causing particular awareness. Most pedestrians crossing a street probably know that their chances of being hit by an automobile are significantly greater than zero. Some apparently choose to ignore or consider insignificant the probability of an accident as they blithely jaywalk through a busy intersection. Others, by virtue of training or awareness of the value of safety, wait for the proper time to cross. A minority, because of an extreme sensitivity to the possibility of an accident, wait until no cars at all are in sight before attempting to cross. With such divergent reactions to danger, it is logical that each of the three groups would place a different value on an insurance policy for pedestrians. The *utility* of the policy would be unique to each.

Not only does the utility vary among parties, it also varies with time. In the pedestrian example, either the jaywalker or the normally cautious street crosser might be converted to the hypercautious category after being an eyewitness to a serious accident. Putting the situation in monetary terms, we would anticipate that the witness would put a higher value on the insurance policy after viewing the accident. If such eyewitnesses could buy protection, they would probably be willing to pay more for it after the accident, even though the probability that an accident will occur to them has not changed.

An individual or corporation places a value on an alternative which achieves a balance between the potential return and the probability of receiving those returns. Under common conditions a gamble may be attractive to one party, while a conservative approach is desirable to another. Later, the positions could be reversed. Each party rates the anticipated returns of an alternative in relation to its chance of successful culmination in a manner that reflects their current economic status.

**Example 15.6 Reaction to Risk**

Suppose each of two university students was offered a proposition of putting up \$100 for a double-or-nothing return on the throw of a single die. Each would win \$100 if a 3, 4, 5, or 6 showed, or would lose the ante if a 1 or a 2 turned up. Assume that both students have the same gambling instincts, but one is working for school fees and the other is the recent recipient of a large inheritance. The alternatives for acceptance or rejection of the proposition are expressed in the payoff table:

| | *Win* (2/3) | *Lose* (1/3) |
|---|---|---|
| *Accept* | $100 | −$100 |
| *Reject* | 0 | 0 |

The two students would undoubtedly consider the proposition with different feelings. The lucky student with ample resources would probably consider it a "good" bet. This student might question the reason behind the odds or the fairness of the die, but if the probabilities appeared true, would likely accept the proposition because of the adventure aspects and because the consequence of a $100 loss would not be disastrous. On the other hand, the working student would quite possibly reject the proposition because the loss of $100 would be a painful experience. The working student would appreciate the gain, but even with favorable odds, the potential reverse in fortunes would outweigh the opportunity of winning. This student might accept the proposition if the odds of winning were upped to 10 to 1, or if the original odds could be applied to a smaller sum, such as $10. The reaction to risk is a function of the degree of that risk relative to the importance of the gain or loss.

**REFERENCE CONTRACT**

The concept of a reference contract is a means of taking into account the unique value that individual parties place on different alternatives. To develop this reference contract, we must determine the amount of money a party would demand or be willing to pay to be relieved of the obligation stated in a proposition. The fact that individuals are often inconsistent in stating their preferences for specified combinations of risk and the associated consequences necessitates a measure of caution in the analysis.

| | *S1* (0.5) | *S2* (0.3) | *S3* (0.2) |
|---|---|---|---|
| *C* | $80 | 0 | −$40 |
| *T* | $100 | $40 | −$100 |
| *N* | 0 | 0 | 0 |

Returning to the two students described in Example 15.6, consider their reactions to the opportunity of investing the money instead in campus services. The three alternatives are to invest in a food catering service $C$, or a tutoring service $T$, or to do nothing $N$. The payoff table shows the returns to be expected. The probability of success $S$ in the ventures is based on enrollment in the university and the general state of the economy. We assume the stated probabilities and the expected rewards are realistic. The outcomes are the returns from an investment of $100 over a year's period.

Let us look first at the preferences of the wealthy student, whom we shall call Mr. Loaded. He is, at the time of the proposition, inclined to be a risk taker. He feels that he will not be hurt by the consequences of a poor investment of this magnitude but will receive disproportionately high satisfaction from a success.

By asking him to relate the value he attaches to wins or losses associated with various probabilities, we hope to establish the cash equivalence of his preferences. We shall consider a range of values from a gain of $100 to a loss of $100, although we could use any scale. The starting question could be, "What amount of cash would you be willing to accept in lieu of a contract that assures you (probability of 1.0) a gain of $100?" Any sensible person would ask at least $100, even though hoping to receive more for it. Next we could ask what amount he would take in place of a contract that gave him an 80 percent chance of winning $100 and a 20 percent chance of losing $100. Since he is a risk taker, he would probably ask for at least $80. This means he places an $80 value on the opportunity to win $100 at a probability of 0.8 with the associated 20 percent chance of losing $100. We could say he is indifferent to the two alternatives. By continuing this procedure through a selected range of discrete probabilities, we would obtain his reference contract or *utility function* as depicted in the following table:

| | | REFERENCE CONTRACT | |
|---|---|---|---|
| *P($100 loss)* | *P($100 gain)* | *Mr. Loaded* | *Ms. Broke* |
| 0.0 | 1.0 | $100 | $100 |
| 0.2 | 0.8 | 80 | 40 |
| 0.4 | 0.6 | 40 | 0 |
| 0.6 | 0.4 | 0 | −40 |
| 0.8 | 0.2 | −40 | −70 |
| 1.0 | 0.0 | −100 | −100 |

The preferences for the working student, denoted as Ms. Broke in the table, could be determined by the same procedure. Because of her tight financial condition, we would expect her to try to avoid risk. Her cautious nature is evident in her response to the offer of an 80 percent chance of a $100 gain versus a 20 percent chance of a $100 loss. She would be willing to accept $40 in place of the alternative. That is, she is indifferent to a sure gain of $40 in place of winning $100 with a probability of 0.8 or losing $100 with a probability of 0.2. The table continues to reflect her reluctance to incur a large debt. She would pay $40 to be relieved of a proposition that gave her only a 40 percent chance of winning and a 60 percent chance of losing $100.

**TRANSLATED CONTRACT**

Now that we have a utility index for each of the two students, we can personalize the original contract to indicate individual preferences. In the first statement of the payoffs the expected values of the alternatives were

$$\text{EV}(C) = \$80 \times 0.5 + 0 \times 0.3 + (-\$40) \times 0.2 = \$32$$
$$\text{EV}(T) = \$100 \times 0.5 + \$40 \times 0.3 + (-\$100) \times 0.2 = \$42$$
$$\text{EV}(N) = 0 + 0 + 0 = \$0$$

An immediate assumption is that a reasonable person would always choose the tutoring investment. However, by expressing the outcome in terms of the utility of each investment as expressed in the reference contract, we can see how conclusions would vary according to individual preferences. This is done by substituting the equivalent probability, or utility rating, of each return for the monetary outcome. Thus, the dollar payoff table is translated to a utility payoff table.

| | S1 (0.5) | S2 (0.3) | S3 (0.2) |
|---|---|---|---|
| C | $80 | 0 | −$40 |
| T | $100 | $40 | −$100 |
| N | 0 | 0 | 0 |

(a) ORIGINAL PAYOFF

| | S1 | S2 | S3 |
|---|---|---|---|
| C | 0.8 | 0.4 | 0.2 |
| T | 1.0 | 0.6 | 0.0 |
| N | 0.4 | 0.4 | 0.4 |

(b) MR. LOADED'S UTILITY PAYOFF

| | S1 | S2 | S3 |
|---|---|---|---|
| C | 0.96 | 0.6 | 0.4 |
| T | 1.0 | 0.8 | 0.0 |
| N | 0.6 | 0.6 | 0.6 |

(c) MS. BROKE'S UTILITY PAYOFF

**FIGURE 15.20**

Then the expectation for the utility of each alternative is:

*Mr. Loaded*

$EV(C) = 0.5 \times 0.8 + 0.3 \times 0.4 + 0.2 \times 0.2 = 0.56$

$\mathrm{EV}(T) = 0.5 \times 1.0 + 0.3 \times 0.6 + 0.2 \times 0.0 = 0.68$

$\mathrm{EV}(N) = 0.40$

*Ms. Broke*

$\mathrm{EV}(C) = 0.5 \times 0.96 + 0.3 \times 0.6 + 0.2 \times 0.4 = 0.74$

$\mathrm{EV}(T) = 0.5 \times 1.0 + 0.3 \times 0.8 + 0.2 \times 0.0 = 0.74$

$\mathrm{EV}(N) = 0.60$

Mr. Loaded, for instance, is indifferent to a sure reward of $80 and the probability of 0.8 of a $100 gain. Therefore we can substitute 0.8 for the $80 in his payoff table. This procedure is repeated for all the outcomes for each student and is shown in Figure 15.20 along with the original payoff table.

If the goal is to maximize the expected utility, Mr. Loaded would reasonably select an investment in the tutoring service, but Ms. Broke would be highly uncertain. Both investments appear to be equal if we assume that Ms. Broke has been consistent in her stated preferences and that neither investment is too much larger than the utility she places on doing nothing. She should obviously investigate the investment potentials further and assess the value of any other influencing factors.

## Example 15.7 Translated Contracts

A small company is faced with the prospect that one of its products will shortly be outmoded by new developments achieved by a competitor. It would be possible to launch a crash program to develop a new version of the product that would be equal to or better than the competition. However, there is a better than even chance that neither the company's own new version nor the competitor's will receive significantly greater acceptance than the current model. A suggested compromise between doing nothing and developing a new model is to retain the present model but increase the sales budget. From a market survey on the likelihood of acceptance of the new product and the plant engineers' estimates of anticipated developmental costs, the following payoff table was obtained:

| | NEW MODEL | | |
|---|---|---|---|
| *Alternative* | *Acceptance (0.4)* | *Rejection (0.6)* | *Expected Value* |
| Develop new model ($A$) | $300,000 | −$200,000 | $ 0 |
| Increase sales budget ($B$) | −50,000 | 100,000 | 40,000 |
| Retain present model ($C$) | −100,000 | 0 | −40,000 |

Regardless of the reaction to risk, it is apparent that alternative $C$ can be eliminated, because $B$ is a more attractive choice in both possible futures. If the new product fails, $B$ earns a reward of $100,000, compared with no gain for $C$; and if the new version succeeds, the expected loss from $B$ will be $50,000 less than that from $C$. We can say that $B$ *dominates* $C$. An alternative that is dominated by another can be eliminated from the decision process.

| *Utility* | *Dollars* |
|---|---|
| 0.00 | −400,000 |
| 0.06 | −200,000 |
| 0.12 | −100,000 |
| 0.16 | −50,000 |
| 0.20 | 0 |
| 0.30 | 100,000 |
| 0.45 | 200,000 |
| 0.65 | 300,000 |
| 0.95 | 400,000 |

The next step is to consider the utility function of the company. It is a small but ambitious firm. It has made rapid growth, and its products are well diversified. A loss of $200,000 would certainly be serious, but other successful products could carry the company without too much hardship. The management feels that it needs a major accomplishment to build its reputation. A significant success would improve prestige and enlarge the horizons of the whole organization. The company has a proven engineering staff and is willing to gamble. The accompanying utility index describes its current attitude.

Judging just from the described attitude, we might guess that the company would decide to choose alternative $A$, even though the expected dollar value is zero. This is confirmed by converting the payoffs to utilities and calculating the expected value.

| | NEW MODEL | | |
|---|---|---|---|
| *Alternative* | *P(A) (0.4)* | *P(R) (0.6)* | *Expected Value (utility)* |
| $A$ | 0.65 | 0.06 | 0.296 |
| $B$ | 0.16 | 0.30 | 0.244 |
| $C$ | 0.12 | 0.20 | 0.168 |

We can make additional insights about utility functions by plotting the utilities of the

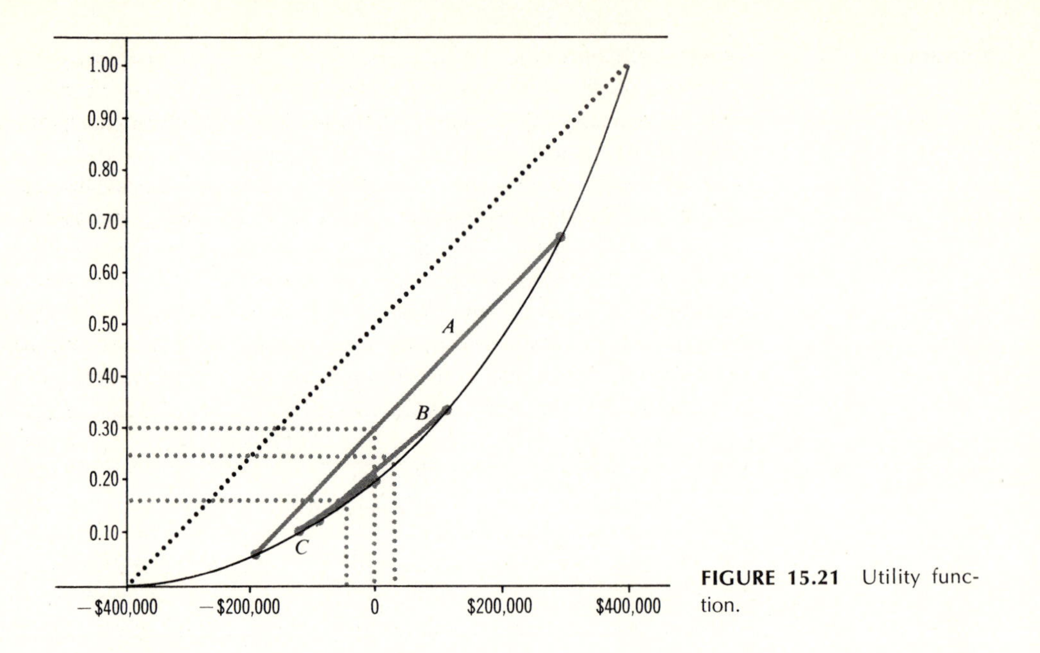


**FIGURE 15.21** Utility function.

money values involved in the decision. The shape of the curve in Figure 15.21 typifies a risk taker; it shows an increasing ***marginal utility*** for money. This means that as the rewards become higher, more value is placed on each additional dollar.

The alternatives can be shown by lines connecting the points on the utility curve which represent the outcomes. The dotted line to each chord shows the expected dollar value of each alternative. For ***A***, the expected dollar value is zero and the expected utility is about 0.3. ***C*** is shown in a similar manner and can be seen to have an expected loss of $40,000, with the corresponding utility rating of 0.17. The dollar expectation of ***B*** is greatest, but its utility is exceeded by that for ***A***.

If the utility function for the company had been linear over its entire range, it would have appeared as the straight dotted line in the figure. Such a function shows the utility in proportion to the monetary value. In this case the expected value calculated from the dollar-payoff table would directly indicate the utility of each alternative.

## PRACTICAL CONSIDERATIONS OF UTILITY

It is easier to manipulate utilities than it is to obtain them. Determining an individual's utility index is often a tedious and time-consuming task, but it can be even more difficult to obtain one that represents an organization's policy. While asking the series of questions necessary to establish a utility scale, side issues arising from the wording of the questions may cause distractions that show up as inconsistent responses. Then there is always the possibility that different responses would have been obtained on another day because of a change in mood or the temporary outlook of the person being questioned.

It should also be apparent that a utility function for a particular set of alternatives is not necessarily valid for another set of alternatives. Many intangible considerations fringe the choice of any specific rating. A manager might indicate a utility function which clearly shows conservative attitudes toward industrial actions, but he or she might have an entirely different set of attitudes for gambling at a gaming table or in the stock market.

Under some conditions it is expedient to employ methods that retain the concepts of utility functions without the requirement of actually enumerating the full range of utilities.

An alternative can be deemed acceptable if it fulfills a minimum level of accomplishment. When applied to management, this is called an *aspiration level*. It is essentially the same concept as that applied to operating systems, where it is called a *standard of performance*. In money decisions it commonly takes the form of a minimum level of profit that is acceptable or a level of cost which is not to be exceeded. Perhaps its most useful application is in decisions where the outcomes vary over periods of time. In a sales situation in which bids on merchandise are received or made intermittently, as in offers made to a used-car dealer, an aspiration level is almost indispensable.

When the range of outcomes is relatively small, the expected dollar values can be used directly without introducing significant errors in preferences. A graph of a utility function is closely approximated by a line over a small range between maximum and minimum dollar returns. The question, "How small is small?" is a matter of judgment. A range of $500 may be trivial to a large corporation, but it could be a meaningful sum to a struggling business.

**QUESTIONS**

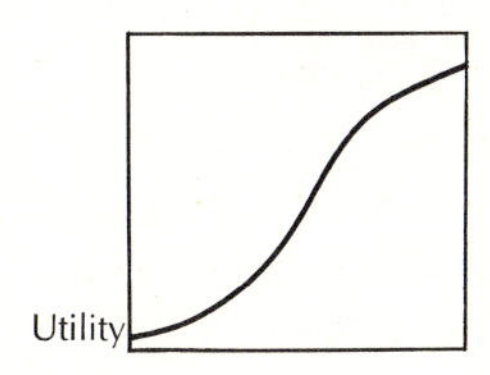


**15.2a** Suppose you were notified that you were to receive a relatively small inheritance from a distant relative, but the exact amount had not yet been determined. Then someone offered to bet you double or nothing for the unknown sum on the flip of a fair coin. Make a payoff table to represent your alternative, letting $I$ be the value of your inheritance. At what value of $I$ would you be willing to accept the bet? What is the significance of this indifference point to your utility index?

**15.2b** What could be said about the individual with the utility function shown?

**15.2c** An individual decides that his utility rating for a loss of $20 is 0.10, and for a gain of $100 is 0.60. In addition, he finds himself indifferent to the two alternatives in the payoff table below.

| | *P* (0.3) | *P* (0.7) |
|---|---|---|
| *A1* | $50 | 50 |
| *A2* | −20 | 100 |

What is his utility rating for $50?

[*U*($50) = 0.45]

**15.2d** The company described in Example 15.7 has determined a new utility index since its decision to develop a refined version of its product.

| *Utility* | *Dollars* |
|---|---|
| 0.00 | −400,000 |
| 0.40 | −200,000 |
| 0.58 | −100,000 |
| 0.65 | −50,000 |
| 0.71 | 0 |
| 0.82 | 100,000 |
| 0.90 | 200,000 |
| 0.95 | 300,000 |
| 0.98 | 400,000 |

1 If this index had been applied to the decision to develop a refined product, what would the decision have been?

*(Option B)*

2 What are some possibilities that could have influenced the altered utility index?
3 At what probability would the expected utility of alternatives $A$ and $B$ be equal under the new utility scale?

[$P(acceptance) = 0.584$]

**15.2e** An investor is deciding whether to buy bonds or stocks. She estimates that the probabilities of inflation and recession are, respectively, 0.7 and 0.3. For her anticipated investment of $10,000, she believes her choice of stocks would gain 10 percent per year during a period of inflation and would lose 5 percent of their value in a recession. Her selection of bonds would show a gain of 4 percent regardless of inflation or recession.

1 Which alternative would she choose if her utility scale were −$1000 = 0.0; −$500 = 0.2; $0 = 0.4; $400 = 0.6; $800 = 0.8; and $1000 = 1.0? The scale shows the dollar values equated to an equivalent utility, and intermediate values can be obtained by straight-line interpolation.

*(Stocks)*

2 Which alternative would the investor choose if her utility scale were directly proportional to dollar values?

*(Stocks)*

3 At what probability of recession would she be indifferent to the two alternatives?

[$P(recession) = 0.4$]

**15.2f** Another way of approaching a decision is temporarily to bypass specific assignments of the probabilities of possible futures by determining at what probability the decision maker would be indifferent to each occurrence. Then, general observations can be made about the future which can narrow the area of consideration. This is accomplished by finding the probabilities at which the utilities of the strategies are equal; that is,

$$\text{EV}(A\,1) = \text{EV}(A\,2) = \text{EV}(A\,3)$$

| | *R* | *N* | *I* |
|---|---|---|---|
| *A1* | 0.9 | 0.5 | 0.2 |
| *A2* | 0.7 | 1.0 | 0.1 |
| *A3* | 0.3 | 0.6 | 0.8 |

The given payoff table shows the utility a firm attaches to three alternative means of replacing obsolete equipment under the possible futures of a recession $R$, normal activity $N$, and inflation $I$.

1 At what probabilities for $R$, $N$, and $I$ would the firm be indifferent to alternatives $A\,1, A\,2$, or $A\,3$?
2 What is the expected value of the utility?
3 How could the firm use this information in narrowing the choice of alternatives?

**15.2g** Would it be accurate to say that one outcome is preferred to another because the utility of one is greater than that of the other?

# APPENDIX A

# SELECTED REFERENCES


Canada, J. R.: *Intermediate Economic Analysis for Management and Engineering,* Prentice-Hall, Englewood Cliffs, N.J., 1971.

DeGarmo, E. P., and J. R. Canada: *Engineering Economy,* 5th ed., Macmillan, New York, 1973.

deNeufville, R., and J. H. Stafford: *Systems Analysis for Engineers and Managers,* McGraw-Hill, New York, 1971.

Emerson, C. R., and W. R. Taylor: *An Introduction to Engineering Economy,* Cardinal, Bozeman, Mont., 1973.

*Engineering Economist, the;* quarterly journal of the Engineering Economy Division of the American Society for Engineering Education, Raymond P. Lutz (ed.), The University of Texas at Dallas, P.O. Box 688, Richardson, Tex.

Fabrycky, W. J., and G. J. Thuesen: *Economic Decision Analysis,* Prentice-Hall, Englewood Cliffs, N.J., 1974.

Fleischer, G. A. (ed.): *Risk and Uncertainty: Non-Deterministic Decision Making in Engineering Economy,* American Institute of Industrial Engineers, Norcross, Ga., 1975.

Grant, E. L., and W. G. Ireson: *Principles of Engineering Economy,* 5th ed., Ronald, New York, 1970.

James, L. D., and R. R. Lee: *Economics of Water Resources Planning,* McGraw-Hill, New York, 1971.

Jelen, F. C.: *Cost and Optimization Engineering,* McGraw-Hill, New York, 1971.

McConnell, C. R.: *Economics,* 6th ed., McGraw-Hill, New York, 1975.

Morris, W. T.: *Engineering Economic Analysis,* Reston Publishing, Reston, Va., 1976.

Newnan, D. G.: *Engineering Economic Analysis,* Engineering Press, San Jose, Calif., 1976.

Ostwald, P. F.: *Cost Estimating for Engineering and Management,* Prentice-Hall, Englewood Cliffs, N.J., 1974.

Park, W. R.: *Cost Engineering Analysis,* Wiley, New York, 1973.

Radford, K. J.: *Managerial Decision Making,* Reston Publishing, Reston, Va., 1975.

Reisman, A.: *Managerial and Engineering Economics,* Allyn and Bacon, Boston, 1971.

Riggs, J. L., and M. S. Inoue: *Introduction to Management Science and Operations Research,* McGraw-Hill, New York, 1975.

———: *Production Systems: Planning, Analysis, and Control,* 2nd ed., Wiley, New York, 1976.

——— and A. J. Kalbaugh: *The Art of Management,* McGraw-Hill, New York, 1974.

Samuelson, P. A.: *Economics,* 9th ed., McGraw-Hill, New York, 1973.

Smith, G. W.: *Engineering Economy: Analysis of Capital Expenditures,* 2d ed., Iowa State University Press, Ames, Iowa, 1973.

Tarquin, A. J., and L. T. Blank: *Engineering Economy,* McGraw-Hill, New York, 1976.

Taylor, G. A.: *Managerial and Engineering Economy,* 2d ed., Van Nostrand, New York, 1975.

Thueson, H. G., W. J. Fabrycky, and G. J. Thuesen: *Engineering Economy,* 4th ed., Prentice-Hall, Englewood Cliffs, N.J., 1971.

Weston, J. F., and E. F. Brigham: *Essentials of Managerial Finance,* 3d ed., Dryden Press, Hinsdale, Ill., 1974.

# APPENDIX B

# INTEREST TABLES

**½% Interest Factors for Discrete Compounding Periods**

| | SINGLE PAYMENT | | UNIFORM SERIES | | | | | |
|---|---|---|---|---|---|---|---|---|
| | Compound Amount Factor | Present Worth Factor | Capital Recovery Factor | Present Worth Factor | Sinking Fund Factor | Compound Amount Factor | Gradient Factor | |
| $N$ | $(F/P, ½, N)$ | $(P/F, ½, N)$ | $(A/P, ½, N)$ | $(P/A, ½, N)$ | $(A/F, ½, N)$ | $(F/A, ½, N)$ | $(A/G, ½, N)$ | $N$ |
| 1 | 1.0050 | .99503 | 1.0051 | .9949 | 1.0001 | .9998 | .0000 | 1 |
| 2 | 1.0100 | .99008 | .50385 | 1.9847 | .49885 | 2.0046 | .4613 | 2 |
| 3 | 1.0150 | .98515 | .33674 | 2.9696 | .33174 | 3.0143 | .9537 | 3 |
| 4 | 1.0201 | .98025 | .25318 | 3.9497 | .24818 | 4.0292 | 1.4531 | 4 |
| 5 | 1.0252 | .97538 | .20305 | 4.9248 | .19805 | 5.0491 | 1.9462 | 5 |
| 6 | 1.0303 | .97052 | .16963 | 5.8951 | .16463 | 6.0741 | 2.4413 | 6 |
| 7 | 1.0355 | .96570 | .14576 | 6.8606 | .14076 | 7.1043 | 2.9364 | 7 |
| 8 | 1.0407 | .96089 | .12786 | 7.8213 | .12286 | 8.1396 | 3.4304 | 8 |
| 9 | 1.0459 | .95611 | .11393 | 8.7772 | .10893 | 9.1800 | 3.9231 | 9 |
| 10 | 1.0511 | .95136 | .10279 | 9.7282 | .09779 | 10.225 | 4.4140 | 10 |
| 11 | 1.0563 | .94663 | .09368 | 10.674 | .08868 | 11.276 | 4.9063 | 11 |
| 12 | 1.0616 | .94192 | .08609 | 11.616 | .08109 | 12.332 | 5.3959 | 12 |
| 13 | 1.0669 | .93723 | .07966 | 12.553 | .07466 | 13.394 | 5.8857 | 13 |
| 14 | 1.0723 | .93257 | .07415 | 13.485 | .06915 | 14.460 | 6.3752 | 14 |
| 15 | 1.0776 | .92793 | .06938 | 14.413 | .06438 | 15.532 | 6.8614 | 15 |
| 16 | 1.0830 | .92332 | .06520 | 15.336 | .06020 | 16.610 | 7.3489 | 16 |
| 17 | 1.0884 | .91872 | .06152 | 16.255 | .05652 | 17.693 | 7.8351 | 17 |
| 18 | 1.0939 | .91415 | .05824 | 17.168 | .05324 | 18.781 | 8.3198 | 18 |
| 19 | 1.0993 | .90961 | .05531 | 18.078 | .05031 | 19.874 | 8.8046 | 19 |
| 20 | 1.1048 | .90508 | .05268 | 18.983 | .04768 | 20.974 | 9.2892 | 20 |
| 21 | 1.1103 | .90058 | .05029 | 19.883 | .04529 | 22.078 | 9.7715 | 21 |
| 22 | 1.1159 | .89610 | .04812 | 20.779 | .04312 | 23.183 | 10.253 | 22 |
| 23 | 1.1215 | .89164 | .04614 | 21.671 | .04114 | 24.304 | 10.735 | 23 |
| 24 | 1.1271 | .88721 | .04433 | 22.558 | .03933 | 25.425 | 11.216 | 24 |
| 25 | 1.1327 | .88280 | .04266 | 23.440 | .03766 | 26.552 | 11.695 | 25 |
| 26 | 1.1384 | .87841 | .04112 | 24.318 | .03612 | 27.685 | 12.173 | 26 |
| 27 | 1.1441 | .87404 | .03969 | 25.192 | .03469 | 28.823 | 12.652 | 27 |
| 28 | 1.1498 | .86969 | .03837 | 26.062 | .03337 | 29.967 | 13.129 | 28 |
| 29 | 1.1555 | .86536 | .03714 | 26.927 | .03214 | 31.116 | 13.605 | 29 |
| 30 | 1.1613 | .86106 | .03599 | 27.788 | .03099 | 32.272 | 14.081 | 30 |
| 31 | 1.1671 | .85678 | .03491 | 28.644 | .02991 | 33.433 | 14.555 | 31 |
| 32 | 1.1730 | .85251 | .03390 | 29.497 | .02890 | 34.600 | 15.029 | 32 |
| 33 | 1.1788 | .84827 | .03295 | 30.345 | .02795 | 35.772 | 15.501 | 33 |
| 34 | 1.1847 | .84405 | .03206 | 31.189 | .02706 | 36.951 | 15.974 | 34 |
| 35 | 1.1906 | .83986 | .03122 | 32.028 | .02622 | 38.135 | 16.446 | 35 |
| 40 | 1.2207 | .81918 | .02765 | 36.164 | .02265 | 44.147 | 18.790 | 40 |
| 45 | 1.2515 | .79901 | .02488 | 40.198 | .01988 | 50.311 | 21.113 | 45 |
| 50 | 1.2831 | .77933 | .02266 | 44.133 | .01766 | 56.630 | 23.416 | 50 |
| 55 | 1.3155 | .76014 | .02085 | 47.971 | .01585 | 63.109 | 25.699 | 55 |
| 60 | 1.3487 | .74142 | .01934 | 51.715 | .01434 | 69.751 | 27.960 | 60 |
| 65 | 1.3828 | .72317 | .01806 | 55.366 | .01306 | 76.561 | 30.201 | 65 |
| 70 | 1.4177 | .70536 | .01697 | 58.928 | .01197 | 83.543 | 32.422 | 70 |
| 75 | 1.4535 | .68799 | .01603 | 62.401 | .01103 | 90.701 | 34.622 | 75 |
| 80 | 1.4902 | .67105 | .01520 | 65.790 | .01020 | 98.040 | 36.802 | 80 |
| 85 | 1.5278 | .65453 | .01447 | 69.094 | .00947 | 105.56 | 38.961 | 85 |
| 90 | 1.5663 | .63841 | .01383 | 72.318 | .00883 | 113.27 | 41.099 | 90 |
| 95 | 1.6059 | .62269 | .01325 | 75.462 | .00825 | 121.18 | 43.218 | 95 |
| 100 | 1.6464 | .60736 | .01273 | 78.528 | .00773 | 129.29 | 45.316 | 100 |

**1% Interest Factors for Discrete Compounding Periods**

| | SINGLE PAYMENT | | UNIFORM SERIES | | | | | |
|---|---|---|---|---|---|---|---|---|
| | Compound Amount Factor | Present Worth Factor | Capital Recovery Factor | Present Worth Factor | Sinking Fund Factor | Compound Amount Factor | Gradient Factor | |
| N | (F/P, 1, N) | (P/F, 1, N) | (A/P, 1, N) | (P/A, 1, N) | (A/F, 1, N) | (F/A, 1, N) | (A/G, 1, N) | N |
| 1 | 1.0100 | .99010 | 1.0100 | .9900 | 1.0000 | .9999 | .0000 | 1 |
| 2 | 1.0201 | .98030 | .50757 | 1.9701 | .49757 | 2.0097 | .4864 | 2 |
| 3 | 1.0303 | .97059 | .34006 | 2.9406 | .33006 | 3.0297 | .9813 | 3 |
| 4 | 1.0406 | .96099 | .25631 | 3.9014 | .24631 | 4.0598 | 1.4751 | 4 |
| 5 | 1.0510 | .95147 | .20606 | 4.8528 | .19607 | 5.1003 | 1.9675 | 5 |
| 6 | 1.0615 | .94205 | .17257 | 5.7947 | .16257 | 6.1512 | 2.4581 | 6 |
| 7 | 1.0721 | .93273 | .14865 | 6.7273 | .13865 | 7.2125 | 2.9469 | 7 |
| 8 | 1.0828 | .92349 | .13071 | 7.6507 | .12071 | 8.2845 | 3.4349 | 8 |
| 9 | 1.0936 | .91435 | .11675 | 8.5649 | .10675 | 9.3672 | 3.9209 | 9 |
| 10 | 1.1046 | .90530 | .10560 | 9.4701 | .09560 | 10.460 | 4.4047 | 10 |
| 11 | 1.1156 | .89634 | .09647 | 10.366 | .08647 | 11.565 | 4.8872 | 11 |
| 12 | 1.1268 | .88746 | .08886 | 11.253 | .07886 | 12.680 | 5.3682 | 12 |
| 13 | 1.1380 | .87868 | .08242 | 12.132 | .07242 | 13.807 | 5.8476 | 13 |
| 14 | 1.1494 | .86998 | .07691 | 13.002 | .06691 | 14.945 | 6.3253 | 14 |
| 15 | 1.1609 | .86137 | .07213 | 13.863 | .06213 | 16.094 | 6.8010 | 15 |
| 16 | 1.1725 | .85284 | .06795 | 14.716 | .05795 | 17.255 | 7.2754 | 16 |
| 17 | 1.1842 | .84440 | .06427 | 15.560 | .05427 | 18.427 | 7.7483 | 17 |
| 18 | 1.1961 | .83604 | .06099 | 16.396 | .05099 | 19.611 | 8.2192 | 18 |
| 19 | 1.2080 | .82776 | .05806 | 17.223 | .04806 | 20.807 | 8.6883 | 19 |
| 20 | 1.2201 | .81957 | .05542 | 18.043 | .04542 | 22.015 | 9.1560 | 20 |
| 21 | 1.2323 | .81145 | .05304 | 18.854 | .04304 | 23.235 | 9.6222 | 21 |
| 22 | 1.2446 | .80342 | .05087 | 19.658 | .04087 | 24.467 | 10.086 | 22 |
| 23 | 1.2571 | .79547 | .04889 | 20.453 | .03889 | 25.712 | 10.549 | 23 |
| 24 | 1.2696 | .78759 | .04708 | 21.240 | .03708 | 26.969 | 11.010 | 24 |
| 25 | 1.2823 | .77979 | .04541 | 22.020 | .03541 | 28.238 | 11.469 | 25 |
| 26 | 1.2952 | .77207 | .04387 | 22.792 | .03387 | 29.521 | 11.927 | 26 |
| 27 | 1.3081 | .76443 | .04245 | 23.556 | .03245 | 30.816 | 12.383 | 27 |
| 28 | 1.3212 | .75686 | .04113 | 24.313 | .03113 | 32.124 | 12.838 | 28 |
| 29 | 1.3344 | .74937 | .03990 | 25.062 | .02990 | 33.445 | 13.291 | 29 |
| 30 | 1.3478 | .74195 | .03875 | 25.804 | .02875 | 34.779 | 13.742 | 30 |
| 31 | 1.3612 | .73461 | .03768 | 26.539 | .02768 | 36.127 | 14.191 | 31 |
| 32 | 1.3748 | .72733 | .03667 | 27.266 | .02667 | 37.488 | 14.640 | 32 |
| 33 | 1.3886 | .72013 | .03573 | 27.986 | .02573 | 38.863 | 15.086 | 33 |
| 34 | 1.4025 | .71301 | .03484 | 28.699 | .02484 | 40.251 | 15.531 | 34 |
| 35 | 1.4165 | .70595 | .03401 | 29.405 | .02401 | 41.653 | 15.973 | 35 |
| 40 | 1.4887 | .67169 | .03046 | 32.831 | .02046 | 48.878 | 18.164 | 40 |
| 45 | 1.5647 | .63909 | .02771 | 36.090 | .01771 | 56.471 | 20.314 | 45 |
| 50 | 1.6445 | .60808 | .02552 | 39.192 | .01552 | 64.452 | 22.423 | 50 |
| 55 | 1.7284 | .57857 | .02373 | 42.142 | .01373 | 72.839 | 24.491 | 55 |
| 60 | 1.8165 | .55049 | .02225 | 44.950 | .01225 | 81.655 | 26.520 | 60 |
| 65 | 1.9092 | .52378 | .02100 | 47.622 | .01100 | 90.920 | 28.508 | 65 |
| 70 | 2.0065 | .49836 | .01993 | 50.163 | .00993 | 100.65 | 30.457 | 70 |
| 75 | 2.1089 | .47418 | .01902 | 52.582 | .00902 | 110.89 | 32.366 | 75 |
| 80 | 2.2164 | .45117 | .01822 | 54.883 | .00822 | 121.64 | 34.236 | 80 |
| 85 | 2.3295 | .42927 | .01752 | 57.072 | .00752 | 132.95 | 36.067 | 85 |
| 90 | 2.4483 | .40844 | .01690 | 59.156 | .00690 | 144.83 | 37.859 | 90 |
| 95 | 2.5732 | .38862 | .01636 | 61.138 | .00636 | 157.32 | 39.614 | 95 |
| 100 | 2.7044 | .36976 | .01587 | 63.024 | .00587 | 170.44 | 41.330 | 100 |

**1½% Interest Factors for Discrete Compounding Periods**

| | SINGLE PAYMENT | | UNIFORM SERIES | | | | | |
|---|---|---|---|---|---|---|---|---|
| | Compound Amount Factor | Present Worth Factor | Capital Recovery Factor | Present Worth Factor | Sinking Fund Factor | Compound Amount Factor | Gradient Factor | |
| *N* | (*F/P*, 1½, *N*) | (*P/F*, 1½, *N*) | (*A/P*, 1½, *N*) | (*P/A*, 1½, *N*) | (*A/F*, 1½, *N*) | (*F/A*, 1½, *N*) | (*A/G*, 1½, *N*) | *N* |
| 1 | 1.0150 | .98522 | 1.0150 | .9852 | 1.0000 | 1.0000 | .0000 | 1 |
| 2 | 1.0302 | .97066 | .51131 | 1.9557 | .49631 | 2.0148 | .4917 | 2 |
| 3 | 1.0456 | .95632 | .34340 | 2.9120 | .32840 | 3.0450 | .9857 | 3 |
| 4 | 1.0613 | .94219 | .25946 | 3.8540 | .24446 | 4.0905 | 1.4760 | 4 |
| 5 | 1.0772 | .92827 | .20910 | 4.7823 | .19410 | 5.1518 | 1.9653 | 5 |
| 6 | 1.0934 | .91455 | .17554 | 5.6967 | .16054 | 6.2290 | 2.4511 | 6 |
| 7 | 1.1098 | .90103 | .15157 | 6.5977 | .13657 | 7.3223 | 2.9351 | 7 |
| 8 | 1.1264 | .88772 | .13359 | 7.4853 | .11859 | 8.4320 | 3.4161 | 8 |
| 9 | 1.1433 | .87460 | .11962 | 8.3598 | .10462 | 9.5585 | 3.8952 | 9 |
| 10 | 1.1605 | .86168 | .10844 | 9.2214 | .09344 | 10.701 | 4.3716 | 10 |
| 11 | 1.1779 | .84894 | .09930 | 10.070 | .08430 | 11.862 | 4.8456 | 11 |
| 12 | 1.1956 | .83640 | .09169 | 10.906 | .07669 | 13.039 | 5.3169 | 12 |
| 13 | 1.2135 | .82404 | .08525 | 11.730 | .07025 | 14.235 | 5.7863 | 13 |
| 14 | 1.2317 | .81186 | .07973 | 12.542 | .06473 | 15.448 | 6.2524 | 14 |
| 15 | 1.2502 | .79987 | .07495 | 13.342 | .05995 | 16.680 | 6.7165 | 15 |
| 16 | 1.2689 | .78805 | .07077 | 14.130 | .05577 | 17.930 | 7.1781 | 16 |
| 17 | 1.2879 | .77640 | .06708 | 14.906 | .05208 | 19.199 | 7.6374 | 17 |
| 18 | 1.3073 | .76493 | .06381 | 15.671 | .04881 | 20.487 | 8.0939 | 18 |
| 19 | 1.3269 | .75363 | .06088 | 16.424 | .04588 | 21.794 | 8.5482 | 19 |
| 20 | 1.3468 | .74249 | .05825 | 17.167 | .04325 | 23.121 | 8.9998 | 20 |
| 21 | 1.3670 | .73152 | .05587 | 17.898 | .04087 | 24.468 | 9.4493 | 21 |
| 22 | 1.3875 | .72071 | .05371 | 18.619 | .03871 | 25.834 | 9.8959 | 22 |
| 23 | 1.4083 | .71006 | .05173 | 19.329 | .03673 | 27.222 | 10.340 | 23 |
| 24 | 1.4294 | .69957 | .04993 | 20.028 | .03493 | 28.630 | 10.782 | 24 |
| 25 | 1.4509 | .68923 | .04827 | 20.718 | .03327 | 30.059 | 11.221 | 25 |
| 26 | 1.4726 | .67904 | .04674 | 21.397 | .03174 | 31.510 | 11.658 | 26 |
| 27 | 1.4947 | .66901 | .04532 | 22.066 | .03032 | 32.983 | 12.093 | 27 |
| 28 | 1.5171 | .65912 | .04400 | 22.725 | .02900 | 34.477 | 12.525 | 28 |
| 29 | 1.5399 | .64938 | .04278 | 23.374 | .02778 | 35.994 | 12.955 | 29 |
| 30 | 1.5630 | .63979 | .04164 | 24.014 | .02664 | 37.534 | 13.382 | 30 |
| 31 | 1.5864 | .63033 | .04058 | 24.644 | .02558 | 39.097 | 13.807 | 31 |
| 32 | 1.6102 | .62102 | .03958 | 25.265 | .02458 | 40.683 | 14.229 | 32 |
| 33 | 1.6344 | .61184 | .03864 | 25.877 | .02364 | 42.293 | 14.649 | 33 |
| 34 | 1.6589 | .60280 | .03776 | 26.479 | .02276 | 43.928 | 15.067 | 34 |
| 35 | 1.6838 | .59389 | .03694 | 27.073 | .02194 | 45.586 | 15.482 | 35 |
| 40 | 1.8139 | .55129 | .03343 | 29.913 | .01843 | 54.261 | 17.522 | 40 |
| 45 | 1.9541 | .51174 | .03072 | 32.550 | .01572 | 63.606 | 19.501 | 45 |
| 50 | 2.1051 | .47504 | .02857 | 34.997 | .01357 | 73.673 | 21.422 | 50 |
| 55 | 2.2677 | .44096 | .02683 | 37.269 | .01183 | 84.518 | 23.283 | 55 |
| 60 | 2.4430 | .40933 | .02539 | 39.378 | .01039 | 96.201 | 25.087 | 60 |
| 65 | 2.6318 | .37997 | .02419 | 41.335 | .00919 | 108.78 | 26.833 | 65 |
| 70 | 2.8351 | .35271 | .02317 | 43.152 | .00817 | 122.34 | 28.523 | 70 |
| 75 | 3.0542 | .32741 | .02230 | 44.839 | .00730 | 136.95 | 30.157 | 75 |
| 80 | 3.2903 | .30392 | .02155 | 46.405 | .00655 | 152.68 | 31.737 | 80 |
| 85 | 3.5445 | .28212 | .02089 | 47.858 | .00589 | 169.63 | 33.262 | 85 |
| 90 | 3.8185 | .26188 | .02032 | 49.207 | .00532 | 187.89 | 34.734 | 90 |
| 95 | 4.1135 | .24310 | .01982 | 50.460 | .00482 | 207.57 | 36.155 | 95 |
| 100 | 4.4314 | .22566 | .01937 | 51.622 | .00437 | 228.76 | 37.524 | 100 |

**2% Interest Factors for Discrete Compounding Periods**

| | SINGLE PAYMENT | | UNIFORM SERIES | | | | | |
|---|---|---|---|---|---|---|---|---|
| | Compound Amount Factor | Present Worth Factor | Capital Recovery Factor | Present Worth Factor | Sinking Fund Factor | Compound Amount Factor | Gradient Factor | |
| *N* | (*F*/*P*, 2, *N*) | (*P*/*F*, 2, *N*) | (*A*/*P*, 2, *N*) | (*P*/*A*, 2, *N*) | (*A*/*F*, 2, *N*) | (*F*/*A*, 2, *N*) | (*A*/*G*, 2, *N*) | *N* |
| 1 | 1.0200 | .98039 | 1.0200 | .9804 | 1.0000 | 1.0000 | .0000 | 1 |
| 2 | 1.0404 | .96117 | .51507 | 1.9415 | .49507 | 2.0199 | .4934 | 2 |
| 3 | 1.0612 | .94232 | .34677 | 2.8837 | .32677 | 3.0603 | .9851 | 3 |
| 4 | 1.0824 | .92385 | .26263 | 3.8075 | .24263 | 4.1214 | 1.4733 | 4 |
| 5 | 1.1040 | .90573 | .21217 | 4.7132 | .19217 | 5.2038 | 1.9584 | 5 |
| 6 | 1.1261 | .88798 | .17853 | 5.6012 | .15853 | 6.3078 | 2.4401 | 6 |
| 7 | 1.1486 | .87056 | .15452 | 6.4717 | .13452 | 7.4339 | 2.9189 | 7 |
| 8 | 1.1716 | .85350 | .13651 | 7.3252 | .11651 | 8.5826 | 3.3940 | 8 |
| 9 | 1.1950 | .83676 | .12252 | 8.1619 | .10252 | 9.7541 | 3.8659 | 9 |
| 10 | 1.2189 | .82035 | .11133 | 8.9822 | .09133 | 10.949 | 4.3347 | 10 |
| 11 | 1.2433 | .80427 | .10218 | 9.7865 | .08218 | 12.168 | 4.8001 | 11 |
| 12 | 1.2682 | .78850 | .09456 | 10.574 | .07456 | 13.411 | 5.2622 | 12 |
| 13 | 1.2935 | .77304 | .08812 | 11.347 | .06812 | 14.679 | 5.7209 | 13 |
| 14 | 1.3194 | .75788 | .08261 | 12.105 | .06261 | 15.973 | 6.1764 | 14 |
| 15 | 1.3458 | .74302 | .07783 | 12.848 | .05783 | 17.292 | 6.6288 | 15 |
| 16 | 1.3727 | .72846 | .07365 | 13.577 | .05365 | 18.638 | 7.0778 | 16 |
| 17 | 1.4002 | .71417 | .06997 | 14.291 | .04997 | 20.011 | 7.5236 | 17 |
| 18 | 1.4282 | .70017 | .06670 | 14.991 | .04670 | 21.411 | 7.9660 | 18 |
| 19 | 1.4567 | .68644 | .06378 | 15.677 | .04378 | 22.839 | 8.4052 | 19 |
| 20 | 1.4859 | .67298 | .06116 | 16.350 | .04116 | 24.296 | 8.8412 | 20 |
| 21 | 1.5156 | .65979 | .05879 | 17.010 | .03879 | 25.781 | 9.2739 | 21 |
| 22 | 1.5459 | .64685 | .05663 | 17.657 | .03663 | 27.297 | 9.7033 | 22 |
| 23 | 1.5768 | .63417 | .05467 | 18.291 | .03467 | 28.843 | 10.129 | 23 |
| 24 | 1.6084 | .62173 | .05287 | 18.913 | .03287 | 30.420 | 10.552 | 24 |
| 25 | 1.6405 | .60954 | .05122 | 19.522 | .03122 | 32.028 | 10.972 | 25 |
| 26 | 1.6733 | .59759 | .04970 | 20.120 | .02970 | 33.669 | 11.388 | 26 |
| 27 | 1.7068 | .58588 | .04829 | 20.706 | .02829 | 35.342 | 11.802 | 27 |
| 28 | 1.7409 | .57439 | .04699 | 21.280 | .02699 | 37.049 | 12.212 | 28 |
| 29 | 1.7758 | .56313 | .04578 | 21.843 | .02578 | 38.790 | 12.619 | 29 |
| 30 | 1.8113 | .55208 | .04465 | 22.395 | .02465 | 40.565 | 13.023 | 30 |
| 31 | 1.8475 | .54126 | .04360 | 22.937 | .02360 | 42.377 | 13.423 | 31 |
| 32 | 1.8844 | .53065 | .04261 | 23.467 | .02261 | 44.224 | 13.821 | 32 |
| 33 | 1.9221 | .52024 | .04169 | 23.987 | .02169 | 46.108 | 14.215 | 33 |
| 34 | 1.9606 | .51004 | .04082 | 24.497 | .02082 | 48.031 | 14.606 | 34 |
| 35 | 1.9998 | .50004 | .04000 | 24.997 | .02000 | 49.991 | 14.994 | 35 |
| 40 | 2.2079 | .45291 | .03656 | 27.354 | .01656 | 60.398 | 16.886 | 40 |
| 45 | 2.4377 | .41021 | .03391 | 29.489 | .01391 | 71.888 | 18.701 | 45 |
| 50 | 2.6914 | .37154 | .03182 | 31.422 | .01182 | 84.573 | 20.440 | 50 |
| 55 | 2.9715 | .33652 | .03014 | 33.174 | .01014 | 98.579 | 22.103 | 55 |
| 60 | 3.2808 | .30480 | .02877 | 34.760 | .00877 | 114.04 | 23.694 | 60 |
| 65 | 3.6223 | .27607 | .02763 | 36.196 | .00763 | 131.11 | 25.212 | 65 |
| 70 | 3.9993 | .25004 | .02667 | 37.497 | .00667 | 149.96 | 26.661 | 70 |
| 75 | 4.4155 | .22647 | .02586 | 38.676 | .00586 | 170.77 | 28.041 | 75 |
| 80 | 4.8751 | .20512 | .02516 | 39.743 | .00516 | 193.75 | 29.355 | 80 |
| 85 | 5.3824 | .18579 | .02456 | 40.710 | .00456 | 219.12 | 30.604 | 85 |
| 90 | 5.9426 | .16827 | .02405 | 41.586 | .00405 | 247.13 | 31.791 | 90 |
| 95 | 6.5611 | .15241 | .02360 | 42.379 | .00360 | 278.05 | 32.917 | 95 |
| 100 | 7.2440 | .13804 | .02320 | 43.097 | .00320 | 312.20 | 33.984 | 100 |

**2½% Interest Factors for Discrete Compounding Periods**

| | SINGLE PAYMENT | | UNIFORM SERIES | | | | | |
|---|---|---|---|---|---|---|---|---|
| | Compound Amount Factor | Present Worth Factor | Capital Recovery Factor | Present Worth Factor | Sinking Fund Factor | Compound Amount Factor | Gradient Factor | |
| $N$ | $(F/P, 2\frac{1}{2}, N)$ | $(P/F, 2\frac{1}{2}, N)$ | $(A/P, 2\frac{1}{2}, N)$ | $(P/A, 2\frac{1}{2}, N)$ | $(A/F, 2\frac{1}{2}, N)$ | $(F/A, 2\frac{1}{2}, N)$ | $(A/G, 2\frac{1}{2}, N)$ | $N$ |
| 1 | 1.0250 | .97561 | 1.0250 | .9756 | 1.0000 | 1.0000 | .0000 | 1 |
| 2 | 1.0506 | .95182 | .51884 | 1.9273 | .49384 | 2.0243 | .4930 | 2 |
| 3 | 1.0768 | .92860 | .35014 | 2.8559 | .32514 | 3.0755 | .9827 | 3 |
| 4 | 1.1038 | .90595 | .26582 | 3.7618 | .24082 | 4.1524 | 1.4681 | 4 |
| 5 | 1.1314 | .88386 | .21525 | 4.6457 | .19025 | 5.2562 | 1.9496 | 5 |
| 6 | 1.1596 | .86230 | .18155 | 5.5079 | .15655 | 6.3875 | 2.4269 | 6 |
| 7 | 1.1886 | .84127 | .15750 | 6.3492 | .13250 | 7.5472 | 2.9002 | 7 |
| 8 | 1.2184 | .82075 | .13947 | 7.1699 | .11447 | 8.7358 | 3.3695 | 8 |
| 9 | 1.2488 | .80073 | .12546 | 7.9707 | .10046 | 9.9542 | 3.8346 | 9 |
| 10 | 1.2800 | .78120 | .11426 | 8.7518 | .08926 | 11.203 | 4.2955 | 10 |
| 11 | 1.3120 | .76215 | .10511 | 9.5140 | .08011 | 12.483 | 4.7524 | 11 |
| 12 | 1.3448 | .74356 | .09749 | 10.257 | .07249 | 13.795 | 5.2052 | 12 |
| 13 | 1.3785 | .72543 | .09105 | 10.982 | .06605 | 15.140 | 5.6539 | 13 |
| 14 | 1.4129 | .70773 | .08554 | 11.690 | .06054 | 16.518 | 6.0985 | 14 |
| 15 | 1.4482 | .69047 | .08077 | 12.381 | .05577 | 17.931 | 6.5391 | 15 |
| 16 | 1.4844 | .67363 | .07660 | 13.054 | .05160 | 19.379 | 6.9756 | 16 |
| 17 | 1.5216 | .65720 | .07293 | 13.711 | .04793 | 20.864 | 7.4081 | 17 |
| 18 | 1.5596 | .64117 | .06967 | 14.353 | .04467 | 22.385 | 7.8365 | 18 |
| 19 | 1.5986 | .62553 | .06676 | 14.978 | .04176 | 23.945 | 8.2609 | 19 |
| 20 | 1.6386 | .61028 | .06415 | 15.588 | .03915 | 25.543 | 8.6813 | 20 |
| 21 | 1.6795 | .59539 | .06179 | 16.184 | .03679 | 27.182 | 9.0976 | 21 |
| 22 | 1.7215 | .58087 | .05965 | 16.765 | .03465 | 28.861 | 9.5100 | 22 |
| 23 | 1.7645 | .56671 | .05770 | 17.331 | .03270 | 30.583 | 9.9183 | 23 |
| 24 | 1.8087 | .55288 | .05591 | 17.884 | .03091 | 32.347 | 10.322 | 24 |
| 25 | 1.8539 | .53940 | .05428 | 18.424 | .02928 | 34.156 | 10.723 | 25 |
| 26 | 1.9002 | .52624 | .05277 | 18.950 | .02777 | 36.010 | 11.119 | 26 |
| 27 | 1.9477 | .51341 | .05138 | 19.463 | .02638 | 37.910 | 11.512 | 27 |
| 28 | 1.9964 | .50089 | .05009 | 19.964 | .02509 | 39.858 | 11.900 | 28 |
| 29 | 2.0463 | .48867 | .04889 | 20.453 | .02389 | 41.854 | 12.285 | 29 |
| 30 | 2.0975 | .47675 | .04778 | 20.929 | .02278 | 43.901 | 12.665 | 30 |
| 31 | 2.1499 | .46512 | .04674 | 21.395 | .02174 | 45.998 | 13.042 | 31 |
| 32 | 2.2037 | .45378 | .04577 | 21.848 | .02077 | 48.148 | 13.415 | 32 |
| 33 | 2.2588 | .44271 | .04486 | 22.291 | .01986 | 50.352 | 13.784 | 33 |
| 34 | 2.3152 | .43191 | .04401 | 22.723 | .01901 | 52.610 | 14.149 | 34 |
| 35 | 2.3731 | .42138 | .04321 | 23.144 | .01821 | 54.926 | 14.511 | 35 |
| 40 | 2.6850 | .37244 | .03984 | 25.102 | .01484 | 67.399 | 16.261 | 40 |
| 45 | 3.0378 | .32918 | .03727 | 26.832 | .01227 | 81.512 | 17.917 | 45 |
| 50 | 3.4370 | .29095 | .03526 | 28.361 | .01026 | 97.480 | 19.483 | 50 |
| 55 | 3.8886 | .25716 | .03365 | 29.713 | .00865 | 115.54 | 20.959 | 55 |
| 60 | 4.3996 | .22729 | .03235 | 30.908 | .00735 | 135.98 | 22.351 | 60 |
| 65 | 4.9777 | .20089 | .03128 | 31.964 | .00628 | 159.11 | 23.659 | 65 |
| 70 | 5.6318 | .17756 | .03040 | 32.897 | .00540 | 185.27 | 24.887 | 70 |
| 75 | 6.3719 | .15694 | .02965 | 33.722 | .00465 | 214.87 | 26.038 | 75 |
| 80 | 7.2092 | .13871 | .02903 | 34.451 | .00403 | 248.36 | 27.115 | 80 |
| 85 | 8.1565 | .12260 | .02849 | 35.095 | .00349 | 286.26 | 28.122 | 85 |
| 90 | 9.2283 | .10836 | .02804 | 35.665 | .00304 | 329.13 | 29.062 | 90 |
| 95 | 10.441 | .09578 | .02765 | 36.168 | .00265 | 377.63 | 29.937 | 95 |
| 100 | 11.813 | .08465 | .02731 | 36.613 | .00231 | 432.51 | 30.751 | 100 |

**3% Interest Factors for Discrete Compounding Periods**

| | SINGLE PAYMENT | | UNIFORM SERIES | | | | | |
|---|---|---|---|---|---|---|---|---|
| | Compound Amount Factor | Present Worth Factor | Capital Recovery Factor | Present Worth Factor | Sinking Fund Factor | Compound Amount Factor | Gradient Factor | |
| $N$ | $(F/P, 3, N)$ | $(P/F, 3, N)$ | $(A/P, 3, N)$ | $(P/A, 3, N)$ | $(A/F, 3, N)$ | $(F/A, 3, N)$ | $(A/G, 3, N)$ | $N$ |
| 1 | 1.0300 | .97087 | 1.0300 | .9709 | 1.0000 | 1.0000 | .0000 | 1 |
| 2 | 1.0609 | .94260 | .52262 | 1.9134 | .49262 | 2.0299 | .4920 | 2 |
| 3 | 1.0927 | .91514 | .35354 | 2.8285 | .32354 | 3.0908 | .9795 | 3 |
| 4 | 1.1255 | .88849 | .26903 | 3.7170 | .23903 | 4.1835 | 1.4622 | 4 |
| 5 | 1.1592 | .86261 | .21836 | 4.5796 | .18836 | 5.3090 | 1.9401 | 5 |
| 6 | 1.1940 | .83749 | .18460 | 5.4170 | .15460 | 6.4682 | 2.4129 | 6 |
| 7 | 1.2298 | .81310 | .16051 | 6.2301 | .13051 | 7.6622 | 2.8809 | 7 |
| 8 | 1.2667 | .78941 | .14246 | 7.0195 | .11246 | 8.8920 | 3.3440 | 8 |
| 9 | 1.3047 | .76642 | .12844 | 7.7859 | .09844 | 10.158 | 3.8022 | 9 |
| 10 | 1.3439 | .74410 | .11723 | 8.5300 | .08723 | 11.463 | 4.2555 | 10 |
| 11 | 1.3842 | .72243 | .10808 | 9.2524 | .07808 | 12.807 | 4.7040 | 11 |
| 12 | 1.4257 | .70139 | .10046 | 9.9537 | .07046 | 14.191 | 5.1475 | 12 |
| 13 | 1.4685 | .68096 | .09403 | 10.634 | .06403 | 15.617 | 5.5863 | 13 |
| 14 | 1.5125 | .66113 | .08853 | 11.295 | .05853 | 17.085 | 6.0201 | 14 |
| 15 | 1.5579 | .64187 | .08377 | 11.937 | .05377 | 18.598 | 6.4491 | 15 |
| 16 | 1.6046 | .62318 | .07961 | 12.560 | .04961 | 20.156 | 6.8732 | 16 |
| 17 | 1.6528 | .60502 | .07595 | 13.165 | .04595 | 21.760 | 7.2926 | 17 |
| 18 | 1.7024 | .58740 | .07271 | 13.753 | .04271 | 23.413 | 7.7072 | 18 |
| 19 | 1.7534 | .57030 | .06982 | 14.323 | .03982 | 25.115 | 8.1169 | 19 |
| 20 | 1.8060 | .55369 | .06722 | 14.877 | .03722 | 26.869 | 8.5219 | 20 |
| 21 | 1.8602 | .53756 | .06487 | 15.414 | .03487 | 28.675 | 8.9221 | 21 |
| 22 | 1.9160 | .52190 | .06275 | 15.936 | .03275 | 30.535 | 9.3176 | 22 |
| 23 | 1.9735 | .50670 | .06082 | 16.443 | .03082 | 32.451 | 9.7084 | 23 |
| 24 | 2.0327 | .49194 | .05905 | 16.935 | .02905 | 34.425 | 10.094 | 24 |
| 25 | 2.0937 | .47762 | .05743 | 17.412 | .02743 | 36.457 | 10.475 | 25 |
| 26 | 2.1565 | .46370 | .05594 | 17.876 | .02594 | 38.551 | 10.852 | 26 |
| 27 | 2.2212 | .45020 | .05457 | 18.326 | .02457 | 40.707 | 11.224 | 27 |
| 28 | 2.2878 | .43709 | .05329 | 18.763 | .02329 | 42.929 | 11.592 | 28 |
| 29 | 2.3565 | .42436 | .05212 | 19.188 | .02212 | 45.217 | 11.954 | 29 |
| 30 | 2.4272 | .41200 | .05102 | 19.600 | .02102 | 47.573 | 12.313 | 30 |
| 31 | 2.5000 | .40000 | .05000 | 20.000 | .02000 | 50.000 | 12.666 | 31 |
| 32 | 2.5750 | .38835 | .04905 | 20.388 | .01905 | 52.500 | 13.016 | 32 |
| 33 | 2.6522 | .37704 | .04816 | 20.765 | .01816 | 55.075 | 13.360 | 33 |
| 34 | 2.7318 | .36606 | .04732 | 21.131 | .01732 | 57.727 | 13.700 | 34 |
| 35 | 2.8137 | .35539 | .04654 | 21.486 | .01654 | 60.459 | 14.036 | 35 |
| 40 | 3.2619 | .30657 | .04326 | 23.114 | .01326 | 75.397 | 15.649 | 40 |
| 45 | 3.7814 | .26445 | .04079 | 24.518 | .01079 | 92.715 | 17.154 | 45 |
| 50 | 4.3837 | .22812 | .03887 | 25.729 | .00887 | 112.79 | 18.556 | 50 |
| 55 | 5.0819 | .19678 | .03735 | 26.774 | .00735 | 136.06 | 19.859 | 55 |
| 60 | 5.8913 | .16974 | .03613 | 27.675 | .00613 | 163.04 | 21.066 | 60 |
| 65 | 6.8296 | .14642 | .03515 | 28.452 | .00515 | 194.32 | 22.183 | 65 |
| 70 | 7.9173 | .12630 | .03434 | 29.123 | .00434 | 230.57 | 23.213 | 70 |
| 75 | 9.1783 | .10895 | .03367 | 29.701 | .00367 | 272.61 | 24.162 | 75 |
| 80 | 10.640 | .09398 | .03311 | 30.200 | .00311 | 321.33 | 25.034 | 80 |
| 85 | 12.334 | .08107 | .03265 | 30.630 | .00265 | 377.82 | 25.834 | 85 |
| 90 | 14.299 | .06993 | .03226 | 31.002 | .00226 | 443.31 | 26.566 | 90 |
| 95 | 16.576 | .06033 | .03193 | 31.322 | .00193 | 519.22 | 27.234 | 95 |
| 100 | 19.217 | .05204 | .03165 | 31.598 | .00165 | 607.23 | 27.843 | 100 |

**4% Interest Factors for Discrete Compounding Periods**

| | SINGLE PAYMENT | | UNIFORM SERIES | | | | | |
|---|---|---|---|---|---|---|---|---|
| | Compound Amount Factor | Present Worth Factor | Capital Recovery Factor | Present Worth Factor | Sinking Fund Factor | Compound Amount Factor | Gradient Factor | |
| *N* | (*F/P*, 4, *N*) | (*P/F*, 4, *N*) | (*A/P*, 4, *N*) | (*P/A*, 4, *N*) | (*A/F*, 4, *N*) | (*F/A*, 4, *N*) | (*A/G*, 4, *N*) | *N* |
| 1 | 1.0400 | .96154 | 1.0400 | .9615 | 1.0000 | 1.0000 | .0000 | 1 |
| 2 | 1.0816 | .92456 | .53020 | 1.8860 | .49020 | 2.0399 | .4900 | 2 |
| 3 | 1.1248 | .88900 | .36035 | 2.7750 | .32035 | 3.1215 | .9736 | 3 |
| 4 | 1.1698 | .85481 | .27549 | 3.6298 | .23549 | 4.2464 | 1.4506 | 4 |
| 5 | 1.2166 | .82193 | .22463 | 4.4517 | .18463 | 5.4162 | 1.9213 | 5 |
| 6 | 1.2653 | .79032 | .19076 | 5.2420 | .15076 | 6.6328 | 2.3853 | 6 |
| 7 | 1.3159 | .75992 | .16661 | 6.0019 | .12661 | 7.8981 | 2.8429 | 7 |
| 8 | 1.3685 | .73069 | .14853 | 6.7326 | .10853 | 9.2140 | 3.2940 | 8 |
| 9 | 1.4233 | .70259 | .13449 | 7.4352 | .09449 | 10.582 | 3.7387 | 9 |
| 10 | 1.4802 | .67557 | .12329 | 8.1108 | .08329 | 12.005 | 4.1769 | 10 |
| 11 | 1.5394 | .64958 | .11415 | 8.7603 | .07415 | 13.486 | 4.6086 | 11 |
| 12 | 1.6010 | .62460 | .10655 | 9.3849 | .06655 | 15.025 | 5.0339 | 12 |
| 13 | 1.6650 | .60058 | .10014 | 9.9855 | .06014 | 16.626 | 5.4529 | 13 |
| 14 | 1.7316 | .57748 | .09467 | 10.563 | .05467 | 18.291 | 5.8655 | 14 |
| 15 | 1.8009 | .55527 | .08994 | 11.118 | .04994 | 20.023 | 6.2717 | 15 |
| 16 | 1.8729 | .53391 | .08582 | 11.652 | .04582 | 21.824 | 6.6716 | 16 |
| 17 | 1.9478 | .51338 | .08220 | 12.165 | .04220 | 23.697 | 7.0652 | 17 |
| 18 | 2.0257 | .49363 | .07899 | 12.659 | .03899 | 25.644 | 7.4526 | 18 |
| 19 | 2.1068 | .47465 | .07614 | 13.133 | .03614 | 27.670 | 7.8338 | 19 |
| 20 | 2.1911 | .45639 | .07358 | 13.590 | .03358 | 29.777 | 8.2087 | 20 |
| 21 | 2.2787 | .43884 | .07128 | 14.029 | .03128 | 31.968 | 8.5775 | 21 |
| 22 | 2.3698 | .42196 | .06920 | 14.450 | .02920 | 34.247 | 8.9402 | 22 |
| 23 | 2.4646 | .40573 | .06731 | 14.856 | .02731 | 36.617 | 9.2969 | 23 |
| 24 | 2.5632 | .39013 | .06559 | 15.246 | .02559 | 39.081 | 9.6475 | 24 |
| 25 | 2.6658 | .37512 | .06401 | 15.621 | .02401 | 41.644 | 9.9921 | 25 |
| 26 | 2.7724 | .36069 | .06257 | 15.982 | .02257 | 44.310 | 10.330 | 26 |
| 27 | 2.8833 | .34682 | .06124 | 16.329 | .02124 | 47.083 | 10.663 | 27 |
| 28 | 2.9986 | .33348 | .06001 | 16.662 | .02001 | 49.966 | 10.990 | 28 |
| 29 | 3.1186 | .32066 | .05888 | 16.983 | .01888 | 52.964 | 11.311 | 29 |
| 30 | 3.2433 | .30832 | .05783 | 17.291 | .01783 | 56.083 | 11.627 | 30 |
| 31 | 3.3730 | .29647 | .05686 | 17.588 | .01686 | 59.326 | 11.936 | 31 |
| 32 | 3.5079 | .28506 | .05595 | 17.873 | .01595 | 62.699 | 12.240 | 32 |
| 33 | 3.6483 | .27410 | .05510 | 18.147 | .01510 | 66.207 | 12.539 | 33 |
| 34 | 3.7942 | .26356 | .05432 | 18.411 | .01432 | 69.855 | 12.832 | 34 |
| 35 | 3.9460 | .25342 | .05358 | 18.664 | .01358 | 73.650 | 13.119 | 35 |
| 40 | 4.8009 | .20829 | .05052 | 19.792 | .01052 | 95.022 | 14.476 | 40 |
| 45 | 5.8410 | .17120 | .04826 | 20.719 | .00826 | 121.02 | 15.704 | 45 |
| 50 | 7.1064 | .14072 | .04655 | 21.482 | .00655 | 152.66 | 16.811 | 50 |
| 55 | 8.6460 | .11566 | .04523 | 22.108 | .00523 | 191.15 | 17.806 | 55 |
| 60 | 10.519 | .09506 | .04420 | 22.623 | .00420 | 237.98 | 18.696 | 60 |
| 65 | 12.798 | .07814 | .04339 | 23.046 | .00339 | 294.95 | 19.490 | 65 |
| 70 | 15.570 | .06422 | .04275 | 23.394 | .00275 | 364.27 | 20.195 | 70 |
| 75 | 18.944 | .05279 | .04223 | 23.680 | .00223 | 448.60 | 20.820 | 75 |
| 80 | 23.048 | .04339 | .04181 | 23.915 | .00181 | 551.21 | 21.371 | 80 |
| 85 | 28.042 | .03566 | .04148 | 24.108 | .00148 | 676.05 | 21.856 | 85 |
| 90 | 34.117 | .02931 | .04121 | 24.267 | .00121 | 827.93 | 22.282 | 90 |
| 95 | 41.508 | .02409 | .04099 | 24.397 | .00099 | 1012.7 | 22.654 | 95 |
| 100 | 50.501 | .01980 | .04081 | 24.504 | .00081 | 1237.5 | 22.979 | 100 |

### 5% Interest Factors for Discrete Compounding Periods

| | SINGLE PAYMENT | | UNIFORM SERIES | | | | | |
|---|---|---|---|---|---|---|---|---|
| | Compound Amount Factor | Present Worth Factor | Capital Recovery Factor | Present Worth Factor | Sinking Fund Factor | Compound Amount Factor | Gradient Factor | |
| N | (F/P, 5, N) | (P/F, 5, N) | (A/P, 5, N) | (P/A, 5, N) | (A/F, 5, N) | (F/A, 5, N) | (A/G, 5, N) | N |
| 1 | 1.0500 | .95238 | 1.0500 | .9524 | 1.0000 | 1.0000 | .0000 | 1 |
| 2 | 1.1025 | .90703 | .53781 | 1.8593 | .48781 | 2.0499 | .4874 | 2 |
| 3 | 1.1576 | .86384 | .36722 | 2.7231 | .31722 | 3.1524 | .9671 | 3 |
| 4 | 1.2155 | .82271 | .28202 | 3.5458 | .23202 | 4.3100 | 1.4386 | 4 |
| 5 | 1.2762 | .78353 | .23098 | 4.3294 | .18098 | 5.5255 | 1.9021 | 5 |
| 6 | 1.3400 | .74622 | .19702 | 5.0756 | .14702 | 6.8017 | 2.3575 | 6 |
| 7 | 1.4070 | .71069 | .17282 | 5.7862 | .12282 | 8.1418 | 2.8048 | 7 |
| 8 | 1.4774 | .67684 | .15472 | 6.4631 | .10472 | 9.5488 | 3.2441 | 8 |
| 9 | 1.5513 | .64461 | .14069 | 7.1077 | .09069 | 11.026 | 3.6753 | 9 |
| 10 | 1.6288 | .61392 | .12951 | 7.7216 | .07951 | 12.577 | 4.0986 | 10 |
| 11 | 1.7103 | .58469 | .12039 | 8.3062 | .07039 | 14.206 | 4.5140 | 11 |
| 12 | 1.7958 | .55684 | .11283 | 8.8631 | .06283 | 15.916 | 4.9214 | 12 |
| 13 | 1.8856 | .53033 | .10646 | 9.3934 | .05646 | 17.712 | 5.3211 | 13 |
| 14 | 1.9799 | .50507 | .10103 | 9.8985 | .05103 | 19.598 | 5.7128 | 14 |
| 15 | 2.0789 | .48102 | .09634 | 10.379 | .04634 | 21.577 | 6.0969 | 15 |
| 16 | 2.1828 | .45812 | .09227 | 10.837 | .04227 | 23.656 | 6.4732 | 16 |
| 17 | 2.2919 | .43630 | .08870 | 11.273 | .03870 | 25.839 | 6.8418 | 17 |
| 18 | 2.4065 | .41553 | .08555 | 11.689 | .03555 | 28.131 | 7.2029 | 18 |
| 19 | 2.5269 | .39574 | .08275 | 12.085 | .03275 | 30.538 | 7.5565 | 19 |
| 20 | 2.6532 | .37690 | .08024 | 12.462 | .03024 | 33.064 | 7.9025 | 20 |
| 21 | 2.7859 | .35895 | .07800 | 12.821 | .02800 | 35.718 | 8.2412 | 21 |
| 22 | 2.9252 | .34186 | .07597 | 13.162 | .02597 | 38.503 | 8.5725 | 22 |
| 23 | 3.0714 | .32558 | .07414 | 13.488 | .02414 | 41.429 | 8.8966 | 23 |
| 24 | 3.2250 | .31008 | .07247 | 13.798 | .02247 | 44.500 | 9.2135 | 24 |
| 25 | 3.3862 | .29531 | .07095 | 14.093 | .02095 | 47.725 | 9.5234 | 25 |
| 26 | 3.5555 | .28125 | .06956 | 14.375 | .01957 | 51.111 | 9.8261 | 26 |
| 27 | 3.7333 | .26786 | .06829 | 14.642 | .01829 | 54.667 | 10.122 | 27 |
| 28 | 3.9200 | .25510 | .06712 | 14.898 | .01712 | 58.400 | 10.411 | 28 |
| 29 | 4.1160 | .24295 | .06605 | 15.140 | .01605 | 62.320 | 10.693 | 29 |
| 30 | 4.3218 | .23138 | .06505 | 15.372 | .01505 | 66.436 | 10.968 | 30 |
| 31 | 4.5379 | .22037 | .06413 | 15.592 | .01413 | 70.757 | 11.237 | 31 |
| 32 | 4.7647 | .20987 | .06328 | 15.802 | .01328 | 75.295 | 11.500 | 32 |
| 33 | 5.0030 | .19988 | .06249 | 16.002 | .01249 | 80.060 | 11.756 | 33 |
| 34 | 5.2531 | .19036 | .06176 | 16.192 | .01176 | 85.063 | 12.005 | 34 |
| 35 | 5.5158 | .18130 | .06107 | 16.374 | .01107 | 90.316 | 12.249 | 35 |
| 40 | 7.0397 | .14205 | .05828 | 17.158 | .00828 | 120.79 | 13.277 | 40 |
| 45 | 8.9846 | .11130 | .05626 | 17.773 | .00626 | 159.69 | 14.364 | 45 |
| 50 | 11.466 | .08721 | .05478 | 18.255 | .00478 | 209.33 | 15.223 | 50 |
| 55 | 14.634 | .06833 | .05367 | 18.633 | .00367 | 272.69 | 15.966 | 55 |
| 60 | 18.678 | .05354 | .05283 | 18.929 | .00283 | 353.56 | 16.605 | 60 |
| 65 | 23.838 | .04195 | .05219 | 19.161 | .00219 | 456.76 | 17.153 | 65 |
| 70 | 30.424 | .03287 | .05170 | 19.342 | .00170 | 588.48 | 17.621 | 70 |
| 75 | 38.829 | .02575 | .05132 | 19.484 | .00132 | 756.59 | 18.017 | 75 |
| 80 | 49.557 | .02018 | .05103 | 19.596 | .00103 | 971.14 | 18.352 | 80 |
| 85 | 63.248 | .01581 | .05080 | 19.683 | .00080 | 1244.9 | 18.634 | 85 |
| 90 | 80.723 | .01239 | .05063 | 19.752 | .00063 | 1594.4 | 18.871 | 90 |
| 95 | 103.02 | .00971 | .05049 | 19.805 | .00049 | 2040.4 | 19.068 | 95 |
| 100 | 131.48 | .00761 | .05038 | 19.847 | .00038 | 2609.7 | 19.233 | 100 |

**6% Interest Factors for Discrete Compounding Factors**

| | SINGLE PAYMENT | | UNIFORM SERIES | | | | | |
|---|---|---|---|---|---|---|---|---|
| | Compound Amount Factor | Present Worth Factor | Capital Recovery Factor | Present Worth Factor | Sinking Fund Factor | Compound Amount Factor | Gradient Factor | |
| $N$ | $(F/P, 6, N)$ | $(P/F, 6, N)$ | $(A/P, 6, N)$ | $(P/A, 6, N)$ | $(A/F, 6, N)$ | $(F/A, 6, N)$ | $(A/G, 6, N)$ | $N$ |
| 1 | 1.0600 | .94340 | 1.0600 | .9434 | 1.0000 | 1.0000 | .0000 | 1 |
| 2 | 1.1236 | .89000 | .54544 | 1.8333 | .48544 | 2.0599 | .4852 | 2 |
| 3 | 1.1910 | .83962 | .37411 | 2.6729 | .31411 | 3.1835 | .9610 | 3 |
| 4 | 1.2624 | .79210 | .28860 | 3.4650 | .22860 | 4.3745 | 1.4269 | 4 |
| 5 | 1.3382 | .74726 | .23740 | 4.2123 | .17740 | 5.6370 | 1.8833 | 5 |
| 6 | 1.4185 | .70496 | .20337 | 4.9172 | .14337 | 6.9751 | 2.3301 | 6 |
| 7 | 1.5036 | .66506 | .17914 | 5.5823 | .11914 | 8.3936 | 2.7673 | 7 |
| 8 | 1.5938 | .62742 | .16104 | 6.2097 | .10104 | 9.8972 | 3.1949 | 8 |
| 9 | 1.6894 | .59190 | .14702 | 6.8016 | .08702 | 11.491 | 3.6130 | 9 |
| 10 | 1.7908 | .55840 | .13587 | 7.3600 | .07587 | 13.180 | 4.0217 | 10 |
| 11 | 1.8982 | .52679 | .12679 | 7.8867 | .06679 | 14.971 | 4.4210 | 11 |
| 12 | 2.0121 | .49698 | .11928 | 8.3837 | .05928 | 16.869 | 4.8109 | 12 |
| 13 | 2.1329 | .46884 | .11296 | 8.8525 | .05296 | 18.881 | 5.1917 | 13 |
| 14 | 2.2608 | .44231 | .10759 | 9.2948 | .04759 | 21.014 | 5.5632 | 14 |
| 15 | 2.3965 | .41727 | .10296 | 9.7121 | .04296 | 23.275 | 5.9257 | 15 |
| 16 | 2.5403 | .39365 | .09895 | 10.105 | .03895 | 25.671 | 6.2791 | 16 |
| 17 | 2.6927 | .37137 | .09545 | 10.477 | .03545 | 28.212 | 6.6237 | 17 |
| 18 | 2.8542 | .35035 | .09236 | 10.827 | .03236 | 30.904 | 6.9594 | 18 |
| 19 | 3.0255 | .33052 | .08962 | 11.158 | .02962 | 33.759 | 7.2864 | 19 |
| 20 | 3.2070 | .31181 | .08719 | 11.469 | .02719 | 36.784 | 7.6048 | 20 |
| 21 | 3.3995 | .29416 | .08501 | 11.763 | .02501 | 39.991 | 7.9148 | 21 |
| 22 | 3.6034 | .27751 | .08305 | 12.041 | .02305 | 43.390 | 8.2163 | 22 |
| 23 | 3.8196 | .26180 | .08128 | 12.303 | .02128 | 46.994 | 8.5096 | 23 |
| 24 | 4.0488 | .24698 | .07968 | 12.550 | .01968 | 50.814 | 8.7948 | 24 |
| 25 | 4.2917 | .23300 | .07823 | 12.783 | .01823 | 54.862 | 9.0719 | 25 |
| 26 | 4.5492 | .21982 | .07690 | 13.003 | .01690 | 59.154 | 9.3412 | 26 |
| 27 | 4.8222 | .20737 | .07570 | 13.210 | .01570 | 63.703 | 9.6027 | 27 |
| 28 | 5.1115 | .19564 | .07459 | 13.406 | .01459 | 68.525 | 9.8565 | 28 |
| 29 | 5.4182 | .18456 | .07358 | 13.590 | .01358 | 73.637 | 10.102 | 29 |
| 30 | 5.7433 | .17412 | .07265 | 13.764 | .01265 | 79.055 | 10.341 | 30 |
| 31 | 6.0879 | .16426 | .07179 | 13.929 | .01179 | 84.798 | 10.573 | 31 |
| 32 | 6.4531 | .15496 | .07100 | 14.083 | .01100 | 90.886 | 10.798 | 32 |
| 33 | 6.8403 | .14619 | .07027 | 14.230 | .01027 | 97.339 | 11.016 | 33 |
| 34 | 7.2507 | .13792 | .06960 | 14.368 | .00960 | 104.17 | 11.227 | 34 |
| 35 | 7.6858 | .13011 | .06897 | 14.498 | .00897 | 111.43 | 11.431 | 35 |
| 40 | 10.285 | .09723 | .06646 | 15.046 | .00646 | 154.75 | 12.358 | 40 |
| 45 | 13.764 | .07265 | .06470 | 15.455 | .00470 | 212.73 | 13.141 | 45 |
| 50 | 18.419 | .05429 | .06344 | 15.761 | .00344 | 290.32 | 13.796 | 50 |
| 55 | 24.649 | .04057 | .06254 | 15.990 | .00254 | 394.14 | 14.340 | 55 |
| 60 | 32.985 | .03032 | .06188 | 16.161 | .00188 | 533.09 | 14.790 | 60 |
| 65 | 44.142 | .02265 | .06139 | 16.289 | .00139 | 719.03 | 15.160 | 65 |
| 70 | 59.071 | .01693 | .06103 | 16.384 | .00103 | 967.86 | 15.461 | 70 |
| 75 | 79.051 | .01265 | .06077 | 16.455 | .00077 | 1300.8 | 15.705 | 75 |
| 80 | 105.78 | .00945 | .06057 | 16.509 | .00057 | 1746.4 | 15.903 | 80 |
| 85 | 141.56 | .00706 | .06043 | 16.548 | .00043 | 2342.7 | 16.061 | 85 |
| 90 | 189.44 | .00528 | .06032 | 16.578 | .00032 | 3140.7 | 16.189 | 90 |
| 95 | 253.52 | .00394 | .06024 | 16.600 | .00024 | 4208.7 | 16.290 | 95 |
| 100 | 339.26 | .00295 | .06018 | 16.617 | .00018 | 5637.8 | 16.371 | 100 |

**7% Interest Factors for Discrete Compounding Periods**

| | SINGLE PAYMENT | | UNIFORM SERIES | | | | | |
|---|---|---|---|---|---|---|---|---|
| | Compound Amount Factor | Present Worth Factor | Capital Recovery Factor | Present Worth Factor | Sinking Fund Factor | Compound Amount Factor | Gradient Factor | |
| *N* | (*F*/*P*, 7, *N*) | (*P*/*F*, 7, *N*) | (*A*/*P*, 7, *N*) | (*P*/*A*, 7, *N*) | (*A*/*F*, 7, *N*) | (*F*/*A*, 7, *N*) | (*A*/*G*, 7, *N*) | *N* |
| 1 | 1.0700 | .93458 | 1.0700 | .9346 | 1.0000 | 1.000 | .0000 | 1 |
| 2 | 1.1449 | .87344 | .55310 | 1.8080 | .48310 | 2.0699 | .4830 | 2 |
| 3 | 1.2250 | .81630 | .38105 | 2.6242 | .31105 | 3.2148 | .9548 | 3 |
| 4 | 1.3107 | .76290 | .29523 | 3.3871 | .22523 | 4.4398 | 1.4153 | 4 |
| 5 | 1.4025 | .71299 | .24389 | 4.1001 | .17389 | 5.7506 | 1.8648 | 5 |
| 6 | 1.5007 | .66635 | .20980 | 4.7665 | .13980 | 7.1531 | 2.3030 | 6 |
| 7 | 1.6057 | .62275 | .18555 | 5.3892 | .11555 | 8.6539 | 2.7302 | 7 |
| 8 | 1.7181 | .58201 | .16747 | 5.9712 | .09747 | 10.259 | 3.1463 | 8 |
| 9 | 1.8384 | .54394 | .15349 | 6.5151 | .08349 | 11.977 | 3.5515 | 9 |
| 10 | 1.9671 | .50835 | .14238 | 7.0235 | .07238 | 13.816 | 3.9459 | 10 |
| 11 | 2.1048 | .47510 | .13336 | 7.4986 | .06336 | 15.783 | 4.3294 | 11 |
| 12 | 2.2521 | .44402 | .12590 | 7.9426 | .05590 | 17.888 | 4.7023 | 12 |
| 13 | 2.4098 | .41497 | .11965 | 8.3576 | .04965 | 20.140 | 5.0647 | 13 |
| 14 | 2.5785 | .38782 | .11435 | 8.7454 | .04435 | 22.550 | 5.4165 | 14 |
| 15 | 2.7590 | .36245 | .10980 | 9.1078 | .03980 | 25.128 | 5.7581 | 15 |
| 16 | 2.9521 | .33874 | .10586 | 9.4466 | .03586 | 27.887 | 6.0895 | 16 |
| 17 | 3.1587 | .31658 | .10243 | 9.7631 | .03243 | 30.839 | 6.4108 | 17 |
| 18 | 3.3798 | .29587 | .09941 | 10.059 | .02941 | 33.998 | 6.7223 | 18 |
| 19 | 3.6164 | .27651 | .09675 | 10.335 | .02675 | 37.378 | 7.0240 | 19 |
| 20 | 3.8696 | .25842 | .09439 | 10.593 | .02439 | 40.994 | 7.3161 | 20 |
| 21 | 4.1404 | .24152 | .09229 | 10.835 | .02229 | 44.864 | 7.5988 | 21 |
| 22 | 4.4303 | .22572 | .09041 | 11.061 | .02041 | 49.004 | 7.8723 | 22 |
| 23 | 4.7404 | .21095 | .08871 | 11.272 | .01871 | 53.434 | 8.1367 | 23 |
| 24 | 5.0722 | .19715 | .08719 | 11.469 | .01719 | 58.175 | 8.3922 | 24 |
| 25 | 5.4273 | .18425 | .08581 | 11.653 | .01581 | 63.247 | 8.6389 | 25 |
| 26 | 5.8072 | .17220 | .08456 | 11.825 | .01456 | 68.674 | 8.8772 | 26 |
| 27 | 6.2137 | .16093 | .08343 | 11.986 | .01343 | 74.481 | 9.1070 | 27 |
| 28 | 6.6486 | .15041 | .08239 | 12.137 | .01239 | 80.695 | 9.3288 | 28 |
| 29 | 7.1140 | .14057 | .08145 | 12.277 | .01145 | 87.344 | 9.5425 | 29 |
| 30 | 7.6120 | .13137 | .08059 | 12.409 | .01059 | 94.458 | 9.7485 | 30 |
| 31 | 8.1449 | .12278 | .07980 | 12.531 | .00980 | 102.07 | 9.9469 | 31 |
| 32 | 8.7150 | .11474 | .07907 | 12.646 | .00907 | 110.21 | 10.137 | 32 |
| 33 | 9.3250 | .10724 | .07841 | 12.753 | .00841 | 118.92 | 10.321 | 33 |
| 34 | 9.9778 | .10022 | .07780 | 12.853 | .00780 | 128.25 | 10.498 | 34 |
| 35 | 10.676 | .09367 | .07723 | 12.947 | .00723 | 138.23 | 10.668 | 35 |
| 40 | 14.973 | .06678 | .07501 | 13.331 | .00501 | 199.62 | 11.423 | 40 |
| 45 | 21.001 | .04762 | .07350 | 13.605 | .00350 | 285.73 | 12.035 | 45 |
| 50 | 29.455 | .03395 | .07246 | 13.800 | .00246 | 406.51 | 12.528 | 50 |
| 55 | 41.313 | .02421 | .07174 | 13.939 | .00174 | 575.90 | 12.921 | 55 |
| 60 | 57.943 | .01726 | .07123 | 14.039 | .00123 | 813.47 | 13.232 | 60 |
| 65 | 81.268 | .01230 | .07087 | 14.109 | .00087 | 1146.6 | 13.475 | 65 |
| 70 | 113.98 | .00877 | .07062 | 14.160 | .00062 | 1614.0 | 13.666 | 70 |
| 75 | 159.86 | .00626 | .07044 | 14.196 | .00044 | 2269.5 | 13.813 | 75 |
| 80 | 224.21 | .00446 | .07031 | 14.222 | .00031 | 3188.8 | 13.927 | 80 |
| 85 | 314.47 | .00318 | .07022 | 14.240 | .00022 | 4478.2 | 14.014 | 85 |
| 90 | 441.06 | .00227 | .07016 | 14.253 | .00016 | 6286.7 | 14.081 | 90 |
| 95 | 618.62 | .00162 | .07011 | 14.262 | .00011 | 8823.1 | 14.131 | 95 |
| 100 | 867.64 | .00115 | .07008 | 14.269 | .00008 | 12381.7 | 14.170 | 100 |

**8% Interest Factors for Discrete Compounding Periods**

| | SINGLE PAYMENT | | UNIFORM SERIES | | | | | |
|---|---|---|---|---|---|---|---|---|
| | Compound Amount Factor | Present Worth Factor | Capital Recovery Factor | Present Worth Factor | Sinking Fund Factor | Compound Amount Factor | Gradient Factor | |
| N | (F/P, 8, N) | (P/F, 8, N) | (A/P, 8, N) | (P/A, 8, N) | (A/F, 8, N) | (F/A, 8, N) | (A/G, 8, N) | N |
| 1 | 1.0800 | .92593 | 1.0800 | .9259 | 1.0000 | 1.0000 | .0000 | 1 |
| 2 | 1.1664 | .85734 | .56077 | 1.7832 | .48077 | 2.0799 | .4807 | 2 |
| 3 | 1.2597 | .79383 | .38803 | 2.5770 | .30804 | 3.2463 | .9487 | 3 |
| 4 | 1.3604 | .73503 | .30192 | 3.3121 | .22192 | 4.5060 | 1.4038 | 4 |
| 5 | 1.4693 | .68059 | .25046 | 3.9926 | .17046 | 5.8665 | 1.8463 | 5 |
| 6 | 1.5868 | .63017 | .21632 | 4.6228 | .13632 | 7.3358 | 2.2762 | 6 |
| 7 | 1.7138 | .58349 | .19207 | 5.2063 | .11207 | 8.9227 | 2.6935 | 7 |
| 8 | 1.8509 | .54027 | .17402 | 5.7466 | .09402 | 10.636 | 3.0984 | 8 |
| 9 | 1.9989 | .50025 | .16008 | 6.2468 | .08008 | 12.487 | 3.4909 | 9 |
| 10 | 2.1589 | .46320 | .14903 | 6.7100 | .06903 | 14.486 | 3.8712 | 10 |
| 11 | 2.3316 | .42889 | .14008 | 7.1389 | .06008 | 16.645 | 4.2394 | 11 |
| 12 | 2.5181 | .39712 | .13270 | 7.5360 | .05270 | 18.976 | 4.5956 | 12 |
| 13 | 2.7196 | .36770 | .12642 | 7.9037 | .04652 | 21.495 | 4.9401 | 13 |
| 14 | 2.9371 | .34046 | .12130 | 8.2442 | .04130 | 24.214 | 5.2729 | 14 |
| 15 | 3.1721 | .31524 | .11683 | 8.5594 | .03683 | 27.151 | 5.5943 | 15 |
| 16 | 3.4259 | .29189 | .11298 | 8.8513 | .03298 | 30.323 | 5.9045 | 16 |
| 17 | 3.6999 | .27027 | .10963 | 9.1216 | .02963 | 33.749 | 6.2036 | 17 |
| 18 | 3.9959 | .25025 | .10670 | 9.3718 | .02670 | 37.449 | 6.4919 | 18 |
| 19 | 4.3156 | .23171 | .10413 | 9.6035 | .02413 | 41.445 | 6.7696 | 19 |
| 20 | 4.6609 | .21455 | .10185 | 9.8181 | .02185 | 45.761 | 7.0368 | 20 |
| 21 | 5.0337 | .19866 | .09983 | 10.016 | .01983 | 50.422 | 7.2939 | 21 |
| 22 | 5.4364 | .18394 | .09803 | 10.200 | .01803 | 55.455 | 7.5411 | 22 |
| 23 | 5.8713 | .17032 | .09642 | 10.371 | .01642 | 60.892 | 7.7785 | 23 |
| 24 | 6.3410 | .15770 | .09498 | 10.528 | .01498 | 66.763 | 8.0065 | 24 |
| 25 | 6.8483 | .14602 | .09368 | 10.674 | .01368 | 73.104 | 8.2253 | 25 |
| 26 | 7.3962 | .13520 | .09251 | 10.809 | .01251 | 79.953 | 8.4351 | 26 |
| 27 | 7.9879 | .12519 | .09145 | 10.935 | .01145 | 87.349 | 8.6362 | 27 |
| 28 | 8.6269 | .11592 | .09049 | 11.051 | .01049 | 95.337 | 8.8288 | 28 |
| 29 | 9.3171 | .10733 | .08962 | 11.158 | .00962 | 103.96 | 9.0132 | 29 |
| 30 | 10.062 | .09938 | .08883 | 11.257 | .00883 | 113.28 | 9.1896 | 30 |
| 31 | 10.867 | .09202 | .08811 | 11.349 | .00811 | 123.34 | 9.3583 | 31 |
| 32 | 11.736 | .08520 | .08745 | 11.434 | .00745 | 134.21 | 9.5196 | 32 |
| 33 | 12.675 | .07889 | .08685 | 11.513 | .00685 | 145.94 | 9.6736 | 33 |
| 34 | 13.689 | .07305 | .08630 | 11.586 | .00630 | 158.62 | 9.8207 | 34 |
| 35 | 14.785 | .06764 | .08580 | 11.654 | .00580 | 172.31 | 9.9610 | 35 |
| 40 | 21.724 | .04603 | .08386 | 11.924 | .00386 | 259.05 | 10.569 | 40 |
| 45 | 31.919 | .03133 | .08259 | 12.108 | .00259 | 386.49 | 11.044 | 45 |
| 50 | 46.900 | .02132 | .08174 | 12.233 | .00174 | 573.75 | 11.410 | 50 |
| 55 | 68.911 | .01451 | .08118 | 12.318 | .00118 | 848.89 | 11.690 | 55 |
| 60 | 101.25 | .00988 | .08080 | 12.376 | .00080 | 1253.1 | 11.901 | 60 |
| 65 | 148.77 | .00672 | .08054 | 12.416 | .00054 | 1847.1 | 12.060 | 65 |
| 70 | 218.59 | .00457 | .08037 | 12.442 | .00037 | 2719.9 | 12.178 | 70 |
| 75 | 321.19 | .00311 | .08025 | 12.461 | .00025 | 4002.3 | 12.265 | 75 |
| 80 | 471.93 | .00212 | .08017 | 12.473 | .00017 | 5886.6 | 12.330 | 80 |
| 85 | 693.42 | .00144 | .08012 | 12.481 | .00012 | 8655.2 | 12.377 | 85 |
| 90 | 1018.8 | .00098 | .08008 | 12.487 | .00008 | 12723.9 | 12.411 | 90 |
| 95 | 1497.0 | .00067 | .08005 | 12.491 | .00005 | 18701.5 | 12.436 | 95 |
| 100 | 2199.6 | .00045 | .08004 | 12.494 | .00004 | 27484.5 | 12.454 | 100 |

**9% Interest Factors for Discrete Compounding Periods**

| | SINGLE PAYMENT | | UNIFORM SERIES | | | | | |
|---|---|---|---|---|---|---|---|---|
| | Compound Amount Factor | Present Worth Factor | Capital Recovery Factor | Present Worth Factor | Sinking Fund Factor | Compound Amount Factor | Gradient Factor | |
| *N* | $(F/P, 9, N)$ | $(P/F, 9, N)$ | $(A/P, 9, N)$ | $(P/A, 9, N)$ | $(A/F, 9, N)$ | $(F/A, 9, N)$ | $(A/G, 9, N)$ | *N* |
| 1 | 1.0900 | .91743 | 1.0900 | .9174 | 1.0000 | 1.0000 | .0000 | 1 |
| 2 | 1.1881 | .84168 | .56847 | 1.7591 | .47847 | 2.0899 | .4784 | 2 |
| 3 | 1.2950 | .77219 | .39506 | 2.5312 | .30506 | 3.2780 | .9425 | 3 |
| 4 | 1.4115 | .70843 | .30867 | 3.2396 | .21867 | 4.5730 | 1.3923 | 4 |
| 5 | 1.5386 | .64993 | .25709 | 3.8896 | .16709 | 5.9846 | 1.8280 | 5 |
| 6 | 1.6770 | .59627 | .22292 | 4.4858 | .13292 | 7.5232 | 2.2496 | 6 |
| 7 | 1.8280 | .54704 | .19869 | 5.0329 | .10869 | 9.2002 | 2.6572 | 7 |
| 8 | 1.9925 | .50187 | .18068 | 5.5347 | .09068 | 11.028 | 3.0510 | 8 |
| 9 | 2.1718 | .46043 | .16680 | 5.9952 | .07680 | 13.020 | 3.4311 | 9 |
| 10 | 2.3673 | .42241 | .15582 | 6.4176 | .06582 | 15.192 | 3.7976 | 10 |
| 11 | 2.5804 | .38754 | .14695 | 6.8051 | .05695 | 17.559 | 4.1508 | 11 |
| 12 | 2.8126 | .35554 | .13965 | 7.1606 | .04965 | 20.140 | 4.4909 | 12 |
| 13 | 3.0657 | .32618 | .13357 | 7.4868 | .04357 | 22.952 | 4.8180 | 13 |
| 14 | 3.3416 | .29925 | .12843 | 7.7861 | .03843 | 26.018 | 5.1325 | 14 |
| 15 | 3.6424 | .27454 | .12406 | 8.0606 | .03406 | 29.360 | 5.4345 | 15 |
| 16 | 3.9702 | .25187 | .12030 | 8.3125 | .03030 | 33.002 | 5.7243 | 16 |
| 17 | 4.3275 | .23108 | .11705 | 8.5435 | .02705 | 36.972 | 6.0022 | 17 |
| 18 | 4.7170 | .21200 | .11421 | 8.7555 | .02421 | 41.300 | 6.2685 | 18 |
| 19 | 5.1415 | .19449 | .11173 | 8.9500 | .02173 | 46.017 | 6.5234 | 19 |
| 20 | 5.6043 | .17843 | .10955 | 9.1285 | .01955 | 51.158 | 6.7673 | 20 |
| 21 | 6.1086 | .16370 | .10762 | 9.2922 | .01762 | 56.763 | 7.0004 | 21 |
| 22 | 6.6584 | .15018 | .10591 | 9.4423 | .01591 | 62.871 | 7.2231 | 22 |
| 23 | 7.2577 | .13778 | .10438 | 9.5801 | .01438 | 69.530 | 7.4356 | 23 |
| 24 | 7.9109 | .12641 | .10302 | 9.7065 | .01302 | 76.787 | 7.6383 | 24 |
| 25 | 8.6228 | .11597 | .10181 | 9.8225 | .01181 | 84.698 | 7.8315 | 25 |
| 26 | 9.3989 | .10640 | .10072 | 9.9289 | .01072 | 93.321 | 8.0154 | 26 |
| 27 | 10.244 | .09761 | .09974 | 10.026 | .00974 | 102.72 | 8.1905 | 27 |
| 28 | 11.166 | .08955 | .09885 | 10.116 | .00885 | 112.96 | 8.3570 | 28 |
| 29 | 12.171 | .08216 | .09806 | 10.198 | .00806 | 124.13 | 8.5153 | 29 |
| 30 | 13.267 | .07537 | .09734 | 10.273 | .00734 | 136.30 | 8.6655 | 30 |
| 31 | 14.461 | .06915 | .09669 | 10.342 | .00669 | 149.57 | 8.8082 | 31 |
| 32 | 15.762 | .06344 | .09610 | 10.406 | .00610 | 164.03 | 8.9435 | 32 |
| 33 | 17.181 | .05820 | .09556 | 10.464 | .00556 | 179.79 | 9.0717 | 33 |
| 34 | 18.727 | .05340 | .09508 | 10.517 | .00508 | 196.97 | 9.1932 | 34 |
| 35 | 20.413 | .04899 | .09464 | 10.566 | .00464 | 215.70 | 9.3082 | 35 |
| 40 | 31.408 | .03184 | .09296 | 10.757 | .00296 | 337.86 | 9.7956 | 40 |
| 45 | 48.325 | .02069 | .09190 | 10.881 | .00190 | 525.83 | 10.160 | 45 |
| 50 | 74.353 | .01345 | .09123 | 10.961 | .00123 | 815.04 | 10.429 | 50 |
| 55 | 114.40 | .00874 | .09079 | 11.014 | .00079 | 1260.0 | 10.626 | 55 |
| 60 | 176.02 | .00568 | .09051 | 11.047 | .00051 | 1944.6 | 10.768 | 60 |
| 65 | 270.82 | .00369 | .09033 | 11.070 | .00033 | 2998.0 | 10.870 | 65 |
| 70 | 416.70 | .00240 | .09022 | 11.084 | .00022 | 4618.9 | 10.942 | 70 |
| 75 | 641.14 | .00156 | .09014 | 11.093 | .00014 | 7112.7 | 10.993 | 75 |
| 80 | 986.47 | .00101 | .09009 | 11.099 | .00009 | 10950.6 | 11.029 | 80 |
| 85 | 1517.8 | .00066 | .09006 | 11.103 | .00006 | 16854.8 | 11.055 | 85 |
| 90 | 2335.3 | .00043 | .09004 | 11.106 | .00004 | 25939.2 | 11.072 | 90 |
| 95 | 3593.1 | .00028 | .09002 | 11.108 | .00003 | 39916.6 | 11.084 | 95 |
| 100 | 5528.4 | .00018 | .09002 | 11.109 | .00002 | 61422.7 | 11.093 | 100 |

**10% Interest Factors for Discrete Compounding Periods**

| | SINGLE PAYMENT | | UNIFORM SERIES | | | | | |
|---|---|---|---|---|---|---|---|---|
| | Compound Amount Factor | Present Worth Factor | Capital Recovery Factor | Present Worth Factor | Sinking Fund Factor | Compound Amount Factor | Gradient Factor | |
| $N$ | $(F/P, 10, N)$ | $(P/F, 10, N)$ | $(A/P, 10, N)$ | $(P/A, 10, N)$ | $(A/F, 10, N)$ | $(F/A, 10, N)$ | $(A/G, 10, N)$ | $N$ |
| 1 | 1.1000 | .90909 | 1.1000 | .9091 | 1.0000 | 1.000 | .0000 | 1 |
| 2 | 1.2100 | .82645 | .57619 | 1.7355 | .47619 | 2.0999 | .4761 | 2 |
| 3 | 1.3310 | .75132 | .40212 | 2.4868 | .30212 | 3.3099 | .9365 | 3 |
| 4 | 1.4641 | .68302 | .31547 | 3.1698 | .21547 | 4.6409 | 1.3810 | 4 |
| 5 | 1.6105 | .62092 | .26380 | 3.7907 | .16380 | 6.1050 | 1.8100 | 5 |
| 6 | 1.7715 | .56448 | .22961 | 4.3552 | .12961 | 7.7155 | 2.2234 | 6 |
| 7 | 1.9487 | .51316 | .20541 | 4.8683 | .10541 | 9.4870 | 2.6215 | 7 |
| 8 | 2.1435 | .46651 | .18745 | 5.3349 | .08745 | 11.435 | 3.0043 | 8 |
| 9 | 2.3579 | .42410 | .17364 | 5.7589 | .07364 | 13.579 | 3.3722 | 9 |
| 10 | 2.5937 | .38555 | .16275 | 6.1445 | .06275 | 15.937 | 3.7253 | 10 |
| 11 | 2.8530 | .35050 | .15396 | 6.4950 | .05396 | 18.530 | 4.0639 | 11 |
| 12 | 3.1384 | .31863 | .14676 | 6.8136 | .04676 | 21.383 | 4.3883 | 12 |
| 13 | 3.4522 | .28967 | .14078 | 7.1033 | .04078 | 24.522 | 4.6987 | 13 |
| 14 | 3.7974 | .26333 | .13575 | 7.3666 | .03575 | 27.974 | 4.9954 | 14 |
| 15 | 4.1771 | .23940 | .13147 | 7.6060 | .03147 | 31.771 | 5.2788 | 15 |
| 16 | 4.5949 | .21763 | .12782 | 7.8236 | .02782 | 35.949 | 5.5492 | 16 |
| 17 | 5.0544 | .19785 | .12466 | 8.0215 | .02466 | 40.543 | 5.8070 | 17 |
| 18 | 5.5598 | .17986 | .12193 | 8.2013 | .02193 | 45.598 | 6.0524 | 18 |
| 19 | 6.1158 | .16351 | .11955 | 8.3649 | .01955 | 51.158 | 6.2860 | 19 |
| 20 | 6.7273 | .14865 | .11746 | 8.5135 | .01746 | 57.273 | 6.5080 | 20 |
| 21 | 7.4001 | .13513 | .11562 | 8.6486 | .01562 | 64.001 | 6.7188 | 21 |
| 22 | 8.1401 | .12285 | .11401 | 8.7715 | .01401 | 71.401 | 6.9188 | 22 |
| 23 | 8.9541 | .11168 | .11257 | 8.8832 | .01257 | 79.541 | 7.1084 | 23 |
| 24 | 9.8495 | .10153 | .11130 | 8.9847 | .01130 | 88.495 | 7.2879 | 24 |
| 25 | 10.834 | .09230 | .11017 | 9.0770 | .01017 | 98.344 | 7.4579 | 25 |
| 26 | 11.917 | .08391 | .10916 | 9.1609 | .00916 | 109.17 | 7.6185 | 26 |
| 27 | 13.109 | .07628 | .10826 | 9.2372 | .00826 | 121.09 | 7.7703 | 27 |
| 28 | 14.420 | .06935 | .10745 | 9.3065 | .00745 | 134.20 | 7.9136 | 28 |
| 29 | 15.862 | .06304 | .10673 | 9.3696 | .00673 | 148.62 | 8.0488 | 29 |
| 30 | 17.448 | .05731 | .10608 | 9.4269 | .00608 | 164.48 | 8.1761 | 30 |
| 31 | 19.193 | .05210 | .10550 | 9.4790 | .00550 | 181.93 | 8.2961 | 31 |
| 32 | 21.113 | .04736 | .10497 | 9.5263 | .00497 | 201.13 | 8.4090 | 32 |
| 33 | 23.224 | .04306 | .10450 | 9.5694 | .00450 | 222.24 | 8.5151 | 33 |
| 34 | 25.546 | .03914 | .10407 | 9.6085 | .00407 | 245.46 | 8.6149 | 34 |
| 35 | 28.101 | .03559 | .10369 | 9.6441 | .00369 | 271.01 | 8.7085 | 35 |
| 40 | 45.257 | .02210 | .10226 | 9.7790 | .00226 | 442.57 | 9.0962 | 40 |
| 45 | 72.887 | .01372 | .10139 | 9.8628 | .00139 | 718.87 | 9.3740 | 45 |
| 50 | 117.38 | .00852 | .10086 | 9.9148 | .00086 | 1163.8 | 9.5704 | 50 |
| 55 | 189.04 | .00529 | .10053 | 9.9471 | .00053 | 1880.4 | 9.7075 | 55 |
| 60 | 304.46 | .00328 | .10033 | 9.9671 | .00033 | 3034.6 | 9.8022 | 60 |
| 65 | 490.34 | .00204 | .10020 | 9.9796 | .00020 | 4893.4 | 9.8671 | 65 |
| 70 | 789.69 | .00127 | .10013 | 9.9873 | .00013 | 7886.9 | 9.9112 | 70 |
| 75 | 1271.8 | .00079 | .10008 | 9.9921 | .00008 | 12709.0 | 9.9409 | 75 |
| 80 | 2048.2 | .00049 | .10005 | 9.9951 | .00005 | 20474.0 | 9.9609 | 80 |
| 85 | 3298.7 | .00030 | .10003 | 9.9969 | .00003 | 32979.7 | 9.9742 | 85 |
| 90 | 5312.5 | .00019 | .10002 | 9.9981 | .00002 | 53120.2 | 9.9830 | 90 |
| 95 | 8555.9 | .00012 | .10001 | 9.9988 | .00001 | 85556.8 | 9.9889 | 95 |
| 100 | 13780.6 | .00007 | .10001 | 9.9992 | .00001 | 137796.1 | 9.9927 | 100 |

**11% Interest Factors for Discrete Compounding Periods**

| | SINGLE PAYMENT | | UNIFORM SERIES | | | | | |
|---|---|---|---|---|---|---|---|---|
| | Compound Amount Factor | Present Worth Factor | Capital Recovery Factor | Present Worth Factor | Sinking Fund Factor | Compound Amount Factor | Gradient Factor | |
| $N$ | $(F/P, 11, N)$ | $(P/F, 11, N)$ | $(A/P, 11, N)$ | $(P/A, 11, N)$ | $(A/F, 11, N)$ | $(F/A, 11, N)$ | $(A/G, 11, N)$ | $N$ |
| 1 | 1.1100 | .90090 | 1.1100 | .9009 | 1.0000 | 1.000 | .0000 | 1 |
| 2 | 1.2321 | .81162 | .58394 | 1.7125 | .47394 | 2.1099 | .4739 | 2 |
| 3 | 1.3676 | .73119 | .40922 | 2.4437 | .29922 | 3.3420 | .9305 | 3 |
| 4 | 1.5180 | .65873 | .32233 | 3.1024 | .21233 | 4.7097 | 1.3698 | 4 |
| 5 | 1.6850 | .59345 | .27057 | 3.6958 | .16057 | 6.2277 | 1.7922 | 5 |
| 6 | 1.8704 | .53464 | .23638 | 4.2305 | .12638 | 7.9128 | 2.1975 | 6 |
| 7 | 2.0761 | .48166 | .21222 | 4.7121 | .10222 | 9.7831 | 2.5862 | 7 |
| 8 | 2.3045 | .43393 | .19432 | 5.1461 | .08432 | 11.859 | 2.9584 | 8 |
| 9 | 2.5580 | .39093 | .18060 | 5.5370 | .07060 | 14.163 | 3.3143 | 9 |
| 10 | 2.8394 | .35219 | .16980 | 5.8892 | .05980 | 16.721 | 3.6543 | 10 |
| 11 | 3.1517 | .31729 | .16112 | 6.2065 | .05112 | 19.561 | 3.9787 | 11 |
| 12 | 3.4984 | .28584 | .15403 | 6.4923 | .04403 | 22.712 | 4.2878 | 12 |
| 13 | 3.8832 | .25752 | .14815 | 6.7498 | .03815 | 26.211 | 4.5821 | 13 |
| 14 | 4.3104 | .23200 | .14323 | 6.9818 | .03323 | 30.094 | 4.8618 | 14 |
| 15 | 4.7845 | .20901 | .13907 | 7.1908 | .02907 | 34.404 | 5.1274 | 15 |
| 16 | 5.3108 | .18829 | .13552 | 7.3791 | .02552 | 39.189 | 5.3793 | 16 |
| 17 | 5.8950 | .16963 | .13247 | 7.5487 | .02247 | 44.500 | 5.6180 | 17 |
| 18 | 6.5434 | .15282 | .12984 | 7.7016 | .01984 | 50.395 | 5.8438 | 18 |
| 19 | 7.2632 | .13768 | .12756 | 7.8392 | .01756 | 56.938 | 6.0573 | 19 |
| 20 | 8.0622 | .12404 | .12558 | 7.9633 | .01558 | 64.201 | 6.2589 | 20 |
| 21 | 8.9490 | .11174 | .12384 | 8.0750 | .01384 | 72.264 | 6.4490 | 21 |
| 22 | 9.9334 | .10067 | .12231 | 8.1757 | .01231 | 81.213 | 6.6282 | 22 |
| 23 | 11.026 | .09069 | .12097 | 8.2664 | .01097 | 91.146 | 6.7969 | 23 |
| 24 | 12.238 | .08171 | .11979 | 8.3481 | .00979 | 102.17 | 6.9554 | 24 |
| 25 | 13.585 | .07361 | .11874 | 8.4217 | .00874 | 114.41 | 7.1044 | 25 |
| 26 | 15.079 | .06631 | .11781 | 8.4880 | .00781 | 127.99 | 7.2442 | 26 |
| 27 | 16.738 | .05974 | .11699 | 8.5478 | .00699 | 143.07 | 7.3753 | 27 |
| 28 | 18.579 | .05382 | .11626 | 8.6016 | .00626 | 159.81 | 7.4981 | 28 |
| 29 | 20.623 | .04849 | .11561 | 8.6501 | .00561 | 178.39 | 7.6130 | 29 |
| 30 | 22.891 | .04368 | .11502 | 8.6937 | .00502 | 199.01 | 7.7205 | 30 |
| 31 | 25.409 | .03935 | .11451 | 8.7331 | .00451 | 221.90 | 7.8209 | 31 |
| 32 | 28.204 | .03545 | .11404 | 8.7686 | .00404 | 247.31 | 7.9146 | 32 |
| 33 | 31.307 | .03194 | .11363 | 8.8005 | .00363 | 275.52 | 8.0020 | 33 |
| 34 | 34.751 | .02878 | .11326 | 8.8293 | .00326 | 306.83 | 8.0835 | 34 |
| 35 | 38.573 | .02592 | .11293 | 8.8552 | .00293 | 341.58 | 8.1594 | 35 |
| 40 | 64.999 | .01538 | .11172 | 8.9510 | .00172 | 581.81 | 8.4659 | 40 |
| 45 | 109.52 | .00913 | .11101 | 9.0079 | .00101 | 986.60 | 8.6762 | 45 |
| 50 | 184.55 | .00542 | .11060 | 9.0416 | .00060 | 1668.7 | 8.8185 | 50 |

**12% Interest Factors for Discrete Compounding Periods**

| | SINGLE PAYMENT | | UNIFORM SERIES | | | | | |
|---|---|---|---|---|---|---|---|---|
| | Compound Amount Factor | Present Worth Factor | Capital Recovery Factor | Present Worth Factor | Sinking Fund Factor | Compound Amount Factor | Gradient Factor | |
| *N* | (*F*/*P*, 12, *N*) | (*P*/*F*, 12, *N*) | (*A*/*P*, 12, *N*) | (*P*/*A*, 12, *N*) | (*A*/*F*, 12, *N*) | (*F*/*A*, 12, *N*) | (*A*/*G*, 12, *N*) | *N* |
| 1 | 1.1200 | .89286 | 1.1200 | .8929 | 1.0000 | 1.0000 | .0000 | 1 |
| 2 | 1.2544 | .79719 | .59170 | 1.6900 | .47170 | 2.1200 | .4717 | 2 |
| 3 | 1.4049 | .71178 | .41635 | 2.4018 | .29635 | 3.3743 | .9246 | 3 |
| 4 | 1.5735 | .63552 | .32924 | 3.0373 | .20924 | 4.7793 | 1.3588 | 4 |
| 5 | 1.7623 | .56743 | .27741 | 3.6047 | .15741 | 6.3528 | 1.7745 | 5 |
| 6 | 1.9738 | .50663 | .24323 | 4.1114 | .12323 | 8.115 | 2.1720 | 6 |
| 7 | 2.2106 | .45235 | .21912 | 4.5637 | .09912 | 10.088 | 2.5514 | 7 |
| 8 | 2.4759 | .40388 | .20130 | 4.9676 | .08130 | 12.299 | 2.9131 | 8 |
| 9 | 2.7730 | .36061 | .18768 | 5.3282 | .06768 | 14.775 | 3.2573 | 9 |
| 10 | 3.1058 | .32197 | .17698 | 5.6502 | .05698 | 17.548 | 3.5846 | 10 |
| 11 | 3.4785 | .28748 | .16842 | 5.9376 | .04842 | 20.654 | 3.8952 | 11 |
| 12 | 3.8959 | .25668 | .16144 | 6.1943 | .04144 | 24.132 | 4.1896 | 12 |
| 13 | 4.3634 | .22918 | .15568 | 6.4235 | .03568 | 28.028 | 4.4682 | 13 |
| 14 | 4.8870 | .20462 | .15087 | 6.6281 | .03087 | 32.392 | 4.7316 | 14 |
| 15 | 5.4735 | .18270 | .14682 | 6.8108 | .02682 | 37.279 | 4.9802 | 15 |
| 16 | 6.1303 | .16312 | .14339 | 6.9739 | .02339 | 42.752 | 5.2146 | 16 |
| 17 | 6.8659 | .14565 | .14046 | 7.1196 | .02046 | 48.883 | 5.4352 | 17 |
| 18 | 7.6899 | .13004 | .13794 | 7.2496 | .01794 | 55.749 | 5.6427 | 18 |
| 19 | 8.6126 | .11611 | .13576 | 7.3657 | .01576 | 63.439 | 5.8375 | 19 |
| 20 | 9.6462 | .10367 | .13388 | 7.4694 | .01388 | 72.051 | 6.0201 | 20 |
| 21 | 10.803 | .09256 | .13224 | 7.5620 | .01224 | 81.698 | 6.1913 | 21 |
| 22 | 12.100 | .08264 | .13081 | 7.6446 | .01081 | 92.501 | 6.3513 | 22 |
| 23 | 13.552 | .07379 | .12956 | 7.7184 | .00956 | 104.60 | 6.5009 | 23 |
| 24 | 15.178 | .06588 | .12846 | 7.7843 | .00846 | 118.15 | 6.6406 | 24 |
| 25 | 16.999 | .05882 | .12750 | 7.8431 | .00750 | 133.33 | 6.7708 | 25 |
| 26 | 19.039 | .05252 | .12665 | 7.8956 | .00665 | 150.33 | 6.8920 | 26 |
| 27 | 21.324 | .04689 | .12590 | 7.9425 | .00590 | 169.37 | 7.0049 | 27 |
| 28 | 23.883 | .04187 | .12524 | 7.9844 | .00524 | 190.69 | 7.1097 | 28 |
| 29 | 26.749 | .03738 | .12466 | 8.0218 | .00466 | 214.58 | 7.2071 | 29 |
| 30 | 29.959 | .03338 | .12414 | 8.0551 | .00414 | 241.32 | 7.2974 | 30 |
| 31 | 33.554 | .02980 | .12369 | 8.0849 | .00369 | 271.28 | 7.3810 | 31 |
| 32 | 37.581 | .02661 | .12328 | 8.1116 | .00328 | 304.84 | 7.4585 | 32 |
| 33 | 42.090 | .02376 | .12292 | 8.1353 | .00292 | 342.42 | 7.5302 | 33 |
| 34 | 47.141 | .02121 | .12260 | 8.1565 | .00260 | 384.51 | 7.5964 | 34 |
| 35 | 52.798 | .01894 | .12232 | 8.1755 | .00232 | 431.65 | 7.6576 | 35 |
| 40 | 93.049 | .01075 | .12130 | 8.2437 | .00130 | 767.07 | 7.8987 | 40 |
| 45 | 163.98 | .00610 | .12074 | 8.2825 | .00074 | 1358.2 | 8.0572 | 45 |
| 50 | 288.99 | .00346 | .12042 | 8.3045 | .00042 | 2399.9 | 8.1597 | 50 |

**13% Interest Factors for Discrete Compounding Periods**

| | SINGLE PAYMENT | | UNIFORM SERIES | | | | | |
|---|---|---|---|---|---|---|---|---|
| | Compound Amount Factor | Present Worth Factor | Capital Recovery Factor | Present Worth Factor | Sinking Fund Factor | Compound Amount Factor | Gradient Factor | |
| N | (F/P, 13, N) | (P/F, 13, N) | (A/P, 13, N) | (P/A, 13, N) | (A/F, 13, N) | (F/A, 13, N) | (A/G, 13, N) | N |
| 1 | 1.1300 | .88496 | 1.1300 | .8850 | 1.0000 | 1.0000 | .0000 | 1 |
| 2 | 1.2769 | .78315 | .59949 | 1.6680 | .46949 | 2.1299 | .4694 | 2 |
| 3 | 1.4428 | .69305 | .42352 | 2.3611 | .29353 | 3.4068 | .9187 | 3 |
| 4 | 1.6304 | .61332 | .33620 | 2.9744 | .20620 | 4.8497 | 1.3478 | 4 |
| 5 | 1.8424 | .54276 | .28432 | 3.5172 | .15432 | 6.4802 | 1.7570 | 5 |
| 6 | 2.0819 | .48032 | .25015 | 3.9975 | .12015 | 8.3226 | 2.1467 | 6 |
| 7 | 2.3525 | .42506 | .22611 | 4.4225 | .09611 | 10.404 | 2.5170 | 7 |
| 8 | 2.6584 | .37616 | .20839 | 4.7987 | .07839 | 12.757 | 2.8684 | 8 |
| 9 | 3.0040 | .33289 | .19487 | 5.1316 | .06487 | 15.415 | 3.2013 | 9 |
| 10 | 3.3945 | .29459 | .18429 | 5.4262 | .05429 | 18.419 | 3.5161 | 10 |
| 11 | 3.8358 | .26070 | .17584 | 5.6869 | .04584 | 21.813 | 3.8133 | 11 |
| 12 | 4.3344 | .23071 | .16899 | 5.9176 | .03899 | 25.649 | 4.0935 | 12 |
| 13 | 4.8979 | .20417 | .16335 | 6.1217 | .03335 | 29.984 | 4.3572 | 13 |
| 14 | 5.5346 | .18068 | .15867 | 6.3024 | .02867 | 34.882 | 4.6049 | 14 |
| 15 | 6.2541 | .15989 | .15474 | 6.4623 | .02474 | 40.416 | 4.8374 | 15 |
| 16 | 7.0672 | .14150 | .15143 | 6.6038 | .02143 | 46.670 | 5.0551 | 16 |
| 17 | 7.9859 | .12522 | .14861 | 6.7290 | .01861 | 53.737 | 5.2588 | 17 |
| 18 | 9.0240 | .11081 | .14620 | 6.8399 | .01620 | 61.723 | 5.4490 | 18 |
| 19 | 10.197 | .09807 | .14413 | 6.9379 | .01413 | 70.747 | 5.6264 | 19 |
| 20 | 11.522 | .08678 | .14235 | 7.0247 | .01235 | 80.944 | 5.7916 | 20 |
| 21 | 13.020 | .07680 | .14081 | 7.1015 | .01081 | 92.467 | 5.9453 | 21 |
| 22 | 14.713 | .06796 | .13948 | 7.1695 | .00948 | 105.48 | 6.0880 | 22 |
| 23 | 16.626 | .06015 | .13832 | 7.2296 | .00832 | 120.20 | 6.2204 | 23 |
| 24 | 18.787 | .05323 | .13731 | 7.2828 | .00731 | 136.82 | 6.3430 | 24 |
| 25 | 21.229 | .04710 | .13643 | 7.3299 | .00643 | 155.61 | 6.4565 | 25 |
| 26 | 23.989 | .04168 | .13565 | 7.3716 | .00565 | 176.84 | 6.5613 | 26 |
| 27 | 27.108 | .03689 | .13498 | 7.4085 | .00498 | 200.83 | 6.6581 | 27 |
| 28 | 30.632 | .03265 | .13439 | 7.4412 | .00439 | 227.94 | 6.7474 | 28 |
| 29 | 34.614 | .02889 | .13387 | 7.4700 | .00387 | 258.57 | 6.8295 | 29 |
| 30 | 39.114 | .02557 | .13341 | 7.4956 | .00341 | 293.18 | 6.9052 | 30 |
| 31 | 44.199 | .02262 | .13301 | 7.5182 | .00301 | 332.30 | 6.9747 | 31 |
| 32 | 49.945 | .02002 | .13266 | 7.5383 | .00266 | 376.50 | 7.0385 | 32 |
| 33 | 56.438 | .01772 | .13234 | 7.5560 | .00234 | 426.44 | 7.0970 | 33 |
| 34 | 63.775 | .01568 | .13207 | 7.5717 | .00207 | 482.88 | 7.1506 | 34 |
| 35 | 72.065 | .01388 | .13183 | 7.5855 | .00183 | 546.65 | 7.1998 | 35 |
| 40 | 132.77 | .00753 | .13099 | 7.6343 | .00099 | 1013.6 | 7.3887 | 40 |
| 45 | 244.62 | .00409 | .13053 | 7.6608 | .00053 | 1874.0 | 7.5076 | 45 |
| 50 | 450.71 | .00222 | .13029 | 7.6752 | .00029 | 3459.3 | 7.5811 | 50 |

**14% Interest Factors for Discrete Compounding Periods**

| | SINGLE PAYMENT | | UNIFORM SERIES | | | | | |
|---|---|---|---|---|---|---|---|---|
| | Compound Amount Factor | Present Worth Factor | Capital Recovery Factor | Present Worth Factor | Sinking Fund Factor | Compound Amount Factor | Gradient Factor | |
| $N$ | $(F/P, 14, N)$ | $(P/F, 14, N)$ | $(A/P, 14, N)$ | $(P/A, 14, N)$ | $(A/F, 14, N)$ | $(F/A, 14, N)$ | $(A/G, 14, N)$ | $N$ |
| 1 | 1.1400 | .87719 | 1.1400 | .8772 | 1.0000 | 1.000 | .0000 | 1 |
| 2 | 1.2996 | .76947 | .60729 | 1.6466 | .46729 | 2.1399 | .4672 | 2 |
| 3 | 1.4815 | .67497 | .43073 | 2.3216 | .29073 | 3.4395 | .9129 | 3 |
| 4 | 1.6889 | .59208 | .34321 | 2.9137 | .20321 | 4.9211 | 1.3369 | 4 |
| 5 | 1.9254 | .51937 | .29128 | 3.4330 | .15128 | 6.6100 | 1.7398 | 5 |
| 6 | 2.1949 | .45559 | .25716 | 3.8886 | .11716 | 8.535 | 2.1217 | 6 |
| 7 | 2.5022 | .39964 | .23319 | 4.2882 | .09319 | 10.730 | 2.4831 | 7 |
| 8 | 2.8525 | .35056 | .21557 | 4.6388 | .07557 | 13.232 | 2.8245 | 8 |
| 9 | 3.2519 | .30751 | .20217 | 4.9463 | .06217 | 16.085 | 3.1462 | 9 |
| 10 | 3.7071 | .26975 | .19171 | 5.2161 | .05171 | 19.337 | 3.4489 | 10 |
| 11 | 4.2261 | .23662 | .18339 | 5.4527 | .04339 | 23.044 | 3.7332 | 11 |
| 12 | 4.8178 | .20756 | .17667 | 5.6602 | .03667 | 27.270 | 3.9997 | 12 |
| 13 | 5.4923 | .18207 | .17116 | 5.8423 | .03116 | 32.088 | 4.2490 | 13 |
| 14 | 6.2612 | .15971 | .16661 | 6.0020 | .02661 | 37.580 | 4.4819 | 14 |
| 15 | 7.1378 | .14010 | .16281 | 6.1421 | .02281 | 43.841 | 4.6990 | 15 |
| 16 | 8.1371 | .12289 | .15962 | 6.2650 | .01962 | 50.979 | 4.9010 | 16 |
| 17 | 9.2763 | .10780 | .15692 | 6.3728 | .01692 | 59.116 | 5.0888 | 17 |
| 18 | 10.574 | .09456 | .15462 | 6.4674 | .01462 | 68.392 | 5.2629 | 18 |
| 19 | 12.055 | .08295 | .15266 | 6.5503 | .01266 | 78.967 | 5.4242 | 19 |
| 20 | 13.743 | .07276 | .15099 | 6.6231 | .01099 | 91.022 | 5.5734 | 20 |
| 21 | 15.667 | .06383 | .14954 | 6.6869 | .00955 | 104.76 | 5.7111 | 21 |
| 22 | 17.860 | .05599 | .14830 | 6.7429 | .00830 | 120.43 | 5.8380 | 22 |
| 23 | 20.361 | .04911 | .14723 | 6.7920 | .00723 | 138.29 | 5.9549 | 23 |
| 24 | 23.211 | .04308 | .14630 | 6.8351 | .00630 | 158.65 | 6.0623 | 24 |
| 25 | 26.461 | .03779 | .14550 | 6.8729 | .00550 | 181.86 | 6.1609 | 25 |
| 26 | 30.165 | .03315 | .14480 | 6.9060 | .00480 | 208.32 | 6.2514 | 26 |
| 27 | 34.388 | .02908 | .14419 | 6.9351 | .00419 | 238.49 | 6.3342 | 27 |
| 28 | 39.203 | .02551 | .14366 | 6.9606 | .00366 | 272.88 | 6.4039 | 28 |
| 29 | 44.691 | .02238 | .14320 | 6.9830 | .00320 | 312.08 | 6.4791 | 29 |
| 30 | 50.948 | .01963 | .14280 | 7.0026 | .00280 | 356.77 | 6.5422 | 30 |
| 31 | 58.081 | .01722 | .14245 | 7.0198 | .00245 | 407.72 | 6.5997 | 31 |
| 32 | 66.212 | .01510 | .14215 | 7.0349 | .00215 | 465.80 | 6.6521 | 32 |
| 33 | 75.482 | .01325 | .14188 | 7.0482 | .00188 | 532.01 | 6.6998 | 33 |
| 34 | 86.049 | .01162 | .14165 | 7.0598 | .00165 | 607.49 | 6.7430 | 34 |
| 35 | 98.096 | .01019 | .14144 | 7.0700 | .00144 | 693.54 | 6.7824 | 35 |
| 40 | 188.87 | .00529 | .14075 | 7.1050 | .00075 | 1341.9 | 6.9299 | 40 |
| 45 | 363.66 | .00275 | .14039 | 7.1232 | .00039 | 2590.4 | 7.0187 | 45 |
| 50 | 700.19 | .00143 | .14020 | 7.1326 | .00020 | 4994.2 | 7.0713 | 50 |

**15% Interest Factors for Discrete Compounding Periods**

| | SINGLE PAYMENT | | UNIFORM SERIES | | | | | |
|---|---|---|---|---|---|---|---|---|
| | Compound Amount Factor | Present Worth Factor | Capital Recovery Factor | Present Worth Factor | Sinking Fund Factor | Compound Amount Factor | Gradient Factor | |
| $N$ | $(F/P, 15, N)$ | $(P/F, 15, N)$ | $(A/P, 15, N)$ | $(P/A, 15, N)$ | $(A/F, 15, N)$ | $(F/A, 15, N)$ | $(A/G, 15, N)$ | $N$ |
| 1 | 1.1500 | .86957 | 1.1500 | .8696 | 1.0000 | 1.000 | .0000 | 1 |
| 2 | 1.3225 | .75614 | .61512 | 1.6257 | .46512 | 2.1499 | .4651 | 2 |
| 3 | 1.5208 | .65752 | .43798 | 2.2832 | .28798 | 3.4724 | .9071 | 3 |
| 4 | 1.7490 | .57175 | .35027 | 2.8549 | .20027 | 4.9933 | 1.3262 | 4 |
| 5 | 2.0113 | .49718 | .29832 | 3.3521 | .14832 | 6.7423 | 1.7227 | 5 |
| 6 | 2.3130 | .43233 | .26424 | 3.7844 | .11424 | 8.7536 | 2.0971 | 6 |
| 7 | 2.6600 | .37594 | .24036 | 4.1604 | .09036 | 11.066 | 2.4498 | 7 |
| 8 | 3.0590 | .32690 | .22285 | 4.4873 | .07285 | 13.726 | 2.7813 | 8 |
| 9 | 3.5178 | .28426 | .20957 | 4.7715 | .05957 | 16.785 | 3.0922 | 9 |
| 10 | 4.0455 | .24719 | .19925 | 5.0187 | .04925 | 20.303 | 3.3831 | 10 |
| 11 | 4.6523 | .21494 | .19107 | 5.2337 | .04107 | 24.349 | 3.6549 | 11 |
| 12 | 5.3502 | .18691 | .18448 | 5.4206 | .03448 | 29.001 | 3.9081 | 12 |
| 13 | 6.1527 | .16253 | .17911 | 5.5831 | .02911 | 34.351 | 4.1437 | 13 |
| 14 | 7.0756 | .14133 | .17469 | 5.7244 | .02469 | 40.504 | 4.3623 | 14 |
| 15 | 8.1369 | .12290 | .17102 | 5.8473 | .02102 | 47.579 | 4.5649 | 15 |
| 16 | 9.3575 | .10687 | .16795 | 5.9542 | .01795 | 55.716 | 4.7522 | 16 |
| 17 | 10.761 | .09293 | .16537 | 6.0471 | .01537 | 65.074 | 4.9250 | 17 |
| 18 | 12.375 | .08081 | .16319 | 6.1279 | .01319 | 75.835 | 5.0842 | 18 |
| 19 | 14.231 | .07027 | .16134 | 6.1982 | .01134 | 88.210 | 5.2307 | 19 |
| 20 | 16.366 | .06110 | .15976 | 6.2593 | .00976 | 102.44 | 5.3651 | 20 |
| 21 | 18.821 | .05313 | .15842 | 6.3124 | .00842 | 118.80 | 5.4883 | 21 |
| 22 | 21.644 | .04620 | .15727 | 6.3586 | .00727 | 137.62 | 5.6010 | 22 |
| 23 | 24.891 | .04018 | .15628 | 6.3988 | .00628 | 159.27 | 5.7039 | 23 |
| 24 | 28.624 | .03493 | .15543 | 6.4337 | .00543 | 184.16 | 5.7978 | 24 |
| 25 | 32.918 | .03038 | .15470 | 6.4641 | .00470 | 212.78 | 5.8834 | 25 |
| 26 | 37.856 | .02642 | .15407 | 6.4905 | .00407 | 245.70 | 5.9612 | 26 |
| 27 | 43.534 | .02297 | .15353 | 6.5135 | .00353 | 283.56 | 6.0318 | 27 |
| 28 | 50.064 | .01997 | .15306 | 6.5335 | .00306 | 327.09 | 6.0959 | 28 |
| 29 | 57.574 | .01737 | .15265 | 6.5508 | .00265 | 377.16 | 6.1540 | 29 |
| 30 | 66.210 | .01510 | .15230 | 6.5659 | .00230 | 434.73 | 6.2066 | 30 |
| 31 | 76.141 | .01313 | .15200 | 6.5791 | .00200 | 500.94 | 6.2541 | 31 |
| 32 | 87.563 | .01142 | .15173 | 6.5905 | .00173 | 577.08 | 6.2970 | 32 |
| 33 | 100.69 | .00993 | .15150 | 6.6004 | .00150 | 664.65 | 6.3356 | 33 |
| 34 | 115.80 | .00864 | .15131 | 6.6091 | .00131 | 765.34 | 6.3705 | 34 |
| 35 | 133.17 | .00751 | .15113 | 6.6166 | .00113 | 881.14 | 6.4018 | 35 |
| 40 | 267.85 | .00373 | .15056 | 6.6417 | .00056 | 1779.0 | 6.5167 | 40 |
| 45 | 538.75 | .00186 | .15028 | 6.6543 | .00028 | 3585.0 | 6.5829 | 45 |
| 50 | 1083.6 | .00092 | .15014 | 6.6605 | .00014 | 7217.4 | 6.8204 | 50 |

**20% Interest Factors for Discrete Compounding Periods**

| | SINGLE PAYMENT | | UNIFORM SERIES | | | | | |
|---|---|---|---|---|---|---|---|---|
| | Compound Amount Factor | Present Worth Factor | Capital Recovery Factor | Present Worth Factor | Sinking Fund Factor | Compound Amount Factor | Gradient Factor | |
| *N* | (*F/P*, 20, *N*) | (*P/F*, 20, *N*) | (*A/P*, 20, *N*) | (*P/A*, 20, *N*) | (*A/F*, 20, *N*) | (*F/A*, 20, *N*) | (*A/G*, 20, *N*) | *N* |
| 1 | 1.2000 | .83333 | 1.2000 | .8333 | 1.0000 | 1.0000 | .0000 | 1 |
| 2 | 1.4400 | .69445 | .65455 | 1.5277 | .45455 | 2.1999 | .4545 | 2 |
| 3 | 1.7280 | .57870 | .47473 | 2.1064 | .27473 | 3.6399 | .8791 | 3 |
| 4 | 2.0736 | .48225 | .38629 | 2.5887 | .18629 | 5.3679 | 1.2742 | 4 |
| 5 | 2.4883 | .40188 | .33438 | 2.9906 | .13438 | 7.4415 | 1.6405 | 5 |
| 6 | 2.9859 | .33490 | .30071 | 3.3255 | .10071 | 9.9298 | 1.9788 | 6 |
| 7 | 3.5831 | .27908 | .27742 | 3.6045 | .07742 | 12.915 | 2.2901 | 7 |
| 8 | 4.2998 | .23257 | .26061 | 3.8371 | .06061 | 16.498 | 2.5756 | 8 |
| 9 | 5.1597 | .19381 | .24808 | 4.0309 | .04808 | 20.798 | 2.8364 | 9 |
| 10 | 6.1917 | .16151 | .23852 | 4.1924 | .03852 | 25.958 | 3.0738 | 10 |
| 11 | 7.4300 | .13459 | .23110 | 4.3270 | .03110 | 32.150 | 3.2892 | 11 |
| 12 | 8.9160 | .11216 | .22527 | 4.4392 | .02527 | 39.580 | 3.4840 | 12 |
| 13 | 10.699 | .09346 | .22062 | 4.5326 | .02062 | 48.496 | 3.6596 | 13 |
| 14 | 12.839 | .07789 | .21689 | 4.6105 | .01689 | 59.195 | 3.8174 | 14 |
| 15 | 15.406 | .06491 | .21388 | 4.6754 | .01388 | 72.034 | 3.9588 | 15 |
| 16 | 18.488 | .05409 | .21144 | 4.7295 | .01144 | 87.441 | 4.0851 | 16 |
| 17 | 22.185 | .04507 | .20944 | 4.7746 | .00944 | 105.92 | 4.1975 | 17 |
| 18 | 26.623 | .03756 | .20781 | 4.8121 | .00781 | 128.11 | 4.2975 | 18 |
| 19 | 31.947 | .03130 | .20646 | 4.8435 | .00646 | 154.73 | 4.3860 | 19 |
| 20 | 38.337 | .02608 | .20536 | 4.8695 | .00536 | 186.68 | 4.4643 | 20 |
| 21 | 46.004 | .02174 | .20444 | 4.8913 | .00444 | 225.02 | 4.5333 | 21 |
| 22 | 55.205 | .01811 | .20369 | 4.9094 | .00369 | 271.02 | 4.5941 | 22 |
| 23 | 66.246 | .01510 | .20307 | 4.9245 | .00307 | 326.23 | 4.6474 | 23 |
| 24 | 79.495 | .01258 | .20255 | 4.9371 | .00255 | 392.47 | 4.6942 | 24 |
| 25 | 95.394 | .01048 | .20212 | 4.9475 | .00212 | 471.97 | 4.7351 | 25 |
| 26 | 114.47 | .00874 | .20176 | 4.9563 | .00176 | 567.36 | 4.7708 | 26 |
| 27 | 137.36 | .00728 | .20147 | 4.9636 | .00147 | 681.84 | 4.8020 | 27 |
| 28 | 164.84 | .00607 | .20122 | 4.9696 | .00122 | 819.21 | 4.8291 | 28 |
| 29 | 197.81 | .00506 | .20102 | 4.9747 | .00102 | 984.05 | 4.8526 | 29 |
| 30 | 237.37 | .00421 | .20085 | 4.9789 | .00085 | 1181.8 | 4.8730 | 30 |
| 31 | 284.84 | .00351 | .20070 | 4.9824 | .00070 | 1419.2 | 4.8907 | 31 |
| 32 | 341.81 | .00293 | .20059 | 4.9853 | .00059 | 1704.0 | 4.9061 | 32 |
| 33 | 410.17 | .00244 | .20049 | 4.9878 | .00049 | 2045.8 | 4.9193 | 33 |
| 34 | 492.21 | .00203 | .20041 | 4.9898 | .00041 | 2456.0 | 4.9307 | 34 |
| 35 | 590.65 | .00169 | .20034 | 4.9915 | .00034 | 2948.2 | 4.9406 | 35 |
| 40 | 1469.7 | .00068 | .20014 | 4.9966 | .00014 | 7343.6 | 4.9727 | 40 |
| 45 | 3657.1 | .00027 | .20005 | 4.9986 | .00005 | 18281.3 | 4.9876 | 45 |
| 50 | 9100.1 | .00011 | .20002 | 4.9994 | .00002 | 45497.2 | 4.9945 | 50 |

**25% Interest Factors for Discrete Compounding Periods**

| | SINGLE PAYMENT | | UNIFORM SERIES | | | | | |
|---|---|---|---|---|---|---|---|---|
| | Compound Amount Factor | Present Worth Factor | Capital Recovery Factor | Present Worth Factor | Sinking Fund Factor | Compound Amount Factor | Gradient Factor | |
| $N$ | $(F/P, 25, N)$ | $(P/F, 25, N)$ | $(A/P, 25, N)$ | $(P/A, 25, N)$ | $(A/F, 25, N)$ | $(F/A, 25, N)$ | $(A/G, 25, N)$ | $N$ |
| 1 | 1.2500 | .80000 | 1.2500 | .8000 | 1.0000 | 1.0000 | .00000 | 1 |
| 2 | 1.5625 | .64000 | .69444 | 1.4400 | .44444 | 2.2500 | .44444 | 2 |
| 3 | 1.9531 | .51200 | .51230 | 1.9520 | .26230 | 3.8125 | .85246 | 3 |
| 4 | 2.4414 | .40960 | .42344 | 2.3616 | .17344 | 5.7656 | 1.2249 | 4 |
| 5 | 3.0518 | .32768 | .37185 | 2.6893 | .12185 | 8.2070 | 1.5631 | 5 |
| 6 | 3.8147 | .26214 | .33882 | 2.9514 | .08882 | 11.259 | 1.8683 | 6 |
| 7 | 4.7684 | .20972 | .31634 | 3.1661 | .06634 | 15.073 | 2.1424 | 7 |
| 8 | 5.9605 | .16777 | .30040 | 3.3289 | .05040 | 19.842 | 2.3872 | 8 |
| 9 | 7.4506 | .13422 | .28876 | 3.4631 | .03876 | 25.802 | 2.6048 | 9 |
| 10 | 9.3132 | .10737 | .28007 | 3.5705 | .03007 | 33.253 | 2.7971 | 10 |
| 11 | 11.642 | .08590 | .27349 | 3.6564 | .02349 | 42.566 | 2.9663 | 11 |
| 12 | 14.552 | .06872 | .26845 | 3.7251 | .01845 | 54.208 | 3.1145 | 12 |
| 13 | 18.190 | .05498 | .26454 | 3.7801 | .01454 | 68.760 | 3.2437 | 13 |
| 14 | 22.737 | .04398 | .26150 | 3.8241 | .01150 | 86.949 | 3.3559 | 14 |
| 15 | 28.422 | .03518 | .25912 | 3.8593 | .00912 | 109.687 | 3.4530 | 15 |
| 16 | 35.527 | .02815 | .25724 | 3.8874 | .00724 | 138.109 | 3.5366 | 16 |
| 17 | 44.409 | .02252 | .25576 | 3.9099 | .00576 | 173.636 | 3.6084 | 17 |
| 18 | 55.511 | .01801 | .25459 | 3.9279 | .00459 | 218.045 | 3.6698 | 18 |
| 19 | 69.389 | .01441 | .25366 | 3.9424 | .00366 | 273.556 | 3.7222 | 19 |
| 20 | 86.736 | .01153 | .25292 | 3.9539 | .00292 | 342.945 | 3.7667 | 20 |
| 21 | 108.420 | .00922 | .25233 | 3.9631 | .00233 | 429.681 | 3.8045 | 21 |
| 22 | 135.525 | .00738 | .25186 | 3.9705 | .00186 | 538.101 | 3.8365 | 22 |
| 23 | 169.407 | .00590 | .25148 | 3.9764 | .00148 | 673.626 | 3.8634 | 23 |
| 24 | 211.758 | .00472 | .25119 | 3.9811 | .00119 | 843.033 | 3.8861 | 24 |
| 25 | 264.698 | .00378 | .25095 | 3.9849 | .00095 | 1054.791 | 3.9052 | 25 |
| 26 | 330.872 | .00302 | .25076 | 3.9879 | .00076 | 1319.489 | 3.9212 | 26 |
| 27 | 413.590 | .00242 | .25061 | 3.9903 | .00061 | 1650.361 | 3.9346 | 27 |
| 28 | 516.988 | .00193 | .25048 | 3.9923 | .00048 | 2063.952 | 3.9457 | 28 |
| 29 | 646.235 | .00155 | .25039 | 3.9938 | .00039 | 2580.939 | 3.9551 | 29 |
| 30 | 807.794 | .00124 | .25031 | 3.9950 | .00031 | 3227.174 | 3.9628 | 30 |
| 31 | 1009.742 | .00099 | .25025 | 3.9960 | .00025 | 4034.968 | 3.9693 | 31 |
| 32 | 1262.177 | .00079 | .25020 | 3.9968 | .00020 | 5044.710 | 3.9746 | 32 |
| 33 | 1577.722 | .00063 | .25016 | 3.9975 | .00016 | 6306.887 | 3.9791 | 33 |
| 34 | 1972.152 | .00051 | .25013 | 3.9980 | .00012 | 7884.609 | 3.9828 | 34 |
| 35 | 2465.190 | .00041 | .25010 | 3.9984 | .00010 | 9856.761 | 3.9858 | 35 |

**30% Interest Factors for Discrete Compounding Periods**

| | SINGLE PAYMENT | | UNIFORM SERIES | | | | | |
|---|---|---|---|---|---|---|---|---|
| | Compound Amount Factor | Present Worth Factor | Capital Recovery Factor | Present Worth Factor | Sinking Fund Factor | Compound Amount Factor | Gradient Factor | |
| *N* | (*F*/*P*, 30, *N*) | (*P*/*F*, 30, *N*) | (*A*/*P*, 30, *N*) | (*P*/*A*, 30, *N*) | (*A*/*F*, 30, *N*) | (*F*/*A*, 30, *N*) | (*A*/*G*, 30, *N*) | *N* |
| 1 | 1.3000 | .76923 | 1.3000 | .7692 | 1.0000 | 1.000 | .0000 | 1 |
| 2 | 1.6900 | .59172 | .73478 | 1.3609 | .43478 | 2.2999 | .4348 | 2 |
| 3 | 2.1969 | .45517 | .55063 | 1.8161 | .25063 | 3.9899 | .8277 | 3 |
| 4 | 2.8560 | .35013 | .46163 | 2.1662 | .16163 | 6.1869 | 1.1782 | 4 |
| 5 | 3.7129 | .26933 | .41058 | 2.4355 | .11058 | 9.0430 | 1.4903 | 5 |
| 6 | 4.8267 | .20718 | .37839 | 2.6427 | .07839 | 12.755 | 1.7654 | 6 |
| 7 | 6.2748 | .15937 | .35687 | 2.8021 | .05687 | 17.582 | 2.0062 | 7 |
| 8 | 8.1572 | .12259 | .34192 | 2.9247 | .04192 | 23.857 | 2.2155 | 8 |
| 9 | 10.604 | .09430 | .33124 | 3.0190 | .03124 | 32.014 | 2.3962 | 9 |
| 10 | 13.785 | .07254 | .32346 | 3.0915 | .02346 | 42.619 | 2.5512 | 10 |
| 11 | 17.921 | .05580 | .31773 | 3.1473 | .01773 | 56.404 | 2.6832 | 11 |
| 12 | 23.297 | .04292 | .31345 | 3.1902 | .01345 | 74.326 | 2.7951 | 12 |
| 13 | 30.287 | .03302 | .31024 | 3.2232 | .01024 | 97.624 | 2.8894 | 13 |
| 14 | 39.373 | .02540 | .30782 | 3.2486 | .00782 | 127.91 | 2.9685 | 14 |
| 15 | 51.185 | .01954 | .30598 | 3.2682 | .00598 | 167.28 | 3.0344 | 15 |
| 16 | 66.540 | .01503 | .30458 | 3.2832 | .00458 | 218.46 | 3.0892 | 16 |
| 17 | 86.503 | .01156 | .30351 | 3.2948 | .00351 | 285.01 | 3.1345 | 17 |
| 18 | 112.45 | .00889 | .30269 | 3.3036 | .00269 | 371.51 | 3.1718 | 18 |
| 19 | 146.18 | .00684 | .30207 | 3.3105 | .00207 | 483.96 | 3.2024 | 19 |
| 20 | 190.04 | .00526 | .30159 | 3.3157 | .00159 | 630.15 | 3.2275 | 20 |
| 21 | 247.06 | .00405 | .30122 | 3.3198 | .00122 | 820.20 | 3.2479 | 21 |
| 22 | 321.17 | .00311 | .30094 | 3.3229 | .00094 | 1067.2 | 3.2646 | 22 |
| 23 | 417.53 | .00240 | .30072 | 3.3253 | .00072 | 1388.4 | 3.2781 | 23 |
| 24 | 542.79 | .00184 | .30055 | 3.3271 | .00055 | 1805.9 | 3.2890 | 24 |
| 25 | 705.62 | .00142 | .30043 | 3.3286 | .00043 | 2348.7 | 3.2978 | 25 |
| 26 | 917.31 | .00109 | .30033 | 3.3297 | .00033 | 3054.3 | 3.3049 | 26 |
| 27 | 1192.5 | .00084 | .30025 | 3.3305 | .00025 | 3971.6 | 3.3106 | 27 |
| 28 | 1550.2 | .00065 | .30019 | 3.3311 | .00019 | 5164.1 | 3.3152 | 28 |
| 29 | 2015.3 | .00050 | .30015 | 3.3316 | .00015 | 6714.4 | 3.3189 | 29 |
| 30 | 2619.9 | .00038 | .30011 | 3.3320 | .00011 | 8729.7 | 3.3218 | 30 |
| 31 | 3405.9 | .00029 | .30009 | 3.3323 | .00009 | 11350.0 | 3.3242 | 31 |
| 32 | 4427.6 | .00023 | .30007 | 3.3325 | .00007 | 14756.0 | 3.3261 | 32 |
| 33 | 5755.9 | .00017 | .30005 | 3.3327 | .00005 | 19184.0 | 3.3276 | 33 |
| 34 | 7482.7 | .00013 | .30004 | 3.3328 | .00004 | 24940.0 | 3.3287 | 34 |
| 35 | 9727.5 | .00010 | .30003 | 3.3329 | .00003 | 32423.0 | 3.3297 | 35 |

**40% Interest Factors for Discrete Compounding Periods**

| | SINGLE PAYMENT | | UNIFORM SERIES | | | | | |
|---|---|---|---|---|---|---|---|---|
| | Compound Amount Factor | Present Worth Factor | Capital Recovery Factor | Present Worth Factor | Sinking Fund Factor | Compound Amount Factor | Gradient Factor | |
| *N* | $(F/P, 40, N)$ | $(P/F, 40, N)$ | $(A/P, 40, N)$ | $(P/A, 40, N)$ | $(A/F, 40, N)$ | $(F/A, 40, N)$ | $(A/G, 40, N)$ | *N* |
| 1 | 1.4000 | .71429 | 1.40000 | .7143 | 1.00000 | 1.0000 | .0000 | 1 |
| 2 | 1.9600 | .51020 | .81667 | 1.2244 | .41667 | 2.3999 | .4167 | 2 |
| 3 | 2.7440 | .36443 | .62936 | 1.5889 | .22936 | 4.3599 | .7798 | 3 |
| 4 | 3.8415 | .26031 | .54077 | 1.8492 | .14077 | 7.1039 | 1.0923 | 4 |
| 5 | 5.3782 | .18593 | .49136 | 2.0351 | .09136 | 10.945 | 1.3579 | 5 |
| 6 | 7.5295 | .13281 | .46126 | 2.1679 | .06126 | 16.323 | 1.5810 | 6 |
| 7 | 10.541 | .09486 | .44192 | 2.2628 | .04192 | 23.853 | 1.7663 | 7 |
| 8 | 14.757 | .06776 | .42907 | 2.3306 | .02907 | 34.394 | 1.9185 | 8 |
| 9 | 20.660 | .04840 | .42034 | 2.3790 | .02034 | 49.152 | 2.0422 | 9 |
| 10 | 28.925 | .03457 | .41432 | 2.4135 | .01432 | 69.813 | 2.1419 | 10 |
| 11 | 40.495 | .02469 | .41013 | 2.4382 | .01013 | 98.738 | 2.2214 | 11 |
| 12 | 56.693 | .01764 | .40718 | 2.4559 | .00718 | 139.23 | 2.2845 | 12 |
| 13 | 79.370 | .01260 | .40510 | 2.4685 | .00510 | 195.92 | 2.3341 | 13 |
| 14 | 111.11 | .00900 | .40363 | 2.4775 | .00363 | 275.29 | 2.3728 | 14 |
| 15 | 155.56 | .00643 | .40259 | 2.4839 | .00259 | 386.41 | 2.4029 | 15 |
| 16 | 217.79 | .00459 | .40184 | 2.4885 | .00185 | 541.98 | 2.4262 | 16 |
| 17 | 304.91 | .00328 | .40132 | 2.4918 | .00132 | 759.77 | 2.4440 | 17 |
| 18 | 426.87 | .00234 | .40094 | 2.4941 | .00094 | 1064.6 | 2.4577 | 18 |
| 19 | 597.62 | .00167 | .40067 | 2.4958 | .00067 | 1491.5 | 2.4681 | 19 |
| 20 | 836.67 | .00120 | .40048 | 2.4970 | .00048 | 2089.1 | 2.4760 | 20 |
| 21 | 1171.3 | .00085 | .40034 | 2.4978 | .00034 | 2925.8 | 2.4820 | 21 |
| 22 | 1639.8 | .00061 | .40024 | 2.4984 | .00024 | 4097.1 | 2.4865 | 22 |
| 23 | 2295.8 | .00044 | .40017 | 2.4989 | .00017 | 5737.0 | 2.4899 | 23 |
| 24 | 3214.1 | .00031 | .40012 | 2.4992 | .00012 | 8032.8 | 2.4925 | 24 |
| 25 | 4499.8 | .00022 | .40009 | 2.4994 | .00009 | 11247.2 | 2.4944 | 25 |

**50% Interest Factors for Discrete Compounding Periods**

| | SINGLE PAYMENT | | UNIFORM SERIES | | | | | |
|---|---|---|---|---|---|---|---|---|
| | Compound Amount Factor | Present Worth Factor | Capital Recovery Factor | Present Worth Factor | Sinking Fund Factor | Compound Amount Factor | Gradient Factor | |
| *N* | (*F/P*, 50, *N*) | (*P/F*, 50, *N*) | (*A/P*, 50, *N*) | (*P/A*, 50, *N*) | (*A/F*, 50, *N*) | (*F/A*, 50, *N*) | (*A/G*, 50, *N*) | *N* |
| 1 | 1.5000 | .66667 | 1.5000 | .6667 | 1.00000 | 1.000 | .0000 | 1 |
| 2 | 2.2500 | .44444 | .90000 | 1.1111 | .40000 | 2.500 | .4000 | 2 |
| 3 | 3.3750 | .29630 | .71053 | 1.4074 | .21053 | 4.750 | .7368 | 3 |
| 4 | 5.0625 | .19753 | .62308 | 1.6049 | .12308 | 8.125 | 1.0153 | 4 |
| 5 | 7.5937 | .13169 | .57583 | 1.7366 | .07583 | 13.187 | 1.2417 | 5 |
| 6 | 11.390 | .08779 | .54812 | 1.8244 | .04812 | 20.781 | 1.4225 | 6 |
| 7 | 17.085 | .05853 | .53108 | 1.8829 | .03108 | 32.171 | 1.5648 | 7 |
| 8 | 25.628 | .03902 | .52030 | 1.9219 | .02030 | 49.257 | 1.6751 | 8 |
| 9 | 38.443 | .02601 | .51335 | 1.9479 | .01335 | 74.886 | 1.7596 | 9 |
| 10 | 57.665 | .01734 | .50882 | 1.9653 | .00882 | 113.33 | 1.8235 | 10 |
| 11 | 86.497 | .01156 | .50585 | 1.9768 | .00585 | 170.99 | 1.8713 | 11 |
| 12 | 129.74 | .00771 | .50388 | 1.9845 | .00388 | 257.49 | 1.9067 | 12 |
| 13 | 194.61 | .00514 | .50258 | 1.9897 | .00258 | 387.23 | 1.9328 | 13 |
| 14 | 291.92 | .00343 | .50172 | 1.9931 | .00172 | 581.85 | 1.9518 | 14 |
| 15 | 437.89 | .00228 | .50114 | 1.9954 | .00114 | 873.78 | 1.9656 | 15 |
| 16 | 656.84 | .00152 | .50076 | 1.9969 | .00076 | 1311.6 | 1.9756 | 16 |
| 17 | 985.26 | .00101 | .50051 | 1.9979 | .00051 | 1968.5 | 1.9827 | 17 |
| 18 | 1477.8 | .00068 | .50034 | 1.9986 | .00034 | 2953.7 | 1.9878 | 18 |
| 19 | 2216.8 | .00045 | .50023 | 1.9991 | .00023 | 4431.6 | 1.9914 | 19 |
| 20 | 3325.2 | .00030 | .50015 | 1.9994 | .00015 | 6648.5 | 1.9939 | 20 |
| 21 | 4987.8 | .00020 | .50010 | 1.9996 | .00010 | 9973.7 | 1.9957 | 21 |
| 22 | 7481.8 | .00013 | .50007 | 1.9997 | .00007 | 14961.7 | 1.9970 | 22 |
| 23 | 11222.7 | .00009 | .50004 | 1.9998 | .00004 | 22443.5 | 1.9979 | 23 |
| 24 | 16834.1 | .00006 | .50003 | 1.9998 | .00003 | 33666.2 | 1.9985 | 24 |
| 25 | 25251.2 | .00004 | .50002 | 1.9999 | .00002 | 50500.3 | 1.9990 | 25 |

**60% Interest Factors for Discrete Compounding Periods**

| | SINGLE PAYMENT | | UNIFORM SERIES | | | | | |
|---|---|---|---|---|---|---|---|---|
| | Compound Amount Factor | Present Worth Factor | Capital Recovery Factor | Present Worth Factor | Sinking Fund Factor | Compound Amount Factor | Gradient Factor | |
| $N$ | $(F/P, 60, N)$ | $(P/F, 60, N)$ | $(A/P, 60, N)$ | $(P/A, 60, N)$ | $(A/F, 60, N)$ | $(F/A, 60, N)$ | $(A/G, 60, N)$ | $N$ |
| 1 | 1.6000 | .62500 | 1.6000 | .6250 | 1.0000 | 1.000 | .0000 | 1 |
| 2 | 2.5600 | .39063 | .98462 | 1.0156 | .38462 | 2.6000 | .3846 | 2 |
| 3 | 4.0959 | .24414 | .79380 | 1.2597 | .19380 | 5.1599 | .6977 | 3 |
| 4 | 6.5535 | .15259 | .70804 | 1.4123 | .10804 | 9.2559 | .9464 | 4 |
| 5 | 10.485 | .09537 | .66325 | 1.5077 | .06325 | 15.809 | 1.1395 | 5 |
| 6 | 16.777 | .05960 | .63803 | 1.5673 | .03803 | 26.295 | 1.2863 | 6 |
| 7 | 26.843 | .03725 | .62322 | 1.6045 | .02322 | 43.072 | 1.3958 | 7 |
| 8 | 42.949 | .02328 | .61430 | 1.6278 | .01430 | 69.915 | 1.4759 | 8 |
| 9 | 68.719 | .01455 | .60886 | 1.6424 | .00886 | 112.86 | 1.5337 | 9 |
| 10 | 109.95 | .00909 | .60551 | 1.6515 | .00551 | 181.58 | 1.5748 | 10 |
| 11 | 175.92 | .00568 | .60343 | 1.6571 | .00343 | 291.53 | 1.6037 | 11 |
| 12 | 281.47 | .00355 | .60214 | 1.6607 | .00214 | 467.45 | 1.6238 | 12 |
| 13 | 450.35 | .00222 | .60134 | 1.6629 | .00134 | 748.92 | 1.6377 | 13 |
| 14 | 720.57 | .00139 | .60083 | 1.6643 | .00083 | 1199.2 | 1.6472 | 14 |
| 15 | 1152.9 | .00087 | .60052 | 1.6652 | .00052 | 1919.8 | 1.6536 | 15 |
| 16 | 1844.6 | .00054 | .60033 | 1.6657 | .00033 | 3072.7 | 1.6579 | 16 |
| 17 | 2951.4 | .00034 | .60020 | 1.6661 | .00020 | 4917.4 | 1.6609 | 17 |
| 18 | 4722.3 | .00021 | .60013 | 1.6663 | .00013 | 7868.8 | 1.6628 | 18 |
| 19 | 7555.7 | .00013 | .60008 | 1.6664 | .00008 | 12591.0 | 1.6641 | 19 |
| 20 | 12089.0 | .00008 | .60005 | 1.6665 | .00005 | 20147.0 | 1.6650 | 20 |

**70% Interest Factors for Discrete Compounding Periods**

| | SINGLE PAYMENT | | UNIFORM SERIES | | | | | |
|---|---|---|---|---|---|---|---|---|
| | Compound Amount Factor | Present Worth Factor | Capital Recovery Factor | Present Worth Factor | Sinking Fund Factor | Compound Amount Factor | Gradient Factor | |
| N | (F/P, 70, N) | (P/F, 70, N) | (A/P, 70, N) | (P/A, 70, N) | (A/F, 70, N) | (F/A, 70, N) | (A/G, 70, N) | N |
| 1 | 1.7000 | .58824 | 1.7000 | .5882 | 1.0000 | 1.000 | .0000 | 1 |
| 2 | 2.8900 | .34602 | 1.0703 | .9343 | .37037 | 2.700 | .3704 | 2 |
| 3 | 4.9130 | .20354 | .87889 | 1.1378 | .17889 | 5.590 | .6619 | 3 |
| 4 | 8.3520 | .11973 | .79521 | 1.2575 | .09521 | 10.502 | .8845 | 4 |
| 5 | 14.198 | .07043 | .75304 | 1.3279 | .05304 | 18.855 | 1.0497 | 5 |
| 6 | 24.137 | .04143 | .73025 | 1.3693 | .03025 | 33.053 | 1.1692 | 6 |
| 7 | 41.033 | .02437 | .71749 | 1.3937 | .01749 | 57.191 | 1.2537 | 7 |
| 8 | 69.757 | .01434 | .71018 | 1.4080 | .01018 | 98.224 | 1.3122 | 8 |
| 9 | 118.58 | .00843 | .70595 | 1.4165 | .00595 | 167.98 | 1.3520 | 9 |
| 10 | 201.59 | .00496 | .70349 | 1.4214 | .00349 | 286.56 | 1.3787 | 10 |
| 11 | 342.71 | .00292 | .70205 | 1.4244 | .00205 | 488.16 | 1.3963 | 11 |
| 12 | 582.62 | .00172 | .70120 | 1.4261 | .00120 | 830.88 | 1.4079 | 12 |
| 13 | 990.45 | .00101 | .70071 | 1.4271 | .00071 | 1413.5 | 1.4154 | 13 |
| 14 | 1683.7 | .00059 | .70042 | 1.4277 | .00042 | 2403.9 | 1.4202 | 14 |
| 15 | 2862.4 | .00035 | .70024 | 1.4280 | .00024 | 4087.7 | 1.4233 | 15 |
| 16 | 4866.0 | .00021 | .70014 | 1.4282 | .00014 | 6950.1 | 1.4252 | 16 |
| 17 | 8272.3 | .00012 | .70008 | 1.4284 | .00008 | 11816.0 | 1.4265 | 17 |
| 18 | 14063.0 | .00007 | .70005 | 1.4284 | .00005 | 20089.0 | 1.4272 | 18 |
| 19 | 23907.0 | .00004 | .70003 | 1.4285 | .00003 | 34152.0 | 1.4277 | 19 |
| 20 | 40642.0 | .00002 | .70002 | 1.4285 | .00002 | 58059.0 | 1.4280 | 20 |

**80% Interest Factors for Discrete Compounding Periods**

| | SINGLE PAYMENT | | UNIFORM SERIES | | | | | |
|---|---|---|---|---|---|---|---|---|
| | Compound Amount Factor | Present Worth Factor | Capital Recovery Factor | Present Worth Factor | Sinking Fund Factor | Compound Amount Factor | Gradient Factor | |
| N | (F/P, 80, N) | (P/F, 80, N) | (A/P, 80, N) | (P/A, 80, N) | (A/F, 80, N) | (F/A, 80, N) | (A/G, 80, N) | N |
| 1 | 1.8000 | .55556 | 1.8000 | .5556 | 1.00000 | 1.0000 | .0000 | 1 |
| 2 | 3.2400 | .30864 | 1.1571 | .8642 | .35714 | 2.8000 | .3571 | 2 |
| 3 | 5.8319 | .17147 | .96556 | 1.0356 | .16556 | 6.0399 | .6291 | 3 |
| 4 | 10.497 | .09526 | .88423 | 1.1309 | .08423 | 11.871 | .8288 | 4 |
| 5 | 18.895 | .05292 | .84470 | 1.1838 | .04470 | 22.369 | .9706 | 5 |
| 6 | 34.012 | .02940 | .82423 | 1.2132 | .02423 | 41.265 | 1.0682 | 6 |
| 7 | 61.221 | .01633 | .81328 | 1.2295 | .01328 | 75.277 | 1.1337 | 7 |
| 8 | 110.19 | .00907 | .80733 | 1.2386 | .00733 | 136.49 | 1.1767 | 8 |
| 9 | 198.35 | .00504 | .80405 | 1.2437 | .00405 | 246.69 | 1.2044 | 9 |
| 10 | 357.04 | .00280 | .80225 | 1.2465 | .00225 | 445.05 | 1.2219 | 10 |
| 11 | 642.68 | .00156 | .80125 | 1.2480 | .00125 | 802.10 | 1.2328 | 11 |
| 12 | 1156.8 | .00086 | .80069 | 1.2489 | .00069 | 1444.7 | 1.2396 | 12 |
| 13 | 2082.2 | .00048 | .80038 | 1.2494 | .00038 | 2601.6 | 1.2437 | 13 |
| 14 | 3748.1 | .00027 | .80021 | 1.2496 | .00021 | 4683.8 | 1.2462 | 14 |
| 15 | 6746.5 | .00015 | .80012 | 1.2498 | .00012 | 8431.9 | 1.2477 | 15 |

**90% Interest Factors for Discrete Compounding Periods**

| | SINGLE PAYMENT | | UNIFORM SERIES | | | | | |
|---|---|---|---|---|---|---|---|---|
| | Compound Amount Factor | Present Worth Factor | Capital Recovery Factor | Present Worth Factor | Sinking Fund Factor | Compound Amount Factor | Gradient Factor | |
| *N* | (*F/P*, 90, *N*) | (*P/F*, 90, *N*) | (*A/P*, 90, *N*) | (*P/A*, 90, *N*) | (*A/F*, 90, *N*) | (*F/A*, 90, *N*) | (*A/G*, 90, *N*) | *N* |
| 1 | 1.9000 | .52632 | 1.9000 | .52632 | 1.00000 | 1.0000 | .00000 | 1 |
| 2 | 3.6100 | .27701 | 1.2448 | .80332 | .34483 | 2.9000 | .34483 | 2 |
| 3 | 6.8589 | .14579 | 1.0536 | .94912 | .15361 | 6.5099 | .59908 | 3 |
| 4 | 13.032 | .07673 | .97480 | 1.0258 | .07480 | 13.368 | .77867 | 4 |
| 5 | 24.760 | .04039 | .93788 | 1.0662 | .03788 | 26.401 | .90068 | 5 |
| 6 | 47.045 | .02126 | .91955 | 1.0874 | .01955 | 51.161 | .98081 | 6 |
| 7 | 89.386 | .01119 | .91018 | 1.0986 | .01018 | 98.207 | 1.0319 | 7 |
| 8 | 169.83 | .00589 | .90533 | 1.1045 | .00533 | 187.59 | 1.0637 | 8 |
| 9 | 322.68 | .00310 | .90280 | 1.1076 | .00280 | 357.42 | 1.0831 | 9 |
| 10 | 613.10 | .00163 | .90147 | 1.1093 | .00147 | 680.11 | 1.0947 | 10 |

**100% Interest Factors for Discrete Compounding Periods**

| | SINGLE PAYMENT | | UNIFORM SERIES | | | | | |
|---|---|---|---|---|---|---|---|---|
| | Compound Amount Factor | Present Worth Factor | Capital Recovery Factor | Present Worth Factor | Sinking Fund Factor | Compound Amount Factor | Gradient Factor | |
| *N* | (*F/P*, 100, *N*) | (*P/F*, 100, *N*) | (*A/P*, 100, *N*) | (*P/A*, 100, *N*) | (*A/F*, 100, *N*) | (*F/A*, 100, *N*) | (*A/G*, 100, *N*) | *N* |
| 1 | 2.000 | .50000 | 2.0000 | .50000 | 1.0000 | 1.000 | .00000 | 1 |
| 2 | 4.000 | .25000 | 1.3333 | .75000 | .33333 | 3.000 | .33333 | 2 |
| 3 | 8.000 | .12500 | 1.1428 | .87500 | .14286 | 7.000 | .57143 | 3 |
| 4 | 16.000 | .06250 | 1.0666 | .93750 | .06667 | 15.000 | .73333 | 4 |
| 5 | 32.000 | .03125 | 1.0322 | .96875 | .03226 | 31.000 | .83871 | 5 |
| 6 | 64.00 | .01562 | 1.0158 | .98438 | .01587 | 63.00 | .90476 | 6 |
| 7 | 128.00 | .00781 | 1.0078 | .99219 | .00787 | 127.00 | .94488 | 7 |
| 8 | 256.00 | .00391 | 1.0039 | .99609 | .00392 | 255.00 | .96863 | 8 |
| 9 | 512.00 | .00195 | 1.0019 | .99805 | .00196 | 511.00 | .98239 | 9 |
| 10 | 1024.0 | .00098 | 1.0009 | .99902 | .00098 | 1023.0 | .99022 | 10 |

# INDEX